D1667360

Encyclopedia of Metalloproteins

Robert H. Kretsinger • Vladimir N. Uversky
Eugene A. Permyakov
Editors

Encyclopedia of Metalloproteins

Volume 2

D–L

With 1109 Figures and 256 Tables

Editors
Robert H. Kretsinger
Department of Biology
University of Virginia
Charlottesville, VA, USA

Vladimir N. Uversky
Department of Molecular Medicine
College of Medicine
University of South Florida
Tampa, FL, USA

Eugene A. Permyakov
Institute for Biological Instrumentation
Russian Academy of Sciences
Pushchino, Moscow Region, Russia

ISBN 978-1-4614-1532-9 ISBN 978-1-4614-1533-6 (eBook)
ISBN 978-1-4614-1534-3 (print and electronic bundle)
DOI 10.1007/ 978-1-4614-1533-6
Springer New York Heidelberg Dordrecht London

Library of Congress Control Number: 2013931183

Printed on acid-free paper

Springer is part of Springer Science+Business Media (www.springer.com)

Preface

Metal ions play an essential role in the functioning of all biological systems. All biological processes occur in a milieu of high concentrations of metal ions, and many of these processes depend on direct participation of metal ions. Metal ions interact with charged and polar groups of all biopolymers; those interactions with proteins play an especially important role.

The study of structural and functional properties of metal binding proteins is an important and ongoing activity area of modern physical and chemical biology. Thirteen metal ions – sodium, potassium, magnesium, calcium, manganese, iron, cobalt, zinc, copper, nickel, vanadium, tungsten, and molybdenum – are known to be essential for at least some organisms. Metallo-proteomics deals with all aspects of the intracellular and extracellular interactions of metals and proteins. Metal cations and metal binding proteins are involved in all crucial cellular activities. Many pathological conditions are correlated with abnormal metal metabolism. Research in metalloproteomics is rapidly growing and is progressively entering curricula at universities, research institutions, and technical high schools.

Encyclopedia of Metalloproteins is a key resource that provides basic, accessible, and comprehensible information about this expanding field. It covers exhaustively all thirteen essential metal ions, discusses other metals that might compete or interfere with them, and also presents information on proteins interacting with other metal ions. *Encyclopedia of Metalloproteins* is an ideal reference for students, teachers, and researchers, as well as the informed public.

Acknowledgements

We extend our sincerest thanks to all of the contributors who have shared their insights into metalloproteins with the broader community of researchers, students, and the informed public.

About the Editors

Robert H. Kretsinger Department of Biology, University of Virginia 395, Charlottesville, VA 22904, USA

Robert H. Kretsinger is Commonwealth Professor of Biology at the University of Virginia in Charlottesville, Virginia, USA. His research has addressed structure, function, and evolution of several different protein families. His group determined the crystal structure of parvalbumin in 1970. The analysis of this calcium binding protein provided the initial characterization of the helix, loop, helix conformation of the EF-hand domain and of the pair of EF-hands that form an EF-lobe. Over seventy distinct subfamilies of EF-hand proteins have been identified, making this domain one of the most widely distributed in eukaryotes.

Dr. Kretsinger has taught courses in protein crystallography, biochemistry, macromolecular structure, and history and philosophy of biology, and has served as chair of his department and of the University Faculty Senate.

His "other" career as a sculptor began, before the advent of computer graphics, with space filling models of the EF-hand (www.virginiastonecarvers.com). He has also been an avid cyclist for many decades.

Vladimir N. Uversky Department of Molecular Medicine, College of Medicine, University of South Florida, Tampa, FL 33612 USA

Vladimir N. Uversky is an Associate Professor at the Department of Molecular Medicine at the University of South Florida (USF). He obtained his academic degrees from Moscow Institute of Physics and Technology (PhD in 1991) and from the Institute of Experimental and Theoretical Biophysics, Russian Academy of Sciences (DSc in 1998). He spent his early career working mostly on protein folding at the Institute of Protein Research and Institute for Biological Instrumentation, Russia. In 1998, he moved to the University of California, Santa Cruz, where for six years he studied protein folding, misfolding, protein conformation diseases, and protein intrinsic disorder phenomenon. In 2004, he was invited to join the Indiana University School of Medicine as a Senior Research Professor to work on intrinsically disordered proteins. Since 2010, Professor Uversky has been with USF, where he continues to study intrinsically disordered proteins and protein folding and misfolding processes. He has authored over 450 scientific publications and edited several books and book series on protein structure, function, folding and misfolding.

Eugene A. Permyakov, Institute for Biological Instrumentation, Russian Academy of Sciences, Pushchino, Moscow Region, Russia

Eugene A. Permyakov received his PhD in physics and mathematics at the Moscow Institute of Physics and Technology in 1976, and defended his Doctor of Sciences dissertation in biology at Moscow State University in 1989. From 1970 to 1994 he worked at the Institute of Theoretical and Experimental Biophysics of the Russian Academy of Sciences. From 1990 to 1991 and in 1993, Dr. Permyakov worked at the Ohio State University, Columbus, Ohio, USA. Since 1994 he has been the Director of the Institute for Biological Instrumentation of the Russian Academy of Sciences. He is a Professor of Biophysics and is known for his work on metal binding proteins and intrinsic luminescence method. He is a member of the Russian Biochemical Society. Dr. Permyakov's primary research focus is the study of physico-chemical and functional properties of metal binding proteins. He is the author of more than 150 articles and 10 books, including *Luminescent Spectroscopy of Proteins* (CRC Press, 1993), *Metalloproteomics* (John Wiley & Sons, 2009), and *Calcium Binding Proteins* (John Wiley & Sons, 2011). He is an Academic Editor of the journals *PLoS ONE* and *PeerJ*, and Editor of the book *Methods in Protein Structure and Stability Analysis* (Nova, 2007).

In his spare time, Dr. Permyakov is an avid jogger, cyclist, and cross country skier.

Section Editors

Sections: Physiological Metals: Ca; Non-Physiological Metals: Pd, Ag

Robert H. Kretsinger Department of Biology, University of Virginia, Charlottesville, VA, USA

Section: Physiological Metals: Ca

Eugene A. Permyakov Institute for Biological Instrumentation, Russian Academy of Sciences, Pushchino, Moscow Region, Russia

Sections: Metalloids; Non-Physiological Metals: Ag, Au, Pt, Be, Sr, Ba, Ra

Vladimir N. Uversky Department of Molecular Medicine, College of Medicine, University of South Florida, Tampa, FL, USA

Sections: Physiological Metals: Co, Ni, Cu

Stefano Ciurli Laboratory of Bioinorganic Chemistry, Department of Pharmacy and Biotechnology, University of Bologna, Italy

Sections: Physiological Metals: Cd, Cr

John B. Vincent Department of Chemistry, The University of Alabama, Tuscaloosa, AL, USA

Section: Physiological Metals: Fe

Elizabeth C. Theil Children's Hospital Oakland Research Institute, Oakland, CA, USA
Department of Molecular and Structural Biochemistry, North Carolina State University, Raleigh, NC, USA

Sections: Physiological Metals: Mo, W, V

Biswajit Mukherjee Department of Pharmaceutical Technology, Jadavpur University, Kolkata, India

Sections: Physiological Metals: Mg, Mn

Andrea Romani Department of Physiology and Biophysics, Case Western Reserve University, Cleveland, OH, USA

Sections: Physiological Metals: Na, K

Sergei Yu. Noskov Institute for BioComplexity and Informatics and Department for Biological Sciences, University of Calgary, Calgary, AB, Canada

Section: Physiological Metals: Zn

David S. Auld Harvard Medical School, Boston, Massachusetts, USA

Sections: Non-Physiological Metals: Li, Rb, Cs, Fr

Sergei E. Permyakov Protein Research Group, Institute for Biological Instrumentation, Russian Academy of Sciences, Pushchino, Moscow Region, Russia

Sections: Non-Physiological Metals: Lanthanides, Actinides

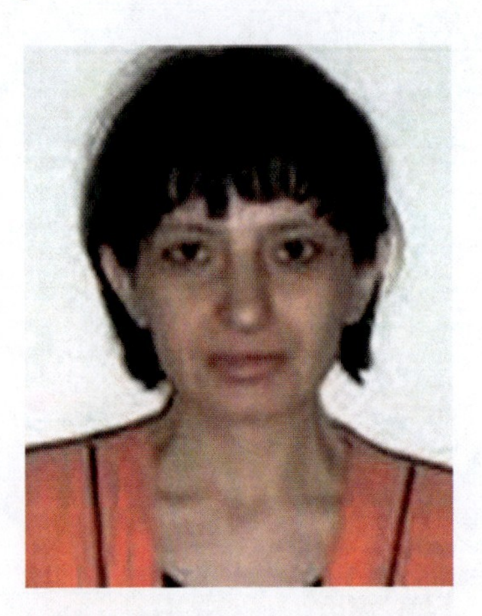

Irena Kostova Department of Chemistry, Faculty of Pharmacy, Medical University, Sofia, Bulgaria

Sections: Non-Physiological Metals: Sc, Y, Ti, Zr, Hf, Rf, Nb, Ta, Tc, Re, Ru, Os, Rh, Ir

Chunying Chen Key Laboratory for Biological Effects of Nanomaterials and Nanosafety of CAS, National Center for Nanoscience and Technology, Beijing, China

Sections: Non-Physiological Metals: Hg, Pb

K. Michael Pollard Department of Molecular and Experimental Medicine, The Scripps Research Institute, La Jolla, CA, USA

Sections: Non-Physiological Metals: Al, Ga, In, Tl, Ge, Sn, Sb, Bi, Po

Sandra V. Verstraeten Department of Biological Chemistry, University of Buenos Aires, Buenos Aires, Argentina

List of Contributors

Satoshi Abe Department of Biomolecular Engineering, Graduate School of Bioscience and Biotechnology, Tokyo Institute of Technology, Yokohama, Japan

Vojtech Adam Department of Chemistry and Biochemistry, Faculty of Agronomy, Mendel University in Brno, Brno, Czech Republic

Central European Institute of Technology, Brno University of Technology, Brno, Czech Republic

Olayiwola A. Adekoya Pharmacology Research group, Department of Pharmacy, Institute of Pharmacy, University of Tromsø, Tromsø, Norway

Paul A. Adlard The Mental Health Research Institute, The University of Melbourne, Parkville, VIC, Australia

Magnus S. Ågren Department of Surgery and Copenhagen Wound Healing Center, Bispebjerg University Hospital, Copenhagen, Denmark

Karin Åkerfeldt Department of Chemistry, Haverford College, Haverford, PA, USA

Takashiro Akitsu Department of Chemistry, Tokyo University of Science, Shinjuku-ku, Tokyo, Japan

Lorenzo Alessio Department of Experimental and Applied Medicine, Section of Occupational Health and Industrial Hygiene, University of Brescia, Brescia, Italy

Mamdouh M. Ali Biochemistry Department, Genetic Engineering and Biotechnology Division, National Research Centre, El Dokki, Cairo, Egypt

James B. Ames Department of Chemistry, University of California, Davis, CA, USA

Olaf S. Andersen Department of Physiology and Biophysics, Weill Cornell Medical College, New York, NY, USA

Gregory J. Anderson Iron Metabolism Laboratory, Queensland Institute of Medical Research, PO Royal Brisbane Hospital, Brisbane, QLD, Australia

Janet S. Anderson Department of Chemistry, Union College, Schenectady, NY, USA

João Paulo André Centro de Química, Universidade do Minho, Braga, Portugal

Claudia Andreini Magnetic Resonance Center (CERM) – University of Florence, Sesto Fiorentino, Italy

Department of Chemistry, University of Florence, Sesto Fiorentino, Italy

Alexey N. Antipov A.N. Bach Institute of Biochemistry Russian Academy of Sciences, Moscow, Russia

Tayze T. Antunes Kidney Research Center, Ottawa Hospital Research Institute, University of Ottawa, Ottawa, ON, Canada

Varun Appanna Department of Chemistry and Biochemistry, Laurentian University, Sudbury, ON, Canada

Vasu D. Appanna Department of Chemistry and Biochemistry, Laurentian University, Sudbury, ON, Canada

Cristina Ariño Department of Analytical Chemistry, University of Barcelona, Barcelona, Spain

Vladimir B. Arion Institute of Inorganic Chemistry, University of Vienna, Vienna, Austria

Farukh Arjmand Department of Chemistry, Aligarh Muslim University, Aligarh, UP, India

Fabio Arnesano Department of Chemistry, University of Bari "Aldo Moro", Bari, Italy

Joan L. Arolas Proteolysis Lab, Department of Structural Biology, Molecular Biology Institute of Barcelona, CSIC Barcelona Science Park, Barcelona, Spain

Afolake T. Arowolo Department of Biochemistry, Microbiology & Biotechnology, Rhodes University, Grahamstown, Eastern Cape, South Africa

Nebojša Arsenijević Faculty of Medical Sciences, University of Kragujevac, Centre for Molecular Medicine, Kragujevac, Serbia

Samuel Ogheneovo Asagba Department of Biochemistry, Delta State University, Abraka, Delta State, Nigeria

Michael Aschner Department of Pediatrics, Division of Pediatric Clinical Pharmacology and Toxicology, Vanderbilt University Medical Center, Nashville, TN, USA

Center in Molecular Toxicology, Vanderbilt University Medical Center, Nashville, TN, USA

Center for Molecular Neuroscience, Vanderbilt University Medical Center, Nashville, TN, USA

The Kennedy Center for Research on Human Development, Vanderbilt University Medical Center, Nashville, TN, USA

Michael Assfalg Department of Biotechnology, University of Verona, Verona, Italy

William D. Atchison Department of Pharmacology and Toxicology, Michigan State University, East Lansing, MI, USA

Bishara S. Atiyeh Division of Plastic Surgery, Department of Surgery, American University of Beirut Medical Center, Beirut, Lebanon

Sílvia Atrian Departament de Genètica, Facultat de Biologia, Universitat de Barcelona, Barcelona, Spain

Christopher Auger Department of Chemistry and Biochemistry, Laurentian University, Sudbury, ON, Canada

David S. Auld Harvard Medical School, Boston, MA, USA

Scott Ayton The Mental Health Research Institute, The University of Melbourne, Parkville, VIC, Australia

Eduard B. Babiychuk Department of Cell Biology, Institute of Anatomy, University of Bern, Bern, Switzerland

Petr Babula Department of Natural Drugs, Faculty of Pharmacy, University of Veterinary and Pharmaceutical Sciences Brno, Brno, Czech Republic

Damjan Balabanič Ecology Department, Pulp and Paper Institute, Ljubljana, Slovenia

Wojciech Bal Institute of Biochemistry and Biophysics, Polish Academy of Sciences, Warsaw, Poland

Graham S. Baldwin Department of Surgery, Austin Health, The University of Melbourne, Heidelberg, VIC, Australia

Cynthia Bamdad Minerva Biotechnologies Corporation, Waltham, MA, USA

Mario Barbagallo Geriatric Unit, Department of Internal Medicine and Medical Specialties (DIMIS), University of Palermo, Palermo, Italy

Juan Barceló Lab. Fisiología Vegetal, Facultad de Biociencias, Universidad Autónoma de Barcelona, Bellaterra, Spain

Khurram Bashir Graduate School of Agricultural and Life Sciences, The University of Tokyo, Tokyo, Japan

Partha Basu Department of Chemistry and Biochemistry, Duquesne University, Pittsburgh, PA, USA

Andrea Battistoni Dipartimento di Biologia, Università di Roma Tor Vergata, Rome, Italy

Mikael Bauer Department of Biochemistry and Structural Biology, Lund University, Chemical Centre, Lund, Sweden

Lukmaan Bawazer School of Chemistry, University of Leeds, Leeds, UK

Carine Bebrone Centre for Protein Engineering, University of Liège, Sart–Tilman, Liège, Belgium

Institute of Molecular Biotechnology, RWTH–Aachen University, c/o Fraunhofer IME, Aachen, Germany

Konstantinos Beis Division of Molecular Biosciences, Imperial College London, London, South Kensington, UK

Membrane Protein Lab, Diamond Light Source, Harwell Science and Innovation Campus, Chilton, Oxfordshire, UK

Research Complex at Harwell, Harwell Oxford, Didcot, Oxforsdhire, UK

Catherine Belle Départment de Chimie Moléculaire, UMR-CNRS 5250, Université Joseph Fourier, ICMG FR-2607, Grenoble, France

Andrea Bellelli Department of Biochemical Sciences, Sapienza University of Rome, Rome, Italy

Gunes Bender Department of Biological Chemistry, University of Michigan Medical School, Ann Arbor, MI, USA

Stefano Benini Faculty of Science and Technology, Free University of Bolzano, Bolzano, Italy

Stéphane L. Benoit Department of Microbiology, The University of Georgia, Athens, GA, USA

Tomas Bergman Department of Medical Biochemistry and Biophysics, Karolinska Institutet, Stockholm, Sweden

Lawrence R. Bernstein Terrametrix, Menlo Park, CA, USA

Marla J. Berry Department of Cell & Molecular Biology, John A. Burns School of Medicine, University of Hawaii at Manoa, Honolulu, HI, USA

Ivano Bertini Magnetic Resonance Center (CERM) – University of Florence, Sesto Fiorentino, Italy

Department of Chemistry, University of Florence, Sesto Fiorentino, Italy

Gerd Patrick Bienert Institut des Sciences de la Vie, Universite catholique de Louvain, Louvain-la-Neuve, Belgium

Andrew N. Bigley Department of Chemistry, Texas A&M University, College Station, TX, USA

Luis M. Bimbo Division of Pharmaceutical Technology, University of Helsinki, Helsinki, Finland

Ohad S. Birk Head, Genetics Institute, Soroka Medical Center Head, Morris Kahn Center for Human Genetics, NIBN and Faculty of Health Sciences, Ben Gurion University, Beer Sheva, Israel

Ruth Birner-Gruenberger Institute of Pathology and Center of Medical Research, Medical University of Graz, Graz, Austria

Cristina Bischin Department of Chemistry and Chemical Engineering, Babes-Bolyai University, Cluj-Napoca, Romania

Florian Bittner Department of Plant Biology, Braunschweig University of Technology, Braunschweig, Germany

Jodi L. Boer Department of Biochemistry and Molecular Biology, Michigan State University, East Lansing, MI, USA

Judith S. Bond Department of Biochemistry and Molecular Biology, Pennsylvania State University College of Medicine, Hershey, PA, USA

Martin D. Bootman Life, Health and Chemical Sciences, The Open University Walton Hall, Milton Keynes, UK

Bhargavi M. Boruah CAS Key Laboratory of Pathogenic Microbiology and Immunology, Institute of Microbiology, Chinese Academy of Sciences, Beijing, China

Graduate University of Chinese Academy of Science, Beijing, China

Sheryl R. Bowley Division of Hemostasis and Thrombosis, Beth Israel Deaconess Medical Center, Harvard Medical School, Boston, MA, USA

Doreen Braun Institute for Experimental Endocrinology, Charité-Universitätsmedizin Berlin, Berlin, Germany

Davorka Breljak Unit of Molecular Toxicology, Institute for Medical Research and Occupational Health, Zagreb, Croatia

Leonid Breydo Department of Molecular Medicine, Morsani College of Medicine, University of South Florida, Tampa, FL, USA

Mickael Briens UPR ARN du CNRS, Université de Strasbourg, Institut de Biologie Moléculaire et Cellulaire, Strasbourg, France

Joan B. Broderick Department of Chemistry and Biochemistry, Montana State University, Bozeman, MT, USA

James E. Bruce Department of Genome Sciences, University of Washington, Seattle, WA, USA

Ernesto Brunet Dept. Química Orgánica, Facultad de Ciencias, Universidad Autónoma de Madrid, Madrid, Spain

Maurizio Brunori Department of Biochemical Sciences, Sapienza University of Rome, Rome, Italy

Susan K. Buchanan Laboratory of Molecular Biology, National Institute of Diabetes and Digestive and Kidney Diseases, US National Institutes of Health, Bethesda, MD, USA

Gabriel E. Büchel Institute of Inorganic Chemistry, University of Vienna, Vienna, Austria

Živadin D. Bugarčić Faculty of Science, Department of Chemistry, University of Kragujevac, Kragujevac, Serbia

Melisa Bunderson-Schelvan Department of Biomedical and Pharmaceutical Sciences, Center for Environmental Health, The University of Montana, Missoula, MT, USA

Jean-Claude G. Bünzli Center for Next Generation Photovoltaic Systems, Korea University, Sejong Campus, Jochiwon–eup, Yeongi–gun, ChungNam–do, Republic of Korea

École Polytechnique Fédérale de Lausanne, Institute of Chemical Sciences and Engineering, Lausanne, Switzerland

John E. Burke Medical Research Council, Laboratory of Molecular Biology, Cambridge, UK

Torsten Burkholz Division of Bioorganic Chemistry, School of Pharmacy, Saarland State University, Saarbruecken, Germany

Bruce S. Burnham Department of Chemistry, Biochemistry, and Physics, Rider University, Lawrenceville, NJ, USA

Ashley I. Bush The Mental Health Research Institute, The University of Melbourne, Parkville, VIC, Australia

Kunzheng Cai Institute of Tropical and Subtropical Ecology, South China Agricultural University, Guangzhou, China

Iván L. Calderón Laboratorio de Microbiología Molecular, Universidad Andrés Bello, Santiago, Chile

Glaucia Callera Kidney Research Center, Ottawa Hospital Research Institute, University of Ottawa, Ottawa, ON, Canada

Marcello Campagna Department of Public Health, Clinical and Molecular Medicine, University of Cagliari, Cagliari, Italy

Mercè Capdevila Departament de Química, Facultat de Ciències, Universitat Autònoma de Barcelona, Cerdanyola del Vallés (Barcelona), Spain

Fernando Cardozo-Pelaez Department of Pharmaceutical Sciences, Center for Environmental Health Sciences, University of Montana, Missoula, MT, USA

Bradley A. Carlson Molecular Biology of Selenium Section, Laboratory of Cancer Prevention, National Cancer Institute, National Institutes of Health, Bethesda, MD, USA

Silvia Castelli Department of Biology, University of Rome Tor Vergata, Rome, Italy

Tommy Cedervall Department of Biochemistry and Structural Biology, Lund University, Chemical Centre, Lund, Sweden

Sudipta Chakraborty Department of Pediatrics and Department of Pharmacology, and the Kennedy Center for Research on Human Development, Vanderbilt University Medical Center, Nashville, TN, USA

Henry Chan Department of Molecular Biology, Division of Biological Sciences, University of California at San Diego, La Jolla, CA, USA

N. Chandrasekaran Centre for Nanobiotechnology, VIT University, Vellore, Tamil Nadu, India

Loïc J. Charbonnière Laboratoire d'Ingénierie Moléculaire Appliquée à l'Analyse, IPHC, UMR 7178 CNRS/UdS ECPM, Strasbourg, France

Malay Chatterjee Division of Biochemistry, Department of Pharmaceutical Technology, Jadavpur University, Kolkata, West Bengal, India

Mary Chatterjee Division of Biochemistry, Department of Pharmaceutical Technology, Jadavpur University, Kolkata, West Bengal, India

François Chaumont Institut des Sciences de la Vie, Universite catholique de Louvain, Louvain-la-Neuve, Belgium

Juan D. Chavez Department of Genome Sciences, University of Washington, Seattle, WA, USA

Chi-Ming Che Department of Chemistry, State Key Laboratory of Synthetic Chemistry and Open Laboratory of Chemical Biology of the Institute of Molecular Technology for Drug Discovery and Synthesis, The University of Hong Kong, Hong Kong, China

Elena Chekmeneva Department of Chemistry, University of Sheffield, Sheffield, UK

Di Chen The Developmental Therapeutics Program, Barbara Ann Karmanos Cancer Institute, and Departments of Oncology, Pharmacology and Pathology, School of Medicine, Wayne State University, Detroit, MI, USA

Hong-Yuan Chen National Key Laboratory of Analytical Chemistry for Life Science, School of Chemistry and Chemical Engineering, Nanjing University, Nanjing, China

Jiugeng Chen Laboratory of Plant Physiology and Molecular Genetics, Université Libre de Bruxelles, Brussels, Belgium

Sai-Juan Chen State Key Laboratory of Medical Genomics, Shanghai Institute of Hematology, Rui Jin Hospital Affiliated to Shanghai Jiao Tong University School of Medicine, Shanghai, China

Zhu Chen State Key Laboratory of Medical Genomics, Shanghai Institute of Hematology, Rui Jin Hospital Affiliated to Shanghai Jiao Tong University School of Medicine, Shanghai, China

Robert A. Cherny The Mental Health Research Institute, The University of Melbourne, Parkville, VIC, Australia

Yana Chervona Department of Environmental Medicine, New York University Medical School, New York, NY, USA

Christopher R. Chitambar Division of Hematology and Oncology, Medical College of Wisconsin, Froedtert and Medical College of Wisconsin Clinical Cancer Center, Milwaukee, WI, USA

Hassanul Ghani Choudhury Division of Molecular Biosciences, Imperial College London, London, South Kensington, UK

Membrane Protein Lab, Diamond Light Source, Harwell Science and Innovation Campus, Chilton, Oxfordshire, UK

Research Complex at Harwell, Harwell Oxford, Didcot, Oxforsdhire, UK

Samrat Roy Chowdhury Department of Pharmaceutical Technology, Jadavpur University, Kolkata, West Bengal, India

Stefano Ciurli Department of Agro-Environmental Science and Technology, University of Bologna, Bologna, Italy

Stephan Clemens Department of Plant Physiology, University of Bayreuth, Bayreuth, Germany

Nansi Jo Colley Department of Ophthalmology and Visual Sciences, UW Eye Research Institute, University of Wisconsin, Madison, WI, USA

Gianni Colotti Institute of Molecular Biology and Pathology, Consiglio Nazionale delle Ricerche, Rome, Italy

Giovanni Corsetti Division of Human Anatomy, Department of Biomedical Sciences and Biotechnologies, Brescia University, Brescia, Italy

Max Costa Department of Environmental Medicine, New York University Medical School, New York, NY, USA

Jos A. Cox Department of Biochemistry, University of Geneva, Geneva, Switzerland

Adam V. Crain Department of Chemistry and Biochemistry, Montana State University, Bozeman, MT, USA

Ann Cuypers Centre for Environmental Sciences, Hasselt University, Diepenbeek, Belgium

Martha S. Cyert Department of Biology, Stanford University, Stanford, USA

Sabato D'Auria National Research Council (CNR), Laboratory for Molecular Sensing, Institute of Protein Biochemistry, Naples, Italy

Verónica Daier Departamento de Química Física/IQUIR-CONICET, Facultad de Ciencias Bioquímicas y Farmacéuticas, Universidad Nacional de Rosario, Rosario, Argentina

Charles T. Dameron Chemistry Department, Saint Francis University, Loretto, PA, USA

Subhadeep Das Division of Biochemistry, Department of Pharmaceutical Technology, Jadavpur University, Kolkata, West Bengal, India

Nilay Kanti Das Department of Dermatology, Medical College, Kolkata, West Bengal, India

Rupali Datta Department of Biological Sciences, Michigan Technological University, Houghton, MI, USA

Benjamin G. Davis Chemistry Research Laboratory, Department of Chemistry, University of Oxford, Oxford, UK

Dennis R. Dean Department of Biochemistry, Virginia Tech University, Blacksburg, VA, USA

Kannan Deepa Biochemical Engineering Laboratory, Department of Chemical Engineering, Indian Institute of Technology Madras, Chennai, Tamil Nadu, India

Claudia Della Corte Unit of Liver Research of Bambino Gesù Children's Hospital, IRCCS, Rome, Italy

Simone Dell'Acqua REQUIMTE/CQFB, Departamento de Química, Faculdade de Ciências e Tecnologia, Universidade Nova de Lisboa, Caparica, Portugal

Dipartimento di Chimica, Università di Pavia, Pavia, Italy

Hakan Demir Department of Nuclear Medicine, School of Medicine, Kocaeli University, Umuttepe, Kocaeli, Turkey

Sumukh Deshpande Institute for Biocomplexity and Informatics, Department of Biological Sciences, University of Calgary, Calgary, AB, Canada

Alessandro Desideri Department of Biology, University of Rome Tor Vergata, Rome, Italy

Interuniversity Consortium, National Institute Biostructure and Biosystem (INBB), Rome, Italy

Patrick C. D'Haese Laboratory of Pathophysiology, University of Antwerp, Wilrijk, Belgium

José Manuel Díaz-Cruz Department of Analytical Chemistry, University of Barcelona, Barcelona, Spain

Saad A. Dibo Division Plastic and Reconstructive Surgery, American University of Beirut Medical Center, Beirut, Lebanon

Pavel Dibrov Department of Microbiology, University of Manitoba, Winnipeg, MB, Canada

Adeleh Divsalar Department of Biological Sciences, Tarbiat Moallem University, Tehran, Iran

Ligia J. Dominguez Geriatric Unit, Department of Internal Medicine and Medical Specialties (DIMIS), University of Palermo, Palermo, Italy

Delfina C. Domínguez College of Health Sciences, The University of Texas at El Paso, El Paso, TX, USA

Rosario Donato Department of Experimental Medicine and Biochemical Sciences, University of Perugia, Perugia, Italy

Elke Dopp Institute of Hygiene and Occupational Medicine, University of Duisburg-Essen, Essen, Germany

Melania D'Orazio Dipartimento di Biologia, Università di Roma Tor Vergata, Rome, Italy

Q. Ping Dou The Developmental Therapeutics Program, Barbara Ann Karmanos Cancer Institute, and Departments of Oncology, Pharmacology and Pathology, School of Medicine, Wayne State University, Detroit, MI, USA

Ross G. Douglas Zinc Metalloprotease Research Group, Division of Medical Biochemistry, Institute of Infectious Disease and Molecular Medicine, University of Cape Town, Cape Town, South Africa

Annette Draeger Department of Cell Biology, Institute of Anatomy, University of Bern, Bern, Switzerland

Gabi Drochioiu Alexandru Ioan Cuza University of Iasi, Iasi, Romania

Elzbieta Dudek Department of Biochemistry, University of Alberta, Edmonton, AB, Canada

Todor Dudev Institute of Biomedical Sciences, Academia Sinica, Taipei, Taiwan

Henry J. Duff Libin Cardiovascular Institute of Alberta, Calgary, AB, Canada

Evert C. Duin Department of Chemistry and Biochemistry, Auburn University, Auburn, AL, USA

R. Scott Duncan Vision Research Center and Departments of Basic Medical Science and Ophthalmology, School of Medicine, University of Missouri, Kansas City, MO, USA

Michael F. Dunn Department of Biochemistry, University of California at Riverside, Riverside, CA, USA

Serdar Durdagi Institute for Biocomplexity and Informatics, Department of Biological Sciences, University of Calgary, Calgary, AB, Canada

Kaitlin S. Duschene Department of Chemistry and Biochemistry, Montana State University, Bozeman, MT, USA

Ankit K. Dutta School of Molecular and Biomedical Science, University of Adelaide, Adelaide, South Australia, Australia

Naba K. Dutta Ian Wark Research Institute, University of South Australia, Mawson Lakes, South Australia, Australia

Paul J. Dyson Institut des Sciences et Ingénierie Chimiques, Ecole Polytechnique Fédérale de Lausanne (EPFL) SB ISIC-Direction, Lausanne, Switzerland

Brian E. Eckenroth Department of Microbiology and Molecular Genetics, University of Vermont, Burlington, VT, USA

Niels Eckstein Federal Institute for Drugs and Medical Devices (BfArM), Bonn, Germany

David J. Eide Department of Nutritional Sciences, University of Wisconsin-Madison, Madison, WI, USA

Thomas Eitinger Institut für Biologie/Mikrobiologie, Humboldt-Universität zu Berlin, Berlin, Germany

Annette Ekblond Cardiology Stem Cell Laboratory, Rigshospitalet University Hospital, Copenhagen, Denmark

Jean-Michel El Hage Chahine ITODYS, Université Paris-Diderot Sorbonne Paris Cité, CNRS UMR 7086, Paris, France

Alex Elías Laboratorio de Microbiología Molecular, Departamento de Biología, Universidad de Santiago de Chile, Santiago, Chile

Jeffrey S. Elmendorf Department of Cellular and Integrative Physiology and Department of Biochemistry and Molecular Biology, and Centers for Diabetes Research, Membrane Biosciences, and Vascular Biology and Medicine, Indiana University School of Medicine, Indianapolis, IN, USA

Sanaz Emami Department of Biophysics, Institute of Biochemistry and Biophysics (IBB), University of Tehran, Tehran, Iran

Vinita Ernest Centre for Nanobiotechnology, VIT University, Vellore, Tamil Nadu, India

Miquel Esteban Department of Analytical Chemistry, University of Barcelona, Barcelona, Spain

Christopher Exley The Birchall Centre, Lennard-Jones Laboratories, Keele University, Staffordshire, UK

Chunhai Fan Laboratory of Physical Biology, Shanghai Institute of Applied Physics, Shanghai, China

Marcelo Farina Departamento de Bioquímica, Centro de Ciências Biológicas, Universidade Federal de Santa Catarina, Florianópolis, SC, Brazil

Nicholas P. Farrell Department of Chemistry, Virginia Commonwealth University, Richmond, VA, USA

Caroline Fauquant iRTSV/LCBM UMR 5249 CEA-CNRS-UJF, CEA/Grenoble, Bât K, Université Grenoble, Grenoble, France

James G. Ferry Department of Biochemistry and Molecular Biology, Eberly College of Science, The Pennsylvania State University, University Park, PA, USA

Ana Maria Figueiredo Instituto de Pesquisas Energeticas e Nucleares, IPEN-CNEN/SP, Sao Paulo, Brazil

David I. Finkelstein The Mental Health Research Institute, The University of Melbourne, Parkville, VIC, Australia

Larry Fliegel Department of Biochemistry, University of Alberta, Edmonton, AB, Canada

Swaran J. S. Flora Division of Pharmacology and Toxicology, Defence Research and Development Establishment, Gwalior, India

Juan C. Fontecilla-Camps Metalloproteins; Institut de Biologie Structurale J.P. Ebel; CEA; CNRS; Université J. Fourier, Grenoble, France

Sara M. Fox Department of Pharmacology and Toxicology, Michigan State University, East Lansing, MI, USA

Ricardo Franco REQUIMTE FCT/UNL, Departamento de Química, Faculdade de Ciências e Tecnologia, Universidade Nova de Lisboa, Caparica, Portugal

Stefan Fränzle Department of Biological and Environmental Sciences, Research Group of Environmental Chemistry, International Graduate School Zittau, Zittau, Germany

Christopher J. Frederickson NeuroBioTex, Inc, Galveston Island, TX, USA

Michael Frezza The Developmental Therapeutics Program, Barbara Ann Karmanos Cancer Institute, and Departments of Oncology, Pharmacology and Pathology, School of Medicine, Wayne State University, Detroit, MI, USA

Barbara C. Furie Division of Hemostasis and Thrombosis, Beth Israel Deaconess Medical Center, Harvard Medical School, Boston, MA, USA

Bruce Furie Division of Hemostasis and Thrombosis, Beth Israel Deaconess Medical Center, Harvard Medical School, Boston, MA, USA

Roland Gaertner Department of Endocrinology, University Hospital, Ludwig-Maximilians University Munich, Munich, Germany

Sonia Galván-Arzate Departamento de Neuroquímica, Instituto Nacional de Neurología y Neurocirugía Manuel Velasco Suárez, Mexico City, DF, Mexico

Livia Garavelli Struttura Semplice Dipartimentale di Genetica Clinica, Dipartimento di Ostetrico-Ginecologico e Pediatrico, Istituto di Ricovero e Cura a Carattere Scientifico, Arcispedale S. Maria Nuova, Reggio Emilia, Italy

Carolyn L. Geczy Inflammation and Infection Research Centre, School of Medical Sciences, University of New South Wales, Sydney, NSW, Australia

Emily Geiger Department of Biological Sciences, Michigan Technological University, Houghton, MI, USA

Alayna M. George Thompson Department of Chemistry and Biochemistry, University of Arizona, Tucson, AZ, USA

Charles P. Gerba Department of Soil, Water and Environmental Science, University of Arizona, Tucson, AZ, USA

Miltu Kumar Ghosh Department of Pharmaceutical Technology, Jadavpur University, Kolkata, West Bengal, India

Pramit Ghosh Department of Community Medicine, Medical College, Kolkata, West Bengal, India

Saikat Ghosh Department of Pharmaceutical Technology, Jadavpur University, Kolkata, West Bengal, India

Hedayatollah Ghourchian Department of Biophysics, Institute of Biochemistry and Biophysics (IBB), University of Tehran, Tehran, Iran

Jessica L. Gifford Department of Biological Sciences, Biochemistry Research Group, University of Calgary, Calgary, AB, Canada

Danuta M. Gillner Department of Chemistry, Silesian University of Technology, Gliwice, Poland

Mario Di Gioacchino Occupational Medicine and Allergy, Head of Allergy and Immunotoxicology Unit (Ce.S.I.), G. d'Annunzio University, Via dei Vestini, Chieti, Italy

Denis Girard Laboratoire de recherche en inflammation et physiologie des granulocytes, Université du Québec, INRS-Institut Armand-Frappier, Laval, QC, Canada

F. Xavier Gomis-Rüth Proteolysis Lab, Department of Structural Biology, Molecular Biology Institute of Barcelona, CSIC Barcelona Science Park, Barcelona, Spain

Harry B. Gray Beckman Institute, California Institute of Technology, Pasadena, CA, USA

Claudia Großkopf Department Chemicals Safety, Federal Institute for Risk Assessment, Berlin, Germany

Thomas E. Gunter Department of Biochemistry and Biophysics, University of Rochester School of Medicine and Dentistry, Rochester, NY, USA

Dharmendra K. Gupta Departamento de Bioquímica, Biología Celular y Molecular de Plantas, Estación Experimental del Zaidin, CSIC, Granada, Spain

Nikolai B. Gusev Department of Biochemistry, School of Biology, Moscow State University, Moscow, Russian Federation

Mandana Haack-Sørensen Cardiology Stem Cell Laboratory, Rigshospitalet University Hospital, Copenhagen, Denmark

Bodo Haas Federal Institute for Drugs and Medical Devices (BfArM), Bonn, Germany

Hajo Haase Institute of Immunology, Medical Faculty, RWTH Aachen University, Aachen, Germany

Fathi Habashi Department of Mining, Metallurgical, and Materials Engineering, Laval University, Quebec City, Canada

Alice Haddy Department of Chemistry and Biochemistry, University of North Carolina, Greensboro, NC, USA

Nguyêt-Thanh Ha-Duong ITODYS, Université Paris-Diderot Sorbonne Paris Cité, CNRS UMR 7086, Paris, France

Jesper Z. Haeggström Department of Medical Biochemistry and Biophysics (MBB), Karolinska Institute, Stockholm, Sweden

James F. Hainfeld Nanoprobes, Incorporated, Yaphank, NY, USA

Sefali Halder Department of Pharmaceutical Technology, Jadavpur University, Kolkata, West Bengal, India

Boyd E. Haley Department of Chemistry, University of Kentucky, Lexington, KY, USA

Raymond F. Hamilton Jr. Department of Biomedical and Pharmaceutical Sciences, Center for Environmental Health, The University of Montana, Missoula, MT, USA

Heidi E. Hannon Department of Pharmacology and Toxicology, Michigan State University, East Lansing, MI, USA

Timothy P. Hanusa Department of Chemistry, Vanderbilt University, Nashville, TN, USA

Edward D. Harris Department of Nutrition and Food Science, Texas A&M University, College Station, TX, USA

Todd C. Harrop Department of Chemistry, University of Georgia, Athens, GA, USA

Andrea Hartwig Department Food Chemistry and Toxicology, Karlsruhe Institute of Technology, Karlsruhe, Germany

Robert P. Hausinger Department of Biochemistry and Molecular Biology, Michigan State University, East Lansing, MI, USA

Department of Microbiology and Molecular Genetics, 6193 Biomedical and Physical Sciences, Michigan State University, East Lansing, MI, USA

Hiroaki Hayashi Department of Dermatology, Kawasaki Medical School, Kurashiki, Japan

Xiao He CAS Key Laboratory for Biomedical Effects of Nanomaterials and Nanosafety & CAS Key Laboratory of Nuclear Analytical Techniques, Institute of High Energy Physics, Chinese Academy of Sciences, Beijing, China

Yao He Institute of Functional Nano & Soft Materials, Soochow University, Jiangsu, China

Kim L. Hein Centre for Molecular Medicine Norway (NCMM), University of Oslo Nordic EMBL Partnership, Oslo, Norway

Claus W. Heizmann Department of Pediatrics, Division of Clinical Chemistry, University of Zurich, Zurich, Switzerland

Michael T. Henzl Department of Biochemistry, University of Missouri, Columbia, MO, USA

Carol M. Herak-Kramberger Unit of Molecular Toxicology, Institute for Medical Research and Occupational Health, Zagreb, Croatia

Christian Hermans Laboratory of Plant Physiology and Molecular Genetics, Université Libre de Bruxelles, Brussels, Belgium

Griselda Hernández New York State Department of Health, Wadsworth Center, Albany, NY, USA

Akon Higuchi Department of Chemical and Materials Engineering, National Central University, Jhongli, Taoyuan, Taiwan

Department of Reproduction, National Research Institute for Child Health and Development, Setagaya–ku, Tokyo, Japan

Cathay Medical Research Institute, Cathay General Hospital, Hsi–Chi City, Taipei, Taiwan

Russ Hille Department of Biochemistry, University of California, Riverside, CA, USA

Alia V. H. Hinz Department of Chemistry, Western Michigan University, Kalamazoo, MI, USA

John Andrew Hitron Graduate Center for Toxicology, University of Kentucky, Lexington, KY, USA

Miryana Hémadi ITODYS, Université Paris-Diderot Sorbonne Paris Cité, CNRS UMR 7086, Paris, France

Christer Hogstrand Metal Metabolism Group, Diabetes and Nutritional Sciences Division, School of Medicine, King's College London, London, UK

Erhard Hohenester Department of Life Sciences, Imperial College London, London, UK

Andrij Holian Department of Biomedical and Pharmaceutical Sciences, Center for Environmental Health, The University of Montana, Missoula, MT, USA

Richard C. Holz Department of Chemistry and Biochemistry, Loyola University Chicago, Chicago, IL, USA

Charles G. Hoogstraten Department of Biochemistry and Molecular Biology, Michigan State University, East Lansing, MI, USA

Ying Hou Key Laboratory for Biomechanics and Mechanobiology of the Ministry of Education, School of Biological Science and Medical Engineering, Beihang University, Beijing, China

Mingdong Huang Division of Hemostasis and Thrombosis, Beth Israel Deaconess Medical Center, Harvard Medical School, Boston, MA, USA

David L. Huffman Department of Chemistry, Western Michigan University, Kalamazoo, MI, USA

Paco Hulpiau Department for Molecular Biomedical Research, VIB, Ghent, Belgium

Amir Ibrahim Plastic and Reconstructive SurgeryBurn Fellow, Massachusetts General Hospital / Harvard Medical School & Shriners Burn Hospital, Boston, USA

Mitsu Ikura Ontario Cancer Institute and Department of Medical Biophysics, University of Toronto, Toronto, Ontario, Canada

Andrea Ilari Institute of Molecular Biology and Pathology, Consiglio Nazionale delle Ricerche, Rome, Italy

Giuseppe Inesi California Pacific Medical Center Research Institute, San Francisco, CA, USA

Hiroaki Ishida Department of Biological Sciences, Biochemistry Research Group, University of Calgary, Calgary, AB, Canada

Vangronsveld Jaco Centre for Environmental Sciences, Hasselt University, Diepenbeek, Belgium

Claus Jacob Division of Bioorganic Chemistry, School of Pharmacy, Saarland State University, Saarbruecken, Germany

Sushil K. Jain Department of Pediatrics, Louisiana State University Health Sciences Center, Shreveport, LA, USA

Peter Jensen Department of Dermato-Allergology, Copenhagen University Hospital Gentofte, Hellerup, Denmark

Klaudia Jomova Department of Chemistry, Faculty of Natural Sciences, Constantine The Philosopher University, Nitra, Slovakia

Raghava Rao Jonnalagadda Chemical Laboratory, Central Leather Research Institute (Council of Scientific and Industrial Research), Chennai, Tamil Nadu, India

Hans Jörnvall Department of Medical Biochemistry and Biophysics, Karolinska Institutet, Stockholm, Sweden

Olga Juanes Dept. Química Orgánica, Facultad de Ciencias, Universidad Autónoma de Madrid, Madrid, Spain

Sreeram Kalarical Janardhanan Chemical Laboratory, Central Leather Research Institute (Council of Scientific and Industrial Research), Chennai, Tamil Nadu, India

Paul C. J. Kamer School of Chemistry, University of St Andrews, St Andrews, UK

Tina Kamčeva Laboratory of Physical Chemistry, Vinča Institute of Nuclear Sciences, University of Belgrade, Belgrade, Serbia

Laboratory of Clinical Biochemistry, Section of Clinical Pharmacology, Haukeland University Hospital, Bergen, Norway

ChulHee Kang Washington State University, Pullman, WA, USA

Kazimierz S. Kasprzak Chemical Biology Laboratory, Frederick National Laboratory for Cancer Research, Frederick, MD, USA

Jane Kasten-Jolly New York State Department of Health, Wadsworth Center, Albany, NY, USA

Jens Kastrup Cardiology Stem Cell Laboratory, Rigshospitalet University Hospital, Copenhagen, Denmark

The Heart Centre, Cardiac Catheterization Laboratory, Rigshospitalet University Hospital, Copenhagen, Denmark

Prafulla Katkar Department of Biology, University of Rome Tor Vergata, Rome, Italy

Fusako Kawai Center for Nanomaterials and Devices, Kyoto Institute of Technology, Kyoto, Japan

Jason D. Kenealey Department of Biomolecular Chemistry, University of Wisconsin, Madison, WI, USA

Bernhard K. Keppler Institute of Inorganic Chemistry, University of Vienna, Vienna, Austria

E. Van Kerkhove Department of Physiology, Centre for Environmental Sciences, Hasselt University, Diepenbeek, Belgium

Kazuya Kikuchi Division of Advanced Science and Biotechnology, Graduate School of Engineering, Osaka University, Suita, Osaka, Japan

Immunology Frontier Research Center, Osaka University, Suita, Osaka, Japan

Michael Kirberger Department of Chemistry, Georgia State University, Atlanta, GA, USA

Masanori Kitamura Department of Molecular Signaling, Interdisciplinary Graduate School of Medicine and Engineering, University of Yamanashi, Chuo, Yamanashi, Japan

Rene Kizek Department of Chemistry and Biochemistry, Faculty of Agronomy, Mendel University in Brno, Brno, Czech Republic

Central European Institute of Technology, Brno University of Technology, Brno, Czech Republic

Nanne Kleefstra Diabetes Centre, Isala clinics, Zwolle, The Netherlands

Department of Internal Medicine, University Medical Center Groningen, Groningen, The Netherlands

Langerhans Medical Research Group, Zwolle, The Netherlands

Judith Klinman Departments of Chemistry and Molecular and Cell Biology, California Institute for Quantitative Biosciences, University of California, Berkeley, Berkeley, CA, USA

Michihiko Kobayashi Graduate School of Life and Environmental Sciences, Institute of Applied Biochemistry, The University of Tsukuba, Tsukuba, Ibaraki, Japan

Ahmet Koc Department of Molecular Biology and Genetics, Izmir Institute of Technology, Urla, İzmir, Turkey

Sergey M. Korotkov Sechenov Institute of Evolutionary Physiology and Biochemistry, The Russian Academy of Sciences, St. Petersburg, Russia

Peter Koulen Vision Research Center and Departments of Basic Medical Science and Ophthalmology, School of Medicine, University of Missouri, Kansas City, MO, USA

Nancy F. Krebs Department of Pediatrics, Section of Nutrition, University of Colorado, School of Medicine, Aurora, CO, USA

Zbigniew Krejpcio Division of Food Toxicology and Hygiene, Department of Human Nutrition and Hygiene, The Poznan University of Life Sciences, Poznan, Poland

The College of Health, Beauty and Education in Poznan, Poznan, Poland

Robert H. Kretsinger Department of Biology, University of Virginia, Charlottesville, VA, USA

Artur Krężel Department of Protein Engineering, Faculty of Biotechnology, University of Wrocław, Wrocław, Poland

Aleksandra Krivograd Klemenčič Faculty of Health Sciences, University of Ljubljana, Ljubljana, Slovenia

Peter M. H. Kroneck Department of Biology, University of Konstanz, Konstanz, Germany

Eugene Kryachko Bogolyubov Institute for Theoretical Physics, Kiev, Ukraine

Naoko Kumagai-Takei Department of Hygiene, Kawasaki Medical School, Okayama, Japan

Anil Kumar CAS Key Laboratory for Biomedical Effects of Nanoparticles and Nanosafety, National Center for Nanoscience and Nanotechnology, Chinese Academy of Sciences, Beijing, China

Graduate University of Chinese Academy of Science, Beijing, China

Thirumananseri Kumarevel RIKEN SPring-8 Center, Harima Institute, Hyogo, Japan

Valery V. Kupriyanov Institute for Biodiagnostics, National Research Council, Winnipeg, MB, Canada

Wouter Laan School of Chemistry, University of St Andrews, St Andrews, UK

James C. K. Lai Department of Biomedical & Pharmaceutical Sciences, College of Pharmacy and Biomedical Research Institute, Idaho State University, Pocatello, ID, USA

Maria José Laires CIPER – Interdisciplinary Centre for the Study of Human Performance, Faculty of Human Kinetics, Technical University of Lisbon, Cruz Quebrada, Portugal

Kyle M. Lancaster Department of Chemistry and Chemical Biology, Cornell University, Ithaca, NY, USA

Daniel Landau Department of Pediatrics, Soroka University Medical Centre, Ben-Gurion University of the Negev, Beer Sheva, Israel

Albert Lang Department of Molecular and Cell Biology, California Institute for Quantitative Biosciences, University of California, Berkeley, Berkeley, CA, USA

Alan B. G. Lansdown Faculty of Medicine, Imperial College, London, UK

Jean-Yves Lapointe Groupe d'étude des protéines membranaires (GÉPROM) and Département de Physique, Université de Montréal, Montréal, QC, Canada

Agnete Larsen Department of Biomedicine/Pharmacology Health, Aarhus University, Aarhus, Denmark

Lawrence H. Lash Department of Pharmacology, Wayne State University School of Medicine, Detroit, MI, USA

David A. Lawrence Department of Biomedical Sciences, School of Public Health, State University of New York, Albany, NY, USA

Laboratory of Clinical and Experimental Endocrinology and Immunology, Wadsworth Center, Albany, NY, USA

Peter A. Lay School of Chemistry, University of Sydney, Sydney, NSW, Australia

Gabriela Ledesma Departamento de Química Física/IQUIR-CONICET, Facultad de Ciencias Bioquímicas y Farmacéuticas, Universidad Nacional de Rosario, Rosario, Argentina

John Lee Department of Biochemistry and Molecular Biology, University of Georgia, Athens, GA, USA

Suni Lee Department of Hygiene, Kawasaki Medical School, Okayama, Japan

Silke Leimkühler From the Institute of Biochemistry and Biology, Department of Molecular Enzymology, University of Potsdam, Potsdam, Germany

Herman Louis Lelie Department of Chemistry and Biochemistry, University of California, Los Angeles, CA, USA

David M. LeMaster New York State Department of Health, Wadsworth Center, Albany, NY, USA

Joseph Lemire Department of Chemistry and Biochemistry, Laurentian University, Sudbury, ON, Canada

Thomas A. Leonard Max F. Perutz Laboratories, Vienna, Austria

Alain Lescure UPR ARN du CNRS, Université de Strasbourg, Institut de Biologie Moléculaire et Cellulaire, Strasbourg, France

Solomon W. Leung Department of Civil & Environmental Engineering, School of Engineering, College of Science and Engineering and Biomedical Research Institute, Idaho State University, Pocatello, ID, USA

Bogdan Lev Institute for Biocomplexity and Informatics, Department of Biological Sciences, University of Calgary, Calgary, AB, Canada

Aviva Levina School of Chemistry, University of Sydney, Sydney, NSW, Australia

Huihui Li School of Chemistry and Material Science, Nanjing Normal University, Nanjing, China

Yang V. Li Department of Biomedical Sciences, Heritage College of Osteopathic Medicine, Ohio University, Athens, OH, USA

Xing-Jie Liang CAS Key Laboratory for Biomedical Effects of Nanoparticles and Nanosafety, National Center for Nanoscience and Nanotechnology, Chinese Academy of Sciences, Beijing, China

Patrycja Libako Faculty of Veterinary Medicine, Wroclaw University of Environmental and Life Sciences, Wrocław, Poland

Carmay Lim Institute of Biomedical Sciences, Academia Sinica, Taipei, Taiwan

Department of Chemistry, National Tsing Hua University, Hsinchu, Taiwan

Sara Linse Department of Biochemistry and Structural Biology, Lund University, Chemical Centre, Lund, Sweden

John D. Lipscomb Department of Biochemistry, Molecular Biology, and Biophysics, University of Minnesota, Minneapolis, MN, USA

Junqiu Liu State Key Laboratory of Supramolecular Structure and Materials, College of Chemistry, Jilin University, Changchun, China

Qiong Liu College of Life Sciences, Shenzhen University, Shenzhen, P. R. China

Zijuan Liu Department of Biological Sciences, Oakland University, Rochester, MI, USA

Marija Ljubojević Unit of Molecular Toxicology, Institute for Medical Research and Occupational Health, Zagreb, Croatia

Mario Lo Bello Department of Biology, University of Rome "Tor Vergata", Rome, Italy

Yan-Chung Lo The Genomics Research Center, Academia Sinica, Taipei, Taiwan

Institute of Biological Chemistry, Academia Sinica, Taipei, Taiwan

Lingli Lu MOE Key Laboratory of Environment Remediation and Ecological Health, College of Environmental & Resource Science, Zhejiang University, Hangzhou, China

Roberto G. Lucchini Department of Experimental and Applied Medicine, Section of Occupational Health and Industrial Hygiene, University of Brescia, Brescia, Italy

Department of Preventive Medicine, Mount Sinai School of Medicine, New York, USA

Bernd Ludwig Institute of Biochemistry, Goethe University, Frankfurt, Germany

Quan Luo State Key Laboratory of Supramolecular Structure and Materials, College of Chemistry, Jilin University, Changchun, China

Jennene A. Lyda Department of Pharmaceutical Sciences, Center for Environmental Health Sciences, University of Montana, Missoula, MT, USA

Charilaos Lygidakis Regional Health Service of Emilia Romagna, AUSL of Bologna, Bologna, Italy

Jiawei Ma Key Laboratory for Biomechanics and Mechanobiology of the Ministry of Education, School of Biological Science and Medical Engineering, Beihang University, Beijing, China

Jian Feng Ma Plant Stress Physiology Group, Institute of Plant Science and Resources, Okayama University, Kurashiki, Japan

Megumi Maeda Department of Biofunctional Chemistry, Division of Bioscience, Okayama University Graduate School of Natural Science and Technology, Okayama, Japan

Axel Magalon Laboratoire de Chimie Bactérienne (UPR9043), Institut de Microbiologie de la Méditerranée, CNRS & Aix-Marseille Université, Marseille, France

Jeanette A. Maier Department of Biomedical and Clinical Sciences L. Sacco, Università di Milano, Medical School, Milano, Italy

Robert J. Maier Department of Microbiology, The University of Georgia, Athens, GA, USA

Masatoshi Maki Department of Applied Molecular Biosciences, Graduate School of Bioagricultural Sciences, Nagoya University, Nagoya, Japan

R. Manasadeepa Department of Pharmaceutical Technology, Jadavpur University, Kolkata, West Bengal, India

David J. Mann Division of Molecular Biosciences, Department of Life Sciences, Imperial College London, South Kensington, London, UK

G. Marangi Istituto di Genetica Medica, Università Cattolica Sacro Cuore, Policlinico A. Gemelli, Rome, Italy

Wolfgang Maret Metal Metabolism Group, Diabetes and Nutritional Sciences Division, School of Medicine, King's College London, London, UK

Bernd Markert Environmental Institute of Scientific Networks, in Constitution, Haren/Erika, Germany

Michael J. Maroney Department of Chemistry, Lederle Graduate Research Center, University of Massachusetts at Amherst, Amherst, MA, USA

Brenda Marrero-Rosado Department of Pharmacology and Toxicology, Michigan State University, East Lansing, MI, USA

Christopher B. Marshall Ontario Cancer Institute and Department of Medical Biophysics, University of Toronto, Toronto, Ontario, Canada

Dwight W. Martin Department of Medicine and the Proteomics Center, Stony Brook University, Stony Brook, NY, USA

Ebany J. Martinez-Finley Department of Pediatrics, Division of Pediatric Clinical Pharmacology and Toxicology, Vanderbilt University Medical Center, Nashville, TN, USA

Center in Molecular Toxicology, Vanderbilt University Medical Center, Nashville, TN, USA

Jacqueline van Marwijk Department of Biochemistry, Microbiology & Biotechnology, Rhodes University, Grahamstown, Eastern Cape, South Africa

Pradip K. Mascharak Department of Chemistry and Biochemistry, University of California, Santa Cruz, CA, USA

Anne B. Mason Department of Biochemistry, University of Vermont, Burlington, VT, USA

Hidenori Matsuzaki Department of Hygiene, Kawasaki Medical School, Okayama, Japan

Jacqueline M. Matthews School of Molecular Bioscience, The University of Sydney, Sydney, Australia

Andrzej Mazur INRA, UMR 1019, UNH, CRNH Auvergne, Clermont Université, Université d'Auvergne, Unité de Nutrition Humaine, Clermont-Ferrand, France

Paulo Mazzafera Departamento de Biologia Vegetal, Universidade Estadual de Campinas/Instituto de Biologia, Cidade Universitária, Campinas, SP, Brazil

Michael M. Mbughuni Department of Biochemistry, Molecular Biology, and Biophysics, University of Minnesota, Minneapolis, MN, USA

Joseph R. McDermott Department of Biological Sciences, Oakland University, Rochester, MI, USA

Megan M. McEvoy Department of Chemistry and Biochemistry, University of Arizona, Tucson, AZ, USA

Astrid van der Meer Interfaculty Reactor Institute, Delft University of Technology, Delft, The Netherlands

Petr Melnikov Department of Clinical Surgery, School of Medicine, Federal University of Mato Grosso do Sul, Campo Grande, MS, Brazil

Gabriele Meloni Division of Chemistry and Chemical Engineering and Howard Hughes Medical Institute, California Institute of Technology, Pasadena, CA, USA

Ralf R. Mendel Department of Plant Biology, Braunschweig University of Technology, Braunschweig, Germany

Mohamed Larbi Merroun Departamento de Microbiología, Facultad de Ciencias, Universidad de Granada, Granada, Spain

Albrecht Messerschmidt Department of Proteomics and Signal Transduction, Max-Planck-Institute of Biochemistry, Martinsried, Germany

Marek Michalak Department of Biochemistry, University of Alberta, Edmonton, AB, Canada

Faculty of Medicine and Dentistry, University of Alberta, Edmonton, AB, Canada

Isabelle Michaud-Soret iRTSV/LCBM UMR 5249 CEA-CNRS-UJF, CEA/Grenoble, Bât K, Université Grenoble, Grenoble, France

Radmila Milačič Department of Environmental Sciences, Jožef Stefan Institute, Ljubljana, Slovenia

Glenn L. Millhauser Department of Chemistry and Biochemistry, University of California, Santa Cruz, Santa Cruz, CA, USA

Marija Milovanovic Faculty of Medical Sciences, University of Kragujevac, Centre for Molecular Medicine, Kragujevac, Serbia

Shin Mizukami Division of Advanced Science and Biotechnology, Graduate School of Engineering, Osaka University, Suita, Osaka, Japan

Immunology Frontier Research Center, Osaka University, Suita, Osaka, Japan

Cristina Paula Monteiro Physiology and Biochemistry Laboratory, Faculty of Human Kinetics, Technical University of Lisbon, Cruz Quebrada, Portugal

Augusto C. Montezano Kidney Research Center, Ottawa Hospital Research Institute, University of Ottawa, Ottawa, ON, Canada

Pablo Morales-Rico Department of Toxicology, Cinvestav-IPN, Mexico city, Mexico

J. Preben Morth Centre for Molecular Medicine Norway (NCMM), Nordic EMBL Partnership, University of Oslo, Oslo, Norway

Jean-Marc Moulis Institut de Recherches en Sciences et Technologies du Vivant, Laboratoire Chimie et Biologie des Métaux (IRTSV/LCBM), CEA–Grenoble, Grenoble, France

CNRS, UMR5249, Grenoble, France

Université Joseph Fourier–Grenoble I, UMR5249, Grenoble, France

Isabel Moura REQUIMTE/CQFB, Departamento de Química, Faculdade de Ciências e Tecnologia, Universidade Nova de Lisboa, Caparica, Portugal

José J. G. Moura REQUIMTE/CQFB, Departamento de Química, Faculdade de Ciências e Tecnologia, Universidade Nova de Lisboa, Caparica, Portugal

Mohamed E. Moustafa Department of Biochemistry, Faculty of Science, Alexandria University, Alexandria, Egypt

Amitava Mukherjee Centre for Nanobiotechnology, VIT University, Vellore, Tamil Nadu, India

Biswajit Mukherjee Department of Pharmaceutical Technology, Jadavpur University, Kolkata, West Bengal, India

Balam Muñoz Department of Toxicology, Cinvestav-IPN, Mexico city, Mexico

Francesco Musiani Department of Agro-Environmental Science and Technology, University of Bologna, Bologna, Italy

Joachim Mutter Naturheilkunde, Umweltmedizin Integrative and Environmental Medicine, Belegarzt Tagesklinik, Constance, Germany

Bonex W. Mwakikunga Council for Scientific and Industrial Research, National Centre for Nano–Structured, Pretoria, South Africa

Department of Physics and Biochemical Sciences, University of Malawi, The Malawi Polytechnic, Chichiri, Blantyre, Malawi

Chandra Shekar Nagar Venkataraman Condensed Matter Physics Division, Materials Science Group, Indira Gandhi Centre for Atomic Research, Kalpakkam, Tamil Nadu, India

Hideaki Nagase Kennedy Institute of Rheumatology, Nuffield Department of Orthopaedics, Rheumatology and Musculoskeletal Sciences, University of Oxford, London, United Kingdom

Sreejayan Nair University of Wyoming, School of Pharmacy, College of Health Sciences and the Center for Cardiovascular Research and Alternative Medicine, Laramie, WY, USA

Manuel F. Navedo Department of Physiology and Biophysics, University of Washington, Seattle, WA, USA

Tim S. Nawrot Centre for Environmental Sciences, Hasselt University, Diepenbeek, Belgium

Karel Nesmerak Department of Analytical Chemistry, Faculty of Science, Charles University in Prague, Prague, Czech Republic

Gerd Ulrich Nienhaus Institute of Applied Physics and Center for Functional Nanostructures (CFN), Karlsruhe Institute of Technology (KIT), Karlsruhe, Germany

Department of Physics, University of Illinois at Urbana–Champaign, Urbana, IL, USA

Crina M. Nimigean Department of Anesthesiology, Weill Cornell Medical College, New York, NY, USA

Department of Physiology and Biophysics, Weill Cornell Medical College, New York, NY, USA

Department of Biochemistry, Weill Cornell Medical College, New York, NY, USA

Yasumitsu Nishimura Department of Hygiene, Kawasaki Medical School, Okayama, Japan

Naoko K. Nishizawa Graduate School of Agricultural and Life Sciences, The University of Tokyo, Tokyo, Japan

Research Institute for Bioresources and Biotechnology, Ishikawa Prefectural University, Ishikawa, Japan

Valerio Nobili Unit of Liver Research of Bambino Gesù Children's Hospital, IRCCS, Rome, Italy

Nicholas Noinaj Laboratory of Molecular Biology, National Institute of Diabetes and Digestive and Kidney Diseases, US National Institutes of Health, Bethesda, MD, USA

Aline M. Nonat Laboratoire d'Ingénierie Moléculaire Appliquée à l'Analyse, IPHC, UMR 7178 CNRS/UdS ECPM, Strasbourg, France

Sergei Yu. Noskov Institute for Biocomplexity and Informatics, Department of Biological Sciences, University of Calgary, Calgary, AB, Canada

Wojciech Nowacki Faculty of Veterinary Medicine, Wroclaw University of Environmental and Life Sciences, Wrocław, Poland

David O'Connell University College Dublin, Conway Institute, Dublin, Ireland

Masafumi Odaka Department of Biotechnology and Life Science, Graduate School of Technology, Tokyo University of Agriculture and Technology, Koganei, Tokyo, Japan

Akira Ono Department of Material & Life Chemistry, Faculty of Engineering, Kanagawa University, Kanagawa-ku, Yokohama, Japan

Laura Osorio-Rico Departamento de Neuroquímica, Instituto Nacional de Neurología y Neurocirugía Manuel Velasco Suárez, Mexico City, DF, Mexico

Patricia Isabel Oteiza Departments of Nutrition and Environmental Toxicology, University of California, Davis, Davis, CA, USA

Takemi Otsuki Department of Hygiene, Kawasaki Medical School, Okayama, Japan

Rabbab Oun Strathclyde Institute of Pharmacy and Biomedical Sciences, University of Strathclyde, Glasgow, UK

Vidhu Pachauri Division of Pharmacology and Toxicology, Defence Research and Development Establishment, Gwalior, India

Òscar Palacios Departament de Química, Facultat de Ciències, Universitat Autònoma de Barcelona, Cerdanyola del Vallès (Barcelona), Spain

Maria E. Palm-Espling Department of Chemistry, Chemical Biological Center, Umeå University, Umeå, Sweden

Claudia Palopoli Departamento de Química Física/IQUIR-CONICET, Facultad de Ciencias Bioquímicas y Farmacéuticas, Universidad Nacional de Rosario, Rosario, Argentina

Tapobrata Panda Biochemical Engineering Laboratory, Department of Chemical Engineering, Indian Institute of Technology Madras, Chennai, Tamil Nadu, India

Lorien J. Parker Biota Structural Biology Laboratory, St. Vincent's Institute of Medical Research, Fitzroy, VIC, Australia

Department of Biochemistry and Molecular Biology, Bio21 Molecular Science and Biotechnology Institute, The University of Melbourne, Parkville, VIC, Australia

Michael W. Parker Biota Structural Biology Laboratory, St. Vincent's Institute of Medical Research, Fitzroy, VIC, Australia

Department of Biochemistry and Molecular Biology, Bio21 Molecular Science and Biotechnology Institute, The University of Melbourne, Parkville, VIC, Australia

Marianna Patrauchan Department of Microbiology and Molecular Genetics, College of Arts and Sciences, Oklahoma State University, Stillwater, OK, USA

Sofia R. Pauleta REQUIMTE/CQFB, Departamento de Química, Faculdade de Ciências e Tecnologia, Universidade Nova de Lisboa, Caparica, Portugal

Evgeny Pavlov Department of Physiology & Biophysics, Faculty of Medicine, Dalhousie University, Halifax, NS, Canada

V. Pennemans Biomedical Institute, Hasselt University, Diepenbeek, Belgium

Harmonie Perdreau Centre for Molecular Medicine Norway (NCMM), Nordic EMBL Partnership, University of Oslo, Oslo, Norway

Alice S. Pereira Departamento de Química, Faculdade de Ciências e Tecnologia, Requimte, Centro de Química Fina e Biotecnologia, Universidade Nova de Lisboa, Caparica, Portugal

Eulália Pereira REQUIMTE, Departamento de Química e Bioquímica, Faculdade de Ciências da Universidade do Porto, Porto, Portugal

Eugene A. Permyakov Institute for Biological Instrumentation, Russian Academy of Sciences, Pushchino, Moscow Region, Russia

Sergei E. Permyakov Protein Research Group, Institute for Biological Instrumentation of the Russian Academy of Sciences, Pushchino, Moscow Region, Russia

Bertil R. R. Persson Department of Medical Radiation Physics, Lund University, Lund, Sweden

John W. Peters Department of Chemistry and Biochemistry, Montana State University, Bozeman, MT, USA

Marijana Petković Laboratory of Physical Chemistry, Vinča Institute of Nuclear Sciences, University of Belgrade, Belgrade, Serbia

Le T. Phung Department of Microbiology and Immunology, University of Illinois, Chicago, IL, USA

Roberta Pierattelli CERM and Department of Chemistry "Ugo Schiff", University of Florence, Sesto Fiorentino, Italy

Elizabeth Pierce Department of Biological Chemistry, University of Michigan Medical School, Ann Arbor, MI, USA

Andrea Pietrobattista Unit of Liver Research of Bambino Gesù Children's Hospital, IRCCS, Rome, Italy

Thomas C. Pochapsky Department of Chemistry, Rosenstiel Basic Medical Sciences Research Center, Brandeis University, Waltham, MA, USA

Ehmke Pohl Biophysical Sciences Institute, Department of Chemistry, School of Biological and Biomedical Sciences, Durham University, Durham, UK

Joe C. Polacco Department of Biochemistry/Interdisciplinary Plant Group, University of Missouri, Columbia, MO, USA

Arthur S. Polans Department of Ophthalmology and Visual Sciences, UW Eye Research Institute, University of Wisconsin, Madison, WI, USA

K. Michael Pollard Department of Molecular and Experimental Medicine, The Scripps Research Institute, La Jolla, CA, USA

Charlotte Poschenrieder Lab. Fisiología Vegetal, Facultad de Biociencias, Universidad Autónoma de Barcelona, Bellaterra, Spain

Thomas L. Poulos Department of Biochemistry & Molecular Biology, Pharmaceutical Science, and Chemistry, University of California, Irivine, Irvine, CA, USA

Richard D. Powell Nanoprobes, Incorporated, Yaphank, NY, USA

Ananda S. Prasad Department of Oncology, Karmanos Cancer Center, Wayne State University, School of Medicine, Detroit, MI, USA

Walter C. Prozialeck Department of Pharmacology, Midwestern University, Downers Grove, IL, USA

Qin Qin Wise Laboratory of Environmental and Genetic Toxicology, Maine Center for Toxicology and Environmental Health, Department of Applied Medical Sciences, University of Southern Maine, Portland, ME, USA

Thierry Rabilloud CNRS, UMR 5249 Laboratory of Chemistry and Biology of Metals, Grenoble, France

CEA, DSV, iRTSV/LCBM, Chemistry and Biology of Metals, Grenoble Cedex 9, France

Université Joseph Fourier, Grenoble, France

Stephen W. Ragsdale Department of Biological Chemistry, University of Michigan Medical School, Ann Arbor, MI, USA

Frank M. Raushel Department of Chemistry, Texas A&M University, College Station, TX, USA

Frank Reith School of Earth and Environmental Sciences, The University of Adelaide, Centre of Tectonics, Resources and Exploration (TRaX) Adelaide, Urrbrae, South Australia, Australia

CSIRO Land and Water, Environmental Biogeochemistry, PMB2, Glen Osmond, Urrbrae, South Australia, Australia

Tony Remans Centre for Environmental Sciences, Hasselt University, Diepenbeek, Belgium

Albert W. Rettenmeier Institute of Hygiene and Occupational Medicine, University of Duisburg-Essen, Essen, Germany

Rita Rezzani Division of Human Anatomy, Department of Biomedical Sciences and Biotechnologies, Brescia University, Brescia, Italy

Marius Réglier Faculté des Sciences et Techniques, ISM2/BiosCiences UMR CNRS 7313, Aix-Marseille Université Campus Scientifique de Saint Jérôme, Marseille, France

Oliver-M. H. Richter Institute of Biochemistry, Goethe University, Frankfurt, Germany

Agnes Rinaldo-Matthis Department of Medical Biochemistry and Biophysics (MBB), Karolinska Institute, Stockholm, Sweden

Lothar Rink Institute of Immunology, Medical Faculty, RWTH Aachen University, Aachen, Germany

Alfonso Rios-Perez Department of Toxicology, Cinvestav-IPN, Mexico city, Mexico

Rasmus Sejersten Ripa Cardiology Stem Cell Laboratory, Rigshospitalet University Hospital, Copenhagen, Denmark

Cluster for Molecular Imaging and Department of Clinical Physiology, Nuclear Medicine and PET, Rigshospitalet University Hospital, Copenhagen, Denmark

Marwan S. Rizk Deptartment of Anesthesiology, American University of Beirut Medical Center, Beirut, Lebanon

Nigel J. Robinson Biophysical Sciences Institute, Department of Chemistry, School of Biological and Biomedical Sciences, Durham University, Durham, UK

João B. T. Rocha Departamento de Química, Centro de Ciências Naturais e Exatas, Universidade Federal de Santa Maria, Santa Maria, RS, Brazil

Juan C. Rodriguez-Ubis Dept. Química Orgánica, Facultad de Ciencias, Universidad Autónoma de Madrid, Madrid, Spain

Harry A. Roels Louvain Centre for Toxicology and Applied Pharmacology, Université catholique de Louvain, Brussels, Belgium

Andrea M. P. Romani Department of Physiology and Biophysics, School of Medicine, Case Western Reserve University, Cleveland, OH, USA

S. Rosato Struttura Semplice Dipartimentale di Genetica Clinica, Dipartimento di Ostetrico-Ginecologico e Pediatrico, Istituto di Ricovero e Cura a Carattere Scientifico, Arcispedale S. Maria Nuova, Reggio Emilia, Italy

Barry P. Rosen Department of Cellular Biology and Pharmacology, Florida International University, Herbert Wertheim College of Medicine, Miami, FL, USA

Erwin Rosenberg Institute of Chemical Technologies and Analytics, Vienna University of Technology, Vienna, Austria

Amy C. Rosenzweig Departments of Molecular Biosciences and of Chemistry, Northwestern University, Evanston, IL, USA

Michael Rother Institut für Mikrobiologie, Technische Universität Dresden, Dresden, Germany

Benoît Roux Department of Pediatrics, Biochemistry and Molecular Biology, The University of Chicago, Chicago, IL, USA

Namita Roy Choudhury Ian Wark Research Institute, University of South Australia, Mawson Lakes, South Australia, Australia

Jagoree Roy Department of Biology, Stanford University, Stanford, USA

Kaushik Roy Division of Biochemistry, Department of Pharmaceutical Technology, Jadavpur University, Kolkata, West Bengal, India

Marian Rucki Centre of Occupational Health, Laboratory of Predictive Toxicology, National Institute of Public Health, Praha 10, Czech Republic

Anandamoy Rudra Department of Pharmaceutical Technology, Jadavpur University, Kolkata, West Bengal, India

Giuseppe Ruggiero National Research Council (CNR), Laboratory for Molecular Sensing, Institute of Protein Biochemistry, Naples, Italy

Kelly C. Ryan Department of Chemistry, Lederle Graduate Research Center, University of Massachusetts at Amherst, Amherst, MA, USA

Lisa K. Ryan New Jersey Medical School, The Public Health Research Institute, University of Medicine and Dentistry of New Jersey, Newark, NJ, USA

Janusz K. Rybakowski Department of Adult Psychiatry, Poznan University of Medical Sciences, Poznan, Poland

Ivan Sabolić Unit of Molecular Toxicology, Institute for Medical Research and Occupational Health, Zagreb, Croatia

Kalyan K. Sadhu Division of Advanced Science and Biotechnology, Graduate School of Engineering, Osaka University, Suita, Osaka, Japan

Anita Sahu Institute of Pathology and Center of Medical Research, Medical University of Graz, Graz, Austria

P. Ch. Sahu Condensed Matter Physics Division, Materials Science Group, Indira Gandhi Centre for Atomic Research, Kalpakkam, Tamil Nadu, India

Milton H. Saier Jr. Department of Molecular Biology, Division of Biological Sciences, University of California at San Diego, La Jolla, CA, USA

Jarno Salonen Laboratory of Industrial Physics, Department of Physics, University of Turku, Turku, Finland

Abel Santamaría Laboratorio de Aminoácidos Excitadores, Instituto Nacional de Neurología y Neurocirugía Manuel Velasco Suárez, Mexico City, DF, Mexico

Luis F. Santana Department of Physiology and Biophysics, University of Washington, Seattle, WA, USA

Hélder A. Santos Division of Pharmaceutical Technology, University of Helsinki, Helsinki, Finland

Dibyendu Sarkar Earth and Environmental Studies Department, Montclair State University, Montclair, NJ, USA

Louis J. Sasseville Groupe d'étude des protéines membranaires (GÉPROM) and Département de Physique, Université de Montréal, Montréal, QC, Canada

R. Gary Sawers Institute for Microbiology, Martin-Luther University Halle-Wittenberg, Halle (Saale), Germany

Janez Ščančar Department of Environmental Sciences, Jožef Stefan Institute, Ljubljana, Slovenia

Marcus C. Schaub Institute of Pharmacology and Toxicology, University of Zurich, Zurich, Switzerland

Sara Schmitt The Developmental Therapeutics Program, Barbara Ann Karmanos Cancer Institute, and Departments of Oncology, Pharmacology and Pathology, School of Medicine, Wayne State University, Detroit, MI, USA

Paul P. M. Schnetkamp Department of Physiology & Pharmacology, Hotchkiss Brain Institute, University of Calgary, Calgary, AB, Canada

Lutz Schomburg Institute for Experimental Endocrinology, Charité – University Medicine Berlin, Berlin, Germany

Gerhard N. Schrauzer Department of Chemistry and Biochemistry, University of California, San Diego, La Jolla, CA, USA

Ruth Schreiber Department of Pediatrics, Soroka University Medical Centre, Ben-Gurion University of the Negev, Beer Sheva, Israel

Ulrich Schweizer Institute for Experimental Endocrinology, Charité-Universitätsmedizin Berlin, Berlin, Germany

Ion Romulus Scorei Department of Biochemistry, University of Craiova, Craiova, DJ, Romania

Lucia A. Seale Department of Cell & Molecular Biology, John A. Burns School of Medicine, University of Hawaii at Manoa, Honolulu, HI, USA

Lance C. Seefeldt Department of Chemistry and Biochemistry, Utah State University, Logan, UT, USA

William Self Molecular Biology & Microbiology, Burnett School of Biomedical Sciences, University of Central Florida, Orlando, FL, USA

Takashi Sera Department of Applied Chemistry and Biotechnology, Graduate School of Natural Science and Technology, Okayama University, Okayama, Japan

Aruna Sharma Laboratory of Cerebrovascular Research, Department of Surgical Sciences, Anesthesiology & Intensive Care medicine, University Hospital, Uppsala University, Uppsala, Sweden

Hari Shanker Sharma Laboratory of Cerebrovascular Research, Department of Surgical Sciences, Anesthesiology & Intensive Care medicine, University Hospital, Uppsala University, Uppsala, Sweden

Honglian Shi Department of Pharmacology and Toxicology, University of Kansas, Lawrence, KS, USA

Xianglin Shi Graduate Center for Toxicology, University of Kentucky, Lexington, KY, USA

Satoshi Shinoda JST, CREST, and Department of Chemistry, Graduate School of Science, Osaka City University, Sumiyoshi-ku, Osaka, Japan

Maksim A. Shlykov Department of Molecular Biology, Division of Biological Sciences, University of California at San Diego, La Jolla, CA, USA

Siddhartha Shrivastava Rensselaer Nanotechnology Center, Rensselaer Polytechnic Institute, Troy, NY, USA

Center for Biotechnology and Interdisciplinary Studies, Rensselaer Polytechnic Institute, Troy, NY, USA

Sandra Signorella Departamento de Química Física/IQUIR-CONICET, Facultad de Ciencias Bioquímicas y Farmacéuticas, Universidad Nacional de Rosario, Rosario, Argentina

Amrita Sil Department of Pharmacology, Burdwan Medical College, Burdwan, West Bengal, India

Radu Silaghi-Dumitrescu Department of Chemistry and Chemical Engineering, Babes-Bolyai University, Cluj-Napoca, Romania

Simon Silver Department of Microbiology and Immunology, University of Illinois, Chicago, IL, USA

Britt-Marie Sjöberg Department of Biochemistry and Biophysics, Stockholm University, Stockholm, SE, Sweden

Karen Smeets Centre for Environmental Sciences, Hasselt University, Diepenbeek, Belgium

Stephen M. Smith Departments of Molecular Biosciences and of Chemistry, Northwestern University, Evanston, IL, USA

Małgorzata Sobieszczańska Department of Pathophysiology, Wroclaw Medical University, Wroclaw, Poland

Young-Ok Son Graduate Center for Toxicology, University of Kentucky, Lexington, KY, USA

Martha E. Sosa Torres Facultad de Quimica, Universidad Nacional Autonoma de Mexico, Ciudad Universitaria, Coyoacan, Mexico DF, Mexico

Jerry W. Spears Department of Animal Science, North Carolina State University, Raleigh, NC, USA

Sarah R. Spell Department of Chemistry, Virginia Commonwealth University, Richmond, VA, USA

Christopher D. Spicer Chemistry Research Laboratory, Department of Chemistry, University of Oxford, Oxford, UK

St. Hilda's College, University of Oxford, Oxford, UK

Alessandra Stacchiotti Division of Human Anatomy, Department of Biomedical Sciences and Biotechnologies, Brescia University, Brescia, Italy

Jan A. Staessen Study Coordinating Centre, Department of Cardiovascular Diseases, KU Leuven, Leuven, Belgium

Unit of Epidemiology, Maastricht University, Maastricht, The Netherlands

Maria Staiano National Research Council (CNR), Laboratory for Molecular Sensing, Institute of Protein Biochemistry, Naples, Italy

Anna Starus Department of Chemistry and Biochemistry, Loyola University Chicago, Chicago, IL, USA

Alexander Stein Hubertus Wald Tumor Center, University Cancer Center Hamburg (UCCH), University Hospital Hamburg-Eppendorf (UKE), Hamburg, Germany

Iryna N. Stepanenko Institute of Inorganic Chemistry, University of Vienna, Vienna, Austria

Martin J. Stillman Department of Biology, The University of Western Ontario, London, ON, Canada

Department of Chemistry, The University of Western Ontario, London, ON, Canada

Walter Stöcker Johannes Gutenberg University Mainz, Institute of Zoology, Cell and Matrix Biology, Mainz, Germany

Barbara J. Stoecker Department of Nutritional Sciences, Oklahoma State University, Stillwater, OK, USA

Edward D. Sturrock Zinc Metalloprotease Research Group, Division of Medical Biochemistry, Institute of Infectious Disease and Molecular Medicine, University of Cape Town, Cape Town, South Africa

Minako Sumita Department of Biochemistry and Molecular Biology, Michigan State University, East Lansing, MI, USA

Kelly L. Summers Department of Biology, The University of Western Ontario, London, ON, Canada

Raymond Wai-Yin Sun Department of Chemistry, State Key Laboratory of Synthetic Chemistry and Open Laboratory of Chemical Biology of the Institute of Molecular Technology for Drug Discovery and Synthesis, The University of Hong Kong, Hong Kong, China

Claudiu T. Supuran Department of Chemistry, University of Florence, Sesto Fiorentino (Florence), Italy

Hiroshi Suzuki Department of Biochemistry, Asahikawa Medical University, Asahikawa, Hokkaido, Japan

Q. Swennen Biomedical Institute, Hasselt University, Diepenbeek, Belgium

Ingebrigt Sylte Medical Pharmacology and Toxicology, Department of Medical Biology, University of Tromsø, Tromsø, Norway

Yoshiyuki Tanaka Graduate School of Pharmaceutical Sciences Tohoku University, Sendai, Miyagi, Japan

Shen Tang Department of Chemistry, Georgia State University, Atlanta, GA, USA

Akio Tani Research Institute of Plant Science and Resources, Okayama University, Kurashiki, Okayama, Japan

Pedro Tavares Departamento de Química, Faculdade de Ciências e Tecnologia, Requimte, Centro de Química Fina e Biotecnologia, Universidade Nova de Lisboa, Caparica, Portugal

Jan Willem Cohen Tervaert Clinical and Experimental Immunology, Maastricht University, Maastricht, The Netherlands

Tiago Tezotto Departamento de Produção Vegetal, Universidade de São Paulo/ Escola Superior de Agricultura Luiz de Queiroz, Piracicaba, SP, Brazil

Elizabeth C. Theil Children's Hospital Oakland Research Institute, Oakland, CA, USA

Department of Molecular and Structural Biochemistry, North Carolina State University, Raleigh, NC, USA

Frank Thévenod Faculty of Health, School of Medicine, Centre for Biomedical Training and Research (ZBAF), Institute of Physiology & Pathophysiology, University of Witten/Herdecke, Witten, Germany

David J. Thomas Pharmacokinetics Branch – Integrated Systems Toxicology Division, National Health and Environmental Research Laboratory, U.S. Environmental Protection Agency, Research Triangle Park, NC, USA

Ameer N. Thompson Department of Anesthesiology, Weill Cornell Medical College, New York, NY, USA

Department of Physiology and Biophysics, Weill Cornell Medical College, New York, NY, USA

Rüdiger Thul School of Mathematical Sciences, University of Nottingham, Nottingham, UK

Jacob P. Thyssen National Allergy Research Centre, Department of Dermato-Allergology, Copenhagen University Hospital Gentofte, Hellerup, Denmark

Milon Tichy Centre of Occupational Health, Laboratory of Predictive Toxicology, National Institute of Public Health, Praha 10, Czech Republic

Dajena Tomco Department of Chemistry, Wayne State University, Detroit, MI, USA

Hidetaka Torigoe Department of Applied Chemistry, Faculty of Science, Tokyo University of Science, Tokyo, Japan

Rhian M. Touyz Kidney Research Center, Ottawa Hospital Research Institute, University of Ottawa, Ottawa, ON, Canada

Institute of Cardiovascular & Medical Sciences, BHF Glasgow Cardiovascular Research Centre, University of Glasgow, Glasgow, UK

Chikashi Toyoshima Institute of Molecular and Cellular Biosciences, The University of Tokyo, Tokyo, Japan

Lennart Treuel Institute of Applied Physics and Center for Functional Nanostructures (CFN), Karlsruhe Institute of Technology (KIT), Karlsruhe, Germany

Institute of Physical Chemistry, University of Duisburg–Essen, Essen, Germany

Shweta Trivedi Department of Animal Science, North Carolina State University, Raleigh, NC, USA

Thierry Tron iSm2/BiosCiences UMR CNRS 7313, Case 342, Aix-Marseille Université, Marseille, France

Chin-Hsiao Tseng Department of Internal Medicine, National Taiwan University College of Medicine, Taipei, Taiwan

Division of Endocrinology and Metabolism, Department of Internal Medicine, National Taiwan University Hospital, Taipei, Taiwan

Tsai-Tien Tseng Center for Cancer Research and Therapeutic Development, Clark Atlanta University, Atlanta, GA, USA

Samantha D. Tsotsoros Department of Chemistry, Virginia Commonwealth University, Richmond, VA, USA

Petra A. Tsuji Department of Biological Sciences, Towson University, Towson, MD, USA

Hiroshi Tsukube JST, CREST, and Department of Chemistry, Graduate School of Science, Osaka City University, Sumiyoshi-ku, Osaka, Japan

Sławomir Tubek Institute of Technology, Opole, Poland

Raymond J. Turner Department of Biological Sciences, University of Calgary, Calgary, AB, Canada

Toshiki Uchihara Laboratory of Structural Neuropathology, Tokyo Metropolitan Institute of Medical Science, Tokyo, Japan

Takafumi Ueno Department of Biomolecular Engineering, Graduate School of Bioscience and Biotechnology, Tokyo Institute of Technology, Yokohama, Japan

Christoph Ufer Institute of Biochemistry, Charité – Universitätsmedizin Berlin, Berlin, Germany

İrem Uluisik Department of Molecular Biology and Genetics, Izmir Institute of Technology, Urla, İzmir, Turkey

Balachandran Unni Nair Chemical Laboratory, Central Leather Research Institute (Council of Scientific and Industrial Research), Chennai, Tamil Nadu, India

Vladimir N. Uversky Department of Molecular Medicine, University of South Florida, College of Medicine, Tampa, FL, USA

Joan Selverstone Valentine Department of Chemistry and Biochemistry, University of California, Los Angeles, CA, USA

Marian Valko Department of Chemistry, Faculty of Natural Sciences, Constantine The Philosopher University, Nitra, Slovakia

Faculty of Chemical and Food Technology, Slovak Technical University, Bratislava, Slovakia

J. David van Horn Department of Chemistry, University of Missouri-Kansas City, Kansas City, MO, USA

Frans van Roy Department for Molecular Biomedical Research, VIB, Ghent, Belgium

Department of Biomedical Molecular Biology, Ghent University, Ghent, Belgium

Marie Vancová Institute of Parasitology, Biology Centre of the Academy of Sciences of the Czech Republic and University of South Bohemia, České Budějovice, Czech Republic

Jaco Vangronsveld Centre for Environmental Sciences, Hasselt University, Diepenbeek, Belgium

Antonio Varriale National Research Council (CNR), Laboratory for Molecular Sensing, Institute of Protein Biochemistry, Naples, Italy

Milan Vašák Department of Inorganic Chemistry, University of Zürich, Zürich, Switzerland

Claudio C. Vásquez Laboratorio de Microbiología Molecular, Departamento de Biología, Universidad de Santiago de Chile, Santiago, Chile

Oscar Vassallo Department of Biology, University of Rome Tor Vergata, Rome, Italy

Claudio N. Verani Department of Chemistry, Wayne State University, Detroit, MI, USA

Nathalie Verbruggen Laboratory of Plant Physiology and Molecular Genetics, Université Libre de Bruxelles, Brussels, Belgium

Sandra Viviana Verstraeten Department of Biological Chemistry, IQUIFIB (UBA-CONICET), School of Pharmacy and Biochemistry, University of Buenos Aires, Argentina, Buenos Aires, Argentina

Ramon Vilar Department of Chemistry, Imperial College London, South Kensington, London, UK

John B. Vincent Department of Chemistry, The University of Alabama, Tuscaloosa, AL, USA

Hans J. Vogel Department of Biological Sciences, Biochemistry Research Group, University of Calgary, Calgary, AB, Canada

Vladislav Volarevic Faculty of Medical Sciences, University of Kragujevac, Centre for Molecular Medicine, Kragujevac, Serbia

Anne Volbeda Metalloproteins; Institut de Biologie Structurale J.P. Ebel; CEA; CNRS; Université J. Fourier, Grenoble, France

Eugene S. Vysotski Photobiology Laboratory, Institute of Biophysics Russian Academy of Sciences, Siberian Branch, Krasnoyarsk, Russia

Anne Walburger Laboratoire de Chimie Bactérienne (UPR9043), Institut de Microbiologie de la Méditerranée, CNRS & Aix-Marseille Université, Marseille, France

Andrew H.-J. Wang Institute of Biological Chemistry, Academia Sinica, Taipei, Taiwan

Jiangxue Wang Key Laboratory for Biomechanics and Mechanobiology of the Ministry of Education, School of Biological Science and Medical Engineering, Beihang University, Beijing, China

Xudong Wang Department of Pathology, St Vincent Hospital, Worcester, MA, USA

John Wataha Department of Restorative Dentistry, University of Washington HSC D779A, School of Dentistry, Seattle, WA, USA

David J. Weber Department of Biochemistry and Molecular Biology, University of Maryland School of Medicine, Baltimore, MD, USA

Nial J. Wheate Faculty of Pharmacy, The University of Sydney, Sydney, NSW, Australia

Chris G. Whiteley Graduate Institute of Applied Science and Technology, National Taiwan University of Science and Technology, Taipei, Taiwan

Roger L. Williams Medical Research Council, Laboratory of Molecular Biology, Cambridge, UK

Judith Winogrodzki Department of Microbiology, University of Manitoba, Winnipeg, MB, Canada

John Pierce Wise Sr. Wise Laboratory of Environmental and Genetic Toxicology, Maine Center for Toxicology and Environmental Health, Department of Applied Medical Sciences, University of Southern Maine, Portland, ME, USA

Pernilla Wittung-Stafshede Department of Chemistry, Chemical Biological Center, Umeå University, Umeå, Sweden

Bert Wolterbeek Interfaculty Reactor Institute, Delft University of Technology, Delft, The Netherlands

Simone Wünschmann Environmental Institute of Scientific Networks in Constitution, Haren/Erika, Germany

Robert Wysocki Institute of Experimental Biology, University of Wroclaw, Wroclaw, Poland

Shenghui Xue Department of Biology, Georgia State University, Atlanta, GA, USA

Xiao-Jing Yan Department of Hematology, The First Hospital of China Medical University, Shenyang, China

Xiaodi Yang School of Chemistry and Material Science, Nanjing Normal University, Nanjing, China

Jenny J. Yang Department of Chemistry, Georgia State University, Atlanta, GA, USA

Natural Science Center, Atlanta, GA, USA

Vladimir Yarov-Yarovoy Department of Physiology and Membrane Biology, Department of Biochemistry and Molecular Medicine, School of Medicine, University of California, Davis, CA, USA

Katsuhiko Yokoi Department of Human Nutrition, Seitoku University Graduate School, Matsudo, Chiba, Japan

Vincenzo Zagà Department of Territorial Pneumotisiology, Italian Society of Tobaccology (SITAB), AUSL of Bologna, Bologna, Italy

Carla M. Zammit School of Earth and Environmental Sciences, The University of Adelaide, Centre of Tectonics, Resources and Exploration (TRaX) Adelaide, Urrbrae, South Australia, Australia

CSIRO Land and Water, Environmental Biogeochemistry, PMB2, Glen Osmond, Urrbrae, South Australia, Australia

Lourdes Zélia Zanoni Department of Pediatrics, School of Medicine, Federal University of Mato Grosso do Sul, Campo Grande, MS, Brazil

Huawei Zeng United States Department of Agriculture, Agricultural Research Service, Grand Forks Human Nutrition Research Center, Grand Forks, ND, USA

Cunxian Zhang Department of Pathology, Women & Infants Hospital of Rhode Island, Kent Memorial Hospital, Warren Alpert Medical School of Brown University, Providence, RI, USA

Chunfeng Zhao Institute for Biocomplexity and Informatics and Department of Biological Sciences, University of Calgary, Calgary, AB, Canada

Anatoly Zhitkovich Department of Pathology and Laboratory Medicine, Brown University, Providence, RI, USA

Boris S. Zhorov Department of Biochemistry and Biomedical Sciences, McMaster University, Hamilton, ON, Canada

Sechenov Institute of Evolutionary Physiology and Biochemistry, Russian Academy of Sciences, St. Petersburg, Russia

Yubin Zhou Department of Chemistry, Georgia State University, Atlanta, GA, USA

Division of Signaling and Gene Expression, La Jolla Institute for Allergy and Immunology, La Jolla, CA, USA

Michael X. Zhu Department of Integrative Biology and Pharmacology, The University of Texas Health Science Center at Houston, Houston, TX, USA

Marcella Zollino Istituto di Genetica Medica, Università Cattolica Sacro Cuore, Policlinico A. Gemelli, Rome, Italy

D

$d(G_2T_2G_2TGTG_2T_2G_2)$

▶ Strontium and DNA Aptamer Folding

D1-3 (Widely Used, but May Be Confused with Dopamine D1 (D2, D3)-Receptors)

▶ Selenium and Iodothyronine Deiodinases

DapE, DapE-Encoded N-Succinyl-L, L-Diaminopimelic Acid Desuccinylase

▶ Zinc Aminopeptidases, Aminopeptidase from Vibrio Proteolyticus (Aeromonas proteolytica) as Prototypical Enzyme

Darinaparsin

▶ Arsenic in Therapy

Deferoxamine, Desferrioxamine B, Desferoxamine B, DFO-B, DFOA, DFB, Desferal

▶ Chromium(VI), Oxidative Cell Damage

Definition of Dissociation Constant

▶ Magnesium in Biological Systems

Degradation of Thyroid Hormones

▶ Selenoproteins and the Biosynthesis and Activity of Thyroid Hormones

Deiodinase 1-3 (Dio1-3, Official Symbols)

▶ Selenium and Iodothyronine Deiodinases

Deoxyribozymes

▶ DNA-Platinum Complexes, Novel Enzymatic Properties

Dermatitis

▶ Chromium and Allergic Reponses

V.N. Uversky et al. (eds.), *Encyclopedia of Metalloproteins*, DOI 10.1007/978-1-4614-1533-6,

Detoxification of Anticancer Metallodrugs

► Platinum- and Ruthenium-Based Anticancer Compounds, Inhibition of Glutathione Transferase P1-1

DFT, Density Functional Theory

► Zinc Aminopeptidases, Aminopeptidase from Vibrio Proteolyticus (Aeromonas proteolytica) as Prototypical Enzyme

5′-DI (Type I-5′-Deiodinase), 5′-DII (Type II-5′-Deiodinase), 5-DIII (Type III-5-Deiodinase)

► Selenium and Iodothyronine Deiodinases

Diabetes

► Chromium and Diabetes

Dication/Divalent Metal Cation

► Calcium Ion Selectivity in Biological Systems

1,2-Dihydroxy-3-keto-5-thiomethylpent-1-ene Dioxygenase

► Acireductone Dioxygenase

Dimeric Unit: Subunit Dimer

► Monovalent Cations in Tryptophan Synthase Catalysis and Substrate Channeling Regulation

Dio1 (Official Gene Symbol)

► Selenium and Iodothyronine Deiodinases

Dioxygenase

► Iron Proteins, Mononuclear (non-heme) iron oxygenases

Dipeptide Hydrolase and Peptidase P

► Angiotensin I-Converting Enzyme

Dipeptidylcarboxypeptidase I

► Angiotensin I-Converting Enzyme

Diseases of Selenoprotein Deficiency

► Selenocysteinopathies

Dissociation Constant: K_d

► Magnesium Binding Sites in Proteins

Distribution: Partition

► Thallium, Distribution in Plants

DMA^{III} – Dimethylarsenite

► Arsenic, Biologically Active Compounds

DMA^V – Dimethylarsenate

► Arsenic, Biologically Active Compounds

DNA Enzymes

► DNA-Platinum Complexes, Novel Enzymatic Properties

DNA Lesions: Adducts

► Hexavalent chromium and DNA, biological implications of interaction

DNA Lesions: Cross-Links

► Hexavalent chromium and DNA, biological implications of interaction

DNA Lesions: DNA Double Strand Breaks

► Hexavalent chromium and DNA, biological implications of interaction

DNA Lesions: DNA Single Strand Breaks

► Hexavalent chromium and DNA, biological implications of interaction

DNA-Platinum Complexes, Novel Enzymatic Properties

Akon Higuchi
Department of Chemical and Materials Engineering, National Central University, Jhongli, Taoyuan, Taiwan
Department of Reproduction, National Research Institute for Child Health and Development, Setagaya–ku, Tokyo, Japan
Cathay Medical Research Institute, Cathay General Hospital, Hsi–Chi City, Taipei, Taiwan

Synonyms

Deoxyribozymes; DNA enzymes; DNAzymes

Definition

DNA-Pt complexes are DNA molecules where Pt atoms or Pt nanoparticles are chemically bound to DNA molecules. DNA-Pt molecules express enzymatic activity similar to peroxidase. When DNA-Pt complexes are prepared using DNA aptamer, DNAzyme-linked aptamer assay can be developed where DNA aptamer can bind to specific substrate with high specificity and high binding affinity similar to those of antibodies or protein enzymes.

Background

RNA and proteins are known to function as biocatalysts, while DNA has not played a role as a biocatalyst in evolutionary history. Nine classes of natural ribozymes are known, which catalyze phosphoester cleavage/formation or peptide-bond-formation reactions (Breaker 2004). The ability of RNA to serve as a catalyst was first shown for the self-splicing group I intron of *Tetrahymena thermophila* and the RNA moiety of RNAse P (Scherer and Rossi 2003). An HIV ribozyme directed to the nucleolar compartment was reported to inhibit HIV replication successfully (Michienzi et al. 2000). Furthermore, double-stranded

RNAs (dsRNAs), which induced targeted degradation of complementary RNA sequences by a process of RNA interference (RNAi), have also been reported (Matsuoka et al. 2007).

It was considered to be impossible to generate the double helix structure of DNA into intricate active enzymatic sites, because the chemical stability of DNA would prevent self-modifying reactions. However, current DNA engineering has created artificial enzymatic ability in DNA (i.e., deoxyribozymes, DNA enzymes, or DNAzymes), and artificial DNAzymes possess various biological activities, such as DNA modification, RNA modification, and RNA cleavage (Matsuoka et al. 2007). An RNA-cleaving phosphodiester-linked deoxyribozyme, targeting Egr-1 (an immediate-early gene inducibly expressed in growth-quiescent fibroblasts exposed to serum), inhibits endothelial expression of fibroblast growth factor (FGF)-2, but not that of vascular endothelial growth factor (VEGF) and consequently inhibits human breast carcinoma growth in nude mice (Matsuoka et al. 2007). A catalytic DNA molecule, Dz13, suppresses vascular permeability and transendothelial emigration of leukocytes by targeting of the transcription factor c-Jun. DNAzyme was also reported as signal amplification for DNA detection, when DNA was complexed with hemin. The DNAzyme expresses peroxidase characteristics (Ito and Hasuda 2004). DNA-hemin complex can catalyze a chemical oxidation and generate a colorimetric output. Therefore, DNA-hemin complex can be used in the detection of specific DNA such as M13 phage single-stranded DNA (Higuchi et al. 2008; Matsuoka et al. 2007).

Preparation and Enzymatic Activity of DNA-Pt Complex

DNA-Pt complexes can be easily prepared by mixing of DNA and potassium tetrachloroplatinate (II) ($K_2[PtCl_4]$) or *cis*-diamminedichloro platinum (cisplatin, $Pt(NH_3)_2Cl_2$) and incubating under shaking at pH 7–11 and 25–90°C for 8–120 h (Higuchi et al. 2008, 2009, 2010; Matsuoka et al. 2007). Typical reaction conditions are reaction temperature of 90°C, pH 9.0, and a reaction time of 8 h. The enzymatic activity site of DNA-Pt complex is the platinum bound to the DNA. Therefore, each DNA-Pt complex has multiple active catalytic sites, whereas a native protein enzyme such as a horseradish peroxidase has only a single active site. Therefore, the enzymatic activity per unit mole of the DNA-Pt complex increases with increasing DNA length.

The enzymatic reaction of DNA-Pt complex obeys Michaelis-Menten kinetics. K_M (Michaelis-Menten constant) for DNA-Pt complex was reported to be the same order as K_M for hemin and hemin-DNA complex, but one or two orders of magnitude higher than that of horseradish peroxidase, as is shown in Table 1 (Higuchi et al. 2008; Ito and Hasuda 2004). The rate of the reaction catalyzed by the DNA-Pt complex, k_{cat}, was found to be of the same order as that of hemin and hemin-DNA complex but two or three orders of magnitude lower than that of horseradish peroxidase (Table 1) (Dos Santos et al. 2005; Higuchi et al. 2008; Sentchouk and Grintsevich 2004). If the optimal DNA sequence where Pt binds to DNA with much efficiency, which is currently unknown, would be included in DNA-Pt complex, the enzymatic activity of DNA-Pt complex could increase dramatically and may become close to that of horseradish peroxidase in future.

DNA-Platinum Complexes, Novel Enzymatic Properties, Table 1 Comparison of kinetic parameters for peroxidation by DNA-Pt complex, hemin, hemin-DNA complex, and horseradish peroxidase (HRP)

	k_{cat} (min^{-1})	K_M (mM)	k_{cat}/K_M	substrate	References
DNA-Pt complex	1.1×10^2	4.88	0.2×10^2	TMB	Higuchi et al. 2008
Hemin	1.3×10^2	1.5	0.8×10^2	ABTS[a]	Ito and Hasuda 2004
Hemin-DNA complex	7.3×10^2	2.2	3.3×10^2	ABTS[a]	Ito and Hasuda 2004
HRP	2.0×10^4	0.033	10^7	ABTS[a]	Ito and Hasuda 2004
HRP	0.8×10^4			TMB	Dos Santos et al. 2005
HRP	3.4×10^5	0.25	1.4×10^6	TMB	Sentchouk and Grintsevich 2004

[a]ABTS indicates 2,2′-azobis(3-ethylbenzothioline)-6-sulfonic acid.

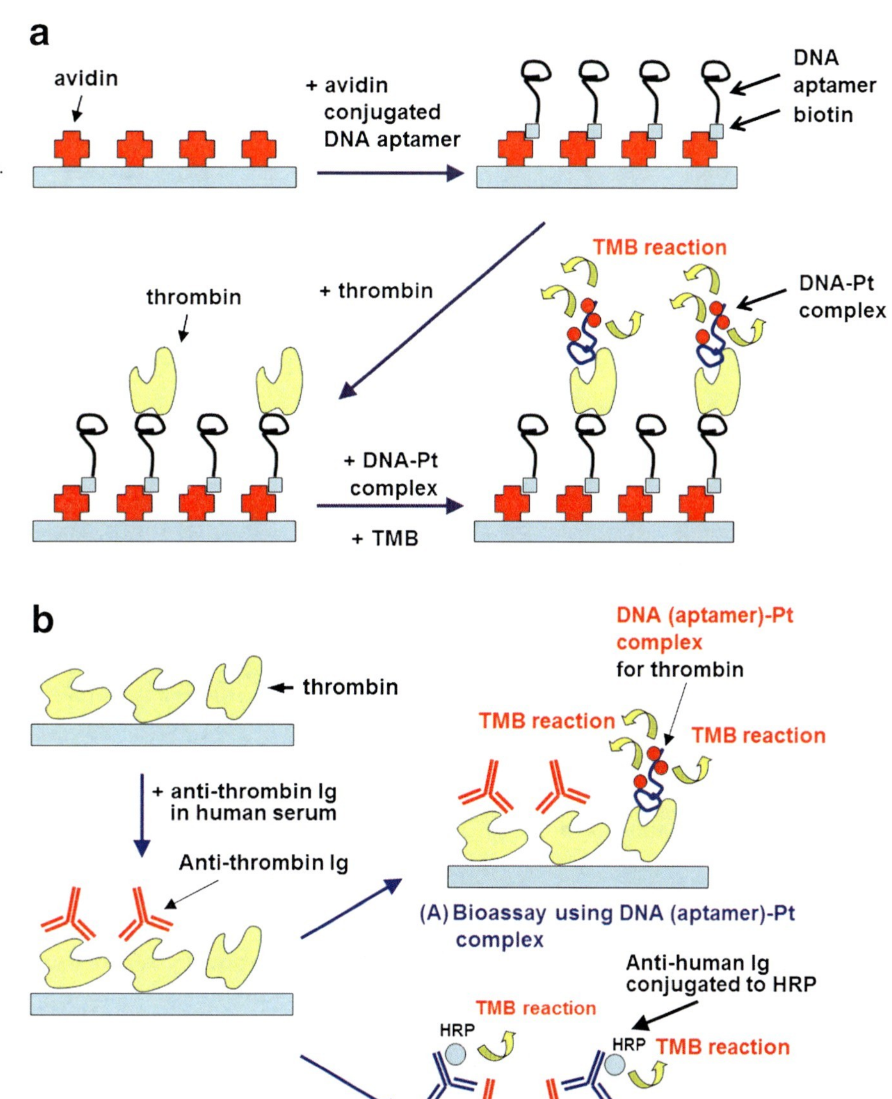

DNA-Platinum Complexes, Novel Enzymatic Properties, Fig. 1 Schematic representation of DNAzyme-linked aptamer assay (DLAA). (**a**) Sandwich method of DLAA targeting thrombin. (**b**) Competitive DLAA targeting antithrombin immunoglobulin (Ig) (**b**)

DNAzyme-Linked Aptamer Assay (DLAA) Using DNA-Pt Complex

Two types of DNAzyme-linked aptamer assay (DLAA) using DNA-Pt complex were recently reported (Higuchi et al. 2008, 2009, 2010). A schematic representation of these two types of DLAA (i.e., sandwich type and competitive type) is shown in Fig. 1. The sandwich DLAA (Fig. 1a) is similar in principle to the conventional sandwich method of enzyme-linked immunosorbent assay (ELISA), which relies on the first antibody and the second antibody conjugated with horseradish peroxidase. In this case, two types of different aptamer are targeting the analyte. One aptamer is conjugated with biotin, and therefore, it can bind to streptavidin-conjugated plates, which are commercially available. Higuchi et al. developed the sandwich type of DLAA

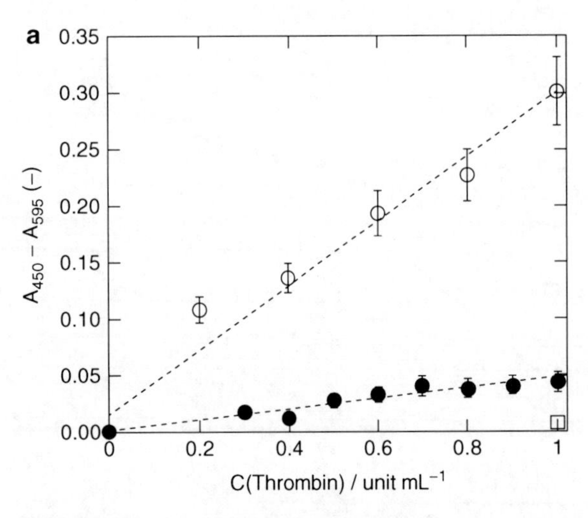

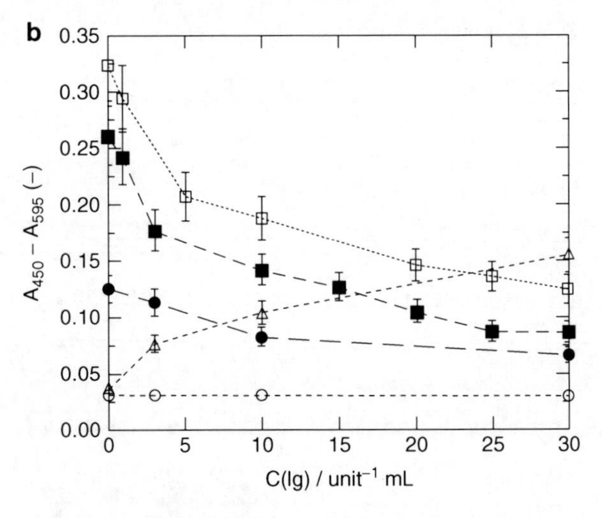

DNA-Platinum Complexes, Novel Enzymatic Properties, Fig. 2 Calibration curves of sandwich type and competitive type of DLAA. (**a**) Calibration curve of the concentration of thrombin detected by sandwich DLAA using 2 (●) and 4 μmol/L (○) DNA-Pt complex prepared from DNA aptamer of thrombin. (**b**) Calibration curve of the concentration of anti-thrombin immunoglobulin (Ig) by competitive DLAA using 2.7 μmol/L DNA-Pt complex. Thrombin-coated plates were incubated with the complex for 15 min at 26°C (○), 1 h at 4°C (●), 1 h at 26°C (□), and 24 h at 26°C (■). The calibration curve for ELISA using a second IgA/G/M conjugated with HRP from a commercial kit is also plotted (Δ). DNA-Pt complexes were prepared at pH 9.0 and a reaction temperature of 90°C, with a reaction time of 8 h. Data are expressed as means ± SD of four independent measurements

targeting thrombin, which was successfully used to analyze the concentration of thrombin (Higuchi et al. 2008). Figure 2a shows the dependence of the absorption difference (A_{450}–A_{595}) measured using the DLAA on the concentration of thrombin in the sample solution. 3,3′,5,5′-Tetramethylbenzidine (TMB)-H_2O_2 solution was used as a substrate of DNAzyme in the DLAA. When TMB is reacted under DNAzyme or horseradish peroxidase in the solution, the color of the solution changes to be yellow, which is analyzed as changes in absorbance at 450 nm (A_{450}) while the absorbance at 595 nm is measured as the baseline (A_{595}). Linear correlation is observed between the absorbance difference and the concentration of thrombin in the concentration range from 0.2 to 1 μmol/L of thrombin in Fig. 2a.

The competitive type of DLAA is shown in Fig. 1b and is based on the principle similar to that of competitive ELISA. Higuchi et al. also developed the competitive DLAA targeting antithrombin antibody (Higuchi et al. 2008). Since the ELISA kit for the antithrombin antibody is commercially available, the performance of the DLAA can be readily compared with that of the commercial kit. Thrombin-coated plates and a standard solution of antithrombin antibodies in serum solution were used in this study. Figure 2b shows the concentration calibration curves for antithrombin antibody detection using the competitive DLAA and the commercial kit. The commercial ELISA kit is based on the sandwich ELISA. In competitive DLAA, the absorbance difference (A_{450}–A_{595}) decreases with increasing concentration of antithrombin antibodies within the concentration range from 1 to 30 unit/mL, whereas the absorbance difference (A_{450}–A_{595}) increases with increasing concentration of antithrombin antibodies for the commercial ELISA kit. In this study, four different sets of binding conditions of DNAzyme (DNA aptamer-Pt complex) were investigated: (a) binding time 15 min at 26°C, (b) binding time 1 h at 26°C, and (c) binding time 24 h at 26°C, and (d) binding time 1 h at 4°C. The 15 min of binding time (a) is not sufficient for the DNA-Pt complex to bind to thrombin and consequently to detect antithrombin antibodies, because no peroxidase activity is observed. The absorbance difference (A_{450}–A_{595}) decreases with increasing concentration of antithrombin antibodies for binding conditions in the above (b)–(d). Optimal temperature (e.g., 26°C) and optimal binding time (e.g., 1 h) allow higher detection of antithrombin antibodies using the competitive DLAA.

One of the drawbacks of DLAA is lower sensitivity to analyze the substrate. This is partially originated from the fact that the free DNA aptamer without immobilization of Pt is contained in the DNA aptamer-Pt complex solution. Another explanation is that the Pt-nanoparticles-binding DNA aptamer, which expresses high enzymatic activity, is not highly abundant in the DNA aptamer-Pt complex solutions. Therefore, Higuchi et al. used DNA-Pt complex after fractionation by size through the ultrafiltration membranes in DLAA (Higuchi et al. 2009). The enzymatic activity was found to be the highest in the filtrate of DNA-Pt complex solutions prepared with cisplatin or $K_2[PtCl_4]$ after ultrafiltration through membranes having molecular weight cutoff (MWCO) of 300,000 Da. These showed 1.2-fold and 1.6-fold higher activity, respectively, than the corresponding unfractioned complexes. Sandwich-type DLAA using unfractioned or fractioned DNA-Pt complexes successfully detected the target protein of thrombin (Higuchi et al. 2009). DLAA incorporating a DNA-Pt ($K_2[PtCl_4]$) complex fractioned through ultrafiltration membranes having MWCO of 300,000 Da showed the highest sensitivity among DLAAs prepared using fractioned DNA-Pt complexes, and was reported to have a 13-fold higher sensitivity than those made with unfractioned DNA-Pt complexes (Higuchi et al. 2009).

Future Direction and Characteristics of DNA-Pt Complex

One drawback of natural enzymes is their denaturation upon heating or pH change. The enzymatic activity of horseradish peroxidase decreased with heat treatment above 60°C, whereas the activity of DNA-Pt complex remained constant after the heat treatment even at higher temperatures, above 80°C (Matsuoka et al. 2007). Although the enzymatic activity of DNA-Pt complex reported currently is one or two order lower than that of horseradish peroxidase, DNA-Pt complex is stable at room temperature and enzymatic activity is stable under heat treatment or pH change. Currently, the optimal Pt-binding of DNA sequence is unknown. The enzymatic activity of DNA-Pt complex might be dramatically increased, when the optimal sequence of DNA for binding of Pt molecules or nanoparticles is introduced into DNA-Pt complex. The fractionation of DNA-Pt complex using ultrafiltration membranes is also important to obtain high enzymatic activity of DNA-Pt complex (Higuchi et al. 2008). DNA-Pt complex has the possibility to replace horseradish peroxidase and to be used as an enzyme unit in ELISA instead of horseradish peroxidase as well as the component of the DNAzyme-linked aptamer assay (DLAA). The DNA-Pt complex can be easily prepared by incubating DNA with the Pt complexes, and the DNA or DNA aptamer can be prepared via the polymerase chain reaction (PCR). Therefore, DLAA using the DNA-Pt complex allows for the development without using antibodies which are produced by animal or plant cells in culture medium.

Cross-References

▶ Deoxyribozymes
▶ DNA Enzymes
▶ DNAzymes
▶ Strontium and DNA Aptamer Folding

References

Breaker RR (2004) Natural and engineered nucleic acids as tools to explore biology. Nature 432:838–845

Dos Santos FJN, Ximenes VF, da Fonseca LM et al (2005) Horseradish peroxidase-catalyzed oxidation of rifampicin: reaction rate enhancement by co-oxidation with anti-inflammatory drugs. Biol Pharm Bull 28:1822–1826

Higuchi A, Siao Y-D, Yang S-T et al (2008) Preparation of a DNA aptamer-Pt complex and its use in the colorimetric sensing of thrombin and anti thrombin antibodies. Anal Chem 80:6580–6586

Higuchi A, Siao Y-D, Hsieh P-V et al (2009) Preparation of fractioned DNA aptamer-Pt complex through ultrafiltration and the colorimetric sensing of thrombin. J Membr Sci 328:97–103

Higuchi A, Yang S-T, Siao Y-D et al (2010) Peroxidase activity of DNA aptamer-Pt complexes prepared with cisplatin. J Biomat Sci Polym Ed 21:67–82

Ito Y, Hasuda H (2004) Immoblization of DNAzyme as a thermo-stable enzyme. Biotechnol Bioeng 86:72–77

Matsuoka Y, Onodera T, Higuchi A et al (2007) Novel enzymatic properties of DNA-Pt complexes. Biomacromolecules 8:2684–2688

Michienzi A, Cagnon L, Bahner I et al (2000) Ribozyme-mediated inhibition of HIV 1 suggests nucleolar trafficking of HIV-1 RNA. Proc Natl Acad Sci USA 97:8955–8960

Scherer LJ, Rossi JJ (2003) Ribozyme-mediated inhibition of HIV 1 suggests nucleolar trafficking of HIV-1 RNA. Nat Biotechnol 21:1457–1465

Sentchouk VV, Grintsevich EE (2004) Oxidation of benzidine and its derivatives by thyroid peroxidase. Biochem (Moscow) 69:201–207

DNAzymes

► DNA-Platinum Complexes, Novel Enzymatic Properties

Docking Motif

► Calcineurin

Double-Stranded DNA

► Mercury and DNA

DppA, D-Aminopeptidase from *Bacillus Subtilis*

► Zinc Aminopeptidases, Aminopeptidase from Vibrio Proteolyticus (Aeromonas proteolytica) as Prototypical Enzyme

DREAM

► Calcium, Neuronal Sensor Proteins

Drug Carrier

► Porous Silicon for Drug Delivery

Dsel

► Selenoprotein K

dSelK

► Selenoprotein K

Dysprosium

Takashiro Akitsu
Department of Chemistry, Tokyo University of Science, Shinjuku-ku, Tokyo, Japan

Definition

A lanthanoid element, the ninth element (yttrium group) of the f-elements block, with the symbol Dy, atomic number 59, and atomic weight 162.500. Electron configuration $[Xe]4f^{10}6s^2$. Dysprosium is composed of stable (^{158}Dy, 0.10%; ^{160}Dy, 2.34%; ^{161}Dy, 18.91%; ^{162}Dy, 25.51%; ^{163}Dy, 24.90%; ^{164}Dy, 28.18%) and two radioactive (^{154}Dy; ^{156}Dy, 0.06%) isotopes. Discovered by L. de Boisbaudran in 1886. Dysprosium exhibits oxidation states III and IV; atomic radii: 178 pm, covalent radii 193 pm; redox potential (acidic solution) Dy^{3+}/Dy −2.353 V, Dy^{3+}/Dy^{2+} −2.6 V; electronegativity (Pauling) 1.22. Ground electronic state of Dy^{3+} is $^6H_{15/2}$ with $S = 5/2$, $L = 5$, $J = 6$ with $\lambda = -380\ cm^{-1}$. Most stable technogenic radionuclide ^{154}Dy (half-life 3×10^6 y). The most common compounds: Dy_2O_3, $Dy(OH)_3$, $DyCl_3$, $Dy_2(SO_4)_3$. Biologically, dysprosium is of moderate toxicity, and it is known that exposure to $DyCl_3$ caused conjunctivitis in rabbits when applied directly to their eyes (Atkins et al. 2006; Cotton et al. 1999; Huheey et al. 1997; Oki et al. 1998; Rayner-Canham and Overton 2006).

Cross-References

► Lanthanide Ions as Luminescent Probes
► Lanthanide Metalloproteins
► Lanthanides and Cancer
► Lanthanides in Biological Labeling, Imaging, and Therapy

► Lanthanides in Nucleic Acid Analysis
► Lanthanides, Physical and Chemical Characteristics

References

Atkins P, Overton T, Rourke J, Weller M, Armstrong F (2006) Shriver and Atkins inorganic chemistry, 4th edn. Oxford University Press, Oxford/New York

Cotton FA, Wilkinson G, Murillo CA, Bochmann M (1999) Advanced inorganic chemistry, 6th edn. Wiley-Interscience, New York

Huheey JE, Keiter EA, Keiter RL (1997) Inorganic chemistry: principles of structure and reactivity, 4th edn. Prentice Hall, New York

Oki M, Osawa T, Tanaka M, Chihara H (1998) Encyclopedic dictionary of chemistry. Tokyo Kagaku Dojin, Tokyo

Rayner-Canham G, Overton T (2006) Descriptive inorganic chemistry, 4th edn. W. H. Freeman, New York

E

E2

▶ Acireductone Dioxygenase

E2′

▶ Acireductone Dioxygenase

EC 1.10.3

▶ Ascorbate Oxidase

EC 1.13.11.54 & 1.13.11.53

▶ Acireductone Dioxygenase

EC 1.15.1.1

▶ Zinc in Superoxide Dismutase

EC 3.6.3.9

▶ Sodium/Potassium-ATPase Structure and Function, Overview

Effect of Nutritional Zinc Deficiency on Hemostasis Dysfunction

▶ Zinc in Hemostasis

Effects

▶ Thallium, Effects on Mitochondria

EF-Hand Calcium-Binding Proteins

▶ Barium Binding to EF-Hand Proteins and Potassium Channels

V.N. Uversky et al. (eds.), *Encyclopedia of Metalloproteins*, DOI 10.1007/978-1-4614-1533-6,

EF-Hand III: the Third EF-Hand

▶ Magnesium Binding Sites in Proteins

EF-Hand Protein

▶ Calcium-Binding Proteins, Overview

EF-Hand Proteins

Hiroaki Ishida and Hans J. Vogel
Department of Biological Sciences, Biochemistry Research Group, University of Calgary, Calgary, AB, Canada

Synonyms

Calcium-binding proteins; Helix-loop-helix calcium-binding poteins; Regulatory calcium binding proteins

Definition

The proteins that contain one or more functional helix-loop-helix EF-hand Ca^{2+}-binding motifs. These proteins can mediate intracellular Ca^{2+} signals and control various physiological events.

Calcium Signaling

In eukaryotic cells, changes in cytosolic calcium (Ca^{2+}) concentration are used as an intracellular signaling mechanism to control cellular functions such as fertilization, cell proliferation, muscle contraction, the immune response, gene transcription, and apoptosis. Upon excitation by various stimuli, the intracellular Ca^{2+} concentration transiently increases from 10^{-7}M in the resting state to 10^{-5}M in the exicited state. This is the result of Ca^{2+} release from internal storage organelles (the endoplasmic or sarcoplasmic reticulum in animal cells, the vacuole in plant cells) and/or Ca^{2+} influx from the extracellular milieu through ion-specific channels found in the plasma membrane. The increase in Ca^{2+} concentration has both spatial and temporal aspects and gives rise to so-called spikes, puffs, and waves, depending on the source and duration of the signal (Berridge et al. 2003). Once formed, the divergent Ca^{2+} signals are translated and further modulated by a wide spectrum of ▶ Ca^{2+}-binding proteins that regulate specific cellular responses.

Intracellular Ca^{2+}-binding proteins are classified into two groups: the EF-hand proteins and the non-EF-hand proteins. This latter groups includes the Ca^{2+}-binding ▶ annexins and ▶ cadherins, two protein families which help form the membrane scaffold and participate in Ca^{2+}-dependent cell adhesion, respectively. The ▶ C2 domain family is another large group of non-EF-hand Ca^{2+}-binding proteins, its members are involved in signal transduction and membrane trafficking. The C2 domain is formed by eight β-strands which can bind two or three Ca^{2+}-ions in the loop regions. In contrast, members of the EF-hand superfamily bind Ca^{2+} through a well-characterized helix-loop-helix motif. With 250 identified members in the human genome, the EF-hand superfamily is the largest family of eukaryotic Ca^{2+}-binding proteins.

The EF-Hand Motif

The original discovery of the EF-hand motif was made by Kretsinger and Nockolds (1973) who described a 30-amino-acid-residue stretch that forms a unique structure and binds the Ca^{2+} ion in the small Ca^{2+}-binding protein ▶ parvalbumin. This motif is made up of two α-helices (the E- and F-helices of parvalbumin) connected by a 9-residue loop. The E- and F-helices are almost perpendicular to each other, resembling the thumb and index finger of a right hand, while the closed middle finger approximates the Ca^{2+}-chelating loop (Fig. 1a). The EF-hand binds Ca^{2+} through a pentagonal bipyramidal coordination geometry, and the chelating residues are indicated either based on their position in the primary sequence (1, 3, 5, 7, 9, and 12) or according to their coordination geometry (X, Y, Z, −Y, −X, and −Z) (Fig. 1b, c). Five of Ca^{2+}'s seven coordinating ligands are either directly or indirectly supplied by the amino acids found in the loop. The first three, 1(X), 3(Y), and 5(Z),

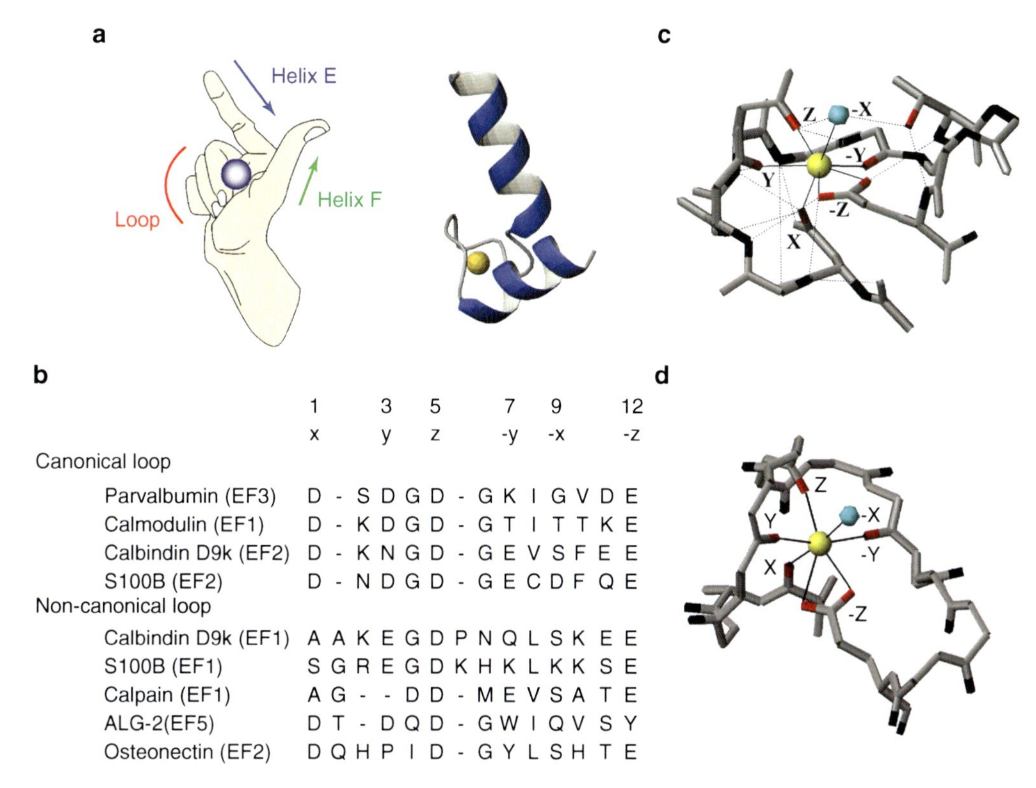

EF-Hand Proteins, Fig. 1 (**a**) The helix-loop-helix EF-hand calcium-binding motif resembles a *right hand*; the index finger and thumb represent the E- and F-helices, respectively, the middle finger mimics the Ca^{2+}-chelating loop of parvalbumin. (**b**) Sequence alignments of canonical and noncanonical EF-hand Ca^{2+}-chelating loops. (**c**) Ca^{2+} coordination by a canonical EF-hand (EF1 of CaM). (**d**) Ca^{2+} coordination by a noncanonical pseudo-EF-hand (EF1 of calbindin D_{9k}). The Ca^{2+} ions and the water molecules are shown as *yellow* and *blue spheres*, respectively, in panels (**c**) and (**d**). The pentagonal bipyramidal coordination positions are labeled

are provided by side-chain oxygen atoms of the amino acid residues found at these positions, typically aspartate residues. Position 7(−Y) is supplied by the main-chain carbonyl oxygen atom of this residue and as such the identity of this amino acid is highly variable. The ligand at the ninth position (−X) is most often indirectly involved in the coordination sphere as it hydrogen bonds a water molecule that directly chelates the bound Ca^{2+} ion. The remaining two coordinating groups come from a bidentate carboxylate ligand supplied by a carboxylate-containing amino acid side chain at the 12th (−Z) position. Although commonly referred to as the EF-loop's 12th residue, this amino acid is not structurally part of the loop, instead it is located in the exiting helix three residues removed from the loop's C-terminal. The identity of the amino acid at the 12th position affects the ability of the EF-hand to distinguish Ca^{2+} from the chemically similar Mg^{2+} cation. Most EF-hands have a glutamate residue at this position, an amino acid which has a side chain long enough to reach the bound Ca^{2+} ion. However, in approximately 10% of EF-hands, an aspartate residue occupies the 12th position. The side chain of aspartate is shorter than that of glutamate and, in many cases, cannot adequately reach the bound Ca^{2+} ion, and as such can only act as a monodentate ligand. Consequently, these so-called Asp12 EF-loops have reduced affinity for Ca^{2+} as well as an increased affinity for the chemically similar but smaller magnesium cation (Mg^{2+}). Finally, some preference is also exhibited by the nonchelating residues of the EF-loop. In particular, position 6 is a highly conserved glycine residue that adopts a unique main-chain conformation (phi = ~60°, psi = ~20°) facilitating the bend in the loop.

In addition to the canonical EF-hand motif, several noncanonical EF-hands are known. The entire subfamily of ► S100 and the related protein calbindin D_{9K}

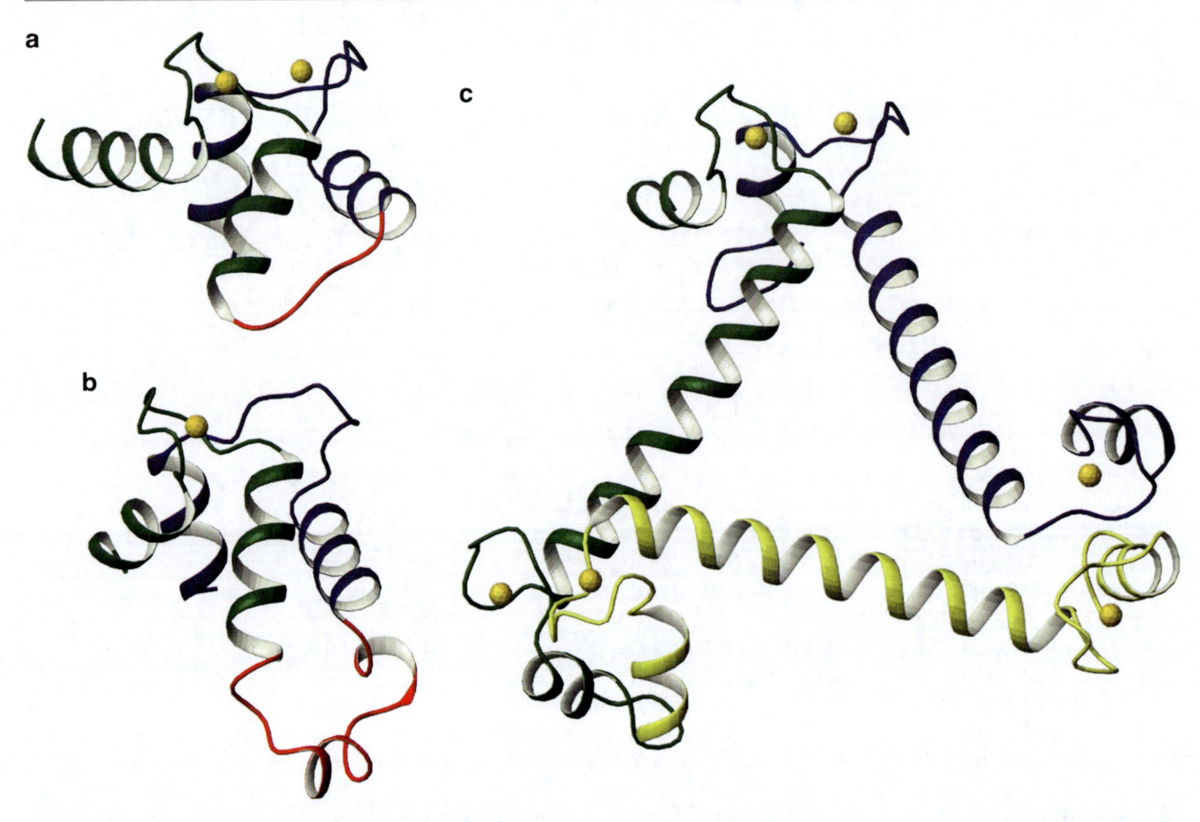

EF-Hand Proteins, Fig. 2 A pair of EF-hand motifs is connected by a linker of varying length. The EF-hand motifs in a pair are shown in different colors, and the linker regions are shown in *red*. (**a**) The N-terminal domain of CaM (1EXR). (**b**) The C-terminal domain of recoverin (1OMR). (**c**) The N-terminal domain of EhCaBP1 (2NXQ), where the EF-hand pairs are formed intermolecularly between three proteins

contain the "pseudo-EF-hand." In this variant, in response to a two-residue insertion, the loop is turned inside out and all Ca^{2+}-chelating groups except the −Z glutamate are provided by main-chain carbonyl oxygen atoms (Fig. 1b, d). Other less common functional EF-hand motifs include an 11-residue Ca^{2+}-binding loop in calpain and a 13-residue Ca^{2+}-binding loop in the extracellular EF-hand protein osteonectin.

The EF-Hand Pair

EF-hand motifs typically occur as a pair packed together to create a four-helix bundle (Fig. 2a). In most cases, the two EF-hand motifs are arranged in tandem in the amino acid sequence, separated by a linker of varying lengths (Fig. 2). This organization is thought to originate from the duplication of a single EF-gene early in evolution, and the paired EF-hands in most cases have subsequently evolved to be not identical. In the well-characterized eukaryotic regulatory Ca^{2+}-binding protein ► calmodulin (CaM), the first (EF1) and second (EF2) EF-hands share roughly 36% sequence identity and are connected by a short five-to-six-residue linker that forms an extended conformation. In contrast, the linker region between the EF3 and EF4 motifs of recoverin, a vision-associated protein, is longer and forms two helical structures (Fig. 2b). Intermolecular EF-hand pairs also occur. In the crystal structure of the N-terminal domain of the EhCaBP1 from the parasitic protozoan *Entamoeba histolytica*, the protein forms a homotrimer in which EF1 of one molecule creates a pair with EF2 from the next protein (Fig. 2c). Similar intermolecular EF-hand pairs are also found for the EF5 motif of the ► penta-EF-hand family proteins (discussed below). The fact that functional EF-hands are frequently paired with an EF-hand that can no longer bind Ca^{2+} highlights the importance of the paired structural unit.

Recoverin, the protein example mentioned above, has four EF-hand motifs, yet EF1 and EF4 which are paired with the functional EF2 and EF3 motifs are themselves incapable of binding Ca^{2+}. Instead, these nonfunctional EF-hands serve to contribute to the overall conformational stability of recoverin and help to create a target-binding interface. Indeed, as seen in Ca^{2+}- and integrin-binding protein 1 (CIB1), two nonfunctional EF-hands can couple together in the same manner as their functional counterparts (Yamniuk and Vogel 2006; PDB access code: 2L4H). The significance of the paired orientation is also seen by the lack of examples of a single EF-hand motif. Although the presence of a single EF-hand has been predicted from the primary sequence of EF-hand proteins (e.g., the ER Ca^{2+}-sensor stromal interaction molecule-1 (STIM1; Stathopulos et al. 2008) and KCBP-interacting Ca^{2+}-binding protein (PDB access code: 3H4S)), subsequent structural analysis of these examples has in both cases revealed a "hidden EF-hand" pairing with the known EF-hand.

The paired, face-to-face organization of EF-hand motifs enables the two Ca^{2+}-binding loops to interact through backbone hydrogen bonds, frequently forming a short antiparallel β-sheet (Fig. 2a). This arrangement also facilitates the positive cooperativity often observed upon Ca^{2+} binding to EF-hand domains. The cooperative phenomenon enables the pair to act as a Ca^{2+}-binding unit as the presence of Ca^{2+} in one EF-hand increases the affinity of the second for this metal ion (Grabarek 2006). It is thought that binding of the first Ca^{2+} ion to the EF-hand pair positions several of the Ca^{2+}-chelating ligands in the second EF-loop into the proper conformation. This effect is mediated by hydrogen bond and dipole interactions between the two Ca^{2+}-chelating loops, in particular the short β-sheet that links residues 7 and 8 of both EF-loops. In addition, the extensive hydrophobic packing between the helices of the two EF-hand motifs also contributes to the cooperativity of Ca^{2+} binding. Through positive cooperativity, functional EF-hand pairs have a much greater affinity for the Ca^{2+} ion than isolated sites.

EF-Hand Protein Arrangements

Two EF-Hands. All EF-hand protein structures determined to date contain 2, 3, 4, 5, or 6 EF-hand motifs (Fig. 3). Calbindin D_{9K}, the smallest member of the EF-hand protein family, consists of only two EF-hand motifs: a pseudo-EF-hand (EF1) and a canonical EF-hand motif (EF2). The S100 protein family, the largest EF-hand subfamily with 25 members in humans, also contains a pseudo-EF-hand/canonical EF-hand pair, forming a structure similar to calbindin D_{9K}. However unlike calbindin D_{9K}, all S100s occur as either stable homo- or heterodimers thereby forming a compact globule containing a total of four EF-hand motifs (Santamaria-Kisiel et al. 2006).

Three EF-Hands. The Ca^{2+}-buffer parvalbumin is a rare example of an EF-hand protein containing only three EF-hand motifs. Parvalbumin has two functional EF-hand motifs that form a pair (EF2 and EF3). The odd number of EF-hands results in the presence of an unpaired EF-hand motif. This EF-hand does not bind Ca^{2+} and, instead, interacts with the EF2/EF3 pair via extensive hydrophobic interactions, resembling a bound target peptide and stabilizing the conformation of this protein.

Four EF-Hands. Proteins that are comprised of four EF-hand motifs are the most abundant group within the EF-hand family. Included are well-known EF-hand proteins such as CaM, the muscle contractile protein ► troponin C (TnC), myosin light chains, centrins, and all members of the ► neuronal Ca^{2+} sensor (NCS) protein family (e.g., frequenin, recoverin, and CaBPs) (Burgoyne et al. 2004). These proteins are made up of two globular domains (the N- and C-terminal domains) connected by a linker sequence of varying length, where each domain contains a pair of EF-hand motifs. The group can be further subdivided into those proteins that are dumbbell-shaped and those that are a compact ellipsoid. CaM and TnC are examples of extended "dumbbell" proteins in which the N- and C-terminal domains act almost as independent Ca^{2+}-sensing and target-binding units. For TnC, the two domains even have separate functions dictated by each domain's affinity and specificity for the Ca^{2+}ion. Due to its high affinity for Ca^{2+} as well as the ability to bind Mg^{2+}, the C-terminal domain is continuously occupied by either metal ion and has a structural role, binding TnC to the remainder of the troponin complex. In contrast, the lower Ca^{2+} affinity of the N-terminal domain allows it to serve as a Ca^{2+} sensor. The members of the NCS protein family form a compact ellipsoidal structure in which the two EF-hand domains interact significantly with each other.

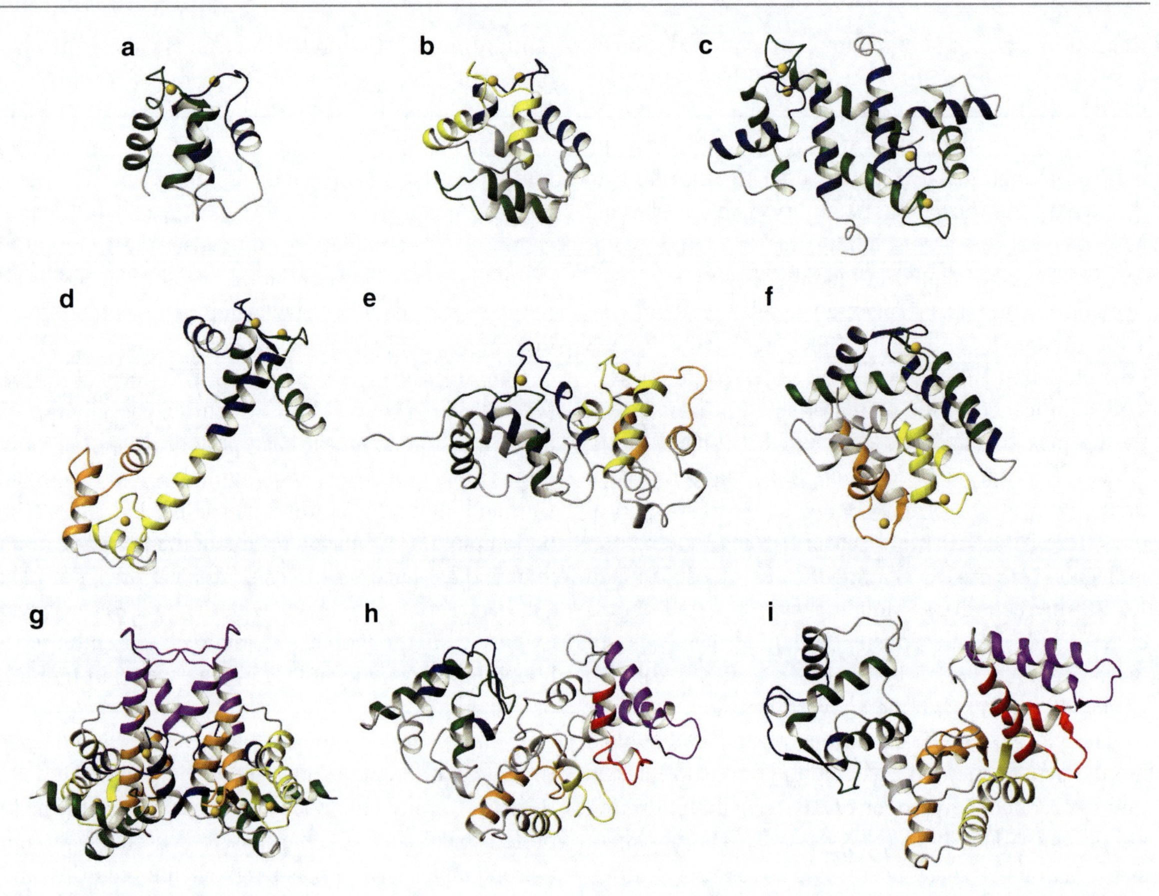

EF-Hand Proteins, Fig. 3 Examples of EF-hand proteins containing 2, 3, 4, 5, or 6 EF-hand motifs. The individual EF-hand motifs are colored differently in *green, navy, yellow, orange, purple,* and *red* starting from the first (N-terminal) EF-hand motif. (**a**) Ca^{2+}-calbindin D_{9k} (1B1G), (**b**) Ca^{2+}-bound parvalbumin (1RWY), (**c**) Ca^{2+}-bound S100B (1QLK), (**d**) Ca^{2+}-bound CaM (1EXR), (**e**) Ca^{2+}-bound recoverin (1JSA), (**f**) Ca^{2+}-bound *Nereis* SCP (2SCP), (**g**) Ca^{2+}-bound calpain domain IV (1DVI) which forms a homodimer via the intermolecular EF-hand pair colored in *purple*, (**h**) Ca^{2+}-bound calbindin D_{28k} (2G9B), and (**i**) Ca^{2+}-free *Danio rerio* secretagogin (2BE4)

This interaction often facilitates cooperative Ca^{2+} and target binding involving both lobes. For example, interlobe interactions between the EF1/EF2 and EF3/EF4 pairs of the globular *Nereis* sarcoplasmic Ca^{2+}-binding protein enable Ca^{2+} to bind EF1 with positive cooperativity despite being paired with nonfunctional EF2. Similarly, the functional EF2 and EF3 hands of recoverin that are paired with the nonfunctional EF1 and EF4 motifs, respectively, bind Ca^{2+} with positive cooperativity, an effect also attributed to the energetics of exposing an internally bound N-terminal myristoyl group in recoverin's Ca^{2+}-bound state.

Five EF-Hands. The penta-EF-hand proteins calpain, apoptosis-linked protein (ALG-2), sorcin, peflin, and grancalcin all have five EF-hand motifs. Like the parvalbumins, the members of this family all contain one unpaired EF-hand motif (EF5). Here the extra EF-hand is used to form an intermolecular EF-hand pair, leading to the formation of protein dimers containing a total of ten EF-hand motifs (Fig. 3g). Although the majority of penta-EF-hand proteins do not bind Ca^{2+} to EF5, the EF5/EF5′ pair of ALG-2 is capable of binding this metal ion albeit through a noncanonical EF-loop motif (Fig. 1b). As mentioned above, similar functional intermolecular EF-hand pairs are observed in EhCaBP1 (Fig. 2c). This process is driven by the apparently "sticky" nature of an unpaired EF-hand, first seen by the spontaneous ability of synthetic peptides encompassing single helix-loop-helix EF-hands to form a homodimer (Shaw et al. 1990).

Six EF-Hands. EF-hand proteins with six EF-hand motifs (hexa-EF-hand proteins) include the neuroprotective protein ► calbindin D_{28k} and the homologous calretinin, reticulocalbindin, and secretagogin. Calbindin D_{28k} contains three EF-hand domains that are arranged in tandem, each of these includes a pair of EF-hand motifs that together create a bulky V-shaped structure. Only four of the six EF-hand motifs (EF1, EF3, EF4, and EF5) possess functional Ca^{2+}-binding sites. The crystal structure of the Ca^{2+}-free form of *Danio rerio* secretagogin shows that it forms similar secondary structures to Ca^{2+}-bound calbindin D_{28k}, but with a large difference in the relative domain orientation (Fig. 3h, i). For secretagogin, the first EF-hand domain appears to be rotated almost 180° relative to the rest of the protein structure. Whether or not this difference reflects a Ca^{2+}-dependent conformational change of all hexa-EF-hand proteins, remains to be determined.

Multidomain Proteins. Most EF-hand proteins consist solely of multiple helix-loop-helix EF-hand motifs and are therefore highly α-helical proteins. However, there are examples of EF-hand motifs as a separate domain in a larger protein. For example, the N-terminal luminal region of ~700-residue STIM proteins contains a single EF-hand pair containing a functional EF-hand (EF1) paired with a nonfunctional "invisible" EF-hand (EF2). A unique feature of the STIM EF-hand pair is that it interacts with an adjacent sterile alpha motif (SAM) domain, an interaction essential to keep the Ca^{2+} affinity of EF1 reasonably low. Furthermore, the ~900-amino-acid-residue cytoskeletal actin-binding protein actinin contains a single pair of EF-hand motifs near its C-terminal end (residues 746–822). Similarly, the 50 kDa DNA-binding protein nucleobindin contains a pair of functional EF-hand motifs in the center of the protein (residues 253–316).

Ca^{2+}-Binding Affinity

Despite extensive similarities in the primary sequence of their Ca^{2+}-chelating loops, EF-hand proteins exhibit Ca^{2+} affinities ranging from nM to mM. Furthermore, the affinity for Ca^{2+} cannot be determined solely from the amino acid sequence of a given Ca^{2+}-chelating loop as both the magnitude of cooperativity between EF-hand motifs as well as free energy contributions caused by the Ca^{2+}-induced conformational change are also contributing determinants (Gifford et al. 2007). Since electrostatics is a long-range effect, negatively charged amino acid side chains in the vicinity of the Ca^{2+}-binding loop also contribute to the overall affinity of a given EF-hand. Several techniques are available for determining the Ca^{2+}-binding constants of EF-hand proteins including a sensitive chromophoric competition assay (Linse 2002). As would be expected, measured Ca^{2+} affinities match the functional role of the EF-hand protein. STIM proteins monitor the fluctuation of ER luminal Ca^{2+} concentration and feature the weakest identified affinity for this metal ion (a few hundred μM). As a result, STIM proteins can act as Ca^{2+} sensors despite the very high Ca^{2+} levels in this storage organelle. CaM and similar cytosolic Ca^{2+}-sensor proteins that translate the change in intracellular cytoplasmic Ca^{2+} concentration into a biochemical response have a Ca^{2+} affinity tuned to sense μM fluctuations in concentration. On the other hand, parvalbumin, calbindin $D_{9K,}$ and other proteins that buffer the cytoplasmic Ca^{2+} concentration both spatially and temporally exhibit the highest Ca^{2+} affinities, in the nM range. Finally, in many cases, the presence of a target protein further modulates the affinity of an EF-hand for Ca^{2+}. For example, the Ca^{2+} affinity for CaM increases by up to two orders of magnitude in the presence of a CaM-binding target peptide.

Consequences of Ca^{2+} Binding

Many Ca^{2+} sensor proteins experience significant conformational changes upon binding this metal ion, changes that alter the biochemical character of these proteins facilitating interaction with their specific downstream target proteins in a Ca^{2+}-dependent manner (Ikura and Ames 2006). The ubiquitous Ca^{2+}-sensor CaM undergoes dramatic conformational changes upon binding four Ca^{2+} ions (Fig. 4a), altering the surface character from hydrophilic in the apo-state to rather hydrophobic in the Ca^{2+}-bound form (see Fig. 4b). In the absence of Ca^{2+}, the two helices of each EF-hand motif in CaM are arranged in an almost antiparallel orientation with interhelical angles of 130–140°. Upon binding Ca^{2+}, the helices open up and the interhelical angles become roughly 90°, exposing a previously buried hydrophobic region (Fig. 4). The backbone root-mean-square deviation between the

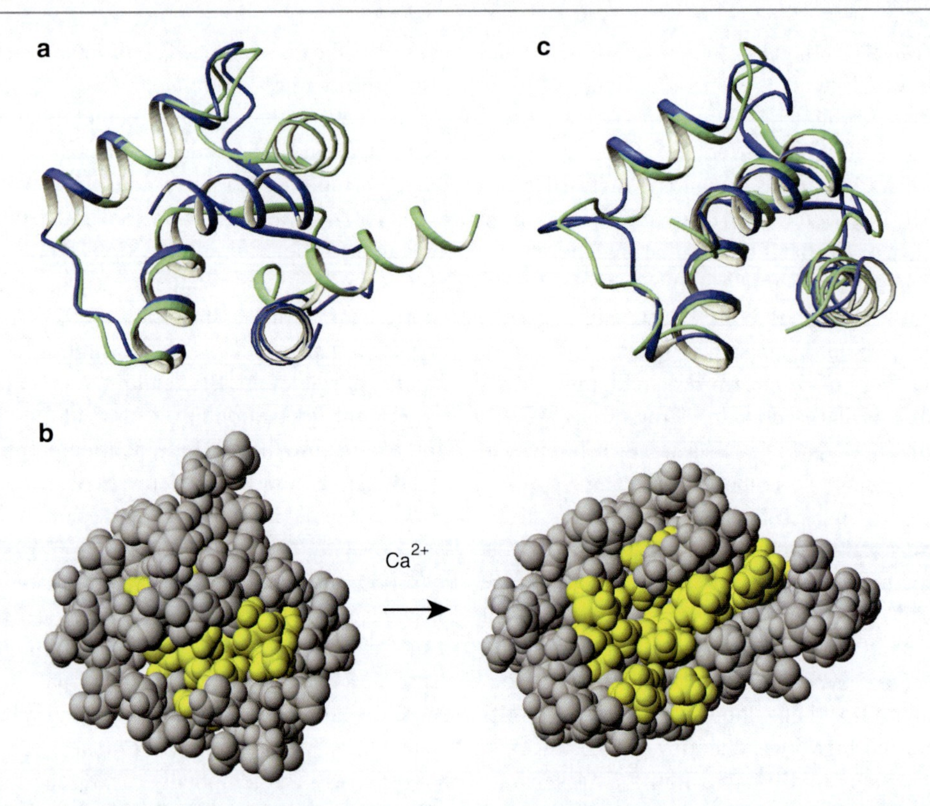

EF-Hand Proteins, Fig. 4 Conformational changes that occur upon Ca^{2+} binding illustrate the difference between Ca^{2+}-sensor and Ca^{2+}-buffer proteins. (**a**) Structures of the N-terminal domain of the Ca^{2+}-sensor CaM with (1EXR) and without Ca^{2+} ions (1F70) are superimposed and show a large change in helix orientation. (**b**) The surface structure of the N-terminal domain of CaM in the absence (*left*) and the presence of Ca^{2+} ions (*right*). The hydrophobic side chains are highlighted in *yellow*. The large increase in the hydrophobic surface allows CaM to act as a Ca^{2+}-sensor protein. Intriguingly, CaM's hydrophobic surface is very rich in methionines. These residues have high flexibility and their sulfur atom gives them the property of polarizability, two features that contribute to CaM's versatility. (**c**) The structures of Ca^{2+}-free (1CLB) and Ca^{2+}-bound calbindin D_{9k} (1B1G) are superimposed, revealing the absence of a structural change for this Ca^{2+}-buffer protein

CaM N-terminal domain structures before and after Ca^{2+} binding is 4.3 Å. Through the exposed hydrophobic surfaces, CaM interacts and regulates over 300 different proteins including protein kinases and phosphatases, receptors, pumps, and channels. This versatility is thought to arise due to the independent nature of CaM's two Ca^{2+}- and target-binding lobes; the highly mobile linker regions enables CaM to quickly reposition its target-binding units to accommodate various target proteins (Fig. 5a) (Ishida and Vogel 2006). As the prototypical Ca^{2+}-sensor, CaM regulates its target proteins through a number of mechanisms: by binding to and removing an autoinhibitory domain from a number of protein kinases (e.g., the Ca^{2+}-CaM-dependent protein kinases and myosin light-chain kinases), inducing dimer formation (e.g., the small conductive Ca^{2+}-CaM-activated potassium channel and plant glutamate decarboxylase), and by facilitating both domain organization (e.g., *Bacillus anthracis* edema factor) and ternary complex formation (e.g., plant MAPK-phosphatase).

The S100s also experience a significant structural change upon Ca^{2+} binding (Fig. 5b). For this family, the EF-hand motifs also serve to facilitate the formation of homo- and heterodimers. The first helix of EF1 (the pseudo-EF-hand) typically forms the interdimer interaction. Potentially due to its role in this interaction, the Ca^{2+} affinity of this EF-hand is ~10 times weaker than canonical EF2 and Ca^{2+} binding to EF1 causes very small structural changes. In contrast, the binding of Ca^{2+} to EF2 reorients the first helix of this EF-hand leading to the opening of a modest hydrophobic target-binding pocket. Despite the high similarity of their structures, the members of the S100 protein family recognize member-specific target proteins. Moreover, individual S100 proteins can interact with

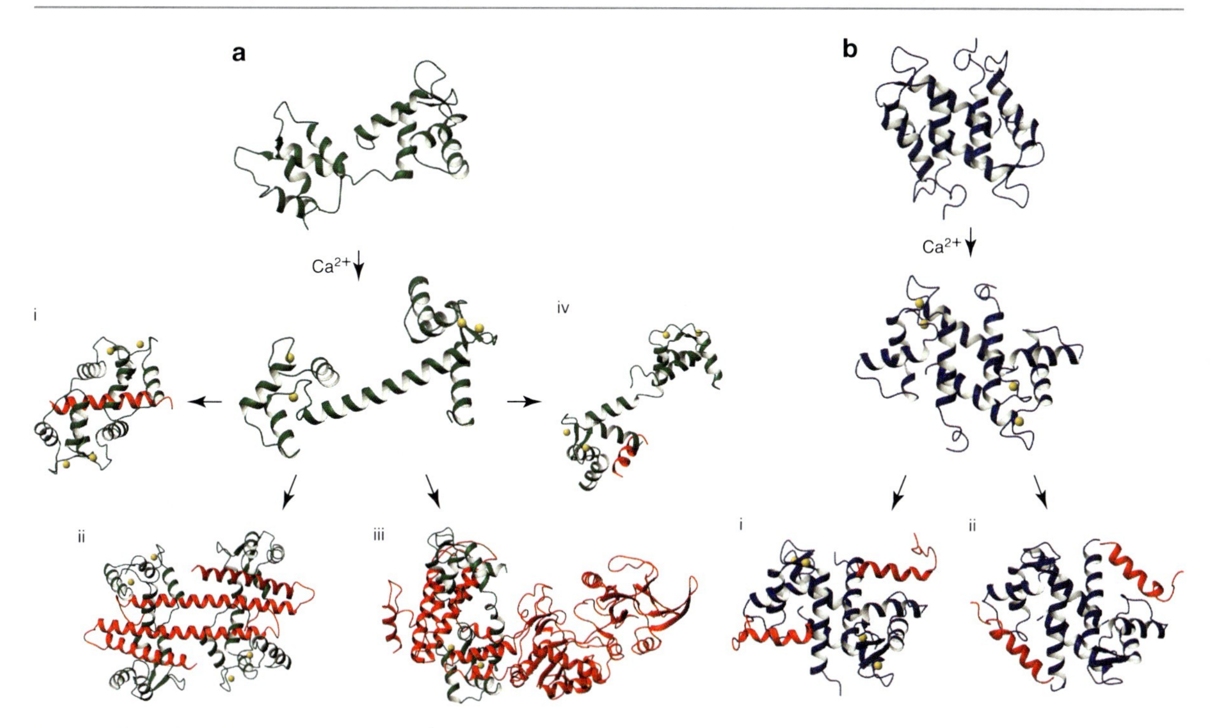

EF-Hand Proteins, Fig. 5 The role of Ca^{2+}-sensor proteins. (**a**), CaM binds four Ca^{2+} ions and alters its conformation. Ca^{2+}-bound CaM interacts and regulates many different proteins and enzymes in a diverse manner. The figures are generated using Ca^{2+}-free CaM (1DMO), Ca^{2+}-bound CaM (1EXR), CaM complexed with CaMKI (*i*, 1MXE), Ca^{2+}-CaM-activated K^+ channel (*ii*, 1G4Y), edema factor (EF) (*iii*, 1K93), and MAPK-phosphatase (*iv*, 2KN2 and 2ROA). (**b**), S100B binds four Ca^{2+} ions and opens its target-binding pocket. Ca^{2+}-bound S100B can bind different target proteins in a different way yet through the same target-binding pocket. The figures represent Ca^{2+}-free S100B (1SYM), Ca^{2+}-bound S100B (1QLK), S100B complexed with p53 (*i*, 1DT7), or with NDR (*ii*, 1PSB)

several different target proteins in different fashions (Bhattacharya et al. 2004). For example, the tumor suppressor protein p53 and the nuclear serine/threonine kinase NDR both bind a homodimer of S100B, but bind differently through the same target-binding pocket (Fig. 5b). Due to its dimer architecture, the S100 proteins can potentially interact with two target proteins simultaneously, thereby bringing two different proteins into close proximity.

Many of the NCS protein family, such as the retinal-specific EF-hand protein recoverin, are myristoylated at their N-terminus. In the Ca^{2+}-free form of recoverin, this myristoyl group is sequestered in the hydrophobic interior of the protein. Binding of two Ca^{2+} ions to EF2 and EF3 induces a rotation of the EF-hand domains, releasing the myristoyl group to the solvent (Fig. 6). In response, recoverin binds to the membrane where it functions as an inhibitor of rhodopsin kinase. This mechanism is commonly referred to as a "Ca^{2+}-myristoyl switch." Interestingly, several other myristoylated NCS proteins do not appear to have such a switch and are constantly anchored to the membrane via their N-terminal myristoyl group. For these proteins, the binding of Ca^{2+} has other effects.

In contrast with these Ca^{2+}-sensor examples, Ca^{2+}-buffer proteins (e.g., parvalbumin and calbindin D_{9K}) experience very small structural changes upon binding Ca^{2+}, consistent with their role in buffering and modulating intracellular Ca^{2+} signals (Schwaller 2009). The three-dimensional structures of calbindin D_{9K} in its Ca^{2+}-free and-bound states are almost identical except for small changes in the Ca^{2+}-chelating loops and the loop connecting the two EF-hands (Fig. 4c). For both states, the two helices of the EF-hands of calbindin D_{9K} are arranged almost perpendicular in the apo-form, and Ca^{2+} binding does not require a major conformational change. This "Ca^{2+}-ready state" is thought to be responsible for the high Ca^{2+} affinity of Ca^{2+}-buffer proteins. Because Ca^{2+} binding induces no changes in the molecular surface of Ca^{2+}-buffers, this class of EF-hand proteins cannot bind targets in a Ca^{2+}-dependent manner.

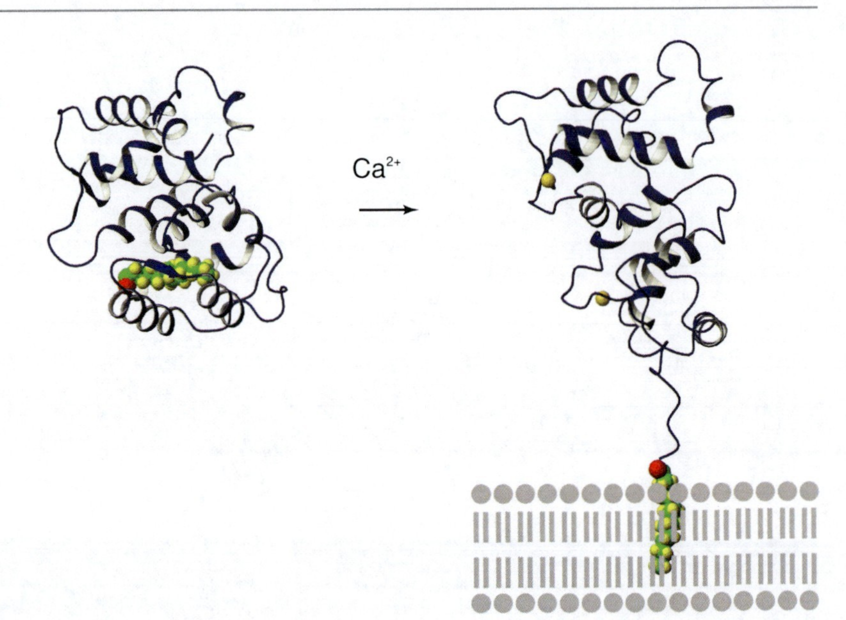

EF-Hand Proteins, Fig. 6 The Ca^{2+}-myristoyl switch of recoverin. The structures of Ca^{2+}-free (*left*, 1IKU) and Ca^{2+}-bound recoverin (*right*, 1JSA). The binding of two Ca^{2+} ions cause a domain rotation that leads to the exposure of the covalently attached myristoyl group and subsequent association of the protein with the membrane. The myristoyl group is indicated as a space-filling model

A survey of available structures for EF-hand proteins reveals that Ca^{2+} binding causes protein-specific conformational changes and that the degree of change in the interhelical angles of a given EF-hand motif can be quite variable. The helix angle in each protein has been tuned to match a specific biological function (Babini et al. 2005).

Automodulation of EF-Hand Proteins

When EF-hand domains are part of a larger multidomain protein, they often interact with other internal domains, much like an endogenous target resembling the CaM-target interaction. This interaction can modulate the protein's activity, stability, target binding, and Ca^{2+}-affinity. The Ca^{2+}-dependent protein kinases (CDPKs) are a group of plant-specific Ca^{2+}-sensor proteins which all contain a CaM-like domain that regulates the activity of the enzyme (DeFalco et al. 2009). Much like how CaM regulates many of its targets, the CaM-like domain controls the activity of the enzyme by binding to an intrinsic autoinhibitory region. Consequently, CDPKs do not require CaM to translate the Ca^{2+} chemical signal. A similar CaM-target-type interaction is seen with STIM1, the ER-based regulator of store-operated Ca^{2+} entry. When the Ca^{2+} level of the ER is high, Ca^{2+} binds to EF1 which together with the "hidden" EF-hand forms a stable intramolecular interaction with the internal SAM domain. However, upon depletion of the luminal ER Ca^{2+}, apo-STIM1 loses the molecular contacts between the EF-hand pair and the SAM domain and oligomerizes as a result. Another example is seen with the NCS family member CIB1. This protein has a C-terminal extension that interacts with its target-binding surface, an interaction that increases the target selectivity to CIB. In contrast, the EF-hand modules that are integrated into actinin and nucleobindin do not appear to interact directly with other portions of these proteins.

EF-Hand Proteins in Plants and Bacteria

Since the original discovery of the EF-hand helix-loop-helix structure, a vast amount of information about Ca^{2+}-signaling and Ca^{2+}-regulatory proteins in mammals has been accumulated. Our knowledge about the role of EF-hand proteins in plants and bacteria is also rapidly expanding. In plants, increased cytoplasmic Ca^{2+} concentration occurs in response to various stresses such as cold, high salt, or invading pathogens. Plants are unique in that they have multiple calmodulin isoforms with slight sequence variations, contrasting with the single extremely conserved calmodulin in vertebrates. Plants also have a group of some 50 calmodulin-like proteins, the functions of which have not yet been defined, as well as numerous CDPKs (DeFalco et al. 2009). Comparatively, little is known about the role of Ca^{2+} in bacteria. The cytosolic Ca^{2+} concentration in bacterial cells is maintained at a low µM level through ► Ca^{2+}-ATPases (Ca^{2+} pumps), similar to eukaryotes. Recently, a total of 390 putative EF-hand or EF-hand-like proteins have been predicted from sequence pattern searches of bacterial genomes. Included in this list are unique EF-hand-like motifs with uncommon secondary structures such as sheet-loop-helix or sheet-loop-sheet (Chen et al. 2010). These putative bacterial Ca^{2+}-binding proteins will need to be tested experimentally for their ability to bind Ca^{2+} as well as to determine their functional significance.

Cross-References

► Annexins
► Bacterial Calcium Binding Proteins
► C2 Domain Proteins
► Calcium ATPase
► Cadherins
► Calbindin D_{28k}
► Calcium in Biological Systems
► Calcium, Local and Global Cell Messenger
► Calcium, Neuronal Sensor Proteins
► Calcium-Binding Proteins
► Calmodulin
► Parvalbumin
► Penta-EF-Hand Calcium-Binding Proteins
► S100 Proteins
► Sarcoplasmic Calcium-Binding Protein Family: SCP, Calerythrin, Aequorin, and Calexcitin
► Troponin

References

Babini E, Bertini I, Capozzi F, Luchinat C, Quattrone A, Turano M (2005) Principal component analysis of the conformational freedom within the EF-hand superfamily. J Proteome Res 4:1961–1971

Berridge MJ, Bootman MD, Roderick HL (2003) Calcium signalling: dynamics, homeostasis and remodelling. Nat Rev Mol Cell Biol 4:517–529

Bhattacharya S, Bunick CG, Chazin WJ (2004) Target selectivity in EF-hand calcium binding proteins. Biochim Biophys Acta 1742:69–79

Burgoyne RD, O'Callaghan DW, Hasdemir B, Haynes LP, Tepikin AV (2004) Neuronal Ca^{2+}-sensor proteins: multitalented regulators of neuronal function. Trends Neurosci 27:203–209

Chen YY, Xue SH, Zhou YB, Yang JJ (2010) Calciomics: prediction and analysis of EF-hand calcium binding protein by protein engineering. Sci China Chem 53:52–60

DeFalco TA, Bender KW, Snedden WA (2009) Breaking the code: Ca^{2+} sensors in plant signalling. Biochem J 425:27–40

Gifford JL, Walsh MP, Vogel HJ (2007) Structures and metal-ion-binding properties of the Ca^{2+}-binding helix-loop-helix EF-hand motifs. Biochem J 405:199–221

Grabarek Z (2006) Structural basis for diversity of the EF-hand calcium-binding proteins. J Mol Biol 359:509–525

Ikura M, Ames JB (2006) Genetic polymorphism and protein conformational plasticity in the calmodulin superfamily: two ways to promote multifunctionality. Proc Natl Acad Sci USA 103:1159–1164

Ishida H, Vogel HJ (2006) Protein-peptide interaction studies demonstrate the versatility of calmodulin target protein binding. Protein Pept Lett 13:455–465

Kretsinger RH, Nockolds CE (1973) Carp muscle calcium-binding protein.II. Structure determination and general description. J Biol Chem 248:3313–3326

Linse S (2002) Calcium binding to proteins studied via competition with chromophoric chelators. Methods Mol Biol 173:15–24

Santamaria-Kisiel L, Rintala-Dempsey AC, Shaw GS (2006) Calcium-dependent and -independent interactions of the S100 protein family. Biochem J 96:201–214

Schwaller B (2009) The continuing disappearance of "pure' Ca^{2+} buffers. Cell Mol Life Sci 66:275–300

Shaw GS, Hodges RS, Sykes BD (1990) Calcium-induced peptide association to form an intact protein domain: ^{1}H NMR structural evidence. Science 249:280–283

Stathopulos PB, Zheng L, Li GY, Plevin MJ, Ikura M (2008) Structural and mechanistic insights into STIM1-mediated initiation of store-operated calcium entry. Cell 135:110–122

Yamniuk AP, Vogel HJ (2006) Insights into the Structure and Function of Calcium- and Integrin-Binding Proteins. Calcium Bind Prot 1:150–155

EF-hand Proteins and Magnesium

Jessica L. Gifford and Hans J. Vogel
Department of Biological Sciences, Biochemistry Research Group, University of Calgary, Calgary, AB, Canada

Synonyms

Ca^{2+}-binding protein; Helix-loop-helix Ca^{2+}-binding protein

Definition

Members of the EF-hand superfamily of regulatory calcium-binding proteins bind the calcium cation through chelating ligands presented by the EF-loop. In an analogous manner, these same ligands bind the chemically similar magnesium cation. Due to the high concentration of magnesium in the cell, this divalent cation often occupies EF-hands particularly under resting conditions at which the calcium concentration of the cytoplasm is low. Competition between calcium and magnesium cations for EF-hands modulates the responsiveness of members of the EF-hand superfamily. Magnesium-bound EF-hands can also have a function independent of the calcium-bound form.

Magnesium, the Eukaryotic Cell, and the EF-hand

In the cell, the divalent magnesium cation (Mg^{2+}) is the second most abundant cation after potassium (K^+) (Romani 2007; Romani and Scarpa 2000). As a cofactor in adenine triphosphate (ATP)-dependent enzymatic reactions, Mg^{2+} plays an important role in numerous biological processes including nucleic acid metabolism, chromatin organization, protein synthesis, metabolic reactions in both the cytosol and the mitochondria, and membrane transport. Furthermore, Mg^{2+} serves as an allosteric regulator modulating enzyme active site structure, can induce structural stabilization of proteins and nucleic acids, and mediate protein-protein interactions. In the majority of mammalian cells, the total Mg^{2+} content ranges between 14 and 20 mM; however, as more than 90% of intracellular Mg^{2+} is bound to ribosomes, phospholipids, or nucleotides (ATP, GTP, etc.), the concentration of intracellular free Mg^{2+} is in the order of 0.5–0.7 mM. The Mg^{2+} content of mammalian cells is tightly regulated by both entry and efflux across the plasma membrane, through organelle compartmentation (in the nucleus, endo(sarco)plasmic reticulum (ER), and mitochondria), and through intracellular buffering by ATP, other phosphonucleotides, and cytosolic proteins (discussed below). Although large fluxes of Mg^{2+} cross the cell membrane in either direction following a variety of hormonal and nonhormonal stimuli, the cytosolic free Mg^{2+} concentration remains relatively static.

In contrast with Mg^{2+}, the intracellular concentration of the calcium cation (Ca^{2+}) fluctuates between 10^{-7} and 10^{-5} M, depending on the activation state of the cell, an oscillation that enables Ca^{2+} to act as a potent secondary messenger (Berridge et al. 2000). A major class of intracellular proteins that intercept this Ca^{2+} signal bind the cation using a helix-loop-helix motif termed the EF-hand (Gifford et al. 2007). Amazingly, EF-hand Ca^{2+}-receptors respond specifically to micromolar concentrations of Ca^{2+} despite the presence of a ~1,000-fold excess of chemically similar Mg^{2+}. As many EF-hand Ca^{2+}-binding sites have Mg^{2+} dissociation constants in the submillimolar range, this excess of Mg^{2+} is sufficient to fully or partially saturate the Ca^{2+}-binding sites of many EF-hand proteins, particularly in the Ca^{2+}-depleted cytosol of the resting cell. Thus, to convey Ca^{2+} signals, EF-hand proteins must respond differently to Ca^{2+} than to Mg^{2+}. This entry will focus on the structural differences between Ca^{2+}-bound and Mg^{2+}-bound EF-hands as well as the functional consequences of this difference.

The EF-hand as a Ca^{2+} Binding Motif

Proteins containing the EF-hand motif constitute a large and functionally diverse family. Composed of two α-helices bridged by a Ca^{2+}-chelation loop, the EF-hand is defined by its secondary structure as well as the ligands presented by the loop (Fig. 1). Due to the chemical properties of Ca^{2+} and its preference for "hard" ligands provided by oxygen atoms, EF-hands and EF-loops in particular are rich in the carboxylate-containing amino acids aspartate and glutamate, the side chains of which provide most coordinating atoms for the bound cation (Falke et al. 1994). The identity of the coordinating side chains is semi-conserved in the most common (canonical) EF-hand and their positioning in the loop satisfies the requirement of Ca^{2+} for seven ligands arranged in a pentagonal bipyramidal geometry (Fig. 2a). The chelating residues are notated in two ways first based on their linear position in the loop and second on their alignment on the axes of the pentagonal bipyramid: 1(+*X*), 3(+*Y*), 5(+*Z*), 7(−*Y*), 9(−*X*), 12(−*Z*). The EF-loop provides either directly (or indirectly through a hydrogen-bonded water molecule) five of the seven Ca^{2+}-coordinating groups. The remaining two ligands originate in a bidentate manner

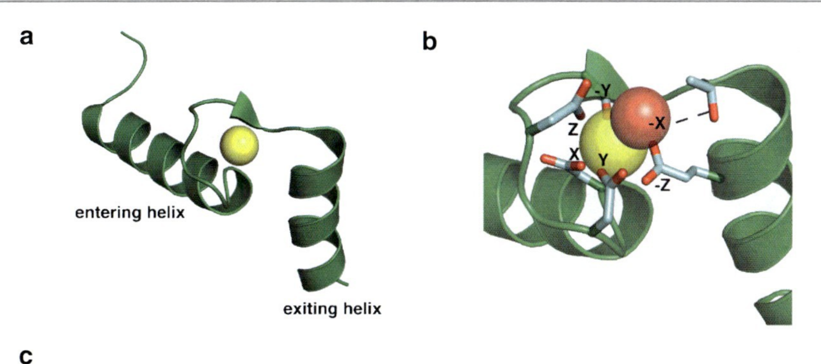

c

EF-loop position	1	2	3	4	5	6	7	8	9	10	11	12
Coordinating ligand	X sc[a]		Y sc		Z sc		−Y bb		−X sc[b]			−Z sc2[c]
Amino Acid	Asp		Asp Asn		Asp Ser Ans	Gly			Asp Ser Thr Glu Asn Gly Gln			Glu Asp[d]

a) Coordinating ligand provided by the side chain (sc) or the backbone (bb)
b) Coordinating ligand typically provided by a water molecule hydrogen bonded to the side chain of this amino acid
c) Side chain provides a bidentate coordinating ligand
d) Presence of Asp in ~10 % of EF-loops increases the affinity and selectivity for Mg^{2+} (see text for details)

EF-hand Proteins and Magnesium, Fig. 1 *Coordination of* Ca^{2+} *by the EF-hand.* (**a**) A single EF-hand from the N-terminal domain of CaM. Extending for 29 amino acid residues, the motif is formed by a nine-residue entering helix, a nine-residue loop including a short β-strand, and an 11-residue exiting helix (PDB code: 1EXR). (**b**) Coordination scheme of the bound Ca^{2+} ion by ligands provided by the EF-loop. The chelating side chains are indicated in *light blue* (carbon atoms) and *red* (oxygens), and the coordinating water molecule and bound Ca^{2+} ion as *red* and *yellow spheres*, respectively. (**c**) The coordination scheme as well as sequence preference of the Ca^{2+}- and Mg^{2+}-binding EF-hand loop

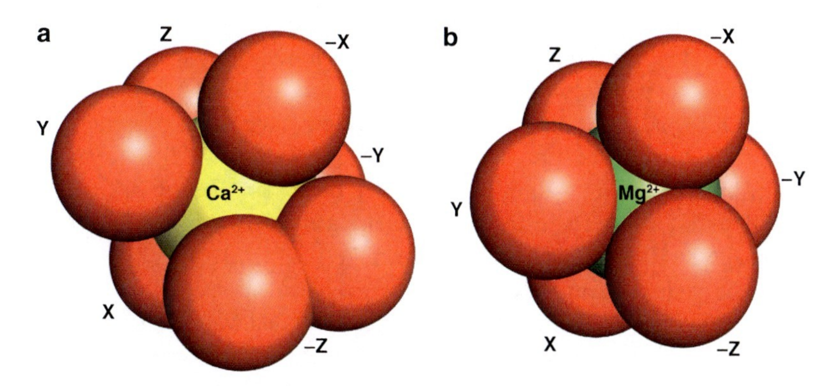

EF-hand Proteins and Magnesium, Fig. 2 *Coordination scheme of (**a**)* Ca^{2+} *and (**b**)* Mg^{2+}. Ca^{2+} requires seven ligands (preferably oxygen atoms) arranged in a pentagonal bipyramid (5 equatorial, 2 axial). Mg^{2+} also has a preference for oxygen ligands; however, due its smaller size, it requires six ligands arranged in an octahedron (4 equatorial, 2 axial). The Ca^{2+} and Mg^{2+} ions and coordinating oxygen ligands are represented by *yellow*, *green*, and *red spheres*, respectively

from the side chain of a Glu or Asp amino acid residue located in the exiting helix, three residues removed from the loop's C-terminus. Not technically part of the EF-loop, this residue is commonly referred to as the loop's twelfth residue and plays a considerable role in the different conformation responses of the EF-hand to either Ca^{2+} or Mg^{2+} (discussed below).

The EF-hand motif almost always occurs as a pair forming a discrete domain (Fig. 3) and members of the EF-hand superfamily typically contain two, four, or six EF-hands (Nelson and Chazin 1998). In the absence of Ca^{2+}, the motifs are stacked against one another in a face-to-face manner, forming a four helix-bundle stabilized by a short antiparallel β-sheet connecting

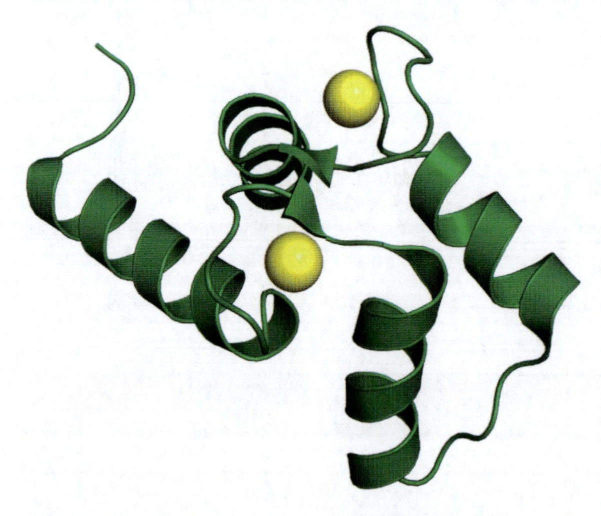

EF-hand Proteins and Magnesium, Fig. 3 *A pair of EF-hands from the N-terminal domain of CaM.* In addition to hydrophobic contacts between the helices, a small β-sheet is formed between the Ca^{2+}-binding loops of the two EF-hands. In most cases, the EF-hand pair is the functional Ca^{2+}-binding unit. The bound Ca^{2+} ion is indicated as a *yellow sphere* (PDB code: 1EXR)

the two Ca^{2+}-binding loops. Extensive hydrophobic contacts among the flanking helices in concert with the antiparallel β-sheet and additional hydrogen bonds that connect the Ca^{2+}-binding loops enable the EF-hand pair to act as a Ca^{2+}-binding unit, often binding this cation with positive cooperativity increasing the responsiveness of the EF-hand to the Ca^{2+} signal.

For most EF-hands, the binding of Ca^{2+} enacts a conformational change. In the CaM-subfamily, the response of the EF-hand to Ca^{2+} has been described as a "closed-to-open" domain transition (Fig. 4a). Characterized by a change in the angle between the entering and exiting helices, this conformational rearrangement is the consequence of Ca^{2+} chelation by the EF-loop (Grabarek 2006). Following initial chelation of the Ca^{2+} ion by ligands in the N-terminal part of the loop, the exiting helix of the EF-hand must be repositioned by ~2 Å in order for the bidentate twelfth (−*Z*) ligand to complete Ca^{2+}'s coordination sphere (Fig. 4b). In 90% of functional EF-hands, the side chain of a glutamate residue (termed Glu12) supplies this bidentate ligand and there is direct coupling between the engagement of these final ligands to the induced conformational change. For the CaM-subfamily, this "off" to "on" transition leads to the exposure of a hydrophobic pocket that serves as a target interaction site. For other EF-hand subfamilies, Ca^{2+}-binding can lead to a rotation of the interhelical angle between the entering and exiting helices, domain rotation, or no conformational change at all (particularly for proteins that buffer the Ca^{2+} concentration of the cytosol) (Yap et al. 1999).

The EF-hand as a Mg^{2+} Binding Motif

In vitro, the EF-hand can bind metal cations from groups Ia, IIa, and IIIa including the lanthanides. With the exception of sodium (Na^{+}), K^{+}, and Mg^{2+}, most cations found in these groups occur at extremely low concentrations in vivo and present little competition for Ca^{2+} binding to EF-hands. Na^{+} and K^{+} exist at concentrations 10^{2}–10^{6}-fold higher than Ca^{2+}. The larger ionic radius of K^{+} (1.33 angstrums vs. 0.99 angstrums for Ca^{2+}) excludes this cation from the EF-loop. In contrast, due to a similar ionic radius (1.02 angstrums), Na^{+} can enter the EF-loop, however, the low charge density of this cation leads it to have a weak affinity for the EF-hand. In contrast. The close chemical properties of Ca^{2+} and Mg^{2+} (both group IIa metals) combined with a 10^{2}–10^{4}-fold concentration excess of Mg^{2+} leads to a complicated case of cation selectivity.

Like Ca^{2+}, Mg^{2+} has a preference for the "hard" ligands provided by the oxygen atoms of the EF-loop and due to its divalent charge can stabilize the O-O repulsion of the EF-hand's coordinating array. Despite these similarities, EF-hands can discriminate between these two cations due to differences in ionic radii and consequentially hydration and coordination geometry. Unlike the larger Ca^{2+} ion which can accommodate five oxygen atoms, only four oxygen atoms fit in a plane around Mg^{2+} (Fig. 2b). As a result, Mg^{2+} has a strict requirement for six ligands arranged in octahedral geometry as opposed to Ca^{2+}'s penchant for seven ligands in a pentagonal bipyramid. By controlling ligand flexibility, EF-hands can exclude Mg^{2+}. Energetically, the hydration state of Mg^{2+} further contributes to the selectivity of the EF-hand for Ca^{2+}. Due to a 30% smaller ionic radius (0.66 Å), the tight ordering of water molecules solvating Mg^{2+} leads the hydrated ion to have an ~400x larger volume than the dehydrated form. In contrast, the volume of the hydrated Ca^{2+} is ~25x that of the bare ion (Maguire and Cowan 2002). This difference in hydration influences both the kinetics and energetics of Mg^{2+} binding to the EF-hand as the kinetic on-rate of this ion is

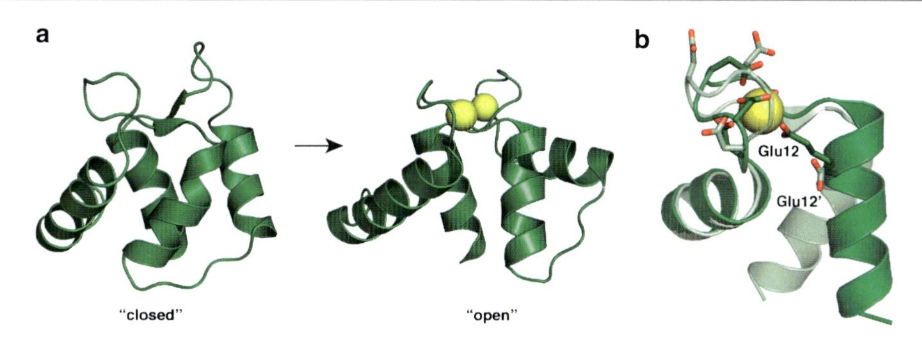

EF-hand Proteins and Magnesium, Fig. 4 *The Ca^{2+}-Induced Conformational Change*. (**a**) The *closed*-to-*open* transition induced by the binding of Ca^{2+} as seen for the N-terminal domain of CaM. The helices move from an antiparallel arrangement in the apo form to being nearly perpendicular in the Ca^{2+}-bound state. (**b**) The engagement of the bidentate ligands provided by Glu12 leads to a repositioning of the exiting helix with respect to the entering helix. Here, the apo form is indicated in light green and the Ca^{2+}-bound in dark green, the bound Ca^{2+} is represented as a *yellow sphere* (PDB codes: 1EXR and 1DMO)

limited by its dehydration rate and binding affinity is weakened by the steep desolvation energy of this cation (Martin 1990). Despite these selectivity mechanisms, many members of the EF-hand superfamily still bind Mg^{2+} to their EF-loops and EF-hands that bind Mg^{2+} fall into two groups: "Ca^{2+}-specific" and "Ca^{2+}/Mg^{2+}" EF-hands (Table 1).

Ca^{2+}-Specific EF-hands

Due to the excess concentration of free Mg^{2+} in the cell, many EF-hands have sufficient affinity for this cation to be partially or even fully saturated with Mg^{2+} at resting Ca^{2+} concentrations. For "Ca^{2+}-specific" EF-hands, key to effective Ca^{2+} signaling is a difference in conformation between the Ca^{2+}- and Mg^{2+}-bound states of the EF-hand. Solution NMR-based studies indicate that the Ca^{2+}-specific EF-hands of calmodulin (CaM) bind Mg^{2+} with a dissociation constant in the millimolar range implying that, at resting Ca^{2+} levels, this protein is almost 50% saturated with Mg^{2+}. Although Mg^{2+} engages the Ca^{2+} ligands of all four of CaM's EF-hands, Mg^{2+} appears to bind ligands only in the N-terminal part of each loop [ligands 1(+*X*), 3(+*Y*), 5(+*Z*), 7(-*Y*)] (Fig. 5). By not engaging the C-terminal ligands of the loop, in particular the critical Glu12, the chelation of Mg^{2+} has no effect on the overall conformation of the EF-hand causing only localized rigidity of the EF-loop, these sites remain in the "closed" or "off" position (Gifford et al. 2007; Grabarek 2011). The crystal structure of Mg^{2+}-calbindin D9K illustrates the coordination scheme of Mg^{2+} when bound to Ca^{2+}-specific EF-hands. In agreement with Mg^{2+}-CaM, Mg^{2+} interacts only with the N-terminal ligands of the EF-loop and the loop is in an extended conformation similar to that of the apo form. The potential ligands provided by Glu12 are too far from the bound cation to contribute to its coordination and instead a water molecule completes the coordination sphere of Mg^{2+}. For Ca^{2+}-specific EF-hands, due to stereochemical constraints imposed by the paired domain structure of

EF-hand Proteins and Magnesium, Table 1 Examples of known Ca^{2+}-specific and Ca^{2+}/Mg^{2+} EF-hand-containing proteins (Gifford et al. 2007)

Ca^{2+}-specific EF-hands	$K_{Mg} = 10^3\ M^{-1}$
CaM	Ubiquitous eukaryotic Ca^{2+}-binding protein
Calbindin D_{9K}	Ca^{2+} buffer, intracellular Ca^{2+} transport
Ca^{2+}/Mg^{2+} EF-hands	$K_{Mg} = 10^4$–$10^5\ M^{-1}$
Parvalbumin	Ca^{2+} buffer
Troponin C	Skeletal and cardiac muscle contraction
Centrin	Cell cycle
Caldendrin	Neuronal protein, modulates Ca^{2+} entry into the cytoplasm
CIB	Platelet aggregation
Myosin RLC	Molluscan muscle contraction; regulatory light chain
Neurocalcin	Endocytosis
GCAPs	Light sensitivity
NCS-1 (frequenin)	Neurotransmission regulation
DREAM	Transcriptional repressor
Aequorin	Bioluminescence
Calbindin D_{28K}	Found in intestinal epithelium and brain
CaV1.2 channel	Ca^{2+} channel, possibly involved in Mg^{2+} regulation

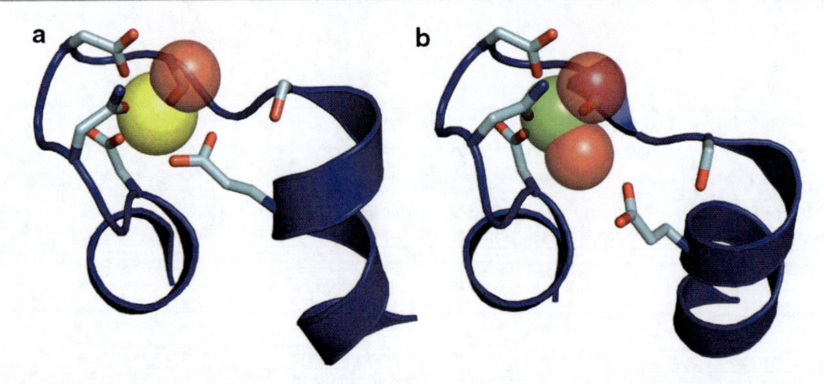

EF-hand Proteins and Magnesium, Fig. 5 *Coordination of Mg^{2+} by the Ca^{2+}-specific EF-hand.* (**a**) Ca^{2+} bound to the Ca^{2+}-specific EF-loop of calbindin D9K. (**b**) Mg^{2+} bound to the same EF-hand. In contrast with Ca^{2+}, Mg^{2+} binds only the N-terminal ligands of the loop. The potential ligand provided by Glu12 is too far from the bound cation to be used and instead a second water molecule completes the coordination sphere. In (**A**/**B**) the Ca^{2+} and Mg^{2+} ions are represented by *yellow* and *green spheres*, respectively; the water ligands(s) as a *red sphere*; portions of the entering and exiting helices as *dark blue*; and the chelating side chains as *light blue* (carbon), *red* (oxygen), and *blue* (nitrogen) (PDB codes: 3ICB and 1IG5, respectively)

EF-hands the smaller Mg^{2+} ion cannot engage the ligands of the EF-loop in the same way as Ca^{2+}, as a result Mg^{2+} can serve to stabilizes the apo-conformation of the EF-hand (Grabarek 2006, 2011).

Ca^{2+}/Mg^{2+} EF-hands

In contrast with the Ca^{2+}-specific EF-hands, "Ca^{2+}/Mg^{2+}" EF-hands have an affinity for Mg^{2+} several fold higher and, at least in the resting cell, functionally bind this cation. Mg^{2+} interacts with the EF-loop of Ca^{2+}/Mg^{2+} EF-hands in almost the same manner as Ca^{2+} including the use of a ligand provided by the loop's twelfth residue. Critical to the cation-interchangeability of many Ca^{2+}/Mg^{2+} EF-hands is the ability of the Glu12 side chain to adapt the number of ligands presented by the EF-loop. Observed in the crystal structure of the Ca^{2+}/Mg^{2+} EF-hands of Mg^{2+}-bound parvalbumin as well as in molecular dynamics simulations of the EF-hands of the C-terminal domain of skeletal troponin (sTnC), a 120° rotation of the Cα-Cβ bond of Glu12 enables this key amino acid to provide a bidentate ligand for Ca^{2+} and a monodentate ligand for Mg^{2+} (PDB codes: 4CPV and 4PAL) (Fig. 6a, b). For Ca^{2+}/Mg^{2+} EF-hands, the flexibility of this side chain is critical: If the side chain is sterically blocked into a bidentate configuration, the EF-hand becomes Ca^{2+} specific as the loop cannot adjust to the smaller ionic radius and octahedral coordination geometry of Mg^{2+}. As the rotation of the Glu12 side chain is limited by the local environment, it is difficult if not impossible from the primary structure to predict if a given EF-hand is a Ca^{2+} or Ca^{2+}/Mg^{2+} site. Because the side chain of Glu12 provides a monodentate ligand for the coordination of Mg^{2+}, the binding of this cation mimics that of Ca^{2+}, and Mg^{2+}-bound EF-hands have a conformation essentially the same as that of the Ca^{2+}-bound form. For Ca^{2+}/Mg^{2+} EF-hands, the binding of Mg^{2+} induces a conformational change and is not conformationally inconsequential as it is for Ca^{2+}-specific EF-hands.

A second class of Ca^{2+}/Mg^{2+} EF-hand contains a noncanonical EF-loop. In approximately 10% of known EF-hands, an aspartate residue instead of a glutamate occupies the loop's 12th position. "Asp12" EF-hands, termed to distinguish them from the Glu12 variety, tend to be smaller and more compact than the canonical loop, a characteristic that has effects on both Ca^{2+} coordination and Mg^{2+} selectivity. As seen in the crystal structure of CIB (*C*a^{2+}- and *i*ntegrin-*b*inding protein), the side chain of an Asp12 EF-hand is too short for both oxygen atoms to coordinate Ca^{2+} and although one does, the second is replaced by a bridging water molecule (Fig. 6c, d). Due to a less favorable binding energy, the inclusion of this water in the coordination sphere of Ca^{2+} decreases the EF-hand's affinity for this cation (Yamniuk et al. 2008). In contrast, the monodentate coordination presented by Asp12 favors the binding of Mg^{2+}. This subtle difference in coordination scheme increases the EF-hand's affinity for Mg^{2+} and reverses the Mg^{2+}/Ca^{2+} specificity of the binding loops. The preference of Mg^{2+} for a shorter side chain in the last position of the loop is

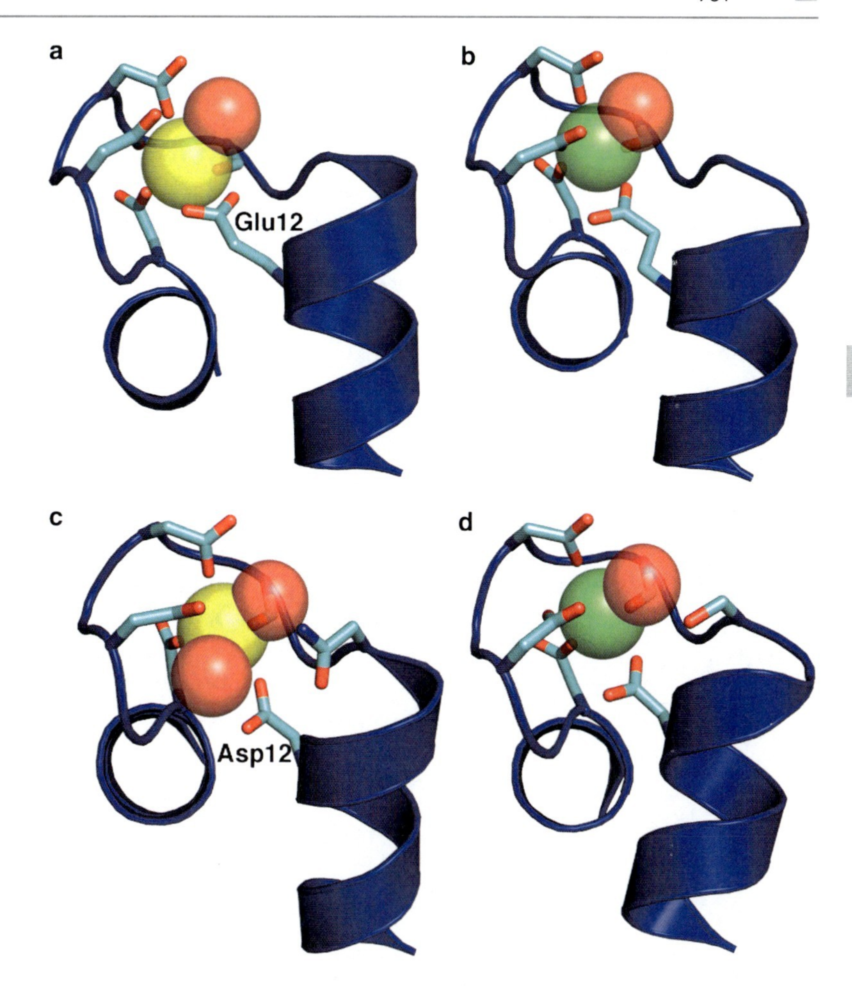

EF-hand Proteins and Magnesium, Fig. 6 Coordination of Mg^{2+} by Ca^{2+}/Mg^{2+} EF-hands. (**a**) Ca^{2+} bound to the Ca^{2+}/Mg^{2+} Glu12 EF-loop of parvalbumin EF-hand 3. (**b**) To accommodate Mg^{2+} in this same EF-hand, the side chain of Glu12 rotates so that only one ligand is provided for Mg^{2+} as opposed to two for Ca^{2+}. (**c**) Ca^{2+} bound to the Ca^{2+}/Mg^{2+} loop of CIB EF-hand 3. Instead of glutamate, an aspartate occupies the 12th position of this EF-hand. The side chain of Asp12 is too short to act as a bidentate ligand for Ca^{2+}, and instead provides one ligand. The coordination sphere of Ca^{2+} is completed by a water molecule. (**d**) Mg^{2+} bound to the Ca^{2+}/Mg^{2+} loop of scallop myosin RLC. The smaller Asp12 side chain provides a monodentate ligand and favors the binding of Mg^{2+} over Ca^{2+}. In (**a-d**) the coloring scheme is the same as in Fig. 5 (PDB codes are 4CPV, 4PAL, 1XO5, and 1WDC, respectively)

counterintuitive as the ionic radius of this cation is smaller than that of Ca^{2+}. It would be expected that an Asp12 loop would have to compress significantly more to coordinate the bound Mg^{2+} ion, requiring a larger shift of the exiting helix compared to when there is a glutamate in this position. However, as seen from the crystal structure of the Ca^{2+}/Mg^{2+} Asp12 EF-hand of scallop myosin RLC (*r*egulatory *l*ight *c*hain), because the Asp12 side chain only provides a monodentate ligand to the octahedral coordination of Mg^{2+}, the conformation of the loop is not significantly constrained.

Functional Roles for Mg^{2+} Binding to EF-hands

Although not traditionally thought of as a Mg^{2+}-binding motif, the chelation of Mg^{2+} by EF-hands is a physiologically important event and serves a number of functions. For several EF-hand-containing proteins, the binding of Ca^{2+} by EF-loops induces structure formation from a poorly defined molten globule apo-state. These "structural" EF-hands, such as those found in the invertebrate SCPs (*s*arcoplasmic *b*inding *pro*teins) and the C-terminal domain of sTnC, tend to be Ca^{2+}/Mg^{2+} EF-hands guaranteeing occupancy by either cation and structure formation at any cytoplasmic Ca^{2+} concentration. As Glu12 structural EF-hands tend to have affinities that ensure Ca^{2+} saturation at resting intracellular Ca^{2+} concentrations, their ability to bind Mg^{2+} is not as significant as it is for Asp12 structural EF-hands. Due to a weakened Ca^{2+} affinity caused by the aspartate amino acid residue in the loop's 12th position, the ability of these EF-hands to bind Mg^{2+} in Ca^{2+}/Mg^{2+} sites is critical as this cation maintains protein structure in the absence, and sometimes presence, of Ca^{2+}. The ability of many EF-hand proteins to be saturated with Mg^{2+} at low cytoplasmic Ca^{2+} concentrations makes their unstructured apo form physiologically irrelevant. Furthermore, as the Mg^{2+}-bound conformation tends to differ in subtle ways from that of the Ca^{2+}-bound state, these proteins which include CIB, caldendrin, and GCAPs (*g*uanylate *c*yclase *a*ctivation *p*roteins) are still able to respond conformationally to the increase in Ca^{2+} concentration that occurs upon cell stimulation.

Due to its ability to effectively compete for EF-hand binding sites, Mg^{2+} can modulate the affinity of EF-hands for Ca^{2+} (Grabarek 2011). In the presence of Mg^{2+}, both Ca^{2+}-specific and Ca^{2+}/Mg^{2+} EF-hands experience a decrease in the kinetic on-rate for Ca^{2+} as the binding of this cation becomes limited by the off-rate of Mg^{2+}. By decreasing the apparent Ca^{2+} affinity and thus sensitivity of an EF-hand, Mg^{2+} competition delays the response of a Ca^{2+}-receptor to an increased cytoplasmic Ca^{2+} concentration. As a result, the presence of Mg^{2+} shifts the target enzyme activation curve of a given Ca^{2+}-receptor to a higher Ca^{2+} concentration. The extent to which Mg^{2+} can decrease the affinity of an EF-hand for Ca^{2+} is represented by the following simplified relation:

$$K'_{\mathrm{Ca}} = K_{\mathrm{Ca}} / \left(1 + K_{\mathrm{Mg}}\left[\mathrm{Mg}^{2+}\right]\right)$$

where K_{Ca} and K'_{Ca} are the binding constants for Ca^{2+} in the absence and presence of Mg^{2+} respectively, and K_{Mg} is the binding constant for Mg^{2+}. This relation becomes more complicated if, as is the case for CaM, TnC, and many others, Ca^{2+} binds with positive cooperativity, or if the target protein is present. Owing to a more complete loop coordination, the effect of Mg^{2+} competition on Ca^{2+} affinity is more significant for Ca^{2+}/Mg^{2+} sites than it is for Ca^{2+}-specific sites. The N- and C-terminal domains of sTnC contain a pair of Ca^{2+}-specific and Ca^{2+}/Mg^{2+} EF-hands, respectively. The Ca^{2+} on-rate for the Ca^{2+}/Mg^{2+} EF-hands of the C-terminal domain is more than 100-fold lower than that of the Ca^{2+}-specific sites found in the N-terminal domain of the protein. Although other factors influence the on-rates of these EF-hands, this decrease is partially attributed to the dissociation rate of Mg^{2+}. In vitro, the presence of a target protein is known to increase the Ca^{2+} affinity of EF-hand-containing proteins, often to the strength that Ca^{2+} would bind at resting cytoplasmic Ca^{2+} concentrations. Thus, the ability of Mg^{2+} to stabilize the "off state" of some EF-hand proteins, particularly those with Ca^{2+}-specific EF-hands, may play an active role in the Ca^{2+}-dependent regulation of cellular processes by facilitating the switching off of their respective target enzymes at resting Ca^{2+} levels (Grabarek 2011). Furthermore, the binding of Mg^{2+} stabilizes the preferred conformation of downstream targets that interact with the apo-state of EF-hand Ca^{2+}receptors, for example, CaM and so-called IQ domain-containing targets (Jurado et al. 1999).

Finally, for some members of the EF-hand superfamily, Mg^{2+}-bound EF-hands have a function distinct from that of Ca^{2+}. It is the Mg^{2+}-bound form of DREAM (*d*ownstream *r*egulatory *e*lement *a*ntagonist *m*odulator) that binds sequence-specific DNA targets; the binding of Ca^{2+} to this protein disrupts this interaction. Furthermore, evidence suggests that the single EF-hand found in the cytoplasmic C-terminal of the cardiac CaV1.2 channel is specific for and regulated by Mg^{2+}. Mg^{2+} would bind this EF-hand under conditions of stress (e.g., ischemia, heart failure) when the ATP levels in the cytoplasm drop and the concentration of free Mg^{2+} correspondingly rises. As a result, the ability of this channel to conduct Ca^{2+} is inhibited and there is an adjustment in the contractile function of the heart (Brunet et al. 2005).

Conclusion

Although the EF-hand helix-loop-helix is traditionally thought of as a Ca^{2+}-binding motif, it also functionally binds the chemically similar Mg^{2+} cation. Differences in identity and number of ligands supplied by the EF-loop create three different EF-hand Mg^{2+} coordination schemes, each having a distinct affinity and selectivity for Mg^{2+}. The ability of some EF-hand-containing proteins to bind Mg^{2+} at low cytoplasmic Ca^{2+} levels brings a structured conformation to these Ca^{2+}-responders that are unstructured in their apo-state. Due to the high cytoplasmic concentration of Mg^{2+} and the complex interplay between Ca^{2+} and this cation, Mg^{2+} modulates the Ca^{2+} sensitivity of many EF-hand-containing proteins. Finally, for some members of the EF-hand superfamily, binding of Mg^{2+} serves a purpose independent of that of Ca^{2+}.

Underscoring the importance of this metal ion-protein interaction is the potential link between the ability of Mg^{2+} to bind EF-hands and some pathological conditions. At times of cellular stress, an increase in cytoplasmic Mg^{2+} concentration due to release from internal stores or the depletion of cellular ATP levels would decrease the Ca^{2+} sensitivity of EF-hands and increase the threshold Ca^{2+} concentration required to activate Ca^{2+}-responders. Mg^{2+} deficiency could have the opposite effect: excessive activation of Ca^{2+}-regulated cellular processes due to a lack of Mg^{2+}-modulated EF-hand Ca^{2+} affinity. The functions of the heart and brain are particularly susceptible to slight changes in the responsiveness of EF-hands as not only are they directly regulated by precise Ca^{2+} transients but they are also modulated through Ca^{2+}-sensitive signaling pathways.

References

Berridge MJ, Lipp P, Bootman MD (2000) The versatility and universality of calcium signalling. Nat Rev Mol Cell Biol 1:11–21

Brunet S, Scheuer T, Klevit R, Catterall WA (2005) Modulation of CaV1.2 channels by Mg2+ acting at an EF-hand motif in the COOH-terminal domain. J Gen Physiol 126:311–323

Falke JJ, Drake SK, Hazard AL, Peersen OB (1994) Molecular tuning of ion binding to calcium signaling proteins. Q Rev Biophys 27:219–290

Gifford JL, Walsh MP, Vogel HJ (2007) Structures and metal-ion-binding properties of the Ca2 + -binding helix-loop-helix EF-hand motifs. Biochem J 405:199–221

Grabarek Z (2006) Structural basis for diversity of the EF-hand calcium-binding proteins. J Mol Biol 359:509–525

Grabarek Z (2011) Insights into modulation of calcium signaling by magnesium in calmodulin, troponin C and related EF-hand proteins. Biochim Biophys Acta 1813:913–921

Jurado LA, Chockalingam PS, Jarrett HW (1999) Apocalmodulin. Physiol Rev 79(3):661–682

Maguire ME, Cowan JA (2002) Magnesium chemistry and biochemistry. Biometals 15:203–210

Martin RB (1990) Bioinorganic chemistry of magnesium. Met Ions Biol Syst 26:1–13

Nelson MR, Chazin WJ (1998) Structures of EF-hand Ca(2+)-binding proteins: diversity in the organization, packing and response to Ca2+ binding. Biometals 11:297–318

Romani AM (2007) Magnesium homeostasis in mammalian cells. Front Biosci 12:308–331

Romani AM, Scarpa A (2000) Regulation of cellular magnesium. Front Biosci 5:D720–D734

Yamniuk AP, Gifford JL, Linse S, Vogel HJ (2008) Effects of metal-binding loop mutations on ligand binding to calcium- and integrin-binding protein 1. Evolution of the EF-hand? Biochemistry 47:1696–1707

Yap KL, Ames JB, Swindells MB, Ikura M (1999) Diversity of conformational states and changes within the EF-hand protein superfamily. Proteins 37:499–507

EF-Hand Superfamily = Calmodulin Superfamily = Troponin Superfamily

▶ Calmodulin

EF-Hand Superfamily=Calmodulin Superfamily=Troponin Superfamily

▶ Calbindin D_{28k}

EGF-Like Domain: Epidermal Growth Factor-Like Domain

▶ Blood Clotting Proteins

Electron Spin Resonance (ESR), Electron Paramagnetic Resonance (EPR)

▶ Chromium(VI), Oxidative Cell Damage

Electron Transfer

▶ Heme Proteins, Heme Peroxidases

Electron Transfer Shuttles

▶ Iron-Sulfur Cluster Proteins, Ferredoxins

4f Elements

▶ Lanthanides, Physical and Chemical Characteristics

Enamelysin

► Zinc Matrix Metalloproteinases and TIMPs

Endocytosis

► C2 Domain Proteins

Endonexins

► Annexins

Endopeptidase 24.18

► Zinc Meprins

Endopeptidase-2

► Zinc Meprins

Endoplasmic Reticulum Stress Response

► Cadmium and Stress Response

Energy Transduction

► Heme Proteins, Cytochrome c Oxidase

Environmental Arsenic Exposure and Diabetes

► Arsenic-Induced Diabetes Mellitus

Environmental Dysregulation of Autoimmunity

► Silica, Immunological Effects

EnzMet

► Enzyme Metallography and Metallographic In Situ Hybridization

Enzymatic Tellurite Reduction

► Catalases as NAD(P)H-Dependent Tellurite Reductases

Enzyme Activation

► Iron-Sulfur Cluster Proteins, Fe/S-S-adenosylmethionine Enzymes and Hydrogenases

Enzyme Inhibition

► Platinum(II) Complexes, Inhibition of Kinases

Enzyme Involved in Phospholipid Metabolism

► Platinum (IV) Complexes, Inhibition of Porcine Pancreatic Phospholipase A2

Enzyme Mechanism

► Heme Proteins, Heme Peroxidases

► Zinc and Iron, Gamma and Beta Class, Carbonic Anhydrases of Domain Archaea

Enzyme Metallography

► Enzyme Metallography and Metallographic In Situ Hybridization

Enzyme Metallography and Metallographic In Situ Hybridization

Richard D. Powell and James F. Hainfeld
Nanoprobes, Incorporated, Yaphank, NY, USA

Synonyms

EnzMet; Enzyme metallography; Silver in situ hybridization; SISH

Definition

Enzyme metallography (EnzMet) is a method for staining, labeling, or detecting specific nucleic acid sequences, proteins, or other biochemical targets for observation by light or electron microscopy, or for conductimetric detection. The enzyme metallographic method consists of the application of a primary, secondary, or tertiary enzymatic probe such as an antibody or streptavidin conjugated to horseradish ► peroxidase, which targets sites of interest. Once bound, this is developed with a substrate containing dissolved metal ions. The enzyme selectively reduces and deposits the metal ions in the form of nanoparticles, which are visualized directly by electron microscopy, or indirectly as a dark, punctuate stain by light microscopy, by eye, or other sensing devices. Current enzyme metallography methods deposit metallic ► silver, yielding a black stain in the bright-field light microscope. The enzyme conjugate is most often used as a secondary probe: The target is first bound with a primary probe which reacts directly with the target of interest, which is then bound by the enzyme-linked secondary probe. The most common application is in situ hybridization (ISH), in which a target nucleic acid sequence within cells or tissues is bound by a complementary oligonucleotide probe linked to a small recognition tag such as biotin, which is then detected using an enzyme-linked secondary probe which binds to the tag, such as streptavidin. Enzyme metallography is also used for immunohistochemistry (IHC), in which a protein of interest is labeled using a primary antibody which binds to the target, followed by an enzyme-linked secondary antibody which binds to the primary. On biochips, tagged oligonucleotide probes with enzyme-linked secondary probes may be used to form conductive bridges between microelectrodes which are then detected electrically.

Principles and Characteristics

Many enzymes act as catalysts for chemical reactions yielding products with intense colors or signal properties which may be detected microscopically, analytically, or spectroscopically with very high sensitivity and specificity. Immunoenzymatic probes, in which these enzymes are linked to antibodies which bind to sites of interest, are used to localize and detect biochemical targets in cells and tissues. Well-known examples include the reaction of horseradish ► peroxidase with chromogenic substrates such as diaminobenzidine (DAB) to yield intensely colored products, and the reaction of alkaline phosphatase with an adamantyl 1,2-dioxetanephenyl phosphate substrate to yield light, termed chemiluminescence. Some enzymes can also catalyze the conversion of soluble metal ions to insoluble metallic deposits, which may be visualized by light microscopy, or in the case of high atomic number elements such as lead, by electron microscopy: These reactions provide a method for localizing enzymes in cells (enzyme histology).

In enzyme metallography, the active enzyme is conjugated to a targeting agent that binds to a site of interest, and revealed by treatment with a metal-containing, or "metallographic" substrate after binding. Thus, the process combines metallographic visualization with immunological or biochemical targeting, constituting a general method for the localization and visualization of targets. Current enzyme metallographic processes use silver-based substrates to deposit metallic silver. While the exact mechanism is not known, it is believed that the enzyme acts to transfer electrons selectively to silver (I) ions in solution to yield silver (0) atoms which are then deposited either on the enzyme itself or in very close proximity, rapidly forming deposits shown by electron microscopy to take the form of nanoparticles in the size range 10–100 nm (Powell et al. 2006), or irregular "nanoflowers" with diameters up to several hundred nanometers (Festag et al. 2008). In the bright-field light microscope, the deposited silver produces a discrete, black stain with sharp edges and a punctuate appearance. This morphology indicates negligible diffusion of the reaction product, and hence very high resolution, and contrasts with the more diffuse staining obtained with organic chromogens such as DAB.

Metallographic In Situ Hybridization

Metallographic in situ hybridization is the detection of specific nucleic acid sequences in cells or tissues using a procedure in which metal is deposited from solution to produce an optical signal, which is usually observed by light microscopy. This method includes two variations: (a) visualization using ▶ gold nanoparticles enlarged using either silver or gold autometallography, which is the selective deposition of silver or gold from solution onto the nanoparticle surface; or (b) visualization using enzyme metallography (EnzMet) in which an enzyme, in this case horseradish ▶ peroxidase, is used to selectively deposit metal from solution. The high sensitivity of the latter method, combined with its high signal definition, visual contrast, and high resolution, makes it highly suited to enumerating copy numbers of genes, and hence detection of conditions in which gene amplification plays a role. This is applicable to a wide variety of conditions, but its initial clinical application has been to assess the status of the *HER2* gene in breast cancer. The *HER2* (ERBB2) proto-oncogene encodes a transmembrane receptor tyrosine kinase of Mr185 kDa (p. 185) which is structurally and functionally analogous to the epidermal growth factor: amplification of this gene, and overexpression of the concomitant oncoprotein, occur in about 20–30% of invasive breast cancers, and correlate with aggressive malignant behavior and poor patient outcomes. The *HER2* oncoprotein is also the target of the humanized monoclonal antibody therapy Herceptin (Trastuzumab). Therefore, amplification of the *HER2* gene is a clinical indicator for Herceptin therapy.

Assessment of *HER2* gene amplification requires accurate copy number enumeration in order to distinguish genuine amplification, or the presence of multiple copies of the gene at its usual position, from polysomy, or duplication of chromosome 17, on which the gene resides. In normal, or non-amplified tissue, in situ hybridization for the *HER2* gene produces two copies (one for each strand of the DNA duplex); chromosome 17 polysomy gives between two and six copies; and *HER2* gene amplification produces more than six copies (Wolff et al. 2007). Each copy produces one spot upon in situ hybridization. Gene copy number is established by averaging the number of spots per cell. This is easiest and most accurate when the signals are sharply defined, highly resolved, and have high visual contrast with the surrounding tissue. The conventional method for determination of gene amplification is fluorescent in situ hybridization (FISH) in which gene copies are detected using a gene probe linked to a fluorescent label; this is observed by fluorescence microscopy where the genes are visualized as bright spots. However, chromogenic methods, which produce colored or dark signals against a bright background, provide the practicing pathologist with several advantages: Bright-field optics and microscopes are simpler, more universal, and more affordable than fluorescent systems; the observer need not wait for dark adaptation before recording copy number data; and the surrounding tissue is visible so that signals appear in morphological context, which may provide additional insights into the condition and help resolve any ambiguities in the results. The metal deposits are stable and do not fade over time or during observation the way fluorescent tags do. This also makes them preferable both for observation and for archiving.

Several chromogenic methods for *HER2* gene evaluation have been developed, using tagged gene probes combined with secondary probes linked to different reporter groups to develop the visible signal. Enzyme-linked probes used with DAB produce brown spots, but the signals are sometimes relatively diffuse and higher contrast and resolution are desirable. As alternatives, metallographic in situ hybridization methods based on gold nanoparticles with silver or gold autometallography were developed, shown schematically in Fig. 1. The first such process to be developed was the use of gold nanoparticle (Nanogold) conjugated secondary or tertiary probes, visualized

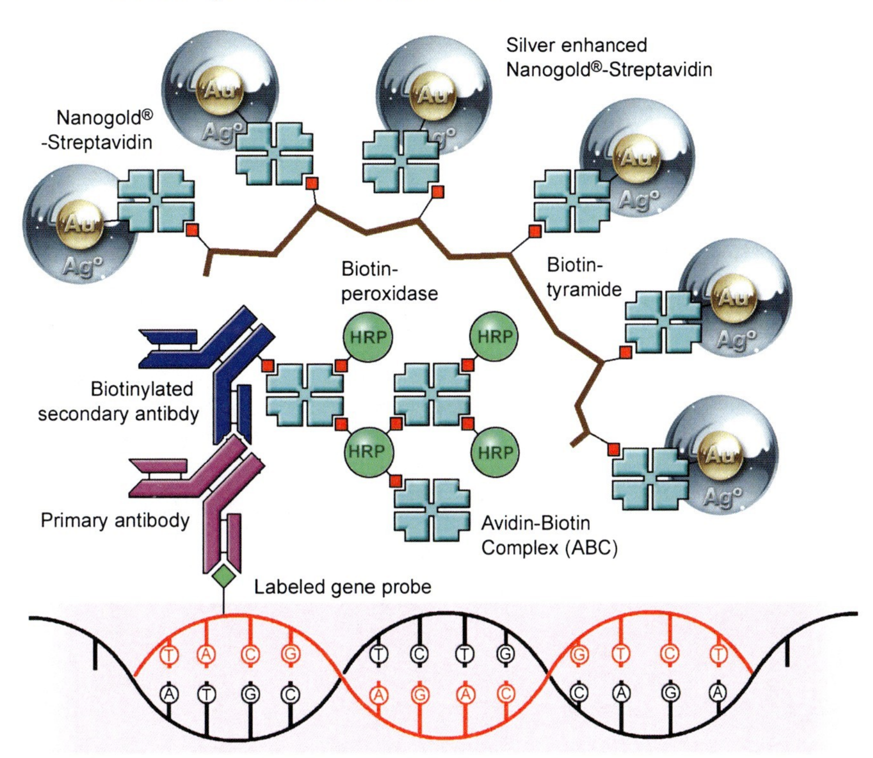

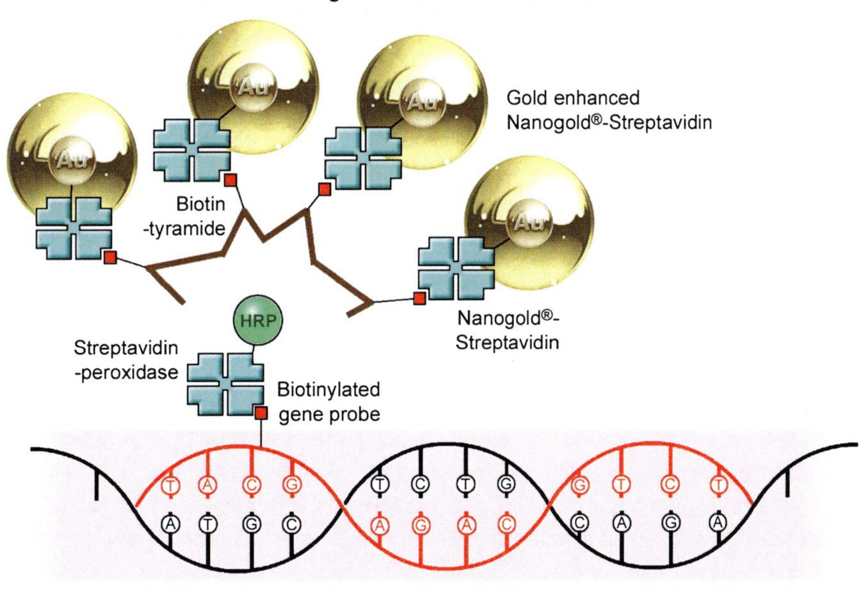

Enzyme Metallography and Metallographic In Situ Hybridization, Fig. 1 Gold nanoparticle (Nanogold) based metallographic in situ hybridization. (**a**) Detection of HPV16 in SiHa cells using a labeled gene probe, biotinylated secondary antibody, avidin-biotin complex followed by tyramide signal amplification producing catalytic deposition of biotin-tyramide, visualized using Nanogold-streptavidin and silver acetate autometallography. (**b**) Detection of *HER2* gene copies in paraffin-embedded human invasive breast carcinoma biopsy specimen using a biotinylated gene probe detected using streptavidin-► peroxidase, tyramide signal amplification giving catalytic biotin deposition, and visualization with Nanogold-streptavidin and gold enhancement (gold based autometallography)

by autometallographic enhancement using silver enhancement. The sensitivity of this procedure was maximized through the use of avidin-biotin complex followed by a tyramide signal amplification step before the metallographic detection, shown schematically in Fig. 1a. Although this procedure was highly sensitive and enabled the visualization of one to a few copies of human papillomavirus 16 in biopsy specimens, it was not developed for *HER2* gene amplification assessment because alternative, simpler protocols yielded

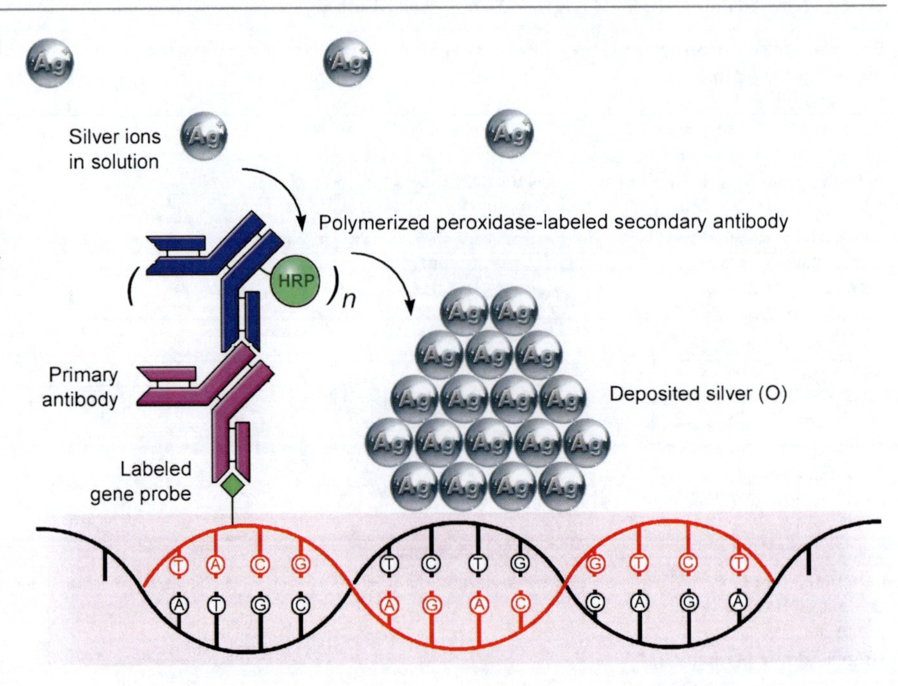

Enzyme Metallography and Metallographic In Situ Hybridization, Fig. 2 Enzyme metallographic (EnzMet) in situ hybridization detection procedure: A labeled gene probe is detected using a polymerized HRP-secondary antibody conjugate developed using enzyme metallography, showing catalytic deposition of metallic silver

superior results. GOLDFISH (Gold-Facilitated In Situ Hybridization: Fig. 1b) provides similar results with fewer steps. In GOLDFISH, Nanogold-conjugated secondary probes were combined with autometallographic amplification using gold enhancement: This yielded large, confluent signals in cells in which the *HER2* gene was amplified and clearly differentiated these regions from normal tissue. However, this method did not provide sufficient differentiation between chromosome 17 polysomy and genuine *HER2* gene amplification for clinical use (Tubbs et al. 2005b).

However, at this time, the enzyme metallographic reaction of horseradish ▶ peroxidase was discovered (Fig. 2). When applied as a detection method for in situ hybridization, this was found to produce smaller, more discrete signals with very high contrast and resolution wapplied as a detection method ith minimal nonspecific background, enabling accurate copy number enumeration (Powell et al. 2007). A comparison of in situ hybridization results obtained by different methodologies is shown in Fig. 3. In order to differentiate cases with *HER2* gene amplification from those where chromosome 17 polysomy results in low-level signal amplification, a second probe, directed against the centromeric region of chromosome 17, is used in conjunction with the *HER2* gene probe and revealed using Fast Red K, which produces a larger, red pot for each copy of chromosome 17. The ratio of black spots, which correspond to copies of the *HER2* gene, to red spots, which correspond to copies of the entire chromosome 17, gives the degree of *HER2* gene amplification (Tubbs et al. 2005b; Powell et al. 2007).

Enzyme metallography has been optimized for use on automated slide staining instruments manufactured by Ventana Medical Systems (part of the Roche Group). An enzyme metallography in situ hybridization assay for *HER2* gene amplification in breast cancer, for use on the Benchmark XT or BenchMark ULTRA automated slide strainer, was introduced in Europe in 2007, and approved by the US Food and Drug Administration (FDA) in the United States in 2011 for clinical use in assessing patients for humanized monoclonal antibody therapy.

Modifications to the procedure include the development of a combined *HER2* gene amplification and *HER2* protein assay, in which copies of the *HER2* gene are revealed using enzyme metallography, and the *HER2* protein is stained immunohistochemically and revealed using Fast Red K. This assay, termed EnzMet GenePro, allows direct comparison of gene amplification with protein overexpression (Downs-Kelly et al. 2005). In a study of 80 cases of invasive breast cancer, excellent agreement was found between the two components; only two discordant cases were present, one where gene amplification determined by

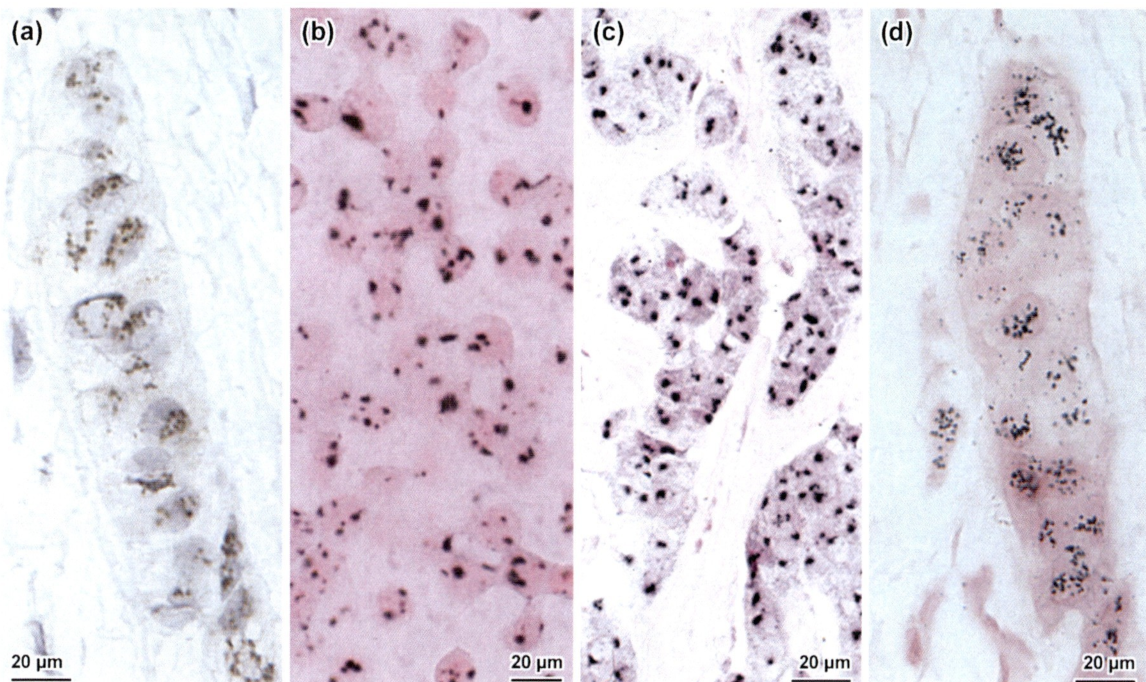

Enzyme Metallography and Metallographic In Situ Hybridization, Fig. 3 Comparison of in situ hybridization detection visualized with enzymatic DAB with that visualized by metallographic methods. (**a**) *HER2* gene amplification in paraffin-embedded human invasive breast carcinoma biopsy, visualized using enzymatic DAB development (Courtesy of Dr. R. R. Tubbs, Case Western University Cleveland Clinic Foundation). (**a–c**) Metallographic in situ hybridization. (**b**) Single copies of HPV-16 in SiHa cells visualized by tyramide signal amplification (TSA; also known as CARD, or catalyzed reporter deposition), followed by detection with streptavidin-Nanogold and silver acetate autometallography, according to the scheme shown in Fig. 1a. Copies of HPV-16 appear as single spots (H & E counterstain). (**c**) *HER2* gene amplification in paraffin-embedded human invasive breast carcinoma biopsy, visualized using Nanogold with gold enhancement (GOLDFISH procedure; Fig. 1b), showing large, confluent nuclear signals from multiple gene copies in close proximity (Nuclear fast *red* counterstain). (**d**) EnzMet™ detection of the amplification of individual *HER2* gene copies in paraffin-embedded human invasive breast carcinoma biopsy, according to the scheme shown in Fig. 2. Normal, non-amplified cells contain two copies of the *HER2* gene, while the infiltrating *HER2*-amplified carcinoma cells show multiple copies (Courtesy of Dr. R. R. Tubbs, Case Western University Cleveland Clinic Foundation) ((**b**) From Powell 2007 and (**c**) from Tubbs 2002; used with permission)

EnzMet was not accompanied by protein overexpression, and one where relatively high protein overexpression occurred in the absence of gene amplification.

Metallographic Immunohistochemistry

Enzyme metallography also provides highly sensitive and specific staining in immunohistochemistry (IHC), the staining of specific proteins or other biomarkers within tissue specimens. Immunohistochemical protocols usually comprise the detection of a target protein by the initial application of a primary antibody that binds specifically to the target protein, followed by a secondary antibody conjugated to a label or reporter group that generates a signal that reveals the protein of interest. Conventional immunohistochemical procedures use enzymatically labeled secondary antibodies with organic chromogens to stain a wide variety of markers. Enzyme metallography has demonstrated significant advantages for IHC in addition to its advantages for ISH, producing higher contrast and a sharper signal that is more easily differentiated from other stains and counterstains (Tubbs et al. 2005a; Liu et al. 2011).

In a study comparing its properties with those of conventional staining procedures using

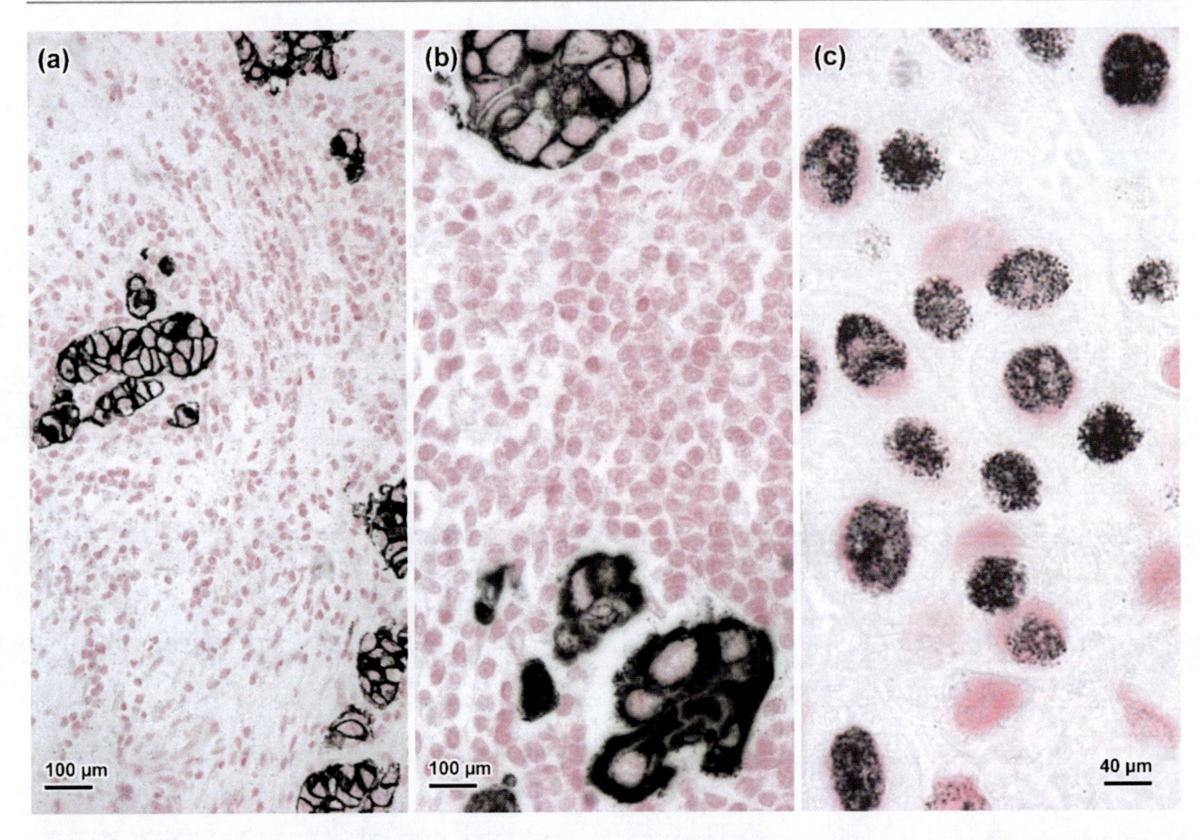

Enzyme Metallography and Metallographic In Situ Hybridization, Fig. 4 Immunohistochemistry using enzyme metallography illustrating the utility of this approach for the detection and staining of different protein targets: (**a**) cell membrane (*HER2*), (**b**) cytoplasmic (cytokeratin AE1/3 cocktail), and (**c**) nuclear (ER) antigens. Nuclear fast red counterstain (Tubbs et al. 2005a; used with permission)

diaminobenzidine (DAB) in high-complexity tissue midiarrays taken from paraffin-embedded biopsy specimens; 88 solid tumors were evaluated by automated EnzMet IHC using primary antibodies specific to the target, followed by labeled secondary antibodies and development using the metallographic substrate. Three specific targets were selected to evaluate the ability of EnzMet IHC to specifically localize antigens in the nucleus (estrogen receptor), cytoplasm (cytokeratins), and cytoplasmic membrane (*HER2*) (Fig. 4). The staining patterns and intensities showed full concordance between EnzMet detection and conventional IHC. The EnzMet staining was sharper and more highly resolved, with many regions assuming a punctuate appearance that defined the staining more clearly. The black color of the EnzMet reaction product is easily differentiated from other stains and counterstains, and the increased detection sensitivity over DAB allowed additional dilution of the primary antibodies (Tubbs et al. 2005a).

Immunoelectron Microscopy with Enzyme Metallography

The metallic silver deposited by enzyme metallography is considerably more electron-dense than biological materials and tissues, and hence provides excellent contrast as a staining method for electron microscopy. ► Peroxidase-linked secondary antibodies have been used with the enzyme metallographic substrate to localize polar tube proteins in microsporidia, a class of protozoans responsible for a number of tropical diseases and opportunistic infections associated with the human immunodeficiency virus, by light microscopy and by pre-embedding and on-section immunoelectron microscopy. In a typical preparation, rabbit kidney (RK-13 cells) were infected with microsporidia, grown on slides, fixed for electron microscopy, then labeled. After initial incubation with primary antibodies against specific polar tube proteins, a universal detection system incorporating a tagged secondary antibody and polymerized ▶ peroxidase-streptavidin tertiary detection system was applied and developed with a modified formulation of the enzyme metallographic reagent. Light microscope examination showed intense staining of the polar tubes and infective sporoplasm. Areas of interest were marked on the back of the slides; after dehydration, infiltration, and embedding, thin sections were cut, counterstained with uranyl acetate and lead citrate, then examined using a transmission electron microscope. Stained structures previously visualized by light microscopy could be identified clearly by electron microscopy, where they were shown to be densely labeled with silver nanoparticles up to 30–100 nm in size (Powell et al. 2006; Liu et al. 2011).

EnzMet immunohistochemistry combined with heavy atom staining has also been used to visualize retrograde labeled neurons within the Western Mosquito fish (*Gambusia affinis*) spinal cord by serial block-face scanning electron microscopy, a serial section method for the electron microscopic analysis of large specimens which is suited to investigating connectivity in neural circuits. Enzyme metallography provides high contrast staining at neuronal boundaries, which facilitates serial reconstruction (Liu et al. 2011).

Other Applications of Enzyme Metallography

The black signal generated by enzyme metallography is ideally suited to optical detection methods. In addition to its use for light microscopic visualization, it is also a sensitive detection reagent for the immunoenzymatic detection of proteins on western blots and immunodot blots (Liu et al. 2011), where the black signal can provide higher contrast than conventional organic chromogens.

Enzyme metallography has also been applied to the development of an electrical detection method for biochips. Electrical detection affords advantages for the detection of multiple targets using biochips because multiple independent electrical contacts can be fabricated and identified on a microscopic scale in a robust, miniaturized, and highly portable format. Detection was conducted on silicon oxide substrate chips photolithographically coated with titanium and gold: Conductivity measurement sites were 10 μm structured electrode gaps, printed with a capture oligonucleotide. This bound the target, which was then detected using a biotin-labeled detection probe, polymeric streptavidin-▶ peroxidase secondary probe, and metallographic development resulting in a conductive link between the two electrodes (Möller et al. 2005). Examination by scanning electron microscopy and atomic force microscopy revealed a layer of deposited silver which took the form of intersecting flat plates, giving the appearance "nanoflowers" up to about 200 nm in diameter (Festag et al. 2008).

Acknowledgments The primary work described in this entry was partly supported by NIH SBIR grants R44 GM064257 and R43 MH086994, and STTR grant R42 CA083618; this entry was partly supported by NIH grant R01 GM085802. We are grateful to Dr. Raymond R. Tubbs (Case Western University Cleveland Clinic Foundation) for sharing results and helpful discussions.

Cross-References

- ▶ Gold and Nucleic Acids
- ▶ Magnesium, Physical and Chemical Properties
- ▶ Peroxidases

References

Downs-Kelly E, Pettay J, Hicks D, Skacel M, Yoder B, Rybicki L, Myles J, Sreenan J, Roche P, Powell R, Hainfeld J, Grogan T, Tubbs R (2005) Analytical validation and interobserver reproducibility of EnzMet GenePro: a second-generation bright-field metallography assay for concomitant detection of HER2 gene status and protein expression in invasive carcinoma of the breast. Am J Surg Pathol 29:1505–1511

Festag G, Schüler T, Möller R, Csáki A, Fritzsche W (2008) Growth and percolation of metal nanostructures in electrode gaps leading to conductive paths for electrical DNA analysis. Nanotechnology 19:125303

Liu W, Mitra D, Joshi V, Powell R, Hainfeld J, Serrano-Velez J, Torres-Vasquez I, Rosa-Molinar E, Takvorian P (2011) EnzMet for versatile, highly sensitive light and electron microscopy staining. Microsc Microanal 17(suppl 2):116–117

Moelans CB, de Weger RA, Van der Wall E, van Diest PJ (2011) Current technologies for HER2 testing in breast cancer. Crit Rev Oncol Hematol 80:380–392

Möller R, Powell RD, Hainfeld JF, Fritzsche W (2005) Enzymatic control of metal deposition as key step for a low-background electrical detection for DNA chips. Nano Lett 5:1475–1482

Powell R, Joshi V, Thelian A, Liu W, Takvorian P, Cali A, Hainfeld J (2006) Light and electron microscopy of

microsporida using enzyme metallography. Microsc Microanal 12(Suppl 2):424CD
Powell RD, Pettay JD, Powell WC, Roche PC, Grogan TM, Hainfeld JF, Tubbs RR (2007) Metallographic in situ hybridization. Hum Pathol 38:1145–1159
Tubbs R, Pettay J, Powell R, Hicks DG, Roche P, Powell W, Grogan T, Hainfeld JF (2005a) High-resolution immunophenotyping of subcellular compartments in tissue microarrays by enzyme metallography. Appl Immunohistochem Mol Morphol 13:371–375
Tubbs RR, Pettay J, Skacel M, Downs-Kelly E, Powell RD, Hicks DG, Hainfeld JF (2005b) Gold and silver-facilitated metallographic in situ hybridization procedures for detection of HER2 gene amplification. In: Hacker GW, Tubbs RR (eds) Molecular morphology in human tissues. CRC Press, Boca Raton, pp 101–106
Wolff AC, Hammond ME, Schwartz JN, Hagerty KL, Allred DC, Cote RJ, Dowsett M, Fitzgibbons PL, Hanna WM, Langer A, McShane LM, Paik S, Pegram MD, Perez EA, Press MF, Rhodes A, Sturgeon C, Taube SE, Tubbs R, Vance GH, van de Vijver M, Wheeler TM, Hayes DF (2007) American Society of Clinical Oncology/College of American Pathologists guideline recommendations for human epidermal growth factor receptor 2 testing in breast cancer. Arch Pathol Lab Med 131:18–43

Enzyme: Catalyst

► Monovalent Cations in Tryptophan Synthase Catalysis and Substrate Channeling Regulation

Enzymes Regulating Thyroid Hormones Synthesis

► Selenoproteins and the Biosynthesis and Activity of Thyroid Hormones

Epidemiology

► Cadmium and Health Risks
► Chromium and Allergic Reponses

Epilysin

► Zinc Matrix Metalloproteinases and TIMPs

Epi-Molecular

► Silver Impregnation Methods in Diagnostics

EPR, Electron Paramagnetic Resonance

► Zinc Aminopeptidases, Aminopeptidase from Vibrio Proteolyticus (Aeromonas proteolytica) as Prototypical Enzyme

Equilibrium

► Zinc Homeostasis, Whole Body

Erbium

Takashiro Akitsu
Department of Chemistry, Tokyo University of Science, Shinjuku-ku, Tokyo, Japan

Definition

A lanthanoid element, the eleventh element (yttrium group) of the f-elements block, with the symbol Er, atomic number 68, and atomic weight 167.259. Electron configuration $[Xe]4f^{12}6s^2$. Erbium is composed of stable (^{162}Er, 0.139%; ^{164}Er, 1.601%; ^{166}Er, 33.503%; ^{167}Er, 22.869%; ^{168}Er, 26.978%; ^{170}Er, 14.910%) and five radioactive (^{160}Er; ^{165}Er; ^{169}Er; ^{171}Er; ^{172}Er) isotopes. Discovered by C. G. Mosander in 1843. Erbium exhibits oxidation state III; atomic radii: 176 pm, covalent radii 190 pm; redox potential (acidic solution) Er^{3+}/Er −2.296 V; electronegativity (Pauling) 1.24. Ground electronic state of $^4I_{15/2}$ with $S = 3/2$, $L = 6$, $J = 15/2$ with $\lambda = -820\ cm^{-1}$. Most stable technogenic radionuclide ^{169}Er (half-life 9.4 days). The most common compounds: Er_2O_3, ErX_3 (X = F, Cl, Br, and I), and $Er(OH)_3$. Biologically, erbium is of low to moderate toxicity, and known that the tonus and contractility of the rabbit ileum in response to acetylcholine or nicotine was decreased dose dependently by $ErCl_3$ (Atkins et al. 2006; Cotton et al. 1999; Huheey et al. 1997; Oki et al. 1998; Rayner-Canham and Overton 2006).

Cross-References

- ▶ Lanthanide Ions as Luminescent Probes
- ▶ Lanthanide Metalloproteins
- ▶ Lanthanides and Cancer
- ▶ Lanthanides in Biological Labeling, Imaging, and therapy
- ▶ Lanthanides in Nucleic Acid Analysis
- ▶ Lanthanides, Physical and Chemical Characteristics

References

Atkins P, Overton T, Rourke J, Weller M, Armstrong F (2006) Shriver and Atkins inorganic chemistry, 4th edn. Oxford University Press, Oxford/New York

Cotton FA, Wilkinson G, Murillo CA, Bochmann M (1999) Advanced inorganic chemistry, 6th edn. Wiley-Interscience, New York

Huheey JE, Keiter EA, Keiter RL (1997) Inorganic chemistry: principles of structure and reactivity, 4th edn. Prentice Hall, New York

Oki M, Osawa T, Tanaka M, Chihara H (1998) Encyclopedic dictionary of chemistry. Tokyo Kagaku Dojin, Tokyo

Rayner-Canham G, Overton T (2006) Descriptive inorganic chemistry, 4th edn. W. H. Freeman, New York

Erythrocuprein

▶ Zinc in Superoxide Dismutase

Erythronium

▶ Vanadium Metal and Compounds, Properties, Interactions, and Applications

Essential Nutrient for Plants

▶ Nickel in Plants

Europium

Takashiro Akitsu
Department of Chemistry, Tokyo University of Science, Shinjuku-ku, Tokyo, Japan

Definition

A lanthanoid element, the sixth element (yttrium group) of the f-elements block, with the symbol Eu, atomic number 63, and atomic weight 151.964. Electron configuration [Xe]$4f^7 6s^2$. Europium is composed of stable (^{153}Eu, 52.2%) and three radioactive (^{150}Eu; ^{151}Eu, 47.8%; ^{152}Eu) isotopes. Discovered by E. A. Demarcay in 1901. Europium exhibits oxidation states II and III; atomic radii: 180 pm, covalent radii 190 pm; redox potential (acidic solution) Eu^{3+}/Eu −2.407 V; Eu^{3+}/Eu^{2+} −0.360 V; electronegativity (Pauling) 1.2. Ground electronic state of Eu^{3+} is $^7F_0$ with S = 3, L = 3, J = 0 with $\lambda = 230\ cm^{-1}$. Most stable technogenic radionuclide ^{151}Eu (half-life 5×10^{18} years). The most common compounds: Eu_2O_3, EuO, and $Eu(OH)_3$. Europium easily forms divalent compounds. Biologically, europium is of low to moderate toxicity, and medically high concentration of Eu-compounds is also detected in the spleen of alcoholic patients (Atkins et al. 2006; Cotton et al. 1999; Huheey et al. 1997; Oki et al. 1998; Rayner-Canham and Overton 2006).

Cross-References

- ▶ Lanthanide Ions as Luminescent Probes
- ▶ Lanthanide Metalloproteins
- ▶ Lanthanides and Cancer
- ▶ Lanthanides in Biological Labeling, Imaging, and Therapy
- ▶ Lanthanides in Nucleic Acid Analysis
- ▶ Lanthanides, Physical and Chemical Characteristics

References

Atkins P, Overton T, Rourke J, Weller M, Armstrong F (2006) Shriver and Atkins inorganic chemistry, 4th edn. Oxford University Press, Oxford/New York

Cotton FA, Wilkinson G, Murillo CA, Bochmann M (1999) Advanced inorganic chemistry, 6th edn. Wiley-Interscience, New York

Huheey JE, Keiter EA, Keiter RL (1997) Inorganic chemistry: principles of structure and reactivity, 4th edn. Prentice Hall, New York

Oki M, Osawa T, Tanaka M, Chihara H (1998) Encyclopedic dictionary of chemistry. Tokyo Kagaku Dojin, Tokyo

Rayner-Canham G, Overton T (2006) Descriptive inorganic chemistry, 4th edn. W. H. Freeman, New York

Evolution

► Zinc and Iron, Gamma and Beta Class, Carbonic Anhydrases of Domain Archaea

Excitation-Contraction Coupling (EC Coupling)

► Calcium Sparklets and Waves

Excitation-Transcription Coupling (ET Coupling)

► Calcium Sparklets and Waves

Exocytosis

► C2 Domain Proteins

Exudative Diathesis

► Selenium and Muscle Function

F

Fe Metabolism

► Iron-Sulfur Cluster Proteins, Nitrogenases

f-Electron Systems – Lanthanide and Actinide Elements

► Actinide and Lanthanide Systems, High Pressure Behavior

Fertilizers

► Polonium and Cancer

Fe-S Clusters

► Iron-Sulfur Cluster Proteins, Nitrogenases

Fibrin

► Nanosilver, Next-Generation Antithrombotic Agent

Fictionalization of Gold Nanomaterials

► Gold Nanomaterials as Prospective Metal-based Delivery Systems for Cancer Treatment

Five-EF-Hand Calcium-Binding Proteins

► Penta-EF-Hand Calcium-Binding Proteins

FK506

► Calcineurin

FlAsH-EDT_2

► Biarsenical Fluorescent Probes

Fluorescent Arsenical Helix/Hairpin Binder

► Biarsenical Fluorescent Probes

Fluorescent Probes for Zn^{2+}

► Zinc, Fluorescent Sensors as Molecular Probes

Fluorescent Protein–Gold Nanoparticle Conjugates

► Gold Nanoparticles and Fluorescent Proteins, Optically Coupled Hybrid Architectures

V.N. Uversky et al. (eds.), *Encyclopedia of Metalloproteins*, DOI 10.1007/978-1-4614-1533-6,

Formate Dehydrogenase

William Self
Molecular Biology & Microbiology, Burnett School of Biomedical Sciences, University of Central Florida, Orlando, FL, USA

Formate dehydrogenase (FDH) catalyzes the two electron oxidation of formic acid to carbon dioxide (Khangulov et al. 1998). The enzyme (from *Escherichia coli*) that is linked to hydrogenase 3 (FDH-H) is the best studied model in this class of enzymes. Two other formate dehydrogenases are expressed in *E. coli*, one linked to nitrate respiration (FDH-N) and a third that is expressed under aerobic conditions (FDH-O), which is poorly studied. FDH-H contains a bis-molybdopterin guanine dinucleotide (MGD) cofactor that positions the catalytic Mo active site near a selenocysteine residue (Boyington et al. 1997). Although FDH-H is the best-studied model for selenoproteins in bacteria, an early study in the 1970s had established that selenium was present in a tungsten-containing clostridial formate dehydrogenase (Andreesen and Ljungdahl 1973). It is now well established that both selenocysteine and cysteine can serve in the active site for this class of Mo/W formate dehydrogenases (Moura et al. 2004).

Formate dehydrogenases are known to oxidize formate that is produced during anaerobic fermentation of sugars produced through the pyruvate formate lyase (PFL) and thus aid in respiration on nitrate or convert formate to hydrogen and carbon dioxide in the absence of alternative electron acceptors (Moura et al. 2004). The study of hydrogen production and analysis of mutants incapable of hydrogen production in *E. coli* led ultimately to the discovery of genes encoding the transport of molybdenum, maturation of the Mo cofactor and the proteins and tRNA necessary for the insertion of selenocysteine into selenoproteins (Self et al. 2001; Stadtman 2002). The role of both sulfur and selenium-dependent formate dehydrogenases have been studied in methanogenic archaea, where these enzymes serve as electron donors for methanogenesis (Costa et al. 2010; Jones and Stadtman 1981).

Based on the widespread occurrence of FDH enzymes in genomes, and the diversity of metals and metalloids at the active site, the question arises as to why some organisms prefer to insert tungsten over molybdenum or selenium over sulfur. A general occurrence of tungsten in archaea was first recognized, but the preference for tungsten may or may not be based on catalytic efficiency (Moura et al. 2004). Given the importance of hydrogen production in biofuels and the current interest in biofuels, much of the recent work centers around applied studies to enhance hydrogen production through formate-hydrogenlyase complexes (Hu and Wood 2010; Maeda et al. 2008; Sanchez-Torres et al. 2009). However, a recent study has also identified the poorly understood FDH-O enzyme as being present in the *E. coli* respiratory chain, possibly playing a direct role in energy metabolism (Sousa et al. 2011). Another recent study demonstrates that both FDH-O and FDH-N can both oxidize hydrogen directly and subsequently reduce benzyl viologen (Soboh et al. 2011). These recent studies show that this class of enzymes still holds many mysteries and thus warrants further research.

References

Andreesen JR, Ljungdahl LG (1973) Formate dehydrogenase of *Clostridium thermoaceticum*: incorporation of selenium-75, and the effects of selenite, molybdate, and tungstate on the enzyme. J Bacteriol 116:867–873

Boyington JC, Gladyshev VN, Khangulov SV, Stadtman TC, Sun PD (1997) Crystal structure of formate dehydrogenase H: catalysis involving Mo, molybdopterin, selenocysteine, and an Fe4S4 cluster. Science 275:1305–1308

Costa KC, Wong PM, Wang T, Lie TJ, Dodsworth JA, Swanson I, Burn JA, Hackett M, Leigh JA (2010) Protein complexing in a methanogen suggests electron bifurcation and electron delivery from formate to heterodisulfide reductase. Proc Natl Acad Sci U S A 107:11050–11055

Hu H, Wood TK (2010) An evolved *Escherichia coli* strain for producing hydrogen and ethanol from glycerol. Biochem Biophys Res Commun 391:1033–1038

Jones JB, Stadtman TC (1981) Selenium-dependent and selenium-independent formate dehydrogenases of *Methanococcus vannielii*. Separation of the two forms and characterization of the purified selenium-independent form. J Biol Chem 256:656–663

Khangulov SV, Gladyshev VN, Dismukes GC, Stadtman TC (1998) Selenium-containing formate dehydrogenase H from *Escherichia coli*: a molybdopterin enzyme that catalyzes

formate oxidation without oxygen transfer. Biochemistry 37:3518–3528
Maeda T, Sanchez-Torres V, Wood TK (2008) Protein engineering of hydrogenase 3 to enhance hydrogen production. Appl Microbiol Biotechnol 79:77–86
Moura JJ, Brondino CD, Trincao J, Romao MJ (2004) Mo and W bis-MGD enzymes: nitrate reductases and formate dehydrogenases. J Biol Inorg Chem 9:791–799
Sanchez-Torres V, Maeda T, Wood TK (2009) Protein engineering of the transcriptional activator FhlA To enhance hydrogen production in *Escherichia coli*. Appl Environ Microbiol 75:5639–5646
Self WT, Grunden AM, Hasona A, Shanmugam KT (2001) Molybdate transport. Res Microbiol 152:311–321
Soboh B, Pinske C, Kuhns M, Waclawek M, Ihling C, Trchounian K, Trchounian A, Sinz A, Sawers G (2011) The respiratory molybdo-selenoprotein formate dehydrogenases of *Escherichia coli* have hydrogen: benzyl viologen oxidoreductase activity. BMC Microbiol 11:173
Sousa PM, Silva ST, Hood BL, Charro N, Carita JN, Vaz F, Penque D, Conrads TP, Melo AM (2011) Supramolecular organizations in the aerobic respiratory chain of *Escherichia coli*. Biochimie 93:418–425
Stadtman TC (2002) Discoveries of vitamin B12 and selenium enzymes. Annu Rev Biochem 71:1–16

Förster Resonance Energy Transfer (FRET)

► Calcium Sparklets and Waves

Francium, Physical and Chemical Properties

Fathi Habashi
Department of Mining, Metallurgical, and Materials Engineering, Laval University, Quebec City, Canada

Element 87 occurs in nature in minute amounts as the radioactive element ^{223}Fr ($t_{1/2} = 21.8$ min) which is the most stable of all 27 isotopes. It decays into astatine, radium, and radon. Francium is only monovalent and shows a very similar behavior to the other heavy alkali elements. It has the most negative standard potential of all elements. Relatively pure ^{223}Fr solutions are obtained by its elution with $NH_4Cl - CrO_3$ from a cation exchanger loaded with ^{227}Ac. The solution is further purified by passing it through a SiO_2 column loaded with $BaSO_4$.

References

Habashi F (ed) (1997) Handbook of extractive metallurgy. WILEY-VCH, Weinheim, p. 1592

Francium, Physiological Effects

Sergei E. Permyakov
Protein Research Group, Institute for Biological Instrumentation of the Russian Academy of Sciences, Pushchino, Moscow Region, Russia

Francium is a highly rare and unstable element. The estimated total amount of Francium in the Earth's crust does not exceed 30 g. Besides, the most stable isotope of Francium (^{223}Fr) has half-life of 22 min. Because of rarity and instability of Francium, it is poorly studied and has no commercial use.

The isotope ^{221}Fr is formed by alpha decay of ^{225}Ac. Since half-life of ^{221}Fr is 5 min, it is able to cause radiation damage, which may be of use in cancer treatment (Perey and Chevallier 1951). ^{221}Fr is accumulated during radioimmunotherapy with ^{225}Ac in kidneys and likely contributes to the long-term renal toxicity observed in mice (Song et al. 2009).

References

Perey M, Chevallier A (1951) Fixation of element 87, francium, in experimental sarcoma of rat. C R Seances Soc Biol Fil 145:1208–1211
Song H, Hobbs RF, Vajravelu R et al (2009) Radioimmunotherapy of breast cancer metastases with alpha-particle emitter ^{225}Ac: comparing efficacy with $^{21}3Bi$ and ^{90}Y. Cancer Res 69:8941–8948

Free Radicals

► Arsenic, Free Radical and Oxidative Stress

Frq1

► Calcium, Neuronal Sensor Proteins

Functional Genomics

► Iron-Sulfur Cluster Proteins, Ferredoxins

Functionalization of AgNPs

► Colloidal Silver Nanoparticles and Bovine Serum Albumin

Funnelins

► Zinc Metallocarboxypeptidases

G

G_4-DNA

► Strontium and DNA Aptamer Folding

Gadolinium

Takashiro Akitsu
Department of Chemistry, Tokyo University of Science, Shinjuku-ku, Tokyo, Japan

Definition

A lanthanoid element, the seventh element of the f-elements block, with the symbol Gd, atomic number 64, and atomic weight 157.25. Electron configuration $[Xe]4f^7 5d^1 6s^2$. Gadolinium is composed of stable (^{154}Gd, 2.18%; ^{155}Gd, 14.80%; ^{156}Gd, 20.47%; ^{157}Gd, 15.65%; ^{158}Gd, 24.84%) and two radioactive (^{152}Gd, 0.20%; ^{160}Gd, 21.86%) isotopes. Discovered by J. C. G. de Marignac in 1880. Gadolinium exhibits oxidation states III (and II); atomic radii: 180 pm, covalent radii 197 pm; redox potential (acidic solution) Gd^{3+}/Gd −2.397 V; Gd^{3+}/Gd^{2+} −3.9 V; electronegativity (Pauling) 1.2. Ground electronic state of Gd^{3+} is $^8S_{7/2}$ with $S = 7/2$, $L = 0, J = 7/2$ with $\lambda = 0\ cm^{-1}$. Most stable technogenic radionuclide ^{160}Gd (half-life $> 1.3 \times 10^{21}$ years). The most common compounds: Gd_2O_3, GdF_3, and $Gd(OH)_3$. Biologically, gadolinium is of high toxicity, and inhalational exposure to high concentrations of Gd_2O_3 results in pneumonitis and acute inflammation in the lung (Atkins et al. 2006; Cotton et al. 1999; Huheey et al. 1997; Oki et al. 1998; Rayner-Canham and Overton 2006).

Cross-References

► Lanthanide Ions as Luminescent Probes
► Lanthanide Metalloproteins
► Lanthanides and Cancer
► Lanthanides in Biological Labeling, Imaging, and Therapy
► Lanthanides in Nucleic Acid Analysis
► Lanthanides, Physical and Chemical Characteristics

References

Atkins P, Overton T, Rourke J, Weller M, Armstrong F (2006) Shriver and Atkins inorganic chemistry, 4th edn. Oxford University Press, Oxford/New York

Cotton FA, Wilkinson G, Murillo CA, Bochmann M (1999) Advanced inorganic chemistry, 6th edn. Wiley-Interscience, New York

Huheey JE, Keiter EA, Keiter RL (1997) Inorganic chemistry: principles of structure and reactivity, 4th edn. Prentice Hall, New York

Oki M, Osawa T, Tanaka M, Chihara H (1998) Encyclopedic dictionary of chemistry. Tokyo Kagaku Dojin, Tokyo

Rayner-Canham G, Overton T (2006) Descriptive inorganic chemistry, 4th edn. W. H. Freeman, New York

Gallium

► Gallium(III) Complexes, Inhibition of Proteasome Activity

V.N. Uversky et al. (eds.), *Encyclopedia of Metalloproteins*, DOI 10.1007/978-1-4614-1533-6,

Gallium and Apoptosis

▶ Gallium, Therapeutic Effects

Gallium Complexes

▶ Gallium(III) Complexes, Inhibition of Proteasome Activity

Gallium Compounds

▶ Gallium(III) Complexes, Inhibition of Proteasome Activity

Gallium Delivery by Transferrin Receptor-1-Mediated Endocytosis

▶ Gallium Uptake and Transport by Transferrin

Gallium Effect on Bacteria

▶ Gallium, Therapeutic Effects

Gallium in Bacteria, Metabolic and Medical Implications

Christopher Auger, Joseph Lemire, Varun Appanna and Vasu D. Appanna
Department of Chemistry and Biochemistry, Laurentian University, Sudbury, ON, Canada

Synonyms

Bacterial response to gallium; The molecular effects of gallium on bacterial physiology

Definition

Chelation therapy: medication involved in the removal of excess metals.

Gallium: metal utilized as a semiconductor.

Pentose phosphate pathway: a metabolic network that produces the antioxidant NADPH and ribose.

Pseudomonas fluorescens: a metabolically versatile microbe.

Tricarboxylic acid cycle: an essential component of aerobic respiration.

Introduction

Due to their ability to survive and grow in a wide range of environmental conditions, bacteria present themselves as ideal models for the study of a variety of biological phenomena, including the interactions of metal toxins with living organisms. Mechanisms by which these prokaryotes dispose of metal pollutants and circumvent their toxic influence have major implications in environmental, medical, and biotechnological domains. This may include the bioremediation of waste sites, chelation therapy, and the production of unique metabolites. *Pseudomonas fluorescens*, a soil microbe, is known to thrive in media supplemented with numerous toxic metals like Ga, aluminum (Al), lead (Pb), zinc (Zn), and yttrium (Y). Owing to its physicochemical similarity to Fe, an essential nutrient in almost all organisms, the interaction of Ga with living organisms has become a subject of ongoing importance.

Ga Occurrence and Toxicity

Ga is a silvery, glass-like semi-metallic element by and large found in its salt form in bauxite and zinc ores (Downs 1993). While normally inert, increasing industrial activity reliant on this metal has rendered Ga accessible to a wide spectrum of organisms. Easily purified from the smelting process, the dominant application of this trivalent metal is in the semiconductors gallium arsenide and gallium nitride (Chitambar 2010). Given the efficiencies of these materials and our reliance on them for computers, high-speed optical communication, and photovoltaic cells, it is clear that the amount of Ga present in the environment will only escalate in the foreseeable future. As a result, the toxicological aspect of this metal cannot be ignored.

Ga, an Fe Mimetic

Ga is not known to have any biological role (Al-Aoukaty et al. 1992). Historically, its low bioavailability has seemingly excluded it from any physiological function. However, it is well recognized that Ga can exert its harmful effect by interfering with the metabolism of Fe. The functions of the latter are quite diverse. Fe is crucial in heme-containing proteins such as hemoglobin, myoglobin, and cytochrome P450. This metal also plays a necessary role in iron-sulfur (Fe-S) cluster containing proteins, which are pervasive in nature, catalyzing a wide array of biochemical reactions. As such, any disruption in the function of these metalloproteins would have consequences on multiple cellular operations. Ga, with an ionic radius of 0.67 Å and a trivalent charge, is sufficiently analogous to Fe^{3+} to displace it from Fe-dependent biomolecules (Downs 1993). However, while Fe is redox active, Ga is not, and the substitution often leads to the inactivation of the target protein. This feature of Ga is exploited in cancer therapy. Radiopharmaceuticals are radioactive tracers that permit the imaging and treatment of many diseases (Chitambar 2010). Cancer cells, due to their rapidly proliferating nature, have an increased need for Fe. By incorporating itself in proteins in lieu of Fe, radiolabeled Ga^{67} allows for the noninvasive visualization of tumors in a patient (Chitambar 2010). This interchange of metals could also contribute minimally to the eradication of cancerous cells, arresting growth via disruption of vital cellular processes and the liberation of Fe (Chitambar 2010).

Free Fe, a Pro-Oxidant: A Consequence of Ga Toxicity

Once displaced from metalloproteins, free Fe further disturbs the operation of the cell via the production of reactive oxygen species (ROS) (Chenier et al. 2008). Fenton chemistry is a series of reactions whereby Fe(II) is initially oxidized by hydrogen peroxide to form Fe(III), hydroxyl radical, and hydroxyl anion. Hydrogen peroxide can then catalyze the reduction of Fe(III) back to Fe(II) with the concomitant production of peroxide radical and a proton (Bériault 2004). These radicals are dangerous, as they can disable DNA, proteins, and lipids. Consequently, the homeostasis of Fe is a strongly regulated process. Proteins, such as transferrin and ferritin, which bind Fe tightly yet reversibly, are indispensable for its storage and delivery (Bériault 2004). Ga therefore exerts its deleterious effects via a two-pronged approach. Firstly, the displacement of Fe from its cognitive protein may subsequently inactivate the protein. Secondly, the release of free Fe into the microenvironment can generate oxidative radicals which, if not properly controlled, can induce cell death.

Physiological Responses of *Pseudomonas fluorescens* to Ga

Although Ga is toxic, some microbial systems have evolved to counter its deleterious effects. *Pseudomonas fluorescens* is an obligate aerobe known to inhabit primarily soil, plants, and water surfaces (Paulsen et al. 2005). This gram-negative, rod-shaped, and nonpathogenic bacterium derived its name from its ability to produce the fluorescent pigment pyoverdin, a siderophore, under iron-limiting conditions (Paulsen et al. 2005). The capability to grow on different carbon sources under a variety of conditions renders *P. fluorescens* ideal for commercial applications (Paulsen et al. 2005). This microbe has been implemented in bioremediation, the production of antifreeze agents as well as biofertilizers (Paulsen et al. 2005). The extreme metabolic versatility of this organism renders it suitable for the study of adaptation in harsh environments. Despite the consequences associated with Ga internalization, *P. fluorescens* is capable of adapting to the toxic effects of this metal. When *P. fluorescens* is grown in a medium containing citrate as the sole carbon source chelated to 1 mM Ga, cells reach their stationary phase within 68 h, compared to the 24 h observed in the control cultures (Al-Aoukaty et al. 1992) (Fig. 1). However, this lag phase is reversed if 20 μM Fe is included in the medium (Al-Aoukaty et al. 1992). The Ga-stressed cells have a pale appearance as opposed to the light brown coloration of the control cells. This may be due to the inability of *P. fluorescens* to access Fe in the Ga-containing medium.

Ga Toxicity: Molecular Insights

The electron transport chain, a series of complexes that is responsible for generating the majority of the universal energy currency adenosine triphosphate (ATP) during aerobic metabolism, is highly susceptible to stress arising from oxidative radicals and metal

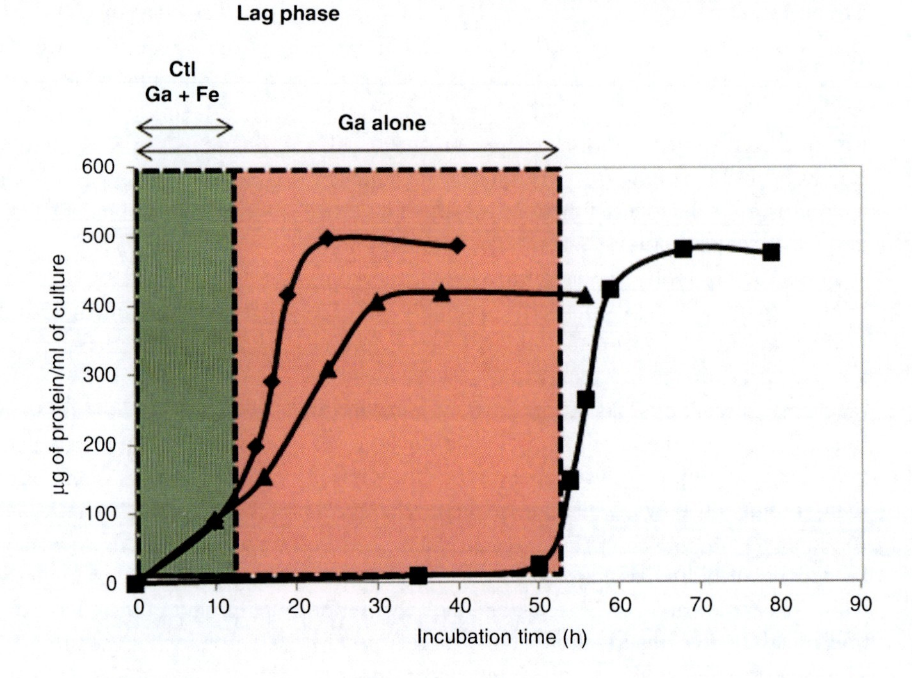

Gallium in Bacteria, Metabolic and Medical Implications, Fig. 1 Biomass curve of *Pseudomonas fluorescens* treated with Ga. When Ga is added to growth media with citrate as the sole carbon source, the lag phase is increased to 50 h (■), as opposed to 13 h in the control cells (◆). Supplementation with exogenous Fe reverses the growth shift observed under Ga stress (▲) (Ctl = control. Adapted from Al-Aoukaty et al. 1992)

pollution (Chenier et al. 2008). Because this machinery is riddled with heme groups, iron-sulfur clusters, and various metal cofactors, Ga is capable of reacting with and disabling the cell's primary source of energy (Fig. 2). In order to assess the activity of these complexes, total membrane protein can first be separated by blue native polyacrylamide gel electrophoresis (BN-PAGE). Secondly, activity staining can be performed in order to determine whether or not these proteins are functional (Bériault et al. 2007). In-gel analysis of complexes I, II, and IV show that they are severely crippled by the addition of gallium to the medium (Chenier et al. 2008).

Ga and Fe-Dependent Metabolism

A key reaction catalyzing the entry of citrate into the TCA cycle is that performed by aconitase (ACN). The latter contains a labile $[Fe_4\text{-}S_4]^{2+}$ cluster and allows for the isomerization of citrate to isocitrate (Chenier et al. 2008). Under oxidative stress, the conventional $[Fe_4\text{-}S_4]^{2+}$ cluster is converted to an inactive $[Fe_3\text{-}S_4]^{+}$ form, thus incapacitating the enzyme (Chenier et al. 2008). As such, ACN acts as a gatekeeper to the TCA cycle. Under normal conditions, citrate catabolism proceeds through the TCA cycle. When *P. fluorescens* is exposed to Ga, ACN activity is diminished (Chenier et al. 2008).

As fumarase (FUM) is also affected by ROS and Ga toxicity, *P. fluorescens* elaborates multiple isoforms of FUM. While FUM A and B rely on iron-sulfur clusters for their activity, FUM C is iron-independent (Chenier et al. 2008). In the presence of Ga, this microbe evokes the synthesis of FUM C in order to counteract the decrease in FUM A activity (Chenier et al. 2008). By doing so, carbon flow proceeds through the TCA cycle rather than grinding to a halt. These metabolic manipulations enable the survival of *P. fluorescens* under Ga stress (Fig. 3).

Metabolic Flux Through the TCA Cycle

As citrate is the sole source of carbon and in the absence of any citrate lyase, *P. fluorescens* activates two enzymes, namely, isocitrate lyase (ICL) and isocitrate dehydrogenase (ICDH), downstream of ACN. The former cleaves isocitrate into succinate and glyoxylate, feeding various anaploretic pathways. ICDH exists in two isoforms; a membrane-bound nicotinamide adenine dinucleotide (NAD)-dependent variant, and a soluble NADP-dependent form (Bériault 2004). Both catalyze the decarboxylation of isocitrate to alpha-ketoglutarate (KG) with the concomitant reduction of their respective cofactors. In *P.fluorescens* subjected to Ga stress, ICL and ICDH-NADP are upregulated (Bériault 2004). Hence, this metabolic

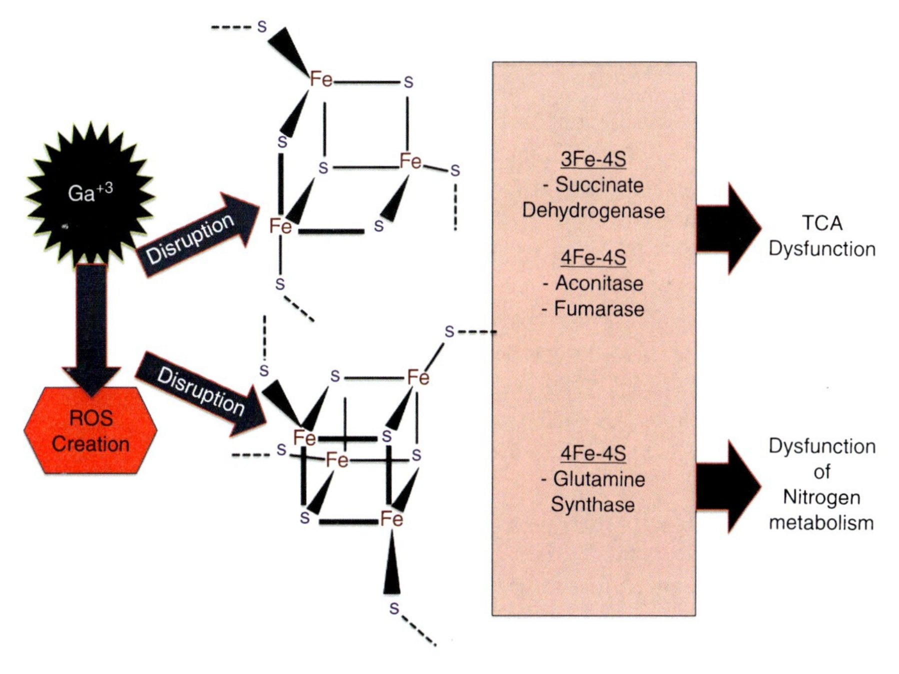

Gallium in Bacteria, Metabolic and Medical Implications, Fig. 2 The disruption of Fe-containing proteins by Ga. Ga disrupts Fe homeostasis and creates ROS, leading to the formation of dysfunctional Fe-S clusters. The Fe-S clusters are crucial to the functioning of numerous key enzymes in the TCA cycle and nitrogen metabolism

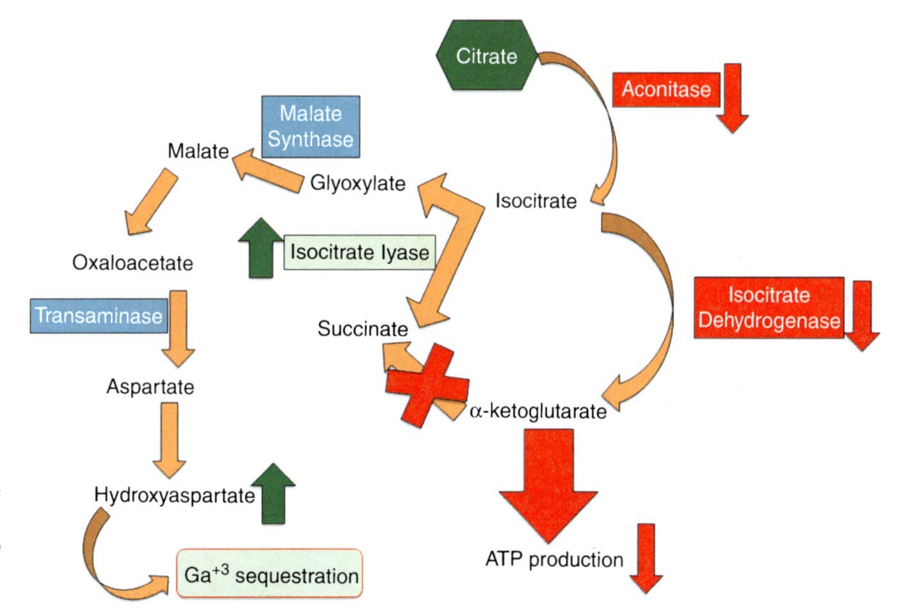

Gallium in Bacteria, Metabolic and Medical Implications, Fig. 3 Metabolic shift induced by Ga toxicity. Citrate is redirected to alternate metabolic pathways in order to ensure the survival of the bacterium

flux generated by ICL and ICDH-NADP acts as a potent vacuum and helps metabolize citrate despite an almost defunct ACN (Fig. 3).

Elimination of NADH

The oxidation of NADH by complex I is the first step of oxidative phosphorylation and is a major site of superoxide formation. It is downregulated in Ga-stressed *P. fluorescens* as are enzymes generating NADH (Chenier et al. 2008). This nicotinamide derivative is also a pro-oxidant. NADH production is attenuated by several mechanisms. Alpha-ketoglutarate dehydrogenase (αKGDH) is an NADH-forming enzyme that is downregulated (Bériault 2004).

The downregulation of αKGDH also serves to pool KG, which can act as a scavenger of ROS. The peroxide-mediated decarboxylation of this alpha-keto acid generates succinate, CO_2, and H_2O, thus providing the cell with an effective means of mopping up ROS in situ. Another stratagem that is utilized to control NADH production and allows the functioning of the TCA cycle, albeit ineffectively, is the expression of NADH oxidase (NOX). *P. fluorescens* employs a water-generating NOX to convert NADH into NAD, which can be utilized to propel other metabolic reactions (Chenier et al. 2008). This Ga-triggered metabolic shift allows the catabolism of citrate, the generation of KG, and the regeneration of NAD.

NADPH Combats Ga Toxicity

While NADH acts as a pro-oxidant, NADPH serves as an antioxidant, acting as fuel for the activity of ROS detoxifying enzymes such as glutathione peroxidase. Hence, its production is favored during Ga stress. The upregulation of ICDH-NADP is one method of accomplishing this task (Bériault et al. 2007). Glucose-6-phosphate dehydrogenase (G6PDH) is the first enzyme catalyzing the entry of glucose-6-phophate into the pentose phosphate pathway. When *P. fluorescens* is exposed to Ga, three different isoforms of G6PDH can be found whereas untreated cells only have one (Bériault et al. 2007). As NADP levels need to be kept elevated for the purpose of NADPH production, two enzymes, NAD kinase and NADP phosphatase are modulated. The former that catalyzes the phosphorylation of NAD to produce NADP is upregulated, while the latter dephosphorylates NADP into NAD and is downregulated (Chenier et al. 2008) (Fig. 4).

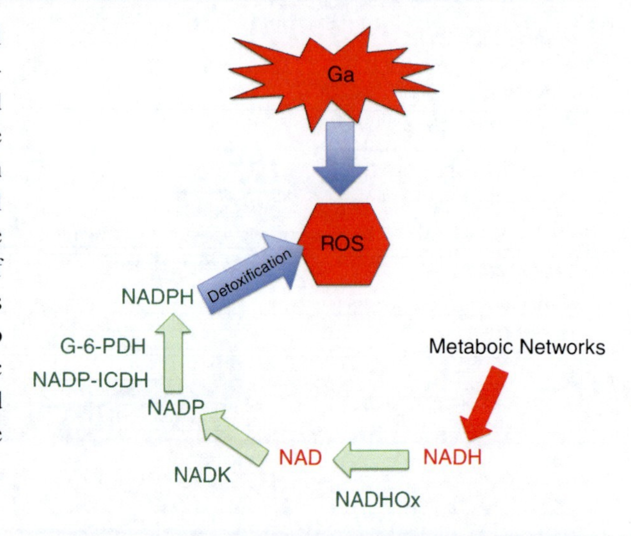

Gallium in Bacteria, Metabolic and Medical Implications, Fig. 4 Ga promotes the reduction in NADH and production of NADPH. Ga forces the microbe to diminish NADH production with a concomitant increase in NADPH synthesis. Simultaneously, Ga promotes the activity of NADH oxidase (NADHOx) leading to the oxidation of NADH to NAD. NAD is diverted to NADP by NAD kinase (NADK). The NADP is pooled and utilized by NADP-dependent isocitrate dehydrogenase (NADP-ICDH) and glucose-6-phosphate dehydrogenase (G6PDH) to create NADPH, the ultimate reductive capacity for ROS detoxification. Arrows in *green* are pathways, which are increased. Those in *red* are those that are reduced in activity

Ga Sequestration and Metal Chelation Therapy

Despite the critical role metabolic adaptation plays in *P. fluorescens* to circumvent Ga toxicity, it does not tackle the root of the problem. If the microbe is to resume business as usual, the source of the toxicity must be dealt with. Siderophores are often deployed by microorganisms in order to chelate and internalize Fe. Numerous siderophores have been isolated and can be used to treat heavy metal toxicity. Desferrioxamine, a chelator produced by *Streptomyces pilosus* is often utilized to treat acute iron poisoning and iron overload (Bériault 2004). *P. fluorescens* treated with Ga elaborates a hydroxyaspartate containing metabolite involved in the chelation of this trivalent metal (Bériault 2004). Hence, enzymes like aspartate transaminase (AST) and malate dehydrogenase (MDH) that play a critical role in the production of aspartate are upregulated (Bériault 2004). Figure 5 depicts the toxicity associated with Ga and the subsequent detoxification evoked by *P. fluorescens* confronted with this metal toxin.

Ga as an Antimicrobial

Inhibition of Biofilm Formation in *Pseudomonas aeruginosa*

Bacterial resistance to commercially available antibiotics is a growing concern. As such, the discovery of alternative therapies for treatment of pathogenic organisms has become a priority. Resistance is thought to occur primarily via transfer of genetic material from the commensal bacteria of a person to the pathogenic

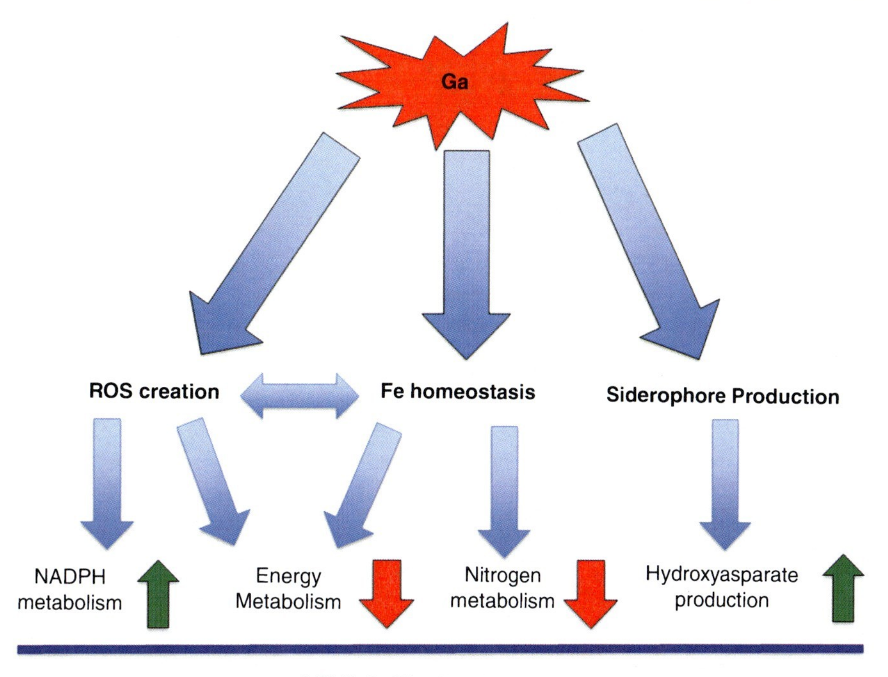

Gallium in Bacteria, Metabolic and Medical Implications, Fig. 5 A global view of Ga toxicity and adaptation in *Pseudomonas fluorescens*. The deprivation of Fe, the generation of ROS, and the production of siderophores evoked by Ga are shown

organism once infection occurs (Rasko and Sperandio 2010). Therefore, it is plausible that drugs targeting virulence factors of the invading bacteria and not the human flora will have a greater lifespan (Rasko and Sperandio 2010). One common virulence factor employed by the gram-negative pathogen *P. aeruginosa* is the establishment of a biofilm. The latter is an aggregate of microorganisms embedded within a protective matrix with increased resistance toward antibiotic treatments.

Ga, because it is FDA approved for intravenous administration and is a Fe mimetic, is a potential candidate for the treatment of pathogenic bacteria in humans (Chitambar 2010). One study has indicated that gallium treatment in mice is capable of blocking acute and chronic *P. aeruginosa* infections (Chitambar 2010). These microbes release siderophores into the environment in order to chelate and internalize available Fe which is required for survival. Ga acts as a "Trojan horse" and is taken up instead. This strategy limits the ability of *P. aeruginosa* to form biofilms in vivo, and the bacteria are quickly eradicated (Chitambar 2010). Indeed, mice subjected to gallium treatment are able to overcome the infection and survive much longer than their untreated counterparts (Chitambar 2010).

Interference with Fe Metabolism in Mycobacteria

Monocytes and macrophages, two subsets of white blood cells (WBCs), form the cornerstone of the human immune response. Their ability to uptake and destroy foreign matter via phagocytosis is indispensable for the clearance of dangerous microorganisms. While this environment is normally too hostile for the survival of the bacteria, some are capable of proliferating within the confines of WBCs. *Mycobacterium tuberculosis* and *Mycobacterium avium* complex (MAC) are pathogenic bacteria that can grow and replicate in monocytes and macrophages (Chitambar 2010). Mycobacteria are known to produce two siderophores, the soluble exochelins and the membrane-bound mycobactins (Olakanmi et al. 2000). These compounds are thought to participate in intraphagosomal Fe acquisition once extracellular Fe is delivered to the WBCs via transferrin.

Ga, because of its similarity to Fe, is well suited to target pathogens which replicate within the phagosome. At sites of infection, WBCs uptake large amounts of Fe to fuel biological processes and impede microbial invasiveness. Gallium nitrate ($Ga(NO_3)_3$) and Ga-transferrin substitute for Fe and inhibit mycobacteria growth in media and within human macrophages, an effect which can be reversed by the addition of exogenous

Fe (Olakanmi et al. 2000). Since Ga is released from transferrin at a higher pH (~6.5) than the release of Fe (~5.5), it is postulated that the pH of the mycobacterial phagosome (≥6.0) renders it an effective environment in which Ga can compete with Fe for intracellular targets (Olakanmi et al. 2000). Suppression of *Mycobacterium* growth by Ga is thought to occur by the inhibition of Fe-dependent antioxidant enzymes and ribonucleotide reductase. The latter plays a key role in mycobacterial proliferation, but has negligible activity in terminally differentiated macrophages (Olakanmi et al. 2000). Furthermore, the concentrations of $Ga(NO_3)_3$ required to impede mycobacterial growth in vitro are approved for human use (Olakanmi et al. 2000).

Improved Cultural Selectivity in Mycological Media

Isolating and culturing clinically relevant fungi, such as the *Aspergillus* spp. and the zygomycetes, prove to be difficult tasks owing to the presence of bacterial contaminants that outgrow the fungus of interest. Sabouraud Dextrose Agar (SDA) is a fungal selective media commonly utilized to culture commensal and pathogenic fungi in clinical microbiology laboratories (Moore et al. 2009). Generally, this medium is supplemented with streptomycin, penicillin, and chloramphenicol to curb the growth of bacteria (Moore et al. 2009). However, the emergence of multi- and pan-drug resistant microorganisms has compromised the effectiveness of these antibiotics.

The ability of Ga salts such as $Ga(NO_3)_3$ and Ga-maltolate to disrupt Fe metabolism in bacteria renders them ideal agents for the suppression of bacterial proliferation in vitro. Supplementation of SDA with 2 mM $Ga(NO_3)_3$ efficaciously inhibits growth of several bacteria, including *Escherichia coli*, *Klebsiella pneumoniae*, *P. aeruginosa*, and *Staphylococcus aureus* (Moore et al. 2009). At this concentration, only the fungi and *Burkholderia cenocepacia* are viable (Moore et al. 2009). It is important to note that the effective concentration of Ga for antimicrobial activity is dependent on the local concentration of Fe, which reverses its toxic effect (Moore et al. 2009).

Conclusion

The ability of bacteria to adapt to and overcome a wide range of environmental pollutants renders them ideal for the study of these agents and their noxious effects. *P. fluorescens*, due to its metabolic versatility, proliferates readily in environments rich in Ga. In order to accomplish such a feat, this organism reworks various metabolic networks, particularly the TCA cycle, in an effort to circumvent and eliminate Ga's adverse influence. The upregulation of enzymes downstream of ACN allows for continuous carbon flux through the TCA cycle, a task also made possible via the switch to a Fe-independent isoform of FUM. A finely tuned metabolic-balancing act exists in order to ensure an adequate ratio of NADPH to NADH and avoid a buildup of oxidative stress. The rerouting of TCA cycle intermediates towards the production of aspartate protects the organism by producing chelating agents to bind and expel Ga. Bacteria which cannot adapt to the noxious effects of Ga exhibit decreased virulence and quickly succumb to death and clearance by the immune system. Thus, a better understanding of bacterial physiology and biochemistry under the influence of Ga is critical in understanding Fe metabolism and designing effective therapeutic approaches to combat bacterial infection.

Cross-References

▶ Gallium, Physical and Chemical Properties
▶ Gallium, Therapeutic Effects
▶ Gallium Uptake and Transport by Transferrin
▶ Iron Homeostasis in Health and Disease
▶ Iron, Physical and Chemical Properties

References

Al-Aoukaty A, Appanna V, Falter H (1992) Gallium toxicity and adaptation in *Pseudomonas fluorescens*. FEMS Microbiol Lett 71:265–272

Bériault R (2004) The metabolic network involved in the survival of *Pseudomonas fluorescens* exposed to gallium, a pro-oxidant and an iron mimetic. M.Sc. thesis, Laurentian University

Bériault R, Hamel R, Chenier D, Mailloux R, Joly H, Appanna V (2007) The overexpression of NADPH-producing enzymes counters the oxidative stress evoked by gallium, an iron mimetic. Biometals 20:165–176

Chenier D, Bériault R, Mailloux R, Baquie M, Abramia G, Lemire J, Appanna V (2008) Involvement of fumarase C and NADH oxidase in metabolic adaptation of *Pseudomonas fluorescens* cells evoked by aluminum and gallium toxicity. Appl Environ Microbiol 74:3977–3984

Chitambar C (2010) Medical applications and toxicities of gallium compounds. Int J Environ Res Public Health 7:2337–2361
Downs A (ed) (1993) Chemistry of aluminum, gallium, indium and thallium. Springer, New York
Moore J, Murphy A, Miller B, Loughrey A, Rooney P, Elborn J, Goldsmith C (2009) Improved cultural selectivity of medically significant fungi by suppression of contaminating bacterial flora employing gallium (III) nitrate. J Microbiol Methods 76:201–203
Olakanmi O, Britigan B, Schlesinger L (2000) Gallium disrupts iron metabolism of mycobacteria residing within human macrophages. Infect Immun 68:5619–5627
Paulsen I, Press C, Ravel J, Kobayashi D, Myers G, Mavrodi D, Deboy R, Seshadri R, Ren Q, Madupu R, Dodson R, Durkin A, Brinkac L, Daugherty S, Sullivan S, Rosovitz M, Gwinn M, Zhou L, Schneider D, Cartinhour S, Nelson W, Weldman J, Watkins K, Tran K, Khouri H, Plerson E, Plerson L, Thomashow L, Loper J (2005) Complete genome sequence of the plant commensal *Pseudomonas fluorescens* Pf-5. Nat Biotechnol 23:873–878
Rasko D, Sperandio V (2010) Anti-virulence strategies to combat bacteria-mediated disease. Nat Rev Drug Discov 9:117–128

Gallium Incorporation by the Iron-Acquisition Pathway

► Gallium Uptake and Transport by Transferrin

Gallium Nitrate, Apoptotic Effects

Christopher R. Chitambar
Division of Hematology and Oncology, Medical College of Wisconsin, Froedtert and Medical College of Wisconsin Clinical Cancer Center, Milwaukee, WI, USA

Synonyms

Metallodrugs in cancer treatment; Therapeutic targeting of iron-dependent tumor growth with iron-mimetic metals

Definition

Apoptosis refers to a form of cell death that occurs through a specific programmed sequence of events in cells. It involves the activation of specific genes and their proapoptotic protein products that act through different pathways to eventually produce proteolytic cleavage of cytoskeletal proteins, nuclear fragmentation, and cell death. The activity of apoptotic proteins is counterbalanced by the activity of antiapoptotic proteins; hence, the balance between pro- and antiapoptotic processes determines whether a cell will live or die. The apoptotic process is often disrupted in cancer cells, thus empowering these cells with a survival advantage over normal cells. Compounds that can trigger apoptosis in malignant cells are of considerable interest since such agents may potentially be developed as drugs to treat cancer.

Background and Introduction

Gallium, atomic number 31, is a shiny, silvery-white-colored group IIIa metal that has found widespread use in the fields of electronics and medicine. Gallium has a melting point of 28.7646°C making it one of the few metals that is near-liquid and can melt when held in the hand. In aqueous solution, Ga^{3+} tends to form chelates through bonds with oxygen and nitrogen atoms present on ligands. In solution, gallium hydrolyzes when the pH is near neutral, leading to the formation of insoluble $Ga(OH)_3$. Gallium shares properties with iron in regard to its ionic radius and trivalent state; however, unlike iron (III), Ga (III) is not reduced to divalent Ga(II) and therefore does not directly participate in redox reactions. In the circulation, gallium binds avidly to the iron transport protein transferrin, albeit slightly more weakly than iron (III).The biochemistry of gallium has been reviewed (Bernstein 1998).

While there appears to be no physiologic function for gallium in the human body, the ability of gallium compounds to interact with certain biologic processes has led to the development of various gallium compounds as potential therapeutic and diagnostic agents. Studies conducted in the 1960s demonstrated that radiogallium (^{67}Ga), when injected into tumor-bearing animals, concentrated primarily in tumor cells. This finding was rapidly advanced to humans leading to the development of the ^{67}Ga scan for imaging and detection of tumors in patients with cancers. While its clinical application was examined in a number of different malignancies, ^{67}Ga scanning proved to be most useful for the detection of viable lymphomatous tumors in Hodgkin's and non-Hodgkin's lymphoma. The reader is referred to a recent review for additional discussion of radiogallium in tumor imaging (Chitambar 2010).

Gallium Nitrate as a Metallodrug for Cancer Treatment

The ability of ^{67}Ga to localize in tumor cells prompted further investigation to evaluate the potential antitumor activity of stable gallium nitrate. Screening studies conducted in the mid-1970s at the National Cancer Institute in the USA compared the antitumor activities and toxicity of the salts of the group IIIa metals gallium, aluminum, and indium in a rodent tumor model. These studies revealed that gallium nitrate had the highest antineoplastic activity and lowest toxicity in mice and rats inoculated with solid tumors including Walker 256 carcinosarcoma, reticulum cell sarcoma A-RCS, mammary carcinoma YMC, lymphosarcoma P1798, and others. These interesting observations led to further studies to determine the toxicity profile of gallium nitrate in larger animals and eventually to its testing in humans as an investigational drug (NSC 15200). In initial clinical trials, to determine its toxicity and antitumor activity in humans, gallium nitrate was administered to patients as either a brief intravenous infusion over 20 min or a continuous intravenous infusion for 5–7 days. The latter was found to be associated with improved clinical outcomes and less toxic side effects; it was therefore adopted as the preferred method of drug administration for gallium nitrate in subsequent clinical studies. The major dose-limiting toxicity of brief intravenous infusion of gallium nitrate was kidney dysfunction resulting from the deposition of precipitates containing gallium, calcium, and phosphate in the renal tubules. In contrast, nausea, vomiting, and diarrhea were the dose-limiting toxicities associated with continuous intravenous infusion of gallium nitrate. Of the various cancers examined, bladder cancer and non-Hodgkin's lymphoma emerged as the two malignancies most sensitive to the antineoplastic activity of gallium nitrate(Chitambar 2004).

Whereas the clinical activity of gallium nitrate as an antineoplastic agent has been convincingly established in several clinical studies, an understanding of the basic mechanisms by which gallium induces tumor cell death has lagged behind its clinical development. Subsequent discussion will focus on our current understanding of the steps involved in the apoptotic action of gallium nitrate.

Mechanisms of Gallium-Induced Cell Death

Iron-Related Processes Involved in Gallium-Induced Cell Death

Exposure of malignant cells to gallium nitrate in vitro results in growth arrest and cell death. In general, leukemia and lymphoma cell lines appear to be more sensitive to the growth-inhibitory effects of gallium nitrate than solid tumor cell lines. Our current understanding of the mechanisms of action of gallium is largely derived from studies conducted in the former cell lines.

Gallium's chemical properties enable it to function as an iron mimetic and, consequently, perturb cellular iron homeostasis. Iron is an essential component of many enzymes involved in cell function and viability. Several studies have shown that cancer cells have a greater requirement for iron than normal cells; certain cancer cells display increased transferrin receptors (responsible for iron uptake), increased ferritin (iron storage protein) levels, and decreased ferroportin levels (iron efflux protein) (Hogemann-Savellano et al. 2003; Arosio et al. 1990; Pinnix et al. 2010). These changes enable malignant cells to acquire increased amounts of iron to support their proliferation relative to nonmalignant cells. Interference with cellular iron utilization by gallium thus has a greater impact on the growth of malignant cells than on normal cells.

Cellular uptake of gallium: The initial step in gallium's cytotoxic action involves its targeting to cancer cells and its entry into these cells. One model of cellular gallium uptake involves the binding of gallium to transferrin and its incorporation into cells via transferrin receptor-mediated endocytosis of transferrin-gallium. Transferrin receptors are present in high numbers on lymphoma and bladder cancer cells. Clues to the similarity between gallium and iron transport into cells were provided by early studies which showed that the uptake of ^{67}Ga by cells in tissue culture could be enhanced by the addition of exogenous transferrin to the culture medium. Evidence for the importance of transferrin in this process was provided by studies which demonstrated (1) that ^{67}Ga in the circulation was bound exclusively to transferrin and (2) that ^{67}Ga was initially incorporated into cells through transferrin receptor-mediated endocytosis of transferrin-gallium in a manner similar to that of ^{59}Fe. The uptake of ^{67}Ga uptake by melanoma cells implanted in a mouse was blocked by a monoclonal antibody against the transferrin receptor, thus providing additional evidence for the role of the transferrin receptor in cellular gallium uptake in vivo. In contrast to the above data, some studies have shown that ^{67}Ga uptake by cells may also occur independently of the transferrin receptor. However, this non-transferrin receptor ^{67}Ga uptake pathway is less well defined; it may be similar to that used by cells for transferrin receptor-independent iron uptake. It is likely that the cellular uptake of ^{67}Ga by transferrin-dependent or transferrin-independent pathways may be dependent on cell type.

The process of cellular uptake of stable gallium nitrate tends to follow that of ^{67}Ga, although there may be some additional steps dictated by the aqueous chemistry of gallium. Under physiologic conditions, approximately one-third of transferrin in the circulation is occupied by iron (III); hence, the remaining metal-binding sites of transferrin are available to bind nonradioactive gallium and facilitate its (receptor-mediated) uptake by cells. However, while high concentrations of gallium favor its avid binding to transferrin in vitro, low concentrations of gallium favor its dissociation from transferrin to form $Ga(OH)_4$. It is probable that following its intravenous administration as gallium nitrate, a variable fraction of gallium exists in the circulation as $Ga(OH)_4$ rather than as transferrin-gallium. Whether $Ga(OH)_4$ is taken up by cells and whether it contributes to the cytotoxicity of gallium nitrate is not known.

Interference of iron transport by gallium: Studies in human leukemic cell lines have shown that transferrin-gallium can compete with transferrin-iron for binding to cell surface transferrin receptors thus inhibiting cellular iron uptake. Moreover, gallium interferes with endosomal acidification necessary for the dissociation of iron from transferrin and its subsequent trafficking from the endosomal to the cytoplasmic compartments. The net consequence of gallium's action at these steps is the development of a state of cellular iron deprivation. The cytotoxicity of gallium in certain malignant cell lines can be reversed in vitro by the addition of iron salts or hemin (iron-protoporphyrin) (Chitambar et al. 1988), suggesting that interference with cellular iron utilization by gallium plays a role in its cytotoxicity. Patients treated with gallium nitrate may develop a microcytic hypochromic anemia and an increase in zinc protoporphyrin levels, findings that are consistent with red cell iron depletion in vivo.

Inhibition of the iron-dependent activity of ribonucleotide reductase: In human leukemic HL60 cells, transferrin-gallium inhibits cell proliferation leading to cell cycle arrest in early S-phase and to a block in DNA synthesis. Studies utilizing electron spin resonance (ESR) spectroscopy and measurements of nucleotide pools in cells showed that blockade of cellular iron uptake by transferrin-gallium results in inhibition of the activity of the R2 subunit of ribonucleotide reductase, the enzyme responsible for deoxyribonucleotide synthesis. This enzyme consists of two dimeric subunits termed R1 and R2. The R1 subunit contains substrate and effector-binding sites, while the R2 subunit contains an essential binuclear iron center and a tyrosyl free radical that has a characteristic signal on ESR spectroscopy. The activity of the R2 subunit increases as cells enter S phase, but since the R2 subunit has a half-life of 3–4 h, a continuous supply of iron is needed for its activity to support DNA synthesis. Disruption of the flow of iron to the R2 subunit will diminish ribonucleotide reductase activity. Hence, inhibition of cellular iron uptake by transferrin-gallium will lead to a block in ribonucleotide reductase and an arrest in DNA synthesis. Iron-containing compounds such as hemin and soluble iron salts reverse the inhibitory effect of gallium on the activity of R2 subunit. Cells exposed to gallium display a decrease in R2 subunit activity (as evidenced by a decrease in the tyrosyl free radical signal of R2). This is due to a conversion of iron-containing R2 to apoR2 (iron-poor R2) that lacks the ESR signal rather than to a loss of R2 protein. The addition of iron salts to cell lysates of gallium-treated cells restores the R2 ESR signal to normal in minutes, thus indicating that active R2 is rapidly regenerated from apoR2 with iron.

Cellular iron deprivation does not appear to be the sole mechanism by which gallium blocks ribonucleotide reductase activity. Gallium nitrate was shown to block the enzymatic activities of CDP and ADP reductase in a cell-free assay system, suggesting that gallium can directly inhibit the activity of ribonucleotide reductase. Although the mechanism by which this occurs has not been elucidated, it appears reasonable to speculate that gallium likely disrupts the iron center of the R2 subunit leading to loss of enzymatic activity. Hence, gallium inhibits ribonucleotide reductase through both an indirect (blockade of cellular iron uptake) and a direct action.

Interaction of gallium with iron metabolism in microbial systems: Further evidence for the interaction of gallium with iron metabolism has been provided by studies in bacteria that show that gallium interferes with iron utilization by certain microorganisms leading to their death. Moreover, gallium promotes a pro-oxidant state and invokes an antioxidant response in *Pseudomonas fluorescens*, suggesting a role for oxidative stress in gallium's mechanisms of antimicrobial action. *Mycobacterium tuberculosis*, *Mycobacterium avium*, and *Pseudomonas aeruginosa* are among the important pathogens in humans that have been reported to be susceptible to gallium compounds in preclinical studies.

Iron-Independent Actions of Gallium That May Produce Cell Death

Gallium has been shown to interfere with cellular processes unrelated to iron. However, the extent to which these effects of gallium contribute to cell death is unclear. DNA polymerases and tyrosine phosphatases can be inhibited by gallium, but a correlation between these effects and cell growth inhibition could not be demonstrated. Other gallium complexes such as gallium chloride can inhibit tubulin polymerization while ligands containing pyridine/4-6-substituted phenolic moieties complexed to gallium can inhibit proteasomal function and the growth of prostate cells in vitro and in animal xenografts. Although the action of gallium nitrate on proteasome function has not been studied, the ability of these newer gallium compounds to act on the proteasome suggests that gallium may also induce apoptotic cell death through mechanisms that are independent of its action on iron metabolism.

Activation of Proapoptotic Proteins by Gallium

Morphologic changes and DNA fragmentation which are typical of apoptosis occur in human lymphoma CCRF-CEM cells incubated with transferrin-gallium. These changes can be prevented by transferrin-iron, again underscoring the interaction of gallium with iron metabolism. It is known that apoptotic cell death occurs by the activation of executioner caspases-3/-7, which, in turn, cleave DNA, lipids, and proteins. These caspases may be activated through two major pathways: (1) an extrinsic pathway in which signals from the cell surface Fas-Associated Death Domain activate caspase-8 which then activates caspase-3 and (2) an intrinsic pathway in which apoptosis occurs though the loss of the mitochondrial membrane potential and the release of cytochrome c and apoptosis-activating factor-1 (APAF-1) from the mitochondria to the cytoplasm. These proteins combine with caspase-9 to form an apoptosome that cleaves procaspase-3 to yield active caspase-3 (Elmore 2007). The activities of cellular proapoptotic and antiapoptotic proteins play a critical role in determining the fate of a cell in response to a potential apoptosis-inducing agent. Studies in human lymphoma CCRF-CEM and DoHH2 cells show that gallium nitrate induces apoptosis via the intrinsic pathway through steps that include the activation of proapoptoticBax, loss of mitochondrial membrane potential, and the release of cytochrome c. These events lead to activation of caspase-3 and cell death. The levels of the antiapoptotic proteins Bcl-2 and Bcl-$X_L$do not appear to be affected by gallium nitrate.

Whereas induction of apoptosis through the initial activation of Bax is frequently seen with a variety of different agents, it is likely not the sole trigger for gallium-induced cell death. With apoptosis induced by gallium maltolate, a novel gallium compound, loss of mitochondrial membrane potential occurs in the absence of Bax activation, thus suggesting that this gallium formulation has a direct action on the mitochondrion (Chitambar et al. 2007). In this regard, it is important to recall that numerous enzymes of the mitochondrial citric acid cycle and electron transport chain contain iron-sulfur clusters that are essential for enzymatic function. Given the interaction of gallium with iron-containing proteins, it is very likely that these iron-sulfur clusters may be disrupted by gallium resulting in loss of mitochondrial function and cell viability. It is anticipated that future studies will elucidate the action of gallium on the mitochondrion.

Cellular Adaptation to Gallium-Induced Apoptosis

Studies utilizing cell lines resistant to the growth-inhibitory effects of gallium nitrate have provided insights into some of the biologic targets of gallium and the adaptive changes cells may undergo to circumvent the cytotoxicity of gallium.

Gallium-resistant CCRF-CEM cells display a decrease in gallium and iron uptake, an increase in the mRNA-binding activity of iron regulatory protein, and a decrease in the content of ferritin. The cytotoxicity of gallium nitrate and the uptake of ^{67}Ga by gallium-resistant cells can be restored by the addition of exogenous transferrin to cells; this suggests that one step by which cells adapt to exposure to gallium nitrate is through decreasing the transferrin receptor-mediated uptake of this metal. Further evidence for this notion is provided by the observation that the growth of CCRF-CEM cells resistant to gallium nitrate can still be inhibited by gallium maltolate, a compound that can be incorporated into cells independent of the transferrin–transferrin receptor pathway.

Recently, studies using gene array technology to identify metal metabolism genes that may be altered by the development of gallium resistance have revealed some unexpected findings. These investigations showed that metallothionein-2A, a protein involved in the binding and intracellular sequestration of divalent metals such as zinc and cadmium (but not iron), was markedly increased in gallium-resistant CCRF-CEM cells. The upregulation of metallothionein-2A in these cells was associated with an increase in the binding of metal-responsive transcription factor-1 to metal-response elements present on the metallothionein gene. An action of gallium on proteins of zinc metabolism would not have been anticipated since gallium is a trivalent metal. A role for metallothionein in modulating gallium's cytotoxicity was further suggested by the observation that increased levels of metallothionein, whether due to endogenous production or by induction through prior exposure of cells to zinc, led to a reduction in the growth-inhibitory effects of gallium nitrate. Exposure of CCRF-CEM cells to gallium nitrate produces an increase in reactive oxygen species (ROS), a decrease in glutathione within 1–4 h, and a subsequent increase in metallothionein-2A and heme oxygenase-1 gene expression. Gallium-induced activation of heme oxygenase-1 involves signaling through the p38 mitogen-activated protein kinase pathway with activation of Nrf-2, a transcription factor for heme oxygenase-1 (Yang and Chitambar 2008). The gallium-induced increase in heme oxygenase-1 can be blocked by p38 MAP kinase inhibitors, while the expression of both metallothionein-2A and heme oxygenase-1 can be reduced by the antioxidant N-acetyl-L-cysteine. These studies suggest a model in which cells exposed to gallium initially generate a cytoprotective response by elevating metallothionein and heme oxygenase-1 levels in response to ROS; however, cell death eventually ensues when this response is overwhelmed.

Conclusions and Future Directions

Gallium nitrate is a metal salt that has displayed important clinical activity in the treatment of certain cancers and is now being evaluated as an antimicrobial agent. Central to the mechanism of gallium nitrate's apoptotic effects is its ability to disrupt iron-dependent processes in cells at several levels including mitochondrial function. Additional processes involved in gallium-induced cell death are likely but these remain to be elucidated. Newer gallium compounds are in development which include gallium maltolate, G4544, Tris(8quinolonato)Ga (III), gallium thiosemicarbazones, and gallium complexes with pyridine and 4-6-substitued phenolic moieties. These compounds may display greater antitumor activity than gallium nitrate and may also possess additional mechanisms of cytotoxic action. Their advancement as therapeutic agents is awaited with anticipation.

Cross-References

▶ Gallium in Bacteria, Metabolic and Medical Implications
▶ Gallium, Therapeutic Effects
▶ Gallium Uptake and Transport by Transferrin
▶ Iron Homeostasis in Health and Disease

References

Arosio P, Levi S, Cairo G, Cazzola M, Fargion S (1990) Ferritin in malignant cells. In: Ponka P, Schulman HM, Woodworth RC (eds) Iron transport and storage. CRC Press, Boca Raton

Bernstein LR (1998) Mechanisms of therapeutic activity for gallium. Pharmacol Rev 50:665–682

Chitambar CR (2004) Gallium compounds as antineoplastic agents. Curr Opin Oncol 16:547–552

Chitambar CR (2010) Medical applications and toxicities of gallium compounds. Int J Environ Res Public Health 7:2337–2361

Chitambar CR, Matthaeus WG, Antholine WE, Graff K, O'Brien WJ (1988) Inhibition of leukemic HL60 cell growth by transferrin-gallium: effects on ribonucleotide reductase

G

and demonstration of drug synergy with hydroxyurea. Blood 72:1930–1936

Chitambar CR, Purpi DP, Woodliff J, Yang M, Wereley JP (2007) Development of gallium compounds for treatment of lymphoma: gallium maltolate, a novel hydroxypyrone gallium compound induces apoptosis and circumvents lymphoma cell resistance to gallium nitrate. J Pharmacol Exp Ther 322:1228–1236

Elmore S (2007) Apoptosis: a review of programmed cell death. Toxicol Pathol 35:495–516

Hogemann-Savellano D, Bos E, Blondet C, Sato F, Abe T, Josephson L, Weissleder R, Gaudet J, Sgroi D, Peters PJ, Basilion JP (2003) The transferrin receptor: a potential molecular imaging marker for human cancer. Neoplasia 5:495–506

Pinnix ZK, Miller LD, Wang W, D'Agostino R Jr, Kute T, Willingham MC, Hatcher H, Tesfay L, Sui G, Di X, Torti SV, Torti FM (2010) Ferroportin and iron regulation in breast cancer progression and prognosis. Sci Transl Med 2:43–56

Yang M, Chitambar CR (2008) Role of oxidative stress in the induction of metallothionein-2A and heme oxygenase-1 gene expression by the antineoplastic agent gallium nitrate in human lymphoma cells. Free Radic Biol Med 45:763–772

Gallium Uptake

▶ Gallium, Therapeutic Effects

Gallium Uptake and Transport by Transferrin

Jean-Michel El Hage Chahine,
Nguyêt-Thanh Ha-Duong and Miryana Hémadi
ITODYS, Université Paris-Diderot Sorbonne Paris Cité, CNRS UMR 7086, Paris, France

Synonyms

Gallium delivery by transferrin receptor-1-mediated endocytosis; Gallium incorporation by the iron-acquisition pathway; Transferrin as a gallium mediator

Definitions

T, human serum transferrin or apotransferrin; TFe, monoferric human serum transferrin; TFe_2, iron-saturated transferrin or holotransferrin; TFR, transferrin receptor; R, transferrin receptor subunit; TGa, transferrin with a gallium-loaded C-lobe; TGa_2, gallium-saturated transferrin in both N- and C-lobes; FeL, iron-nitrilotriacetate; GaL, gallium-nitrilotriacetate.

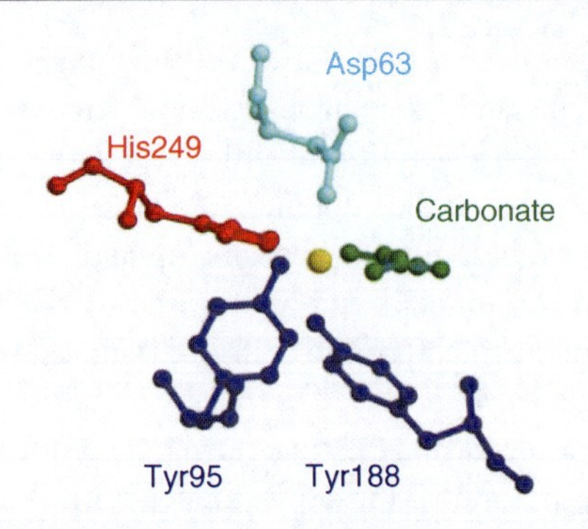

Gallium Uptake and Transport by Transferrin, Scheme 1 The N-binding site of human serum-transferrin (constructed with the RasTop shareware with the Protein Data Bank coordinates: 1A8F)

Transferrins are the most important iron conveyors in vertebrates and invertebrates and also in some microorganisms and bacteria (Aisen 1998). Human serum transferrin (T) is a glycoprotein composed of a single chain of about 700 amino acids organized in two semi-equivalent lobes: the C-lobe and the N-lobe (Aisen 1998). Both lobes consist of two domains. Each contains one iron-binding site, where the metal is coordinated to two phenolates of two tyrosines, an imidazol of a histidine and a carboxylate of an aspartate. It is also coordinated to a synergistic carbonate adjacent to an arginine (Scheme 1). When in the iron-free apo state the two domains of each lobe are in an open conformation in which the protein ligands are in contact with the biological fluid. When iron-loaded, the two domains act as a pair of jaws that engorge Fe(III) in a closed conformation, where iron is buried about 10 Å beneath the protein surface (Scheme 2) (Aisen 1998; Ha-Duong et al. 2008). Human serum transferrin (T) is a major partner in the iron-acquisition pathway. When iron-loaded, holotransferrin (TFe_2) interacts with transferrin receptor 1 (TFR), which is anchored in the plasma membrane. The protein/protein adduct is then internalized in the cytoplasm in an endosome, the acidification of which leads to iron release in the endosome. The protein-protein adduct, consisting of iron-free apo-serum transferrin in interaction with the receptor, is recycled back to the plasma membrane,

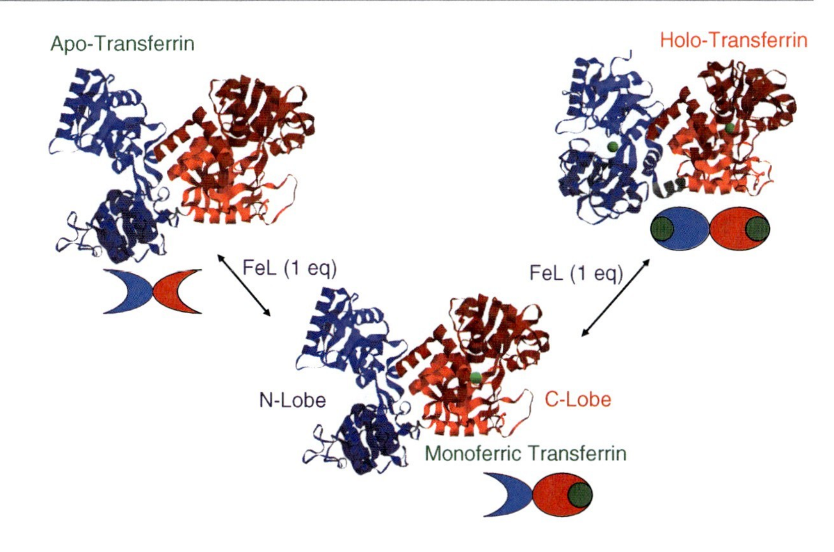

Gallium Uptake and Transport by Transferrin, Scheme 2 Iron uptake by transferrin constructed by RasTop 2.2 (PDB 1CB6 and 1B0L)

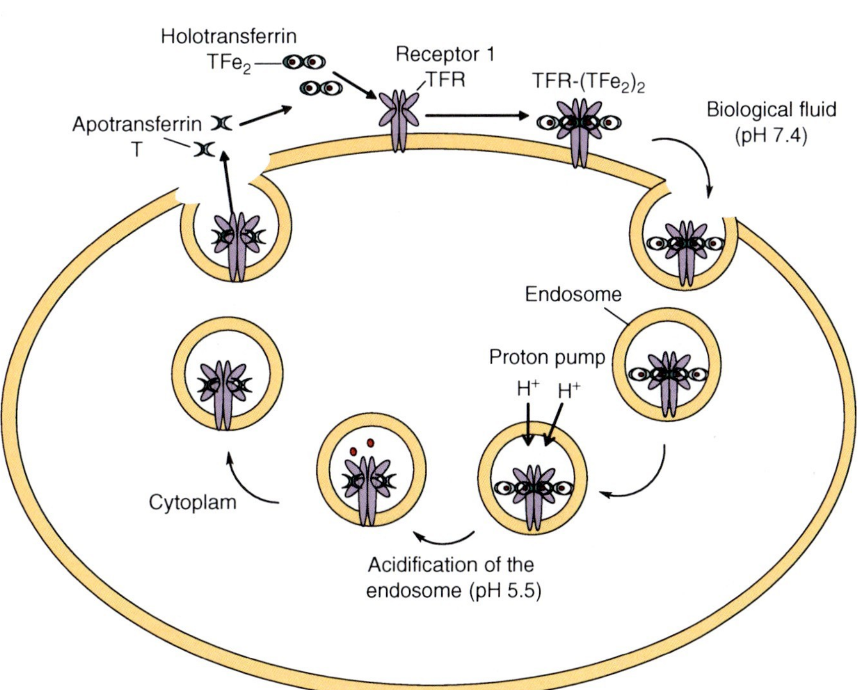

Gallium Uptake and Transport by Transferrin, Fig. 1 Iron-acquisition cycle: step 1, holotransferrin TFe2 interacts with transferrin receptor 1 TFR; step 2, the protein-protein adduct is internalized in the cytosol by receptor-mediated endocytosis; step 3, the endosome is acidified down to pH~5.6; step 4, iron is released from the protein-protein adduct and is actively externalized from the endosome; step 4, the apotransferrin (T) in interaction with TFR is recycled back to the plasma membrane, where it is released in the bloodstream ready for a new iron-acquisition cycle (Aisen 1998; Ha-Duong et al. 2008)

where T is readily released in the plasma to achieve another iron-delivery cycle (Fig. 1) (Aisen 1998; Ha-Duong et al. 2008).

Human serum transferrin can also form very stable complexes with about 40 metals other than iron. These include practically all transition metals, actinides, metals of Group 13 of the periodic table (Harris and Messori 2002). This fact led to assume that transferrin is involved in the incorporation of these metals in the cell and their transport across the blood–brain barrier (Ha-Duong et al. 2008; Harris and Messori 2002). The toxicity of aluminum, or the efficiency of gallium or bismuth in some specific therapies, were explained, for example, by a competition with iron toward metal uptake by transferrin and metal delivery by receptor-mediated endocytosis (Ha-Duong et al. 2008). However, this delivery requires at least two conditions: (1) a stable complex should be formed between the metal and T; (2) this complex must be recognized by the transferrin receptor (Ha-Duong et al. 2008).

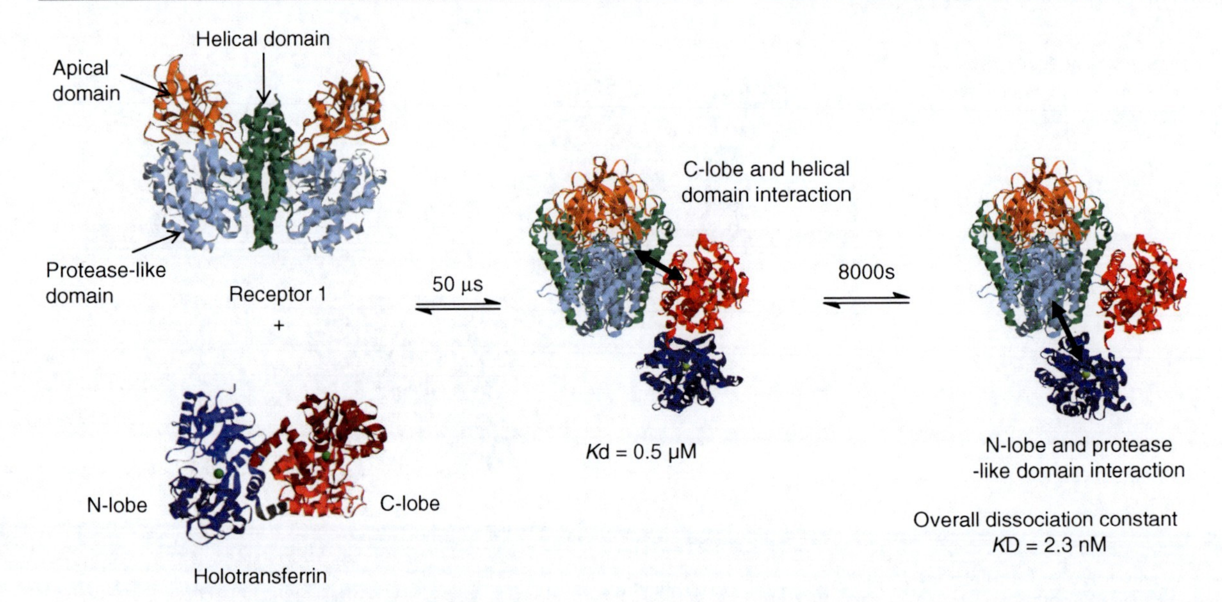

Gallium Uptake and Transport by Transferrin, Fig. 2 The C-lobe of iron-saturated transferrin interacts very rapidly (50 μs) with the helical domain of transferrin receptor 1 (TFR) to yield a first protein-adduct with a dissociation constant $K_d = 0.5$ μM. This interaction is followed by a very slow (~2 h) change in the conformation of the protein-protein adduct to yield the final product in which the N-lobe of the holotransferrin becomes in interaction with the protease-like domain of the receptor with an overall dissociation constant $K_D = 2.3$ nM (Chikh et al. 2007). The structures are constructed with RASTOP 2.2 open-share program by the use of the coordinates available at the Protein Data bank (1CX8 and 1SUV)

In addition, even when the metal is released in the endosome after acidification, it would still require means of delivery to its potential targets in the cell outside this endosome.

Transferrin receptor 1 is a 190 kDa homodimeric protein arranged in two subunits (R) linked by two disulfide bridges. R contains a transmembrane domain, a cytoplasmic endodomain of about 15 kDa and a soluble ectodomain directed toward the biological fluid. The receptor is arranged in four domains: the helical, the apical, the protease-like, which forms with the plasma membrane a pseudo-cavity of about 10 Å and, finally, the endodomain (Lawrence et al. 1999). The C-lobe of holotransferrin (TFe_2) or that of a C-lobe only iron-loaded transferrin interacts primarily with the helical domain of the receptor. This interaction is followed by that of the N-lobe with the protease-like domain (Fig. 2) (Cheng et al. 2004). The first interaction is extremely fast and occurs in the 50 μs range whereas the second occurs in about 2 h (Ha-Duong et al. 2008).

The mechanisms for iron, aluminum, bismuth, cobalt, uranium, and gallium uptake by serum transferrin were established recently (Ha-Duong et al. 2008). It was also shown that, besides aluminum, the cobalt-, bismuth-, uranium-, and gallium-loaded transferrins interact with the transferrin receptor. This renders the incorporation of these metals possible by the iron-acquisition pathway (Ha-Duong et al. 2008).

Gallium is a nonphysiological metal, which belongs to Group 13 of the periodic table. Ga(III) has practically the same ionic radius as Fe(III) and is diamagnetic. Gallium has a very low redox potential which implies that in neutral media, contrary to iron, this metal exists exclusively as a complex of Ga(III) (Harris and Messori 2002). Ga(III) is, therefore, used to mimic Fe(III) without the redox and paramagnetic inconveniences. Moreover, gallium can be used in chemotherapy, medical imagery, has effects on bone metabolism, and an antineoplastic activity (Bernstein 1998; Chitambar and Purpi 2010). Part of this action is explained by the fact that Ga(III) forms a stable complex with transferrin and can, therefore, interfere with the metabolism of iron (Bernstein 1998).

Gallium Uptake by Transferrin: Methods and Results

The transferrin gallium complex possesses specific absorption and emission spectra (Chikh et al. 2007).

These spectra associated with the fast-mixing techniques, such as stopped-flow (mixing time ~1 ms), permit to follow with time the formation of the transferrin-gallium complex (Chikh et al. 2007). When transferrin is rapidly mixed with a gallium donor, such as nitrilotriacetatoGa(III), gallium is transferred from the chelate (GaL) to transferrin in three differentiated kinetic steps (Fig. 3). The use of the techniques and methods of chemical relaxation (fast kinetics) allows attributing each step to a series of chemical processes involved in gallium uptake by transferrin (Chikh et al. 2007; Eigen 1967).

In the C-lobe of apotransferrin, the phenols of the tyrosine ligands are in contact with bulk aqueous medium and are, therefore, in the protonated form. Furthermore, under natural conditions, only the C-lobe of apotransferrin is in interaction with the synergistic hydrogenocarbonate, whereas the N-lobe is not. A first gallium transfer occurs in about 50 s from the gallium chelate to the C-lobe in interaction with HCO_3^- to produce a first transferrin-gallium intermediate, in which the tyrosine ligands are still in the protonated form. This intermediate undergoes a series of proton losses from the protein ligands and conformation changes to produce, in the second kinetic process, an intermediate species, in which gallium is coordinated to the two phenolates of the tyrosines and a carbonate. This leads to a drop in the apparent pK_as of the phenols from about 9 in T down to 8 in the kinetic intermediate. This latter undergoes, then, a series of very slow conformation changes, which allow the N-lobe of transferrin to acquire a second gallium and to reach the thermodynamic equilibrium (Fig. 3) (Chikh et al. 2007).

Interaction of Gallium-Loaded T with R: Methods and Results

Transferrin receptor 1 possesses a typical fluorescence emission spectrum. Adding TGa_2 to TFR results in a increase in fluorescence intensity accompanied by a small red shift in the emission maximum (Chikh et al. 2007). Similar spectral modifications were also observed with iron-loaded and other metals–loaded transferrins (Ha-Duong et al. 2008). These spectra allowed the titration of R by TGa_2 and, subsequently, the determination of the dissociation constant of the TGa_2/receptor 1 protein/protein adduct ($K_{DGa(III)} = 1.10 \pm 0.12$ μM) (Chikh et al. 2007). $K_{DGa(III)}$ is very high as compared to that reported for the interactions of holotransferrin with R in the final equilibrated state (2.3 nM) (Ha-Duong et al. 2008).

This interaction was confirmed by a T-jump kinetic approach. The T-jump technique is based on Joule heating by the discharge of a capacitor in the solution. Heating times can be as low as 200 ns and heating amplitudes can be as high as 10°C. This technique allows to measure kinetic runs occurring in the microseconds to the milliseconds range (Ha-Duong et al. 2008). In the present case the detection was based on the variations in the emission spectra. When a solution of R in the presence of holotransferrin is submitted to a fast T-jump, a single kinetic process corresponding to the interaction of the C-lobe of TGa_2 with R occurs in the 150 μs range as an exponential increase in the fluorescence emission (Fig. 4).

Can Gallium Follow the Iron Acquisition Pathway?

Gallium, iron, aluminum, and bismuth are all considered as hard metals (Harris and Messori 2002) and are complexed by the same protein ligands (Aisen 1998). These include the two phenolates of the two tyrosines, the carboxylate of an aspartate and the imidazol of a histidine. Upon complex formation with T, shifts occur in the apparent pK_a of the ligands. These are related to the nature of the metal. Indeed, the binding site undergoes a chelation process and a change in its environment when the protein transits from an open structure, where all ligands are in contact with the bulk, to the closed structure, where the ligands are engaged in the complex and protected from the external medium (Aisen 1998). The discrepancies in these apparent pK_as shifts reported for iron, bismuth, and cobalt when compared to gallium and aluminum imply that the local environment of aluminum and gallium are at variance with those of Fe(III), Bi(III), and cobalt (Ha-Duong et al. 2008). As with iron, gallium gain by the C-site triggers a series of proton dissociation-dependent changes in the conformation of the protein, which lead to the activation of the N-binding site, thus allowing the second metal uptake by the N-lobe (Chikh et al. 2007).

When gallium-saturated transferrin is mixed with FeL, iron replaces gallium. The affinity of transferrin

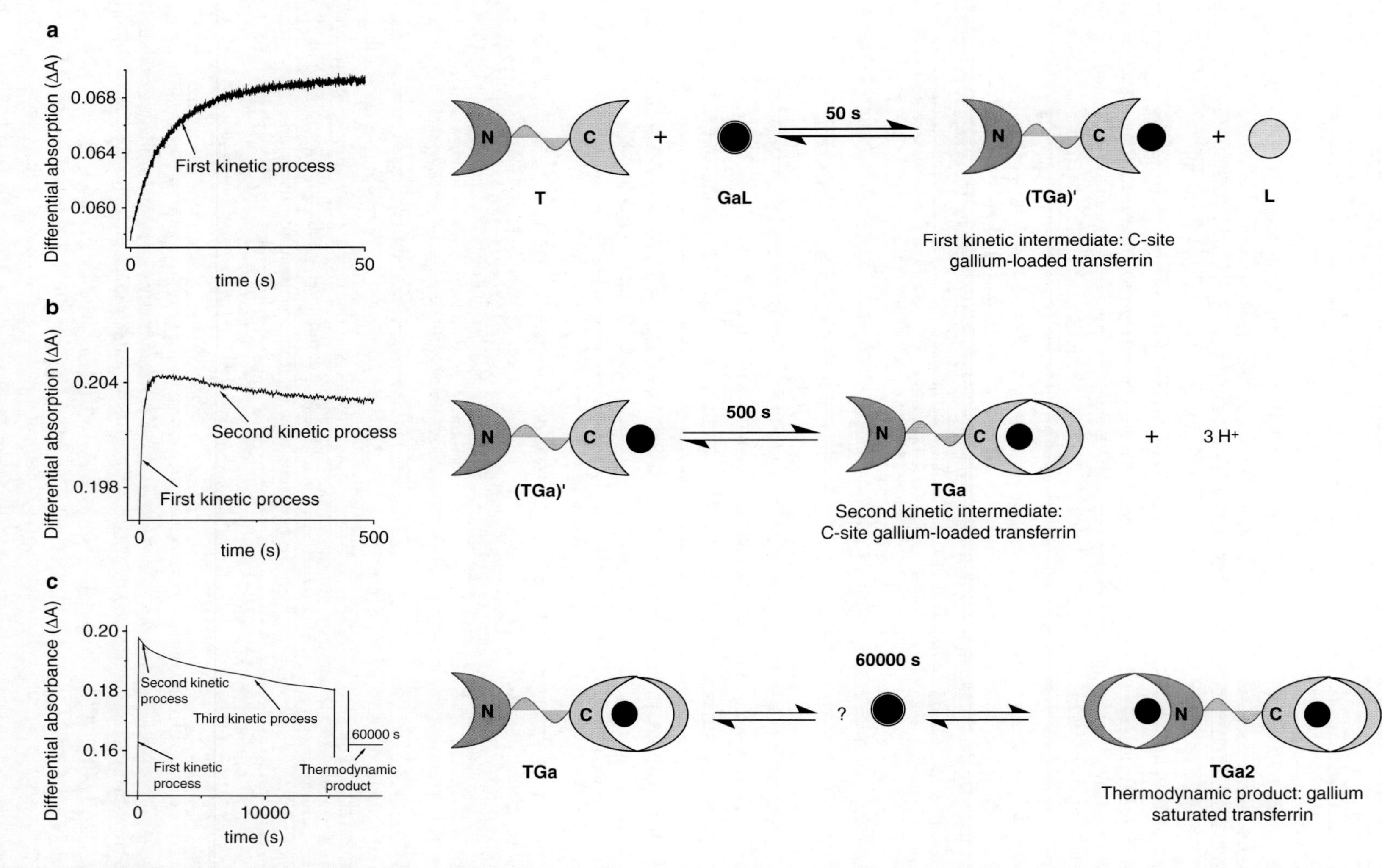

Gallium Uptake and Transport by Transferrin, Fig. 3 (**a**) Ga(III) is transferred from gallium nitrilotriacetate to the C-lobe of transferrin in interaction with bicarbonate to yield a first kinetic product, as shown by the 50 s change in the absorption of the protein when mixed with the chelate. (**b**) A second slow absorption change (500 s) of the kinetic product describes a conformation change accompanied by the loss of three protons from the protein ligands. (**c**) A ultimate very slow absorption change (~17 h) shows the final change in the conformation, which allow a second gallium transfer to the N-lobe of the protein (Chikh et al. 2007)

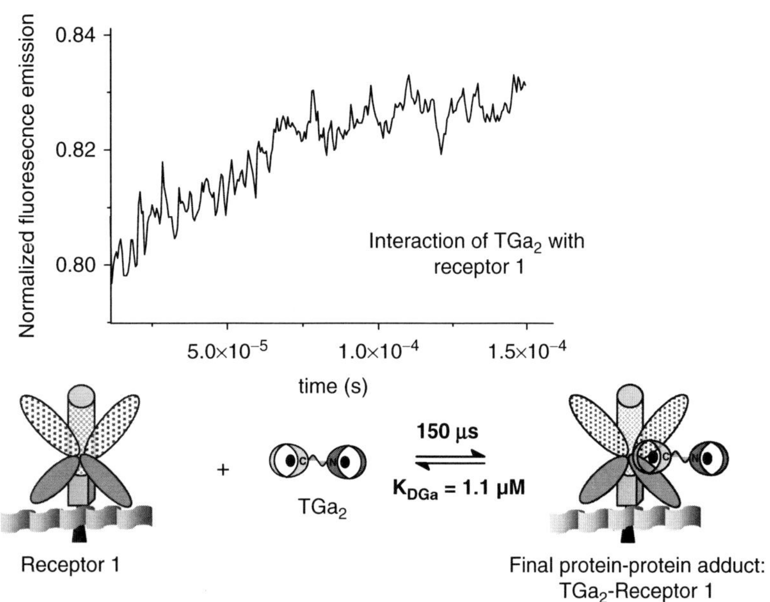

Gallium Uptake and Transport by Transferrin, Fig. 4 Emission growth in 150 μs after an instantaneous T-jump performed in a solution of gallium-saturated transferrin in the presence of receptor 1. This corresponds to the interaction of the C-lobe of the metal-loaded transferrin with the helical domain of R with a dissociation constant $K_{DGa(III)} = 1.1$ μM. No interaction of the N-lobe with the protease-like domain of R is observed (Chikh et al. 2007)

Gallium Uptake and Transport by Transferrin, Table 1 Summary of the main results and interpretations (Chikh et al. 2007)

	Affinity of T for the metal	Dissociation constant (C-Lobe)	Average time of interaction with the C-lobe	Overall dissociation constant (K_D)	Average time of interaction with the N-lobe	Average time for receptor recycling	Internalization with metal-loaded C-lobe
Iron(III)	10^{21}	0.5 μM	50 μs	2.3 nM	8,000 s	Few minutes	Yes
Gallium(III)	10^{19}	1.1 μM	150 μs	1.1 μM	No interaction	Few minutes	Probable

for gallium is about 10^{19} M^{-1} (Harris and Messori 2002), whereas that for iron is in the 10^{21} M^{-1}range (Table 1) (Aisen 1998). Furthermore, the first step in iron uptake by transferrin occurs in less than 0.1 s, whereas the first step in gallium uptake occurs in the 50 s range. Therefore, as compared to gallium, iron uptake by transferrin is both thermodynamically and kinetically favored. This suggests that there cannot be a competition between iron and gallium toward metal uptake by transferrin. However, only 40% of the transferrin circulating in the blood stream is iron-loaded and there is still plenty of protein available for complex formation with other metals (Aisen 1998). Is this availability for gallium sufficient to imply that this metal follows the iron-acquisition pathway by receptor-mediated endocytosis? The answer is negative, because in order to follow the iron-acquisition pathway, the gallium-loaded transferrin should also be recognized by R.

There are two binding sites on a receptor subunit, the first is in the helical domain, common to the C-site of TFe_2 and HFE. The second is in the protease-like domain and is specific to the N-lobe of TFe_2 (Cheng et al. 2004). The N-lobe interacts with the protease-like domain where it occupies the cavity formed by the plasma membrane and the protease-like domain (Cheng et al. 2004; Chikh et al. 2007; Giannetti et al. 2003). The interaction of TFe_2 with R occurs in two differentiated steps (Fig. 2, Table 1). The first corresponds to that of the C-lobe with the helical domain of R. It is extremely fast and occurs with a dissociation constant of 0.5 μM (Fig. 2, Table 1). It is followed by the slow interaction of the N-lobe with the protease-like domain of R. This process lasts about 2 h and stabilizes the protein/protein adduct 300-fold to lead to an overall dissociation constant, $K_{DFe(III)} = 2.3$ nM (Fig. 2, Table 1) (Ha-Duong et al. 2008). With TGa_2, only the interaction of the C-lobe with the helical domain is detected with a dissociation constant $K_{DGa(III)} = 1.1\ \mu M$ (Fig. 4, Table 1). This value is higher than that reported for the interaction of TFe_2 with R and should, therefore, imply that there is no possible competition between holotransferrin and TGa_2 toward the interaction with R in the final equilibrated state. However, the dissociation constant of the R-C lobe intermediate at the end of the first fast step is 0.5 μM (Table 1) (Ha-Duong et al. 2008). This value is of the same order of magnitude as that reported for TGa_2 (1.1 μM, Table 1) (Chikh et al. 2007). Therefore, at the end of the first fast process of the interaction of C-lobe of holotransferrin with R and before the very slow interaction of the N-lobe with the protease-like domain, TGa_2 can compete with holotransferrin for recognition by R. Receptor-mediated endocytosis and iron release in the endosome occur in a few minutes (Chikh et al. 2007). This lapse of time is sufficient to permit the internalization of TGa_2 long before thermodynamic equilibrium between TFe_2 and R is attained. The same case was reported for the interaction of bismuth-loaded transferrin, which also occurs in a single fast step with a dissociation constant of 4 μM (Ha-Duong et al. 2008). Subsequently, TGa_2 can compete with holotransferrin for the interaction with the helical domain of R leading, eventually, to gallium uptake by the iron-acquisition pathway. This was shown for HFE which interacts with only the helical domain of the receptor with a lower K_d value than that of the overall holotransferrin-R interaction. HFE competes with holotransferrin for the recognition by R and perturbs iron acquisition by the cell (Giannetti et al. 2003).

Conclusion

Receptor-mediated iron-acquisition pathway is assumed to be involved in the transport and delivery of several metals from the blood stream to the cell and across the blood–brain barrier. For gallium, the affinity of R for the metal-loaded transferrin is about two orders of magnitude lower than that for holotransferrin at equilibrium. Despite this fact, gallium can be internalized by receptor 1-mediated endocytosis. Indeed, the main responsibility in this transport lies in the fast interaction of the C-lobe of metal-loaded transferrin with the helical domain of R. This process is very fast for both iron-loaded and gallium-loaded transferrins, and the affinities involved are of the same order of magnitude. Both processes are extremely fast compared to endocytosis, which lasts a few minutes. Likewise, endocytosis is extremely fast when compared to the interaction of the N-lobe of holotransferrin with the protease-like domain of R which is controlled by a change in conformation lasting several hours. This implies a dynamic metal transport based more on kinetics than on thermodynamics.

Cross-References

- ▶ Bismuth in Brain, Distribution
- ▶ Calsequestrin
- ▶ Cobalt Transporters
- ▶ Manganese, Interrelation with Other Metal Ions in Health and Disease
- ▶ Nickel Carcinogenesis
- ▶ Vanadium Ions and Proteins, Distribution, Metabolism, and Biological Significance

References

Aisen P (1998) Transferrin, the transferrin receptor, and the uptake of iron by cells. Met Ions Biol Syst 35:585–631

Bernstein LR (1998) Mechanisms of therapeutic activity for gallium. Pharmacol Rev 50(4):665–682

Cheng Y et al (2004) Structure of the human transferrin receptor-transferrin complex. Cell 116(4):565–576

Chikh Z et al (2007) Gallium uptake by transferrin and interaction with receptor 1. J Biol Inorg Chem 12(1):90–100

Chitambar CR, Purpi DP (2010) A novel gallium compound synergistically enhances bortezomib-induced apoptosis in mantle cell lymphoma cells. Leuk Res 34(7):950–953

Eigen M (1967) Nobel lecture

Giannetti AM et al (2003) Mechanism for multiple ligand recognition by the human transferrin receptor. PLoS Biol 1(3):E51

Ha-Duong NT et al (2008) Kinetics and thermodynamics of metal-loaded transferrins: transferrin receptor 1 interactions. Biochem Soc Trans 36(Pt 6):1422–1426

Harris WR, Messori L (2002) A comparative study of aluminum(III), Gallium(III), indium(III) and Thallium(III) binding to human serum transferrin. Coord Chem Rev 228:237–262

Lawrence CM et al (1999) Crystal structure of the ectodomain of human transferrin receptor. Science 286(5440): 779–782

Gallium(III) Complexes, Inhibition of Proteasome Activity

Michael Frezza[1], Sara Schmitt[1], Dajena Tomco[2], Di Chen[1], Claudio N. Verani[2] and Q. Ping Dou[1]
[1]The Developmental Therapeutics Program, Barbara Ann Karmanos Cancer Institute, and Departments of Oncology, Pharmacology and Pathology, School of Medicine, Wayne State University, Detroit, MI, USA
[2]Department of Chemistry, Wayne State University, Detroit, MI, USA

Synonyms

Gallium; Gallium(III) ion; Gallium Complexes; Gallium Compounds

Definition

Gallium complexes were characterized by elemental, spectroscopic, and spectrometric methods. Structure **1** and **3** were elucidated by X-ray crystallography. Biological activity was investigated by growth suppression, proteasome inhibition, and apoptosis induction in vitro and in vivo.

Basic Characteristics

Chemical Properties of Gallium

Gallium is described as a gray semimetal. The most stable cation is the trivalent Ga(III) with an electronic configuration $[Ar]3d^{10}$ formed by the loss of two $4s$ and one $4p$ electrons. The solubility and chemical parameters of gallium have major implications on its therapeutic utility. The trivalent form of gallium is redox-inactive, often being used as a probe for electroactive ligands (Frezza et al. 2007).

Biochemical Properties of Gallium

Investigation into the therapeutic applications of gallium dates to the 1970s where gallium presented an optimal therapeutic and toxicity profile compared to other Group 13 (previously IIIA) elements. Gallium was subsequently found to be active against accelerated bone turnover, tumor growth, and immune diseases. Gallium nitrate and other gallium compounds have been investigated in clinical trials as single agents or in combination with pre-existing chemotherapeutic agents and have been found to be active, especially toward lymphomas and bladder cancer (Frezza et al. 2007).

The therapeutic application of the gallium(III) ion appears to be underscored by its chemical similarity to the trivalent ferric ion, where comparable charge, lability, and ionic radii (0.76 Å vs. 0.78 Å) are observed. Therefore, gallium is commonly recognized as iron(III) upon cellular uptake and interacts with iron specific sites in enzymes and proteins. Gallium also concentrates in regions that are highly dependent of iron, including inflammation and malignancy (Chen et al. 2009). Gallium, unlike iron, is redox inactive. However, unlike iron ions, the gallium(III) ion is redox inactive precluding its participation in biologically relevant redox reactions such as oxygen transport. One of the critical targets of intracellular gallium appears to be the enzyme ribonucleotide reductase, which catalyzes the rate-limiting step in DNA synthesis and is upregulated in highly proliferating cancer cells. This gallium-dependent inhibition of ribonucleotide reductase is in part related to the inhibition of cellular proliferation (Chitamber et al. 1988).

Several gallium complexes have been synthesized and evaluated in clinical trials (Collery et al. 1996). Such complexes have shown an increase in bioavailability and superior anticancer activity compared to gallium salts. These findings provide compelling evidence for the development of novel gallium-based complexes that may show alternative mechanisms of action leading to durable anticancer activity and decreased toxicity.

Cellular 26S Proteasome as a Target for Gallium(III) Complexes

A series of gallium complexes described as $[Ga^{III}(L^X)_2]ClO_4^-$ with asymmetric NN′O-containing pyridine amino phenolate ligands have been developed as potential cellular proteasome inhibitors. The phenolate moiety groups were appended with electron withdrawing and donating groups such as methoxy (**1**), nitro (**2**), chloro (**3**), bromo (**4**), and iodo (**5**) positioned at the 4th and 6th positions (Fig. 1). Due to the flexibility of the ligands, facial coordination takes place and the overall geometry of the complexes is distorted octahedrally (Fig. 1).

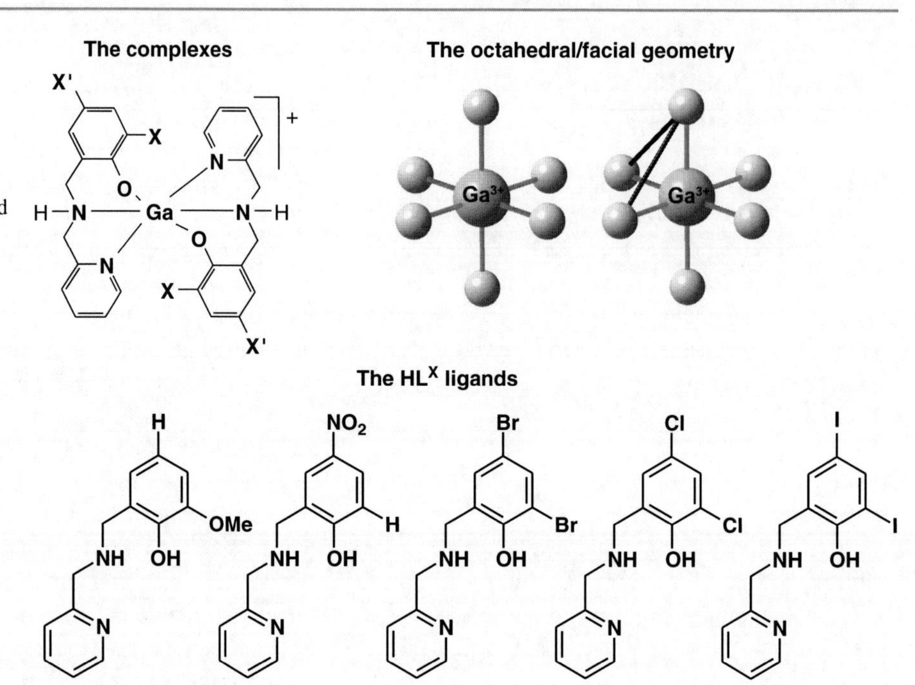

Gallium(III) Complexes, Inhibition of Proteasome Activity, Fig. 1 Gallium complexes. $[Ga^{III}(L^X)_2]ClO_4$ and their ligands HL^X, with X = H, methoxide (**1**), nitro (**2**), chloro (**3**), bromo (**4**), and iodo (**5**). All compounds exhibit facially coordinated ligands and a perchlorate (ClO_4^-) counterion

These complexes were first tested against cisplatin-resistant neuroblastoma cells. It was shown in ranking order, methoxy (**1**) = nitro (**2**) < chloro(**3**) < bromo (**4**) < iodo (**5**), that the species containing halogen substituents showed preferential growth inhibition in human neuroblastoma cells with activity superior to cisplatin. In addition, neuroblastoma cells treated with gallium chloride rendered cells viable, with results similar to the less active methoxy-substituted complex. Furthermore, this series of Ga(III) complexes was tested against normal human fibroblasts and were found to be nontoxic up to 25 μM treatment (Shakya et al. 2006).

Promising results acquired in the laboratory highlight a novel approach utilizing metal-based complexes as tumor-specific proteasome inhibitors. The proteasome is a multicatalytic protease responsible for the degradation of the majority of cellular proteins involved in various cellular processes such as cell cycle progression, angiogenesis, and apoptosis (Dou and Li 1999). The ubiquitin-proteasome pathway in eukaryotes led to the realization that specific molecular machinery is available in the cell to eliminate defective proteins and enzymes (Hershko et al. 1982). This discovery led to the 2004 Nobel Prize in Chemistry to Hershko, Ciechanover, and Rose. An increased proteasome activity has been identified in malignant cells, making its inhibition a venue for cancer therapy of unprecedented relevance (Li and Dou 2000). Additionally, inhibition of the proteasomal chymotrypsin-like activity was found to be associated with induction of tumor cell apoptosis (An et al. 1998). In 2003, bortezomib became the first proteasome inhibitor to be approved as an anticancer drug (Dou and Goldfarb 2002). However, drug resistance and toxicity coupled with the lack of activity toward solid tumors, and interactions with some natural products has limited its clinical development.

A possible molecular mechanism for these gallium (III) complexes responsible for their growth-inhibitory effects was investigated. Whether these gallium complexes could target and inhibit the proteasomal activity of purified 20S and cellular 26S proteasome in intact prostate cancer cells was tested. The studies revealed that complexes **3**, **4**, and **5** were able to target and inhibit proteasomal activity associated with apoptosis induction in various prostate tumor cells including LNCaP (androgen-dependent) and C4-2B (androgen-independent). Specifically, complexes **3**, **4**, and **5** inhibited the chymotrypsin-like activity of the proteasome by 39%, 62%, and 79% in LNCaP cells and 38%, 58%, and 81% in C4-2B cells at 50 μM treatment for 18 h. The ability to inhibit proteasome activity was indicated by accumulation of ubiquitinated proteins and the proteasomal target protein p27. Inhibition of proteasomal activity by these gallium(III) complexes was associated with apoptosis induction as measured by poly(ADP-ribose) polymerase (PARP) cleavage and nuclear staining. One interesting observation was that treatment of androgen receptor (AR)-containing prostate cancer cells with these gallium(III) complexes causes downregulation of AR protein, associated with tumor cell apoptosis.

As observed with cisplatin-resistant neuroblastoma cells, the iodo-substituted complex **5** achieved superior therapeutic efficacy in a human prostate cancer model. Accordingly, complex **5** exhibited potent proteasome inhibitory activity against both intact 26S proteasome ($IC_{50} = 17$ μmol) and purified 20S proteasome ($IC_{50} = 16$ μmol). It is notable to point out that treatment with the iodo-substituted ligand alone showed only minimal cytotoxicity. Furthermore, complex **5** (20 mg/kg/day for 30 days) was able to exert the same effect in vivo by inhibiting the growth of prostate PC-3 tumors grown in mice (by 66%), associated with proteasome inhibition and apoptosis induction. Immunohistochemistry studies showed higher levels of p27 and TUNEL-positive cells in tumors treated with complex **5,** when compared to tumors treated with vehicle control or ligand alone (Chen et al. 2007). The superior biological activity of complex **5** suggests that the iodo-substituted ligand attains certain physiochemical properties that impart significant therapeutic efficacy. Since the active complexes from this study consistently contain halogen-substituted ligands, the heightened efficacy may be governed by the strong π-donating groups (I > Br > Cl). For example, iodine harbors weak electron withdrawing capability, but retains strong π-donating effects. Although not conclusive, this property might play a role in activating the ring system of the complex and thus influencing coordination to the proteasome.

In order to gain insight into a possible mechanism of action for these gallium complexes, the biological role of other bivalent transition metals with the same NN′O ligand platform was then investigated. It was found that the 2:1 metal complex is unable to coordinate to available terminal threonine residues available in the active centers of the 20S proteasome without the loss of one ligand. It was therefore hypothesized that the primary pharmacophore responsible for the biological activity of these metal complexes is an [ML] species with a 1:1 metal-to-ligand ratio. In support of this hypothesis, it was shown that both a 1:1 and 1:2 complexes with copper(II) showed similar levels of pharmacological activity both in vitro and under cell culture conditions (Hindo et al. 2009). To test this working hypothesis further, similar 1:2 metal-to-ligand complexes using divalent nickel(II) and zinc(II) were designed. The nickel(II) ion has a $3d^8$ configuration, which leads to nonzero ligand-field stabilization energies (LFS) and, as a result, fosters slower ligand exchange. On the other hand, the zinc(II) ion exhibits a $3d^{10}$ configuration, is predominantly ionic, and displays rapid ligand dissociation and exchange. Consistent with the above hypothesis, it was found that the zinc(II) complex, but not the nickel(II) complex was able to inhibit the proteasomal chymotrypsin-like activity under cell-free conditions and in cultured human prostate cancer cells (Frezza et al. 2009). Therefore, liberation of one ligand from the complex should facilitate potent proteasome-inhibitory activity, allowing the 1:1 metal-to-ligand species to coordinate to amino acid residues in the proteasome. Thus, gaining a better understanding of the detailed nature of the pharmacophore and its role in proteasome inhibition should aid in the development of metal complexes containing NN′O tridentate ligands with gallium and other transition metals. In conclusion, these metal complexes could serve as platforms for development of novel proteasome inhibitor anticancer drugs for the treatment of prostate and other human cancers.

References

An B, Goldfarb RH, Siman R et al (1998) Novel dipeptidyl proteasome inhibitors overcome Bcl-2 protective function and selectively accumulate the cyclin-dependent kinase inhibitor p27 and induce apoptosis in transformed, but not normal, human fibroblasts. Cell Death Differ 5:1062–1075

Chen D, Frezza M, Shakya R et al (2007) Inhibition of the proteasome activity by gallium(III) complexes contributes to their anti-prostate tumor effects. Cancer Res 67:9258–9265

Chen D, Milacic V, Frezza M et al (2009) Metal complexes, their cellular targets, and potential in cancer therapy. Curr Pharm Des 15:777–791

Chitamber CR, Matthaeus WG, Antholine WE et al (1988) Inhibition of leukemic HL60 cell growth by transferrin-gallium: effects on ribonucleotide reductase and demonstration of drug synergy with hydroxyurea. Blood 72:1930–1936

Collery P, Domingo JL, Keppler BK (1996) Preclinical toxicology and tissue distribution of a novel antitumour gallium compound: tris 8(quinolinolato)gallium(III). Anticancer Res 16:687–691

Dou QP, Goldfarb RH (2002) Bortezomib (millennium pharmaceuticals). IDrugs 5:828–834
Dou QP, Li B (1999) Proteasome inhibitors as potential novel anticancer agents. Drug Resist Updat 2:215–223
Frezza M, Verani CN, Chen D et al (2007) The therapeutic potential of gallium-based complexes in anti-tumor drug design. Lett Drug Des Disc 4:311–317
Frezza M, Hindo SH, Tomco D et al (2009) Comparative activitities of nickel(II) and zinc(II) Complexes of asymmetric [NN'O] ligands as 26S proteasome inhibitors. Inorg Chem 48:5928–5937
Hershko A, Ciechanover H et al (1982) Mechanisms of intracellular protein breakdown. Annu Rev Biochem 51:335–364
Hindo SH, Frezza M, Tomco D et al (2009) Metals in anticancer therapy: copper(II) complexes as inhibitors of the 20S proteasome. Eur J Med Chem 44:4353–4361
Li B, Dou QP (2000) Bax degradation by the ubiquitin/proteasome pathway-dependent pathway: involvement in tumor survival and progression. Proc Natl Acad Sci USA 97:3850–3855
Shakya R, Peng F, Liu J et al (2006) Synthesis, structure, and anticancer activity of gallium(III) complexes with asymmetric tridentate ligands: growth inhibition and apoptosis induction of cisplatin-resistant neuroblastoma cells. Inorg Chem 45:6263–6268

Gallium(III) Ion

▶ Gallium(III) Complexes, Inhibition of Proteasome Activity

Gallium, Physical and Chemical Properties

Fathi Habashi
Department of Mining, Metallurgical, and Materials Engineering, Laval University, Quebec City, Canada

Gallium is a silvery white low-melting point metal (Fig. 1) belonging to the less typical metals group, i.e., when it loses its outermost electrons it will not have the electronic structure of inert gases. There is a volume expansion of 3.2% on solidification of the metal. Gallium is as a by-product of the production of aluminum oxide. Its concentration in bauxite is 0.003% and 0.008%. The largest reserves of gallium are contained in phosphate ores and in coal. When phosphorus is produced electrothermally from phosphate rock, the gallium is concentrated in the flue dust.

Gallium, Physical and Chemical Properties, Fig. 1 Low melting of gallium

Likewise, gallium may be obtained from the fly ash of coal. Gallium contained in zinc ore (sphalerite) has lost economic importance due to the change over to hydrometallurgical methods of zinc extraction. The main use of gallium is as a raw material in the manufacture of semiconductors, for example, the arsenide or the phosphide for the production of light-emitting diodes.

Physical Properties

Atomic number	31
Atomic weight	69.72
Relative abundance in the Earth's crust,%	1.5×10^{-3}
Melting point, °C	29.78
Boiling point, °C	2,403
Density, g/cm^3	
Solid at 24.6°C	5.904
Liquid at 29.8°C	6.095
Crystal form	Orthorhombic
Lattice constants, nm	$a = 0.4523$
	$b = 0.45198$
	$c = 0.76602$
Thermal expansion coefficients, K^{-1}	
Along the a axis	1.65×10^{-5}
Along b axis	1.13×10^{-5}
Along c axis	3.1×10^{-5}

Chemical Properties

Metallic gallium dissolves slowly in dilute mineral acids but rapidly in aqua regia and concentrated sodium hydroxide. In its compounds, the valency is +3. Its oxygen compounds resemble those of aluminum in that there are high- and low-temperature forms of Ga_2O_3 and two hydroxides, $Ga(OH)_3$ and GaO·OH. Substances which lower the hydrogen ion concentration throw down a white gelatinous precipitate from Ga(III) salt solutions. This precipitate is amorphous to x-rays and has a variable water content (gallium oxide hydrate, $Ga_2O_3 \cdot xH_2O$). It dissolves both in acids and in strong bases and differs from aluminum oxide hydrate in being soluble in ammonia solutions. As the precipitate ages, its solubility in caustic alkalis diminishes.

Gallium halides have covalent character and therefore have good solubility in many nonpolar solvents in which they exist in dimeric form. Like aluminum, gallium is precipitated as the white oxide hydrate from solutions of its salts by reducing the hydrogen ion concentration. Gallium sulfate, $Ga_2(SO_4)_3$, crystallizes from aqueous solutions as an 18-hydrate similar to $Al_2(SO_4)_3 \cdot 18H_2O$. It can be dehydrated by heating and decomposes above 520°C with loss of SO_3. With ammonium sulfate, it forms the double salt $(NH_4)Ga(SO_4)_2 \cdot 12H_2O$, ammonium gallium alum.

The crude metal contains 99–99.9% gallium. The material for semiconductor manufacture is used in purities from 99.9999% (6 N) to 99.999999% (8 N). The binary compounds, GaP and GaAs, are of great industrial importance. They may be prepared by direct combination of the elements at high temperature.

References

Greber JF (1997) Gallium. In: Habashi F (ed) Handbook of extractive metallurgy. Wiley, Weinheim, pp 1523–1530

Habashi F (2006) Gallium Update. In: Travaux du Comité international pour l''étude de la bauxite, de l''alumine et de l''aluminium (ICSOBA), vol 33, No. 37. Hungarian National Committee of ICSOBA, Budapest, pp 141–153. Hungarian translation in Kohászat 140(3):29–34

Gallium, Therapeutic Effects

Lawrence R. Bernstein
Terrametrix, Menlo Park, CA, USA

Synonyms

Gallium and apoptosis; Gallium effect on bacteria; Gallium uptake

Definition

Gallium: Gallium is an element (atomic number 31; atomic weight 69.723) in Group XIII of the periodic table (below aluminum, above indium), classed as a semimetal or poor metal. It occurs in the Earth's crust at an average abundance of about 15–19 parts per million (ppm) (similar to the abundance of nitrogen and about ten times that of tin or arsenic), widely distributed in soils and rocks. The element is extracted mainly from bauxite (in which it occurs at an average concentration of roughly 50 ppm) as a byproduct of aluminum refining, and to a lesser extent from zinc ores.

Introduction

Gallium is not known to have any essential biological function, but it exerts a variety of therapeutically useful biological activities. Many of these activities derive from gallium's ability to act as an irreducible mimic of ferric iron (Fe^{3+}). Gallium's ability to compete with other essential metals, particularly zinc, is likely responsible for some of its other biological activities.

The ionic radii, electronegativity, ionization potentials, and coordination chemistry of Ga^{3+} are remarkably similar to those of Fe^{3+}. In the body, Ga^{3+} can be transported in the blood serum by the iron transport protein transferrin, and appears to follow many of the uptake and transport pathways observed for iron (Bernstein 1998).

It is, however, the differences between Ga^{3+} and Fe^{3+} that allow gallium to be therapeutically useful. By being irreducible under physiological conditions, gallium cannot participate in redox reactions, such as the

Fenton-type reactions that make free iron (mainly as Fe^{2+} in solution) highly toxic. Gallium also is not observed to enter Fe^{2+}-bearing molecules such as heme, and so does not interfere with oxygen transport (via hemoglobin) or with cytochrome-mediated reactions. Furthermore, at pH 7.4 and 25°C, Ga^{3+} has a solubility of about 1 μM (98.4% as $[Ga(OH)_4]^-$ and 1.6% as $Ga(OH)_3$) whereas Fe^{3+} has a solubility of only about 10^{-18} M. Thus, small amounts of nonprotein-bound gallium can exist in solution at physiological conditions, versus insignificant amounts of nonprotein-bound Fe^{3+}, permitting biological interactions for Ga^{3+} that would not be possible for Fe^{3+}.

Gallium metalloproteins: Gallium is known to substitute for iron in several human proteins. Crucial to many of gallium's activities in the body, it binds to the two metal sites on transferrin (TF). Typically, serum TF is about 33% saturated with Fe, leaving 67% available for binding by Ga (about 2.7 μg/mL). Under usual physiologic conditions, the binding constants for gallium are $\log K_1 = 20.3$ and $\log K_2 = 19.3$; the corresponding binding constants for Fe^{3+} are $\log K_1 = 22.8$ and $\log K_2 = 21.5$. However, whereas Fe^{3+} remains bound to TF down to a pH of about 5.5, Ga^{3+} starts to dissociate from TF at pH < 6.8; it is $>50\%$ dissociated at pH 6 (McGregor and Brock 1992). This difference in stability may provide a therapeutic advantage to TF-bound Ga, as it may be released sooner than Fe in endosomes of cancer cells and phagosomes of some pathogens, helping it to compete against Fe (Bernstein et al. 2011). In addition to binding to TF, gallium binds even more avidly to the related protein lactoferrin as well as to the iron storage protein ferritin.

The substitution of Ga^{3+} for Fe^{3+} is also observed in bacterial proteins, including ferric-binding protein (FbpA), an iron transporter in the same protein superfamily as TF (Weaver et al. 2008). The presence of such substitutions could inhibit bacterial growth. Ga^{3+} can also substitute for Fe^{3+} in non-ribosomal peptide microbial siderophores.

Ionic gallium is observed to dose-dependently inhibit alkaline phosphatase (Boskey et al. 1993) and matrix metalloproteinase activity (Panagakos et al. 2000). This effect is hypothesized to be caused by the substitution of gallium for zinc in these proteins. Gallium and zinc are chemically similar, with gallium found substituting for zinc in minerals (e.g., gallium was first discovered replacing zinc in sphalerite, cubic ZnS).

Therapeutic gallium compounds: Numerous gallium compounds have shown evidence of therapeutic activity (Table 1). These compounds include inorganic and organic salts, metal-organic complexes, metalloproteins, and organometallics. As with all metals and semimetals, the chemical form in which gallium occurs is crucial to its absorption, distribution, excretion, and activity. The biochemical properties of gallium are thus dependent on the chemical form in which it is delivered. Organometallic gallium compounds (having Ga–C bonds), as well as some highly stable complexes (including Ga porphyrins), tend to remain fairly intact under physiological conditions, so their molecular conformation helps determine their activity. Other compounds, such as some orally administered gallium salts and complexes, are prodrugs in that they dissociate before reaching the bloodstream, with the gallium becoming bound mainly to transferrin and other proteins. Depending on their route of administration and chemical nature, gallium compounds can undergo various degrees of metabolic alteration in the body, resulting in varying contributions from the gallium itself versus its molecular host.

Currently, the only gallium compounds reported to have been tested in humans are citrated gallium nitrate for injection (Ganite®; approved for use in the USA to treat cancer-related hypercalcemia); gallium maltolate for oral administration (Phase I trials); gallium 8-quinolinolate for oral administration (Phase I trials); and gallium chloride for oral administration (in a few cancer patients).

Therapeutic activities: The known therapeutic activities in which gallium may play a major role include: (1) Activity against pathological hyperproliferation, particularly against some aggressive cancers; (2) Anti-inflammatory and immunomodulating activity, as observed in animal models of rheumatoid arthritis, multiple sclerosis, and lupus; (3) Anti-bone-resorptive and anti-hypercalcemic activity, plus possible anabolic bone activity; (4) Activity against some pathogenic microbes, including *Pseudomonas aeruginosa* and some intracellular bacteria; and (5) Possible analgesic activity, including against neuropathic pain.

Gallium in the Treatment of Cancer

Following the discovery of cisplatin's (*cis*-diamminedichloroplatinum(II)) anticancer activity in 1969, there

Gallium, Therapeutic Effects, Table 1 Therapeutic gallium compounds

Compound	Therapeutic activities	References
Inorganic salts and glasses		
$GaCl_3$	In vitro *and animal studies*: Anticancer, particularly in aggressive tumors; active against *Pseudomonas aeruginosa* *Human clinical studies*: Oral bioavailability low. Possible potentiation of cisplatin and etoposide therapy in lung cancer patients.	Bernstein (2005); Banin et al. (2008)
$Ga(NO_3)_3$ (In many of the in vitro and animal studies, and in all of the human clinical studies, citrate was added to solutions of gallium nitrate in order to neutralize the pH and to increase stability. The molar concentration of the citrate was generally the same as that of the Ga)	In vitro *and animal studies*: Anti-bone-resorptive; possible anabolic activity on bone; immunomodulating: effective in animal models of rheumatoid arthritis, multiple sclerosis, and others; anticancer; antimicrobial, including against *Mycobacteria*, *Pseudomonas*, *Staphylococcus*, *Francisella*, and others; antibiofilm activity *Human clinical studies*: Low oral bioavailability. Intravenous formulation approved in the USA for cancer-related hypercalcemia. High efficacy observed for Paget's disease of bone, lymphoma, multiple myeloma; low to moderate efficacy observed for urothelial carcinoma, ovarian cancer, metastatic prostate cancer	Bernstein (2005)
$Ga_2(SO_4)_3$	Suppressed adjuvant arthritis in rats	Bernstein (1998)
Ga_2O_3-doped phosphate-based glasses	Activity against *Staphylococcus aureus*, *Escherichia coli*, and *Clostridium difficile*	Valappil et al. (2008)
Organic salts and complexes		
Ga citrate	Activity against *Pseudomonas fluorescens*	Al-Aoukaty et al. (1992)
Ga tartrate	Eliminated experimental syphilis in rabbits at a single dose of 30–45 mg Ga/kg intramuscularly or 15 mg Ga/kg intravenously. Eliminated *Trypanosoma evansi* infection in mice at 225 mg/kg	Levaditi et al. (1931)
Ga maltolate (tris(3-hydroxy-2-methyl-4 H-pyran-4-onato)gallium(III))	In vitro *and animal studies*: Anticancer, including against hepatocellular carcinoma; effective in animal models of rheumatoid arthritis; antimicrobial: effective against *P. aeruginosa*, *Rhodococcus equi*, *S. aureus*, *Staphylococcus epidermidis* and other bacteria; antibiofilm *Human clinical studies*: Oral formulation safe in Phase I trials; anecdotal efficacy observed in hepatocellular carcinoma, metastatic colon cancer, lymphoma. Topical formulation has shown anecdotal efficacy in pain, including postherpetic neuralgia, inflammation, actinic keratosis, and psoriasis	Bernstein et al. (2000); Bernstein (2005); Chua et al. (2006)
Ga 8-quinolinolate (tris(8-quinolonato)Ga (III))	In vitro *and animal studies*: Anticancer, including against lung cancer, melanoma; anti-bone-resorptive *Human clinical studies*: Oral formulation safe in Phase I trials; possible efficacy in renal cell carcinoma.	Bernstein (2005)
Ga complex of pyridoxal isonicotinoyl hydrazone	Potently antiproliferative in T-lymphoblastic leukemic CCRF-CEM cells	Bernstein (2005)
Ga protoporphyrin IX	Activity against *Yersinia enterocolitica*, *S. aureus*, *Mycobacterium smegmatis* and others	Stojiljkovic et al. (1999); Bozja et al. (2004)
$[GaL_2]ClO_4$; L = 2,4-diiodo-6-{[(pyridin-2-ylmethyl)amino]methyl}phenol	Inhibited PC-3 prostate cancer xenographs in mice; inhibited proteosome activity	Chen et al. (2007)
Ga-deferoxamine	Active against *P. aeruginosa* in vitro and in rabbit cornea together with gentamicin	Banin et al. (2008)

(continued)

Gallium, Therapeutic Effects, Table 1 (continued)

Compound	Therapeutic activities	References
Ga complexes of thiosemicarbazones	Activity against several cancer cell lines; activity against *Cryptococcus*	Kowol et al. (2009); Bastos et al. (2010); Mendes et al. (2009)
$(LH)_2[GaCl_4]Cl$; L = 1-methyl-4,5-diphenylimidazole	Modest activity against several cancer cell lines	Zanias et al. (2010)
$[Ga\text{-}3\text{-}Madd]^+$, $[Ga\text{-}5\text{-}Madd]^+$, $[Ga\text{-}3\text{-}Eadd]^+$, $[Ga\text{-}3\text{-}M\text{-}5\text{-}Quadd]^{+a}$	Activity against *Plasmodium falciparum*	Sharma et al. (1997); Ocheskey et al. (2003, 2005); Harpstrite et al. (2003)
Ga complexes of pyrazoles, indazoles, and benzopyrazoles	Moderate activity against human immunodeficiency virus	Kratz et al. (1992)
Ga complexes of pyrazole-imine-phenols and pyrazole-amine-phenols	Activity against MCF-7 breast cancer and PC-3 prostate cancer cell lines	Silva et al. (2010)
Ga-curcumin	Activity against mouse lymphoma L1210 cell line	Mohammadi et al. (2005)
Diphenyl Ga chloride	Moderate activity against *Candida albicans*, *Cryptococcus neoformans*, *Bacillus subtilis*, and *S. aureus*	Srivastava et al. (1973)
Tris(*N*-methylthioacetohydroxamato) Ga(III)	Moderate activity against several species of bacteria and fungi	Abu-Dari and Mahasneh (1993)
Protein complexes		
Ga transferrin	Active against MCF-7 breast cancer and HeLa cervical carcinoma cells in vitro	Jiang et al. (2002); Head et al. (1997); Olakanmi et al. (2010)
	Active against *Francisella tularensis* and *Francisella novicida* in vitro	
Ga lactoferrin	Active against *F. tularensis* and *F. novicida* in vitro and protected against *F. novicida* nasally introduced to mice	Olakanmi et al. (2010)
Organometallics, organometallic complexes		
$[((CH_3)_2 Ga)(5\text{-phenyl-1,3,4-oxadiazole-2-thio})]_4$	Active against several cancer cell lines	Gallego et al. (2011)
Dimeric methyl-Ga complexes of some carboxylates	Active against several cancer cell lines	Kaluđerović et al. (2010)
α-dimethylamino-cyclohexoxyl-dimethyl-gallium	Active against *P. falciparum*	Yan et al. (1991)
Miscellaneous		
Doxorubicin-Ga-transferrin conjugate	Active in vitro against MCF-7 breast cancer cells, including those resistant to doxorubicin; anecdotal efficacy in breast cancer patients	Wang et al. (2000); Bernstein (2005)
Sulfonated Ga corrole bound to a heregulin-modified protein	Active against HER-positive cancer in rats	Agadjanian et al. (2009)
Liposomal $Ga(NO_3)_3$ and gentamicin	Very active against *P. aeruginosa* biofilms	Halwani et al. (2008)
Yeast-incorporated Ga	Antiresorptive and anabolic to bone in rats	Ma and Fu (2010)

[a]$[Ga\text{-}3\text{-}Madd]^+$ = [{1,12-bis(2-hydroxy-3-methoxybenzyl)-1,5,8,12-tetraazadodecane}gallium(III)]$^+$;
$[Ga\text{-}5\text{-}Madd]^+$ = [{1,12-bis(2-hydroxy-5-methoxybenzyl)-1,5,8,12-tetraazadodecane}gallium(III)]$^+$;
$[Ga\text{-}3\text{-}Eadd]^+$ = [{1,12-bis(2-hydroxy-3-ethyl-benzyl)-1,5,8,12-tetraazadodecane}gallium(III)]$^+$;
$[Ga\text{-}3\text{-}M\text{-}5\text{-}Quadd]^+$ = [{1,12-bis(2-hydroxy-3-methoxy-5-(quinolin-3-yl)-benzyl)-1,5,8,12-tetraazadodecane}-gallium(III)]$^+$

was a systematic effort to determine the anticancer efficacy of other metal compounds. By 1971, after an initial round of in vitro and animal screening, gallium emerged as being particularly promising. Specifically, parenterally administered gallium nitrate showed efficacy in several animal cancer models, including Walker 256 ascites carcinosarcoma in rats and implanted human medulloblastoma in mice. In vitro activity of gallium compounds has been observed in a number of cancer cell lines, including those for lymphoma (Chitambar 2010), breast cancer (Wang et al. 2000), leukemia (Bernstein 1998), and hepatocellular carcinoma (Chua et al. 2006) (see Table 1 for more examples of in vitro and in vivo anticancer activity).

The discovery of gallium's anticancer activity coincided with the introduction of ^{67}Ga-scans to diagnose and locate a variety of cancers in the body. ^{67}Ga-scans were found to be particularly sensitive to lymphomas, some sarcomas, and bone tumors, but much less sensitive to many other types of cancer. For Ga-sensitive cancers, ^{67}Ga-scans were noted to detect growing cancer tissue rather than cancer tissue that was necrotic or otherwise less active. Many of the ^{67}Ga-avid cancers soon proved to be most susceptible to therapeutic gallium treatment, though this correlation was apparently not documented or utilized.

Mechanisms of antiproliferative activity: The sensitivity of multiplying cancer cells to gallium appears due mainly to their high need for iron. This need stems primarily from the requirement for iron in the active site of ribonucleotide reductase, an enzyme essential for the synthesis of DNA. To obtain sufficient iron, most cancer cells highly overexpress TF-receptor (though some cancer cells can also take up iron by TF-receptor-independent mechanisms). If Ga is present on the available TF, the Ga-TF will be taken into the cell by endocytosis and compete with iron. Ga then hinders acidification of the TF-bearing endosome, possibly inhibiting release of Fe (which occurs at pH < 5.5). Gallium thus interferes with the uptake and utilization of iron by cancer cells; in addition, it has direct inhibitory activity on ribonucleotide reductase (Bernstein 1998). Cancer cells that are consequently unable to make DNA and multiply ultimately undergo apoptosis (see "► Gallium Nitrate, Apoptotic Effects" for further information regarding apoptosis).

Other antiproliferative mechanisms of action have been proposed for gallium, including inhibition of protein tyrosine phosphatases and DNA polymerases, but these mechanisms have not been further substantiated.

The natural targeting of gallium to multiplying cancer cells, as is observed in gallium scans, contributes greatly to the efficacy and low toxicity of therapeutically administered gallium. Most healthy cells, even those that are rapidly proliferating, generally do not take up significant amounts of gallium. The inability of Ga to enter heme likely accounts for its lack of accumulation in healthy proliferating hematopoietic cells of the bone marrow. The reasons for the low Ga-avidity of other proliferating healthy cells, such as gastrointestinal mucosal cells and the transient cells of hair follicles, are not known, but may be due to efficient local iron recycling.

Clinical experience: Gallium nitrate became the first preferred gallium compound for experimental and therapeutic use in the USA, due to its relative ease of synthesis and handling. Because aqueous solutions of gallium nitrate are acidic and tend to precipitate gallium hydroxides over time, citrate was added to the solutions; the citrate chelated the gallium, thus neutralizing the pH and promoting stability. The US National Cancer Institute sponsored animal toxicology studies with gallium nitrate, and clinical studies in cancer patients were launched in about 1974 (Adamson et al. 1975).

Small clinical trials found that intravenously administered citrated gallium nitrate (CGN) is effective in non-Hodgkin's lymphoma (43% response) and bladder carcinoma (40% response), with lower response levels in urothelial carcinoma, cervical carcinoma, ovarian carcinoma, squamous cell carcinoma, and metastatic prostate carcinoma (Bernstein 2005). Intravenous CGN must be administered as a slow infusion over several days to avoid renal toxicity. Typical anticancer doses for CGN are 200–500 mg/m^2/day for $\geq$5 days.

CGN administered at 30 mg/m^2/day during alternate 2-week periods, combined with a bimonthly 5-day infusion at 100 mg/m^2/day, together with the M-2 chemotherapy protocol, was highly effective in a study of 13 patients with advanced multiple myeloma. These patients had significantly reduced pain, increased total body calcium, stable bone density, and reduction in vertebral fractures relative to a matched group of 167 patients who received only the M-2 protocol. Most significantly, there was a marked increase in survival in the CGN group (mean survival of 87+ months with several long-term survivors) versus the M-2 only group (mean survival of 48 months with no long-term survivors).

To avoid the inconvenience and renal toxicity of intravenous CGN, gallium maltolate was developed as an orally active gallium compound. Phase I clinical trials have found no renal toxicity or other serious or dose-limiting toxicity (Bernstein et al. 2000). The lack of renal toxicity is due to the differences in blood speciation between parenterally administered CGN and orally administered gallium maltolate. When CGN is introduced into the blood, much of the gallium forms [Ga $(OH)_4]^-$ (gallate) in serum. As a small anionic group, gallate is rapidly concentrated and excreted by the kidney, where it can reach toxically high levels, sometimes forming gallium-calcium-phosphate precipitates in the renal tubules. Gallium from orally administered gallium maltolate, however, becomes almost entirely TF-bound in the blood, with very low renal excretion and no renal toxicity. Anecdotal cases have shown responses to gallium maltolate in advanced hepatocellular carcinoma (Bernstein et al. 2011), lymphoma, metastatic breast cancer, metastatic prostate cancer, and metastatic colorectal cancer.

Gallium 8-quinolinolate is also being developed as an orally active form of gallium. Animal testing has shown activity against some cancers as well as anti-bone-resorptive activity. Phase I clinical trials to date have demonstrated safety and apparent responses in renal cell carcinoma.

Antimicrobial Activity of Gallium

The first biological activity observed for gallium was against pathogenic microbes. In 1931, Levaditi et al. reported that a single dose of gallium tartrate, at 30–45 mg Ga/kg intramuscularly or 15 mg Ga/kg intravenously, eliminated experimental syphilis in rabbits. They also reported that this compound eliminated *Trypanosoma evansi* infection in mice at 225 mg/kg (750 mg/kg was tolerated). Despite these encouraging results, no further antimicrobial studies on gallium were published for several decades.

Emery (1971) found that Ga^{3+} could bind to the fungal siderophore ferrichrome and then be transported into the organism. Subsequently, it was found that other microbial siderophores (low molecular weight Fe^{3+} transporters), including pyoverdine (from *P. aeruginosa*) and deferoxamine, could also bind Ga^{3+}. Suggestions arose for the possible use of Ga^{3+}-substituted siderophores as antimicrobials, but little progress was apparently made in this direction until the late 2000s.

As seen in Table 2, there has been increased research on gallium's antimicrobial properties since the late 1990s, with a number of gallium compounds showing antimicrobial activity at levels that are therapeutically promising. Stojiljkovic et al. (1999) described Ga porphyrins that exploit the heme uptake systems of some bacteria, resulting in highly potent activity against them. Olakanmi et al. (2000) reported that gallium nitrate and gallium transferrin were effective against *Mycobacterium tuberculosis* and *Mycobacterium avium complex*, including those organisms living within human macrophages. Orally administered gallium maltolate at 3–30 mg/kg/day was then found to significantly reduce the number of colony forming units and liver tubercles in guinea pigs infected with *M. tuberculosis* (L.S. Schlesinger 2003, unpublished data).

Gallium appears to be particularly effective against *P. aeruginosa*. This Gram-negative, aerobic bacterium is ubiquitous and highly adaptable, living in soils, plants, and animals throughout a wide variety of environments. It is an opportunistic infector of humans, colonizing open wounds and burns, as well the bladder and lungs. In the presence of sufficient iron, it can form biofilms (commonly in conjunction with other bacterial species) that are highly resistant to immunological attack and to antibiotics. Such biofilms are particularly dangerous when they form in the lungs, as commonly occurs in patients with cystic fibrosis or AIDS. Gallium nitrate was found to inhibit the growth and kill the planktonic form of *P. aeruginosa* dose-dependently at concentrations greater than 1 μM; at 0.5 μM biofilm growth was prevented, and at 100 μM established biofilms were destroyed (Kaneko et al. 2007). In addition, gallium nitrate was effective in treating two murine models of *P. aeruginosa* lung infection. Expression of the transcriptional regulator pvdS, which regulates the synthesis of pyoverdine and other proteins related to iron transport, was suppressed by Ga, contributing to its antibacterial effect (Kaneko et al. 2007).

Gallium maltolate locally administered by subcutaneous injection at 25 mg/kg to burned mouse skin infected with *P. aeruginosa* resulted in 100% survival, versus 0% survival in untreated mice and those treated with an equivalent dose of gallium nitrate (DeLeon et al. 2009). Treatment with gallium maltolate also prevented systemic spread of the bacteria from already colonized wounds. The higher efficacy of gallium maltolate relative to gallium nitrate may be due to its higher lipophilicity (octanol partition coefficient of 0.4), resulting in higher penetration of

Gallium, Therapeutic Effects, Table 2 Antimicrobial activity of gallium compounds

Microorganism	Gallium compound	Activity	References
Bacteria			
Acinetobacter baumannii	Ga maltolate	Local administration inhibited growth in burned mouse skin	DeLeon et al. (2009)
Bacillus cereus	Ga-MTAH[a]	ZOI: 100 μg/mL: 0 mm; 800 μg/mL: 24 mm	Abu-Dari and Mahasneh (1993)
B. subtilis	Diphenyl Ga chloride	MIC: 12.5 μg/mL	Srivastava et al. (1973)
	Ga protoporphyrin IX	MIC: 0.2 μg/mL	Stojiljkovic et al. (1999)
Burkholderia cepacia complex	Ga nitrate	MIC: 64 μg/mL (250 μM Ga)	Peeters et al. (2008)
Citrobacter freundii	Ga protoporphyrin IX	MIC: 1–2 μg/mL (growth medium Fe-restricted w/dipyridyl)	Stojiljkovic et al. (1999)
Escherichia coli	Diphenyl Ga chloride	MIC: >100 μg/mL	Srivastava et al. (1973)
	Ga_2O_3-doped PO_4 glass	Significant activity in disk diffusion assay	Valappil et al. (2008)
	Ga protoporphyrin IX	MIC: <0.5 μg/mL (growth medium Fe-restricted w/dipyridyl)	Stojiljkovic et al. (1999)
	Ga-MTAH[a]	ZOI: 100 μg/mL: 8 mm; 800 μg/mL: 42 mm	Abu-Dari and Mahasneh (1993)
Francisella novicida	Ga lactoferrin	IC50: 1 μM; protected against *F. novicida* nasally introduced to mice	Olakanmi et al. (2010)
	Ga transferrin	IC50: 10 μM	Olakanmi et al. (2010)
F. tularensis	Ga lactoferrin	IC50: 3 μM	Olakanmi et al. (2010)
	Ga transferrin	IC50: 3 μM	Olakanmi et al. (2010)
Haemophilus ducreyi	Ga protoporphyrin IX	MGIC: 32 μg/mL	Bozja et al. (2004)
Helicobacter pylori	Ga protoporphyrin IX	MIC: 0.19 μg/mL	Stojiljkovic et al. (1999)
Klebsiella pneumoniae	Ga protoporphyrin IX	MIC: 2 μg/mL	Stojiljkovic et al. (1999)
Listeria monocytogenes	Ga protoporphyrin IX	MIC: 0.2 μg/mL	Stojiljkovic et al. (1999)
Mycobacterium smegmatis	Ga protoporphyrin IX	MIC: 0.4 μg/mL	Stojiljkovic et al. (1999)
M. tuberculosis	Ga nitrate	IC50: 25–100 μM for bacteria within human macrophages	Olakanmi et al. (2000)
	Ga transferrin	Inhibited iron uptake by the bacteria	Olakanmi et al. (2000)
	Ga maltolate	Orally administered drug showed efficacy in infected guinea pigs at 3.3–30 mg/kg/day	Schlesinger et al. (2003), unpublished data
M. bovis	Ga protoporphyrin IX	MIC: 0.4 μg/mL	Stojiljkovic et al. (1999)
Neisseria gonorrhoeae	Ga protoporphyrin IX	MIC: 0.2 μg/mL	Stojiljkovic et al. (1999); Bozja et al. (2004)
		MGIC: 8 μg/mL; effective in murine vaginal model	
N. meningitidis	Ga protoporphyrin IX	MIC: 0.2 μg/mL	Stojiljkovic et al. (1999)
Proteus mirabilis	Ga protoporphyrin IX	MIC: 2 μg/mL (growth medium Fe-restricted w/ dipyridyl)	Stojiljkovic et al. (1999)
Pseudomonas aeruginosa	Ga nitrate	MIC: 2 μg/mL; effective against biofilms; effective in mouse lung infection	Kaneko et al. (2007)
	Ga chloride	Planktonic phase MIC: 2 μg/mL (32 μM)	Banin et al. (2008)
		0.07 μg/mL (1 μM) prevented biofilm formation	
	Ga maltolate	Oral drug showed dose-dependant efficacy against mouse urinary tract infection	Wirtz et al. (2006)
		Local administration was highly effective in mouse burn/infection model	DeLeon et al. (2009)

(*continued*)

Gallium, Therapeutic Effects, Table 2 (continued)

Microorganism	Gallium compound	Activity	References
	Ga-deferoxamine	Planktonic phase MIC: 2 μg/mL (32 μM)	Banin et al. (2008)
		1 μM (0.07 μg/mL) prevented biofilm formation	
		Active together with gentamicin in rabbit corneal infection	
	Ga_2O_3-doped PO_4 glass	Significant activity in disk diffusion assay	Valappil et al. (2008)
	Liposomal $Ga(NO_3)_3$ and gentamicin	Formulations with 0.16 or 0.3 μM Ga more potent than either gentamicin or $Ga(NO_3)_3$ alone against drug-resistant biofilms	Halwani et al. (2008)
P. fluorescens	Ga citrate	Growth lag of 40 h at 1 mM in PO_4-rich medium [note: PO_4 may remove Ga from solution as insoluble Ga phosphate]	Al-Aoukaty et al. (1992)
Rhodococcus equi	Ga nitrate	Growth inhibition observed at 50 μM Ga	Harrington et al. (2006)
	Ga maltolate	MIC: 0.6 μg/mL (8 μM Ga)	Coleman et al. (2010)
Salmonella typhimurium	Ga-MTAH[a]	ZOI: 100 μg/mL: 0 mm; 800 μg/mL: 18 mm	Abu-Dari and Mahasneh (1993)
Staphylococcus aureus	Diphenyl Ga chloride	MIC: 12.5 μg/mL	Srivastava et al. (1973)
	Ga_2O_3-doped PO_4 glass	Significant activity in disk diffusion assay	Valappil et al. (2008)
	Ga protoporphyrin IX	MIC: 1–2.5 μg/mL	Stojiljkovic et al. (1999)
	Ga maltolate	MIC: 375–2,000 μg/mL (0.84–4.5 mM Ga); active against biofilms	Baldoni et al. (2010)
	Ga-MTAH[a]	ZOI: 100 μg/mL: 4 mm; 800 μg/mL: 32 mm	Abu-Dari and Mahasneh (1993)
S. epidermidis	Ga maltolate	MIC: 94–200 μg/mL (210–450 μM Ga)	Baldoni et al. (2010)
Streptococcus pyogenes	Ga protoporphyrin IX	MIC: >4 μg/mL	Stojiljkovic et al. (1999)
Treponema pallidum	Ga tartrate	Eliminated experimental syphilis in rabbits at single dose of 30–45 mg Ga/kg intramuscularly or 15 mg Ga/kg intravenously	Levaditi et al. (1931)
Yersinia enterocolitica	Ga protoporphyrin IX	MIC: 0.4 μg/mL (growth medium Fe-restricted w/dipyridyl)	Stojiljkovic et al. (1999)
Y. pseudotuberculosis	Ga protoporphyrin IX	MIC: 0.2–0.4 μg/mL	Stojiljkovic et al. (1999)
Fungi			
Aspergillus parasiticus	Ga-MTAH[a]	ZOI: 100 μg/mL: 0 mm; 800 μg/mL: 16 mm	Abu-Dari and Mahasneh (1993)
Candida albicans	Diphenyl Ga chloride	MIC: 25 μg/mL	Srivastava et al. (1973)
C. tropicalis	Ga-MTAH[a]	ZOI: 100 μg/mL: 8 mm; 800 μg/mL: 24 mm	Abu-Dari and Mahasneh (1993)
C. neoformans	Diphenyl Ga chloride	MIC: 25 μg/mL	Srivastava et al. (1973)
Fusarium moniliforme	Ga-MTAH[a]	ZOI: 100 μg/mL: 4 mm; 800 μg/mL: 20 mm	Abu-Dari and Mahasneh (1993)
F. solami	Ga-MTAH[a]	ZOI: 100 μg/mL: 8 mm; 800 μg/mL: 34 mm	Abu-Dari and Mahasneh (1993)
Protozoa			
Plasmodium falciparum	$[Ga\text{-}3\text{-}Madd]^{+b}$	IC50 (chloroquine sensitive): ≫20 μM	Sharma et al. (1997)
		IC50 (chloroquine resistant): 0.5-0.6 μM	
	$[Ga\text{-}5\text{-}Madd]^{+c}$	IC50 (chloroquine sensitive): 2 μM	Ocheskey et al. (2003)
		IC50 (chloroquine resistant): ≫15 μM	
	$[Ga\text{-}3\text{-}Eadd]^{+d}$	IC50 (chloroquine sensitive): 0.086 μM	Harpstrite et al. (2003)
		IC50 (chloroquine resistant): 0.8 μM	

(*continued*)

Gallium, Therapeutic Effects, Table 2 (continued)

Microorganism	Gallium compound	Activity	References
	[Ga-3-M-5-Quadd]$^{+e}$	IC50 (chloroquine sensitive): 0.6 μM	Ocheskey et al. (2005)
		IC50 (chloroquine resistant): 1.4 μM	
	Ga protoporphyrin IX	IC50: 127 μM	Begum et al. (2003)
	Ga maltolate	IC50 (chloroquine sensitive or resistant): 15 μM	Goldberg et al. (1998), unpublished data
	Ga nitrate	IC50 (chloroquine sensitive or resistant): 90 μM	
	α-dimethylamino-cyclohexoxyl-dimethyl-Ga	Single dose of 1–3 mg/kg to mice rapidly killed both sexual and asexual forms of the parasite in blood	Yan et al. (1991)
Trypanosoma evansi	Ga tartrate	Eliminated infection in mice at 225 mg/kg	Levaditi et al. (1931)
Viruses			
Human immunodeficiency virus (HIV)	*trans*-dichlorotetrakis (benzimidazole) gallium(III) chloride	IC50: 5 μM	Kratz et al. (1992)
	Ga nitrate	IC50: 7 μM	Stapleton et al. (1999)

Abbreviations: *IC50* half maximal inhibitory concentration, *MGIC* minimal growth inhibitory concentration, *MIC* minimal inhibitory concentration, *ZOI* zone of inhibition
[a]Tris(*N*-methylthioacetohydroxamato)gallium(III)
[b][{1,12-bis(2-hydroxy-3-methoxybenzyl)-1,5,8,12-tetraazadodecane}gallium(III)]$^+$
[c][{1,12-bis(2-hydroxy-5-methoxybenzyl)-1,5,8,12-tetraazadodecane}gallium(III)]$^+$
[d][{1,12-bis(2-hydroxy-3-ethyl-benzyl)-1,5,8,12-tetraazadodecane}gallium(III)]$^+$
[e][{1,12-bis(2-hydroxy-3-methoxy-5-(quinolin-3-yl)-benzyl)-1,5,8,12-tetraazadodecane}-gallium(III)]$^+$

tissues and bacteria. Gallium maltolate was also effective when administered orally in a murine urinary tract infection model using *P. aeruginosa* (Wirtz et al. 2006). Several other Ga compounds and formulations have also shown activity against *P. aeruginosa* and biofilm formation (Table 2).

A number of Ga compounds have shown in vitro efficacy against *Plasmodium falciparum* (the major causative agent of malaria), including chloroquine-resistant strains (see Table 2), but no in vivo studies have been published.

Due to its inhibition of ribonucleotide reductase, gallium may have antiviral activity. Stapleton et al. (1999) and Kratz et al. (1992) reported in vitro activity of Ga compounds against human immunodeficiency virus (HIV). Because HIV is a retrovirus that is dependent upon host ribonucleotide reductase to synthesize DNA, a strategy that targets host ribonucleotide reductase may avoid the development of drug resistance.

Gallium compounds have also shown promise against other pathogenic organisms, including *Staphylococcus aureus*, *Staphylococcus epidermidis*, *Francisella* sp., *Acinetobacter baumannii*, and *Rhodococcus equi* (Table 2). Due to its novel mechanism of action, low toxicity, and activity against drug-resistant strains, there is justification for further exploration of gallium as a potential antimicrobial. Some gallium compounds may be particularly effective when topically applied to the skin, eyes, lungs (by inhalation), bladder (by instillation), or elsewhere, where they can rapidly achieve high local concentrations and can prevent or treat biofilms. Gallium compounds may also be useful as antimicrobial coatings on medical devices.

Anti-inflammatory and Immunomodulating Activity of Gallium

Immunomodulating activity for gallium was noted at about the same time as the early anticancer studies were getting underway, in the early 1970s. Since then, a number of in vitro and animal studies have shown that gallium can act to suppress inflammation and some pathological immunological responses without being generally immunosuppressive.

Gallium appears particularly effective at inhibiting abnormal T-cell-mediated immunological reactions. Ga-TF greatly suppresses alloantigen-induced proliferation of mixed lymphocytes, also reducing the amounts of IL-2 receptor and increasing the amount of TF-receptor on activated T-cells. It does not, however, inhibit IL-2 secretion or the induction of IL-2-stimulated cytokine-activated killer T-cell activity. Gallium nitrate inhibited both antigen-specific and mitogenic proliferative responses in purified-protein-derivative-specific rat T-cells. While gallium nitrate was found to suppress T-cell activation and some interferon-gamma (IFN-γ) secretion in cell cultures, it did not directly interfere with the normal growth and repair response of gonadal vein endothelial cells to IFN-γ and TNF-α (which may have been enhanced by Ga) (Bernstein 1998).

Intravenously administered gallium nitrate, generally at doses of 10–45 mg Ga/kg/day, has shown efficacy in a number of animal models of T-cell-mediated autoimmune disease. Efficacy has been observed in adjuvant-induced arthritis in rats (reduced synovitis, pannus, subchondral resorption, cartilage degeneration, and periosteal new bone formation), experimental autoimmune encephalomyelitis in rats (a model for demyelinating diseases such as multiple sclerosis, caused by exposure to myelin basic protein (MBP); the proliferative response to MBP was suppressed in T-cells from animals treated with Ga at certain times); experimental autoimmune uveitis in rats (retinal and choroidal inflammation prevented; lymphocyte proliferative responses and humoral immune response decreased); and mouse models of systemic lupus erythematosus and Type 1 diabetes. Some activity in mouse models of asthma and endotoxic shock has also been seen (Bernstein 1998).

Orally administered gallium maltolate showed efficacy in two models of inflammatory arthritis in rats: adjuvant-induced arthritis and streptococcus cell wall–induced chronic arthritis (Schwendner et al. 2005). In both models, oral gallium maltolate dose-dependently reduced joint inflammation, bone degradation, liver and spleen enlargement, and other measures of inflammation. No toxicity was observed.

Gallium nitrate has demonstrated direct immunological effects on macrophages. The effects include transient inhibition of major histocompatibility complex (MHC) class II by murine macrophages; inhibition of inflammatory cytokine and NO secretion by activated murine macrophage-like RAW 264 cells; and inhibition of NO secretion from activated murine ANA-1 macrophages, without inhibition of TNF-α secretion.

In some strains of mice, gallium nitrate was effective at inhibiting acute allograft rejection and prolonging survival. Chronic rejection, however, was not suppressed.

Many of these immunomodulating effects are again likely related to Ga^{3+} being an irreducible mimic of Fe^{3+}. Pro-inflammatory T-helper type 1 (Th-1) cells are much more sensitive to inactivation by iron deprivation than are anti-inflammatory, pro-antibody Th-2 cells (Thorson et al. 1991). Furthermore, Fe^{3+} chelates tend to be highly pro-inflammatory (as they may be siderophores from pathogens); the irreducibility of Ga^{3+} may help to suppress this reaction. The antiproliferative activity of gallium, in this case on certain lymphocytes, may also contribute to gallium's immunomodulating activity.

Effects of Gallium on Bone and on Serum Calcium Levels

A period of intense animal and clinical investigations into gallium's biological activities during the late 1940s and early 1950s revealed the tendency of gallium to concentrate in bone, particularly at sites of bone growth, healing, or tumors. Later, during the early clinical trials of CGN in cancer patients, in the 1970s and 1980s, it was observed that serum calcium levels became normalized in many hypercalcemic patients who were treated with CGN. Hypercalcemia (abnormally high serum calcium, which can be life-threatening) occurs fairly commonly in cancer patients, particularly those with breast and lung cancers. Further clinical and animal studies showed that the reduction in serum calcium was due to inhibited bone mineral resorption rather than increased calcium excretion.

The mechanisms for preferred uptake of Ga at sites of bone remodeling remain poorly understood, though it is known that the uptake is at least partially independent of TF and TF-receptor mediation. Animals or humans lacking TF, that have Fe-saturated TF, or that have TF-receptor blocked by an antibody, show skeletal Ga uptake at the same or greater levels than those with normal TF and TF-receptor. Ga^{3+} in aqueous solution is known to adsorb strongly to calcium phosphates, and gallium phosphates are highly insoluble. Limited experimental evidence (Ga and Ca absorption edge spectroscopy of bone from Ga-treated rats) suggests that about half of the Ga incorporated into bone is in the form of phosphates. The high phosphate concentrations at sites of bone remodeling may cause the precipitation of Ga-phosphates at these sites.

The discovery of gallium's anti-bone-resorptive effect led to numerous preclinical and clinical studies that explored gallium's mechanisms of action as well as its potential in the treatment of cancer-related hypercalcemia and metabolic bone diseases. These diseases included Paget's disease of bone (characterized by abnormally rapid turnover of bone, often accompanied by pain) and osteoporosis. Ga is observed to dose-dependently inhibit the resorption of bone by osteoclasts (bone resorbing cells of macrophage lineage), without being toxic to the osteoclasts at antiresorptive concentrations. This is in contrast to bisphosphonates, whose antiresorptive activity derives in part from their toxicity to osteoclasts. Gallium was found to directly inhibit acid production by vacuolar-class ATPase in osteoclasts. In patients with Paget's disease treated with CGN, Ga was found concentrated almost exclusively in osteoclast nuclei, presenting the possibility for Ga to act on DNA transcription and expression.

Several controlled clinical studies found CGN (administered as a continuous infusion at 200 mg/m^2 for 5 days) to be effective in the treatment of cancer-related hypercalcemia; comparative studies found CGN to be superior to etidronate or calcitonin, and at least as effective as pamidronate. The US Food and Drug Administration approved this drug in 1991 for the treatment of cancer-related hypercalcemia.

Paget's disease has been successfully treated with low doses of CGN (0.25 or 0.5 mg/kg/day administered by subcutaneous injection). The patients had significantly reduced markers of bone turnover, including serum alkaline phosphatase, urinary hydroxyproline, and *N*-telopeptide collagen crosslinks excretion (Chitambar 2010).

Considerable data exist that suggest anabolic (bone mineral increase) activity for gallium. Evidence includes increased observed bone formation and bone calcium content in Ga-treated rats, and elevated serum alkaline phosphatase in Ga-treated postmenopausal women. CGN dose-dependently (from 5 to 100 μM) decreased constitutive and vitamin D3-stimulated osteocalcin (OC) and OC mRNA levels in rat osteogenic sarcoma osteoblast-like 17/2.8 cells and in normal rat osteoblasts (osteocalcin being an inhibitor of bone formation) (Bernstein 1998).

Effects on Pain

In a brief (7-day) clinical trial of gallium in patients with metastatic prostate cancer, a rapid and significant reduction in pain was noted (Scher et al. 1987). Several years later, in a study of multiple myeloma patients, a significant reduction of pain was recorded for those patients who received gallium nitrate plus a standard form of chemotherapy versus those who received chemotherapy alone (Warrell et al. 1993). In both of these cases, the analgesic effect may have been due to the activity of gallium against the cancer, bone resorption, and/or associated inflammation. Neither of these observations regarding pain reduction were discussed or followed up.

Starting in 2006, a topical cream formulation (0.25 or 0.5 wt.% gallium maltolate in 50% water and 50% hydrophilic petrolatum (Aquaphor®)) has been administered cutaneously to a few individuals suffering from various types of pain (Bernstein 2012). The first case was a 100-year-old woman who had severe facial (trigeminal) postherpetic neuralgia for 4 years and who had responded poorly or not at all to a large variety of systemic and locally administered narcotics, anesthetics, analgesics, antiepileptics, antipsychotics, and other medications. Topically applied, low-dose gallium maltolate has provided nearly complete pain relief that lasts about 6–8 h; the treatment has been used everyday for more than 5 years. The topical formulation has also been found effective in other individuals against postherpetic neuralgia and against pain and itching due to insect bites and stings, spider bites, infections, allergic reactions, inflammation, and burns. The mechanisms for the anti-pain activity are not known; they may relate to gallium's anti-inflammatory activity, plus possible interference with neuropeptides, most of which are Zn-dependent. It is also likely that Ga is acting on one or more presently unknown pain pathways.

Cross-References

- ▶ Gallium in Bacteria, Metabolic and Medical Implications
- ▶ Gallium Nitrate, Apoptotic Effects
- ▶ Gallium Uptake and Transport by Transferrin

References

Abu-Dari K, Mahasneh AN (1993) Antimicrobial activity of thiohydroxamic acids and their metal complexes: II. The synthesis and antimicrobial activity of N-methyllthioacetohydroxamic acid and its Zn, Cu, Fe and Ga complexes. Dirasat Nat Sci 20B:7–15

Adamson RH, Canellos GP, Sieber SM (1975) Studies on the antitumor activity of gallium nitrate (NSC-15200) and other group IIIa metal salts. Cancer Chemother Rep 1 59:599–610

Agadjanian H, Ma J, Rentsendorj A et al (2009) Tumor detection and elimination by a targeted gallium corrole. Proc Natl Acad Sci USA 106:6105–6110

Al-Aoukaty A, Appanna VD, Falter H (1992) Gallium toxicity and adaptation in *Pseudomonas fluorescens*. FEMS Microbiol Lett 92:265–272

Baldoni D, Steinhuber A, Zimmerli W et al (2010) In vitro activity of gallium maltolate against Staphylococci in logarithmic, stationary, and biofilm growth phases: comparison of conventional and calorimetric susceptibility testing methods. Antimicrob Agents Chemother 54: 157–163

Banin E, Lozinski A, Brady KM et al (2008) The potential of desferrioxamine-gallium as an anti-*Pseudomonas* therapeutic agent. Proc Natl Acad Sci USA 105: 16761–16766

Bastos TO, Soares BM, Cisalpino PS et al (2010) Coordination to gallium(III) strongly enhances the potency of 2-pyridineformamide thiosemicarbazones against Cryptococcus opportunistic fungi. Microbiol Res 165:573–577

Begum K, Kim HS, Kumar V et al (2003) In vitro antimalarial activity of metalloporphyrins against *Plasmodium falciparum*. Parasitol Res 90:221–224

Bernstein LR (1998) Mechanisms of therapeutic activity for gallium. Pharmacol Rev 50:665–682

Bernstein LR (2005) Therapeutic gallium compounds. In: Gielen M, Tiekink ERT (eds) Metallotherapeutic drugs and metal-based diagnostic agents: the use of metals in medicine. Wiley, New York

Bernstein LR (2012) Successful treatment of refractory postherpetic neuralgia with topical gallium maltolate: case study. Pain Med 13:915–918

Bernstein LR, Tanner T, Godfrey C, Noll B (2000) Chemistry and pharmacokinetics of gallium maltolate, a compound with high oral gallium bioavailability. Metal Based Drugs 7:33–48

Bernstein LR, van der Hoeven JJM, Boer RO (2011) Hepatocellular carcinoma detection by gallium scan and subsequent treatment by gallium maltolate: rationale and case study. Anticancer Agents Med Chem 11:585–590

Boskey AL, Ziecheck W, Guidon P et al (1993) Gallium nitrate inhibits alkaline phosphatase activity in a differentiating mesenchymal cell culture. Bone Miner 20:179–192

Bozja J, Yi K, Shafer WM, Stojiljkovic I (2004) Porphyrin-based compounds exert antibacterial action against the sexually transmitted pathogens *Neisseria gonorrhoeae* and *Haemophilus ducreyi*. Int J Antimicrob Agents 24:578–584

Chen D, Frezza M, Shakya R et al (2007) Inhibition of the proteasome activity by gallium(III) complexes contributes to their anti prostate tumor effects. Cancer Res 67: 9258–9265

Chitambar CR (2010) Medical applications and toxicities of gallium compounds. Int J Environ Res Public Health 7:2337–2361

Chua MS, Bernstein LR, Li R, So SKS (2006) Gallium maltolate is a promising chemotherapeutic agent for the treatment of hepatocellular carcinoma. Anticancer Res 26:1739–1744

Coleman M, Kuskie K, Liu M et al (2010) In vitro antimicrobial activity of gallium maltolate against virulent *Rhodococcus equi*. Vet Microbiol 146(1–2):175–178

DeLeon K, Balldin F, Watters C et al (2009) Gallium maltolate treatment eradicates *Pseudomonas aeruginosa* infection in thermally injured mice. Antimicrob Agents Chemother 53:1331–1337

Emery T (1971) Role of ferrichrome as a ferric ionophore in *Ustilago sphaerogena*. Biochemistry 10:1483–1488

Gallego B, Kaluđerović MR, Kommera H et al (2011) Cytotoxicity, apoptosis and study of the DNA-binding properties of bi- and tetranuclear gallium(III) complexes with heterocyclic thiolato ligands. Invest New Drugs 29(5):932–944

Halwani M, Yebio B, Suntres ZE et al (2008) Co-encapsulation of gallium with gentamicin in liposomes enhances antimicrobial activity of gentamicin against *Pseudomonas aeruginosa*. J Antimicrob Chemother 62:1291–1297

Harpstrite SE, Beatty AA, Collins SD et al (2003) Metalloantimalarials: targeting of *P. falciparum* strains with novel iron (III) and gallium(III) complexes of an amine phenol ligand. Inorg Chem 42:2294–2300

Harrington JR, Martens RJ, Cohen ND et al (2006) Antimicrobial activity of gallium against virulent Rhodococcus equi in vitro and in vivo. J Vet Pharmacol Ther 29:121–127

Head JF, Wang F, Elliott RL (1997) Antineoplastic drugs that interfere with iron metabolism in cancer cells. Adv Enzyme Regul 37:147–169

Jiang XP, Wang F, Yang DC et al (2002) Induction of apoptosis by iron depletion in the human breast cancer MCF-7 cell line and the 13762NF rat mammary adenocarcinoma in vivo. Anticancer Res 22:2685–2692

Kaluđerović MR, Gómez-Ruiz S, Gallego B et al (2010) Anticancer activity of dinuclear gallium(III) carboxylate complexes. Eur J Med Chem 45:519–525

Kaneko Y, Thoendel M, Olakanmi O, Britigan BE, Singh PK (2007) The transition metal gallium disrupts *Pseudomonas aeruginosa* iron metabolism and has antimicrobial and antibiofilm activity. J Clin Invest 117:877–888

Kowol CR, Trondl R, Heffeter P et al (2009) Impact of metal coordination on cytotoxicity of 3-aminopyridine-2-carboxaldehyde thiosemicarbazone (triapine) and novel insights into terminal dimethylation. J Med Chem 52:5032–5043

Kratz F, Nuber B, Weiss J (1992) Synthesis and characterization of potential antitumor and antiviral gallium(III) complexes of N-heterocycles. Polyhedron 11:487–498

Levaditi C, Bardet J, Tchakirian A et al (1931) Le gallium, propriétés thérapeutiques dans la syphilis et les

trypanosomiases expérimentales. C R Acad Sci Hebd Seances Acad Sci D Sci Nat 192:1142–1143
Ma Z, Fu Q (2010) Therapeutic effect of organic gallium on ovariectomized osteopenic rats by decreased serum minerals and increased bone mineral content. Biol Trace Elem Res 133:342–349
McGregor SJ, Brock JH (1992) Effect of pH and citrate on binding of iron and gallium by transferrin in serum. Clin Chem 38:1883–1885
Mendes IC, Soares MA, Dos Santos RG et al (2009) Gallium(III) complexes of 2-pyridineformamide thiosemicarbazones: cytotoxic activity against malignant glioblastoma. Eur J Med Chem 44:1870–1877
Mohammadi K, Thompson KH, Patrick BO et al (2005) Synthesis and characterization of dual function vanadyl, gallium and indium curcumin complexes for medicinal applications. J Inorg Biochem 99:2217–2225
Ocheskey JA, Polyakov VR, Harpstrite SE et al (2003) Synthesis, characterization, and molecular structure of a gallium(III) complex of an amine-phenol ligand with activity against chloroquine-sensitive *Plasmodium falciparum* strains. J Inorg Biochem 93:265–270
Ocheskey JA, Harpstrite SE, Oksman A et al (2005) Metalloantimalarials: synthesis and characterization of a novel agent possessing activity against *Plasmodium falciparum*. Chem Commun (Camb) 2005:1622–1624
Olakanmi O, Britigan BE, Schlesinger LS (2000) Gallium disrupts iron metabolism of mycobacteria residing within human macrophages. Infect Immun 68:5619–5627
Olakanmi O, Gunn JS, Su S, Soni S, Hassett DJ, Britigan BE (2010) Gallium disrupts iron uptake by intracellular and extracellular *Francisella* strains and exhibits therapeutic efficacy in a murine pulmonary infection model. Antimicrob Agents Chemother 54:244–253
Panagakos FS, Kumar E, Venescar C et al (2000) The effect of gallium nitrate on synoviocyte MMP activity. Biochimie 82:147–151
Peeters E, Nelis HJ, Coenye T (2008) Resistance of planktonic and biofilm-grown *Burkholderia cepacia* complex isolates to the transition metal gallium. J Antimicrob Chemother 61:1062–1065
Scher HI, Curley T, Geller N et al (1987) Gallium nitrate in prostatic cancer: evaluation of antitumor activity and effects on bone turnover. Cancer Treat Rep 71:887–893
Schwendner SW, Allamneni KP, Bendele A (2005) Oral gallium maltolate is efficacious in acute and chronic models of rheumatoid arthritis. Ann Rheum Dis 64(Suppl III):168
Sharma V, Beatty A, Goldberg DE, Piwnica-Worms D (1997) Structure of a novel antimalarial gallium(III) complex with selective activity against chloroquine-resistant *Plasmodium falciparum*. Chem Commun 1997:2223–2224
Silva F, Marques F, Santos IC, Paulo A, Rodrigues AS, Rueff J, Santos I (2010) Synthesis, characterization and cytotoxic activity of gallium(III) complexes anchored by tridentate pyrazole-based ligands. J Inorg Biochem 104:523–532
Srivastava TN, Bajpai KK, Singh K (1973) Anti-microbial activities of diaryl gallium, indium and thallium compounds. Ind J Agric Sci 43:88–93
Stapleton JT, Klinzman D, Olakanmi O et al (1999) Gallium nitrate: a potent inhibitor of HIV-1 infection in vitro. Abs Intersci Conf Antimicrob Agents Chemother 39:74
Stojiljkovic I, Kumar V, Srinivasan N (1999) Non-iron metalloporphyrins: potent antibacterial compounds that exploit haem/Hb uptake systems of pathogenic bacteria. Mol Microbiol 31:429–442
Thorson JA, Smith KM, Gomez F, Naumann PW, Kemp JD (1991) Role of iron in T cell activation: TH1 clones differ from TH2 clones in their sensitivity to inhibition of DNA synthesis caused by IgG Mabs against the transferrin receptor and the iron chelator deferoxamine. Cell Immunol 134:126–137
Valappil SP, Ready D, Abou Neel EA et al (2008) Antimicrobial gallium-doped phosphate-based glasses. Adv Funct Mater 18:732–741
Wang F, Jiang X, Yang DC et al (2000) Doxorubicin-gallium-transferrin conjugate overcomes multidrug resistance: evidence for drug accumulation in the nucleus of drug resistant MCF-7/ADR cells. Anticancer Res 20:799–808
Warrell RP Jr, Lovett D, Dilmanian FA et al (1993) Low-dose gallium nitrate for prevention of osteolysis in myeloma: results of a pilot randomized study. J Clin Oncol 11: 2443–2450
Weaver KD, Heymann JJ, Mehta A, Roulhac PL, Anderson DS, Nowalk AJ, Adhikari P, Mietzner TA, Fitzgerald MC, Crumbliss AL (2008) Ga^{3+} as a mechanistic probe in Fe^{3+} transport: characterization of Ga^{3+} interaction with FbpA. J Biol Inorg Chem 13:887–898
Wirtz UF, Kadurugamuwa J, Bucalo LR et al (2006) Efficacy of gallium maltolate in a mouse model for *Pseudomonas aeruginosa* chronic urinary tract infection. In: Abstracts: American society for microbiology, 106th general meeting 2006, Orlando (abstract A-074)
Yan G, Wang G, Li Y (1991) Effects of a-dimethylamino-cyclohexoxyl-dimethyl gallium on ultrastructure of erythrocytic stage of *Plasmodium berghei* and *P. yoelii*. Acta Pharmacol Sin 12:530–533
Zanias S, Papaefstathiou GS, Raptopoulou CP et al (2010) Synthesis, structure, and antiproliferative activity of three gallium(III) azole complexes. Bioinorg Chem Appl 2010:168030

Gating

▶ Potassium Channel Diversity, Regulation of Potassium Flux across Pores
▶ Potassium Channels, Structure and Function

GCAP

▶ Calcium, Neuronal Sensor Proteins

GCP-II, Glutamate Carboxypeptidase II

► Zinc Aminopeptidases, Aminopeptidase from Vibrio Proteolyticus (Aeromonas proteolytica) as Prototypical Enzyme

Ge-132

► Germanium-Containing Compounds, Current Knowledge and Applications

Gelatinase

► Zinc Matrix Metalloproteinases and TIMPs

Gene Name: CALB1=CAB27

► Calbindin D_{28k}

Genetic/Hereditary and Nonhereditary Diseases of Selenoprotein Deficiency

► Selenocysteinopathies

Germanium

► Germanium-Containing Compounds, Current Knowledge and Applications

Germanium Complex as Antioxidant

► Germanium l-Cysteine Alpha-Tocopherol Complex as Stimulator to Antioxidant Defense System

Germanium l-Cysteine Alpha-Tocopherol Complex as Stimulator to Antioxidant Defense System

Mamdouh M. Ali
Biochemistry Department, Genetic Engineering and Biotechnology Division, National Research Centre, El Dokki, Cairo, Egypt

Synonyms

Germanium complex as antioxidant

Definition

Germanium L-cysteine α-tocopherol complex [germanium dichloro tetrakis (L-cysteinyl-α-tocopherol amide) dichloride] acts as a protective agent against γ-irradiation-induced free radical's production and liver toxicity.

Metal is not only an important raw material of industrial construction, but is also beneficial to human health. Germanium is a chemical element with the symbol Ge and atomic number 32. It is a lustrous, hard, grayish-white metalloid in the carbon group. Germanium has five naturally occurring isotopes ranging in atomic mass number from 70 to 76. It forms a large number of organometallic compounds.

Germanium normalizes many physiological functions such as lowering of high blood pressure in humans and rats. It also restores deviant blood

characteristics to their normal range including pH, glucose, the minerals sodium, potassium, calcium and chlorides, triglycerides, cholesterol, uric acid, hemoglobin, and leucocytes (white blood cells) (Kidd 1987).

Organic germanium appears to be remarkably safe. Germanium sesquioxide (Ge-132) is highly safe, even at dosages as high as 10 g per day. The nutritional substance has practically no toxicity and no influence on reproductive functions. In fact, no trace of germanium can be found in the body after 20–30 h. Therefore, it cannot accumulate in the tissue. Dr. Rinehardt observed that a few patients who are very immunodeficient may experience slight afternoon fevers. Rather than being a cause for alarm, he believes these very low-grade temperatures indicate that the germanium is stimulating the patient's compromised immune system.

Organic germanium has found to be an impressive immunostimulant. Several studies have demonstrated its marked antitumor effects (Badawi and Hafiz 2007) interferon-inducing activity (Xin et al. 1996) and restoring the immune function in immune-depressed animals. Organic germanium compounds demonstrate their anticancer activity in laboratory animals infected with a wide range of different cancers as immunostimulant, which has been shown to be mediated by activation and stimulation of T cells, natural killer cells, lymphokine, and macrophage activity (Suzuki et al. 1986). Organic germanium's immune-enhancing properties were also noted in many studies, particularly in the case of cancer and arthritis (Fukazawa et al. 1994).

Organogermanium complexes have been found to have impressive antitumor activity (Morsy et al. 2008). Germanium–amino acid complexes revealed cytotoxic activity against Ehrlich ascitic carcinoma (EAC) cell lines in which germanium–histidine complex exhibited very high cytotoxicity. L-cysteine administration prevented liver fibrosis by suppressing hepatic stellate cell proliferation and activation. L-cysteine and its derivatives modulate lymphocyte functions and immune responses (Badawi and Hafiz 2007).

Organic Germanium Enriches Oxygen Supply

Organic germanium lowers the requirement for oxygen consumption by organs in culture and increases the life span of animals under oxygen stress. Investigating the effects of organic germanium upon oxygen consumption in the liver and diaphragm indicated a decline in oxygen consumption. The organic germanium plays the same role as oxygen in the body, thereby increasing the body's oxygen supply. There is a relationship between oxygen supply, blood viscosity, and blood flow. When more oxygen is available, blood viscosity decreases, thereby increasing the blood flow to all organs. Organic germanium protects against carbon monoxide asphyxiation, stroke, and Raynaud's disease, conditions linked with oxygen starvation. Upon taking therapeutic doses of organic germanium, there is often a warm, glowing, even a tingling feeling, that has been attributed by to its oxygenation effect. Individuals suffering from diseases of the circulatory system, such as Raynaud's disease, which may lead to gangrene and limb amputation, have shown significant improvement in their condition after taking organic germanium (Asai 1980).

The structure of organic germanium, a crystalline lattice network extensively bonded with negative oxygen ions, is said to actually substitute for oxygen, and to enable the attraction and elimination of acidifying hydrogen ions, which detoxifies the blood. In the electron transport scheme during oxidative metabolism, electrons are transferred along a set of electron acceptors, ending up ultimately with the combination of hydrogen and oxygen to form water. However, when there is an oxygen deficiency, the loss of electrons can result in the accumulation of positive hydrogen ions, which leads to blood acidification. Ge-132 has negatively charged oxygen ions, which can clear away these hydrogen ions, and thus detoxify the blood. Organic germanium's ability to facilitate transfer electrons permits it to act as an electron sink during oxidative metabolism, thus enhancing the body's generation of energy without the intake of extra oxygen. The electron transport system can be likened to a fire bucket brigade. If there is a shortage of electron acceptors, the entire process can grind to a halt, just as if one person is not in position, the bucket of water cannot get passed to put out the fire. Organic germanium has been shown to be an excellent electron conductor, and thus can significantly contribute to the efficiency of the entire process of oxidative metabolism, which ultimately generates energy for the body (Goodman 1988).

Germanium and Protection from Radiation

There is growing evidence that radiation injury to living cells is, to large extent, due to oxidative stress (Wallace 1998). The interaction of ionizing radiation with living cells induces a variety of reaction products and a complex chain reaction in which many macromolecules and their degradation products participate. Major biomarkers of oxidative damage to living cells are: lipid peroxidation (LPx) products, DNA hydroxylation and protein hydroxylation products (De Zwart et al. 1999). Efficient defense and repair mechanisms exist in living cells to protect against oxidant species (Deger et al. 2003). The antioxidative defense system is composed of methods to transfer sensitive material to compartments better protected from the action of reactive species, complex transition metals, a potential source of electrons, thereby rendering them unreactive, inhibit vulnerable processes such as DNA replication, repair damaged molecules, initiate apoptosis, and, possibly the most important considering the liability to internal and external modifying factors, activate antioxidant enzymes, and finally use a variety of direct free radical scavengers (Riley 1994). Enzymes involved in antioxidative defense, particularly well documented, are the antioxidative properties of the superoxide dismutase, glutathione peroxidases, and catalase (De Zwart et al. 1999). A need exists for nontoxic and inexpensive drugs for clinical radiation protection. Antioxidant agents may help to protect against chemically induced or radiation-induced toxicity in rats (Sagrista et al. 2002). Previous studies have indicated that some commonly used antioxidants of plant origin include vitamin E, vitamin C, selenium, phenolic compounds, carotenoids, and flavonoids (Jagetia and Reddy 2005).

Numerous biochemical compounds such as cysteine (Badaloo et al. 2002), germanium (Yang and Kim 1999), and α-tocopherol (Chow 1991) have been used individually to target oxygen and oxygen-free radicals in an attempt to reduce radiation-induced damage. Various studies have suggested that germanium compounds may have a protective effect against liver injury and have a similar oxygen-enriching properties and rigorously documented antioxidant effects (Goodman 1988; Chow 1991; Yang and Kim 1999). Ge-132 administered to patients undergoing radiotherapy for cancer offers protection against radiation-induced killing of white and red blood cells (Goodman 1988). The precise mechanisms describing how organic germanium can protect cells from radiation damage have not been elucidated. However, according to Asai (1980), atoms of Ge-132 securely fasten to red blood cells and shelter these cells from oncoming electrons by diverting them around the atom. It is postulated that organic germanium could have an affinity for blood cells: possibly electronic attraction, given germanium's extended network shape and its ability to conduct electrons efficiently (Goodman 1988). α-tocopherol is the most important lipid-soluble chain-breaking antioxidant in tissues and blood. The vitamin might protect cellular components against peroxidation damage via free radical–scavenging mechanism (Chow 1991). In addition, cysteine is a potent antioxidant that has been shown to protect from oxidative stress. In particular, cysteine is known to increase the intracellular stores of glutathione, thereby enhancing endogenous antioxidant levels (Jayalakshmi et al. 2005).

Ali et al. (2007) examined the potency of new prepared germanium L-cysteine α-tocopherol complex [germanium dichloro tetrakis (L-cysteinyl-α-tocopherol amide) dichloride] as a protective agent against γ-irradiation-induced free radicals' production and liver toxicity and gave the result: The treatment of rats with germanium L-cysteine α-tocopherol prior to whole body γ-irradiation seems to exert a beneficial antioxidant protective effect by maintaining high blood levels of reduced glutathione and by enhancing the activities of antioxidant enzymes (catalase, superoxide dismutase, and glutathione peroxidase). It also offers a considerable protection against radiation-induced liver injury by inhibiting the chain reaction of membrane lipid peroxidation that follows irradiation, thus protecting the membrane lipids from peroxidation damage. According to foregoing results, germanium L-cysteine α-tocopherol could be possibly used as radioprotector against treatment side effects induced by radiotherapy after further experiments that can be extended to other organs and the results could be used for planning a more effective therapy program combining radiation therapy and radioprotection.

Even in substances high in germanium such as garlic, ginseng, and aloe, the amount of Ge-132 is extremely small. For this reason, germanium supplements may be more costly than some nutritional compounds, but naturally synthesized germanium sesquioxide is your best source of this vital element. As a preventative aid, most people like 25–100 mg per day.

Germanium and Heart Disease

Two of the major causes of heart disease are hypertension or high blood pressure and arteriosclerosis, clogged arteries. Unfortunately, many people do not realize they suffer from this "silent killer" and risk a sudden death heart attack or stroke. Due to its ability to normalize metabolic functions, germanium appeared to lower blood pressure. At the Japan Experimental Medical Research Institute in Tokyo, test animals with normal blood pressure showed no change when researchers administered organic germanium. However, lab animals with high blood pressure returned to normal levels after 7–10 days of germanium use. Arteriosclerosis also affects thousands of people. Again, most individuals are unaware that their arteries are gradually constricting as cholesterol builds up on the walls. Preliminary results suggest that germanium may be beneficial in reducing cholesterol levels as it is in regulating high blood pressure.

In an examination of 30 surgery patients, in which all test subjects had elevated cholesterol and triglycerides prior to receiving germanium, the following observations were made: After 4 weeks of treatment, a definite decrease in cholesterol was observed which was attributed to the germanium. The average drop was from 360 to 260 mg. Many doctors are urging their patients to reduce their cholesterol below 250 mg. Treating a patient who suffered from severe coronary artery disease. Cholesterol-lowering medication and a low cholesterol diet had failed to bring his serum cholesterol level below 435 mg. Without a change in diet or medication, but with a daily dose of germanium, his level dropped to 234 mg.

Germanium Boosts the Immune System

Previous researches believe that germanium had the ability to stimulate the body's natural defense mechanisms which makes it an ideal candidate for combating viruses, fungi, and bacteria, especially *Candida albicans*, a yeast infection.

Excellent results were obtained in using germanium in a 53-year-old patient who had been suffering from recurrent diverticulitis, an inflamed colon, for 5 years. Despite treatment with antibiotics, the patient still experienced fevers, night sweats, headaches, and fatigue. Within 10 days of receiving 500 mg per day of germanium, the patient experienced a complete regression of his symptoms. Germanium does not directly kill viruses and bacteria. Some researchers believe that germanium plays an active role in returning the body's defenses to normal so that the body can fight off the invading germs. This is the prime reason why the International AIDS Treatment Conference approved germanium's use for clinical testing on AIDS patients.

A virus is a cleverly disguised enemy which invades your cells and uses the cell's own reproductive processes to clone itself. To destroy this enemy, your body mounts an attack using its warrior cells. These T and B lymphocytes, macrophages, and natural killer cells devour foreign cells and defend the body against viruses and cancer. However, none of these cells can function properly if the person is deficient in interferon. Germanium appears to significantly enhance the body's production of interferon, a protein which blocks viruses from infecting cells.

In animal experiments, germanium has reduced the harmful effects of influenza. At the Germanium Research Institute, germanium substantially prolonged the life expectancy of mice injected with influenza virus. In the control group, all the mice who were not given germanium died by day 14. In the groups treated with 100 mg of Ge-132, 60% of the animals survived to day 20.

Dr. Rinehardt has found that germanium's antiviral powers have been particularly potent in one patient, against the Epstein Barr virus, a very debilitating infection related to herpes, which, like herpes, has no known cure.

Germanium and Cancer

Not surprisingly, many cancer patients are immunodeficient. Preliminary research suggests that germanium appears to exhibit an ability to help normalize the body's defenses in cancer patients. A report by (Fukazawa et al. 1994) states that, studies in immunosuppressed animals and patients with malignancies (cancer) or rheumatoid arthritis suggest that Ge-132 restores the normal function of T cells, B lymphocytes, and antibody-forming cells.

In a report given at the International Symposium at Osaka in July of 1981, Japanese doctors observed that

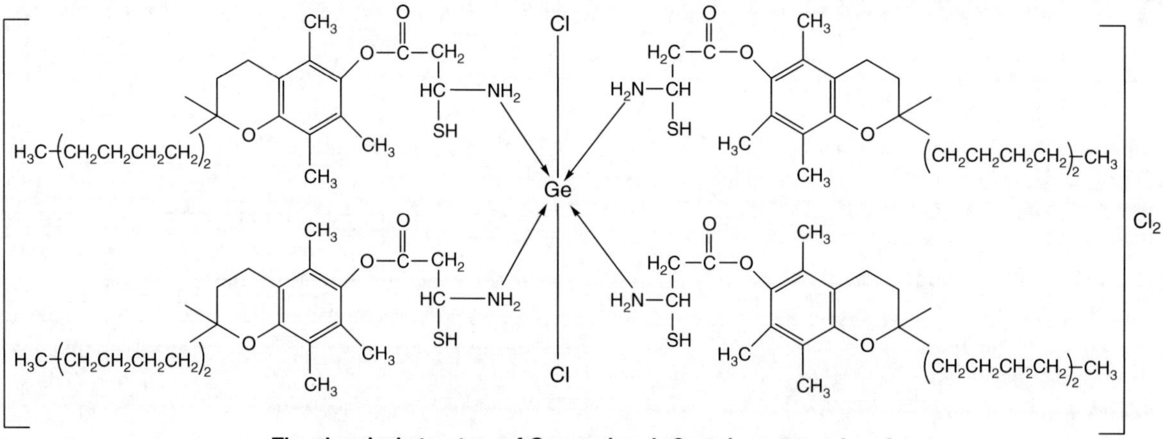

The chemical structure of Germanium L-Cysteine α-tocopherol

Germanium l-Cysteine Alpha-Tocopherol Complex as Stimulator to Antioxidant Defense System, Fig. 1

Ge-132 normalized activity in immune cells such as T lymphocytes and NK cells and increased interferon production in seven cancer patients. The doctors concluded that Ge-132 had increased the patient's resistance in both malignancies and rheumatic disorders. According to this study, one patient's abdomen cancer disappeared after 16 months of Ge-132 treatment and intermittent therapy with a chemotherapy agent. Another patient's postoperative condition was controlled well on Ge-132 alone (Fig. 1).

Cross-References

- ► Antioxidant Enzymes
- ► Germanium

References

Ali MM, Eman N, Sherein K et al (2007) Role of germanium L-cysteine α-tocopherol complex as stimulator to some antioxidant defense system in gamma-irradiated rats. Acta Pharmaceutica 57:1–12

Asai K (1980) Mechanism of action. In: Miracle cure: organic germanium. Japan Publications Inc, Tokyo

Badaloo A, Reid M, Forrester T et al (2002) Cysteine supplementation improves the erythrocyte glutathione synthesis rate in children with severe edematous malnutrition. Am J Clin Nutr 76:646–652

Badawi AM, Hafiz AA (2007) Synthesis and immunomodulatory activity of some novel amino acid germinates. J Iran Chem Sci 4(1):107–113

Chow CK (1991) Vitamin E and oxidative stress. Free Rad Biol Med 11:215–232

De Zwart LL, Meerman JHN, Commandeur JNM et al (1999) Biomarkers of free radical damage applications in experimental animals and in humans. Free Rad Biol Med 26:202–226

Deger Y, Dede S, Belge A et al (2003) Effects of X-ray radiation on lipid peroxidation and antioxidant systems in rabbits treated with antioxidant compounds. Biol Trace Elem Res 94:149–156

Fukazawa H, Ohashi Y, Sekiyama S et al (1994) Multidisciplinary treatment of head and neck cancer using BCG, OK-432, and GE-132 as biologic response modifiers. Head Neck 16:30–38

Goodman S (1988) Therapeutic effect of organic germanium. Med Hypotheses 26:207–215

Jagetia GC, Reddy TK (2005) Modulation of radiation-induced alteration in the antioxidant status of mice by naringin. Life Sci 77:780–794

Jayalakshmi K, Sairam M, Singh SB et al (2005) Neuroprotective effect of N-acetyl cysteine on hypoxia-induced oxidative stress in primary hippocampal culture. Brain Res 1046:97–104

Kidd PM (1987) Germanium-I 32 (Ge-I 32): Homeostatic normalizer and immunostimulant. A review of its preventive efficacy. Intern Clin Nutr Rev 7:11–19

Morsy I, Salwa MI, Badawi AM et al (2008) Metal ions in biology and medicine. Bastia, France

Riley PA (1994) Free radicals in biology: Oxidative stress and the effects of ionizing radiation. Int J Radiat Biol 65:27–33

Sagrista ML, Garcia AE, Africa De Madariaga M et al (2002) Antioxidant and pro-oxidant effect of the thiolic compounds N-acetyl-L-cysteine and glutathione against free radical-induced lipid peroxidation. Free Radical Res 36:329–340

Suzuki F, Brutkiewicz R, Pollard R (1986) Cooperation of lymphokine(s) and macrophages in expression of antitumor activity of carboxyethylgermanium sesquioxide (Ge-132). Anticancer Res 6:177–182

Wallace SS (1998) Enzymatic processing of radiation-induced free radical damage in DNA. Rad Res 150(Suppl. 5):S60–S79
Xin H, Han T, Gong S et al (1996) Experimental studies on effects of zinc and germanium on immune function and antioxidation in mice. Zhonghua Yu Fang Yi Xue Za Zhi 30:221–224
Yang MK, Kim YG (1999) Protective role of Ge-132 against paraquat-induced oxidative stress in the livers of senescence-accelerated mice. J Toxicol Environ Health A 58:289–297

Germanium Oxide

▶ Germanium-Containing Compounds, Current Knowledge and Applications

Germanium Poisoning

▶ Germanium, Toxicity

Germanium(+IV)

▶ Germanium-Containing Compounds, Current Knowledge and Applications

Germanium, Physical and Chemical Properties

Fathi Habashi
Department of Mining, Metallurgical, and Materials Engineering, Laval University, Quebec City, Canada

Germanium is a hard, grayish-white element, has a metallic luster but is brittle like glass. It is classified as a metalloid having the same crystal structure as diamond. It does not form minerals but is present in the lattices of certain sulfides such as sphalerite, ZnS. It is a semiconducting element. Its first industrial application was in 1947 with the development of the transistor. Until the early 1970s, it was used extensively in the solid-state electronics industry, although it has been replaced almost entirely by the cheaper silicon. The main use of metallic germanium currently is for infrared optics. Germanium tetrachloride is used in optical fiber production. Germanium dioxide is used mainly as a catalyst in the production of polyester and synthetic textile fiber. In 2010, the world's total production of germanium was between 100 and 120 t. This comprised germanium recovered from zinc concentrates, fly ash from burning coal, and recycled material. The main producers of germanium end products are located in the United States, Belgium, France, Germany, and Japan.

Physical Properties

Atomic number	32
Atomic weight	72.61
Relative abundance in the Earth's crust, %	7×10^{-4}
Density, g/cm^3	5.3234
Melting point, °C	938.4
Boiling point, °C	2,833
Heat of fusion, kJ mol^{-1}	36.94
Heat of vaporization, kJ mol^{-1}	334
Molar heat capacity, J	23.222
Atomic radius, pm	122
Lattice parameter, nm	0.565790
Ionization energies, kJ mol^{-1}	
First	762
Second	1537.5
Third	3302.1
Microhardness (Vickers-ASTM E384), kg/mm^2	780
Tensile fracture strength, MPa	100
Band gap, eV	0.67
Intrinsic resistivity, W cm	47.6
Transmission wavelength range, mm	2–16
Absorption coefficient at l = 10.6 mm	0.02 cm^{-1}
Refractive index at l = 10.6 mm	4.0027
Thermal conductivity, W cm^{-1} K^{-1}	0.586
Specific heat, J kg^{-1} K^{-1}	310
Linear thermal expansion, K^{-1}	5.9×10^{-6}

Chemical Properties

Elemental germanium is stable in air. At temperatures above 400°C in oxygen, a passivating oxide layer is formed. This layer is destroyed by water vapor. Germanium resists concentrated hydrochloric and hydrofluoric acids even at their boiling points, but reacts slowly with hot sulfuric acid. It dissolves at room temperature in nitric acid to form nitrate and in alkaline media in the presence of an oxidizing agent, e.g., hydrogen peroxide to form germanates. Germanium and silicon are miscible and form a continuous range of alloys. The system with tin has an eutectic.

Germanium can have valences of two or four. The divalent compounds are less stable and oxidize easily to the tetravalent compounds, which are similar to their silicon analogues. Germanium reacts readily with halogens to form tetrahalides. Germanium tetrachloride is obtained by reaction of chlorine gas with elemental germanium and of hydrochloric acid with GeO_2 or germanates. Germanium tetrachloride is soluble in carbon tetrachloride. All the tetrahalides are readily hydrolyzed to hydrated germanium dioxide.

Germanides can be produced by melting metals with germanium and freezing the melt, e.g., Mg_2Ge. Germanes (germanium hydrides) are evolved when germanides react with acids: mono-germane, GeH_4, di-germane, Ge_2H_6, and tri-germane, Ge_3H_8, can be separated by fractional distillation. Mono-germane is similar in structure to methane is a colorless, highly inflammable, toxic gas of low stability with a characteristic unpleasant odor. Polygermanes are similar to alkanes with formula Ge_nH_{2n+2} containing up to five germanium atoms are known. The germanes are less volatile and less reactive than their corresponding silicon analogues. The germanium hydrohalides with one, two, and three halogen atoms are colorless reactive liquids.

Most germanium compounds are comparatively low in toxicity because of pharmacological inertness, diffusibility, and rapid excretion. However, some exceptions exist, the most important being germanium hydrides. Germanium tetrafluoride and tetrachloride vapors irritate the eyes and mucous membranes of the respiratory tract. Gaseous germanium hydride is the most toxic of germanium compounds when inhaled. Like other metal hydrides such as AsH_3, it shows hemolytic action in animals. The lethal concentration in air is 150 ppm.

References

http://en.wikipedia.org/wiki/Germanium

Scoyer J, Wolf H-U (1997) Germanium. In: Habashi F (ed) Handbook of extractive metallurgy. Wiley, Weinheim, pp 1505–1524

Germanium, Toxicity

Katsuhiko Yokoi
Department of Human Nutrition, Seitoku University Graduate School, Matsudo, Chiba, Japan

Synonyms

Germanium poisoning

Definition

Germanium compounds are relatively less toxic compared to other metalloids and metals. However, relatively high doses of germanium dioxide and other inorganic germanium compounds caused severe poisoning including fatal cases via so-called germanium-containing health foods. Impairments in kidney, nerves, muscles, and bone marrows are conspicuous. Milligram order germanium per 1-g tissue was found from kidney, blood, brain, and other tissues of the poisoning cases.

Introduction

Germanium is an element with an atomic number of 32, an atomic weight of 72.63, and a valence of +2 or +4 (Lide 1998). It belongs to a chemical group IVA in the periodic table that includes C, Si, Ge, Sn, and Pb. Its chemical properties are similar to arsenic, silicon, tin, and antimony. Germanium is an important metalloid element to make semiconductors, and its oxide is amphoteric. Elemental germanium is a metalloid that is slowly oxidized in the air.

Inorganic germanium compounds include germanium dioxide (GeO_2), germanium lactate-citrate salt, germanium tetrachloride and other tetrahalides, and germanium hydride. Some "dietary supplements" contain GeO_2 or germanium lactate-citrate salt, and germanate ions (Tao and Bolger 1997). Because of the amphoteric nature of GeO_2, once GeO_2 is hydrolyzed by acid or alkaline and neutralized by its counterpart, the obtained germanate ion is freely soluble in water. The solubilized GeO_2 is colorless, odorless, and tasteless and impossible to distinguish from plain water by senses. Organic germanium compounds include germanium sesquioxides like bis(2-carboxyethylgermanium) sesquioxide or bis(β-carboxyethylgermanium) sesquioxide, usually known as Ge-132. Many organic germanium compounds were tested for possible antitumor and immune-stimulating effects and are sometimes contained in some "dietary supplements."

According to the 2006 UK Total Diet Study, germanium was detected from only the offal and meat categories at the concentrations of 2 and 1 μg/kg, respectively. The population dietary exposure is 0.1–1.5 μg Ge/day (UK-FSA 2009).

As per the author's knowledge, there is no clear evidence that germanium is an essential element for life, including microorganisms to higher plants and animals. However, the consistent occurrence of germanium in coal ash may indicate the absorption of germanium by prehistoric plants (Furst 1987). Altered bone and liver mineral composition in rats fed low germanium diet was reported (Uthus and Seaborn 1996). The question arises of whether germanium may have a physiological role but is still unanswered.

Absorption, Distribution, Metabolism, and Excretion of Inorganic Germanium

Rosenfeld (Rosenfeld 1954) studied the absorption, distribution, metabolism, and excretion of inorganic germanium. When neutralized GeO_2 was given to rats via a stomach tube (50 mg Ge/kg body weight), 96.4% of germanium was absorbed without causing diarrhea and gastrointestinal tract irritations. A 93.8% of intramuscularly administered dose of GeO_2 suspension was absorbed within an hour after administration without any signs of muscular damage. While in the blood, germanium distributes in plasma and blood cells. The plasma protein binding was not found for germanium. Inorganic germanium is considered to be delivered as free germanate ions in blood stream (Rosenfeld 1954). Rosenfeld found that under equilibrium conditions, the distribution of soluble germanium, i.e., germanate ions between erythrocytes and surrounding fluid, roughly approximates the relative water content of the intracellular and extracellular fluid compartments. Absorbed soluble germanium distributes throughout the entire bodies, and the highest concentration occurs in the kidneys. Intraperitoneally injected germanate to rats was excreted into urine (78.8% of the injected dose) and feces (13.0%) within 5 days (Rosenfeld 1954). Kidney is a major excretory organ for soluble inorganic germanium or germanate and an important target organ for inorganic germanium toxicity.

From the above conspicuous metabolic behavior, Rosenfeld stated that with respect to the type of distribution, soluble inorganic germanium compounds resemble substances like alcohol and urea, rather than inorganic bromides and iodides, which do not penetrate cells, or arsenicals, which are selectively fixed by them. From the dilute aqueous solution, at least three germanate ions, i.e., $[GeO(OH)_3]^-$, $[GeO_2(OH)]^{2-}$, and $\{[Ge(OH)_4]_8(OH)_3\}^{3-}$, have been identified. One form of germanium hydroxide $GeO(OH)_2$ is planar and electronically analogous to H_2CO_3 (Qin et al. 2006). This chemical resemblance to carbonic acid may explain the metabolic behavior of neutralized GeO_2.

Recently, dermal absorption of inorganic germanium was investigated using rats (Yokoi et al. 2008). Hydrophilic ointment containing neutralized GeO_2 (0, 0.21 and 0.42 mg/g) spread on a 2 cm × 2 cm polyvinylidene chloride sheet was topically applied to rat skin. When the topical dose was doubled, the germanium concentrations in plasma and kidney were increased more than double. The excessive rise of renal germanium concentration indicates a breakdown of excretion system of inorganic germanium.

Toxicity of Germanium Dioxide and Germanate

Acute oral toxicity of inorganic germanium is relatively low. The reported median lethal doses (LD_{50}) of germanium compounds in rodents are summarized in Table 1. When Rosenfeld and Wallace's acute toxicity data of germanium in rats (Table 1 of their article) was read (Rosenfeld and Wallace 1953), LD_{50} by intraperitoneal injection of neutralized GeO_2, i.e., sodium germanate, was estimated to be 750 mg Ge/kg body weight. Germanic acid, like boric acid, forms complex with polyvalent alcohol such as mannitol and glycerin. It is evident that germanate is more toxic than germanium–sugar alcohol complexes such as Ge–mannitol, or CMC suspension of GeO_2. Hypothermia was conspicuous in inorganic germanium intoxication of experimental animals in acute and subacute studies (Rosenfeld 1954).

Germanium, Toxicity, Table 1 LD_{50} for germanium compounds (mg compounds/kg) in rodents

Compounds	Medium	Species	Route	Male	Female
Germanium dioxide	1–10% mannitol (pH 5.4–5.6)	Mouse	p.o.	6,300	5,400
			i.p.	2,250	2,025
			s.c.	2,750	2,550
		Rat	p.o.	3,700	3,700
			i.p.	1,800	1,620
			s.c.	2,025	1,910
Germanium dioxide	0.5% CMC suspension	Mouse	p.o.	>10,000	–
			i.p.	1,550	
Neutralized germanium dioxide	Solubilized by sodium hydroxide and neutralized by hydrochloric acid (pH 7.3)	Rat	i.p.	750	–
Triethylgermanium acetate	Arachis oil	Rat	p.o.	125–250	–
tri-n-Butylgermanium	Arachis oil	Rat	p.o.	> 375	–
bis-β-Carboxyethylgermanium sesquioxide	0.5% CMC suspension	Mouse	p.o.	12,500	11,400
			i.p.	2,110	2,230
			s.c.	7,550	8,050
		Rat	p.o.	11,700	11,000
			i.p.	3,500	3,200
bis-β-Carboxyethylgermanium sesquioxide	PH 7 solution (neutralized by sodium bicarbonate)	Mouse	p.o.	11,500	11,700
			i.v.	5,720	5,720
			s.c.	10,800	12,900
		Rat	p.o.	10,600	9,500
			i.v.	4,800	4,380
			s.c.	16,300	17,800
β-Carboxyethylgermanium sesquioxide	0.5% CMC suspension	Mouse	p.o.	>10,000	–
			i.p.	2,820	
β-Carbamoylethyl germanium sesquioxide	0.5% CMC suspension	Mouse	p.o.	>10,000	–
			i.p.	7,100	
Propagermanium	5% Gum arabic suspension	Mouse	p.o.	5,600	5,800
			i.p.	1,250	1,300
			s.c.	>4,000	>4,000
		Rat	p.o.	7,700	7,050
			i.p.	1,750	1,670
			s.c.	>4,000	>4,000
Propagermanium	Saline	Mouse	i.v.	>200	>200
		Rat	i.v.	>200	>200
Propagermanium	10% Sodium bicarbonate	Mouse	i.v.	>3,000	>3,000
		Rat	i.v.	>3,000	>3,000

Rosenfeld and Wallace (1953); Cremer and Aldridge (1964); Hatano et al. (1981); Kanda et al. (1990).

In spite of the relatively low acute oral toxicity, human toxicity of inorganic germanium, especially GeO_2 or sodium germanate-containing so-called health foods, is well known because of many poisoning cases. Germanium has been expected as an alternative drug for cancer, immune impairments, viral infections, and health promotion. GeO_2 powder or solubilized and neutralized GeO_2 solution is added to some so-called health foods and drinks as an elixir of health. Tao and Bolger summarized reported cases of germanium poisoning and found that nine persons died of germanium poisoning out of 31 reported cases (Tao and Bolger 1997). Most of cases of germanium poisoning were reported from Japan. They were caused by inorganic germanium (unidentified forms), GeO_2, or unidentified germanium compounds.

Major signs and symptoms of GeO_2 or germanate intoxication in Japanese cases were renal damage, anemia, myopathy, peripheral neuropathy, weight loss, fatigue, and gastrointestinal symptoms (nausea, vomiting, and anorexia). Renal failure was characterized by tubuloepithelial degeneration with swollen mitochondria containing electron-dense granules. The similar electron-dense deposits were also found from mitochondria of hepatocytes and myocytes of germanium poisoning patients. Urinalysis was usually normal (absence of proteinuria and hematuria). Extrarenal organs involved were muscles (both skeletal and cardiac), nerves (both peripheral and central), bone marrow (characterized by hypoplasia), and liver (characterized by liver dysfunction with severe steatosis) (Schauss 1991; Tao and Bolger 1997). Milligram order germanium per 1-g tissue was found from kidney, blood, brain, and other tissues of the poisoning cases.

Four European cases were poisoning due to germanium lactate-citrate. The renal pathology of European cases was similar to the Japanese cases of GeO_2 intoxication. Different from the Japanese cases, hepatic involvements were conspicuous in the European cases. The typical form of the toxicity was lactic acidosis that was not seen in the Japanese cases. Whether there were any cases of genuine organic germanium poisoning by so-called health foods is uncertain because GeO_2 was detected from the ingested preparations that were reported to be organic germanium once (Tao and Bolger 1997). Sometimes, inorganic germanium was erroneously designated as "organogermanium" or "organic germanium" (Schauss 1991; Tao and Bolger 1997).

Human poisoning due to GeO_2 was reproduced in experimental animals (Tao and Bolger 1997). Independent researchers demonstrated the characteristic renal dysfunction with remarkable histological changes in kidneys, i.e., tubuloepithelial degeneration with mitochondrial electron-dense deposits after subchronic and chronic doses of GeO_2. The mitochondrial electron-dense deposits were revealed to contain germanium, iron, and sulfur by electron energy-dispersive X-ray microanalysis (Sanai et al. 1991). The mitochondrial degeneration due to accumulated germanium deposits that associates iron and sulfur could partly explain the germanium-induced tissue damage. Germanium-induced myopathy, associating mitochondrial electron-dense inclusions in degenerated myocytes, was also reproduced in experimental rats orally administered GeO_2. Germanium-induced neuropathy characterized by segmental demyelination and axonal degeneration was also demonstrated in rats orally given GeO_2.

Tumorigenicity and carcinogenicity were not shown by lifetime low-dose exposure to 5 mg Ge/L in drinking water as sodium germanate (Tao and Bolger 1997). Kamildzhanov et al. found decreased spermatozoa motility in rats exposed to GeO_2 by inhalation (Kamildzhanov et al. 1986). They also reported the synergistic toxicity of GeO_2 and arsenous anhydride (i.e., arsenic trioxide) in an inhalation study.

Toxicity of Gaseous Germanium Compounds

Gaseous germanium compounds are synthetic and do not occur naturally. Germanium hydride is a gas at the ambient temperature and the boiling point of germanium tetrachloride is 86.55 °C (Lide 1998). Exposure to gaseous germanium compounds occurs in the work place. Germanium hydride exposure causes neurological abnormalities and lung damage, while germanium tetrachloride can irritate skin (Furst 1987).

Toxicity of Organic Germanium Compounds

Organic germanium is defined as the compounds that contain covalent bonds between carbon and germanium. Organic acid salts of germanium such as germanium lactate-citrate are not included in organic germanium compounds. Toxicological data are limited for organic germanium. The LD_{50}s for selected organic germanium compounds are also listed in Table 1.

The Japanese Ministry of Health, Labour and Welfare has approved propagermanium or proxigermanium, a hydrophilic polymer of 3-oxygermyl propionates, as a drug for the treatment of chronic B hepatitis in 1994. Since 1994 to the reevaluation in 2003, several adverse effects of propagermanium (Serocion®) were reported. According to Hirayama, since 1994, moderate to severe liver damage has been reported in about 4% of 32,700 recipients after propagermanium treatments.

Among them, six patients died of acute liver failure, four with severe chronic hepatitis, one with flare of chronic hepatitis, and one with chronic hepatitis with hepatocellular carcinoma (Hirayama et al. 2003). Because of these severe adverse effects, the Japanese Ministry warned possible acute exacerbation of hepatitis and fatal cases in the course of propagermanium treatment of chronic B hepatitis.

The LD_{50}s of propagermanium in rodents are shown in Table 1. In beagle dogs, no significant changes were observed with intravenous administration of 125–500 mg propagermanium/kg/day for 6 months.

An azaspiran-germanium compound (2-aza-8-germanspiro[4,5]-decane-2-propamine-8,8-diethyl-N, N-dimethyl dichloride) (NSC 192965; spirogermanium) has cytotoxic effects and tested chemotherapeutic effects in phase I and phase II studies that were inconclusive. During the clinical trials, neurotoxic adverse effects, including dizziness, vertigo, nystagmus, ataxia, and paresthesia and pulmonary toxicity were observed (Schauss 1991).

In contrast to GeO_2 (52 mg Ge/kg body weight/day) that induced severe renal toxicity, the similar dose of bis-β-carboxyethylgermanium sesquioxide as an element (51 mg Ge/kg body weight/day) did not cause significant abnormalities in kidney and blood biochemistry or tissue germanium accumulation after a 24-week oral administration of the preparation mixed into the powdered rat chow (Sanai et al. 1991). A French group found moderate renal dysfunction of bis-β-carboxyethylgermanium sesquioxide by subchronic study (6 months) in rats with a sixfold oral dose (306 mg Ge/kg body weight/day) (Anger et al. 1992).

Cross-References

- ▶ Arsenic
- ▶ Cadmium and Health Risks
- ▶ Germanium L-Cysteine Alpha-Tocopherol Complex as Stimulator to Antioxidant Defense System
- ▶ Germanium, Physical and Chemical Properties
- ▶ Lead Nephrotoxicity
- ▶ Mercury Nephrotoxicity
- ▶ Mercury Neurotoxicity
- ▶ Rubredoxin, Interaction with Germanium
- ▶ Silver, Neurotoxicity

References

Anger F, Anger JP, Guillou L, Papillon A, Janson C, Sublet Y (1992) Subchronic oral toxicity (six months) of carboxyethylgermanium sesquioxide $[(HOOCCH_2CH_2Ge)_2O_3]_n$ in rats. Appl Organomet Chem 6:267–272

Cremer JE, Aldridge WN (1964) Toxicological and biochemical studies on some trialkylgermanium compounds. Br J Ind Med 21(3):214–217

Furst A (1987) Biological testing of germanium. Toxicol Ind Health 3:167–204

Hatano M, Ishimura K, Fuchigami K, Ito I, Hongo Y, Hosokawa Y, Azuma I (1981) Toxicological studies on germanium dioxide (GeO_2) (1). Acute, subacute, chronic toxicity and successive irritation to the eye. Oyo Yakuri 21:773–796 (in Japanese)

Hirayama C, Suzuki H, Ito M et al (2003) Propagermanium: a nonspecific immune modulator for chronic hepatitis B. J Gastroenterol 38:525–532

Kamildzhanov A, Ubaidullaev RU, Akbarov AA (1986) Combined effects of arsenous anhydride and germanium dioxide after their hygienic regulation in atmospheric air. Gig Sanit 8:82–84 (in Russian)

Kanda K, Igawa E, Shinoda M, Asaeda N, Yoshiyasu T, Iwai H, Nagai N, Tamano S, Ichikawa K, Koide M (1990) Acute toxicity test of proxigermanium (SK-818) in mice and rats. Kiso To Rinsho 24:5–18 (in Japanese)

Lide DR (1998) CRC handbook of chemistry and physics, 79th edn. CRC Press, New York

Qin C, Gao L, Wang E (2006) Germanium: inorganic chemistry. In: King RB (ed) Encyclopedia of inorganic chemistry. Wiley, Hoboken, NJ. doi:10.1002/0470862106.ia079

Rosenfeld G (1954) Studies of metabolism of germanium. Arch Biochem Biophys 48:84–94

Rosenfeld G, Wallace EJ (1953) Studies of the acute and chronic toxicity of germanium. AMA Arch Ind Hyg Occup Med 8:466–479

Sanai T, Okuda S, Onoyama K et al (1991) Chronic tubulointerstitial changes induced by germanium dioxide in comparison with carboxyethylgermanium sesquioxide. Kidney Int 40:882–890

Schauss AG (1991) Nephrotoxicity and neurotoxicity in humans from organogermanium compounds and germanium dioxide. Biol Trace Elem Res 29:267–280

Tao SH, Bolger PM (1997) Hazard assessment of germanium supplements. Regul Toxicol Pharmacol 25:211–219

UK-FSA (2009) Measurement of the concentrations of metals and other elements from the 2006 UK total diet study. FSA, London. http://www.food.gov.uk/multimedia/pdfs/fsis0109metals.pdf

Uthus EO, Seaborn CD (1996) Deliberations and evaluations of the approaches, endpoints and paradigms for dietary recommendations of the other trace elements. J Nutr 126:2452S–2459S

Yokoi K, Kawaai T, Konomi A et al (2008) Dermal absorption of inorganic germanium in rats. Regul Toxicol Pharmacol 52:169–173

Germanium-Containing Compounds, Current Knowledge and Applications

Erwin Rosenberg
Institute of Chemical Technologies and Analytics, Vienna University of Technology, Vienna, Austria

Synonyms

Carboxyethyl germanium sesquioxide; Ge-132; Germanium oxide; Germanium(+IV); Germanium; Organogermanium compounds; Propagermanium; Spirogermanium

Definition

Germanium (Ge) is the element with atomic number 32. Being a member of the group 14 of the periodic table, it stands in many of its physical and chemical properties between the elements silicon and tin. While the element has important applications in the electronics, semiconductor, and optics industries, the organic compounds of germanium have been investigated as anticancer, anti-inflammatory, and immunostimulating drugs.

General Information and Historical Outline

Germanium is a main group IV (group 14) element that has atomic number 32 and relative atomic mass 72.61. Its specific gravity is 5.323 g/cm^3, melting point 937.4°C (1210.6 K), and boiling point 2,830°C (3103.2 K). Ge has five stable isotopes (^{70}Ge, 20.45%; ^{72}Ge, 27.41%; ^{73}Ge, 7.77%; ^{74}Ge, 36.58%; ^{76}Ge, 7.79%), and 13 further radioactive isotopes of mass numbers 65–69, 71, 75, 77, and 78 whose half-lives range between 20 ms and 287 days.

Germanium is a brittle, relatively hard element (Mohs hardness of 6.0) and has a refractive index of 4.0. With its good transparency in the mid-infrared range and its low dispersion, it is highly suitable for the construction of infrared optics (windows, lenses), which, particularly when designed for rugged field use, benefit from its surface hardness and mechanical strength. The principal region of transparency extends throughout the entire mid-IR range from around 550–5,500 cm^{-1}, corresponding to approximately 1.8–18 μm. The short wavelength cutoff corresponds to an energy gap of 0.68 eV at 300 K to 0.83 eV at 50 K, respectively, and a transmission level of approximately 47% (Scoyer et al. 2000).

Ge is a relatively inert semimetal and is stable in air at ambient temperature. It is oxidized only at temperatures above approximately 1,000 K to form GeO_2. The element is insoluble in water, hydrochloric acid, and dilute alkali, but can readily be dissolved in the oxidizing media, such as nitric or sulfuric acid, and by basic peroxides, nitrates, and carbonates. According to the position in the periodic table of elements, Ge occurs in addition to its elemental form in the oxidation states + II and + IV, where the latter possesses the higher stability and is the one in which Ge compounds typically occur in the environment. In minerals, Ge often appears in the form of the oxide (GeO_2) or the sulfide (GeS_2), and in solution as Germanic acid, $Ge(OH)_4$. Divalent compounds of germanium are mainly produced synthetically and have importance only in the chemical industries.

The concentration of Ge related to the entire earth mass is estimated at 7.3 ppm (w/w); in the surface layer of the earth, it is estimated at 5.6 ppm (w/w), in the continental earth crust 1.6 ppm (w/w), and in the oceans 5.10^{-5} mg/L. Related to the mass of the entire earth, it ranks 22nd in relative abundance, while based on the mass of the surface layer of the earth it is 46th and based on its concentration in the continental earth crust it is only 55th (Rosenberg 2009). Germanium is thus not a very rare element, but it is found in nature only in widely dispersed form. There are but few minerals which contain Ge at significant concentrations, the best known being mixed hydroxides, such as Stottite, $FeGe(OH)_6$, or Schauerteite, $Ca_3Ge(SO_4)_2(OH)_6.3H_2O$, and mixed sulfides, such as Germanite, $Cu_3(Ge, Fe)S_4$, and Argyrodite, $4Ag_2S.GeS_2$, the latter being the form that led to the discovery of this element by Winkler in 1885 (Winkler 1886). None of these minerals occurs in amounts that would justify its industrial mining. Economically more feasible is the production of Ge from silver, zinc, and copper refineries where Ge can be recovered as a by-product. Technical production usually involves a pre-concentration process in which the concentrate from the Zn production, containing only about 0.02% Ge is enriched to a content of 6–8%. The second step of the production of Ge relies on the volatility of $GeCl_4$ which is formed from GeO_2 in the presence of HCl and Cl_2. $GeCl_4$ can thus be separated from the concentrate solution by distillation, is then hydrolyzed again to (pure) GeO_2, and then further reduced to elemental Ge (Scoyer et al. 2000).

From the organic compounds of the elements of group 14 of the periodic table of elements, organogermanium compounds were the least investigated among its homologues up to the middle of the twentieth century. This is all the more surprising since the first organogermanium compound, tetraethylgermane, was synthesized as early as 1887 by Winkler by the reaction of tetrachlorogermane and diethylzinc which was only about a quarter century later than the first organic compounds of silicon, tin and lead were obtained.

The synthesis of Et_4Ge was an additional and significant proof of the assumption that the germanium discovered by Winkler belongs to Group IV of the Periodic Table and that it was identical to Mendeleev's *eca*-silicon.

During the period between 1887 and 1925, no new organogermanium compounds were reported. These 40 years of apparent disinterest resulted mainly from the very limited availability and the high prices of germanium and its simplest inorganic derivatives. This reflected the low natural reserves of Argyrodite, the only mineral source of germanium known at that time.

With the discovery of new sources of germanium in 1922, this picture changed dramatically. In particular, 0.1–0.2% of Ge was found in a residue of American zinc ore after zinc removal (Clark 1951). This was followed by the discovery of two germanium-rich minerals: Germanite (in 1924), a mineral from Southwestern Africa which contained about 5.1% Ge, and Rhenierite, a mineral from the Belgian Congo, containing 6–8% of Ge, which became another source of germanium. In the following decade, the 1930s, processing wastes of coal ashes and sulfide ores became the main sources of germanium (Scoyer et al. 2000). The gradually improving availability of Ge allowed American, English, and German chemists to start fundamental investigations of organogermanium compounds from the mid-1920s, although germanium was still very expensive at that time.

Organogermanium chemistry was stimulated by the rapid development of the chemistry of other organometallic compounds, in particular of the elements silicon and tin. A further motivation was the significant role that this element and its organic derivatives were starting to have in the electronics industry (Haller 2006). Also, first biological effects of organogermanium compounds (including anticancer, hypotensive, immunomodulating, and other kinds of physiological action) were described (Lukevics and Ignatovich 2002). In addition, the eventual decrease in the prices of elemental germanium and its derivatives expanded their production and helped their growth.

The rapid expansion of organogermanium chemistry is clearly illustrated by the continuous increase in the number of compounds described in literature: In 1951, 230 organogermanium compounds were known; in 1961, there were 260; and in 1963, there were more than 700 (Voronkov and Abzaeva 2002).

Current Use of Germanium

Ge has a number of technical uses which make it, despite the relatively low annual production of approximately 60 t/year, an element of strategic importance. For this reason the United States of America's National Defense Stockpile held up to the late 1990s tens of tons of this semi-metal in order to not be dependent on the international production and availability of this technologically and strategically important element. Key germanium applications include infrared optics, night vision instruments, optical fibers, catalyst in PET (polyethylene-terephthalate) production, and semiconductor and photovoltaic cell industries (Rosenberg 2009).

The use of Ge for the production of organogermanium compounds is only of minor importance in comparison to its other technical uses. Most organogermanium compounds are tetravalent. Being an element of Main Group IV (or Group 14 according to the new IUPAC numbering) of the periodic table of the elements, both + II and + IV oxidation states are possible, although the higher oxidation state is more stable. The electronegativity of Ge is 2.01 according to Pauling. It is thus closer to carbon (EN: 2.50) and hydrogen (EN: 2.20) than the electronegativity of the other elements of the same group (silicon: 1.74, tin: 1.72, and lead: 1.55). Consequently, both germanium-carbon and germanium-hydrogen bonds can be considered as almost apolar when compared to the bonds of Si–H (ΔEN 0.46), Sn–H (ΔEN 0.48), and Pb–H (ΔEN 0.65). Being practically apolar in the GeH_4 molecule, the polarity of the Ge–H bond depends mainly on the substitution of Ge with electron providing or withdrawing groups, such as alkyl groups R (weakly electron-pushing) or halogens X (strongly electron-withdrawing), resulting in positively or negatively polarized Ge:

$$\overset{\delta+}{R_3Ge} - \overset{\delta-}{H} \qquad \overset{\delta-}{X_3Ge} - \overset{\delta+}{H}$$

This fact is important as it explains the reactivity of organogermanium compounds.

In order to form organogermanium compounds, three generic routes exist (Voronkov and Abzaeva 2002):

(a) *Direct reaction of Ge with alkyl halogenides* using a copper catalyst. This reaction proceeds with very good yields; however, statistical mixtures of the various alkylgermanium halides are produced, with the dialkylated R_2GeX_2 typically being the favored product:

$$n\ RX + Ge \rightarrow R_nGeX_{4-n} \qquad (1)$$

(R = alky or aryl group; X = halogen; $n = 1, \ldots, 4$)

(b) *Reaction of germanium halide with an organometallic compound* M – R (a form of the Wurtz synthesis, a typical metathesis reaction):

$$\equiv Ge-X + M-R \rightarrow \equiv Ge-R + M-X \qquad (2)$$

(R = alky or aryl group; M = Li, Na, K, Mg, Zn, Hg, Al; X = halogen)

This reaction proceeds particularly well with Grignard reagents (compounds of the general form RMgX, with R being an alkyl or aryl moiety, and X a halogen) and can be used for the synthesis of tetrasubstituted organogermanium compounds. As an example, $GeCl_4$ can be reacted with the methyl-Grignard MeMgI to give Me_4Ge with 93% yield.

(c) *The hydrogermylation reaction*. In this synthetic route, organogermanium compounds are prepared by forming trihalogen germanium hydride in a first step, followed by the addition of this compound to unsaturated organic compounds:

$$Ge + 3HCl \rightarrow HGeCl_3 + H_2 \qquad (3)$$

$$HGeCl_3 + CH_2 = CH - R \rightarrow Cl_3GeCH_2CH_2-R \qquad (4)$$

This latter reaction is of particular importance, since it is the most widely employed route for the synthesis of the pharmaceutically relevant germanium sesquioxide (also called *bis*-carboxyethylgermanium sesquioxide or Ge-132).

Organogermanium Compounds with Therapeutic Use

Germanium sesquioxide (Scheme 1), IUPAC systematic name: 3-[(2-Carboxyethyl-oxogermyl) oxy-oxogermyl]propanoic acid, CAS-No. 12758-40-6, also known under the synonyms propagermanium, proxigermanium, Ge-132, 2-carboxyethylgermasesquioxane, SK-818, or *bis*-(2-carboxyethylgermanium) sesquioxide, is the best known organogermanium drug. Its therapeutic uses were first described by Asai and coworkers of the Asai Germanium Research Institute (Tokyo, Japan) in 1968 who had screened a large number of germanium compounds for their biological effects. The drug was patented in Japan in 1971 and in the USA in 1972, after which date Asai and coworkers strongly promoted the use of this compound in the treatment of various diseases. Ge sesquioxide has an unusual planar structure in which each Ge atom binds to the three surrounding oxygen atoms, forming a 12-membered ring. The carboxylate chains are arranged alternately above and below the plane defined by the germanium and oxygen atoms. This particular three-dimensional structure explains why Ge sesquioxide is a crystalline substance which does not dissolve in organic solvents (except ethylene glycol), but is soluble in water upon heating; and also why it does not melt nor decompose at temperatures below 320°C.

Ge sesquioxide has been claimed to be a potent antioxidant and a useful drug in anticancer and antiviral treatments. This was based on reports that claimed the induction of interferon γ (IFN-γ) and the induction of immunological responses such as

Germanium-Containing Compounds, Current Knowledge and Applications, Scheme 1 Structure of germanium sesquioxide (=*bis*-(2-carboxyethylgermanium) sesquioxide, Ge-132)

increased natural killer (NK) cell, macrophage and T-cell production after oral administration of Ge sesquioxide (Kaplan et al. 2004).

Ge-132, as well as spirogermanium (= 8,8-diethyl-2-[3-(*N*-dimethylamino)-propyl]-2-aza-8-germaspiro [4.5]decane) and other related synthetic derivatives have demonstrated significant antitumor activity in vitro in cancer cell lines and in vivo in animal models of some cancers.

Their antitumor activities have been established against various experimental tumors, such as Ehrlich ascites tumor and Lewis lung carcinoma. In clinical studies, limited value was revealed for antitumor germanium complexes.

The exact anticancer mechanism still remains relatively unclear, though induction of IFN-γ may be the key. Ge-132 may also possess DNA binding specificity like cisplatin, to inhibit cancer cell proliferation. A similar mode of action is reported for some Ge-132 derivatives (Shangguan et al. 2005).

Several cases of Ge-related toxicity have been reported; however, these were linked to large amounts of Ge (15–426 g) being ingested over prolonged periods up to 36 months. Studies have narrowed acute and chronic Ge-related toxicity specifically to inorganic germanium dioxide (GeO_2) and metallic Ge which can accumulate in the liver, kidney and spleen, peripheral nerves, lungs and muscle. Since Ge-132 is also synthesized from GeO_2, it is possible that residual GeO_2 can contaminate formulations claiming to be pure organogermanium. This most likely was the case in the few documented cases of anticancer treatment with Ge-132 with fatal consequences attributed to Ge intake (Tao and Bolger 1997).

In contrast to inorganic germanium, organogermanium compounds including Ge-132 have characteristic low toxicity and are readily excreted via the kidney with very low accumulation in major organs and tissues.

There are but few clinical trials reported in the literature which have assessed the efficacy in the treatment of the various forms of cancer. It appears that most of these studies are either inadequately documented, or they have been conducted under insufficiently controlled conditions, particularly as concerns the dose and the purity of the administered drug. It is thus hardly possible to arrive to a firm and experimentally underpinned conclusion on the efficiency of Ge-132 in anticancer treatment, and on the potential toxicity of Ge-132 formulations (Horneber and Wolf 2011).

Along this line, a review from 1997 reported on 31 cases of organ toxicity related to different germanium compounds, including GeO_2, germanium-lactate-citrate, and Ge-132, which were ingested over a period from 2 to 36 months. In the two cases of Ge-132 intake, contamination with GeO_2 was detected. All cases showed renal impairment in terms of chronic or acute renal failure, which caused death in nine cases. Other clinical findings were anemia, which occurred in all cases, gastrointestinal disturbances, weight loss, myopathy, and liver dysfunction. If at all, renal function improved very slowly and remained impaired in some patients observed as long as 40 months. As a consequence, governmental institutions of several countries imposed alerts because of possible injury to health (Tao and Bolger 1997).

In most countries, Ge-132 falls under the regulations of dietary supplements. An import alert on germanium products was imposed by the U.S. FDA in 1988, because of possible health hazard. In Germany, governmental institutions warned consumers of possibly fatal kidney damage. In the UK, supplements containing germanium were voluntarily withdrawn by the industry. In Japan, Ge-132 is approved for the treatment of chronic hepatitis B.

Spirogermanium. Spirogermanium was first synthesized in 1974 and represents an azaspiran-germanium compound which was tested in various phase I/II trials to examine its anti-tumor effect (Scheme 2). Due to a markedly negative risk-benefit profile, in particular

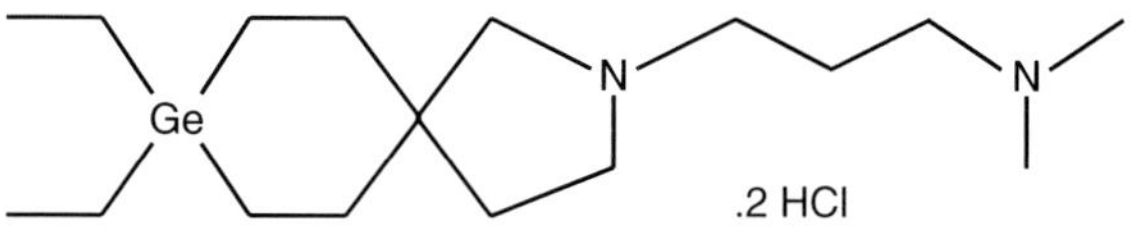

Germanium-Containing Compounds, Current Knowledge and Applications, Scheme 2 Structure of spirogermanium (=N-(3-dimethylaminopropyl)-2-aza-8,8-diethyl-8-germaspiro [4.5]decane)

Germanium-Containing Compounds, Current Knowledge and Applications, Scheme 3 R^1–R^4 = H, alkyl, aryl; Z = O, S; Y = OH, OR, OM, NH_2, NRR′. Ge-132 (*bis*-(2-carboxyethyl-germanium) sesquioxide) is a particular representative of this compound class with R^1–R^4 = H; Y = OH; and Z = O

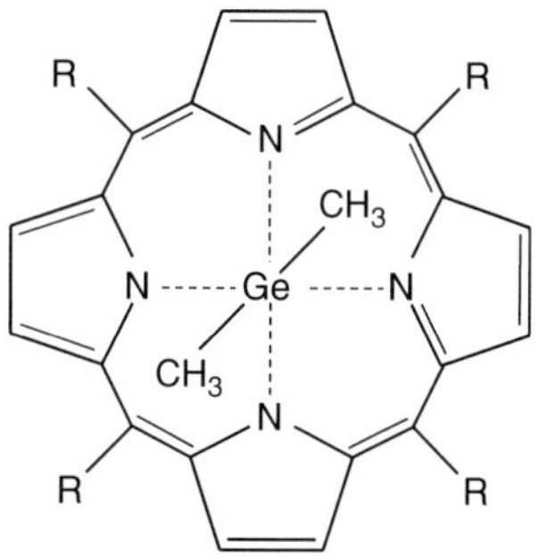

neurologic toxicity, spirogermanium was abandoned for therapeutic use.

Spirogermanium inhibits DNA, RNA, and protein synthesis in vitro and decreases cell survival after exposure to as little as 1.0 μg/mL for 24 h. Quiescent cells seem more resistant. At higher concentrations, cytolysis is observed (Gerber and Léonard 1997).

Other Organogermanium Compounds

In addition to the two best studied organogermanium drugs, a number of other organic germanium derivatives have been tested in anticancer protocols. Many of these studies were led by the Asai Germanium Research Institute in Japan. Starting from the sesquioxide of 2-carboxyethylgermanium, derivatives with similar structures have also been studied as antitumor drugs but also as antibacterial agents (see Scheme 3).

Another germanium compound shown to have an anticancer activity for rats and mice is a porphyrine dimethylgermanium complex (Scheme 4).

The biological activity of trialkyl germanium on fungi, yeasts, and bacteria was also studied. It was found that some organogermanium compounds present a rather strong activity against *Streptococcus lactis*. Similarly, $Me_3GeCH_2CH_2NHCH_2Ph$ has shown a high antibacterial activity with a wide spectrum of targets. Also the organogermanium sesquisulfur has an excellent antibacterial activity (see Scheme 5).

Other germanium compounds that have been shown to exhibit biological effects are the germatranes and trithiagermatranes (showing psychotropic and antitumor activities, Scheme 6) and germathiolactones which are active as biological antioxidants (Scheme 7).

$(Et_3Ge)_2S$ is an extremely strong fungicide, with an efficiency that exceeds many of the best fungicides on the market.

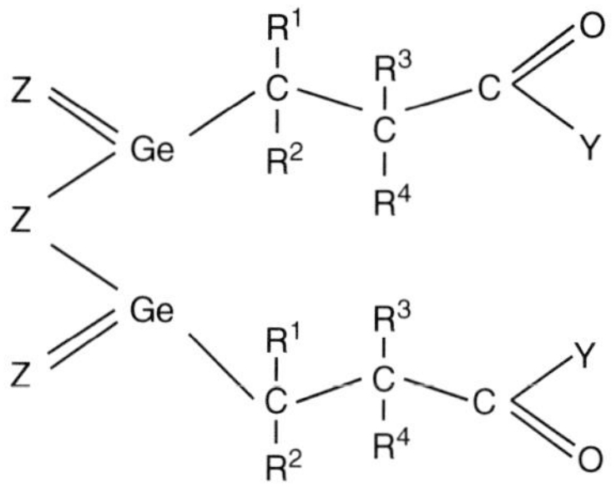

R = 3,5-di-*t*-butylphenyl

Germanium-Containing Compounds, Current Knowledge and Applications, Scheme 4 Structure of the porphyrine dimethylgermanium complex

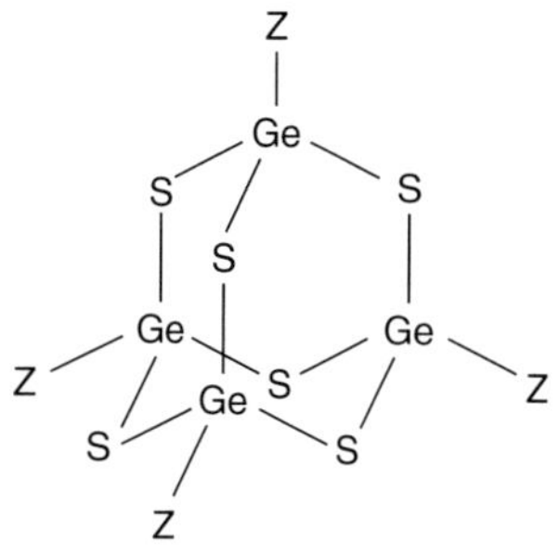

Z = CH_2CH_2COOH

Germanium-Containing Compounds, Current Knowledge and Applications, Scheme 5 Structure of the organogermanium sesquisulfur

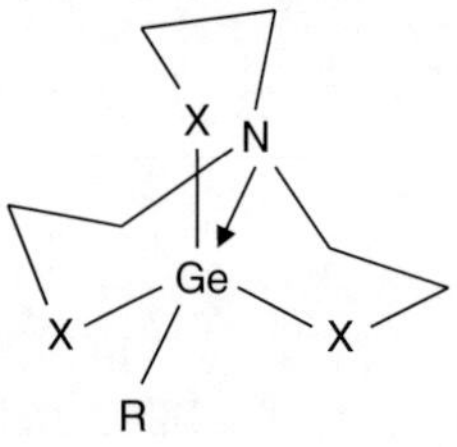

X = O or S

Germanium-Containing Compounds, Current Knowledge and Applications, Scheme 6 Structure of the germatranes and trithiagermatranes

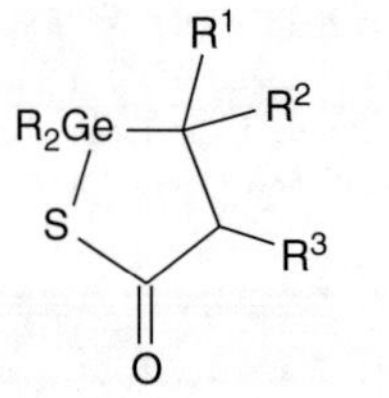

R^1, R^2, R^3 = H, alkyl group

Germanium-Containing Compounds, Current Knowledge and Applications, Scheme 7 Structure of the germathiolactones

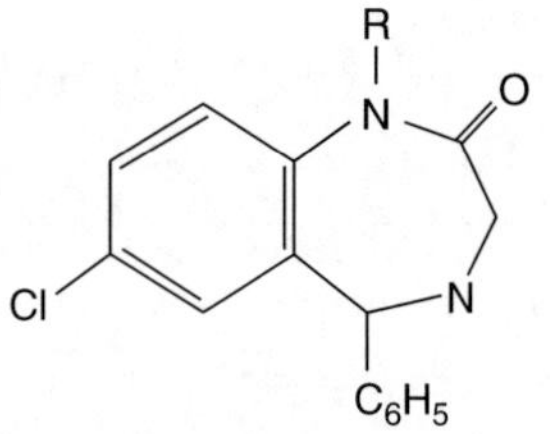

R = CH_3: valium
R = $Ge(CH_3)_3$: trimethylgermanium analogue of valium

Germanium-Containing Compounds, Current Knowledge and Applications, Scheme 8 Structure of valium and its trimethylgermanium analogue

In the valium molecule, substitution of the methyl group connected to the nitrogen atom by a trimethylgermyl group leads to a nontoxic organogermylated valium whose psycholeptic activity is higher than that for valium (Scheme 8).

Numerous organic germanium derivatives have shown an important radioprotective activity. Many of the compounds that were studied for this purpose were organogermanium compounds containing sulfur, such as germathiazolidines, germadithioacetals, germatranes, and germanium sulfides. More than one hundred of these derivatives have been tested and characterized by their dose reduction factor (DRF). The DRF is the ratio between the LD50/30 days of treated mice and the LD50/30 days of non-treated mice. With irradiations in the range 7.5–8.5 Gy (grays), DRFs in the range 1.2–1.7 were obtained.

These organometallic derivatives usually have a lower toxicity and a radioprotective activity greater than that for the corresponding organic derivatives, namely, cysteamine, methylcysteamine, and N-substituted cysteamine.

It should be emphasized that this strong radioprotective activity was obtained in many cases with organogermylated derivatives injected in lower doses than typically required for the corresponding organic compounds. The organogermylated derivatives seem to exhibit this pronounced radioprotective activity due to their amphiphilic properties, where the presence of organometallic groups improves the aqueous solubility, while the presence of organic ligands increases the lipophilic solubility, thereby favoring their transfer through the cellular membranes (Satge 2004).

Biological Effects of Germanium Compounds

Mutagenicity of Germanium. Dibutylgermanium dichloride exhibited an LC50 of about 600 μg/ml (2.33 mM) in a mutation assay on hypoxanthine guanine phosphoribosyl transferase (HGPRT) carried out on Chinese hamster ovarian (CHO) cells; the incidence of mutants increased with dose up to 400 μg/ml (1.55 mM). It is assumed that this electrophilic compound acts by binding to DNA bases, thereby forming penta- and hexa-coordination compounds. Also the immunosuppressive potential of this compound in lymphocytes was evaluated on the basis of cytotoxicity and impairment of antibody formation. No toxic action was observed in lymphocytes at concentrations up to 64 μg/ml (0.25 mM), but the number of antibody-producing cells was reduced to one half of the control.

Ge-132 is a potent antimutagen when added to agar at concentrations of 5–30 μg/ml agar, markedly reducing the number of revertant mutants in B/r WP2 Trp − *Escherichia coli* which are induced by gamma irradiation at a dose of 10 kRad (~100 Gy). Germanium oxide (GeO_2) was found to have a dose-dependent antimutagenic activity for Trp-P-2 frameshift reverse mutations induced by 3-amino-1-methyl-5H-pyrido-[3,4-b]indole in *Salmonella typhimurium* strains TA 98 and TA 1538. Doses of

600 μg/plate of germanium dioxide reduced the mutagenic potential of benzo[a]pyrene in three of four *S. typhimurium* strains tested.

Germanium dioxide (GeO_2) was also reported to exhibit an antimutagenic effect after administration of cadmium chloride ($CdCl_2$) to laboratory mice. Administration of 0.05 mg/kg germanium oxide significantly reduced the action of 1.35 mg cadmium chloride in the DNA synthesis inhibition test when the $CdCl_2$ was injected 1 h later. However, at other concentration ratios of the two compounds, a less pronounced antagonistic effect was observed. Germanium dioxide at doses of 0.1 or 0.5 mg/ml/kg decreased markedly the number of micronuclei or chromosome aberrations induced by 0.7, 1.4, or 2.7 mg/kg of $CdCl_2$ in vivo in mouse bone marrow cells. Germanium dioxide also protected in vivo against sperm head changes in morphology and in vitro from the sister chromatid exchanges (SCEs) in human peripheral blood lymphocytes caused by $CdCl_2$. It should be noted that GeO_2 given alone had no detectable effect on all these tests.

Carcinogenicity of Germanium. No carcinogenic effects have been reported so far for organogermanium compounds. Some, especially the organic compounds of germanium, have been reported to exhibit antineoplastic properties in experimental animals and man. In one of the studies performed with Long-Evans rats, the group that received sodium germanate in drinking water at a concentration of 5 ppm throughout their entire lifetime showed significantly lower incidence of tumors than in the control group.

A few inorganic and organic germanium compounds have been reported to have antineoplastic activity in man. However, in a further study, inorganic germanium appeared not to alter the rate of dimethylhydrazine-induced colon cancer in Western rats (Gerber and Léonard 1997).

As early as 1979, a Japanese patent claimed, on the basis of preliminary experiments, an antitumor activity of Ge-132 on mouse tumors which might be related to the ability of this compound to induce interferon (Brutkiewicz and Suzuki 1987). Ge-132 (administered at 8 mg/day) and "natural organic germanium" (at 70 mg/kg/day, obtained as an extract from the plant *Hydrangea macrophylla* with an original content of 600 ppm Ge) significantly decrease the incidence of intestinal cancer induced by 1,2-dimethylhydrazine in male Sprague–Dawley rats; inorganic germanium (GeO_2) (at 27 mg/kg/day) did not exhibit such anticancer effect. Similar antitumor activities of germanium compounds have also been reported for other systems, such as on Lewis lung tumor, or murine ascites tumor.

This rather heterogeneous and uncertain information may be summarized in that germanium (and its organic compounds) very likely appears not to be carcinogenic; its anticarcinogenic properties, however, appear more doubtful and, thus, do not justify the risk of using germanium compounds in anticancer therapy or "bio-oxidative therapy."

Teratogenicity of germanium. The organic compound dimethyl germanium oxide, $(CH_3)_2GeO$, has been reported to be teratogenic in chick embryos causing limb abnormalities, umbilical hernias, and anophthalmia. Doses of 40 and 100 mg/kg of germanium dioxide (in the form of sodium metagermanate; $Na_2GeO_3.7H_2O$) injected intravenously into pregnant hamsters on day 8 of gestation had some embryopathic effects as indicated by an increased embryonic resorption rate, but did not produce obvious malformations.

Physiological Importance of Germanium

Germanium is not considered an essential element. There is no known biological requirement for germanium, germanates, or any organogermanium compound. Ge deficiency has not been demonstrated in any animal. The estimated average dietary intake of Ge in humans is 1.5 mg/day. Ge is widely distributed in edible foods, all of which, with few exceptions, contain less than 5 ppm Ge, since higher levels are toxic to most plants.

It has been suggested that germanium can interact with silicon in bone metabolism. It can interfere with the action of loop diuretic drugs and inhibit the activity of a number of enzymes including lactate and alcohol dehydrogenase. Hexobarbital-induced sleeping times are increased in mice treated with germanium compounds suggesting that inhibition of cytochrome P450 activity may also occur. Organic germanium compounds have been reported to inhibit the detoxification enzyme glutathione-S-transferase.

Germanium compounds are readily absorbed following oral exposure. Germanium is distributed throughout the body tissues, with some enrichment in the kidney and thyroid. Organic germanium is thought not to accumulate to the same extent as inorganic germanium compounds but few data on germanium metabolism are available.

Germanium is largely excreted in the urine. Some biliary and fecal excretion also occurs.

In plants, germanium is taken up by the same routes as silicon, owing to its chemical similarity. Uptake of larger amounts of Ge from the cultivation medium, however, inhibits plant growth in comparison to a control group grown without Ge. This inhibition is purely competitive and is only dependent on the Ge/Si ratio in the growing medium. In the case of boron deficiency, Ge administration, however, may to a certain degree substitute for the demand of this element, thus alleviating the negative effects of boron undersupplementation.

Analysis of Ge Compounds

The analysis of Ge in inorganic and biological matrices is mainly performed by atomic spectrometric techniques, mostly along with other relevant elements. Both optical and mass spectrometric techniques are used. Among the optical spectrometric detection techniques, atomic absorption spectrometry (AAS) and optical emission spectrometry (OES) are used, and are capable of detecting the presence of this element at the low and sub-ppb (ng/ml) level. Sensitive detection of Ge by AAS requires either electrothermal evaporization in the graphite furnace, or the formation of volatile germanium hydride (GeH_4) by the hydride technique.

Inductively coupled plasma-mass spectrometry (ICP-MS) reaches even higher sensitivity than the two former techniques. Either of the atomic spectrometric techniques requires a complete digestion of the organic or inorganic matrix, achieved by the digestion with nitric and/or hydrochloric acid in open or closed vessels, often with the aid of microwave irradiation. It is evident that such an analytical procedure provides only total element concentration information.

If a more comprehensive picture on the various Ge species present in a sample (speciation analysis) is required, then hyphenated techniques are often used. Hyphenated techniques combine the separation power of high-resolution chromatographic (or, more rarely, electrophoretic) separation techniques with the high detection power of spectrometric detection. An instrumental setup that has been used for the analysis of volatile organogermanium compounds is gas chromatography with microwave-induced plasma-atomic emission spectrometry (GC-MIP-AES). This technique allows the sensitive and selective analysis of organogermanium compounds, with as little as few pg (abs.) of Ge being detectable. The combination of gas chromatography with mass spectrometry with an inductively coupled plasma as atomization and ionization device (GC-ICP-MS) can be used with similar sensitivity. As hyphenated techniques with atomic spectrometric detection do not provide molecule-specific information, this information has to come either from the comparison of chromatographic retention times with authentic standards, or from the complementary use of GC with molecular mass spectrometry (GC-MS). This analytical technique is available in many laboratories, and can equally be used for the speciation analysis of volatile germanium compounds. The rich fragmentation of organogermanium compounds provides increased structural information, however, at the cost of reduced sensitivity.

All gas chromatographic approaches require that the germanium compounds either be volatile, or can be volatilized by an appropriate derivatization reaction. This disadvantage can be overcome by using separation in liquid phase, particularly high-performance liquid chromatography (HPLC) or ion chromatography (IC). The latter has been used with ICP-MS detection to achieve the important separation between inorganic germanium and germanium sesquioxide to control the purity of this drug. The combination of ion chromatography with organic mass spectrometry is possible, although more difficult to achieve due to the limited compatibility of ion chromatographic separation conditions with the requirements of atmospheric pressure ionization sources from molecular mass spectrometry.

Cross-References

▶ Boron, Biologically Active Compounds
▶ Germanium L-Cysteine Alpha-Tocopherol Complex as Stimulator to Antioxidant Defense System
▶ Germanium, Toxicity
▶ Platinum Anticancer Drugs
▶ Rubredoxin, Interaction with Germanium
▶ Silicon, Biologically Active Compounds

References

Brutkiewicz RR, Suzuki F (1987) Biological activities and antitumor mechanism of an immunopotentiating organogermanium compound, Ge-132 (review). In Vivo 1:189–203

Clark JW (1951) Minor metals: Germanium. In: U. S. Bureau of Mines, Minerals Yearbook 1949, U.S. Department of the Interior, Washington, pp 1311–1313

Gerber GB, Léonard A (1997) Mutagenicity, carcinogenicity and teratogenicity of germanium compounds. Mutat Res Rev Mutat Res 387:141–146

Haller EE (2006) Germanium: from its discovery to SiGe devices. Mater Sci Semicond Process 9:408–422

Horneber M, Wolf E, CAM-Cancer Consortium (2011) Propagermanium. Available online: http://www.cam-cancer.org/CAM-Summaries/Dietary-approaches/Propagermanium. Accessed 1 Aug 2012

Kaplan BJ, Parish WW, Andrus GM, Simpson JS, Field CJ (2004) Germane facts about germanium sesquioxide: I. Chemistry and anticancer properties. J Altern Complement Med 10:337–344

Lukevics E, Ignatovich L (2002) Biological activity of organogermanium compounds. In: Rappoport Z (ed) The chemistry of organic germanium, tin and lead compounds, Wiley, Chichester, pp 1653–1683

Rosenberg E (2009) Germanium: environmental occurrence, importance and speciation. Rev Environ Sci Biotechnol 8:29–57

Satge J (2004) Some applications of germanium and its derivatives. Main Group Met Chem 27:301–307

Scoyer J, Guislain H, Wolf HU (2000) Germanium and germanium compounds. In: Ullmann's encyclopedia of industrial chemistry, vol 16. Wiley-VCH, Weinheim, pp 629–641. doi:10.1002/14356007.a12_351

Shangguan G, Xing F, Qu X, Mao J, Zhao D, Zhao X, Ren J (2005) DNA binding specificity and cytotoxicity of novel antitumor agent Ge132 derivatives. Bioorg Med Chem Lett 15:2962–2965

Tao SH, Bolger PM (1997) Hazard assessment of germanium supplements. Regul Toxicol Pharmacol 25:211–219

Voronkov MG, Abzaeva KA (2002) Genesis and evolution in the chemistry of organogermanium, organotin and organolead compounds. In: Rappoport Z (ed) The chemistry of organic germanium, tin and lead compounds, vol 2. Wiley, New York, pp 1–130

Winkler CA (1886) Germanium, Ge, a new nonmetallic element. Chem Ber 19:210–211

Gla: Γ-Carboxyglutamic Acid

▶ Blood Clotting Proteins

Glucose Tolerance Factor

▶ Chromium and Diabetes
▶ Chromium and Glucose Tolerance Factor
▶ Chromium and Membrane Cholesterol

Glucose Transporters

▶ Sodium/Glucose Co-transporters, Structure and Function

Glutamatic Acid: Glutamate, Glu, E

▶ Magnesium Binding Sites in Proteins

Glutathione Peroxidase, GPx

▶ Selenoproteins and Thyroid Gland

Glutathione, Gamma-Glutamyl-Cysteinyl-Glycine, GSH

▶ Magnesium in Health and Disease

Glycerol Channel Fps1

▶ Arsenic and Yeast Aquaglyceroporin

Glycerol Facilitator Fps1

▶ Arsenic and Yeast Aquaglyceroporin

Gold

▶ Gold Nanomaterials as Prospective Metal-based Delivery Systems for Cancer Treatment

Gold and Nucleic Acids

Eugene Kryachko
Bogolyubov Institute for Theoretical Physics, Kiev, Ukraine

Synonyms

Apoferritin, Activation by Gold, Silver, and Platinum Nanoparticles; Biomineralization of gold nanoparticles from gold complexes in cupriavidus metallidurans CH34; Colloidal silver nanoparticles and bovine serum albumin; Gold complexes as prospective metal-based anticancer drugs; Gold nanomaterials as prospective metal-based delivery systems for cancer treatment; Gold nanoparticle platform for protein-protein interactions and drug discovery; Gold nanoparticles and proteins, interaction; Gold, ultrasmall nanoclusters and proteins, interaction; Mercury and DNA

Definition

Gold is one of the seven "coinage" metals that the ancients related to certain gods and to certain stellar objects as well. For instance, gold was naturally linked to the Sun due to its bright yellow. Alchemists even considered gold as "condensed sunbeams" (Newton (1951)) and gave gold the symbol of a circle, the hallmark of mathematical perfection.

Alchemy aimed at transmuting basic metals into gold with the help of the hypothetical substance called the Philosophers' Stone. It emerged from the encounter of the mysticism and practical arts of Egypt (Alexandria) with the rational spirit of the Greek philosophers (see, e.g., Pearsall (1986) and references therein). Among the "coinage" metals, silver was linked to the Moon, iron to Mars, copper to Venus, tin to Jupiter, lead to Saturn, and mercury obviously to Mercury. The 7 days of the week were also associated in the same way with these seven metals: Sunday to gold, Monday to silver, etc.

One of the greatest alchemists, Zosimos of Panopolis, who, as reported, wrote 28 books on alchemy, alludes to a tincture that transforms silver into gold. The famous Arabian chemist Geber (Abu Musa Jabir ibn Hayyan) thought that gold is composed of sulfur and mercury. On the other hand, Robert Boyle (1627–1691), who is likely to be the first to have introduced the concept of a cluster: "There are Clusters wherein the Particles stick not so close together" (Boyle (1661), p. 153), believed that sulfur and mercury cannot be extracted from gold: "I can easily enough sublime Gold into the form of red Crystals of a considerable length; and many other ways may Gold be disguised and help to constitute Bodies of very different Natures both from It and from one another, Yellow, Fixt, Ponderous and Malleable Gold it was before its commixture" (Boyle (1661), p. 429).

The word "gold" is believed to have been derived from Sanskrit and means "to shine" (Newton (1951)). The atom of gold, Au, has the atomic number 79 in the Periodic Table of the Elements with an atomic mass equal to 196.96654 a.u.. Its ground-state electronic configuration is $(Xe)4f^{14}5d^{10}6s^{1}$. Gold has always been considered as the noblest atom. What is the origin of its *nobleness*? It might primarily be a historical reason: gold was found in some river sands in its native metallic form and attracted man's attention due to its color and luster and its resilience to tarnish and corrosion (Newton (1951)). In his Natural History (Translation by Bostock and Riley, Bonn, 1857; Book 33, Chapter 19), Pliny emphasized the latter property of gold by writing that "Those persons are manifestly in error who think that it is the resemblance of its colour to the stars that is so prized in gold." Gold was mined in more than 95 sites (such as, e.g., Um Rus, Barramiya, Samut, Hamash, El Sid, Atud, El Sukari) in the basement rocks of the Eastern Desert of Egypt, under the dynasty of the Roman rulers of Egypt. Gold is not subject to rust, verdigris, or emanation. In addition, gold steadily resists the corrosive action of salt and vinegar. This was probably the reason why gold was used as a medium of exchange for nearly 3,000 years. The first gold coins appeared in Lydia in 700 B.C.

The word "gold" has been used in many different fields, mostly to evoke perfection, like, for example, the famous "gold section" in geometry. In chemistry, the concept of a "gold number" is used to represent the protective action of colloid (Partington (1964), p. 738). This concept was introduced by R. A. Zsigmondy (Z. Anal. Chem. XL, 697 (1901)) after the works by J. B. Richter and M. Faraday (see, e.g., Phil. Trans. CXLVII, 145 (1857)) who recognized the protective action of gelatin on the precipitation of colloidal gold. They found that colloidal gold contains finely divided metallic gold and observed the light scattering by them, the so called "Tyndall effect" (Partington (1964), p. 729).

Another side of the *nobleness* of gold has been discovered in the last two decades, though gold clusters were apparently first studied experimentally in the mid-seventies (Louis 2012).

Surprisingly High Catalytic Activity of Gold Nanoclusters

Gold is the least reactive of the coinage metals. In its bulk form, gold is essentially inert and has a yellowish color. Gold has for a long time been considered as a poor catalyst (see, e.g., Kryachko and Remacle (2006)). However, in the 1980s Haruta (see Haruta (1997) and references therein) discovered that oxide supported by nanosized gold particles, particularly gold, dispersed on various metal oxides as well as nanosized islands on titania oxide, demonstrate a surprisingly high catalytic activity that make them very well suited for use as chemical catalysts in many reactions like combustion of hydrocarbons, reduction of nitrogen oxide, propylene epoxidation, and low-temperature oxidation of carbon monoxide (for review see Kryachko and Remacle (2006)).

Gold exhibits very strong relativistic effects (Pyykkö (1988)) and this is likely to be a key reason for the puzzling and unique behavior or "misbehavior" (Remacle and Kryachko (2005b)) of gold clusters which can be summarized as:

1. Neutral and anionic gold clusters favor 2D structures to unusually large sizes (Remacle and Kryachko (2005b); see also Berry and Smirnov (2011)): for Au_n^- the 2D-3D transition occurs for n = 12, 13, 14 (see Kryachko and Remacle (2010)); the size threshold for the 2D-3D coexistence is lower, n = 7, for cationic than neutral gold clusters: the 2D-3D coexistence develops for Au_5^+ and Au_7^+on the cationic potential energy surfaces while only for Au_9 on the neutral.
2. Small gold clusters exhibit gas-phase and catalytic reactivity with, for example, O_2 (Mills et al. (2003)) and CO (Wallace and Whetten (2000)), which is primarily due to the existence of a band gap.
3. Gold possesses a high activation barrier for the H_2 chemisorption (Hammer and Nørskov (1995)).
4. Finite gold clusters are a very common building "bricks" for nanostructured materials, electronic devices, and novel nanocatalytic systems.
5. Neutral gold clusters do not follow the typical patterns inherent to Lennard-Jones, Morse, or Gupta systems. For example, the ground-state cluster Au_6 does not have an octahedral structure in the gas phase, although it takes such a geometry for MgO-supported gold nanoclusters.

Golden Fullerenes

For larger $n > 12$, gold clusters can possess, in addition to the icosahedral structure, tetrahedral, cage-like, and tubular structures. The discovery, in the 1980s, of the buckyball C_{60} and larger fullerenes (Kroto et al. (1985)) enables to form, by trapping guest atoms, ions, and molecules into their nanosized voids or nanovoids, so-called endohedral or @-fullerenes which manifest a variety of remarkable features and raised the question of existence of similar fullerene- or cage-like structures or hollow cages for other chemical elements. Among the latter, gold is a particular element – definitely "noblesse oblige" – because of its rather unique properties which are mostly dictated by the strong relativistic effects (Pyykkö (1988)) which tend to stabilize the 6*s* orbital and, on the contrary, destabilize the 5*d* one, and thus favor their hybridization.

The first fullerene-type cluster of gold – a "golden fullerene" – is the smallest hollow icosahedral cage $Au_{12}(I_h)$ that was predicted in 2002 and is only stable endohedrally, that is, as $M@Au_{12}$ with a metal atom M = W, V, Nb, and Ta (see, e.g., Kryachko and Remacle (2010, 2011)). The cage Au_{14} is similar: it is bound as the endohedral golden fullerene $M@Au_{14}$ and unbound in the gas phase (see, e.g., Kryachko and Remacle (2010, 2011)). Larger hollow cages $Au_{N=16-18}$ in the anionic charge state have recently been observed in photoelectron spectroscopy experiments (see, e.g., Kryachko and Remacle (2010)). Their voids are characterized by a diameter that determines a spatial confinement of ca. 5.5 Å, that is, these voids are large enough to accommodate some guest atoms. The concept of golden fullerenes has recently been extended to $N > 18$ and already reached 55 (see, e.g., Kryachko and Remacle (2010)). Among them is, for instance, the ground-state and highly stable icosahedral golden fullerene $Au_{32}(I_h)$ whose HOMO-LUMO gap $\Delta := \varepsilon_{LUMO} - \varepsilon_{HOMO}$, defined as the difference of the energy eigenvalues of the HOMO

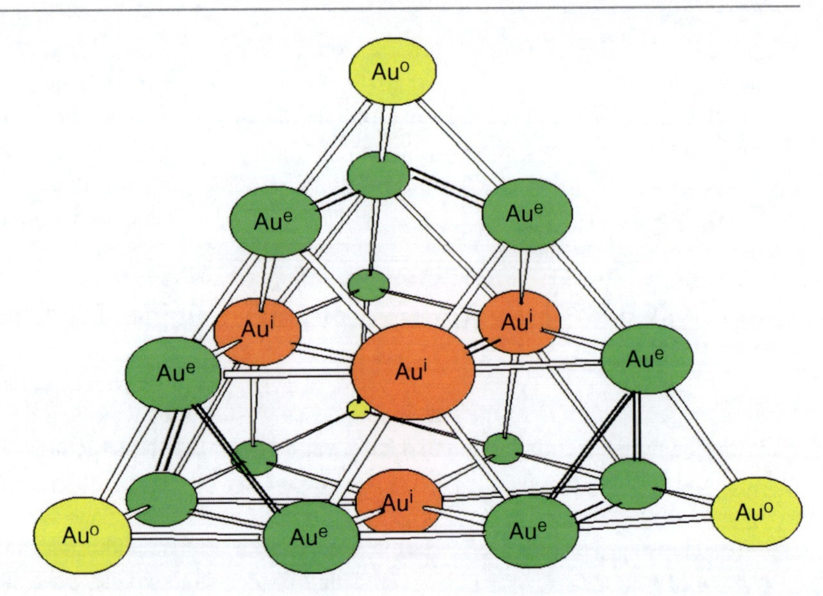

Gold and Nucleic Acids, Fig. 1 The ground-state gold $Au_{20}(T_d)$ cluster with a 20-vertex tetrahedron structure. Its four 3-coordinated vertex atoms Au^o are indicated by the yellow circles, twelve 6-coordinated edge atoms Au^e by the green, and four 9-coordinated inner atoms Au^i by the brown ones. The calculated bond angle $\angle Au^oAu^oAu^o = 60.0°$ (Copyright (2012) Wiley. Used with permission from Figure 1 of Kryachko and Remacle (2007))

(highest occupied molecular orbital) and of the LUMO (lowest unoccupied molecular orbital), falls into the interval of 1.5–2.5 eV, being thus close to $\Delta \in$ (1.57,1.80 eV) of C_{60} (see, e.g., Kryachko and Remacle (2007)). Generally speaking, an arbitrary 3D molecular structure is either space filled, compact, without any void or possesses some voids (emptiness) that result in a hollow cage shape. The space-filled structures, such as the magic gold cluster $Au_{20}(T_d)$ (Kryachko and Remacle (2007)), shown in Fig. 1, are expected to be more energetically stable, at least in the neutral charge state. The 20-nanogold cluster Au_{20} exhibits a large variety of 2D and 3D isomeric forms. A peculiarity of the Au_{20} cluster originates from that is referred to as a so-called magic cluster, characterized by an unusual stability; a large LUMO-HOMO energy gap of 1.77 eV, and for this reason may be perceived as "superatoms"; and a high ionization energy, and should be, by all canons of the jellium model, highly chemically inert.

The Concept of Void Reactivity

When molecules interact with one another, chemical bonding patterns arise between their atoms (Kryachko and Remacle (2010)). Typically, these patterns are spread over the outer space of each interacting partner. In this sense, cages are different – by definition, they possess atoms which are assumed to be capable of forming chemical bonds with molecules from the outer space as well as from the inner one (inside the cage, in voids), that is, with those which are encapsulated or confined within the cage. This hence implies a bifunctionality of the chemical reactivity – the outer or exo-reactivity and the inner, void, or endo-reactivity – of some atoms which compose cages and thus suggests potential routes to control both kinds of chemical reactivities. The features, such as the ionization energies and electron affinities that indicate how a given cage overall reorganizes either upon removal or upon addition of an electron, which are examined in the previous subsection are those which are usually invoked to characterize chemical reactivity. They are, however, the global characteristics of a cage that cannot be partitioned into those which may be solely ascribed either to the outer or to the inner void.

The void regions of the isomeric forms of 20-nanogold cluster Au_{20} defined as cage I, II, and III (Kryachko and Remacle (2010)) are in fact spatially confined areas which can accommodate some heteroatom M or heteroatoms (dopant). A remarkable feature of these cages is that they all have only one-atom-layer that separates the void from the outer surface which may thus facilitate the direct control of the outer reactivity from the inner one. A typical point of view that often prevails is that the void reactivity of a cage is a direct consequence of the spatial confinement, that is, the dopant feels the cage's boundary, and therefore, the size of dopant plays a decisive role for a stable encapsulation. Put in other words, if the size of the void is of the same order of magnitude or comparable with that

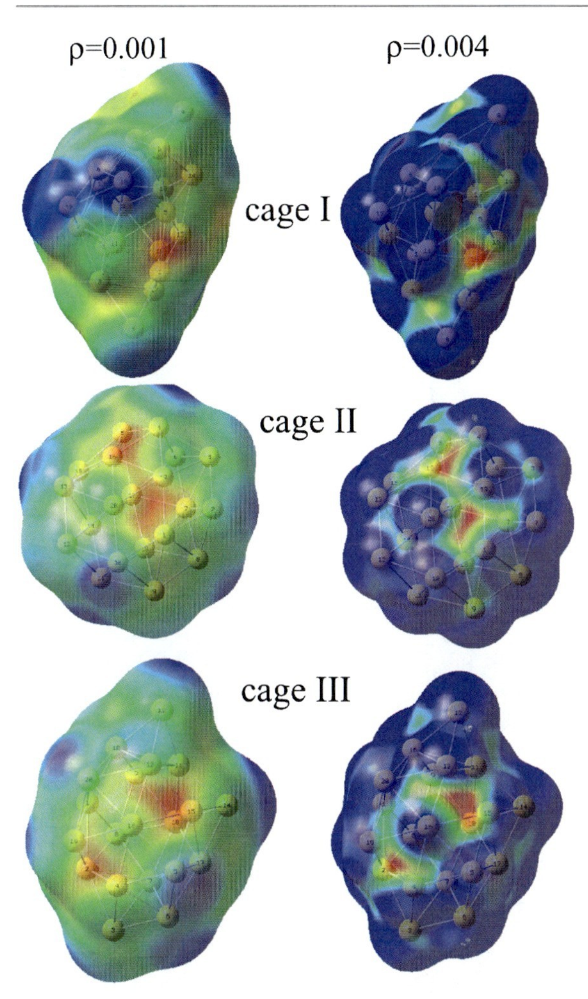

Gold and Nucleic Acids, Fig. 2 The MEPs of cage I, II, and III mapped from −0.016 (*red*) to +0.016 (*blue*) $|e|/(4\pi\varepsilon_0 a_0)$ onto 0.001 $|e|\cdot Å^{-3}$ isosurface of the one-electron density ρ(r) and from −0.01 to +0.01 $|e|/(4\pi\varepsilon_0 a_0)$ onto 0.004 (the *right* column, correspondingly, for cages I, II, and III, respectively) $|e|\cdot Å^{-3}$ isosurface of the one-electron density ρ(**r**). ρ(**r**) is computed at the B3LYP computational level (Copyright (2012) Wiley. Used with permission from Figure 2 of Kryachko and Remacle (2010))

of the dopant, one may anticipate that doping is stable. Obviously, doping influences the outer reactivity. It is well known that the patterns of the molecular orbitals, particularly of the HOMO and LUMO, and the concept of the molecular electrostatic potential (MEP) are crucial for understanding of both facets of chemical reactivity. For example, the buckyball fullerene C_{60} exhibits two different behaviors of the MEP. It is positive in the entire void region, where C_{60} is capable to encage atoms, ions, and some molecules, and reaches the outer central regions of the pentagon rings. On the contrary, it is negative in the outer region and the most negative at the midpoints of the bonds linked two hexagonal rings.

The MEP patterns of the studied golden fullerenes I, II, and III are plotted in Fig. 2.

Gold Nanoparticles and Nanotechnology

The "bottom-up" strategy in molecular electronics and biosensor technology often utilizes biohybrid complexes where biological molecules, DNA and peptides in particular, serve as templates. It is therefore of current interest to assemble the DNA molecule directly on gold nanoparticles. In addition, the mechanism of interaction between the DNA and gold is by itself an important issue, both theoretically and experimentally. Recent experimental studies showed that DNA bases, adenine (A), thymine (T), guanine (G), and cytosine (C) interact with Au surfaces in a specific and sequence-dependent manner. The relative binding affinities of these nucleobases for adsorption on polycrystalline Au films obey the following order: $A > C \geq G > T$ (see, e.g., Kryachko (2009)). The computational models of the interaction of DNA bases with bare Au_3 and Au_4 gold clusters as simple catalytic models of Au particles are demonstrated in Fig. 3 (Remacle and Kryachko (2005a, b); Kryachko (2009)).

Therefore, as follows from Fig. 3, two key bonding ingredients underlie the base-gold and base pair-gold hybridizations: the *anchoring*, either of the Au-N or Au-O type, and the *nonconventional* N-H···Au *hydrogen bonding*. The former is the leading bonding factor and results into stronger binding and coplanar coordination when the ring nitrogen atoms of the nucleobases are involved. The anchor bond predetermines the formation of the nonconventional H-bonding via prearranging the charge distribution within the entire interacting system and "galvanizing" an unanchored atom of the gold cluster to act as a nonconventional proton acceptor, through its lone-pair-like $5d_{\pm 2}$ and $6s$ orbitals. Solid computational evidence has been provided to show that the nonconventional hydrogen bonding is of a new, nonconventional type, and that it sustains and even reinforces the anchoring bond by a cooperative 'back donation' mechanism.

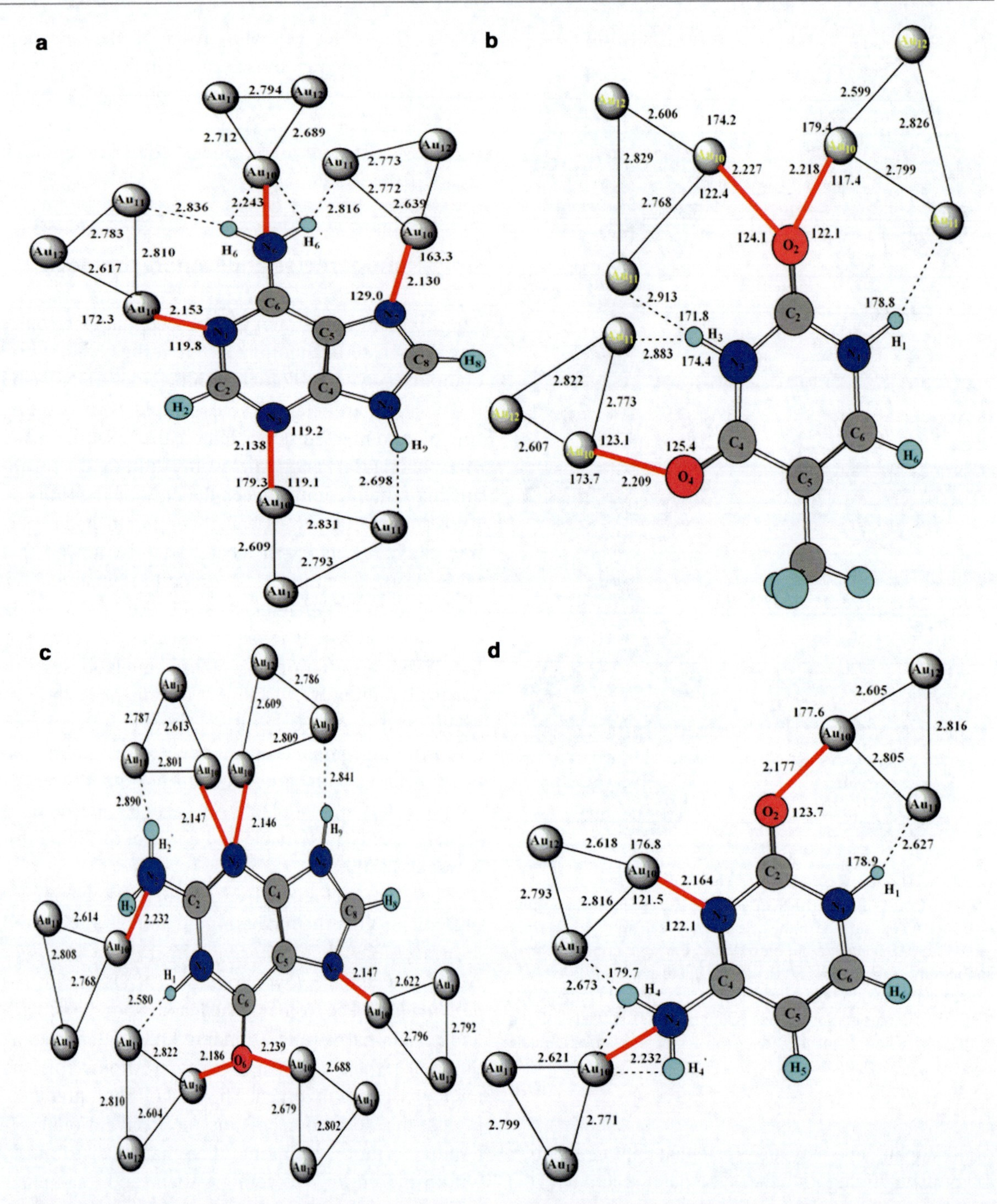

Gold and Nucleic Acids, Fig. 3 (**a**) The four possible planar (N_1, N_3, N_7) and nonplanar (N_6) binding sites of the gold cluster Au_3 to adenine. Also shown is the NH_2 anchored complex $A \cdot Au_3$ (N_6). (**b**) The three possible planar ($O_2(N_1)$, $O_2(N_3)$, O_4) binding sites of the gold cluster Au_3 to thymine. (**c**) The six possible planar ($N_3(N_2)$, $N_3(N_9)$, $O_6(N_1)$, $O_6(N_7)$, N_7) and nonplanar(N_2) binding sites of the gold cluster Au_3 to guanine. The anchoring in N_2 is to the amino group. (**d**) The three possible planar ($O_2(N_1)$, N_3) and nonplanar (N_4) binding sites of the gold cluster Au_3 to cytosine. For the binding site N_4, the anchor bond is to the NH_2 group. For each complex, the anchor bond is drawn as a *thick* (*red*) line and the nonconventional H-bond as a *dotted line*. The bond lengths are given in Å and bond angles in deg (Reprinted with permission from Figures 1–4 of Remacle and Kryachko (2005b). Copyright (2012) The American Chemical Society)

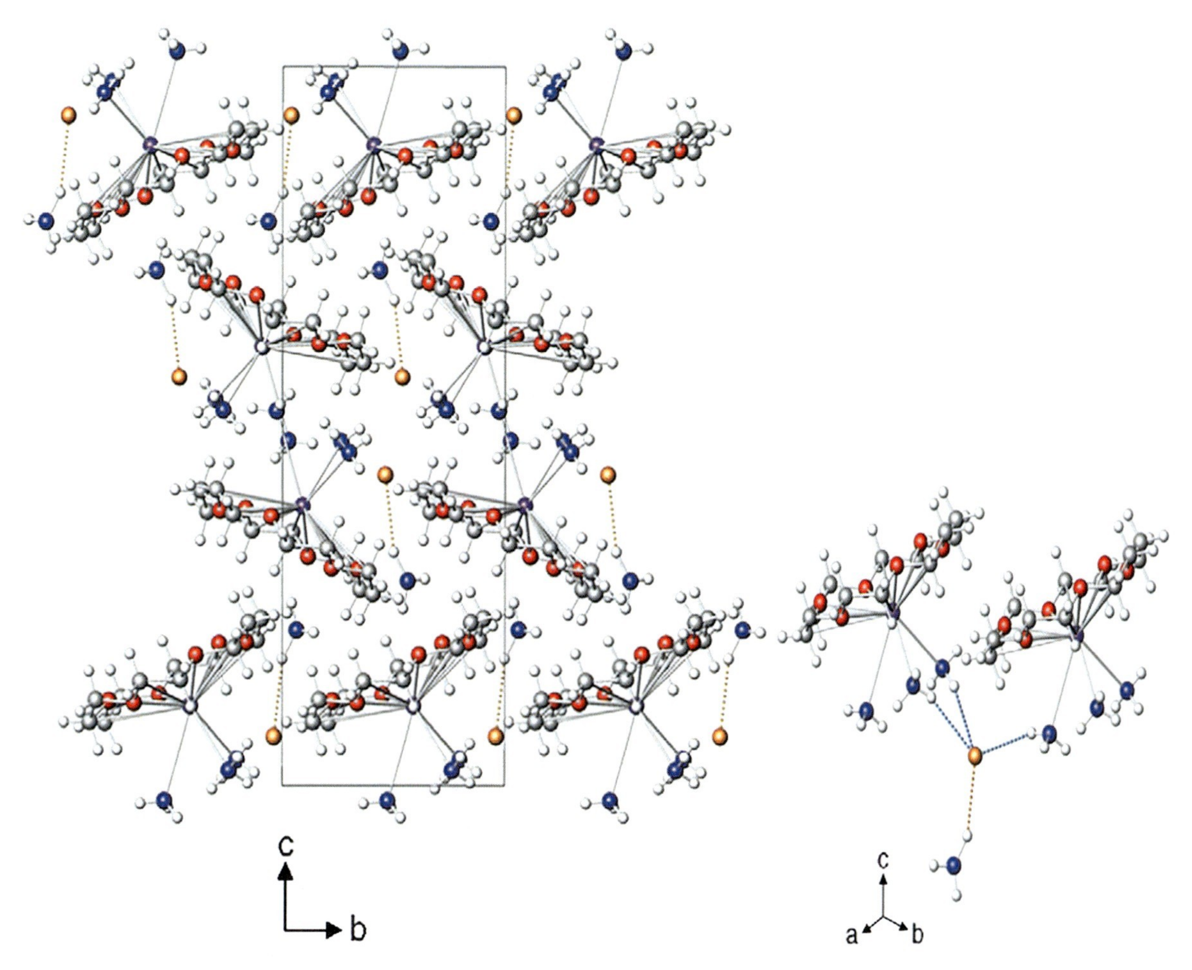

Gold and Nucleic Acids, Fig. 4 *Left* panel: The structure of the crown compound $[Rb([18]crown\text{-}6)(NH_3)_3]Au\text{-}NH_3$: carbon atoms are shown in *gray*, H *white*, N *blue*, O *red*, Au *orange*, Rb *violet*. *Right* panel: Environment of the auride anions in $[Rb([18]crown\text{-}6)(NH_3)_3]Au\text{-}NH_3$ (Copyright (2012) Wiley. Used with permission from Figures 1 and 3 of Nuss and Jansen (2006))

Gold as Nonconventional Hydrogen Bond Acceptor

This feature of gold clusters was discovered in 2005 when it was demonstrated (Kryachko and Remacle (2005)) that the gold clusters Au_n are prone to play, interacting with conventional proton donors such as the O-H and N-H groups, a role of a proton acceptor and hence, to participate in the formation of nonconventional hydrogen bonds. This work showed a strong computational evidence that a triangular gold cluster Au_3 behaves as a proton acceptor with the O-H group of formic acid and with the N-H one of formamide. The bonding involves two ingredients: the anchoring bond between the gold atom and the carbonyl oxygen and the N-H···Au or O-H···Au contacts between Au and the amino group of formamide or the hydroxyl group of formic acid. It was proved (Kryachko and Remacle (2005)) that these contacts share all the common features of the conventional hydrogen bonds and they can therefore be treated as their nonconventional analogs. Since this work, the existence of the X-H···Au_n nonconventional hydrogen bond was computationally demonstrated for a wide variety of molecules in different charge states (Kryachko (2009)): $Z = 0, \pm 1$, ranging from the Au_n-DNA bases and Au_n-DNA duplexes, to $Au_n\text{-}(HF)_m$, $[Au_n^Z\text{-}(H_2O)_m]^Z$, and $[Au_n\text{-}(NH_3)_m]^Z$ complexes. The latter family includes the smallest nanosized tetrahedral gold cluster $Au_{20}^Z(T_d)$. The unveiled charge-state specificity of the bonding ingredients of the $[Au_n\text{-}(NH_3)_m]^Z$ complexes has recently been explored (see, e.g., Kryachko (2009)) to formulate the bonding encoding approach for molecular logic. Recently, on the experimental side, the hydrogen acceptor propensity of the gold atom and some its clusters has been experimentally and computationally examined in the variety of complexes, such as $[Au(H_2O)]^-$-Ar_n, $[Au(H_2O)_{n=1,2}]^-$, $[Au(H_2O)]^-$, the crown compound $[Rb([18]crown\text{-}6)(NH_3)_3]Au\text{-}NH_3$ (Nuss and Jansen (2006)) (see Fig. 4), the complexes of small gold clusters with acetone and with amino acids, the gold (III) antitumor complex, the amine complexes of Au (I), the gold(I) complexes with thione ligands, and $[cytosine\text{-}Au]^-$ and $[uracil\text{-}Au]^-$.

References

Berry RS, Smirnov BM (2011) Charge separation in CO oxidation involving supported gold clusters. JETP 140, No.6(12):1043–1050

Boyle R (1661) The sceptical chymist: or chymico-physical doubts & paradoxes, touching the spagyrist's principles commonly call'd hypostatical, as they are wont to be propos'd and defended by the generality of alchymists (Reprinted in the Everyman Series), London

Hammer B, Nørskov JK (1995) Why gold is the noblest of all the metals. Nature (London) 376:238–240

Haruta M (1997) Size- and support-dependency in the catalysis of gold. Catal Today 36:153

Kroto HW, Heath JR, O'Brien SC, Curl RF, Smalley RE (1985) C_{60}: buckminsterfullerene. Nature (London) 318:162

Kryachko ES (2009) To nano-biochemistry: picture of interactions of DNA with gold. In: Matta C (ed) Quantum biochemistry. Wiley-VCH, Weinheim, pp 245–306, Ch. 8

Kryachko ES, Remacle F (2005) Three-gold clusters form nonconventional hydrogen bonds O-H···Au and N-H···Au with formamide and formic acid. Chem Phys Lett 404(1):142–149

Kryachko ES, Remacle F (2006) Small gold clusters form nonconventional hydrogen bonds X-H···Au: gold-water clusters as example, Ch 11. In: Torro-Labbe A (ed) Theoretical aspects of chemical reactivity. Vol. 16 of Theoretical and computational chemistry, Politzer P (ed). Elsevier, Amsterdam, pp 219–250

Kryachko ES, Remacle F (2007) The magic gold cluster Au_{20}. Int J Quantum Chem 107(14):2922–2934

Kryachko ES, Remacle F (2010) 20-nanogold $Au_{20}(T_d)$ cluster and its hollow cage isomers: structural and energetic properties. J Phys: Conf Ser 248:012026–1–012026–8

Kryachko ES, Remacle F (2011) 20-Nanogold $Au_{20}(T_d)$ and low-energy hollow cages: void reactivity. In: Hoggan P, Maruani J, Piecuch P, Delgado-Barrio G, Brändas EJ (eds) Advances in the theory of quantum systems in chemistry and physics, vol 22, Progress in theoretical chemistry and physics. Springer, Berlin, pp 573–612, Ch. 30

Louis C, Pluchery O (2012) Gold nanoparticles for physics, chemistry and biology. World Scientific.

Mills G, Gordon MS, Metiu H (2003) Oxygen adsorption on Au clusters and a rough Au(111) surface: the role of surface flatness, electron confinement, excess electrons, and band gap. J Chem Phys 118:4198

Newton J (1951) Friend, man and the chemical elements from stone-age hearth to the cyclotron. Ch. Griffin, London

Nuss H, Jansen M (2006) Cs5([12]crown-4)2(O3)5: a supramolecular compound containing the confined ozonide partial structure $^1_\infty$\{Cs8(O3)10\}. Angew Chem Int Ed 45:4369

Partington JR (1964) A history of chemistry, vol 4. Macmillan, London, 4

Pearsall R (1986) The alchemists. Weidenfeld and Nicolson, London

Pyykkö P (1988) Chem Rev 88:563

Remacle F, Kryachko ES (2005a) Complexes of DNA bases and gold clusters Au_3 and Au_4 involving nonconventional N-H···Au hydrogen bonding. Nano Lett 5:735–739

Remacle F, Kryachko ES (2005b) Structure and energetics of two- and three-dimensional neutral, cationic and anionic gold clusters $Au_{5 \le n \le 8}{}^Z$ (Z = 0, ±1). J Chem Phys 122(4):044304–1–0443041–4

Wallace WT, Whetten RL (2000) Carbon monoxide adsorption on selected gold clusters: highly size-dependent activity and saturation compositions. J Phys Chem B 104:10964

Gold Biomineralization in Bacterium Cupriavidus Metallidurans

Carla M. Zammit and Frank Reith
School of Earth and Environmental Sciences, The University of Adelaide, Centre of Tectonics, Resources and Exploration (TRaX) Adelaide, Urrbrae, South Australia, Australia
CSIRO Land and Water, Environmental Biogeochemistry, PMB2, Glen Osmond, Urrbrae, South Australia, Australia

Synonyms

Biomineralization of gold nanoparticles from gold complexes in *Cupriavidus metallidurans* CH34; Interactions of *Cupriavidus metallidurans* CH34 with mobile, ionic gold

Definition

Cupriavidus metallidurans CH34 is a facultative chemolithoautotrophic β-proteobacterium. It facilitates resistance to more than 20 mobile heavy metals via chromosome- and megaplasmid-encoded metal resistance factors. *C. metallidurans* CH34 occurring on gold nuggets mediates the (trans)formation of gold in placer environments. This is due to the ability of *C. metallidurans* CH34 to biomineralize gold via energy-dependent reductive precipitation of highly toxic gold complexes.

Historical Background

Cupriavidus (formerly also known as *Wautersia, Ralstonia*, or *Alcaligenes*) *metallidurans* CH34 was originally isolated from a decantation tank at a metal processing factory in Belgium (Mergeay et al. 1985). It serves as a model organism for the study of metal resistance in microorganisms due to its ability to withstand millimolar concentrations of more than 20 different heavy metal ions. *Cupriavidus metallidurans* CH34 is a facultative chemolithoautotrophic β-proteobacterium that contains two chromosomes and two megaplasmids (Mergeay et al. 1985). Its metal resistance factors are located throughout the genome on both chromosomes and the megaplasmids. These resistance genes have been acquired, rearranged, and recombined from numerous sources, amounting to the many genetic alterations that give this bacterium the ability to adapt and persist in environments high in mobile heavy metals (Janssen et al. 2010).

This ability to adapt to high metal concentrations has made it possible for *Cupriavidus* species to persist in some of the most hostile environments on Earth, e.g., it has been isolated from a range of industrial sites across Belgium, Germany, and the Congo. Species from the genus *Cupriavidus* and its close relative *Ralstonia* have also been detected in ultra-clean spacecraft assembly rooms, spacecraft atmosphere, and spent nuclear fuel pools. Besides anthropogenic contaminated environments, three *Cupriavidus* species have been isolated from volcanic mudflow deposits from the Pinatubo volcano in the Philippines. Additionally, *Cupriavidus* species have also been isolated from the surface of gold grains in Australia (Fig. 1; Reith et al. 2006).

Gold Biomineralization in Bacterium Cupriavidus Metallidurans, Fig. 1 Gold nugget from Western Australia

Metal-Resistant Systems in *C. metallidurans* CH34

Metals make up more than 75% of all elements, and since the industrial revolution, they have undergone drastic redistribution in Earth surface environments. Some metals are essential to microbial life, others are used to gain cellular energy; however, all metals, above threshold concentrations, are toxic to microorganisms. Therefore, microorganisms have developed biochemical systems to deal with toxic levels of metals. Most of these mechanisms are based on changing the speciation of metals via active/passive oxidation/reduction and complex formation, which affect metal mobility in intra- and extracellular aqueous solutions. Whereas passive mechanisms, e.g., passive extracellular biomineralization, involve the interactions between metals with different components of cells, active detoxification and biomineralization are commonly mediated by metal-binding (poly)-peptides and proteins.

To date, *C. metallidurans* CH34 has been identified as the microorganism containing the highest number of genes involved in heavy metals resistance (Mergeay et al. 2003; Monsieurs et al. 2011). Metal resistance in *C. metallidurans* CH34 includes a complex and highly diverse range of mechanisms conveying resistance to Ag^{+}, AsO^{-}, Au^{+}, Au^{3+}, Bi^{3+}, Co^{2+}, Cd^{2+}, CrO_4^{2-}, Cs^{+}, Cu^{+}, Cu^{2+}, $HAsO_4^{2-}$, Hg^{2+}, Ni^{2+}, Pb^{2+}, SeO_3^{2-}, SeO_4^{2-}, Sr^{2+}, Tl^{+}, and Zn^{2+} (Janssen et al. 2010). Many of the active mechanisms that microorganisms in general and *C. metallidurans* CH34 specifically have in place to deal with elevated levels of metals are not specific to one particular metal, but convey resistance to different metals. For example, in a recent study, the *copH* gene in *C. metallidurans* CH34 was upregulated by 13 out of 16 different metal ions (Monsieurs et al. 2011). In this study,

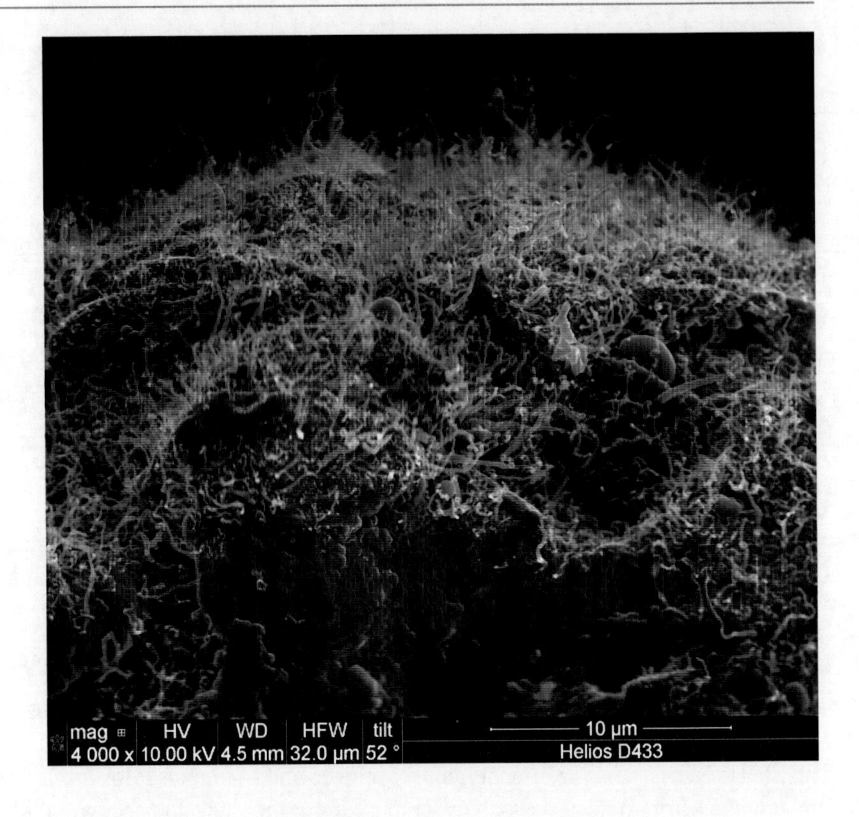

Gold Biomineralization in Bacterium Cupriavidus Metallidurans, Fig. 2 Secondary scanning electron micrograph of a biofilm growing on the surface of a gold nugget from the Flinders Ranges, South Australia

more than 43% of upregulated metal-response genes could be classified as "metal specific," varying between 4% metal-specific response genes for Sr^{2+} and 57% metal-specific response genes for Ni^{2+} (Monsieurs et al. 2011). *Cupriavidus metallidurans* CH34 was shown to have a similar transcriptomic response to metals which clustered in the following categories: Group I – Cd^{2+}, Pb^{2+}, and Zn^{2+}; Group II – Ag^{1+}, Au^{3+}, Cs^{1+}, Hg^{2+}, Sr^{2+}, and Tl^{+}; and Group III – $CrO_4{}^{2+}$, As^{3+}, and Mn^{2+} (Monsieurs et al. 2011). These groupings appeared to not be the result of the physicochemical properties of the metals, but rather to be a function of the location of specific metal resistance genes on chromosomes and plasmids. Having these types of cross-reactive transcriptional networks is highly advantageous for *C. metallidurans* CH34, because it allows the organisms to maintain metal homeostasis when exposed to elevated levels of a range of metals, while being able to conserve metabolic energy required for the production of specific proteins. For example, when *C. metallidurans* CH34 was exposed to equal concentrations of Co^{2+}, Ni^{2+}, Cu^{2+}, Zn^{2+}, and Cd^{2+} combined, it was shown that the metal content within the cell was the same as if only a single metal had been added (Kirsten et al. 2011).

Interactions of *C. metallidurans* CH34 and Gold

The interaction between *C. metallidurans* CH34 and gold has attracted much interest among researchers, because *C. metallidurans* CH34 was shown to be involved in the transformation of gold nuggets (Fig. 2). Gold is a Group IB metal that occurs in aqueous solutions under surface conditions as metallic gold (Au0) as well as the highly toxic aurous, Au(I), and auric, Au(III), complexes. Hence, along with its concentration, its oxidative state and type of complex, as determined by the available ligands, play an important role in its toxicity to cells. Because of their high affinity for organic ligands, Au(III)-complexes are rapidly absorbed by bacterial cells (Reith et al. 2007). Once in contact with a cell membrane, Au(III) is transformed to Au(I), which is taken up into the cytoplasm (Reith et al. 2009). Here it may disturb the redox equilibrium, increase cell permeability, or disrupt metabolic pathways. However, microorganisms such as *C. metallidurans* CH34 and *Salmonella typhimurium* have developed systems to deal with these increased levels of toxic gold (Checa et al. 2007; Reith et al. 2009).

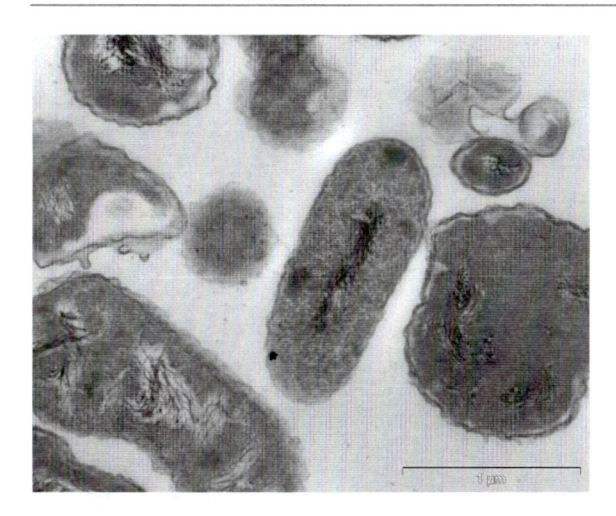

Gold Biomineralization in Bacterium Cupriavidus Metallidurans, Fig. 3 Transmission electron micrograph of an ultrathin section of a *C. metallidurans* cells harboring nanoparticulate gold (*arrow*) in the periplasm

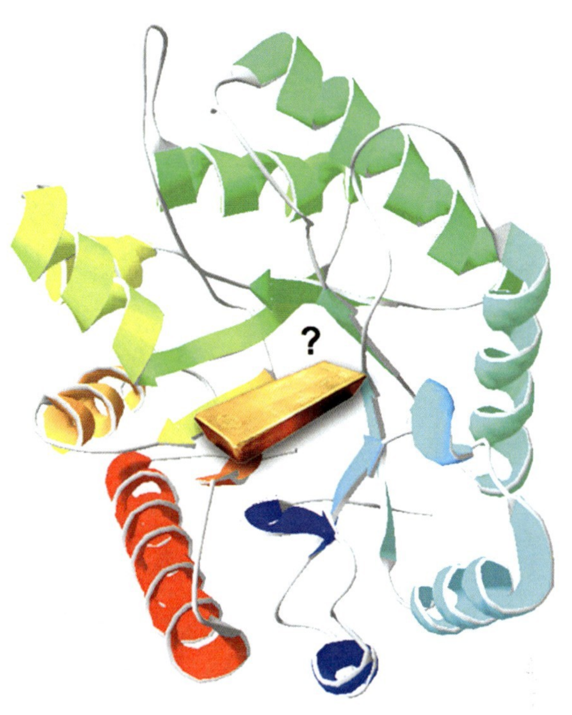

Gold Biomineralization in Bacterium Cupriavidus Metallidurans, Fig. 4 Protein model of Rmet_4684, a gene predicted to encode for a hypothetical protein possibly involved in gold biomineralization by *C. metallidurans*

Gold-Resistant Systems

Two mechanisms employed by microorganisms to tolerate gold have been postulated, the first involves the efflux of Au(I)-complexes through metal transporters, the second involves the active precipitation of Au(I)-complexes (Reith et al. 2007). However, as with other metal resistance systems, deciphering the molecular mechanisms involved in Au detoxification is complex. Actively growing *C. metallidurans* CH34 cells rapidly absorb Au(III)-complexes from solution and reduce them to Au(I)-S complexes on the surface of the cell (Reith et al. 2009). Gold is then distributed across the cell surface and upregulation of genes belonging to oxidative stress and metal detoxification occurs, followed by the appearance of metallic Au nanoparticles in the periplasm (Figs. 3 and 4; Reith et al. 2009). The genes that are most strongly upregulated during this process belong to the *gig* (gold-induced genes) operon, which includes genes for the proteins: GigA and GigB, both cytoplasmic proteins; GigP, a putative periplasmic protein; GigT, a protein with three predicted transmembrane spans, related to a subunit of terminal quinol oxidase found in an archaeal membrane (Mohr 2011). These genes are predicted to be important to the movement of Au(III) into the cytoplasm. A MerR family protein, CupR, specifically responds to gold stress in *C. metallidurans* CH34 and could act as a reporter of Au(I) in the cytoplasm. Next Au(I) needs to be transported into the periplasm where it forms Au(0), this could be achieved with P_{IB1}-type ATPases, such as CupA or CopF. However, the employment of these systems is yet to be experimentally validated and with the two-fold upregulation of 52–303 transcripts upon the addition of Au(III), it is unlikely to be so simple (Reith et al. 2009).

Biomineralization of Gold

Organisms are able to accumulate and precipitate minerals under a range of conditions. This process is termed biomineralization and a variety of different minerals are formed via this process. These include calcium carbonates and phosphates in vertebrates, calcium and magnesium carbonates in invertebrates, silicates in algae and diatoms, as well as iron-oxides and copper and gold minerals in microorganisms. Biomineralization contributes to the natural movement of minerals in the environment and there is a growing amount of evidence that certain microorganisms play

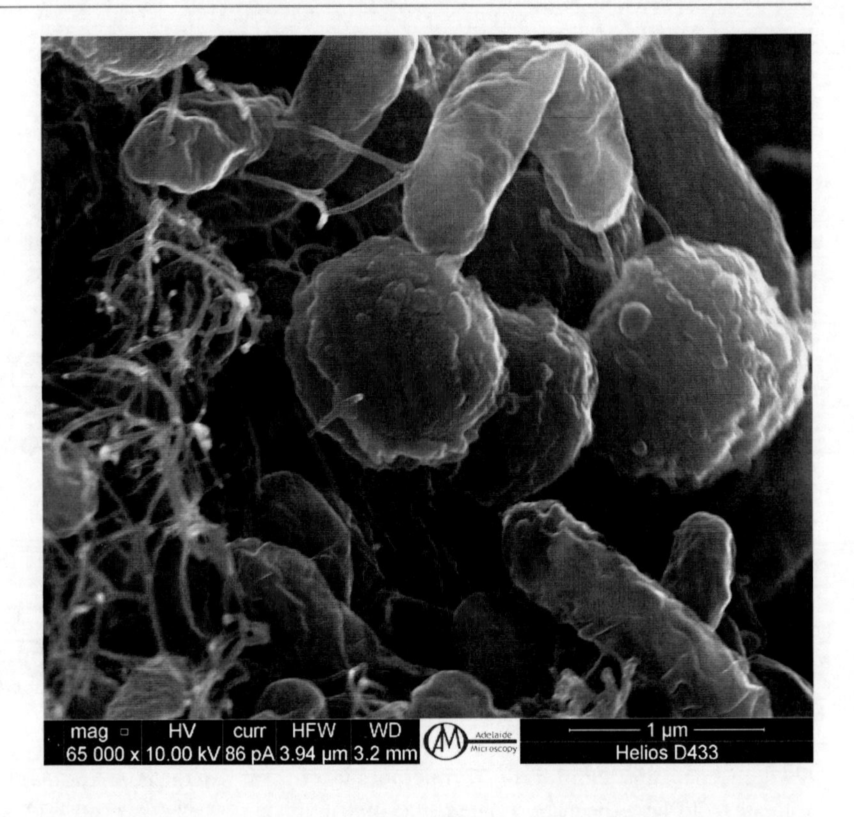

Gold Biomineralization in Bacterium Cupriavidus Metallidurans, Fig. 5 Secondary scanning electron micrograph of spherical gold microparticles formed as a result of biomineralization by *C. metallidurans*

an important role in the movement of gold within the environment, despite gold being seen as an inert substance under surface conditions. *C. metallidurans* cells rapidly accumulate Au(III)-complexes from solution, and promote the biomineralization of Au nanoparticles via energy-dependent reductive precipitation of the highly toxic complexes. These particles can vary in size from a few nanometers to several micrometers (Figs. 3 and 5). The industrial applications of such nanoparticles and microparticles is only just beginning to be explored but includes a broad range of applications, such as bio-imaging, pollution control, water purification, components for technical equipment (e.g., computer components), and uses as catalysts.

Implications for Gold Cycling

It has long been accepted that microorganisms play an important role in the biogeochemical cycling of carbon, oxygen, and nitrogen. The role of microorganisms in the movement of metals within the environment was at first unknown, then overlooked. However, presently, scientists are gaining widespread appreciation for the contribution that microorganisms have in metal solubilization, dispersion, and precipitation. Microorganisms both directly and indirectly influence the movement of metals in the environment. The solubilization of gold occurs via iron and/or reduced inorganic sulfur-oxidizing microorganisms via the production of acid, a by-product of their metabolic reactions; the microbial production of thiosulfate; the sequestration of Au by Au-complexing ligands; the production of cyanide. Once gold is solubilized, it can be distributed in the environment. Finally, the gold is deposited back into the environment via the biomineralization mechanisms discussed above.

Understanding how gold is transported in the environment has many industry-based applications. Technologies derived from fundamental knowledge of how Au is solubilized, dispersed, and precipitated will have a great impact on the mining industry. Mine exploration techniques can be adapted to account for the microbial movement of metals, brining greater accuracy. Processing technologies, based on microbial-metal interactions could be improved (i.e., biooxidation and bioleaching) or further developed (microbially mediated in situ leaching). Bioremediation techniques could be developed for the treatment of metal contaminated sites. The use of nanoparticle technologies is only just in its infancy, and could be greatly explored. In conclusion, although understanding the molecular mechanisms involved in the biomineralization of Au in *C. metallidurans* CH34 is a complex endeavor, the potential applications of such knowledge is wide spanning and of great commercial value.

Cross-References

- ▶ Aluminum, Physical and Chemical Properties
- ▶ Biomineralization of Gold Nanoparticles from Gold Complexes in Cupriavidus Metallidurans CH34
- ▶ Gold and Nucleic Acids
- ▶ Gold Nanomaterials as Prospective Metal-based Delivery Systems for Cancer Treatment
- ▶ Gold Nanoparticle Platform for Protein-Protein Interactions and Drug Discovery
- ▶ Gold Nanoparticles and Fluorescent Proteins, Optically Coupled Hybrid Architectures
- ▶ Gold Nanoparticles and Proteins, Interaction

References

Checa SK, Espariz M, Perez Audero ME et al (2007) Bacterial sensing of and resistance to gold salts. Mol Microbiol 63(5):1307–1318

Janssen PJ, Van Houdt R, Moors H et al (2010) The complete genome sequence of *Cupriavidus metallidurans* strain CH34, a master survivalist in harsh and anthropogenic environments. PLoS One 5(5):e10433

Kirsten A, Herzberg M, Voigt A et al (2011) Contributions of five secondary metal uptake systems to metal homeostasis of *Cupriavidus metallidurans* CH34. J Bacteriol 193(18):4652–4663

Mergeay M, Nies D, Schlegel HG et al (1985) *Alcaligenes eutrophus* CH34 is a facultative chemolithotroph with plasmid-bound resistance to heavy metals. J Bacteriol 162(1):328–334

Mergeay M, Monchy S, Vallaeys T et al (2003) *Ralstonia metallidurans*, a bacterium specifically adapted to toxic metals: towards a catalogue of metal-responsive genes. FEMS Microbiol Rev 27(2–3):385–410

Mohr J (2011) Goldresistenz von *Cupriavidus metallidurans*, Naturwissenschaftlich Fakultaet. Universitate Halle-Wittenberg, Wittenberg

Monsieurs P, Moors H, Van Houdt R et al (2011) Heavy metal resistance in *Cupriavidus metallidurans* CH34 is governed by an intricate transcriptional network. Biometals 24(6):1133–1151

Reith F, Rogers SL, McPhail DC et al (2006) Biomineralization of gold: biofilms on bacterioform gold. Science 313:233–236

Reith F, Lengke MF, Falconer D et al (2007) The geomicrobiology of gold. ISME J 1:567–584

Reith F, Etchmann B, Grosse C (2009) Mechanisms of gold biomineralization in the bacterium *Cupriavidus metallidurans*. PNAS 106(42):17757–17762

Gold Complexes and Their Anticancer Properties

▶ Gold Complexes as Prospective Metal-Based Anticancer Drugs

Gold Complexes as Prospective Metal-Based Anticancer Drugs

Anil Kumar[1,3], Bhargavi M. Boruah[2,3] and Xing-Jie Liang[1]

[1]CAS Key Laboratory for Biomedical Effects of Nanoparticles and Nanosafety, National Center for Nanoscience and Nanotechnology, Chinese Academy of Sciences, Beijing, China

[2]CAS Key Laboratory of Pathogenic Microbiology and Immunology, Institute of Microbiology, Chinese Academy of Sciences, Beijing, China

[3]Graduate University of Chinese Academy of Science, Beijing, China

Synonyms

Gold complexes and their anticancer properties; Gold for health – past and present; Gold physiochemical properties; Mode of action; Overcome cisplatin resistance; Targeted mechanism of gold complex; Toxicity of gold complex

Definition

Gold (*Au*) is a metallic element with an atomic number of 79. It is soft, shiny, dense, as well as an extremely malleable and ductile metal. *Au* is yellow color when in mass but may give black, ruby, or purple color when finely divided. It is a good conductor of electricity and is unaffected by heat, moisture, oxygen, and most reagents except when comes in contact with chlorine, fluorine, and aqua regia. *Au* may be alloyed with various other metals like silver (*Ag*), copper (*Cu*), zinc (*Zn*), nickel (*Ni*), platinum (*Pt*), palladium (*Pd*), tellurium (*Te*), and iron (*Fe*) to give it special properties and various color hues ranging from silver-white to green and orange-red. Due to its relatively uncommon chemical and physical properties, *Au* is indispensible for a wide variety of applications for diagnosis and treatment of cancer. The most advantageous attributes of *Au* is the relative inertness. Listed as one of the "noble" metals, *Au* does not oxidize under normal conditions and thus, have resistance to tarnishing. However, it dissolves in the presence of solutions containing halogens or aqua regia as well as in mixtures of hydrochloric and nitric acids (3:1).

Gold for Health: Past and Present

Besides use as ornaments and as currency, gold (*Au*) has remarkable medical and therapeutic values that have been recognized thousands of years ago. According to the earliest records, use of *Au* for medical purposes can be traced back to the Chinese civilization in 2500 B.C., and after that, several ancient cultures have utilized *Au*-based materials for medicinal purpose for the treatment of a variety of diseases such as smallpox, skin ulcers, measles, fainting, fevers, and falling sickness. The rational use of *Au* in medicine started in the nineteenth century whereby mixture of *Au* chloride and sodium chloride $Na[AuCl_4]$ was used for the treatment of syphilis. Initial research on *Au* complexes could be credited to bacteriologist Robert Koch who reported the use of $K[Au(CN)_2]$ in the treatment of tuberculosis, later replaced by *Au*(I)-thiolated complexes on account of toxicity issues. With the advent of *Au* therapy and following its success in treating rheumatoid arthritis (Fig. 1), *Au* complexes found further applications in the treatment of other rheumatic diseases including psoriatic arthritis, juvenile arthritis, palindromic rheumatism, and discoid lupus erythematosus. With rather more success in recent decades, colloidal *Au* has been used in research applications in medicine, biology, and material sciences. Some *Au*-salts show the anti-inflammatory properties and are used as therapeutics in the treatment of inflammatory skin disorders namely pemphigus, urticaria, and psoriasis. Treatment using *Au*-based drugs is known as chrysotherapy.

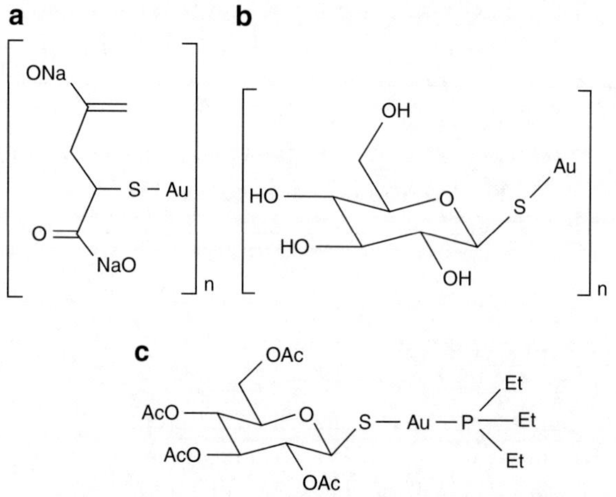

Gold Complexes as Prospective Metal-Based Anticancer Drugs, Fig. 1 Structure of gold(I) complex drugs used for the treatment of rheumatoid arthritis (**a**) sodium aurothiomalate (Myocrisin), (**b**) aurothioglucose (Solganol), and (**c**) tetraacetyl-b-D-thioglucose gold(I) triethyl- phosphine also known as (auranofin) (**a**, Reprinted with permission from Dreaden EC, Mwakwari SC, Sodji QH, Oyelere AK, El-Sayed MA (2009) Tamoxifen−poly(ethylene glycol)−thiol gold nanoparticle conjugates: enhanced potency and selective delivery for breast cancer treatment. Bioconjugate Chem. Copyright (2009 American Chemical Society)

Considerable research has gone into the potential anticancer and antimicrobial activity of *Au* complexes, and the recent decades have seen interesting advances in the pharmacology of *Au*-based drugs on their mechanism of action and biological chemistry related to these findings (Milacic et al. 2008).

Current Cancer Treatment and Challenges

Cancer is one of the most serious health hazard worldwide with about 12 million new cancer cases diagnosed each year which is expected to increase to 25 million per year in 2020 (Cancer Facts and Figures 2012 Annual Publications, *American Cancer Society*). Improvement in understanding, diagnostics, treatment, and prevention has been tremendously achieved through investments in biological and chemical research. With the advancement in cancer surgery, chemotherapy, and radiotherapy, half of the total cancer cases can be cured, while the other half can have the benefit of prolong survival or in worst cases, no survival at all. Cancer chemotherapy was initiated with the discovery of the cytotoxic effects of *N*-mustards (Chabner and Roberts 2005), capable of curbing certain cancer types. Later, ▶ cisplatin (anticancer drug) achieved wide acceptance as a metal-based antineoplastic agent for the treatment of testicular and ovarian cancer, followed by the success of cisplatin analogues like ▶ carboplatin and ▶ oxaliplatin designed on the basis of chemical and biological advantages as anticancer-based drugs. However, undesirable side effects complied with drug resistance have narrowed their applications to treat diverse sorts of cancer, prompting a search for other more effective metal-based drugs (Garza 2008).

Gold Complexes as Potential Anticancer Therapeutics

Metal-based compounds have properties like ligand exchange rates, redox properties, oxidation states, coordination affinities, solubility, biodisponibility, and ▶ biodistribution, which can be tuned for optimal cytotoxic effects with reduced side effects. *Au*-salts are called second-line drugs due to the fact that, besides their use in the treatment of arthritis, *Au* complexes have been found to have potencies to treat cancer. Use of radioactive *Au*-198 dates back to the last century in the treatment of malignancies, which have now been replaced by other radioisotopes, notably iridium-125 and iodine-125, as potent neoplastic suppressants (Sheppard et al. 1947).

Besides *Au*, other metal-based compounds like titanium (*Ti*), germanium (*Ge*), rhodium (*Rh*), rhenium (*Re*), gallium (*Ga*), ruthenium (*Ru*), tin (*Sn*), cobalt (*Co*), and copper (*Cu*) have been studied, and many of them have shown promising results and have even been included in clinical trials. Several *Ru* and *Au* complexes have shown potential applications as anticancer agents, and study of their chemical and biological properties will facilitate the elucidation of a clear structure-activity relationship that in the near future may be used for the design, synthesis, and characterization of more effective anticancer agents with reduced side effects.

Different Oxidation States of Gold and Their Application in Cancer Treatment

Au can exist in multiple oxidation states -I, 0, I, II, III, IV, and V, but only *Au* 0, I. and III are stable in aqueous systems, and, therefore, more applicable for biological environment. In contrast, the oxidation states -I, II, IV, and V are less common, due to their redox properties, stability of I and V states in water is improbable, and hence have limited applications in biological systems. *Au*(I) is thermodynamically more stable than *Au*(III); however, *Au*(I) and *Au*(III) are unstable with respect to *Au*(0) and are readily reduced by mild reducing agents. Many of *Au*(III) complexes are strong oxidizing agents, being reduced to *Au*(I). Nevertheless, considering the body environment as reducing, the appropriate choice of ligand-donor set can stabilize the higher oxidation state of *Au*, which in turn, can control the relative instability, light sensibility, and reduction to metallic *Au* under physiological conditions. Thus, *Au* (III) complexes have been increasingly evaluated for their potential antitumor activity (Garza 2008).

The properties of *Au* complexes for antitumor activity over the past several decades have been based on four underlying principles:

(a) Analogies between square-planar complexes of *Pt*(II) and *Au*(III), both of which are d^8 ions state, which represent the isoelectronic and isostructural differences with that of *Pt*(II) complex.
(b) This analogy explains the immunomodulatory effects of *Au*(I) as potent antiarthritic agents.
(c) Coordination of *Au*(I) and *Au*(III) with known antitumor agents to form new compounds with enhanced activity.
(d) Some *Au*(III) complexes present stability, good enough in physiological environments, thus making them more applicable for this purpose.

Gold(I) Complexes with Anti-Tumor Activity

Auranofin, a *Au*(I) complex, was found to have activity against HeLa cells in vitro and P388 (leukemia cells) in vivo. Subsequent screening led to the discovery of several auranofin analogues (Fig. 1c), but their activity was found to be limited chlorido(triethylphosphane)*Au* (I) (Fig. 2a) showed potent cytotoxic activity in vitro but less antitumor activity in vivo compared to auranofin. *Au*-phosphane complexes such as $[(AuCl)_2\ dppe]$ (*dppe* - bis (diphenylphosphane) ethane) have shown more promising results (Fig. 2b, c) (Berners Price and Filipovska 2011). Another three-coordinate analogues, of *Au*(I) (Fig. 2d), incorporating both mono- and bidentate phosphine ligands, have demonstrated exciting in vitro cytotoxicity against a range of cancer cell lines with formidable potency against the

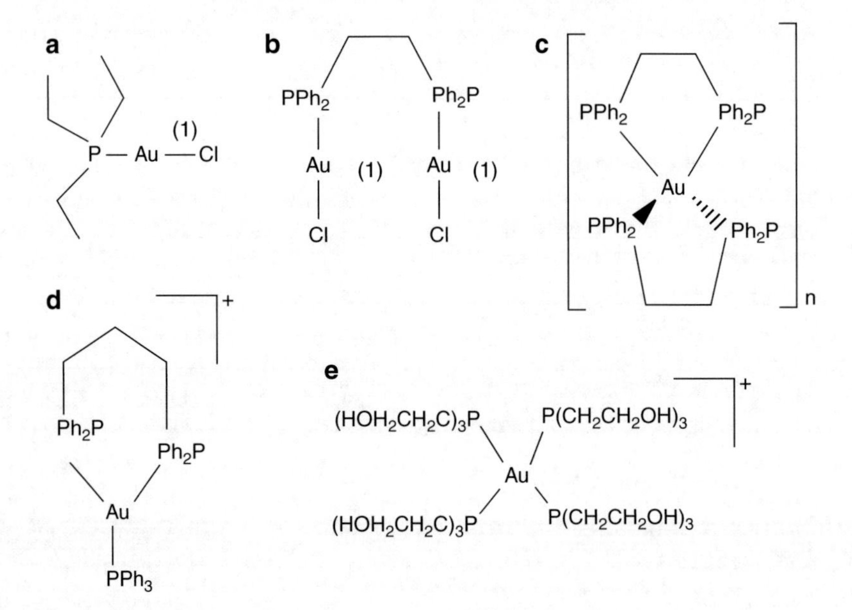

Gold Complexes as Prospective Metal-Based Anticancer Drugs, Fig. 2 Structure of cytotoxic gold(I) compounds (**a**) chlorido(triethylphosphane) gold(I), (**b** and **c**) bis (diphenylphosphane)ethane gold(I) analogues, (**d**) three-coordinate analogues, and (**e**) hydrophilic tetrahedral compound

breast tumor cell line, MCF-7 (Caruso et al. 2003). Similarly, the hydrophilic tetrahedral compound, (Fig. 2e) having all-monodentate phosphine ligands, showed promising activity against a range of prostrate, colon carcinoma, and gastric carcinomas (Pillarsetty et al. 2003).

The *dppe* ligand exhibits antitumor activity by itself, and *Au* protects the ligand from oxidation and aids in the delivery of the active species. Similar experimental results have revealed the use of *Au* as "prodrugs" acting as a platform for delivering anticancer agents, which when administered alone, have fewer efficacies to penetrate into tumors. Such coordination drugs have been reported to alter the normal metabolic pathways and release mechanisms through rearrangement (in solution and biological media) of the diphosphane compounds to produce a rare coordination geometry of *Au*(I) complexes such as $[Au(dppe)_2]Cl$ and $[Au(d_2pypp)_2]Cl$ favoring efficiency for penetration into tumors (Fig. 3).

The mechanism of cytotoxicity mediated by *Au* complexes differs from that of cisplatin, another eminent metal-based anticancer agent. In contrast to cisplatin which targets the DNA, *Au* complexes have the ability to slowdown mitochondrial function and thus inhibit protein synthesis (Krishnamurthy et al. 2008). The effect of structural variation in bis(diphosphane) gold(I) complexes, in general, are active against various types of cancer and kill cells via damage to mitochondria. The coordination of *Au*

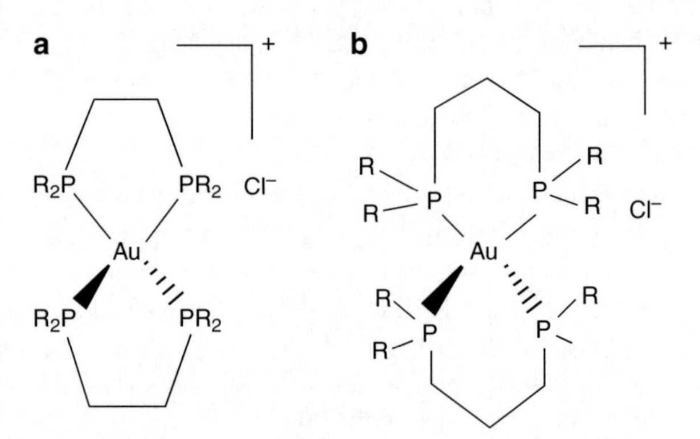

Gold Complexes as Prospective Metal-Based Anticancer Drugs, Fig. 3 Structure of lipophilic cationic gold(I) complex (**a**) $[Au(dppe)_2]Cl$ and (**b**)$[Au(d_2pypp)_2]Cl$ anticancer compounds, where R = 2, 3, or 4 pyridyl

by phosphane ligands with the three different phosphorus-bonds substituents leads to chiral phosphorous coordination compounds, which has been reported to increase potency with increasing number of coordinated phosphorus atoms. However, higher potency related to chirality was not observed. In spite of their efficacy, *Au*(I) complexes have been prevented from clinical use by far due to reports of heart toxicity highlighted during preclinical studies (Hoke et al. 1989). But, toxicities can be circumvented by careful selection of substituents on the phosphane moiety and by tuning the lipophilicity of the cation.

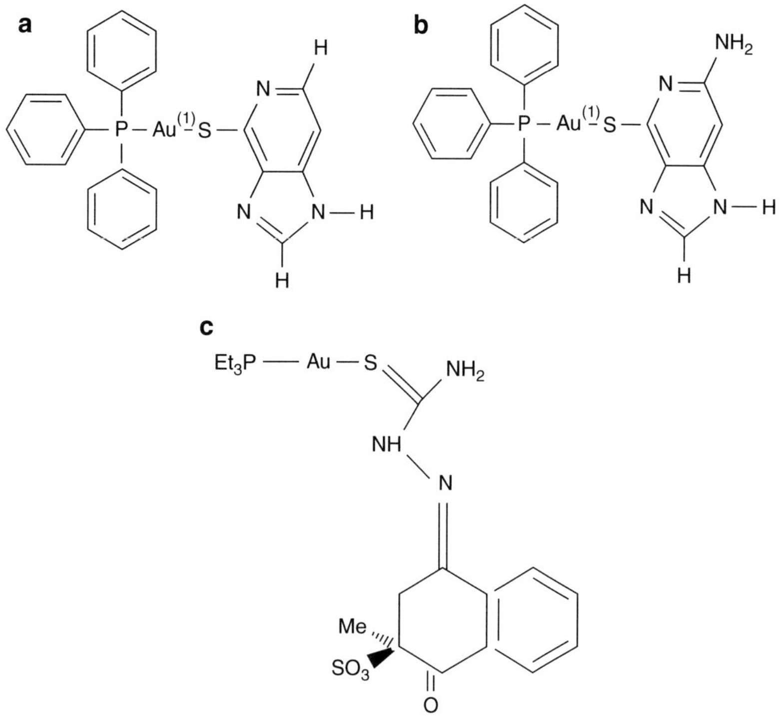

Gold Complexes as Prospective Metal-Based Anticancer Drugs, Fig. 4 Schematic structure of cytotoxic gold(I) (**a**) compounds, 6-mercaptopurine anion derivative, (**b**) 6-thioguanine anion derivative, and (**c**) *N*-heterocyclic carbene ligands

Coupling of phosphane *Au*(I) species to biologically active thiols such as 6-mercaptopurine and 6-thioguanine (Fig. 4a, b) revealed that the presence of phosphane *Au*(I) entity enhanced the potency of the biologically active free thiols, which are already known for anticancer activity in leukemia patients. Although *Au*(I)-drugs have proved their effectiveness in many diseases, knowledge related to the mechanism of action has not been completely understood even after 80 years of clinical use. This lack of knowledge in part is a result of the wide dispersion of *Au* compounds in the body and the absence of effective high-affinity target sites of action.

Gold(III) Complexes as Anticancer Agents

Recent research based on *Au*(III) complexes (Fig. 5) have revealed outstanding cytotoxic properties and are presently being evaluated as potential antitumor agents exhibiting promising in vivo activity against several tumor models in rats and mice (Messori et al. 2003, 2005). In contrast to the *Au*(I) complexes defined by "soft" (easily polarisable) sulfur or phosphorus atom donors, *Au*(III) complexes are characterized by "hard" atom donors like nitrogen, oxygen, and carbon. Four-coordinated *Au*(III) resemble with that of cisplatin having a square-planar geometry. *Au*(III) compounds with imine donors have also been prepared and biologically tested. Several *Au*(III) (Fig. 6a–c) complexes have been reported to show systemic toxic effects and significant differences in the spectrum of action have been noticed compared to cisplatin (Berners Price and Filipovska 2011). Some of the most interesting cytotoxic effects were reported for the complexes containing ligand donors derived from one element or the combination of chloride, nitrogen, oxygen, and carbon. One approach followed was the synthesis of complexes with the mononegative bidentate ligand, damp, (2- [(dimethylamino) methyl] phenyl), and two monodentate anionic ligands, chloride (Cl), or acetate (O_2CCH_3). The damp ligand forms part of a five-membered chelate ring in which the nitrogen of the amine group and the carbon of the aryl ring bind to the metal (Fig. 6i, j) making them readily hydrolyzed and available for substitution.

Initial in vitro studies against a panel of human tumor cell lines indicated that the breast carcinoma cell line (ZR-75-1) is sensitive to the compounds. Further tests in vivo against a xenograft of the same tumor cells demonstrate modest antitumor activity. Analogues of this complex have

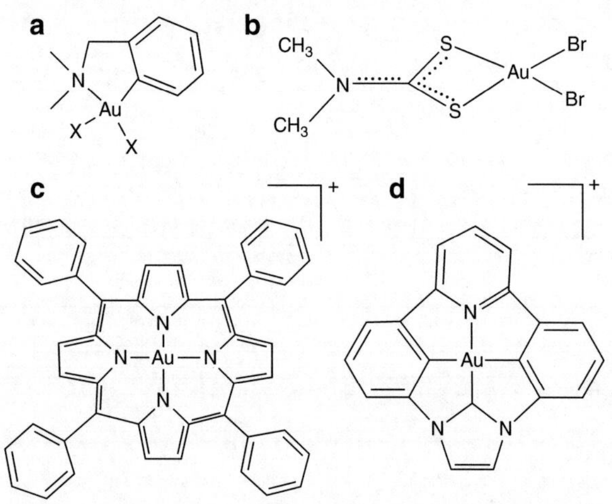

Gold Complexes as Prospective Metal-Based Anticancer Drugs, Fig. 5 Structure of *gold*(III) compounds that have demonstrated its antitumor activity in vivo (**a**) [AuX_2(damap)], (where, X = acetate and malonate) (**b**) [Au(DMDT)Br_2], (**c**) $[Au(III)(TPP)]^+$, and (**d**) $[Au(C\text{^}N\text{^}C\ (NHC)]^+$. The complexes (**a** and **b**) [Au(DMDT)X_2] and [Au(ESDT)X_2] (where DMDT = N, *N*-dimethylthiocarbamate and ESDT = ethylsarcosinedithiocarbamate; X = Cl, Br) are more toxic in vitro than cisplatin and to human tumor cell lines intrinsically resistant to cisplatin

been evaluated in the same way. Also in vitro cytotoxicity studies showed promising activity of two *Au*(III) complexes with bipyridyl-related ligands (Fig. 7), namely [*Au*(bpy)$(OH)_2$]PF6(bpy = 2,2′-dypyridyl) and [*Au*(bpyc-1H)(OH)]PF69 (bpyc = 6-(1,1-dimethylbenzyl)-2,2′-bypiridine) (Marcon et al. 2002).

Gold(I) and (III) Complexes to Overcome Cisplatin Resistance

Many of the *Au*(I) and (III) compounds are able to overcome to a large extent resistance to cisplatin, suggesting that a different mechanism of action is taking part. Carbene ligands of *Au*(I) complex (a derivative of vitamin K3) (Fig. 4c) displayed significant cytotoxicity in the cisplatin sensitive and resistant cancer lines, proving to be relatively more cytotoxic in the latter (Casas et al. 2006). A complex of *Au*(III), [*Au*(phen)Cl_2]Cl (phen = 1,10-phenanthroline) (Fig. 6f), and [*Au*(tpy)Cl]Cl_2 (tpy = 2,2′:6′,2″-terpyridine) (Fig. 6h) exhibited cytotoxicity profile which is similar to that of cisplatin (used at a concentration which is three times higher than the *Au* complexes) on the sensitive line. Moreover,

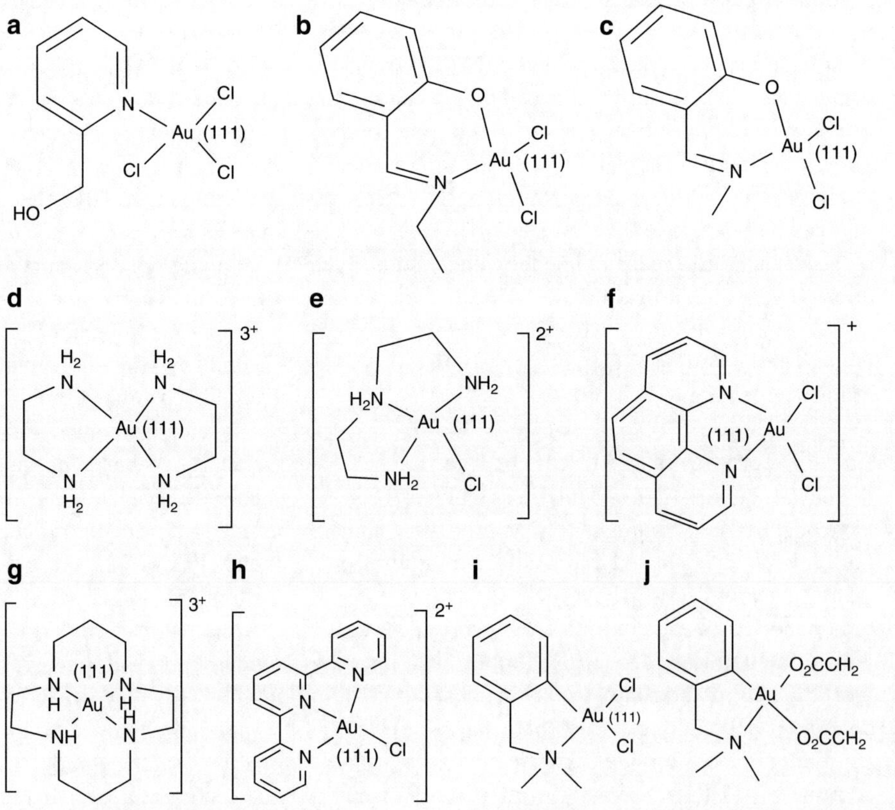

Gold Complexes as Prospective Metal-Based Anticancer Drugs, Fig. 6 Structure of cytotoxic gold(III) compounds and cationic species which show anticancer properties. (Reprinted with permission from Bardhan R, Lal S, Joshi A, Halas NJ (2011) Theranostic nanoshells: from probe design to imaging and treatment of cancer. Acc Chem Res. Copyright (YEAR) American Chemical Society)

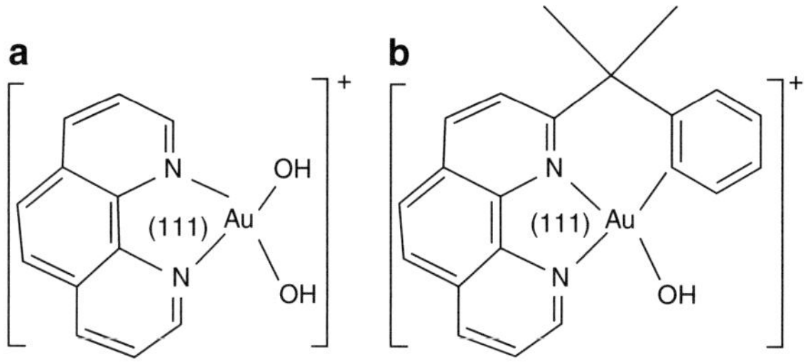

Gold Complexes as Prospective Metal-Based Anticancer Drugs, Fig. 7 Structure of cytotoxic gold(III) complexes with bipyridyl-related ligands, namely, (**a**) [*Au*(bpy)$(OH)_2$]PF6 (bpy = 2,2′-dypyridyl), (**b**) [*Au*(bpyc-1H)(OH)]PF69 (bpyc = 6-(1,1-dimethylbenzyl)-2,2′-bypiridine) compounds, and cationic species which show anticancer properties

[*Au*(tpy)Cl]Cl_2 largely overcomes resistance to cisplatin as it is at least three times more effective than cisplatin itself on the resistant line (Messori et al. 2000). In case of cisplatin-resistant tumor cell lines, *Au*(III) complexes could significantly retain their antitumor potency against CCRF-CEM/R leukemia and A2780/R ovarian carcinoma cell line having only minimal cross-resistance with cisplatin. In addition, the studies suggest that the in vitro interactions of *Au*(III) complexes with calf thymus DNA are weak and to produce only modest modifications of the double helix, whereas significant binding to model proteins takes place, confirming that different mechanisms, compared to cisplatin, could occur (Calamai et al. 1998).

Mode of Action

The cellular and molecular mechanisms of *Au*-salts are still not completely understood. However, *Au* may have diverse inhibitory as well as activating effects on different cellular functions emphasizing its several mechanisms of action. In case of *Au*(I) complexes, *Au* functions as prodrugs which metabolize with both cleavages because of ligand substitution (Champion et al. 1990; Nobili et al. 2010). *Au*(I) complexes have low activation energies and proceed via three-coordinate intermediates. *Au*(I) can undergo ligand exchange with other biological proteins such as albumin and also with ▶ metallothioneins to form aurothioneins. Thiol-exchange reactions are particularly significant as most of the *Au*(I) in blood is carried by the thiol in cysteine-34 of albumin. *Au* concentration in blood can increase in the range of 20–40 μM after injection of *Au*-drugs. Although half-life for excretion is about 5–31 days, *Au* can also remain in the body for many years (Rau 2005). The major area for *Au* deposits in the cells is the lysosome, hence, also termed as auranosome. The ability of *Au* to inhibit enzymes responsible for destroying the joint tissues may be the key mechanism of antiarthritic activity of *Au*. Interestingly, patients who smoke and treated with *Au*-drugs accumulate much higher concentrations of *Au* in their red blood cells (RBCs) compared to nonsmokers. The reason may be traced back to the high-affinity reaction between *Au* and cyanide (readily present in tobacco smoke as HCN) resulting in the formation of [*Au*$(CN)_2$], which can easily pass through cell membranes. In the case of nonsmokers, *Au* reacts with trace amounts of cyanide (formed from SCN-) forming [*Au* (CN) $_2$] which can be detected in the levels of 5–560 nM in their urine (Shaw 1999).

Preliminary, pharmacological investigation through flow cytometry analysis revealed induction of apoptosis after 48-h incubation of the *Au* compounds and the cells although modifications related to cell cycle remained insignificant (Nobili et al. 2010; Coronnello et al. 2005). Novel experimental evidences have indicated *Au*(III) compounds causing cytotoxic effects primarily by direct mitochondrial damage through modification or inhibition of specific proteins or enzymes such as selenoenzyme thioredoxin reductase (TxR). Mammalian TrxR isoenzymes are selenoproteins with a redox active selenocysteine (Sec) residue in their active sites and belong to the pyridine nucleotide-disulfide oxidoreductase family of proteins that catalyze NADPH-dependent reduction of their native Trx substrates. Thus, TxR is a crucial enzyme in the cells for protection against the oxidative stress damages, and its inhibition by *Au*(III) complex can cause big perturbation in the mitochondrial functions (Berners Price and Filipovska 2011; Coronnello et al. 2005).

In case of *Au*(III) porphyrins, a direct mitochondrial mechanism involves the influence of some apoptotic proteins like MAPKs and Bcl-2-, while for *Au*(III) dithiocarbamates, inhibition of proteasomal chymotrypsin-like activity occurs (Casini and Messori 2011; Milacic et al. 2008; Milacic and Dou 2009). Several possible mechanisms have been hypothesized, but the exact mechanism by which *Au*(I)-derivatives inflict cytotoxicity is not completely understood. In vivo as well as in vitro studies have also indicated *Au* internalization in the lymphocytic cells, thus, altering their normal functions. From immunological point of view, suppression of the B lymphocyte function by *Au* compounds could prevent the formation of blocking antibodies that protect tumors and thereby facilitating tumor destruction by T cells.

Toxicity of Gold Complexes

Underlying the potential benefits of using *Au*-based drugs for treating rheumatoid arthritis, inflammatory, and other autoimmune diseases, the risks associated with toxicity of *Au* on organ systems cannot be ruled out. However, the most adverse cases of *Au* toxicity is restricted to skin and mucous membranes and reported in case of blind clinical trials which have no possibility of further adjustment of the dose (Rau 2005).

Analysis of cytotoxicity data of *Au* complex has revealed significant insights due to their structure/function relationship (Messori et al. 2005):

(a) The cytotoxicity of *Au*(III) complexes is strictly related to the presence of the *Au*(III) center. The *Au* complexes, namely, $[Au(en)2]^{3+}$ (en = ethylenediamine) and $[Au(dien)Cl]^{2+}$ (dien = diethylenetriamine) (Fig. 6d, e), are significantly more cytotoxic than the corresponding platinum compounds.
(b) The presence of hydrolysable chloride atoms in the *Au*(III) center or, in general, of important leaving groups does not essentially represent a requirement for cytotoxicity.
(c) Excessive stabilization of the *Au*(III) center results in loss of biological activity as was already observed in case of $[Au(cyclam)]^{3+}$ (cyclam = 1, 4, 8, 11-tetraazacyclotetradecane) (Fig. 6g).
(d) The amount of *Au*(III) that enters in the cells is proportional to the exposure time, at least during the first hour.

The formation of *Au*(III) may be responsible for some of the toxic side effects of *Au*-drugs. Although most of the *Au*, in vivo, is present in the form of *Au*(I) oxidation state, powerful oxidants such as hypochlorous acid (HClO), which can oxidize *Au*(I) to *Au*(III), are generated at sites of inflammation so white blood cells from patients treated with *Au*-drugs become sensitive to *Au*(III).

Recent research related to *Au*(III)-dithiocarbamato complexes has focused on their strong tumor cell growth-inhibitory effects achieved through non-cisplatin-like mechanisms of action. A similar *Au* (III) compound, namely, $[Au(III)Br_2$ (ESDT)] (*AU*L12), exhibited encouraging antitumor activity in vitro as well as in vivo in three transplantable murine tumor models following 80% inhibition of tumor growth. Moreover, the complex showed low acute toxicity levels (lethal dose, LD_{50} 30 mg kg^{-1}) and reduced nephrotoxicity (Marzano et al. 2011). Thus, these results suggest *Au* (III)-dithiocarbamate to be an attractive candidate in clinical trials for tumor-growth inhibition. To sum up, further understanding of the redox cycling of *Au* may lead to a better understanding of the side effects resulting due to *Au*(III) complexes and discover ways to overcome them.

Acknowledgment This work was supported in part by Chinese Natural Science Foundation project (No.30970784 and 81171455), National Key Basic Research Program of China (2009CB930200), Chinese Academy of Sciences (CAS) "Hundred Talents Program" (07165111ZX), and CAS Knowledge Innovation Program. We sincerely thank Prof. Chunying Chen, National Center for Nanoscience and Technology, Chinese Academy of Sciences, for her valuable suggestions for preparing the chapter.

Cross-References

- ▶ Apoptosis
- ▶ Biodistribution of Gold Nanomaterials
- ▶ Carboplatin
- ▶ Magnesium and Cell Cycle
- ▶ Gold and Nucleic Acids
- ▶ Metallothioneins
- ▶ Oxaliplatin

References

Berners Price SJ, Filipovska A (2011) Gold compounds as therapeutic agents for human diseases. Metallomics 3:863–873

Calamai P, Carotti S, Guerri A et al (1998) Cytotoxic effects of *Au*(III) complexes on established human tumor cell lines sensitive and resistant to cisplatin. Anticancer Drug Des 13:67–80

Cancer Facts & Figures (2012) Is an annual publication of the American Cancer Society, Atlanta. http://www.cancer.org/acs/groups/content/@epidemiologysurveilance/documents/document/acspc-031941.pdf. Accessed January 2012

Caruso F, Rossi M, Tanski J et al (2003) Antitumor Activity of the mixed phosphine gold specie chlorotriphenylphosphine-1,3bis(diphenylphosphino)propane gold(I). Med Chem 46:1737–1742

Casas J, Castellano S, Couce EE et al (2006) A gold(I) complex with a vitamin K3 derivative: characterization and antitumoral activity. J Inorg Biochem 100:1858–1860
Casini A, Messori L (2011) Molecular mechanisms and proposed targets for selected anticancer *Au* compounds. Curr Top Med Chem 11:2647–2660
Chabner BA, Roberts TG (2005) Chemotherapy and the war on cancer. Nat Rev Cancer 5:65–72
Champion GD, Graham GG, Ziegler JB (1990) The gold complexes. Baillieres Clin Rheumatol 4:491–534
Coronnello M, Mini E, Caciagli B et al (2005) Mechanisms of cytotoxicity of selected organogold(III) compounds. J Med Chem 48:6761–6765
Garza OA (2008) Design synthesis characterization and, biological studies of ruthenium and gold compounds with anticancer properties. Doctoral thesis, Leiden Institute of Chemistry, Faculty of Science, Leiden University
Hoke GD, Macia RA, Meunier PC et al (1989) In vivo and in vitro cardio-toxicity of a gold-containing antineoplastic drug candidate in the rabbit. Toxicol Appl Pharmacol 100:293–306
Krishnamurthy D, Karver MR, Fiorillo E et al (2008) Gold(I)-mediated inhibition of protein tyrosine phosphatases: a detailed in vitro and cellular study. J Med Chem 51:4790–4795
Marcon G, Carott S, Coronnello M et al (2002) Gold(III) complexes with bipyridyl ligands: solution chemistry, cytotoxicity, and DNA binding properties. J Med Chem 45:1672–1677
Marzano C, Ronconi L, Chiara F et al (2011) Gold(III)-dithiocarbamato anticancer agents: activity, toxicology and histopathological studies in rodents. Int J Cancer 129:487–496
Messori L, Abbate F, Marcon G et al (2000) Gold(III) complexes as potential antitumor agents: solution chemistry and cytotoxic properties of some selected gold(III) compounds. J Med Chem 43:3541–3548
Messori L, Marco G, Orioli P (2003) Gold(III) compounds as new family of anticancer drugs. Bioinorg Chem Appl 1:177–187
Messori L, Marcon G, Innocenti A et al (2005) Molecular recognition of metal complexes by dna: a comparative study of the interactions of the parent complexes [PtCl(TERPY)]Cl and [AuCl(TERPY)]Cl2 with double stranded DNA. Bioinorg Chem Appl 3:239–253
Milacic V, Dou QP (2009) The tumor proteasome as a novel target for gold(III) complexes: implications for breast cancer therapy. Coord Chem Rev 253:1649–1660
Milacic V, Fregona D, Dou QP (2008) *Au* complexes as prospective metal-based anticancer drugs. Histol Histopathol 23:101–108
Nobili S, Mini E, Landini I et al (2010) Gold compounds as anticancer agents: chemistry, cellular pharmacology, and preclinical studies. Med Res Rev 30:550–580
Pillarsetty N, Katti KK, Hoffma TJ et al (2003) In vitro and in vivo antitumor properties of tetrakis((trishydroxy- methyl) phosphine)gold(I) chloride. J Med Chem 46:1130–1132
Rau R (2005) Have traditional DMARDs had their day? Effectiveness of parenteral gold compared to biologic agents. Clin Rheumatol 24:189–202
Shaw CF (1999) Gold-based therapeutic agents. Chem Rev 99:2589–2600
Sheppard CW, Goodell JP, Hahn PF (1947) Colloidal gold containing the radioactive isotope Au198 in the selective internal radiation therapy of diseases of the lymphoid system. J Lab Clin Med 32:1437–1441

Gold for Health – Past and Present

▶ Gold Complexes as Prospective Metal-Based Anticancer Drugs

Gold Nanomaterials

▶ Gold Nanomaterials as Prospective Metal-based Delivery Systems for Cancer Treatment

Gold Nanomaterials as Prospective Metal-based Delivery Systems for Cancer Treatment

Anil Kumar[1,2] and Xing-Jie Liang[1]
[1]CAS Key Laboratory for Biomedical Effects of Nanoparticles and Nanosafety, National Center for Nanoscience and Nanotechnology, Chinese Academy of Sciences, Beijing, China
[2]Graduate University of Chinese Academy of Sciences, Beijing, China

Synonyms

Biodistribution of gold nanomaterials; Biomedical application; Cancer diagnosis and treatment; Fictionalization of gold nanomaterials; Gold; Gold nanomaterials; Gold-based nanomedicine; Method of synthesis; Photothermal therapy; Targeted drug delivery; Toxicity of gold nanomaterials

Definition

Gold nanomaterials (*Au*NMs) are one of the precious nanomaterials that are important in biomedical field because of their application used in our lives to cure disease. *Au*NMs with controlled geometrical, optical, and surface chemical properties are the subject of intensive studies and applications in biology and medicine. *Au*NMs are used in drug carriers, diagnosis, imaging, and for thermotherapy of biological targets. *Au*NMs, such as nanorods (NRs), nanoshells (NSs), nanocages (NCs), and nanostar (aspherical), have the extensive potential to deliver the drug and image the integral part of our imaging toolbox and are useful in the fight against various cancer. Surface of *Au*NMs can be functionalized by using various ligand approaches to conjugate some biological molecules (directly or indirectly) onto their surface for various applications. The toxicity of nanomaterials can be monitored by using difference biocompatible molecules to make them more applicable for biomedical applications. Due to unique optical properties of *Au*NMs, they are being incorporated into new generations of drug-delivery vehicles, contrast agents, and diagnostic devices, some of which are currently undergoing clinical investigation or have been approved by the Food and Drug Administration (FDA) for use in humans.

Gold Nanomaterials (*Au*NMs)

After the great success in clinical application of gold complexes for cancer treatment, the next generation of gold-based materials comprises of the nano-world, where these materials can be synthesized in the nanoscale with different shapes, and sizes meant for different applications in the biomedical field (Dreaden et al. 2012; Dykman and Khlebtsov 2012). The word "nano," derived from the Greek word *"nanos,"* meaning dwarf or extremely small, is used to describe any material or property which occurs with dimensions on the nanometer scale (1–100 nm) (http://en.wikipedia.org/wiki/Nanotechnology). Nanomaterials can be metals, ceramics, polymeric materials, or composite materials. Among these, gold-based nanomaterials (*Au*NMs) are an attractive platform to design nanomedicine for cancer treatment (Dreaden et al. 2011; Jelveh and Chithrani 2011). Colloidal gold is a suspension (or colloid) of sub-micrometer-sized particles of gold in a fluid condition such as water. The color of *Au*NMs suspension varies from size of nanoparticles, it is usually intense red color (for particles less than 100 nm), or a dirty yellowish color (for larger particles) (Fig. 1) (http://en.wikipedia.org/wiki/Colloidalgold). Beyond their beauty, *Au*NMs exhibit properties which are fundamentally different from all others. Due to the unique optical and electronic properties of gold nanoparticles, they are the subject of substantial research, with applications in a wide variety of areas including electron microscopy, electronics, nanotechnology, nanomedicine, and also in material sciences. Properties and applications of *Au*NMs depend upon shape, size, surface coating, and so on. For example, rod-like particles have both transverse and longitudinal absorption peaks, and anisotropy of the shape affects their self-assembly process. In contrast to the bulk materials (macroscale materials), "*Au*NMs" in the nanoscale (1–100 nm) exhibit unique fundamental properties which are particularly influenced by size, shape, diameter, and surface coating of the materials (Huang and El-Sayed. 2010). Due to their nanoscale size, these materials preferentially accumulate at the sites of tumor growth/inflammation and enter into the cells by mechanisms very different and much more rapid than those of small molecules (Huang et al. 2012; Connor et al. 2005).

Their unique optical or photophysical properties allow for their use in bio-diagnostic assays for several kinds of diseases (cancer, bacterial, and HIV-AIDS), it is also used to interpret pregnancy test (such product has already been marketed for the benefit of patients). Due to their facile surface chemistry of *Au*NMs, it can act as artificial antibodies whose binding affinity can be precisely tuned by varying the concentration of binding ligands on their surfaces. The efficient conversion of light into heat by *Au*NRs can allow for the highly specific thermal ablation (► photothermal Therapy) of diseased or infected tissues. *Au*NMs can be used to enhance cancer radiation therapy or increase imaging contrast in diagnostic CT scans (computed tomography) by exploiting their ability to absorb abundant amounts of X-ray radiation. The functional configuration of *Au*NMs can be exploited to facilitating efficient delivery of poorly soluble drug, flouroscence imaging and MRI (contrast agents) (Dykman and Khlebtsov 2012).

Methods for the Synthesis of Gold Nanomaterials (*Au*NMs)

*Au*NMs in various shapes, such as spherical, nanorods, nanoshells, nanocages, were synthesized (Fig. 1a).

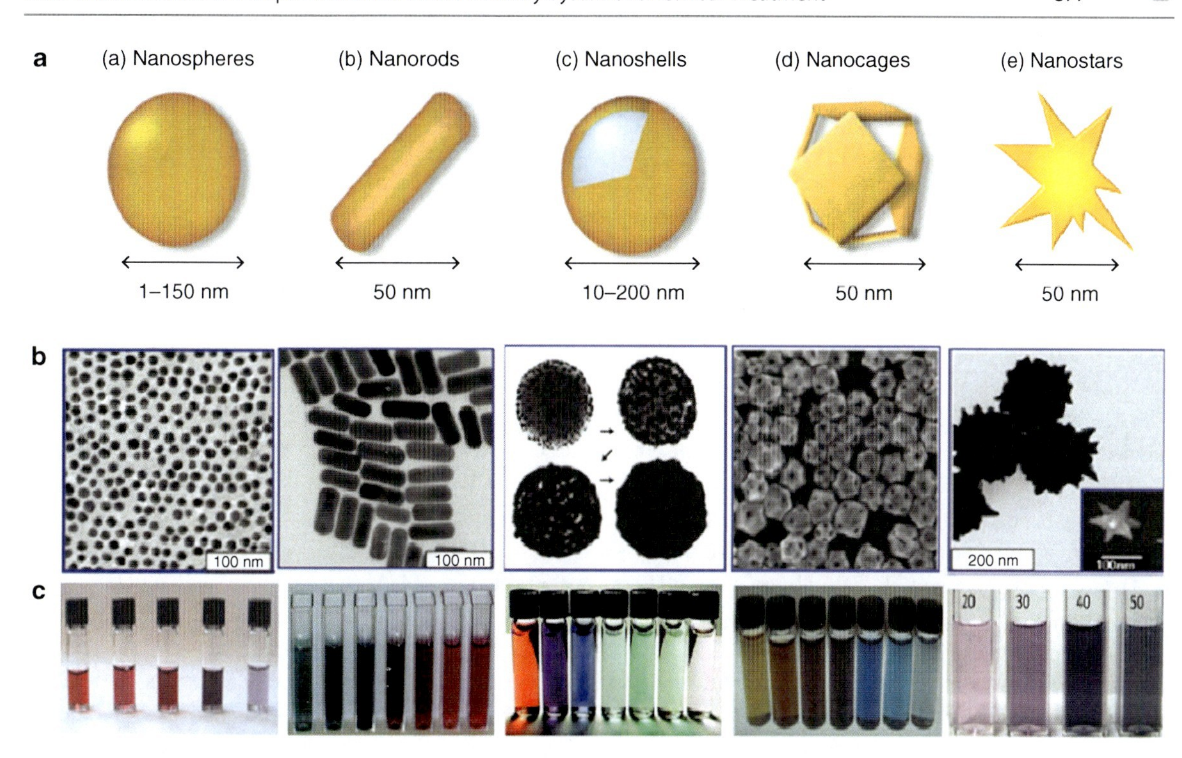

Gold Nanomaterials as Prospective Metal-based Delivery Systems for Cancer Treatment, Fig. 1 (**a**) Different types of gold nanomaterials with different shapes and sizes dependingon the synthesis procedure. (**b**) Image of different types of gold nanomaterials in TEM (Transmission electron microscopy). (**c**) Tunable optical properties of gold nanomaterials by changing the aspect ratios different color was observed during synthesis process (Reprinted with permission American Chemical Society, Elsevier, Future Science Ltd, Royal Society Chemistry, Reference from Jiang et al. 2012; Dreaden et al. 2012; Dykman and Khlebtsov 2012; Huang and El-Sayed 2010)

Among these nanostructures, spherical and gold nanorods are more attractive for nanomedicine application. The synthesis of colloidal gold dates back to 1857 when Michael Faraday discovered the "fine particles" through stabilizing a reduction reaction of gold chloride by phosphorus, with addition of carbon disulfide. Present synthetic methods employ similar strategies with the added advantage of using surface capping ligands to prevent aggregation of the particles by electrostatic and/or physical repulsion. There are three main important routes to synthesize *Au*NMs which include the electrochemical method, template method, and the seed-mediated growth method. Electrochemical method consists of electrochemical synthesis of transition metal clusters within reverse micelles in organic solvent systems; through this synthetic process, the bulk gold metal anode is consumed to form $AuBr_4$ and then the anions, complexed with the cationic surfactants, migrate to the cathode where reduction occurs. This method is more applicable to prepare high yields of gold nanorods. Template method consists of electrochemical deposition of gold within the nanopores of alumina or polycarbonate as template membranes. Seed-mediated growth method is one of the oldest having the longest history among all the three methods, wherein monodispersed *Au*NMs of various sizes and shapes could be synthesized using simple and straightforward synthesis procedure consisting of two basal steps including nucleation and growth. Firstly, gold nanoseeds are prepared by reducing metal salts with a strong reducing agent, and then the growth steps are followed by adding metal salts with a weak reducing agent in the presence of a surfactant. Using this method, monodisperse spherical *Au*NPs of diameter in the range of 10–150-nm diameters could be synthesized with citrate as a reducing agent (Jiang et al. 2012).

The mechanism following formation of these NPs is related to spontaneous nucleation and isotropic growth or LaMer growth (LaMer et al. 1995). In this method, the size of the *Au*NPs can be varied by changing concentrations of sodium citrate. Likewise, particle size can be adjusted by varying proportion of gold ion with that of reducing agent or stabilizer ratio.

Method for the synthesis of gold nanostars was carried by mild reduction of Gold(III) chloride trihydrate ($HAuCl_4.3H_2O$) with ascorbic acid (AA) in the presence of DES (deep-eutectic solvents) which is prepared from mixture of choline and urea (1:2). The exact method to obtain gold nanostar depends upon the variation of the reactant molar ratio.

The key step for synthesis of *Au*NMs is to control their physicochemical properties to obtain well-defined materials, which is carried out by controlling the different parameters like concentration, surfactant, pH, and temperature in the growth of nanoscale process (Dreaden et al. 2012; Jiang et al. 2012).

Optical Properties of Gold nanomaterials for Biomedical Applications

Surface Plasmon Resonance (SPR)

Surface plasmon resonance is a phenomenon occurring at the metal surface when a beam of light is incident on the surface of the molecules at a particular angle and distance (typically in case of gold (*Au*) and silver (*Ag*) metallic NPs). It is well known that diameter and thickness of materials play an important role at the metal surface; the SPR phenomenon results in a gradual reduction in intensity of the reflected light when the size and shape of materials differs. By measuring the appropriate exquisite sensitivity of SPR to the refractive index of the surrounding medium onto the metal surface, it is possible to measure accurately the adsorption and scattering of molecules on the metal surface and their targeted specific ligands.

The photophysical properties *Au*NMs like absorbance and light scattering can be tuned in accordance with the size. It is well known that particles with size of 20 nm exhibit Surface Plasmon Resonance (SPR) and can be used for colorimetric detection of analytes as well as for biological analysis (Kumar et al. 2012). On the other hand, *Au*NPs of large sizes (20–80 nm) have scattering properties that tend to increase with increase in size and diameter, making them more applicable for biomedical applications (Fig. 2).

Importantly, the SPR spectrum depends on the size, shape, and chemical composition of *Au*NMs, as well as the external properties of the nanoparticles surface. However, near-infrared light (800–1,200 nm) is preferred in biomedical applications due to its deeper penetration (both blood and soft tissues are highly transparent within this range).

Surface-Enhanced Raman Scattering (SERS)

Surface-enhanced Raman scattering is also an important property of *Au*NMs that enhances Raman scattering by molecules adsorbed on gold metal surfaces. This surface-enhanced absorption or surface-enhanced Raman scattering is most attractive, because of the huge enhancement of the SERS signal, by a factor of (ca. 10^{14}–10^{15}) which means the technique may detect single molecules.

Due to excitation, a considerable enhancement occurs at the *Au*NMs surface, this enhancement also helps in origination of energy transfer. The electronic coupling between adsorbed molecules and the *Au*NMs surface result in charge transfer between the *Au*NMs metal surface and adsorbed molecules. The elastically scattered visible light by the *Au*NMs themselves it can be used as a imagining by using a dark-field optical microscope. Therefore, inelastic SERS effect which adsorbed by molecules providing a Raman spectrum that leads to the identification of biological molecules (Fig. 2) for diagnosis and treatment of several diseases (Kumar et al. 2012).

Chemical Modifications

Gold nanomaterials can be functionalized by a wide variety of functionalizing agents (ligands, surfactants, polymers, dendrimers, biomolecules, etc.) onto their surface (Fig. 3), allowing them to be used in sensing of biomolecules and cells, diagnosis of diseases, and also for intracellular delivery. It is important that nanoparticles functionalized with ligands exhibiting differential affinity toward proteins and cell surface molecules have been employed for their identification to particular target.

However, the most robust method for stabilizing is by thiolates using the strong *Au–S* bond between the soft acid *Au* and the soft thiolate base. The cross-linking agents contain thiol group on one side and carboxyl or amine group on the other side that help in the conjugation process. The thiol group binds to the gold surface, and the terminal

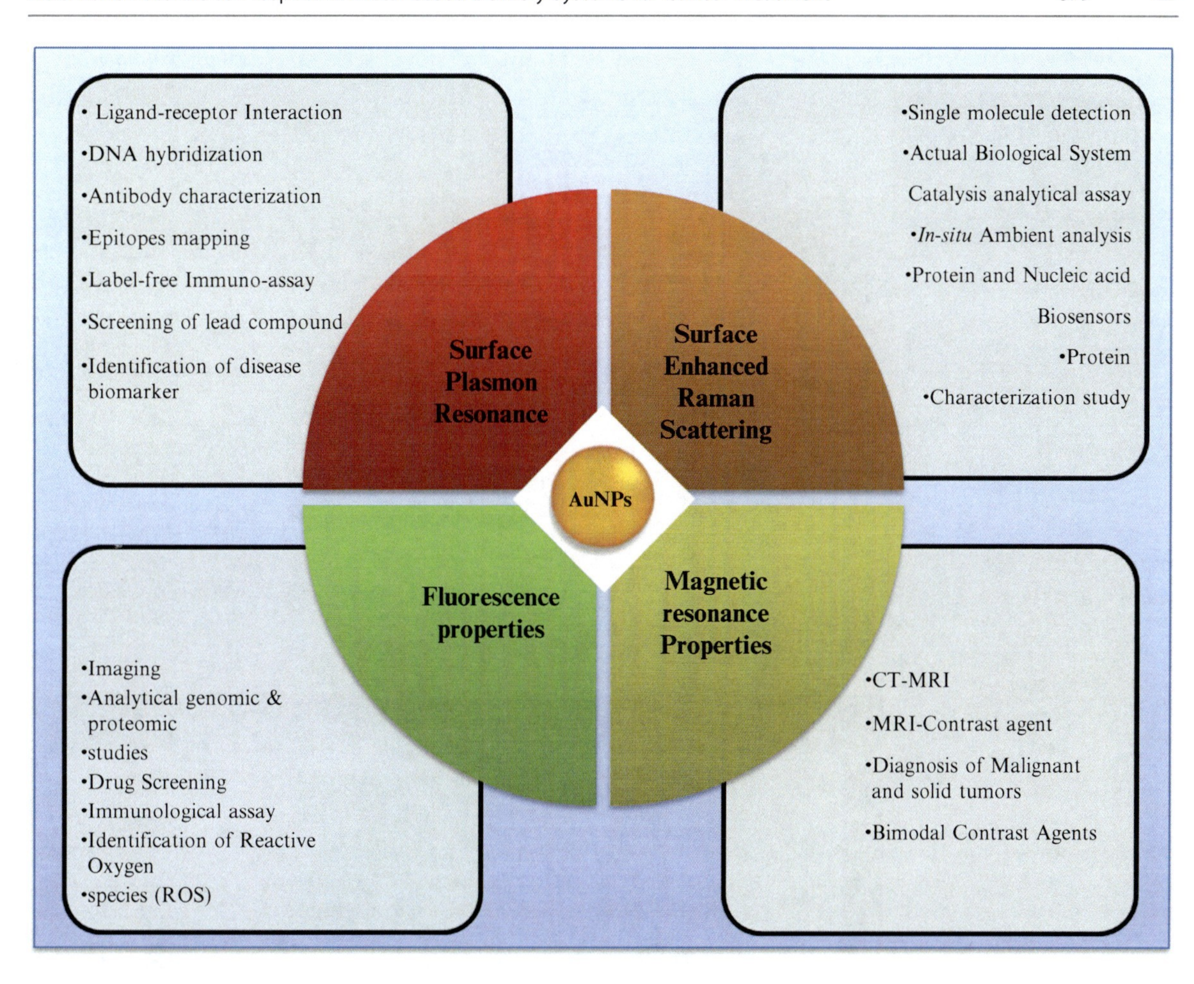

Gold Nanomaterials as Prospective Metal-based Delivery Systems for Cancer Treatment, Fig. 2 Optical properties of gold nanomaterials and its various application used in biomedical field for human welfare

carboxyl or amine group serves as a binding site to conjugate functional biomolecules such as proteins, peptides, antibodies, DNA, or RNA, using 1-ethyl-3(3-dimethylaminopropyl)-carbodiimide-HCl (EDC), NHS (*N*-Hydroxysuccinimide) coupling agents (Kanaras and Bartczak 2011).

Applications of Gold Nanomaterials (*Au*NMs) in Biomedical Field

Due to their comparative size related to proteins and other biomolecules, *Au*NMs can be exploited to alter subcellular processes which are normally not feasible with proteins or small molecules. The diverse advantages of gold nanotechnology can be compiled for effective biomedical applications comprising of simultaneous targeting, diagnostic, and therapeutic functionality, according to patient's need (Fig. 4).

Spherical Gold Nanoparticles

The fabrication of spherical *Au*NPs with varying core sizes makes them an attractive platform for drug-delivery systems (DDSs). A variety of synthetic methods exist for the fabrication of spherical *Au*NPs, core size ranging from 1 to 150 nm in diameter. The single most popular method of Brust and Schiffrin et al. allows for the rapid and scalable synthesis of monolayer protected *Au*NPs (Brust et al. 1994), in which salts $AuCl_4^-$ are reduced by $NaBH_4$ in solution with the desired ligands. By changing the ligand to gold, stoichiometry core sizes from 1.5 to 6 nm can be fabricated which is also known as ultra-small *Au*NPs. Ultrasmall *Au*NPs can

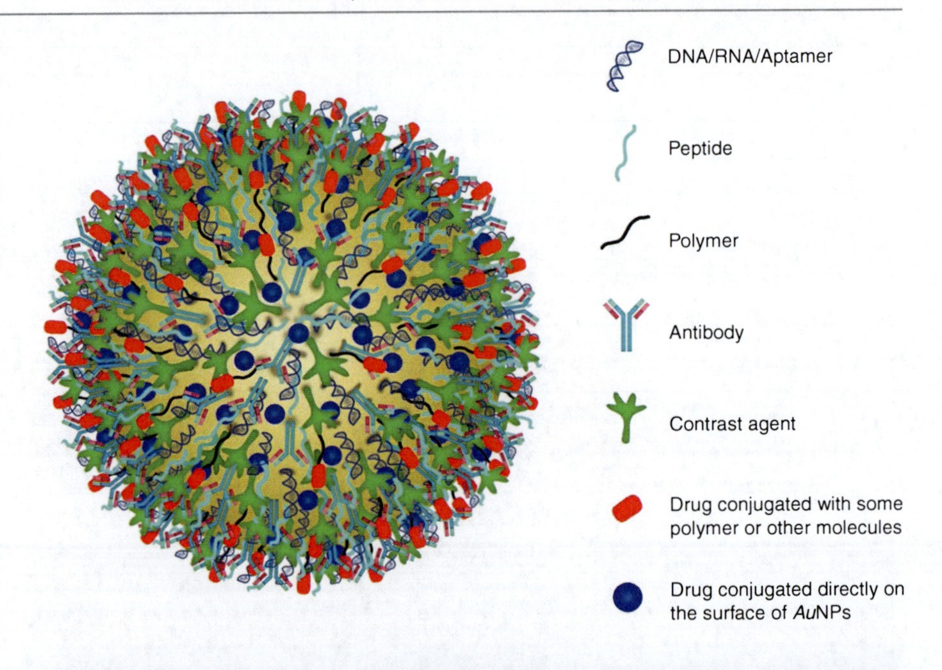

Gold Nanomaterials as Prospective Metal-based Delivery Systems for Cancer Treatment, Fig. 3 Functionalization of gold nanomaterials by using different biological and chemical molecules on the surface of nanomaterials for a wide range of clinical application

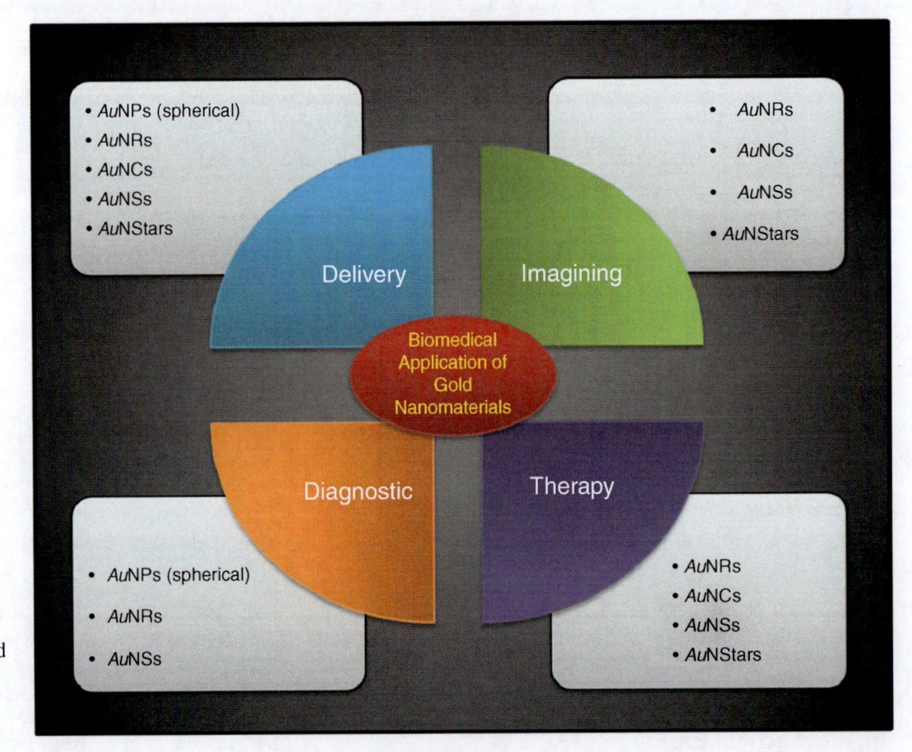

Gold Nanomaterials as Prospective Metal-based Delivery Systems for Cancer Treatment, Fig. 4 Important role of gold nanomaterials in biomedical filed for diagnosis and treatment of cancer disease

also be prepared by using different stabilizing agents, e.g., Tiopronin or GSH (Glutathione) to prepare less than 10 nm *Au*NPs.

The diversity of *Au*NPs is achieved by synthesizing a mixed monolayer by post-functionalization or directly functionalizing with appropriate ligands onto *Au*NPs surface. A widely used method for the synthesis of mixed monolayer protected *Au*NPs is using a place-exchange reaction developed by Murray et al. In this method, ligands with thiol functionality exchange in equilibrium with the bound ligands to *Au*NPs surface. The tunability of the ligand structure allows for enhanced delivery applications, such as the use of polyethylene glycol (PEG) and oligo (ethylene glycol) moieties to create a more biocompatible nanoparticle for biological application. The spherical *Au*NPs ranging from 2 to 10 nm in size play an important role for drug delivery into nucleus for the treatment of diseases like cancer (Kumar et al. 2012), It can be functionalized by using various biological molecules (DNA, RNA Peptides, and various hydrophobic drugs) onto the surface of *Au*NPs for diagnostic and therapeutic applications. It is very well known that *Au*NPs interact between cells and successfully deliver the cargo molecules into the cell cytoplasm or nucleus. The size, charge, and surface functionality affect the intracellular fate and cellular uptake of *Au*NPs. Hydrophobicity and size of *Au*NPs contributed to determining the cellular uptake. Size is an important parameter to characterize the nanomaterials in biological environment for its further application in biological complex (Huang et al. 2012).

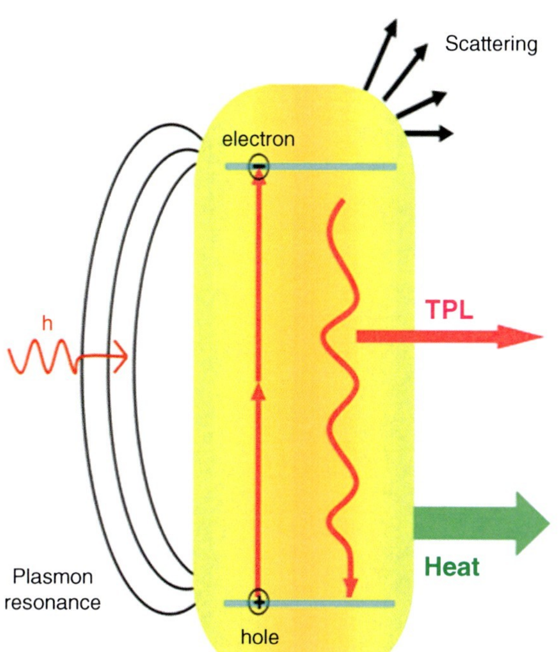

Gold Nanomaterials as Prospective Metal-based Delivery Systems for Cancer Treatment, Fig. 5 Photophysical properties of gold nanorods (*Au*NRs). Near-infrared irradiation induces the excitation of a longitudinal plasmon resonance mode, resulting mostly in absorption but also some resonant light scattering. Through this process an electronic transition from the d-band to sp-band occurs with two-photon absorption, generating an electron–hole pair; recombination of separated charges results in two-photon luminescence (TPL) emission. Heat is also generated as a consequence of electron-photon collisions and that heat is important to damage cancer cells (Reprinted with permission (John Wiley and Sons), Reference from Tong et al. 2009)

Gold Nanorods

Gold nanorods (*Au*NRs) are nanomaterials with structures like cylindroid or rod in shape which generate heat, during excitation (Fig. 5). *Au*NRs can be prepared by using CTAB, a cationic surfactant. While the exact conditions depend upon the requirements, NRs can be reproduced easily through these syntheses and also can yield NRs with different surface modifications, to produce stable NRs suspensions with longer circulation lifetime and specific targeting to diseased cells. Cytotoxicity issues can be monitored by sterically stabilizing surfactants, particularly by using nonionic polymers such as polyethylene glycol (PEG) chains.

There are numerous examples for conjugating *Au*NRs with biomolecules for cell targeting or intracellular delivery, such as folic acid, antibodies, DNA, and therapeutic drug), introduced by various surface functionalization methods similar as spherical *Au*NPs illustrated in (Fig. 4). Most *Au*NRs are used as multiplex biosensors as well as detection method based on localized SPR shifts, using 11-mercaptoundecanoic acid as a linker for protein conjugation. *Au*NRs can be functionalized by antibodies (Abs) such as anti-epidermal growth factor receptor Abs (anti-EGFR) onto PSS-coated NRs, and use these to selectively identify cancer cells using conventional and dark-field microscopy. Anti-EGFR Ab-coated NRs (anti-EGFR-NRs) can also be used as labels for cancer cells based on their polarized Raman spectral emissions (Tong et al. 2009). Recently, antibody-conjugated NRs have also been used as photothermal agents to selectively damage cancer cells by NIR laser irradiation. A NIR laser pulse was used to irradiate *Au*NPs; due to strong light scattering from *Au*NRs in the dark field, it was easy to visualize malignant cells and distinguish them from the nonmalignant cells by a microscope. During the study, it was known that only half of the power density of the laser is enough to destroy malignant cells than that for nonmalignant cells, which provides a good chance for killing cancer cells efficiently and selectively. *Au*NRs can also generate two-photon ▶ luminescence at sufficient intensities for single-particle detection and in vivo imaging. Due to all these characteristic properties,

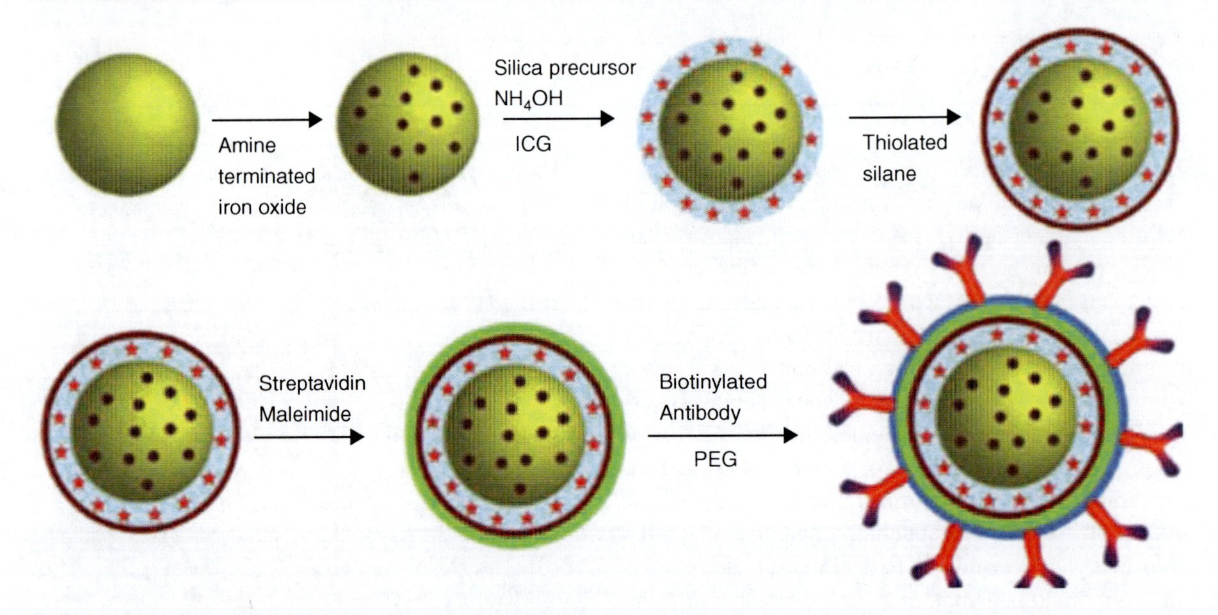

Gold Nanomaterials as Prospective Metal-based Delivery Systems for Cancer Treatment, Fig. 6 Schematic design and fabrication of gold nanoshells (*Au*NSs) with antibodies for theranostic application (Reprinted with permission (Journal of American Chemical Society), Reference from Bardhan et al. 2011)

*Au*NRs have provided a chance for the real-time imaging system as well as for photothermal therapy. On the other hand, *Au*NRs can also mediate tumor cell death by compromising membrane integrity in thermal therapy. In the cell membrane, blabbing was observed due to the disruption of actin filaments, resulting in Ca^{2+} influx and the depolymerization of the intracellular actins network in cytoskeleton (Tong et al. 2009).

Gold Nanoshells

Gold nanoshells (*Au*NSs) are a novel class of optically tunable nanoparticles that consist of a dielectric core covered by a thin metallic gold shell. Based on the relative dimensions of the shell thickness and core radius, nanoshells may be designed to scatter and/or absorb light over a broad spectral range including the near-infrared (NIR), due to this behavior gold nanoshell is also called nanoshell plasmon. By adjusting the size of the nanoparticle core, relative to the thickness of the gold shell, the optical resonance of nanoshells can be precisely and systematically diverse over a broad spectrum including the near-infrared (NIR) region where it provides maximal penetration of light through tissue. The ability to control both wavelength-dependent scattering and absorption of nanoshells offers the opportunity to design nanoshells (Fig. 6) for a theranostic application (Bardhan et al. 2011).

The surface of *Au*NS is a chemically inert material well known for its biocompatibility. In addition, polymers such as PEG can be grafted to nanoshell surfaces through self-assembly process to improve their biocompatibility.

Theranostic probes, used for the detection and treatment of cancer in a single treatment, are NPs containing binding sites on their shell that allow them to attach to a desired location (typically cancerous cells) that can be imaged through dual modality imagery (an imaging strategy that uses x-rays and radionuclide imaging) and through near-infrared fluorescence (http://en.wikipedia.org/wiki/Nanoshell). Through these NPs, treatment of cancer is possible only because of the scattering and absorption that occurs for plasmonics. During such scattering condition, the gold-plated NPs become visible to imaging processes that are tuned to the correct wavelength which is dependent upon the size and geometry of the particles due to plasmonic properties of *Au*NSs has been used in a wide variety of biomedical application (Bardhan et al. 2011).

1. Photothermal Effects

   Excitation of *Au*NSs in the presence of resonant light (due to Photoexcitation of the electron gas) results in rapid nonequilibrium state which results in heating. The scattering electron rapidly increases the surface temperature of the metal; this initial rapid heating is followed by cooling to equilibrium by energy exchange between the electrons and the lattice. Through this process very fast rates of energy dissipation (a few picoseconds) take place relative to lattice cooling, which result in intense photothermal heating. When *Au*NSs are in biological medium, a significant temperature difference between the hot *Au*NSs surface and the cooler surrounding biological medium occurs, this rapid local temperature increase or decrease is responsible for cell death.

2. Fluorescence Enhancement
   When a metal surface is illuminated with resonant light, nanoshell plasmons decay nonradiatively, resulting in absorption, and decay radiatively, resulting in light scattering. These characteristic properties can modify the emission properties of fluorophores in their proximity. This metal fluorophore interaction occurs due to (1) modification of the electromagnetic field and (2) the photonic mode density near the fluorophores; the damping of molecular oscillators by processes such as electron tunneling between the metal and the fluorophore and image interactions between the two systems typically lead to strong quenching of the molecular fluorescence.
3. Optical Imaging and Therapy
   The *Au*NSs (60 nm core radius, 10–12 nm) shell thickness can absorb and scatter light at 800 nm, therefore serving as imaging contrast agents and photothermal therapy. The light scattering properties of *Au*NSs scattered in the NIR physiological "water window" can be employed as contrast agents for dark-field scattering, photoacoustic imaging, and optical coherence tomography (OCT). It is well known that *Au*NSs bound to anti-HER2 antibodies have been used for dark-field imaging and therapy of SKBR3 breast carcinoma cells and also the work related to *Au*NSs has been also examined in vivo for enhancement of optical coherence tomography (OCT) and for inducing photothermal cell death.
4. Multimodal Imaging and Therapy
   Multimodal imaging and therapy system based upon *Au*NSs when the complexes system were constructed for two diagnostic capabilities, such as MRI and near-infrared fluorescence imaging, in addition to photothermal therapy. In some current studies, *Au*NS complexes were constructed by encapsulating *Au*NSs in a silica epilayer doped with ~10-nm iron oxide (Fe_3O_4) NPs, this complex (contrast agents) is now in clinical use for patients (Bardhan et al. 2011).
5. Gene Therapy
   The plasmonic properties of *Au*NSs help in the photothermal ablation to treat solid tumors, this can also be used to light-trigger release of short DNA strands conjugated to the surface of *Au*NSs. The light-triggered release of the DNA from *Au*NSs also allows precise temporal control over DNA/RNA release to treat cancer by controlling their genetic machinery.
6. Drug Delivery
   *Au*NSs play an important role to increase drug accumulation into tumors site, because during the heating process, *Au*NSs generate tumor perfusion and through that perfusion helps to increases the drug transport into the tumor environment. Mice, when treated with nanoshells at 24 h before laser treatment, perfusion was monitored using MRI (Magnetic Resonance Imaging) study, during low heating process (0.8 W cm^{-2}) and high laser intensities (4 W cm^{-2}) decreased contrast uptake, while heating with an intensity of (2 W cm^{-2}) almost doubled the uptake was observed. Thus nanoshells have potential to improve drug delivery for cancer treatment.

Gold Nanocage

Gold nanocages (*Au*NCs) represent a novel class of nanostructures prepared by using a remarkably simple synthetic route that involves galvanic replacement between Ag nanocubes and $HAuCl_4$ in an aqueous solution. *Au*NCs are cubes of gold, with sides about 50-billionths of a meter long and holes at each corner (Fig. 1Bd), are made using silver particles as a mold, and then coated with strands of a smart polymer. In normal conditions, the polymer strands remain extended and bushy, covering the holes in the cube. But when heated, the strands collapse, leaving the holes open and allowing the drug inside to escape.

*Au*NCs have hollow interiors and porous walls, besides having a range of hidden qualities that make them unique for theranostic applications (Xia et al. 2011)

1. AuNCs are single crystals with unique mechanical flexibility and stability, and possess flat surfaces at the atomic level.

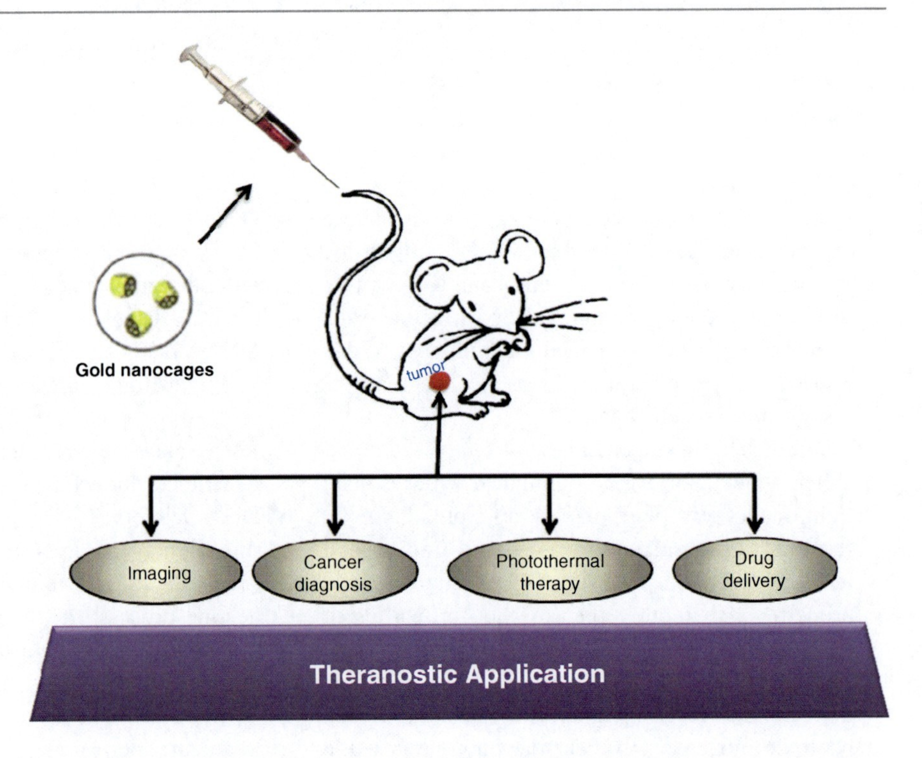

Gold Nanomaterials as Prospective Metal-based Delivery Systems for Cancer Treatment, Fig. 7 Theranostic application of gold nanocages (*Au*NCs) and their significant role in treatment of cancer disease

2. They can be synthesized in large scale, with wall thicknesses tunable in the range of 2–10 nm with an accuracy of 0.5 nm.
3. Their LSPR peaks can be tuned to any wavelength of interest in the range of 600–1,200 nm with precision, by simply controlling the amount of $HAuCl_4$ added to the reaction.
4. Their hollow interiors can be used for encapsulation of drugs and, thus, are suitable drug-delivery systems through the porous walls, with the release being controlled by various stimuli.
5. Their sizes can be readily varied from 20 to 500 nm to optimize biodistribution, facilitate particle permeation through epithelial tissues, or increase drug loading.
6. Their LSPR peaks can be dominated by absorption or scattering to adapt to different imaging modalities.
7. Other noble metals such as ▶ Pd (Palladium) and Pt (Platinum) can be incorporated into the walls during synthesis to maneuver their optical properties.

*Au*NCs have extraordinary large scattering and absorption cross sections, which make them superb optical tracers or contrast agents for imaging modalities such as dark-field microscopy, optical coherence tomography (OCT), photoacoustic tomography (PAT), and multi-photon luminescence-based detection.

The potential biomedical applications of *Au*NCs have been demonstrated both in vitro and in vivo studies for theranostic application (Fig. 7). Due to their high absorption cross section in NIR region, most incident photons are absorbed by gold nanocages and converted into phonons or vibrations of the lattice when resonance frequency is applied. The subsequent generation of heat can cause rise in temperature as well as thermo-elastic expansion, which, in turn, can be used in bio-imaging and photothermal therapy (Fig. 8). Besides their use as drug-delivery system, the surface of *Au*NCs could be readily functionalized with biomolecules or polymers to enhance biocompatibility without significant shift in their SPR peaks.

Gold Nanostar

The star-shaped gold nanoparticle is approximately 25-nm wide and has around 5–10 points. The points are highly spiky (Fig.1Ae), and it is also known as aspherical nanoparticles because of their structure. The use of highly pointed, spiky particles has shown a huge enhancement in electrical conduction, since the charges accumulated at the particle protuberances generate a very large electric local field. It is believed that when the composite material is compressed, the electrons tunnel between the sharp tips at even higher distances with respect to the spherical particles, which

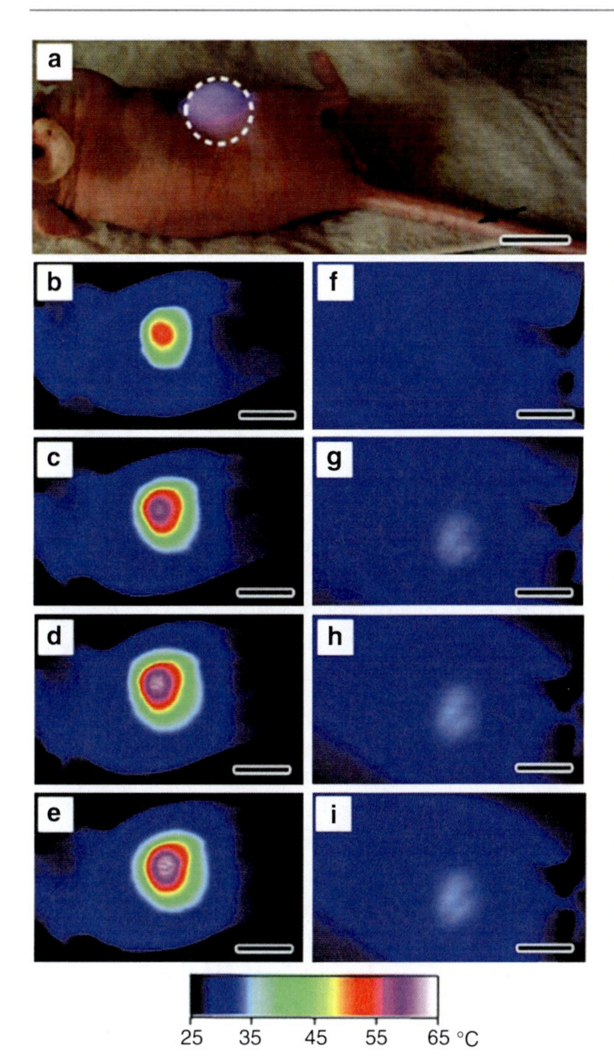

Gold Nanomaterials as Prospective Metal-based Delivery Systems for Cancer Treatment, Fig. 8 Photograph of a tumor-bearing mouse undergoing photothermal treatment using gold nanocages (*Au*NCs). (**a**) PEGylated nanocages were administrated intravenously through the tail vein as indicated by an arrow (in tail region) After the nanocages had been cleared from the circulation (72 h after injection), the tumor on the right flank was irradiated by the diode laser at 0.7 W cm^{-2} with a beam size indicated by the dashed circle in figure. Figure (**b–g**) Thermographic images of nanocage-injected and (**f–i**) saline-injected tumor-bearing mice at different time point intervals: (**b**, **e**) 1 min; (**c**, **f**) 3 min; (**d**, **g**) 5 min; and (E, I) 10 min (Reproduced with permission (John Wiley and Sons), Literature from Chen et al. 2010)

results in a larger reduction of the electrical resistance. Indeed, the local charge density enhancement is well known for these highly irregular metal spiky particles, which are widely used for the surface-enhanced Raman scattering (SERS) effect.

Because the plasmonic behavior of *Au*NPs is largely determined by their geometry and size, aspherical *Au*NPs typically exhibit plasmons in the near-infrared (NIR) region. This region is called the "tissue therapeutic window" because of the lower optical attenuation from water and blood in this spectral range (650–900 nm). At the plasmon resonance, the absorption cross section of aspherical *Au*NPs is strongly enhanced to several orders of magnitude higher than an organic dye; aspherical *Au*NPs absorb NIR photoenergy strongly, thereby efficiently transducing the incident light into heat (Yuan et al. 2012). Moreover, because of their unique plasmonic property, aspherical *Au*NPs emit strong two-photon photoluminescence (TPL), allowing direct particle visualization under multi-photon microscopy. Combining a preferential tumor site accumulation, efficient NIR photothermal transduction, and particle optical traceability, aspherical *Au*NPs are ideal candidates for various biomedical arenas, including in vitro and in vivo photothermal therapy (PTT), surface-enhanced Raman scattering spectroscopy, photoacoustic imaging, biosensing, photodynamic therapy, and also magnetomotive imaging. Gold nanostars are also able to release the drugs easily and that their star shape allowed concentration at specific points with cells. (http://www.medicalnewstoday.com/articles/243856.php/).

Biodistribution of Gold Nanomaterials

Biodistribution study of *Au*NMs depends upon the various aspects such as size, shape, charge, functionalization methods; cells and animal types; particle administration doses and methods; and so on. Most of the studies revealed that *Au*NMs of all sizes were mainly accumulated in various parts of organs like liver, lung, and spleen. The exact accumulation of *Au*NMs in various tissues was dependent on particle size. 15-nm *Au*NPs revealed higher accumulation in all tissues including blood, liver, lung, spleen, kidney, brain, heart, and stomach. On the other hand, 15- and 50-nm *Au*NPs were able to pass blood–brain barrier as evident from gold concentration in brain and around 200-nm *Au*NPs showed very minute presence in different organs including blood, brain, stomach, and pancreas (Sonavane et al. 2008).

Gold Nanomaterials as Prospective Metal-based Delivery Systems for Cancer Treatment, Table 1 Examples of gold-based nanomaterials in therapeutics use (with FDA approval status)

Product/brand name	Component/active ingredient	Delivery route	Target	Company	Current status
Verigene	Gold	In vitro diagnostic	Genetics	Nanosphere	FDA approved
Aurimmune	Colloidal gold nanoparticles coupled to TNF-α and PEG-Thiol(~27 nm)	Intravenous	Solid tumor	Cyt-Immune Sciences	Phase-II
AuroShell	Gold-coated silica nanoparticles (~150 nm)	Intravenous	Solid tumor	Nanospectra Biosciences	Phase-I

It is also known that PEGylation of *Au*NPs influences the biokinetics after IV injection into mice, especially regarding the uptake in liver and spleen. Therefore, PEGylation process presents an important tool to prolong the blood circulation times. Sometimes total translocation of the NPs in circulation and accumulation in secondary organs is not influenced by the PEGylation (PEG-750 or PEG-10 k) process. Therefore, other surface modifications (e.g., carboxylation) play an important role to influence the translocation from the lungs to the circulation.

There is a large number of scatter data published in different journals on the levels of kinetics, biodistribution, and on toxicity of *Au*NMs estimates. Most of the data collected pertained to the effects of the shape, structure, charge, and surface chemical modification of *Au*NMs on their pharmacokinetics in the blood stream, their kinetics of accumulation and elimination from the organism, and their short- and long-term toxicities (Khlebtsov and Dykman. 2011) However, this field is quite unclear and more deep research is needed to understand the behavior of *Au*NMs in massive environment.

Toxicity of Gold Nanomaterials

Potential applications of *Au*NMs have been already investigated in biomedical field including delivery, chemical sensing, and imaging applications while bulk *Au*NMs have been deemed "safe." However, more research needs to be done to study the biocompatibility and environmental impact of *Au*NMs provided they are to be used on a large-scale basis (in vivo) (Khlebtsov and Dykman 2011). It is well known that *Au*NMs internalized by the cells depend upon various aspects such as size, shape, surface coating, and incubation time and so on. The potential concentrations at which *Au*NMs might be used for treatment should be well validated. It has been reported that spherical *Au*NPs are nontoxic to the human leukemia cells in concentrations up to ~100 μM. Surface modification of nanomaterials included a wide range of anionic, neutral, and cationic groups like citrate, cysteine, glucose, biotin (aka vitamin B7 or H), and cetyltrimethylammonium bromide (CTAB). CTAB is the structure-directing agent that can be used to control shape in *Au*NRs, and it appears to form a tightly bound cationic bilayer on *Au*NPs during synthesis. Due to this cationic trimethylammonium headgroup, *Au*NRs show cytotoxicity in cells. At certain concentration, (~10 nm) CTAB and gold salt $HAuCl_4$ were found to be toxic to the cells that can break the cell membrane after exposure. Nevertheless, these materials are still used in several biomedical applications after proper purification and also functionalization of surface by using polymer (e.g., PEG) to make it more relevant for such applications. The cationic *Au*NPs are generally toxic at much lower concentrations than anionic particles, which they relate to the electrostatic interaction between the cationic nanoparticles and the negatively charged cell membranes (Kim et al. 2010). Cytotoxicity of the nanomaterials also depends upon the type of cells, receptor expression, and so on because the cells have very unique natural properties for their selection.

Gold-Based Nanomedicine

The biomedical application of gold stretches back to almost 5,000 years. In today's era of nano-age, nanoparticles have opened new doors for diagnostic and therapeutic applications and many nano-based drugs are already in clinical trials (Table 1). Special progress has been made in the field of tumor targeting whereby *Au*NPs can specifically target and deliver anticancer drugs directly into cancerous tumors, with the added advantage of being simple and cost-effective. Currently, significant research is going on how to improve the therapeutic effects of pharmaceutical drug based upon designing of these materials and at the same time how to control the toxicity and environmental impacts to make these materials more applicable for biomedical applications.

Acknowledgment This work was supported in part by Chinese Natural Science Foundation project (No.30970784 and 81171455), National Key Basic Research Program of China (2009CB930200), Chinese Academy of Sciences (CAS) "Hundred Talents Program" (07165111ZX), and CAS Knowledge Innovation Program. We sincerely thank Bhargavi M Boruah, Laboratory of Pathogenic Microbiology and Immunology, Institute of Microbiology, Chinese Academy of Sciences, Beijing-100101, China Chinese Academy of Sciences, for her help and valuable suggestions in preparing this entry and also Mr. Zhang Xu for his help in preparing the graphics of this manuscript. Author would like to thank every scientific community and people who are involved in cancer research and are making an effort in the fight against cancer.

Cross-References

▶ Gold and Nucleic Acids
▶ Luminescent Tag
▶ Peptide Toxins
▶ Photothermal Therapy

References

Bardhan R, Lal S, Joshi A et al (2011) Theranostic nanoshells: from probe design to imaging and treatment of cancer. Acc Chem Res 44:936–946

Brust M, Walker M, Bethell D et al (1994) Synthesis of thiol-derivatised gold nanoparticles in a two-phase liquid–liquid system. J Chem Soc Chem Commun 1994:801–802

Connor EE, Mwamuka J, Gole A (2005) Gold nanoparticles are taken up by human cells but do not cause acute cytotoxicity. Small 1:325–327

Dreaden EC, Mackey MA, Huang X et al (2011) Beating cancer in multiple ways using nanogold. Chem Soc Rev 40:3391–3404

Dreaden EC, Alkilany AM, Huang X et al (2012) The golden age: gold nanoparticles for biomedicine. Chem Soc Rev 41:2740–2779

Dykman L, Khlebtsov N (2012) Gold nanoparticles in biomedical applications: recent advances and perspectives. Chem Soc Rev 41:2256–2282

Huang X, El-Sayed MA (2010) Gold nanoparticles: optical properties and implementations in cancer diagnosis and photothermal therapy. J Adv Res 1:13–28

Huang K, Ma H, Liu J et al (2012) Size-dependent localization and penetration of ultrasmall gold nanoparticles in cancer cells, multicellular spheroids, and tumors in vivo. ACS Nano 6:4483–4493

Jelveh S, Chithrani DB (2011) Gold nanostructures as a platform for combinational therapy in future cancer therapeutics. Cancer 3:1081–1110

Jiang X-M, Wang L-M, Wang J et al (2012) Gold nanomaterials: preparation, chemical modification, biomedical applications and potential risk assessment. Appl Biochem Biotechnol 166:1533–1551

Kanaras AG, Bartczak D (2011) Preparation of peptide-functionalized gold nanoparticles using one pot EDC/Sulfo-NHS coupling. Langmuir 27:10119–10123

Khlebtsov N, Dykman L (2011) Biodistribution and toxicity of engineered gold nanoparticles: a review of in vitro and in vivo studies. Chem Soc Rev 40:1647–1671

Kim C, Agasti SS, Zhu ZJ et al (2010) Recognition-mediated activation of therapeutic gold nanoparticles inside living cells. Nat Chem 2:962–966

Kumar A, Bhargavi MB, Liang XJ (2011) Gold nanoparticles: promising nanomaterials for the diagnosis of cancer and HIV/AIDS. J Nanomater. doi:10.1155/2011/202187

Kumar A, Ma H, Zhang X et al (2012) Gold nanoparticles functionalized with therapeutic and targeted peptides for cancer treatment. Biomaterials 33:1180–1189

LaMer VK, Dinegar RH (1950) Theory, production and mechanism of formation of monodispersed hydrosols. J Am Chem Soc 72:4847–4854

Sonavane G, Tomoda K, Makino K (2008) Biodistribution of colloidal gold nanoparticles after intravenous administration: effect of particle size. Colloids Surf B Biointerfaces 66:274–280

Tong L, Wei QS, Wei A et al (2009) Gold nanorods as contrast agents for biological imaging: optical properties, surface conjugation and photothermal effects. Photochem Photobiol 85:21–32

Xia Y, Li W, Cobley CM et al (2011) Gold nanocages: from synthesis to theranostic applications. Acc Chem Res 44:914–924

Yuan H, Khoury CG, Wilson CM et al (2012) In vivo particle tracking and photothermal ablation using plasmon-resonant gold nanostars. Nanomed Nanotechnol Biol Med 8:1355–1363

Gold Nanoparticle Platform for Protein-Protein Interactions and Drug Discovery

Cynthia Bamdad
Minerva Biotechnologies Corporation, Waltham, MA, USA

Synonyms

AuNPs; Colloids; Gold nanoparticles; Nanoparticles; NPs; Sol

Definition

Nanoparticles are generally formed from metal or semiconductor material and their diameters range from 10 to 100 nm. Metallic particles of diameter less than ~4 nm are generally referred to as metal clusters, comprised of just a few or up to 300 metal atoms. One reason for the distinction between nanoparticles and metal clusters is that metal clusters do not share the same optical properties as the larger nanoparticles. For example, gold clusters are not characterized by the red color of colloidal gold and cannot produce a color change when aggregated. The most commonly used nanoparticles are gold nanoparticles, which appear red when in a homogeneous suspension and turn blue/gray when aggregated and those made of CdSe, such as Quantum Dots™, which self-fluoresce.

Background

Gold nanoparticles (AuNPs) have been used in biological assays for decades in agglutination assays, radioactive assays, and in pregnancy tests (Geoghegan et al. 1980). In these earlier colloidal gold assays, antibodies were typically adsorbed directly onto the nanoparticles. Upon antigen binding, agglutination or a solution color change would occur. One problem with these earlier approaches was that antibodies could be nonspecifically adsorbed onto bare nanoparticles without significant denaturation, but most other proteins could not. Therefore, early nanoparticle assays were limited to the detection of antibody-antigen interactions. Another problem was that even antibody-coated colloids are unstable when stored in solution for long periods, so the particles either had to be embedded in an absorbent strip or coated with antibody just prior to use. Thus, applications of these nanoparticles were very limited.

Technical advances in nanoparticle synthesis and in nanoscale surface chemistries that were made in the 1990s greatly expanded the applicability of nanoparticles to biomedical testing. Methods were developed to generate nanoparticles out of semiconductor materials (Brus 1991), which were later marketed as Quantum Dots™. These self-fluorescing nanoparticles enabled real-time visualization of, for example, a ligand's journey through a cell. Others attached DNA oligos to gold particles to detect the presence of target DNA (Mirkin et al. 1996). Perhaps the greatest advance that enabled today's nanoparticle technology was the development of methods of molecular self-assembly (Nuzzo and Allara 1983; Pale-Grosdemange et al. 1991). Under certain conditions, thiols line up or "self-assemble" to form single molecule thick layers (monolayers) on gold substrates: self-assembled monolayers (SAMs). These thiols can be terminated with a variety of headgroups, imparting several functionalities to a planar gold surface (Sigal et al. 1996; Bamdad 1998a, b). Improvements to these techniques enabled the formation of SAMs on gold nanoparticles, which stabilized the particles and provided user-friendly surfaces to which proteins could be readily attached (Bamdad 2000; Thompson et al. 2011).

Advantages of Gold Nanoparticles

Nanoparticles are on the same size scale as proteins and about 500–1,000-times smaller than the average cell in the human body, so do little to perturb the native environment during a biological experiment. An important characteristic of gold nanoparticles is that they can self-signal when they find their target. Gold nanoparticles have the intrinsic optical property that when they are in a homogeneous suspension, the solution looks pink/red, but when the particles are aggregated or drawn close together, the solution turns blue/gray. Therefore, in a bioassay, protein-binding partners, immobilized on separate particles and then mixed together, signal the binding event by bringing the particles in close proximity to each other, which induces the pink-to-blue color change. However, bare gold nanoparticles are relatively unstable in solution, due to electromagnetic properties of colloidal particles. In the presence of salt and over a wide range of pHs, changes to the surface charge of the particles cause gold nanoparticles to irreversibly precipitate out of solution. Nanoparticles can be somewhat stabilized by adding detergent to the solution, or as in the early assays, nonspecifically adsorbing proteins to the particles. However, detergents interfere with binding assays and many proteins denature when adsorbed directly onto nanoparticles. A better approach is to coat or "cap" the nanoparticles with molecules that shield surface charge or carry their own charge, which causes the nanoparticles to repel each other. Importantly, sulfur binds to gold, so thiol chemistry

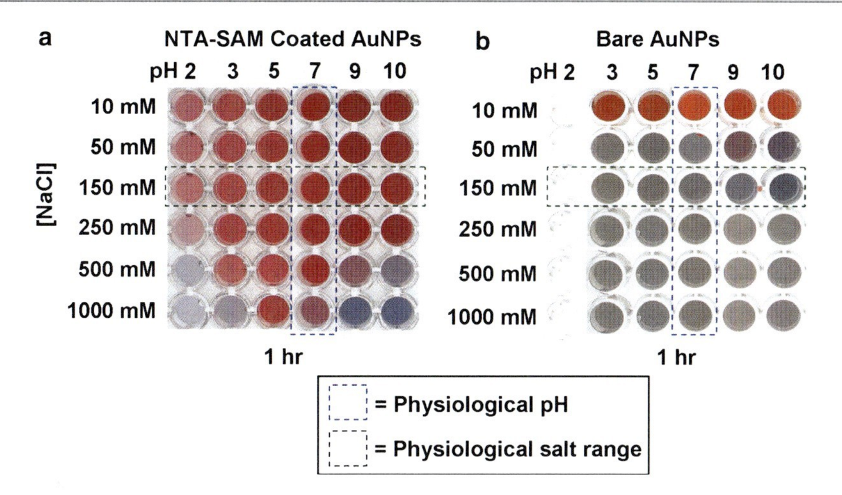

Gold Nanoparticle Platform for Protein-Protein Interactions and Drug Discovery, Fig. 1 *Stability of bare* versus *NTA-SAM-coated nanoparticles*. Salts mask surface charges on gold nanoparticles (AuNPs), which causes them to aggregate and makes the solution color change from *red* to *blue/gray*. Similarly, changes in the protonation state of surface atoms can cause the AuNPs to precipitate out of solution. AuNPs of 25 nm were either *left* bare or coated with NTA-SAMs (nitrilo triacetic acid-self-assembled monolayers) and then incubated in varying concentrations of NaCl as indicated. Resistance to salt-induced aggregation was determined by observing solution color change after 1 h. *Dashed boxes* represent physiological NaCl concentrations and pH. As can be seen, NTA-SAMs (**a**) make AuNPs stable over a wide range of salt concentrations and over a range of pHs. Bare AuNPs (**b**) precipitate out of solution at physiological pH and salt concentrations

can be used to derivatize gold nanoparticles. Thiols are saturated carbon chains terminated with a sulfur atom. The simple addition of thiols to gold nanoparticles will temporarily stabilize the particles, but thiols desorb over time, rendering the particles unstable again. Gold nanoparticles coated with SAMs provide the most stable particles because the thiols are integrated into a robust monolayer that is difficult to disrupt. The reason for the increased stability of SAMs is that sulfur chemisorbs onto gold in a specific tiling pattern, wherein the distance between sulfurs is 4.97 Å (Strong and Whitesides 1988), which is approximately the van der Waals distance between the carbons of the alkyl chain. This means that there is a lot of energy generated between neighboring carbons, as the alkyl chains line up into a uniform, 1-molecule thick layer that is difficult to disrupt. Gold nanoparticles are sometimes coated with carboxy-terminated SAMs. The negative charge of the carboxylates makes the nanoparticles repel each other and thus prevents nanoparticle precipitation. However, SAMs comprised of a single-charged species are unstable as the thiols will rearrange to better distribute charge. In addition, nanoparticles that carry a net charge can repel a target protein of the same charge even if the nanoparticle bears its cognate ligand. Heterologous SAMs comprised of mixed thiols and especially those that incorporate ethylene glycol-terminated thiols, to resist nonspecific binding, are preferred. Heterologous SAMs render gold nanoparticles stable under a multitude of assay conditions, including over a wide range of salt concentrations, pH, and temperatures. For example, 25 nm gold nanoparticles derivatized with SAMs comprised of thiols terminated with triethylene glycol, carboxylates, and nitrilo-tri-acetic acid (NTA-Ni for the specific capture of histidine-tagged proteins) are stable at physiological pH 7 in sodium chloride concentrations from 10 to 500 mM compared to bare gold nanoparticles, which are only stable in 10 nm salt at pH 7 (Fig. 1). Bioassays are typically performed at pH 7.4 in 150 mM NaCl.

Factors that Affect the Accuracy of Nanoparticle Bioassays

Stability

Typically, derivatized nanoparticles are used to detect protein-protein interactions or to screen for drugs that disrupt a targeted interaction. The color change produced by nanoparticle instability is indistinguishable from the color change produced when two particle-immobilized proteins interact and draw the particles close together. Accordingly, one must be sure that under particular assay conditions, the nanoparticles do not precipitate out of solution in the absence of a real protein-protein interaction. Additives, such as Tween-20 or BSA, are often added to stabilize nanoparticles and prevent spontaneous precipitation. However, it should be noted that these additives greatly decrease the sensitivity of the assay. In addition to salt- and pH-induced particle aggregation, many organic solvents cause gold nanoparticles to precipitate. DMSO, which is a solvent that many drug candidates are dissolved into, causes nanoparticle aggregation, so users are cautioned to dissolve drug candidates in DMSO but then dilute as much as possible in the assay buffer.

Particle Size

As discussed above, a solution of gold nanoparticles changes in color from red/pink to blue/gray when the particles are drawn close together. Gold nanoparticles between 10 and 50 nm are characterized by a brilliant red color and absorb light in the visible region with peak wavelengths around 520 nm (van der Hulst 1957). Decreasing the distance between gold nanoparticles causes a visible color change from red to blue as the absorption peak broadens and shifts toward longer wavelengths of about 620 nm (Kreibig and Grenzel 1985; Dusemund et al. 1991; Elghanian et al. 1997). While the effect of inter-particle distance on color change has been well established (Rechberger et al. 2003), the effect of nanoparticle size on detecting the color change has been underappreciated. Color change of gold nanoparticle solutions is also determined by the diameter of the nanoparticle itself (Link and El-Sayed 1999). For example, nanoparticles with diameters less than 4 nm do not emit color whether in a homogeneous suspension or aggregated, so they cannot be used in a self-signaling color change assay. The reverse is also true; the larger the nanoparticle the faster and more intense the color change. In fact, the extinction coefficient of gold nanoparticles increases exponentially with diameter (Nath and Chilkoti 2002), which means that a few interactions among large particles will produce the same color change as many interactions among small particles. Thus, nanoparticles of large diameter are more sensitive for detecting protein-protein interactions. However, particles that are too large are likely to perturb the assay or sediment out of solution. In general, gold nanoparticles between 20 and 40 nm in diameter are optimal for use in bioassays.

To demonstrate the effect of particle size on assay sensitivity, a typical antibody-antigen binding assay was performed in which a histidine-tagged peptide antigen was immobilized onto NTA-Ni-SAM-coated nanoparticles and the cognate antibody was added into the nanoparticle solution. Antibodies are bivalent; so one antibody can simultaneously bind to two different peptides that are immobilized on two different nanoparticles, which causes a color change in the solution as the distance between nanoparticles is reduced. The same antibody-antigen assay was performed in parallel using five different nanoparticle batches with diameters of 15 nm, 20 nm, 25 nm, 30 nm, or 35 nm. Antibody-induced particle aggregation was detected at concentrations as low as 6.7 nm for 30–35-nm diameter nanoparticles, but could not be detected using 25 nm nanoparticles until the antibody concentration was increased by four times. Nanoparticles of 15 nm could not detect the antibody even at concentrations five times greater than the concentration at which nanoparticles of 30 nm detected it (Fig. 2).

Detection Method

Detecting nanoparticle color change by eye is convenient but less sensitive than detecting by a spectrophotometer. Using a spectrophotometer, the characteristic red color of gold nanoparticles can be detected as a peak at wavelength ~520 nm. Although the aggregation-induced color change from pink to blue shifts the peak wavelength to ~620 nm, complete aggregation results in near complete loss of color. A more reliable method for detecting nanoparticle color change with a spectrophotometer is detecting the decrease in the peak at 521 nm. In this method, a known set of interacting proteins is immobilized onto separate nanoparticles and the interaction is allowed to proceed to completion. The absorbance at 521 nm at t_∞ is recorded and used in (1) to calculate K, which is a percentage of maximal aggregation. Comparison of Fig. 2a, b shows the increase in sensitivity when color change is detected by a spectrophotometer measuring the decrease in absorbance at 521 nm.

$$K = \frac{A_{521}^{t} - A_{521}^{t_0}}{A_{521}^{t_\infty} - A_{521}^{t_0}} * 100 \qquad (1)$$

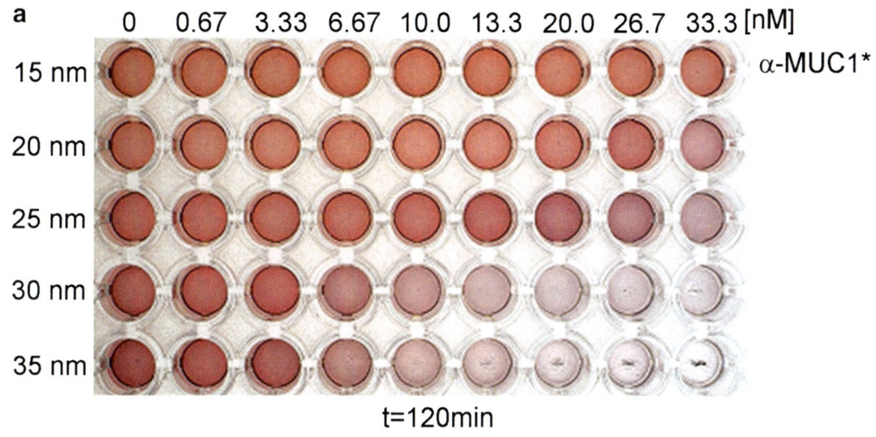

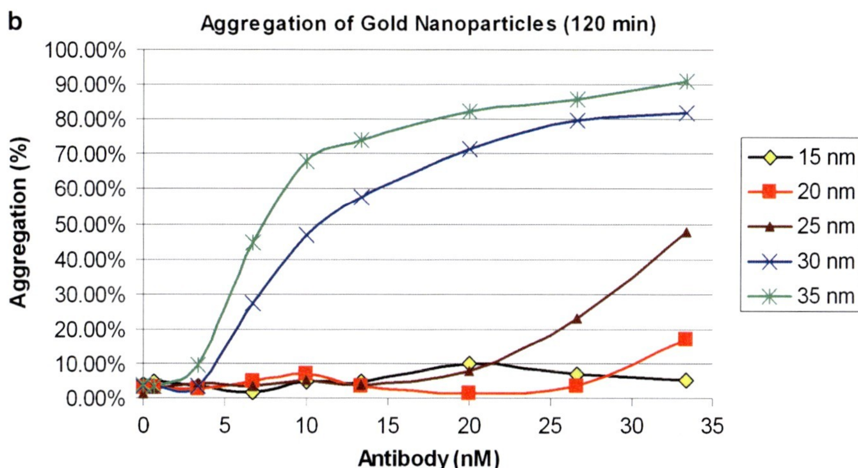

Gold Nanoparticle Platform for Protein-Protein Interactions and Drug Discovery, Fig. 2 *Correlation between nanoparticle size and sensitivity*. The size of gold nanoparticles in solution is a determinant of nanoparticle color change. The same binding assay was performed in parallel using five different nanoparticle batches with diameters of 15 nm, 20 nm, 25 nm, 30 nm, or 35 nm. Nanoparticles were coated with NTA-Ni-SAMs designed to capture and present histidine-tagged species. A histidine-tagged MUC1* sequence peptide was immobilized onto the nanoparticles and the cognate antibody was added over a range of concentrations. Antibodies are bivalent so one antibody can simultaneously bind to two peptides immobilized on two different nanoparticles, which causes a solution color change as the distance between nanoparticles is reduced. (**a**) Nanoparticle color change is clearly visible to the naked eye and shows that the larger the nanoparticle, the more sensitive is the detection. (**b**) A spectrophotometer was used to quantify nanoparticle color change. Equation 1 was used to calculate percentage loss of absorbance at 521 nm. In this method, a known set of interacting proteins is immobilized on nanoparticles and the interaction is allowed to go to completion. The absorbance measured at 521 nm is defined as maximum aggregation, $A^{t\infty}$ and a percentage of aggregation after 120 min is plotted in function of the antibody concentration. As can be seen by comparing (**a**) to (**b**), detection using a spectrophotometer is more sensitive than inspection by eye

Method of Protein Attachment

The fourth factor that contributes to the veracity of any nanoparticle assay is how the proteins are attached to the nanoparticle. As discussed above, proteins can be nonspecifically adsorbed directly onto bare gold nanoparticles. Antibodies may be rigid enough that they are not significantly denatured by this process, however, this method of protein or peptide immobilization is not recommended. In addition to the problem of protein denaturation, nonspecifically bound proteins or peptides will desorb over time, thus compromising the integrity of the assay. Nanoparticle surface coatings, ideally SAMs, that incorporate an entity that enables protein attachment are preferred. Surface coatings with exposed carboxylates can be used to covalently couple proteins using standard

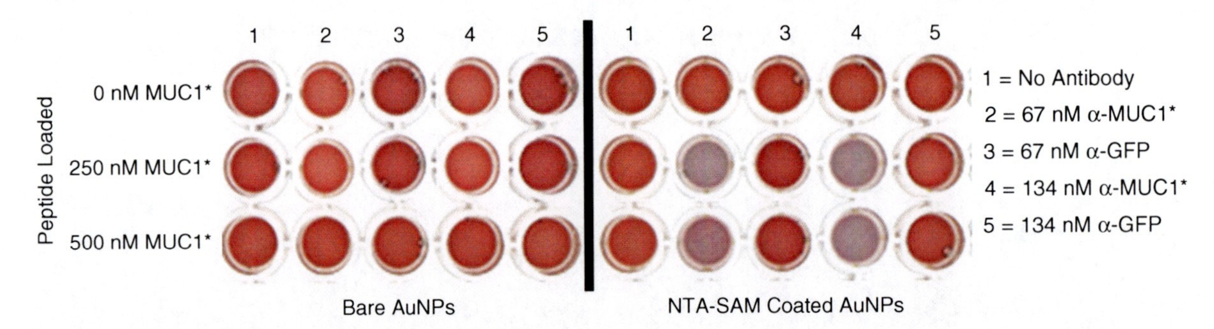

Gold Nanoparticle Platform for Protein-Protein Interactions and Drug Discovery, Fig. 3 *Comparison of peptide loading onto bare gold nanoparticles compared to SAM-coated nanoparticles*. Protein-protein interactions are easily detected when probes are loaded onto NTA-Ni-SAM-coated gold nanoparticles (AuNPs) but undetectable when nonspecifically adsorbed onto bare AuNPs. The color of AuNPs changes from *red* to *blue/gray* as distance between nanoparticles is reduced, for example, when peptides on nanoparticles interact with each other or with a common target. Bare AuNPs and NTA-Ni-SAM-coated AuNPs were loaded with three concentrations of a histidine-tagged MUC1* peptide, then exposed to either the cognate antibody (α-MUC1*) or an irrelevant control antibody (α-GFP). There was no color change for any of the wells in which the MUC1* peptide had been nonspecifically adsorbed onto bare AuNPs (*left*). In stark contrast, when the MUC1* peptides immobilized on NTA-Ni-SAM-coated AuNPs were exposed to α-MUC1*, the color of the solution changed from *red* to *blue/gray*, indicating an interaction, but not when they were exposed to the control antibody, indicating color change is due to specific interaction between binding pairs

EDC/NHS chemistry that couples a primary amine on the protein to the activated carboxylate on the particle. If the concentration of the protein is high relative to the concentration of the particles, one can avoid coupling one protein to many particles. Dynamic light scattering analysis, which measures an effective particle diameter, is informative for determining the optimal protein concentration at which a high percentage of proteins become coupled to a single nanoparticle, rather than one protein coupled to two or more nanoparticles. The latter case renders the protein inaccessible to interactions with binding partners, yet the faulty immobilization mimics real interactions. Another method of attaching proteins to nanoparticles is via interaction between a moiety incorporated into the nanoparticle surface coating and an affinity tag on the protein. Although this technique is mainly used for recombinant proteins expressed with an affinity tag, such as a histidine-tag, it is the easiest method of protein attachment that presents the probe protein in a consistent orientation without denaturing the protein. For example, nanoparticles coated with SAMs that incorporate NTA-Ni, triethylene glycol and carboxy-terminated thiols instantly capture any histidine-tagged protein or peptide that is added to the nanoparticle solution. A simple experiment highlights the pitfalls of directly adsorbing peptides onto bare nanoparticles and the advantage of immobilizing affinity tagged peptides to SAM-coated nanoparticles bearing NTA-Ni. The same peptide is either nonspecifically adsorbed onto bare nanoparticles or immobilized onto NTA-Ni-SAMs via the affinity tag-metal chelate interaction. Peptides nonspecifically adsorbed are unable to bind their cognate antibody or an irrelevant antibody, yet the same peptide immobilized via a SAM is available to bind to its cognate antibody and does not bind the irrelevant antibody (Fig. 3).

Applications

Detecting Protein-Protein Interactions

Generally, to see if two proteins interact, each protein is separately immobilized on two different sets of nanoparticles and the particles are then mixed together. If a color change from red/pink to blue/gray occurs, then the proteins have interacted (Fig. 4). If proteins are immobilized onto nanoparticles via interaction between an affinity tag on the protein and a moiety on the nanoparticle, then proteins need not be immobilized on two different sets of nanoparticles, which are then mixed together. Statistically, enough proteins will immobilize on different nanoparticles that the particles will become cross-linked as the binding partners interact.

As in any bioassay, proper controls must be included in the experimental setup. However, several perhaps unanticipated precautions must be taken when

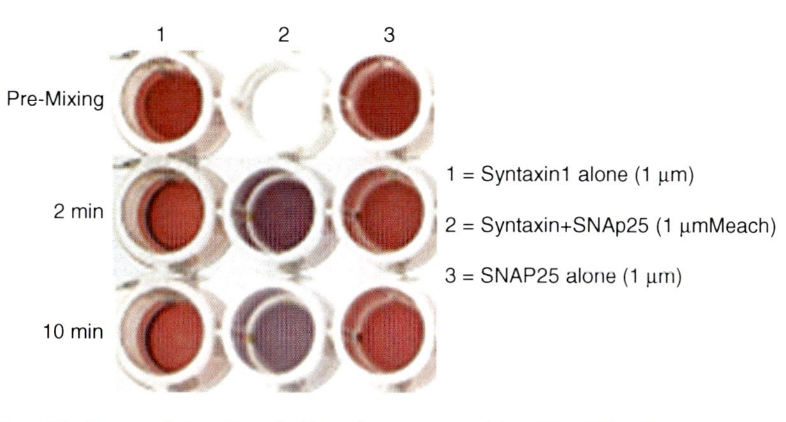

Gold Nanoparticle Platform for Protein-Protein Interactions and Drug Discovery, Fig. 4 *Detecting protein-protein interactions.* Known binding partners SNAP25 and Syntaxin1a were expressed with histidine tags and separately immobilized onto different sets of NTA-Ni-SAM-coated nanoparticles. The NTA-Ni-SAM-coated AuNPs were loaded with equal concentrations of either histidine-tagged SNAP25 or histidine-tagged Syntaxin1a. The two nanoparticle solutions were mixed together and color change indicating their interaction was observed within 2 min

detecting protein-protein interactions wherein one or both are dimeric. If the ligand that is being detected dimerizes its peptide receptor, then the peptide should be immobilized on the nanoparticle at low density. If the peptide is immobilized on the nanoparticle at high density, then the ligand(s) can dimerize two peptides on the same particle and no color change will result. Similarly, if the ligand is added at too high a concentration, then one ligand dimer can bind to each peptide receptor instead of one ligand dimerizing two receptors.

In some cases, one may want to identify an unknown ligand that binds to an orphan receptor. Many growth factor receptors function by ligand-induced dimerization. The extracellular domain of a transmembrane receptor can be attached to nanoparticles and used in a screen of cell lysates and supernatants, suspected of containing the dimerizing growth factor. If the sample added to the receptor-bearing nanoparticles contains the dimerizing ligand, then the characteristic color change will occur. As long as the nanoparticles are coated with properly formed SAMs, optimally bearing ethylene glycol species, the whole cell lysates, supernatants, or fractions thereof can be added to peptide-bearing nanoparticles without being compromised by nonspecific binding. Although detecting interactions among multimers is more complicated than detecting single binding pairs, they are biologically relevant and of great interest in the area of drug discovery.

Drug Screening

If one is able to detect an interaction between proteins in a nanoparticle assay, then screening for drugs that disrupt that interaction is immediately available. A candidate drug is added to a solution containing the two interacting proteins and if the drug candidate inhibits the interaction, the nanoparticle color change will not occur. In nanoparticle assays, it is easier to inhibit an interaction than it is to disrupt an interaction that has already taken place. One reason for this is that proteins on different nanoparticles interact and quickly form a protein-nanoparticle reticulum that sterically hinders a drug candidate's access to the proteins of interest. Therefore, nanoparticle drug screens are most successful when the drug candidates are added first to a multi-well plate, followed by the addition of one set of protein-immobilized particles then the other. To screen for drugs that inhibit the binding of a particle-immobilized peptide to a dimeric ligand, free in solution, one should separately preincubate both the peptide-bearing nanoparticles and the ligand in solution with the candidate drug and then mix the two solutions together. Otherwise, the avidity of the peptide-particle reticulum could mask the inhibitory effect of bona fide inhibitors.

Drug screening for inhibitors of Class I growth factor receptors, which are activated by ligands that dimerize the extracellular domain, has been successfully done even when the activating ligand was unknown. The extracellular domain of the

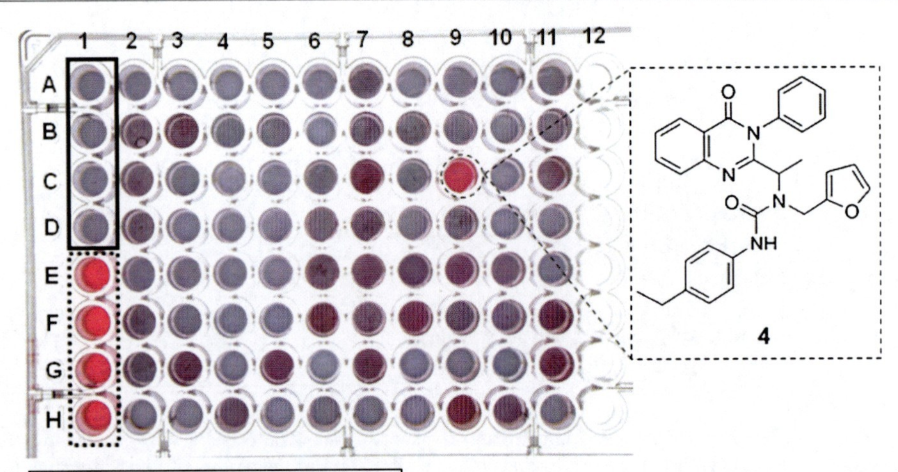

Gold Nanoparticle Platform for Protein-Protein Interactions and Drug Discovery, Fig. 5 *Demonstration of a high-throughput drug screen.* NTA-Ni-SAM-coated gold nanoparticles are a platform for high-throughput drug screening. A small molecule drug library was screened for the ability to disrupt an interaction between MUC1* peptide and its ligand, dimeric NM23. All wells of a 96-well plate contain MUC1* peptide loaded onto NTA-Ni-SAM-coated AuNPs. Column 1, rows A–D are positive controls that contain MUC1*-bearing AuNPs, while E–H are negative controls that contain AuNPs bearing an irrelevant peptide. All other wells contain an equal weight amount of compound of approximately the same molecular weight. The endogenous MUC1* ligand that dimerizes MUC1* (NM23) was then added to every well. The compound in well C9 inhibited the nanoparticle color change and is identified as a potent inhibitor of the MUC1*-NM23 interaction. The drug "hit" of well C9 is shown in the inset to the *right*

transmembrane receptor of MUC1* was synthesized with a histidine tag to facilitate immobilization onto NTA-Ni-SAM-coated nanoparticles. Cell lysates or supernatants from MUC1*-positive or negative cancer cells were separately added to the nanoparticle solutions in search of a supernatant fraction that contained a ligand that dimerized the MUC1* extracellular domain peptides and caused the nanoparticle color change. It was observed that only supernatants from MUC1*-positive cancer cells secreted the dimerizing ligand and cells that had high MUC1* expression secreted more of the dimerizing ligand. Using the strategy outlined above, small molecules from a drug library were screened and several drug hits were identified that inhibited the MUC1*-ligand interaction (Fig. 5). In cell-based assays, these drug hits greatly inhibited the growth of MUC1*-positive cancer cells. As long as proper controls are incorporated, derivatized gold nanoparticles can be an ideal technology for detecting and studying protein-protein interactions and for identifying small molecule or natural product antagonists.

Cross-References

- ▶ Aluminum, Physical and Chemical Properties
- ▶ Apoferritin, Activation by Gold, Silver, and Platinum Nanoparticles
- ▶ Biomineralization of Gold Nanoparticles from Gold Complexes in Cupriavidus Metallidurans CH34
- ▶ Colloidal Silver Nanoparticles and Bovine Serum Albumin
- ▶ Gold and Nucleic Acids
- ▶ Gold Nanomaterials as Prospective Metal-based Delivery Systems for Cancer Treatment
- ▶ Gold Nanoparticles and Fluorescent Proteins, Optically Coupled Hybrid Architectures
- ▶ Gold Nanoparticles and Proteins, Interaction
- ▶ Gold, Ultrasmall Nanoclusters and Proteins, Interaction
- ▶ Nickel, Physical and Chemical Properties
- ▶ Silicon Nanowires
- ▶ Silver in Protein Detection Methods in Proteomics Research

References

Bamdad C (1998a) A DNA self-assembled monolayer for the specific attachment of unmodified double or single stranded DNA. Biophys J 75(4):1997–2003
Bamdad C (1998b) The use of variable density self-assembled monolayers to probe the structure of a target molecule. Biophys J 75(4):1989–1996
Bamdad C (2000) Interaction of colloid-immobilized species with species on non-colloidal structures. WO 00/43783
Brus LE (1991) Quantum crystallites and nonlinear optics. Appl Phys A 53:465–474
Dusemund B, Hoffmann A, Salzmann T, Kreibig U, Schmid G (1991) Cluster matter – the transition of optical elastic-scattering to regular reflection. Z Phys D 20: 305–308
Elghanian R, Storhoff JJ, Mucic RC, Letsinger RL, Mirkin CA (1997) Selective colorimetric detection of polynucleotides based on the distance-dependent optical properties of gold nanoparticles. Science 277:1078–1081
Geoghegan WD, Ambegaonkar S, Calvanico NJ (1980) Passive gold agglutination. An alternative to passive hemagglutination. J Immunol Methods 34:11–21
Kreibig U, Grenzel L (1985) Optical absorption of small metallic nanoparticles. Surf Sci 156:678–700
Link S, El-Sayed MA (1999) Size and temperature dependence of the plasmon absorption of colloidal gold nanoparticles. J Phys Chem B 103:4212–4217
Mirkin CA, Letsinger RL, Mucic RC, Storhoff JJ (1996) A DNA-based method for rationally assembling nanoparticles into macroscopic materials. Nature 382:607–609
Nath N, Chilkoti A (2002) A colorimetric gold nanoparticle sensor to interrogate biomolecular interactions in real time on a surface. Anal Chem 74:504–509
Nuzzo RG, Allara DL (1983) Adsorption of bifunctional organic disulfides on gold surfaces. J Am Chem Soc 105: 4481–4483
Pale-Grosdemange C, Simons ES, Prime KL, Whitesides GM (1991) Formation of self-assembled monolayers by chemisorption of derivatives of oligo(ethylene glycol) of structure $HS(CH_2)11(OCH_2CH_2)mOH$ on gold. J Am Chem Soc 113:12–20
Rechberger W et al (2003) Optical properties of two interacting gold nanoparticles. Opt Commun 220:137–141
Sigal GB, Bamdad C, Barberis A, Strominger J, Whitesides GM (1996) A self-assembled monolayer for the binding and study of histidine-tagged proteins by surface plasmon resonance. Anal Chem 68(3):490–497
Strong L, Whitesides GM (1988) Structures of self-assembled monolayer films of organosulfur compounds adsorbed on gold single crystals: electron diffraction studies. Langmuir 4:546–558
Thompson AB, Calhoun AK, Bamdad C et al (2011) A gold nanoparticle platform for protein-protein interactions and drug discovery. ACS Appl Mater Interfaces 3:2979–2987
Van der Hulst HC (1957) Light scattering by small metal particles. Wiley, New York

Gold Nanoparticles

▶ Gold Nanoparticle Platform for Protein-Protein Interactions and Drug Discovery

Gold Nanoparticles and Fluorescent Proteins, Optically Coupled Hybrid Architectures

Naba K. Dutta[1], Ankit K. Dutta[2] and
Namita Roy Choudhury[1]
[1]Ian Wark Research Institute, University of South Australia, Mawson Lakes, South Australia, Australia
[2]School of Molecular and Biomedical Science, University of Adelaide, Adelaide, South Australia, Australia

Synonyms

Binding of fluorescent proteins at gold nanoparticles; Fluorescent protein–gold nanoparticle conjugates; Protein–gold nanoparticle conjugation

Definition

Gold Nanoparticles (AuNPs) and Gold Nanoclusters (AuNCs)

Gold nanoparticles (AuNPs) are individual metallic gold entities that have at least one dimension in the length scale of approximately 1–100 nm. On the other hand, gold nanoclusters (AuNCs) are ultrasmall gold particles (composed of a few to a hundred gold atoms) with sub-nanometer sizes (<1 nm).

Fluorescent Proteins

Proteins are biopolymers that consist of linear polymers chain built from a series of up to 20 different L-α-amino acid residues in different combination and sequence, which are bonded together by peptide bonds between the carboxyl and amino groups of adjacent amino acid residues. Fluorescent proteins are proteins that exhibit fluorescent characteristics. Normally, the presence of three aromatic amino acid residues, tryptophan, tyrosine, phenylalanine, respectively, in the protein structure enables them to be fluorescent.

Hybrid Architecture

Hybrid architectures are engineered nanomaterials made from two or more constituent entities interacting at the molecular level and exhibit new/novel characteristics, which are significantly different from the two individual constituents.

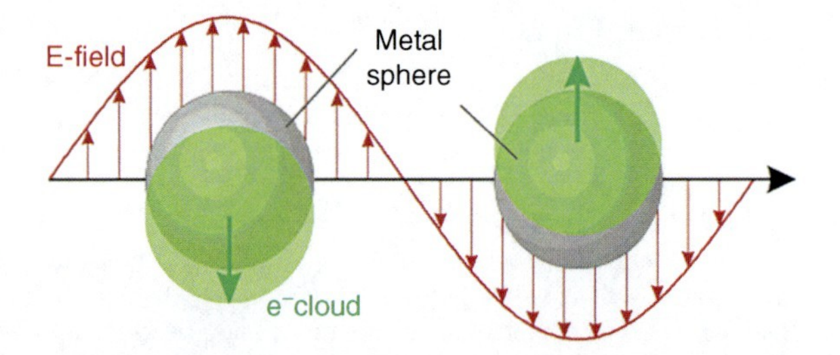

Gold Nanoparticles and Fluorescent Proteins, Optically Coupled Hybrid Architectures, Scheme 1 Schematic representation of plasmon oscillation for a sphere, showing the displacement of the conduction electron charge cloud relative to the nuclei (Reproduced with permission from The Journal of Physical Chemistry B 107,668-677 Copyright © 2010 American Chemical Society)

Bulk Gold, Gold Nanoparticle (AuNP), and Gold Nanoclusters (AuNC): Size Tunable Electronic Transitions and Optical Properties

Dazzling color, inertness, and stability of bulk gold has given it a perpetual attraction and is the material of desire since time immemorial. However, in recent years, AuNPs and AuNCs have attracted great attention due to their unusual properties, high level catalytic activities, and responsiveness. The electronic, optical, electrochemical, magnetic, and catalytic properties displayed by AuNPs are dramatically different from bulk metallic gold (size $>$ μm) and significantly dependent on the size, shape, environment, and the capping layer used to stabilize the AuNPs. The properties of materials change significantly as their size approaches nanoscale, as a consequence of the dramatic increase in the percentage of atoms at the surface of a material. The display of unique characteristics by AuNPs is related to the presence of localized surface plasmon resonance (LSPR). LSPR is initiated when AuNPs are irradiated with electromagnetic wave (e.g., light) and the imposed oscillating electric field of the electromagnetic wave causes the conduction electrons to oscillate coherently, as shown schematically in Scheme 1. AuNPs reveal LSPR by confining conduction electrons in both ground and excited states to dimension smaller than the electron mean free path, ($\sim$20 nm) (Kelly and Coronado 2003).

The resulting frequency of oscillation of LSPR is determined mainly by four factors: (1) the density of electrons, (2) the effective electron mass, (3) the shape, and (4) size of the charge distribution. In AuNPs, LSPR occurs with direct optical excitement that induces visible red color to AuNP colloids. The electric field of photons induces collective oscillation of the free electrons of the AuNP surface only, but not the electrons on the AuNP core. Consequently, on electromagnetic wave exposure, the valence bond electrons of AuNPs and bulk gold (particle size $>$ μm) experience different energies and distributions, and bulk gold exhibits no LSPR. The facile synthesis protocol including new lithographic techniques as well as improved classical wet chemical methods has provided opportunities to synthesize AuNPs with a wide range of sizes, shapes, and dielectric environments; and LSPR is size and shape tunable (Kelly and Coronado 2003). Mie in 1908 (Mie 1908) advanced a theoretical treatment that provides the basis to explain the LSPR of spherical nanoparticles of arbitrary size, though theoretical treatment of the new generation of the complex metal nanoparticles has proved to be of significant challenge.

This is worthy to note that confinement of free electrons in AuNPs below a critical size, when their sizes are comparable to or smaller than the Fermi wavelength ($\sim$0.7 nm) of conductive electrons, they display optical, electronic, and chemical properties of metal clusters dramatically different from bulk gold and AuNPs. These ultrasmall gold nanoparticles are generally known as gold clusters – AuNC. AuNCs with sub-nanometer sizes ($<$1 nm composed of a few to a hundred gold atoms) and demonstrate molecular-like electronic transitions between highest occupied molecular orbital and lowest unoccupied molecular orbital (HOMO/LUMO) energy levels (Sinkeldam et al. 2010) (Scheme 2).

Due to this unique electronic characteristic, AuNCs exhibit photoluminescent properties and size-dependent fluorescence characteristics. AuNCs exhibit discrete absorption and fluorescence that are size tunable from UV to near IR as illustrated in Fig. 1.

These molecular metals exhibit highly polarizable transitions and scale in size according to the simple relation $E_{Fermi}/N^{1/3}$ (E_{Fermi} is the Fermi energy of bulk Au, "N" is the number of gold atoms in AuNC), predicted by the spherical jellium model (Zheng et al. 2004, 2007).

Gold Nanoparticles and Fluorescent Proteins, Optically Coupled Hybrid Architectures, Scheme 2 A simplified Jablonski diagram of a typical photophysical process (Reproduced with permission from Chemical Reviews, 110, 2620–2640, Copyright © 2010 American Chemical Society)

Such AuNCs display attractive set of features, including ultrasmall size, good biocompatibility, and excellent photostability, and have received significant attention as ideal fluorescent labels for multitude of promising applications in advancing fields including biosensors and molecular imaging (Shanga et al. 2011).

Historical Background and Importance of Colloidal Gold Particles

The brilliant red color of colloidal AuNPs has been of significant interest for centuries and used since antiquity to decorate stained glass windows, ceramics, and other decorative items to assure dazzling effect, and in medical treatments including unstable mental and emotional states, arthritis, blood circulation, and tuberculosis (Shaw 1999). In 1857, Faraday (1857) was first to report a scientific approach for gold colloid synthesis and its optical properties and published a complete treaty on colloidal gold. Traditionally, the wet chemical methods for synthesis of AuNPs are generally based on the reduction of gold salts in solution. In 1951, Turkevitch (Turkevitch et al. 1951) demonstrated a facile method to yield AuNPs with an average

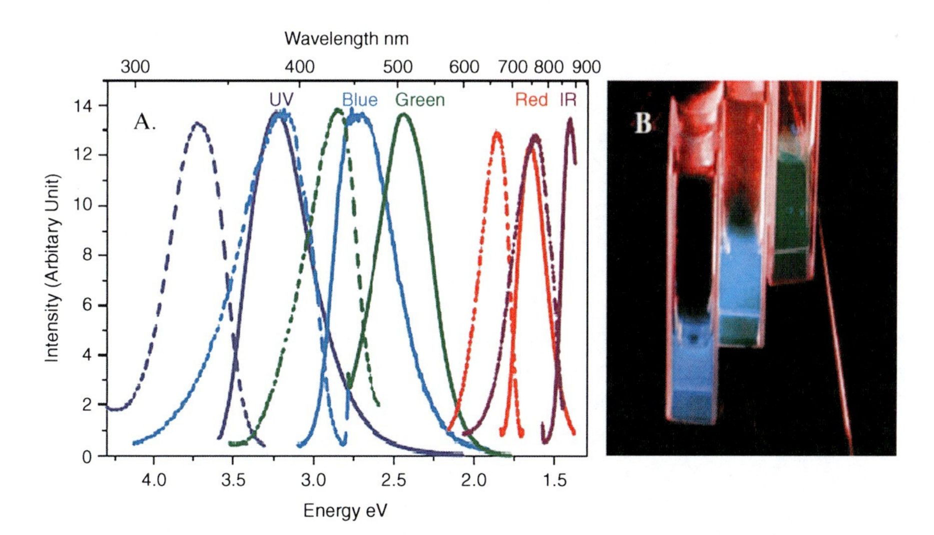

Gold Nanoparticles and Fluorescent Proteins, Optically Coupled Hybrid Architectures, Fig. 1 Excitation (*dashed*) and emission (*solid*) spectra of different gold nanoclusters. Excitation and emission maxima shift to longer wavelength with increasing nanocluster size leads to lower energy emission. (**b**) Emission from the three shortest wavelength emitting gold nanocluster solutions (Reproduced with permission from Physical Review Letters, 2004, 93:077402-1-4, Copyright © 2004 American Physical Society)

diameter of ~20 nm, which was stabilized with citrate. Citrates form a negatively charged electrical double layer on the surface of AuNPs and provide them with the required stability. However, citrate moieties do not form very strong bonds with gold atoms, and such AuNPs are not very stable. Brust et al. in 1994 (Brust et al. 1994) introduced another protocol to synthesize highly stable thiol-capped AuNPs. The covalent stabilization of the AuNPs through Au–thiol interactions on the surface leads to highly stable AuNPs. It is now been well established that many functionalities can bind to metallic gold, for example, sulfur-containing moieties including thiols, thioethers, thioesters, disulfides, isothiocyanates, phosphines, amines, citrate, and carboxylates to yield stable AuNPs. In all the cases, a passivation layer on the surface of AuNPs prevents their coalescence and stabilizes nanoparticles. Further modification of the procedure has also been advanced that includes the preparation of AuNPs/AuNCs using weak ligands and consequently their exchange with stronger ones for long-term stability and function. Based on their capability of sequestering metal ions from solution, dendrimers, and a variety of polymers, biopolymers and block copolymers have also been utilized as templates or directing agent to prepare AuNPs and AuNCs (Daniel and Astruc 2010).

Use of Fluorescent Proteins as Templates/Directing Agent to Prepare AuNPs/AuNCs

In general synthesis of AuNPs using wet chemical methods discussed above often involves the use of harsh chemicals (e.g., surfactant) and sometimes organic solvents (e.g., toluene) to control the size, shape, and morphology of AuNPs. The presence of such impurities and harsh chemicals in the final product is unacceptable for many sensitive biomedical applications including biosensing, bioimaging, drug and gene delivery, and cancer photothermal therapy. This demand has recently prompted the use of proteins and/or peptides as the template and reducing agent, which is inspired by the roles of such molecules in natural biomineralization process. Such biomaterials allow synthesis of AuNPs and AuNCs of controlled shape, size, and organization and function in aqueous solution at room temperature under mild condition that also ensure biocompatibility (Tan et al. 2010).

Most proteins are biopolymers consisting of linear polymer chain amino acid residues, and the combined effect of all of the amino acid side chains in a protein determines the physical, physicochemical, and chemical reactivity including three-dimensional structure. Several independent studies have confirmed that these functionalities are related to the presence of certain amino acid residues in a peptide sequence. Moreover, the large diversity of proteins and/or peptides available for use through the traditional top-down approach or the modern recombinant DNA technology and combinatorial approach offers a unique opportunity for exploration of the structure/property relationship in peptides/proteins. Very recently, intrinsically fluorescent proteins (IFPs) have also been utilized as templates for synthesizing highly fluorescent hybrid architectures. AuNCs that have opened the possibilities of incorporation of biological fluorophores into nanoparticle based devices with unique photophysical properties.

Protein–Gold Nanoparticle Interactions: Exploration of the Key Chemistry

The formation of colloidal metal particles by proteins/peptides requires two basic steps as well known for conventional metal nanoparticle synthesis, namely, (a) the reduction capability for metal ions and (b) capping capability for the nanoparticles formed. Some proteins and IFPs could be used as both the reductant and the capping agent in metal nanoparticle synthesis; and recently research has been focused on developing a guiding principle through exploration of the key chemistry of protein amino acid segments. Due to the complexity in proteins, a variety of short peptide sequences have been developed and used to identify the specific functions of the 20 natural amino acid residues in reduction of gold precursor such as chloroaurate and binding to elemental gold particles thus formed (Tan et al. 2010). Among the 20 standard amino acids, tryptophan (Trp) has been identified as the fastest reducing agent. However, amino acid residues with heterocycles (histidine-His, Trp) and sulfur (cysteine-Cys, methionine-Met) are strong binders and form strong complexes with metal ions (Lippard and Berg 1994); however, they lower the redox potential of gold and hence the reducibility of the metal ions. The amino acid residues with acidic and

Gold Nanoparticles and Fluorescent Proteins, Optically Coupled Hybrid Architectures, Scheme 3 Chemical structure of the fluorescent aromatic amino acid residues

Gold Nanoparticles and Fluorescent Proteins, Optically Coupled Hybrid Architectures, Table 1 Fluorescent characteristics of the aromatic amino acids[a]

	Absorption		Fluorescence	
Amino acid residue	Wavelength (nm)	Absorptivity	Wavelength (nm)	Quantum yield
Tryptophan	280	5,600	348	0.20
Tyrosine	274	1,400	303	0.14
Phenylalanine	257	200	282	0.04

[a]Adopted from http://www.biotek.com/resources/docs/Synergy_HT_Quantitation_of_Peptides_and_Amino_Acids.pdf

hydrophobic side chains without nitrogen are recognized as weak binders. Furthermore, binding affinity also increased with increase in peptide conformational flexibility (i.e., glycine-Gly>glutamine-Gln>proline-Pro). Therefore, the reduction capability of a peptide is not a linear sum of the capabilities of its amino acid constituents. The binding affinity of an individual amino acid residue to AuNPs in a peptide sequence is governed by its reactive side group as well as by its contribution to the peptide's adaptability to conformations favorable for adsorption on gold. The increased binding strength improves the reduction reaction up to a critical point, beyond which a further increase in binding actually suppresses gold nanoparticle formation (Tan et al. 2010). The in-depth investigations using short peptides of various composition and microstructures reveal the possibility of precise kinetic control at the reaction microenvironment level by dialing in different amino acid sequences for the right combination of reduction, capping, and morphogenic functions. Indeed, the regulation of crystal morphology by proteins is often observed in biology, and repeating polypeptide sequence could be found that regulate the morphology of other inorganic crystals. It has been identified that two amino acid sequences, respectively, (a) serine–glutamic acid–lysine–leucine and (b) glycine–alanine–serine–leucine, containing repeats, in the gold-binding polypeptides from Escherichia coli could introduce shape-directing function in gold crystal growth (Tan et al. 2010). However, it is noteworthy that as the peptide chain becomes longer, the increasing complexity of the peptide sequence and the increasing possibility for structure and conformation variations dramatically increase, and formulating any predictive rule is complex.

Origin of Fluorescence in Protein: Molecular Level Understanding

Among the 20 α-amino acid residues, three aromatic amino acid residues, tryptophan, tyrosine, and phenylalanine (Scheme 3), exhibit fluorescent characteristics.

These three amino acid residues have distinct photon absorption and emission wavelengths and differ greatly in their quantum yields and lifetimes with tryptophan having much stronger fluorescence and higher quantum yield than the other two aromatic amino acids (Table 1).

The intensity, quantum yield, and wavelength of maximum fluorescence emission are also dependent on the local environment. Green fluorescent protein (GFP) (Fig. 2) isolated from coelenterates, such as the Pacific jellyfish, Aequorea victoria, or from the sea pansy, Renilla reniformis, is the most well-known intrinsically fluorescent protein (IFP) and was discovered by Shimomura et al. (1962) as a companion protein to aequorin (Tsien 2009).

In GFP, a special case of fluorescence occurs, where the fluorophore originates from an internal serine–tyrosine–glycine sequence, which is posttranslationally

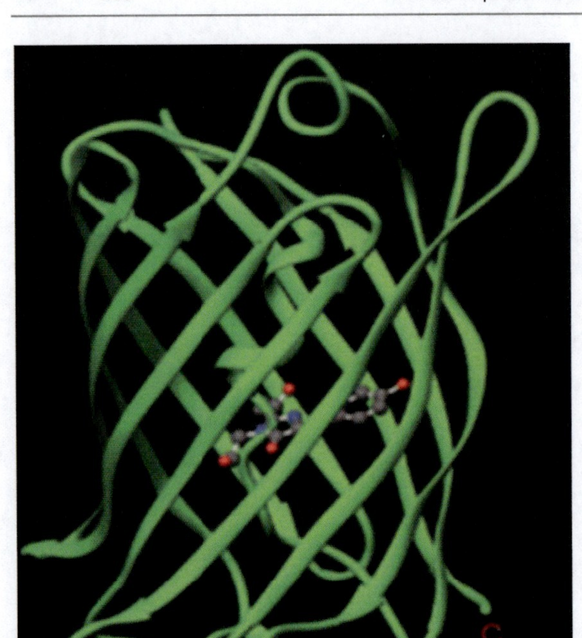

Gold Nanoparticles and Fluorescent Proteins, Optically Coupled Hybrid Architectures, Fig. 2 3D structure of the archetypical Aequorea victoria GFP.(α-Helices and β-strands are shown as *ribbons*, connecting segments as *tubes*, and the chromophore in *ball*-and-*stick* representation). (Reproduced with permission from Science 1996, 273, 1392–1395, Copyright © 1996 American Association for the Advancement of Science)

modified to a 4-(p-hydroxybenzylidene)- imidazolidin-5-one structure The chromophore is a p-hydroxybenzylideneimidazolinone formed from residues 65–67, which are Ser–Tyr–Gly in the native protein. Scheme 4 shows the currently accepted mechanism for chromophore formation.

First, GFP folds into a nearly native conformation, then the imidazolinone is formed by nucleophilic attack of the amide of Gly67 on the carbonyl group of residue 65, followed by dehydration. Finally, molecular oxygen (the process would stop in the absence of O_2) dehydrogenates the α-β bond of residue 66 to put its aromatic group into conjugation with the imidazolinone, which would complete the fluorophore (Heim et al. 1994). A variety of mutants of GFP, creating new variants of GFP that shone more strongly and in different colors – from blue to red and yellow – have also been developed (Fig. 3).

These highly fluorescent IFPs have revolutionized how the dynamics of cellular events can be imaged and open up tremendous possibilities for continuously monitoring gene expression, cell developmental fates, and protein trafficking in living, minimally perturbed cells, tissues, and organisms. Such IFPs have also been used as reporters to monitor transcriptional regulation, as targeted markers for organelles and subcellular structures, and in sensors designed to show changes in cellular environments ranging from pH to protein kinase activity (Tsien 2009; Wang et al. 2004; Orm et al. 1996). For the discovery and development of GFP, the 2008 Nobel Prize in Chemistry has been awarded jointly to O. Shimomura, M. Chalfie, and R. Tsien (http://www.nobelprize.org/nobel_prizes/chemistry/laureates/2008). In view of its enormous potential, fluorescence-based detection strategies that use quenched emission of fluorescence by IFP ligand functionalized AuNPs have generated significant recent interest in biotechnology for fluorescence immunoassay and biosensors.

Fundamental of Fluorescence Process and Design of Optically Coupled Hybrid Architectures

Photo-induced excitation of a chromophore through absorption of a photon of appropriate energy induces the vibronic transition (simultaneous change of vibrational and electronic quantum number) in a fluorophore – commonly known as Franck–Condon state – extremely rapidly (within 10^{-15} s) (Scheme 2). The efficiency of this process is related to the chromophore's absorption cross section (σ), which is proportional to its extinction coefficient (ε) (Sinkeldam et al. 2010). The Jablonski diagram for a typical fluorophore is presented in Scheme 2, which illustrates the possible chain of photophysical events that include the following: (1) internal conversion or vibrational relaxation, (2) fluorescence, intersystem crossing (from singlet state to a triplet state), and (3) phosphorescence. The vibrational relaxation (within 10^{-12}–10^{-10} s) quickly populates the lowest vibronic state of the chromophore's excited state (loss of energy without light emission). This relaxation process generally accounts for the observed Stokes shift – energy of emission wave is lower (longer wavelength) when compared to the excitation energy. Each of the processes occurs

Gold Nanoparticles and Fluorescent Proteins, Optically Coupled Hybrid Architectures, Scheme 4 A tentative molecular interpretation posttranslational autoxidation of the chromophore of *green* fluorescent protein, which is considered to be responsible for the fluorescence in GFP (Tsien 1998)

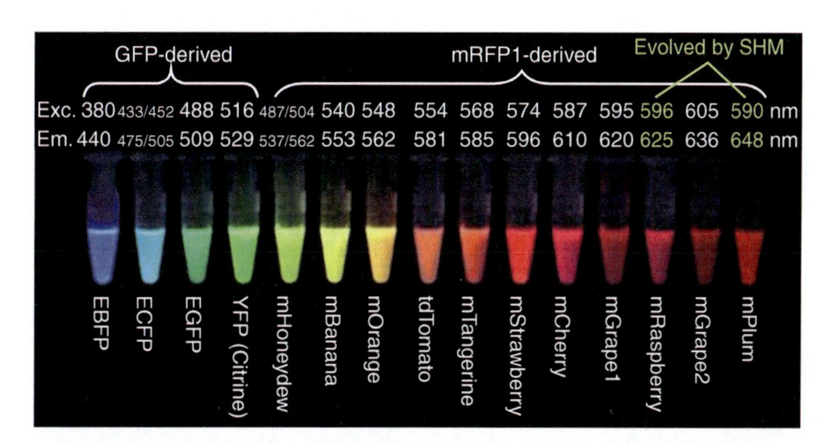

Gold Nanoparticles and Fluorescent Proteins, Optically Coupled Hybrid Architectures, Fig. 3 Monomeric and tandem dimeric fluorescent proteins derived from Aequorea GFP or Discosoma RFP, expressed in bacteria, and purified. This photo is a time exposure of fluorescences excited at different wavelengths and viewed through different cutoff filters. SHM = somatic hypermutation (Reproduced with permission from Angewandte Chemie International Edition **48**, 5612–5626, Copyright © 2009, Wiley-VCH)

with a certain probability, characterized by the decay rate constants (k). The fluorescence lifetime of a fluorophore is the characteristic time required by a population of excited fluorophore to decaying back to the ground state (τ_0) via the loss of energy through fluorescence and other non-radiative processes. Typical organic fluorophores reside in their excited state for a very short period of $\sim(0.5–20) \times 10^{-9}$ s. The fluorescence lifetime (τ) of a fluorophore reflects the emission quantum yield of the chromophores ($Q = \Phi = \tau/\tau_0$). The product of the molar absorptivity (ε) and the fluorescence quantum yield (Φ) of a fluorophore ($\varepsilon\Phi$) is recognized as the brightness (Valeur and Nuno Berberan-Santos 2012). Fluorescence-based techniques have gained enormous popularity for their simplicity, versatility, and sensitivity (up to a 1,000-fold higher than absorption spectrophotometry). Moreover, creative probe design with selectivity to various environmental parameters (pH, polarity, viscosity, presence of quenchers, etc.) makes molecular fluorescence an extremely effective tool for in vitro biophysical and biochemical analyses, as well as in vivo cellular imaging, capable of providing both spatial and temporal information.

Nevertheless, the major limitations of the fluorescence probe include low photostability, high background fluorescence and nonspecific background signals. To achieve these specific goals, activatable fluorescence probes have been designed. Such activatable probes are designed with at least two components, respectively, (1) the fluorophore that acts as the donor and (2) the quencher that acts as the acceptor. Such probes offer high specificity to the target of interest and with superior signal to background ratio in the imaging signal. Inorganic AuNPs show the highest quenching efficiency (up to 99 %) and therefore can offer the highest sensitivity in the development of activatable probes.

Optically Coupled Architectures Based on AuNP–Fluorescent Protein Hybrid: Activatable Probes

The fluorescence-based assays, detection, and imaging techniques are among the most sensitive and popular biological tests. AuNPs are the most commonly used optical sensing nanomaterials because of their strong LSPR absorptions in the visible region, quenching efficiency, facile preparation, high stability and biocompatibility. In addition, the surface chemistry of AuNPs is versatile, allowing the linking of various biofunctional groups (e.g., amphiphilic polymers, silanols, sugars, nucleic acids, and proteins) through strong Au–S or Au–N bonding or through physical adsorption. AuNPs are considered as super quenchers for fluorophores, because they have much stronger molar absorptivity relative to the available organic dyes. For example, the molar absorptivity of 13-nm AuNPs at 518 nm is ca. 2×10^8 cm^{-1} mol^{-1} L. AuNPs have Stern–Volmer quenching constants (KSV) that are more than five orders of magnitude greater than those of typical small-molecule dye quencher pairs. The absorbance of AuNPs is dependent on the size and geometry, for example, as their diameters increase, their plasmon band red shifts. Hence, AuNPs are highly tunable absorbing agents. AuNP surfaces can be chemically conjugated with various biomolecules and targeting agents using facile and straightforward manner. Typically, effective quenching results when the distance between the fluorophore and the AuNP surface is less than 5 nm. In addition, size, shape, and the surface characteristics of the AuNPs are all important parameters that affect fluorescence quenching. The past few years have witnessed the development of many successful strategies for the preparation of high-quality AuNPs and of numerous AuNP-based sensing systems, and activatable probes appear to be the most reliable molecular probe design architecture.

The fluorescence activatable probes, also referred to as molecular beacons, are designed from a donor–quencher conjugates, and the efficiency of quenching is dependent on the distance between these two components. In the native state, these two components are in close proximity, and quenching occurs efficiently; however, with increase of distance between the donor and the quencher in response to the external stimuli/analyte, fluorescence is effectively restored (Fig. 4).

Activatable probes have been designed with various quenchers, fluorophores, and biomolecules; however, fluorescence-based detection strategies that use quenched emission of fluorescence by IFP ligand functionalized AuNPs have generated significant recent interest in biotechnology for fluorescence immunoassay and biosensors (Swierczewska et al. 2011). Hazarika et al. (2006) reported reversible binding of yellow fluorescent protein (EYFP) – a mutant derivative of the naturally occurring green fluorescent protein from the North Pacific jellyfish Aequorea victoria at AuNP hybrids without compromising their biological activity. Dutta and coworkers (Mayavan et al. 2011) investigated the synthesis of optically coupled hybrid architectures based on a new biomimetic fluorescent protein *rec1-resilin* and nanometer-scale AuNPs in one-step method using a non-covalent mode of binding protocol. The presence of uniformly distributed fluorophore amino acid sequences, serine (threonine)–tyrosine–glycine, along the molecular structure of *rec1-resilin* provides significant opportunity to synthesize fluorophore-modified AuNPs bioconjugates with unique photophysical properties. The final properties of the protein–AuNP hybrids are very sensitive not only to the chemical composition of the protein and the size of AuNPs but also on their topological organization including the distance of the fluorophores from AuNP

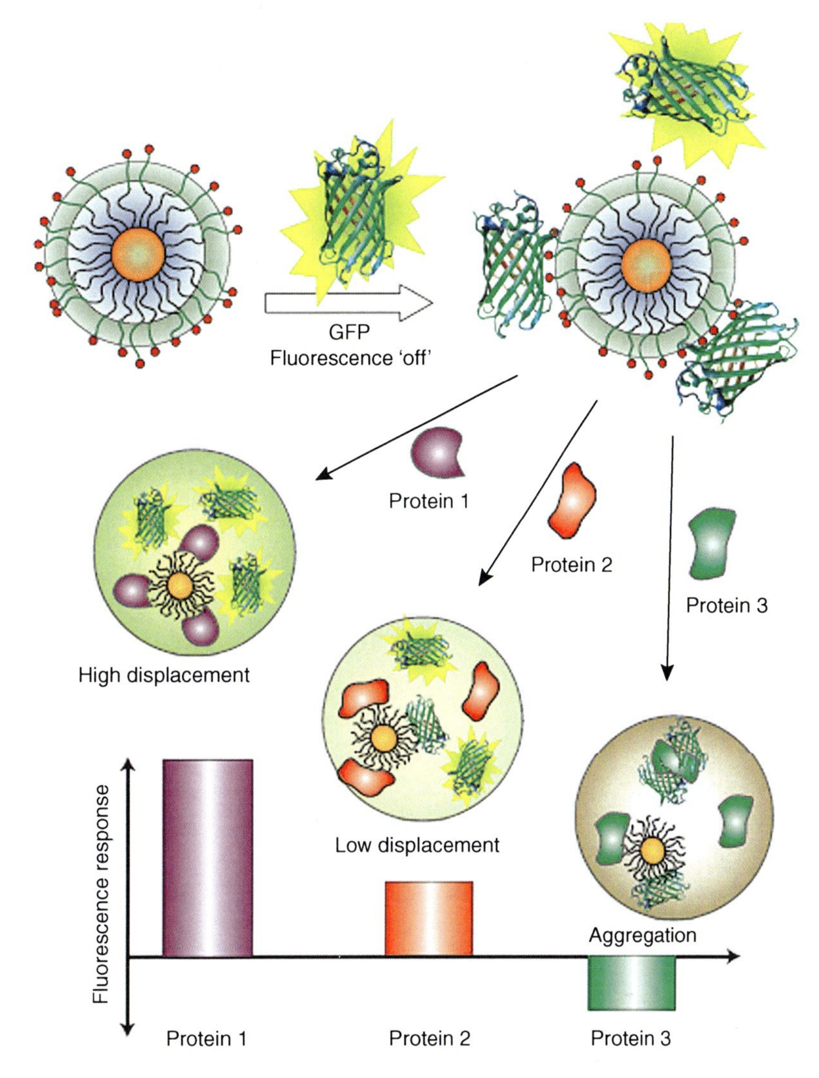

Gold Nanoparticles and Fluorescent Proteins, Optically Coupled Hybrid Architectures, Fig. 4 Schematic of competitive binding of green fluorescent protein (GFP) onto monolayer-protected AuNPs. The competitive binding of GFP protein to the AuNP is affected by specific types of proteins, which, in turn, affects the efficacy of fluorescence quenching. Different proteins can therefore be identified based on the shifts in fluorescence intensity (Reproduced with permission from Nature Chemistry, 1: 461–465. Copyright © 2009 Macmillan Publishers Limited)

surfaces as discussed before. In DNA molecular beacons, AuNPs have been used as an effective proximal quencher (Kobayashi et al. 2010).

Further advancement has also been made to synthesize AuNPs using engineered fluorescent protein as reductant and/or capping agent applied to bio- and chemosensing. Rotello and co-workers (De et al. 2009) have demonstrated the power of protein activatable fluorescent probe from conjugates of AuNPs and green fluorescent protein in rapid and efficient sensing of proteins in human serum with specificity. Figure 4 illustrates the use of competitive binding of GFP to the AuNP, which is affected by the presence of specific types of proteins in the environment that in turn affects the efficacy of fluorescence quenching. Such specific recognition and quantification is attractive, because there is a direct correlation between protein levels and disease states in human serum, but challenging (serum features more than 20,000 proteins, with an overall protein content greater than 1 mM). Using such probe, reproducible fluorescence response patterns with an identification accuracy of 100 % in buffer and 97 % in human serum have been reported from five major serum proteins (human serum albumin, immunoglobulin G, transferrin, fibrinogen, and α-antitrypsin). The arrays have been demonstrated to be enabled to discriminate between different concentrations of the same protein as well as a mixture of different proteins in human serum. Such chemical noses can sense proteins with over 97 % accuracy using in-depth understanding of the fluorescence quenching characteristics of the

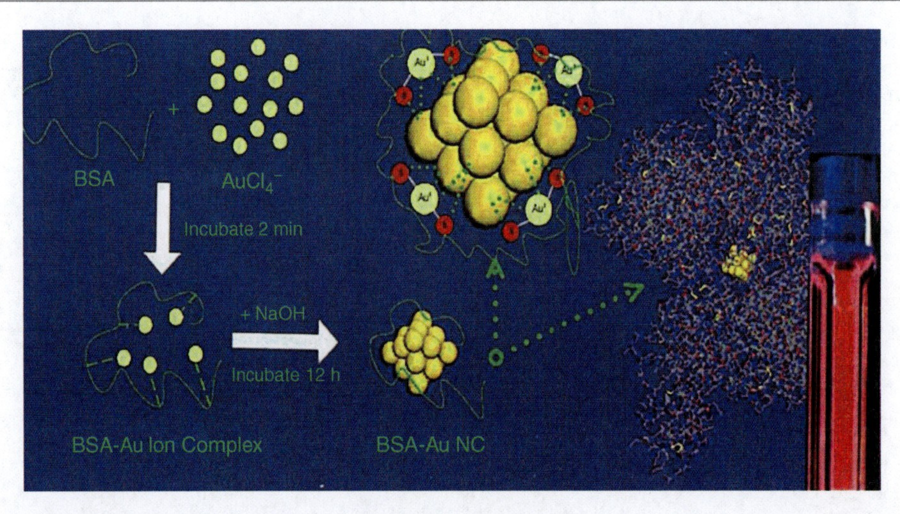

Gold Nanoparticles and Fluorescent Proteins, Optically Coupled Hybrid Architectures, Fig. 5 Schematic representation of the formation of AuNCs in BSA solution. Upon addition of Au(III) ions to the aqueous BSA solution, the protein molecules sequestered Au ions and entrapped them. The reducing capability of BSA is subsequently activated by adjusting the pH to about 12 with NaOH, so the entrapped ions underwent progressive reduction to form AuNCs in situ (Reproduced with permission from Journal of the American Chemical Society **131**, 888–889, Copyright © 2009 American Chemical Society)

molecular probe along with interplay of electrostatic and hydrophobic interactions. So far, AuNP activatable probe studies have mainly been fixed to microtiter plates to use them as diagnostic tools. Nevertheless, AuNPs are widely accepted to be noncytotoxic, biocompatible, and exhibit good cellular uptake and could be extended to in vivo study.

Optically Coupled Hybrid Architectures Based on AuNC–Fluorescent Protein: Ratiometric Fluorescent Probe

Optically coupled hybrid architectures based on AuNC–fluorescent protein bioconjugates have the unique potential to display dual fluorescence emissions – one from the AuNC and the other from the IFP that could be further modulated through the control of AuNC size and morphology. This dual fluorescence offers a distinct advantage over the most fluorescence-AuNP sensors as discussed above, which normally exhibit a single signal fluorescence output on the basis of quenching or recovery. Two well-separated emission bands with comparable intensities from AuNC–IFP have two distinct advantages: (1) the large shift (~100 nm) between excitation and emission peak of AuNC, and (2) presence of two well-separated emission bands with comparable intensities ensures accuracy in determining their intensities and ratios. Ratiometric fluorescence methods can eliminate most or all the ambiguities by using the ratio of two emission bands to demonstrate the quantitative analysis ability independent of fluorophore concentration and environmental conditions.

The power of common protein bovine serum albumin (BSA) in both sequestration and reduction of Au precursors to generate highly fluorescent AuNCs has been clearly demonstrated (Xie et al. 2009). AuNC–BSA was prepared by reduction of chloroauric acid with bovine serum albumin (BSA) based on the capability of BSA to sequester and reduce Au precursors in situ (Fig. 5).

The presence of 21 Tyr residues in BSA is considered to be responsible for the reduction reaction; Fig. 5 presents a schematic representation of the formation of AuNCs in BSA solution. Compared with short peptides, large and complicated proteins possess abundant binding sites that can potentially bind and further reduce metal ions, thus offering better scaffolds for template-driven formation of small metal NCs. The BSA–AuNC coupled hybrids exhibit dual emissions: the blue at ~425 nm from the oxides of BSA and the red at ~635 nm from AuNCs. Such BSA–AuNCs colloidal solutions are highly stable both in solutions (aqueous or buffer) and in the solid form. The power of metal-ion modulated ratiometric fluorescence probe

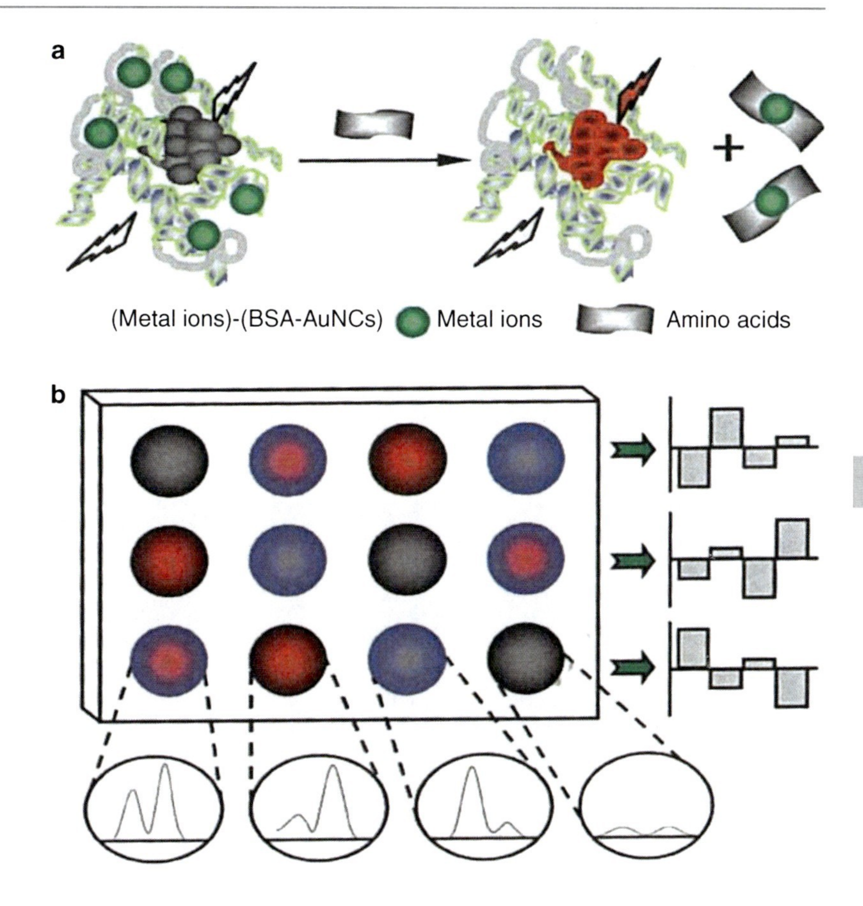

Gold Nanoparticles and Fluorescent Proteins, Optically Coupled Hybrid Architectures, Fig. 6 (**a**) Illustration of The fluorescence ratiometric response to amino acids by the use of metal ion-modulated AuNCs. (**b**) Schematic drawings for the identification of amino acids by ratiometric fluorescence probe array (Reproduced with permission from Analyst, 137, 1618–1623, Copyright © 2012 RSC Publishing)

array constructed from such BSA–AuNCs hybrids for the detection and identification of amino acids has been recently demonstrated (Wang et al. 2012) as illustrated in Fig. 6.

Metal ions easily bind with BSA–AuNCs and thus greatly affect the two emissions (~425 nm and ~635 nm, respectively) in different ways by fluorescence enhancement or quenching. Upon the addition of amino acids, the metal ion-modulated BSA–AuNCs exhibit fluorescence ratiometric responses. The environment and type of interaction of the specific metal ions with the BSA modulate the intensities of the two fluorescence peaks of AuNC–BSA differently. Figure 6a illustrates the typical fluorescence ratiometric response to amino acids by the use of metal ion-modulated AuNCs. For example, the presence of Ni^{2+} in the metal ion-modulated BSA–AuNCs quenches both the fluorescence emissions, while the addition of Zn^{2+} only enhances the red fluorescence. Following the addition of amino acids into the bioconjugates, stronger interaction occurs between the amino acids and the metal ions, compared with BSA that may induce either the disassociation of metal ions from BSA–AuNCs or the aggregation of AuNCs. Consequently, the two fluorescence emissions from BSA–AuNCs independently change with different amino acids, and the resultant evolution of fluorescence ratios can be employed as an efficient indicator of amino acids (Wang et al. 2012). The dual emissions of BSA–AuNCs modulated by a specific metal ion offer four different possible fluorescence responses: (1) both fluorescence quenching, (2) both enhancements, (3) one quenching and another enhancement, and (4) one enhancement and another quenching. Therefore, a modulated BSA–AuNCs ratiometric fluorescent probe array designed with integration of four different metal ions, respectively, Ni^{2+}, Pb^{2+}, Zn^{2+}, and Cd^{2+} and designated as BSA–AuNC–Ni^{2+}, BSA–AuNCs–Pb^{2+}, BSA–AuNCs–Zn^{2+}, and BSA–AuNCs–Cd^{2+} can essentially display 256 kinds of different signals, theoretically. Using such an intelligent probe, a specific fluorescence pattern obtained for a specific amino acid can be easily recognized as the characteristic

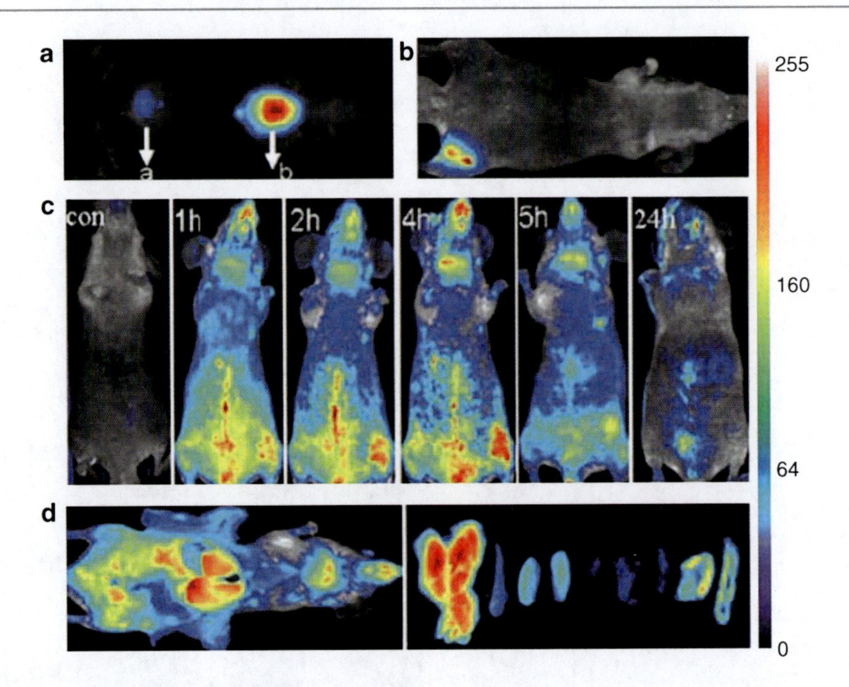

Gold Nanoparticles and Fluorescent Proteins, Optically Coupled Hybrid Architectures, Fig. 7 In vivo fluorescence image of 100 ml. (**a**) AuNCs injected subcutaneously and (**b**) intramuscularly into the mice. (**c**) Real-time in vivo abdomen imaging of intravenously injected with 200 ml of AuNCs at different time points post-injection. (**d**) Ex vivo optical imaging of anatomized mice with injection of 200 ml of AuNCs and some dissected organs during necropsy. The organs are liver, spleen, left kidney, right kidney, heart, lung, muscle, skin, and intestine from *left* to *right* (Reproduced with permission from Nanoscale, 2, 2244–2249. Copyright © 2010 RSC Publishing)

fingerprint. Various amino acids were identified by the fluorescence ratiometric array composed of such four ratiometric probes with each amino acid added into the four probes to obtain the distinct pattern on the changes of fluorescence ratio. Figure 6b represents the likely mechanism for the detection of amino acids. It is well understood that larger complex protein BSA easily binds to various metal ions such as Ni^{2+}, Pb^{2+}, Zn^{2+}, and Cd^{2+} due to the presence of large amount of residual amino, carboxylic and mercapto groups, and disulfide bridges at the peptide chains. This work is remarkable not only for providing an alternative approach of synthesizing highly fluorescent AuNCs but also because the proposed strategy may be extended to other proteins and metals.

The use of BSA–AuNCs coupled hybrids has also been explored as novel contrast imaging agents for tumor fluorescence imaging in vivo (Wu et al. 2010) as illustrated in Fig. 7.

It was clearly demonstrated that the light level produced by the BSA–AuNCs hybrid is sufficient to provide a localised signal under a few millimeters of tissues with a subcutaneous injection. Furthermore, using MDA-MB-45 and Hela tumor xenograft models, in vivo and ex vivo imaging studies show that the ultrasmall NIR AuNCs are able to be highly accumulated in the tumor areas due to enhanced permeability and retention (EPR) effects and yield excellent contrast from the surrounding tissues. The results indicate that the ultrasmall NIR AuNCs appear as very promising contrast imaging agents for in vivo fluorescence tumor imaging.

Future Outlook

The dramatically changed characteristics of colloidal gold particles based on the size, shape, geometry,morphology and interparticle distance, and the environment are currently most intensely investigated. We envisage that these modulation in characteristics –particularly the absorption, plasmonic, and fluorescence characteristics – will enable them for many potential applications in near future for a wide range of areas, including nanotechnology, catalysis, medicine, immunohistochemistry biomarkers, and sensors. The gold-fluorescent protein-based hybrid colloidal gold nanoparticles with range in size from 2 to 100 nm will find wider applications as labeling, delivery, heating, and sensors. The molecular beacons and activatable probes from AuNP–fluorescent protein will find applications in new and exciting ways for fluorescence-based supersensitive assays and detections. Fluorescence-based activatable probes can be used in vitro and in vivo as cellular and clinical bioimaging agents. The fluorescence characteristics of ultrasmall size of AuNCs can provide extraordinary advantages over quantum dots and conventional fluorescent dyes owing to their excellent biocompatibility, robust resistance to photobleaching, and facile synthesis. They are able to sustain a longer blood circulation time than other classical PEGylated nanomaterials and enjoy superior binding efficiency towards tumor through "enhanced permeability and retention." Dual fluorescence emissions from protein–AuNCs-based probes can be greatly modulated in presence of other metal ions and compounds, and the resultant fluorescence ratiometric responses will be identified as a novel sensory method for identification and quantification of different analytes including heavy metals as pollutant, proteins in blood serum. Such AuNC hybrid with further inclusion of vector and drug molecules specific to cancer cells has the potential to make significant in road in tumor diagnostic and therapeutic markets in the continued fight against disease.

Acknowledgment This work has been financially supported by the Australian Research Council.

Cross-References

- ▶ Biomineralization of Gold Nanoparticles from Gold Complexes in Cupriavidus Metallidurans CH34
- ▶ Gold Nanomaterials as Prospective Metal-based Delivery Systems for Cancer Treatment
- ▶ Gold Nanoparticle Platform for Protein-Protein Interactions and Drug Discovery
- ▶ Gold Nanoparticles and Proteins, Interaction
- ▶ Gold, Ultrasmall Nanoclusters and Proteins, Interaction

References

Brust M, Walker M et al (1994) Synthesis of thiol-derivatized gold nanoparticles in a two-phase liquid–liquid system. J Chem Soc, Chem Commun 7:801–802

Daniel M-C, Astruc D (2004) Quantum-size-related properties and applications toward biology, catalysis and nanotechnology. Chem Rev 104:293–346

De M, Rana S et al (2009) Sensing of proteins in human serum using conjugates of nanoparticles and green fluorescent protein. Nat Chem 1:461–465

Faraday M (1857) Experimental relations of gold (and other metals) to Light. Philos Trans 147:145–181

Hazarika P, Kukolka F et al (2006) Reversible binding of fluorescent proteins at DNA-gold nanoparticles. Angew Chem Int Ed 45:6827–6830

Heim R, Prashert DC et al (1994) Wavelength mutations and posttranslational autoxidation of green fluorescent protein. Proc Natl Acad Sci USA 91:12501–12504

Kelly KL, Coronado E (2003) The optical properties of metal nanoparticles: the influence of size, shape, and dielectric environment. J Phys Chem B 107:668–677

Kobayashi H, Ogawa M et al (2010) New strategies for fluorescent probe design in medical diagnostic imaging. Chem Rev 110:2620–2640

Lippard SJ, Berg JM (1994) Principles of bioinorganic chemistry. University Science Books, Mill Valley

Mayavan S, Dutta NK et al (2011) Self-organization, interfacial interaction and photophysical properties of gold nanoparticle complexes derived from resilin-mimetic fluorescent protein rec1-resilin. Biomaterials 32:2786–2796

Mie G (1908) Beiträge zur Optik trüber Medien, speziell kolloidaler Metallösungen. Ann Phys 25:377–445

Orm M, Cubitt AB et al (1996) Crystal structure of the *Aequorea victoria* green fluorescent protein. Science 273:1392–1395

Shanga SL, Dongb S et al (2011) Ultra-small fluorescent metal nanoclusters: synthesis and biological applications. Nano Today 6:401–418

Shaw CF (1999) Gold-based medicinal agents. Chem Rev 99:2589–2600

Sinkeldam RW, Greco NJ et al (2010) Fluorescent analogs of biomolecular building blocks: design, properties, and applications. Chem Rev 110:2579–2619

Swierczewska M, Lee S et al (2011) The design and application of fluorophore–gold nanoparticle activatable probes. Phys Chem Chem Phys 13:9929–9941

Tan YN, Lee JY et al (2010) Uncovering the design rules for peptide synthesis of metal nanoparticles. J Am Chem Soc 132:5677–5686

Tsien RY (1998) Annu Rev Biochem 67:509–44

Tsien RY (2009) Constructing and exploiting the fluorescent protein paintbox (Nobel Lecture). Angew Chem Int Ed 48:5612–5626

Turkevitch J, Stevenson PC et al (1951) Nucleation and growth process in the synthesis of colloidal gold. Discus Faraday Soc 11:55–75

Valeur B, Nuno Berberan-Santos M (2012) Molecular fluorescence: principles and applications, 2nd edn. Wiley-VCH, Weinheim

Wang LWC, Jackson WC et al (2004) Evolution of new nonantibody proteins via. iterative somatic hypermutation. Proc Natl Acad Sci USA 101:16745–16749

Wang M, Mei Q et al (2012) Protein-gold nanoclusters for identification of amino acids by metal ions modulated ratiometric fluorescence. Analyst 137:1618–1623

Wu X, He X et al (2010) Ultrasmall near-infrared gold nanoclusters for tumor fluorescence imaging in vivo. Nanoscale 2:2244–2249

Xie J, Zheng Y et al (2009) Protein-directed synthesis of highly fluorescent gold nanoclusters. J Am Chem Soc 131:888–889

Zheng J, Zhang C et al (2004) Highly fluorescent, water-soluble, size-tunable gold quantum dots. Phys Rev Lett 93:077402-1–077402-4

Zheng J, Nicovich PR et al (2007) Highly fluorescent noble-metal quantum dots. Annu Rev Phys Chem 58:409–431

Gold Nanoparticles and Proteins, Interaction

Ricardo Franco[1] and Eulália Pereira[2]
[1]REQUIMTE FCT/UNL, Departamento de Química, Faculdade de Ciências e Tecnologia, Universidade Nova de Lisboa, Caparica, Portugal
[2]REQUIMTE, Departamento de Química e Bioquímica, Faculdade de Ciências da Universidade do Porto, Porto, Portugal

Synonyms

Gold nanoparticles interaction with proteins; Protein-gold nanoparticle conjugation

Definition

Gold nanoparticles are metallic gold entities with a diameter of 1–100 nm. They are surrounded by a capping layer that imparts chemical functionality to the nanoparticle. In the presence of proteins, gold nanoparticles interact with these biomolecules, and the nature of this interaction mainly depends on local chemical and physical effects.

Historical Background

Gold nanoparticles (AuNPs) have been used as radioactive labels in vivo since the 1950s. Since the 1980s, AuNPs have been conjugated with targeting antibodies and used as density probes for biological transmission electron microscopy. The interest in the study of the interaction of AuNPs with proteins started when research in nanotechnology found that the unique nanoscale properties of AuNPs and their similarity in size to biomolecules and organelles, make them ideal partners for the development of sensors, drug delivery vehicles, and in vivo imaging agents. In fact, from the viewpoint of biological and medical applications, the primary interest in nanoparticles (NPs) stems from the fact that they are small enough to interact with cellular machinery and to potentially reach previously inaccessible targets, such as the brain (Mahmoudi et al. 2011). For in vivo applications, AuNPs are practically nontoxic and easy to functionalize with appropriate capping agents. These functionalized AuNPs are robust delivery/imaging systems, in which the gold core imparts stability to the assembly, while the capping monolayer allows tuning of surface properties, such as charge, hydrophobicity, and binding affinity.

Protein-Gold Nanoparticle Interactions at the Molecular Level

When proteins come in contact with AuNPs, they become adsorbed at the surface of the NP due to van der Waals forces. The initial adsorption depends on the nature of the surfaces of both the AuNP and the protein. In order to better understand this process it is, thus, necessary to first have a closer look at the AuNP surface. In a bare metal, surface atoms are less stable than those within the core. This instability of the surface atoms is primarily due to the lower number of neighbors relative to core atoms. This lower number of neighbors in surface atoms can be partially compensated by adsorption of other species. This lower stability of surface atoms is also the underlying reason for spontaneous growth of nanostructures to macroscopic structures, since the ratio between the number of surface atoms and the number of total atoms decreases as the diameter of the NP increases. For spherical or cubic particles this fraction is inversely proportional to the cubic root of the total number of atoms. In order to prevent the growth of NP into macroscopic materials, it is, therefore, necessary to use an agent that can lower the surface energy of the NP by adsorbing to surface atoms. These agents, commonly called capping agents, can be small molecules or polymers, and are frequently used as co-reagents in NP synthesis. In addition to their role as stabilizers of the surface atoms, preventing undesired growth, the capping agent also improves dispersion of the NPs in the solvent used for the synthesis (Sperling and Parak 2010). For AuNPs in aqueous solution, the most typical capping agent is citrate that confers a strongly negative surface charge to AuNPs. This high charge not only hinders the close contact between AuNPs in solution, thus avoiding aggregation, but also provides a strong interaction with the surrounding water molecules, promoting colloidal stability. Other capping agents commonly used for AuNPs are small molecules with a terminal thiolate functional group. The most representative molecule in this group is 11-mercaptoundecanoic acid (MUA). Thiolates are known to have a very high affinity for gold surfaces. In addition, MUA has a 11 carbon chain that forms a packed layer around the NP, reminiscent of the structure of a micelle, which imparts slow desorption of MUA and hinders diffusion of other species to the AuNP surface. Finally, the carboxylate functional group is directed toward the surface, conveying a highly negative charge to the surface of the NPs. The capped AuNPs may, thus, be viewed as a composite NP made up of a gold metal core and a monolayer of the organic capping agent, the latter forming the surface in direct contact with the solvent. Other thiolate functionalized capping agents imparting characteristic surface properties to the AuNPs are commonly used, namely, to obtain specific hydrophilic/hydrophobic surfaces, surface charge, and/or binding properties. The adsorption of the capping agent is a dynamic process, that is, the capping agent can be exchanged with solvent molecules, ions, and other molecules in solution, potentially decreasing the colloidal stability of the AuNPs in solution. In this respect, citrate is considered a labile capping agent, because the exchange rate is fast, whereas MUA and other thiolate capping agents show much slower exchange rates.

The adsorption process of the protein molecule is controlled by van der Waals forces between the NP surface and the protein. These can be electrostatic attraction forces between the charged surface of the AuNP (more commonly negatively charged) and charged residues on the protein; hydrogen bonding between the capping molecules and the protein; acid-base adduct formation; dipole-dipole interactions; and London (dispersion) forces. In addition, when adsorption takes place, the protein replaces ions, water molecules, and/or capping agent molecules on the AuNP surface, while the AuNP replaces the local solvent shell at the protein surface. The combined effect of all these interactions can be strong enough to partly overcome the intramolecular interactions responsible for the tertiary structure of the protein, eventually leading to conformational changes. The extent of these conformational changes depends not only on the chemical properties of the AuNPs surface, mostly imparted by the capping agent, but also on the robustness of the tertiary structure of the protein. Moreover, it must be emphasized that the conformation stability is not usually the same for each of the protein domains, and thus, adsorption may or may not significantly affect the structure/function of the protein. Experimentally, both types of behavior have been observed.

Often, the adsorption of protein molecules takes place in a two-step process: a first fast adsorption occurs due to the electrostatic interaction between the charged surface and opposite charges in the protein surface; then slow rearrangement may occur, that may or may not lead to replacement of the capping agent and cause direct interaction of protein molecules with the metal surface. The latter slow reorganization step can involve interactions with internal residues of the protein (hydrophobic interactions; cysteine adsorption at the gold surface, etc.), leading to extensive protein conformational changes.

Many proteins have a prolate ellipsoid shape giving rise to two possible orientations relative to the metal surface: "side-on" and "end-on." Side-on adsorption is usually stronger than end-on adsorption, mainly because in the side-on orientation, the contact area between the two nano-objects is larger (Fig. 1). This may lead to irreversible adsorption, whereas end-on adsorption is more usually reversible. It is also to be expected that conformational changes induced by the nanoparticle should be more severe in side-on adsorption.

Many concepts given above are common to the adsorption of proteins to macroscopic surfaces. Nevertheless, other effects arise when the size of the particles is similar to the size of the proteins adsorbed. Figure 2 shows the relative sizes of selected proteins and of AuNPs at the most commonly used sizes.

For instance, the conformational changes induced by adsorption of proteins at flat macroscopic surfaces are commonly more severe than at NPs. Moreover, the distortion is also more intense at larger NPs (low curvature) than at smaller NPs (high curvature).

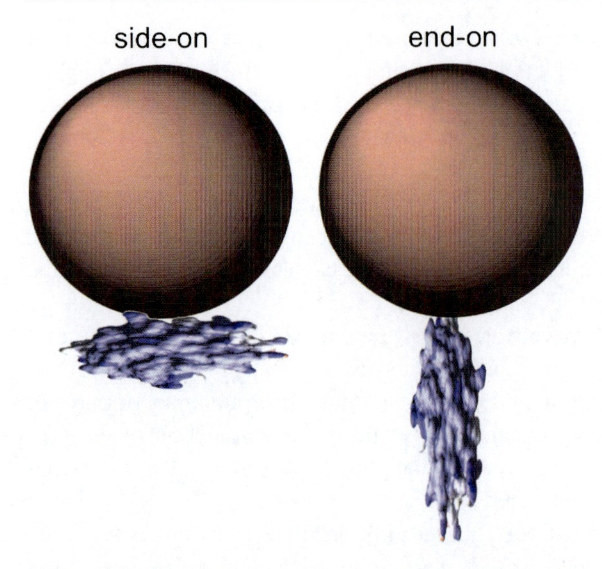

Gold Nanoparticles and Proteins, Interaction, Fig. 1 Possible orientations of a prolate ellipsoid-shaped protein relative to the metal surface, upon conjugation with AuNPs. The side-on adsorption orientation has a larger contact surface area compared to the end-on adsorption orientation

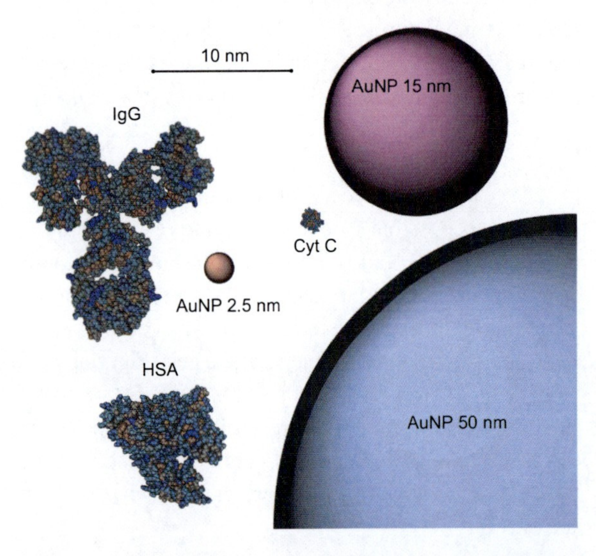

Gold Nanoparticles and Proteins, Interaction, Fig. 2 Relative sizes of AuNPs (most common) and representative proteins. Proteins are IgG Monoclonal Antibody (IgG) (PDB ID: 1IGY); Human Serum Albumin (HSA) (PDB ID: 1AO6); and Horse Heart Cytochrome c (Cyt c) (PDB ID: 1HRC)

This curvature effects are related to the decrease in contact area available for interaction when going from a flat surface to a curved surface (Fig. 3). As the size of NPs approaches the size of the proteins, the curvature effect becomes extremely important. These curvature effects are most commonly observed for globular proteins. Fibrillar proteins, on the other hand, often show large conformational changes in high curvature surfaces. This may be explained by side-on adsorption and wrapping of the protein around the nanoparticle.

Another important size effect arises from the lateral protein-protein interactions that often cause cooperative or anti-cooperative effects in proteins adsorbed at flat surfaces (crowding effect) (Fig. 3). The high curvature of NPs decreases the lateral interactions, with obvious advantages for maintaining biological activity by exposing the protein to the solvent and thus, keeping the protein available for biological interactions (e.g., with substrate in the case of enzymes, or to antigen binding for antibody-AuNP conjugates). In some cases, however, lateral protein-protein interactions are found to be necessary for structure stabilization and in these cases, the decrease in size, and hence the increase in the curvature of the surface, can be detrimental to the biological functions of adsorbed proteins. The increase in NP curvature and the resultant increased surface area determine an increased number of protein motifs on the "outer" surface of protein exposed to the medium. This may in turn lead to avidity effects arising from a dense location of actively interacting motifs on the NP surface (Shemetov et al. 2012).

In conclusion, adsorption of protein molecules at AuNP is a complex process that, depending on the conformational stability of the protein and the surface properties and morphology of the AuNPs, can lead to more or less severe changes in the protein structure. For a specific protein, the key factor for the adsorption process is the chemical composition of the surface of the NP, usually imparted by the capping agent directly surrounding the AuNP. A current common strategy for improving the properties of protein-AuNP conjugates is to alter the surface chemistry of the AuNP, by changing the capping agent, thus modulating the affinity and denaturing properties of the AuNPs.

Characterization Techniques

Conjugates between AuNPs and proteins can be formed by incubation of both components, namely, AuNPs with an appropriate capping agent and the protein containing solution. Alternatively, complex protein-containing mixtures, such as plasma, can be incubated with AuNPs and further characterized for

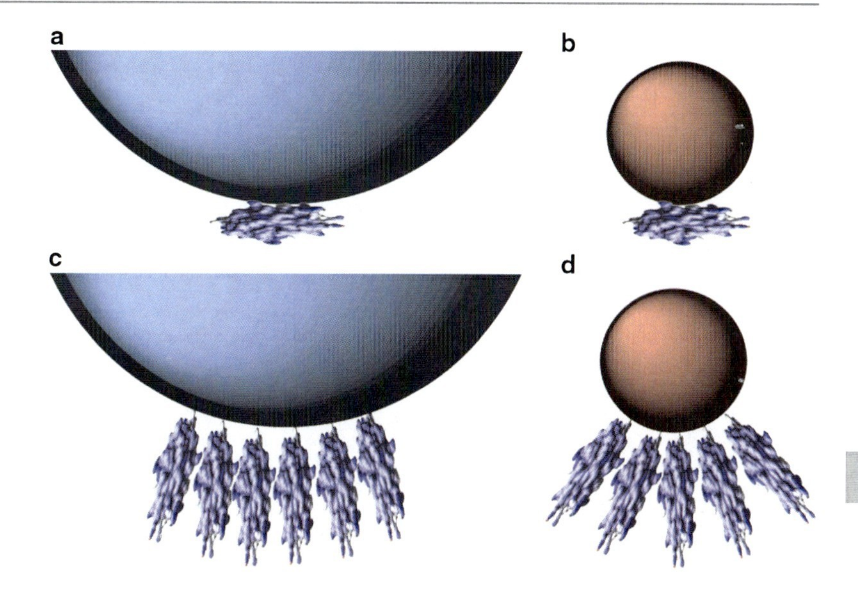

Gold Nanoparticles and Proteins, Interaction, Fig. 3 AuNP curvature effects on protein adsorption modes. The contact area between protein and AuNP is larger in low curvature NPs (**a**) than in high curvature NPs (**b**); lateral protein-protein interaction ("crowding effect") is more extensive in low curvature NPs (**c**) than in high curvature NPs (**d**)

determination of the quality and amount of AuNP-bound proteins (Dobrovolskaia et al. 2009; Mahmoudi et al. 2011). Protein to AuNP binding can occur by electrostatic-hydrophobic interaction, or covalent strategies can be utilized. The latter are of two main types: (a) covalent binding to the AuNP of a thiol from the side chain of a cysteine residue present in the protein; or (b) cross-linking chemistry by formation of an amide bond between carboxylate surface groups, functionalizing the AuNPs, and available amine groups in the protein, using EDC/NHS (1-Ethyl-3-(3-dimethylaminopropyl)-carbodiimide hydrochloride/ N-hydroxysuccinimide). Covalent binding strategies have to be considered with great care, as they often induce enzyme activity loss and deleterious protein structural alterations. In the case of conjugate formation from protein mixtures, these covalent strategies should be altogether avoided due to their lack of specificity. The cross-linking chemistry approach for covalent binding has been used for AuNP-antibody conjugates without appreciable loss of antigen affinity by the antibody, and sometimes with improved results in immunoassays.

AuNP-protein conjugate isolation is a complicated process. A common method to achieve the goal of obtaining pure conjugates is centrifugation followed by washing to remove unbound or weakly bound protein(s). Other common biochemical separation techniques that can be used to isolate AuNP-protein conjugates are gel exclusion chromatography and membrane ultrafiltration. In the case of protein mixtures, it is important that separation techniques have the capability to separate proteins with various affinities, without any disruption of the conjugates, as well as avoiding the creation of new binding opportunities (Mahmoudi et al. 2011).

Agarose gel electrophoresis is a powerful tool toward identification of protein-AuNP conjugates. Protein adsorption at the surface of a capped AuNP leads to reduced electrophoretic mobility, mainly due to reduction of the superficial charge of the AuNPs. This variation of electrophoretic mobility can in some cases be directly related to the degree of protein coverage of the AuNPs. For example, if increasing concentrations of a protein are incubated with citrate-capped AuNP, the resulting conjugates will present lower electrophoretic mobility toward the anode. Electrophoretic mobility data can be plotted versus protein to AuNP concentration ratios and fitted to a Langmuir-type behavior (Fig. 4).

The highest electrophoretic mobility values correspond to full protein coverage of the AuNP that can be attained at known protein concentrations in solution. Agarose gel electrophoresis of AuNP-protein conjugates is still lacking the quantitative detail that could already be achieved for nucleic acid-AuNP conjugates, in which single oligonucleotide discrimination is possible (Sperling and Parak 2010). Experimental optimization can, nevertheless, lead to qualitative results in terms of compactness of the protein corona, and quantitative results in terms of maximal protein coverage of the AuNPs.

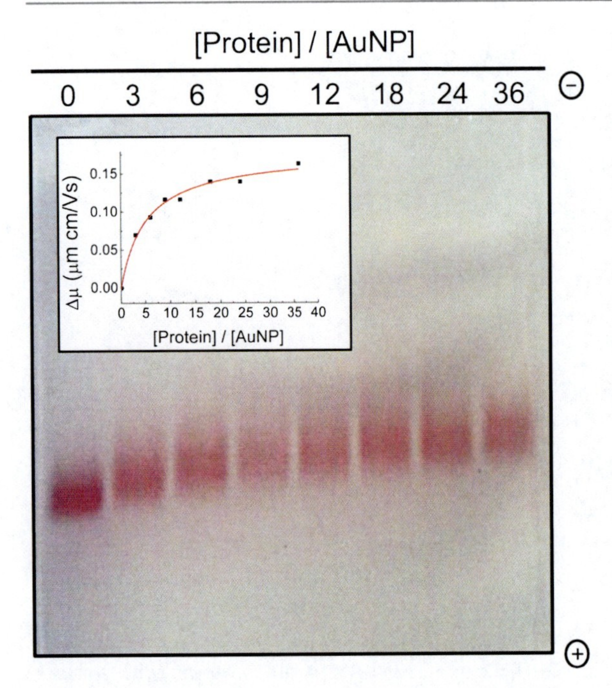

Gold Nanoparticles and Proteins, Interaction, Fig. 4 Agarose gel electrophoresis of tyrosinase-AuNP conjugates, prepared by incubation of the protein with pentapeptide-capped AuNPs, at the indicated concentration ratios. The inset presents a Langmuir fitting to electrophoretic mobility values determined for each lane

Spectroscopic techniques for conjugate characterization are complementary, and full characterization often requires two or more of the following.

Light scattering spectroscopies: Spectroscopic techniques based on light scattering include (a) Dynamic Light Scattering (DLS) also named Photon Correlation Spectroscopy (PCS), (b) Scattering Correlation Spectroscopy (SCS), and (c) Zeta(ζ)-Potential. Light scattering spectroscopic techniques detect protein binding to NPs as an increase in the nanoparticle hydrodynamic radius (DLS and SCS), or as a variation of the surface electric potential of the NPs upon protein adsorption. Equation 1 is a modified Langmuir isothermal adsorption curve which can be adjusted for experimental data of the specific property (S = hydrodynamic radius for DLS; or surface potential for ζ-potential measurements) for the free AuNP and for conjugates obtained for varying concentrations of protein (p):

$$S_{Conj} = S_{AuNP} \times \sqrt[3]{1 + \frac{V_p}{V_{AuNP}} \frac{N}{\left[1 + \left(\frac{K_D}{[p]}\right)^n\right]}} \quad (1)$$

In formula Eq. 1, V_p and V_{AuNP} are the volumes of the protein and the AuNPs, respectively, and their ratio is used as the correction factor for the finite size of the NPs in contrast to the macroscopic solid surfaces for which the equation was originally derived. N is the number of proteins bound, K_D is the dissociation constant, and n is the Hill coefficient, which measures cooperativity of binding. In the traditional Langmuir model, $n = 1$, with proteins binding to the NP independently, with no interaction between free and bound proteins, or between bound proteins. Hill coefficient values above 1 indicate positive cooperativity (the binding of one protein molecule promotes the binding of the next), and values below 1 indicate anticooperativity (strong repulsion between bound and free protein molecules, increasing as more surface binding sites are filled). In most reported cases, an anticooperative binding model is observed (Dominguez-Medina et al. 2012).

UV/visible spectroscopy: Typically, a solution of spherical AuNPs is red due to the collective oscillation of electrons in the conduction band known as the surface plasmon resonance (SPR). Color change of the solution can be induced by AuNP aggregation, with bathochromic shifts as high as 50 nm, a phenomenon that has been advantageously utilized in biodetection assays in several different formats (Baptista et al. 2011).

This effect can also be explored by UV/visible spectroscopy to characterize protein-AuNP conjugates. In fact, aggregation can be promoted by pH changes of the solution. The pH that onsets aggregation is directly related to the protonation state of chemical functionalities at the surface of the AuNPs or of the conjugates. For negatively charged AuNPs, when all functionalities are protonated, the surface potential approaches zero and aggregation begins. The pH of aggregation onset can then be an indirect measure of the degree of protein coverage of the AuNPs. For example, the pH of aggregation onset changes from 3 (corresponding to the pK_a of citrate) to 6, as citrate-capped AuNPs are gradually covered with a cytochrome c corona (Gomes et al. 2008). Figure 5 presents AuNP-cytochrome conjugates before and after pH-induced aggregation, namely, at pH values below and above pH 6, corresponding to aggregation onset.

Fluorescence spectroscopy: AuNPs present interesting photoluminescent properties, causing

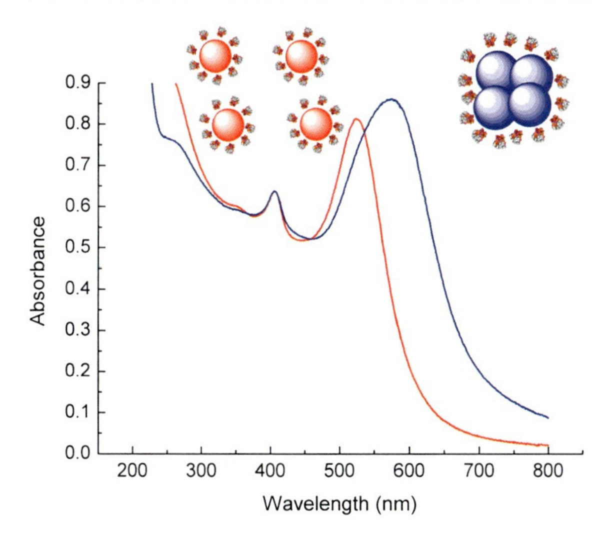

Gold Nanoparticles and Proteins, Interaction, Fig. 5 Visible spectra for cytochrome-AuNP conjugates at pH 8 (non-aggregated conjugates; *red line*), and at pH 4 (aggregated conjugates; *blue line*)

fluorescence enhancement or fluorescence quenching of chromophores in its vicinity, depending on the distance between the metal core and the fluorophore. Chromophores within ca. 5 nm of the surface of the AuNP have their fluorescence quenched, while fluorophores at distances of ca. 10 nm have their fluorescence enhanced up to 100-fold. The enhancement decreases again for longer distances. These effects have been used in the development of molecular diagnostic applications (Baptista et al. 2011). A particularly interesting example of an array-sensor platform for human serum proteins identification is based on specifically functionalized AuNPs-fluorophore constructs. The constructs are made of AuNPs capped with positively charged ligands bearing different terminal functionalities (alkyl, cycloalkyl, aromatic, polar), working as the recognition element, and the anionic green fluorescent protein (GFP) as the transducer element (Moyano and Rotello 2011). The discerning capabilities of the method toward a protein rely on the choice of surface functionalization of the AuNP, responsible for direct interaction with the bioanalyte. This approach is well suited for the detection of proteins present in high levels, but not of biomarkers.

CD spectroscopy: Circular dichroism (CD) spectroscopy in the far-UV region is commonly used for determination of protein secondary structures in solution. Different protein secondary structures (α-helix, β-sheet, etc.) have their own characteristic CD spectra in the UV region. AuNPs are not usually chiral in nature and thus, do not affect the protein CD spectra, although, if their size is large, they can give a scattering contribution which can affect the CD signal. Although CD cannot be applied on complex protein mixtures, it can provide useful information on protein structure changes resulting from adsorption to a NP surface (Mahmoudi et al. 2011).

FTIR and Raman: Vibrational spectroscopies, Raman, and Fourier transform infrared (FTIR) are used to detect conformational changes in proteins upon AuNP attachment, by alterations in the characteristic vibration bands. Also, protein attachment onto NPs can be confirmed through the appearance of additional characteristic bands. FTIR has been used to monitor the structure of NP-bound proteins, with protein secondary structures estimated on the basis of the absorption of amide bonds. Among the amide I, II, and III bands, the amide I vibrational band (1,700–1,600 cm^{-1}) is the most sensitive and is frequently used to determine protein conformation. Raman spectroscopy has an important advantage over FTIR, as spectra are free of interference from water vibrational bands. Raman spectra of proteins consist of bands associated with the protein backbone, and aromatic or sulfur-containing amino acid side chains (Mahmoudi et al. 2011).

Examples

Peptides and Amyloid conjugates: Seminal work by Lévy and collaborators (Shaw et al. 2011) established a rational approach for AuNPs capping by peptides. The proposed pentapeptide, Cys-Ala-Leu-X^1X^2, contains a strong affinity for gold, provided by the thiol side chain of the N-terminal cysteine, and the ability to self-assemble into a dense layer that excludes water, given by the hydrophobic side chains of alanine and leucine. A hydrophilic terminus ensures colloidal stability and water solubility ($X^{1,2}$ = Asn, Lys, etc.). This type of capping agent rapidly forms a compact self-assembled monolayer, which proved to be extremely useful as a bio-friendly surface for enzyme and antibody conjugation, with or without cross-linking agents. This peptide functionalization concept is rather flexible, allowing for specific recognition sequences to be designed based on the exposed amino acids. The exposed peptide recognition sequences can find use in bioassays and analytical devices, taking advantage of the specificity imparted by the peptide sequence; and of the avidity effect given by AuNP functionalization, for increased detection sensitivity. The AuNP-peptide interaction can be further developed in relation to amyloid peptides, implicated in the pathogenesis of Alzheimer's disease. These peptides suffer secondary structure transformation to a beta-sheet-dominated conformation, leading to protein aggregation. Here the approaches can be related to the amyloid-peptide aggregation process, as it can be induced or slowed down by the presence of AuNPs, depending on their surface functionalization, with special relevance for the influence of AuNPs curvature.

Control of enzyme activity on conjugates: Proteins, such as α-chymotrypsin (ChT), adsorb electrostatically to the surface of AuNPs functionalized with tetra(ethylene glycol) carboxylate ligands (AuTCOOH), with retention of protein structure and enzymatic activity. Two major mechanisms have been identified for this stabilization: (a) restriction of ChT unfolding through multivalent electrostatic interactions, and (b) preferential localization of AuTCOOH at the air-water interface, shielding ChT from interfacial denaturation. Interactions of the anionic monolayer surface of the NP with charged substrates and reaction products lead to a dramatic alteration of the substrate selectivity for ChT/Au-TCOOH conjugates in relation to the enzyme alone in solution (Moyano and Rotello 2011). To explore the effects of hydrophobicity on ChT affinity for AuNPs and on the stability of the protein in the resulting conjugates, AuNPs with amino acid cappings were used. Different amino acids provided an extensive range of hydrophobicities to the AuNPs, with increasing AuNP surface hydrophobicity resulting in increased protein binding. Smaller denaturation rate constants were obtained with increasing hydrophobicity, contradicting the general belief that hydrophobic surfaces are detrimental to protein stability.

Cytochrome conjugates: Cytochrome c (Cyt c) has been used as a model protein in several types of studies, including AuNP-protein interaction. The protein's structure and folding are extensively characterized, both experimentally and computationally. Cyt c presents a "front face lysine patch" in its three-dimensional structures, a protein region that is positively charged at neutral pH and is involved in the formation of physiological complexes of Cyt c with other proteins. This highly charged region has been advantageously used for studying adsorption to negatively charged surfaces in electrochemical systems. Studies of Cyt c adsorption to AuNPs with an anionic functionalized surface have shown Cyt c binding over a large surface and with a high binding constant. Conversely, AuNPs with a hydrophobic functionalized surface bind a much smaller surface of Cyt c, spanning from Tyr^{67} to Phe^{82}, suggesting increased selectivity of the functionalized AuNPs for this specific protein surface. A combination of electrostatics and hydrophobicity seems then to be necessary for specific recognition of the different faces of Cyt c (Moyano and Rotello 2011).

Non-covalent AuNP-Cyt c interaction studies have relied in Cyt c from horse heart, whereas interaction studies on covalent AuNP-Cyt c conjugates are based on the protein from *Saccharomyces cerevisiae*. This Cyt c has a unique exposed Cys^{102} that can be linked covalently to AuNPs, inducing protein conformational changes (Gomes et al. 2008). This effect can be explored as a tool for studying protein structure and folding at the AuNP surface, by creating site-directed mutants in which Cys residues are replaced for amino acid residues strategically placed in the protein three-dimensional structure (Aubin-Tam et al. 2009).

Interaction with blood proteins: AuNPs can access the bloodstream either by intentional (medical use) or unintentional (pollution) delivery. It has been recognized that protein adsorption is the initial event when NPs come into contact with a bodily fluid. The so-formed protein corona changes the biological identity of the NPs and can have profound consequences in biodistribution, cell uptake, and other toxicological issues. This initial protein corona is kinetically labile, and its specific composition at a given instant after contact strongly depends on the concentration of blood proteins and their relative affinity for the AuNP (Lynch and Dawson 2008). This initial layer consisting mainly of the more abundant proteins loosely bound to the AuNP surface (soft corona) slowly transforms into a more densely packed layer of higher affinity proteins (hard corona) existing in blood at lower concentrations.

The composition of the protein corona depends primarily on the surface properties of the AuNPs, but size and curvature also seem to be an important factor. Blood has a complex composition of proteins with ca. 10^4 proteins, most of which at very low concentrations.

Albumin is the most abundant protein in the blood, followed by fibrinogen and gamma-globulins. Albumin, apolipoprotein, immunoglobulins, and fibrinogen were found to be the most abundantly AuNP-bound proteins. AuNPs were also found to induce conformational changes in these proteins. This can have profound implications on the biological impact of AuNPs since conformational changes may expose internal protein residues, leading to new epitopes on the protein surface and disturbance of biological mechanisms, for example, inflammatory responses (Shemetov et al. 2012).

Future Directions

Protein and AuNPs interact through complex mechanisms, which are presently the subject of active research. The number of available studies is still modest taking into consideration the complexity of the problem. The characterization of the protein corona adsorbed to the NP is crucial, as this dynamic layer of proteins determines how a NP interacts with living systems and thereby can modify the cellular responses to the NPs. Improved characterization methods should be developed for the quantitative and qualitative analysis of the adsorbed protein layer. In this respect, great promise lies in technological advanced methodologies, such as Differential Centrifugal Sedimentation. On the other hand, established methods, such as Agarose Gel Electrophoresis, should now be explored under the perspective of a NP system.

Another determining step toward medical and nonmedical applications of AuNPs is related to nanotoxicological issues that must be addressed before implementing the technology. It is worth mentioning that the risk assessment procedure should include numerous types of molecular, cell culture, and animal model tests to identify the potential hazards of nanoparticles before they could be used in biomedical applications or released into the environment.

Cross-References

- ▶ Biomineralization of Gold Nanoparticles from Gold Complexes in Cupriavidus Metallidurans CH34
- ▶ Gold and Nucleic Acids
- ▶ Gold Nanomaterials as Prospective Metal-based Delivery Systems for Cancer Treatment
- ▶ Gold Nanoparticle Platform for Protein-Protein Interactions and Drug Discovery
- ▶ Gold Nanoparticles and Fluorescent Proteins, Optically Coupled Hybrid Architectures
- ▶ Gold, Ultrasmall Nanoclusters and Proteins, Interaction

References

Aubin-Tam M-E, Hwang W et al (2009) Site-directed nanoparticle labeling of cytochrome c. Proc Natl Acad Sci 106(11):4095–4100

Baptista PV, Doria G, et al (2011) Chapter 11 – Nanoparticles in molecular diagnostics. In Antonio V (ed) Progress in molecular biology and translational science, vol 104. Amsterdam: Academic Press, pp 427–488

Dobrovolskaia MA, Patri AK et al (2009) Interaction of colloidal gold nanoparticles with human blood: effects on particle size and analysis of plasma protein binding profiles. Nanomed Nanotechnol Biol Med 5(2):106–117

Dominguez-Medina S, McDonough S et al (2012) In situ measurement of bovine serum albumin interaction with gold nanospheres. Langmuir 28(24):9131–9139

Gomes I, Santos NC et al (2008) Probing surface properties of cytochrome c at Au bionanoconjugates. J Phys Chem C112(42):16340–16347

Lynch I, Dawson KA (2008) Protein-nanoparticle interactions. Nano Today 3(1–2):40–47

Mahmoudi M, Lynch I et al (2011) Protein – nanoparticle interactions: opportunities and challenges. Chem Rev 111(9):5610–5637

Moyano DF, Rotello VM (2011) Nano meets biology: structure and function at the nanoparticle interface. Langmuir 27(17):10376–10385

Shaw CP, Fernig DG et al (2011) Gold nanoparticles as advanced building blocks for nanoscale self-assembled systems. J Mater Chem 21(33):12181–12187

Shemetov AA, Nabiev I et al (2012) Molecular interaction of proteins and peptides with nanoparticles. ACS Nano 6(6):4585–4602

Sperling RA, Parak WJ (2010) Surface modification, functionalization and bioconjugation of colloidal inorganic nanoparticles. Philos Trans R Soc A Math Phys Eng Sci 368(1915):1333–1383

Gold Nanoparticles Interaction with Proteins

▶ Gold Nanoparticles and Proteins, Interaction

Gold Nanoparticles, Biosynthesis

Kannan Deepa and Tapobrata Panda
Biochemical Engineering Laboratory, Department of Chemical Engineering, Indian Institute of Technology Madras, Chennai, Tamil Nadu, India

Synonyms

Clusters; Colloidal gold; Monodispersity; SPR

Definition

Nanoparticles are entities that have at least one of their dimensions less than 100 nm. The properties of nanoparticles are markedly different from that of their bulk counterparts. Nanomaterials are known to have widespread applications ranging from consumer goods such as cosmetics and textiles to highly specialized core research fields such as diagnostics, imaging, catalysis, electronics, optics, etc., Though several techniques exist for the synthesis of nanoparticles, they suffer from limitations such as polydispersity and instability of the nanoparticles and the toxic nature of the chemicals involved. Hence, it is necessary to look upon alternative benign routes for the synthesis of nanoparticles. Synthesis of nanoparticles from biological sources is not only an eco-friendly alternative, but also confers additional advantages like water solubility, biocompatibility, and surface functionalization.

Introduction

The vital interactions between biological systems and inorganic materials are implicated in several instances such as the ability of organisms to produce highly structured bioinorganic materials, removal of toxic metals (bioremediation), and the recovery of precious metals such as gold. Oft-cited examples of lower organisms producing inorganic materials include the magnetotactic bacteria synthesizing magnetite nanoparticles that act as magnetic sensors; diatoms and radiolarians possessing highly organized and geometrically symmetric siliceous nanoarchitectures (Sarikaya 1999). Nanoparticle research mainly focuses on synthesis procedures since it is required to synthesize nanoparticles of desired size, shape, and monodispersity. Another hot area of adequate interest is the surface modification of the synthesized nanoparticles (Eustis and El-Sayed 2006). The past decade has witnessed a tremendous increase in the number of publications in the field of "biosynthesis of nanoparticles" of which gold and silver were the most studied. Extensive literature available on the biosynthesis of nanoparticles showcases the wide variety of biological sources (ranging from bacteria to higher fungi and plants) used for the synthesis of nanoparticles. Relatively little information is available on the mode of interaction between the gold ions and the biomolecules leading to the formation of nanoparticles, though several reports suggest the possible role of proteins (enzymes in particular), sometimes with the assistance of cofactors (Durán et al. 2011).

Process Outline

The production of metal nanoparticles by wet synthesis techniques involves two significant steps – (1) reduction of the precursor salt to form metal clusters followed by (2) capping/stabilizing the clusters to reduce their growth rate and to prevent aggregation of the clusters into larger particles. Stabilization is necessary since the preformed clusters are subject to attractive forces (such as van der Waals forces) at short interparticle distances which if not counteracted by equivalent repulsive forces would render the solution unstable. Stabilization is brought about either electrostatically, sterically or by a combination of both (electrosteric stabilization) (Schmid 2004). Electrostatic/charge stabilization involves the formation of an electrical double layer of adsorbed ions and their counter ions. These electrostatic Coulombic repulsive forces between the clusters overcome the attractive forces resulting in the formation of stable colloids. Steric stabilization is achieved by enclosing the clusters in a protective layer consisting of molecules of polymers, surfactants, or ligands. Hence, individual clusters are not within the range of the attractive forces of neighboring clusters, thereby preventing aggregation. Figure 1 shows the scheme for the formation of gold nanoparticles.

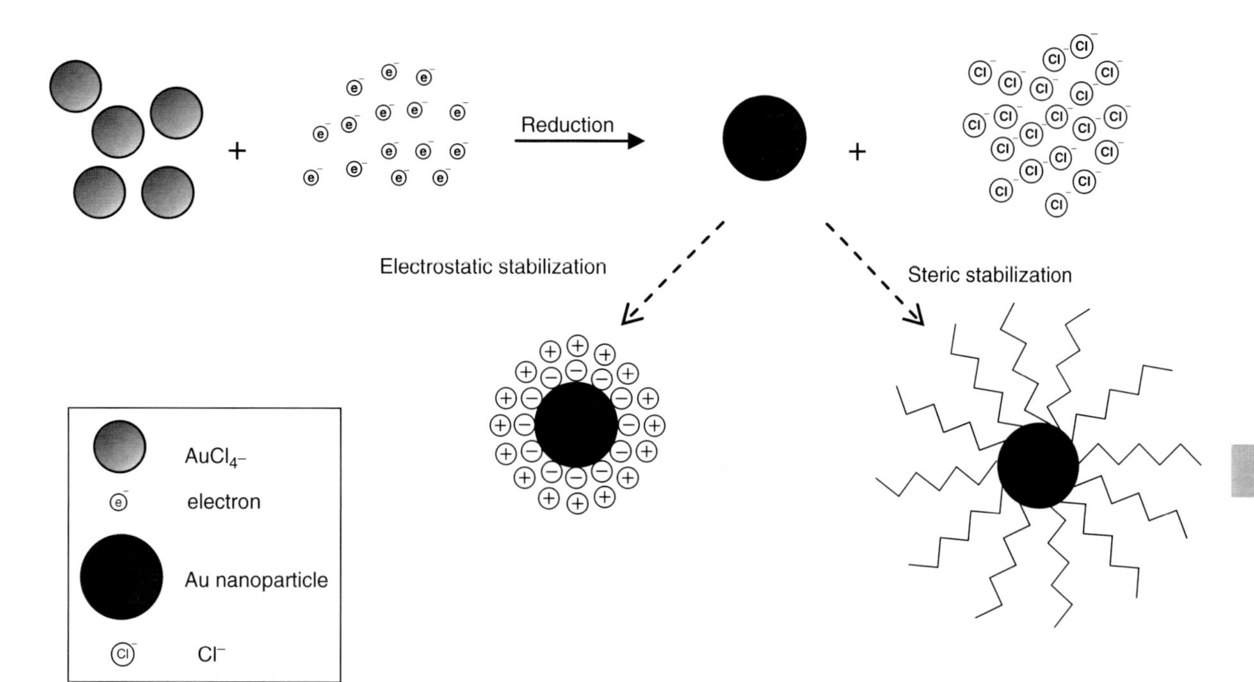

Gold Nanoparticles, Biosynthesis, Fig. 1 Scheme for the formation of gold nanoparticles. $AuCl_4^-$ ions are reduced to Au^0 clusters. The clusters are then stabilized by providing an electrical double layer (electrostatic stabilization) or coated using polymers, surfactants, or ligands (steric stabilization)

In the biosynthesis of gold nanoparticles, the most commonly used precursor salt is Gold (III) chloride ($HAuCl_4$). Reduction of the $AuCl_4^-$ ions to metallic Au proceeds according to the following reaction.

$$\left[\begin{array}{c} Cl \quad\quad Cl \\ Au(III) \\ Cl \quad\quad Cl \end{array}\right]^- + 3e^- \longrightarrow Au^0 + 4Cl^-$$

Reduction and/or stabilization are believed to be mediated by protein molecules.

Challenges

A variety of methods are available for synthesizing gold nanoparticles but they suffer from limitations such as extent of size and shape control, poor degree of monodispersity, and instability. Also, eco-friendly methods that do not employ toxic reducing agents, stabilizers, or solvents are mandatory. The process should not be energy-intensive, i.e., use of high temperatures, radiation, etc., are not recommended.

Sources Explored

A broad range of biological systems have been studied for the synthesis of gold nanoparticles. Live cells or extracts from microorganisms, plants, and even human cells have been used to synthesize gold nanoparticles. The most commonly used protocol for microbial synthesis of gold nanoparticles requires that the washed biomass be suspended in a solution of the gold salt. Cell culture filtrates can also be used instead of the biomass. However, methods for synthesis of gold nanoparticles from plants has variations that are mentioned under the "plants" section. For the purpose of brevity, it would be better to look into a few unique examples.

Bacteria

The interaction between $AuCl_4^-$ ions and cell walls of *Bacillus subtilis* was reported by Beveridge and Murray (1976) as a part of an ensemble study involving the uptake and retention of several metals by this bacterium. It was observed that *microscopic* elemental gold (Au^0) crystals were deposited within the cell walls. The formation of *nanoscale* gold by

Bacillus subtilis was however observed later when gold nanoparticles in the size range of 5–50 nm were precipitated intracellularly (Southam and Beveridge 1994). Gold nanoparticles of two varying size ranges, i.e., clusters (20–50 nm) and crystals ($>$ 100 nm) were synthesized by *Lactobacillus* sp. commonly found in buttermilk (Nair and Pradeep 2002). Other bacteria that are known to synthesize gold nanoparticles include *Escherichia coli* DH5α, *Plectonema boryanum* UTEX 485, *Shewanella algae, Desulfovibrio desulfuricans, Rhodopseudomonas capsulata* or *Rhodobacter capsulatus, Actinobacter* sp., *Bacillus megaterium* D01, *Bacillus licheniformis*, and *Stenotrophomonas maltophilia* among several others (Panda and Deepa 2011).

Plectonema boryanum UTEX 485 reacted differently to $Au(S_2O_3)_2^{3-}$ and $AuCl_4^-$ ions. With $AuCl_4^-$ ions, octahedral platelets of size lesser than 10 nm were observed intracellularly while extracellular particles where in the range of 1–10 μm. With $Au(S_2O_3)_2^{3-}$, nanoparticles of size 10–25 nm were present in the membrane vesicles along with gold sulfide both intracellularly and also encrusted on the cyanobacteria. The anaerobic bacterium *Shewanella algae* showed differences in the size and localization of the nanoparticles based on varying pH conditions. At pH 7, 10–20 nm sized nanoparticles were found in the periplasm; at pH 2.8, particles in the size range of 15–200 nm were found, some of which were present extracellularly; at pH 2, 20 nm sized particles were deposited on cells while 350 nm sized particles were deposited extracellularly. The extracellular reduction of gold ions was attributed to an enzyme which got released from the periplasmic space into the solution.

Actinomycetes and Fungi

Among the actinomycetes, the alkalothermophilic *Thermonospora* and alkalotolerant *Rhodococcus* sp. have been known to synthesize gold nanoparticles. The marine yeast, *Yarrowia lipolytica* was able to synthesize hexagonal and triangular gold nanoparticles on the cell surface within 12 h, and later got dislodged from the cells after 24 h. Fungi were considered as useful candidates for synthesizing nanoparticles owing to their ability to secrete large amount of enzymes (Mandal et al. 2006).

Several fungi such as *Verticillium* sp., *Colletotrichum* sp., *Trichothecium* sp., *Helminthosporium solani*, *Fusarium oxysporum, Fusarium semitectum, Aspergillus niger*, and *Penicillium* sp. are known to synthesize gold nanoparticles (Panda and Deepa 2011). The alkalotolerant fungus, *Trichothecium* sp. produced gold nanoparticles of sizes between 5 and 200 nm extracellularly under stationary condition. Varied shapes of the nanoparticles were also observed such as triangular, hexagonal, spherical, and rod shapes. However, under shaking condition, spherical nanoparticles in the size range of 10–25 nm were observed on the cell wall and cytoplasmic membrane. The effect of temperature on the synthesis of gold nanoparticles by *Penicillium* sp. was studied. Nanoparticles of size 8 ($\pm$2) nm were obtained at 4°C, while at 28°C, nanoparticles of size 30 ($\pm$5) nm were obtained. At fluctuating temperatures between 20°C and 30°C, even bigger particles of size 75 ($\pm$35) nm were obtained indicating higher degree of polydispersity. The extract of *Volvariella volvacea* exhibited variation in nanoparticle shapes at different temperatures. At 40°C, hexagonal structures were formed while dendritic or fractal structures were obtained at 80°C.

Plants

The usage of plants for the synthesis of gold nanoparticles has several advantages such as nonpathogenicity, low toxicity, low cost, and the process is generally faster than when employing microorganisms. Most importantly, the elaborate process of maintaining cell cultures is avoided since *phytosynthesis* makes use of either the extracts derived from plant tissues such as leaves and fruits or isolated compounds such as phyllanthin and apiin. Growing the seedlings in metal-rich soil or agar media are other alternative methods (Rai et al. 2008). Plants that were used for synthesizing gold nanoparticles include *Emblica officinalis* (fruit extract), *Sesbania drummondii, Camellia sinensis, Cymbopogon flexuosus, Pelargonium graveolens, Psidium guajava, Medicago sativa, Cinnamomum camphora, Cinnamomum zeylanicum, Triticum aestivum* (shoots and leaves), *Azadirachta indica, Aloe vera, Cicer arietinum* L. (seeds), *Magnolia kobus, Diopyros kaki*, etc.

By varying the ratio of $HAuCl_4$ to tea extract (*Camellia sinensis*), nanoparticles of different sizes were obtained. Shape control of the biosynthesized nanoparticles was observed with *Cymbopogon flexuosus* wherein predominantly hexagonal particles were obtained at lower concentrations of the extract, while mostly spherical particles were obtained at higher concentrations. Nanoparticles of varying morphologies were produced as an effect of varying temperature in the case of *Magnolia kobus* leaf extract. At 25°C and 60°C, triangular, pentagonal, and hexagonal plates of dimensions 250–300 nm were obtained. However at 95°C, mostly spherical particles were obtained.

Characterization

Physical, chemical, and electrochemical methods of synthesizing nanoparticles clearly state the factors responsible for the formation of nanoparticles. Hence, only the properties of the nanoparticles such as their shape, size, size distribution, degree of polydispersity, stability, etc., need to be characterized. However, a biological process is several times complex due to the innumerable components which are produced by the cell under different conditions. The greatest challenge lies in exploring the actual biological factors responsible for the formation of the nanoparticles apart from characterizing the nanoparticles themselves. Given below are a few techniques used for characterizing the nanoparticles.

Gold in its colloidal state is known to exhibit vivid colors ranging from red, purple, violet, pink, and blue. These colors arise due to the Surface Plasmon Resonance (SPR) of metal nanoparticles. Hence, a change in color of the reaction solution to one of the above-mentioned colors is a primary indication for the formation of colloidal gold. This could be followed by a UV-visible spectrophotometric analysis. The presence of a maximum absorption peak in the visible spectrum around 540 nm indicates the presence of gold nanoparticles in solution. X-ray diffraction (XRD) gives information about the crystallinity of the nanoparticles. Also, the Full Width at Half Maximum (FWHM) of the maximum refraction peak in the diffractogram can be used to calculate the average size of the nanoparticles using the Scherrer equation. Electron microscopy is used to visualize the nanoparticles in order to study their size, shape, alignment, and even the localization of the nanoparticles within cells. Most electron microscopy instruments are equipped with Energy Dispersive X-ray (EDX) analyzer which gives information about the elemental composition of the sample. Selected Area Electron Diffraction (SAED) gives information about the crystal structure of the nanoparticles which can be correlated with XRD data to establish the crystallinity of the particles.

Mechanisms

Several possible mechanisms for the formation of gold nanoparticles using biological systems have been reported (Durán et al. 2011). Tyrosine residues present in the C-terminus of the peptides and are capable of reducing the gold ions and also stabilizing the gold nanoparticles through interactions of the metal surface with the free N-terminus of the peptide. Tryptophan residues of peptides get converted to a transient tryptophyl radical donating an electron in the process, thereby acting as a reducing agent in the formation of metal nanoparticles.

In the microbial synthesis of gold nanoparticles, it is speculated that proteins are involved in reduction and stabilization of the nanoparticles. The role of NADH-dependent reductases is the most widely reported factor responsible for the synthesis of metal nanoparticles by several organisms. NADH acts as the electron carrier. More specifically, NADH-dependent nitrate reductases which got induced by nitrate ions were reported to mediate the process of bioreduction of metal ions. Though both *Fusarium oxysporum* and *Fusarium moniliforme* were able to secrete nitrate reductase, nanoparticles were not obtained with the latter. Electron shuttling compounds such as quinones were necessary in addition to nitrate reductase activity. The absence of such an electron shuttle mechanism in *Fusarium moniliforme* could be responsible for the inability of the organism to produce nanoparticles. The stabilization of nanoparticles by polypeptides or proteins is supported by FTIR analyses showing the presence of amide groups and the disappearance of carboxyl groups as in the synthesis of gold nanoparticles by *Rhizopus oryzae*. The intracellular reducing sugar content of *Penicillium* sp. was responsible for the formation of gold nanoparticles, while carotenoids present in the plasma membrane of *Rhodobacter capsulatus* mediates the reduction of gold ions. A biosurfactant, surfactin produced by *Bacillus subtilis* promotes the stabilization of the gold nanoparticles.

Studies on the synthesis of gold nanoparticles using plant extracts suggest the role of functional groups of the cell walls of plant tissues in the reduction process. Reducing sugars (aldoses), aldehydes, and ketones,

Gold Nanoparticles, Biosynthesis, Table 1 Applications of biosynthesized gold nanoparticles

S. No.	Organism/Compound	Type/Origin	Application	References
1.	*Lactobacillus* sp.	Bacterium	In situ investigations of biomolecules using Surface Enhanced Raman Spectroscopy (SERS)	Nair and Pradeep 2002
2.	*Escherichia coli* DH5α	Bacterium	Study of electron transfer mechanisms of redox-active proteins	Du et al. 2007
3.	*Rhizopus oryzae*	Fungus	Biosorption of organophosphorus pesticides and antimicrobial activity	Das et al. 2009
4.	*Helminthosporum solani*	Phytopathogenic fungus	Enhancement of uptake of anticancer drug, doxorubicin by HEK-293 cells	Kumar et al. 2008
5.	*Cymbopogon flexuosus*	Lemongrass plant	Infrared-absorbing optical coatings on glass	Shankar et al. 2005
6.	Tea leaves	–	Cellular probes in imaging of tumor cells	Nune et al. 2009
7.	*Sesbania drummondii*	Leguminous shrub	Catalytic role in the conversion of toxic pollutant 4-nitrophenol to 4-aminophenol	Sharma et al. 2007
8.	*Scutellaria barbata* D. Don	Herb	-do-	Wang et al. 2009
9.	HEK-293 (human embryonic kidney), HeLa (human cervical cancer), SiHa (human cervical cancer), and SKNSH (human neuroblastoma) cells	Non-malignant and malignant cell lines	Cancer diagnostics	Anshup et al. 2005

hydroxyl groups, amino groups, carboxylic groups, sulfhydryl groups, methoxy groups, and proteins have been reported as the factors mediating phytosynthesis of nanoparticles. Polyol components and water-soluble heterocyclic components are also implicated in the synthesis of metal nanoparticles. Theaflavins and thearubigins, the polyphenols present in tea leaf extract are involved in reduction of the gold ions. Flavonoids and terpenoids present in the leaf extract of *Psidium guajava* and *Pelargonium graveolens*, respectively, were found to be responsible for the formation of gold nanoparticles.

Applications

Gold nanoparticles are known to have maximum utility in the fields of biology and medicine in areas of cellular imaging, targeting tumor cells in cancer therapy, drug delivery, etc. In this context, it is essential to synthesize gold nanoparticles that are free of toxic unreacted reducing agents or surfactants that are commonly used for stabilization in the chemical syntheses protocols. It is possible that gold nanoparticles synthesized through benign routes such as biosynthesis would be most suited for in vivo studies than their chemically prepared counterparts. Gold nanoparticles synthesized using biological systems have been studied for their applications in spectroscopy, electrochemistry, catalysis, environmental biotechnology, and diagnostics. Table 1 lists the potential applications of biosynthesized gold nanoparticles.

Future Directions

In the past decade, several investigations have been carried out on the biosynthesis of gold nanoparticles with much emphasis on the search for novel biological sources for synthesizing them. Researchers have also explored the biological factors and the possible mechanisms involved in the formation of gold nanoparticles. The scope of future studies in this area could be directed on the analysis of extraction procedures of the biogenic nanoparticles. Another exciting area of study would be to genetically modify the organism to overexpress the specific proteins responsible for the formation of nanoparticles. As a consequence, the protein could be extracted in significant amounts and used for large-scale synthesis of the nanoparticles. And finally, the studies on applications of biosynthesized nanoparticles require much more attention.

Cross-References

- ▶ Aluminum, Physical and Chemical Properties
- ▶ Gold and Nucleic Acids
- ▶ Gold Complexes as Prospective Metal-Based Anticancer Drugs
- ▶ Gold Nanomaterials as Prospective Metal-based Delivery Systems for Cancer Treatment
- ▶ Gold Nanoparticle Platform for Protein-Protein Interactions and Drug Discovery
- ▶ Gold Nanoparticles and Fluorescent Proteins, Optically Coupled Hybrid Architectures
- ▶ Gold Nanoparticles and Proteins, Interaction
- ▶ Gold, Ultrasmall Nanoclusters and Proteins, Interaction
- ▶ Gold(III) Complexes, Cytotoxic Effects
- ▶ Metals and the Periodic Table

References

Anshup A, Venkataraman JS, Subramaniam C et al (2005) Growth of gold nanoparticles in human cells. Langmuir 21:11562–11567

Beveridge TJ, Murray RGE (1976) Uptake and retention of metals by cell walls of *Bacillus subtilis*. J Bacteriol 127:1502–1518

Das SK, Das AR, Guha AK (2009) Gold nanoparticles: microbial synthesis and application in water hygiene management. Langmuir 25:8192–8199

Du L, Jiang H, Liu X et al (2007) Biosynthesis of gold nanoparticles assisted by *Escherichia coli* DH5α and its application on direct electrochemistry of hemoglobin. Electrochem Commun 9:1165–1170

Durán N, Marcato PD, Durán M et al (2011) Mechanistic aspects in the biogenic synthesis of extracellular metal nanoparticles by peptides, bacteria, fungi, and plants. Appl Microbiol Biotechnol 90:1609–1624

Eustis S, El-Sayed MA (2006) Why gold nanoparticles are more precious than pretty gold: noble metal surface plasmon resonance and its enhancement of the radiative and nonradiative properties of nanocrystals of different shapes. Chem Soc Rev 35:209–217

Kumar SA, Peter Y, Nadeau JL (2008) Facile biosynthesis, separation and conjugation of gold nanoparticles to doxorubicin. Nanotechnology 19:495101–195110

Mandal D, Bolander ME, Mukhopadhyay D et al (2006) The use of microorganisms for the formation of metal nanoparticles and their application. Appl Microbiol Biotechnol 69:485–492

Nair B, Pradeep T (2002) Coalescence of nanoclusters and formation of submicron crystallites assisted by Lactobacillus strains. Cryst Growth Des 2:293–298

Nune SK, Chanda N, Shukla R et al (2009) Green nanotechnology from tea: phytochemicals in tea as building blocks for production of biocompatible gold nanoparticles. J Mater Chem 19:2912–2920

Panda T, Deepa K (2011) Biosynthesis of gold nanoparticles. J Nanosci Nanotechnol 11:10279–10294

Rai M, Yadav A, Gade A (2008) Current trends in phytosynthesis of metal nanoparticles. Crit Rev Biotechnol 28:277–284

Sarikaya M (1999) Biomimetics: materials fabrication through biology. Proc Natl Acad Sci 96:14183–14185

Schmid G (2004) Nanoparticles – from theory to application. Wiley-VCH, Weinheim

Shankar SS, Rai A, Ahmad A et al (2005) Controlling the optical properties of lemongrass extract synthesized gold nanotriangles and potential application in infrared-absorbing optical coatings. Chem Mater 17:566–572

Sharma NC, Sahi SV, Nath S et al (2007) Synthesis of plant-mediated gold nanoparticles and catalytic role of biomatrix-embedded nanomaterials. Environ Sci Technol 41:5137–5142

Southam G, Beveridge TJ (1994) The in vitro formation of placer gold by bacteria. Geochim Cosmochim Acta 58:4527–4530

Wang Y, He X, Wang K et al (2009) Barbated Skullcup herb extract-mediated biosynthesis of gold nanoparticles and its primary application in electrochemistry. Colloids Surf B 73:75–79

Gold Physiochemical Properties

▶ Gold Complexes as Prospective Metal-Based Anticancer Drugs

Gold(III) Complexes as Anticancer Agents

▶ Gold(III) Complexes, Cytotoxic Effects

Gold(III) Complexes, Cytotoxic Effects

Nebojša Arsenijević[1], Vladislav Volarevic[1], Marija Milovanovic[1] and Živadin D. Bugarčić[2]
[1]Faculty of Medical Sciences, University of Kragujevac, Centre for Molecular Medicine, Kragujevac, Serbia
[2]Faculty of Science, Department of Chemistry, University of Kragujevac, Kragujevac, Serbia

Synonyms

Anticancer characteristics of gold(III) complexes; Antiproliferative characteristics of gold(III) compounds; Antitumor effects of metal-based drugs; Cytotoxic potential of gold(III) complexes; Gold(III) complexes as anticancer agents; Mechanisms of action of gold(III) compounds; Toxicity of gold(III) complexes

The Definition and Importance of the Subject

Gold(III) is isoelectronic with platinum(II), and tetracoordinate gold(III) complexes have the same square-planar geometries as cisplatin, the drug that is currently used for the treatment of several solid tumors. However, effectiveness of cisplatin is limited by toxic side effects and tumor resistance that often leads to the occurrence of secondary malignancies. Several recently published studies that tested anticancer characteristics of gold(III) complexes showed promising results: high cytotoxic effect in vitro and in vivo was noticed against representative tumor cell lines with reduced, or even no, systemic and renal toxicity, suggesting some of the newly synthesized gold(III) compounds as novel, effective anticancer agents. The results achieved so far are summarized here, focusing on the latest in-depth mechanistic studies that have recently provided insights into gold(III) compound mechanism of action and their antitumor, cytotoxic potential, thus opening up new prospects for further pharmacological testing and, hopefully, to enter clinical trials.

Introduction

Gold compounds have a long tradition in applications in medicine as drugs. One of the major goals of modern bioinorganic and bio-organometallic medicinal chemistry research is the development of novel metallo-drugs with a pharmacological activity different from platinum drugs (Casini and Messori 2011). Among the new non-platinum antitumour drugs, gold complexes have recently gained considerable attention as a class of compounds with different pharmacodynamic and pharmacokinetic properties than cisplatin, but with strong cell growth–inhibiting effects. In many cases, the cell growth–inhibiting effects could be related to anti-mitochondrial effects that make the gold complexes interesting drugs (Nobili et al. 2010). Some gold complexes described as antiproliferative comprise a broad variety of different species including phosphine complexes, as well as gold complexes in different oxidation states (Nobili et al. 2010). Also, some gold(I) compounds used for the treatment of rheumatoid arthritis, such as gold thiolates, were considered for possible antitumor activity. It has been shown that gold(I) compounds inhibit tumor cell proliferation in vitro, but unfortunately their in vivo effectiveness as well as significant nephrotoxicity and poor chemical stability limited their use (Casini and Messori 2011). However, the unique chemistry of the gold center, a great propensity to form strong gold–gold bonds (the so-called aurophilic interactions), and rich redox chemistry were further exploited and a variety of gold-based pharmacologically active compounds were synthesized so far (Djeković et al. 2012).

Gold(III) complexes have been synthesized and characterized and show greater chemical stability and a far better toxicological profile than gold(I) complexes. During the early 1990s, there was a revival of interest toward gold(III)-based antitumor compounds in an attempt to obtain pharmaceutically useful substances with an even better stability profile. As a result, several square planar gold(III) complexes, isoelectronic and isostructural with Pt(II), were prepared and tested as potentially new anti-tumor drugs (Che and Sun 2011).

Classification of Gold(III) Complexes

All gold(III) complexes can be grouped into four categories:

1. Classical mononuclear gold(III) complexes
2. Gold(III) porphyrins
3. Organogold(III) compounds
4. Dinuclear gold(III) complexes

Classical Mononuclear Gold(III) Complexes

These compounds are square planar gold(III) compounds accompanied with nitrogen or halide ligands: $[AuCl_4]^-$, $[Au(dien)Cl]Cl_2$ (Audien), $[Au(en)_2]Cl_3$ (Auen), $[Au(cyclam)](ClO_4)_2Cl$ (Aucyclam), $[Au(terpy)Cl]Cl_2$ (Auterpy), and $[Au(phen)Cl_2]Cl$ (Auphen).

Nitrogen ligands are less labile than chloride ligands, while chloride ligands undergo far more facile aquation reactions.

Gold(III) Porphyrins

In these complexes, gold(III) is stabilized with the porphyrin ligand that led to stabilization of gold(III) center and reduces its redox and oxidizing character. Further, reduction of gold(III) to gold(I) or elemental gold is very rare.

Organogold(III) Compounds

These complexes have at least one direct carbon–gold (III) bond which stabilizes oxidation state of gold. Organogold(III) compounds are stable under physiological conditions and have a limited tendency to be reduced to gold(I). They are significantly cytotoxic to human tumor cell lines.

Dinuclear Gold(III) Complexes

Dinuclear gold(III) compounds have been recently synthesized. These complexes are very stable under physiological-like conditions and have significant antiproliferative effects toward different human tumor cell lines. All these compounds contain a common structural motif consisting of an Au_2O_2 "diamond core" linked to two bipyridine ligands in a roughly planar arrangement. Importantly, the introduction of different alkyl or aryl substituents on the 6 (and 60) position(s) of the bipyridine ligand leads to small structural changes that greatly affect the reactivity of the metal centers.

Biological Effects of Gold(III) Complexes

A lot of recently published studies that tested anticancer characteristics of gold(III) complexes showed promising results: high cytotoxic effect in vitro and in vivo was noticed against representative tumor cell lines, suggesting some of the newly synthesized gold (III) compounds as novel, effective anticancer agents (Nobili et al. 2010). Although number of specific hypotheses has been formulated concerning their possible mode of action, the precise mechanism of their biological effects is still largely unknown. However, after detailed analysis, it is possible to define four major intracellular targets for gold(III) complexes: DNA, thioredoxin reductase (TrxR), protein kinase C (PKC), and proteasome (Casini and Messori 2011).

DNA: Target or Not?

Although all gold compounds have a lower affinity for DNA than platinum complexes, and their mode of action is so-called DNA independent, gold(III) center has high propensity to react with nucleobase nitrogens and is able to make stronger interactions with DNA and DNA nucleobases than gold(I) drugs which suggest that DNA could be one of the possible targets for gold(III) compounds (Casini and Messori 2011). Some gold(III) complexes (bipyridyl gold(III) complexes such as$[Au(bipy)(OH)_2][PF_6]$ and $[Au(bipyc\text{-}H)(OH)][PF_6]$) and polyamine complexes (such as $[Au(en)_2]Cl_3$ and $[Au(dien)Cl]Cl_2$) show a very weak interaction with calf thymus DNA and induce slight modifications of the double helix. On the contrary, chloroglycylhistidine gold(III) compounds and $[Au_2(6,6'\text{-dimethyl-}2,2'\text{-bipyridine})(m\text{-}O)_2][PF_6]_2$ complex are able to bind DNA and modify its conformation.

It was recently revealed that for gold(III) complexes bearing a different positive charge, a higher positive charge enhances the DNA binding affinity. The main interactions of gold(III) complexes with DNA depend on the intercalation of the square planar gold(III) chromophore into the DNA double helix. Gold(III) complexes $[Au(DMDT)X_2]$ where DMDT is N, *N*-dimethyldithiocarbamate and $[Au(ESDT)X_2]$ where ESDT is ethylsarcosinedithiocarbamate; X=Cl or Br) possess an appreciable affinity for the DNA double helix, efficiently induce interstrand cross-links, and successfully kill cancer cells. Interestingly, in vitro studies showed that gold(III) porphyrins interact strongly and directly with DNA, but in vivo experiments revealed that these gold(III) compounds cause DNA fragmentation rather than cross-linkage suggesting that their interactions with DNA are noncovalent and reversible.

According to all the above discussed results, it can be concluded that DNA is not the primary or the exclusive target for gold(III) complexes. This observation led a lot of scientists to postulate the existence of some other potential targets for gold complexes, such as thioredoxin reductase (TrxR) (Casini and Messori 2011).

Thioredoxin Redox System: A Principal Target for Gold(III)-Dithiocarbamate Derivatives

The thioredoxin redox system comprises thioredoxin reductase (TrxR), a homodimeric selenium containing flavoprotein, and thioredoxin (Trx), a ubiquitously expressed small protein, with a conserved Cys-Gly-Pro-Cys redox catalytic site capable of reducing a variety of substrates (Nobili et al. 2010). Both Trx and TrxR in mammals are expressed in two isoforms: cytosolic (Trx1 and TrxR1) or mitochondrial (Trx2 and TrxR2). Trx/TrxR system plays multiple functions in the cell as follows: provides reducing equivalents for DNA synthesis through ribonucleotide reductase (RR) enzyme and for reactive oxygen species (ROS) scavenging through the peroxiredoxins (Prxs), and activates several transcription factors involved in the regulation of cell growth and survival, including NF-kB, AP-1, SP-1, and p53 (Casini and Messori 2011).

It was noticed that the expression of cytosolic thioredoxin isoform Trx1 is increased in several human carcinoma and correlates to tumor aggressiveness (Nobili et al. 2010).

Aubipy and Aubipyxil gold(III) compounds inhibit mitochondrial TrxR2 and trigger mitochondrial swelling in cancer cells due to the interaction with selenol moiety, cysteine, and histidine residues present in the active site of the TrxR2 carboxy terminus.

TrxR is overexpressed in prostate cancer and is associated with the onset of resistance to cisplatin. Gold(III)-dithiocarbamate derivatives $[AuCl_2(DMDT)]$ and $[AuBr_2(ESDT)]$ strongly inhibit the activity of the TrxR and were shown to be highly effective against prostate cancer cells (Cattaruzza et al. 2011).

Gold(III)-dithiocarbamate complexes $[Au(DMDT)X_2]$ and $[Au(ESDT)X_2]$ have been shown to trigger ROS generation and ultimately increase the levels of phosphorylated ERK1/2 through the inhibition of both cytosolic and mitochondrial TrxR2, in human uterine cervical carcinoma HeLa cells. Continuous activation of ERK1/2 is supposed to be responsible for gold(III) dithiocarbamate–induced HeLa cell death (Nobili et al. 2010).

The inhibition of TrxR activity was also noticed in Jurkat human leukemia cells and MCF-7 human breast cancer cells treated by gold(III) complexes with thiosemicarbazones further confirming TrxR as a principal target for gold(III)-dithiocarbamate derivatives.

In summing up, as Trx expression is associated with aggressive tumor growth and significantly increased in human leukemia and breast cancer cells, gold(III)-dithiocarbamate-induced inhibition of TrxR activity could further be explored as a new therapeutic approach for the treatment of leukemia and breast cancer (Casini and Messori 2011).

Proteasome: Cellular Target for Gold(III) Complexes

As the ubiquitin–proteasome pathway is essential for cell cycle regulation, apoptosis, angiogenesis, and differentiation, it has recently been investigated as an intracellular target for gold(III) compound–induced cytotoxicity. A gold(III) dithiocarbamate compound showed chymotrypsin-like, trypsin-like, and Peptidyl-glutamyl peptide-hydrolyzing (PGPH)-like peptidase activity, inhibited proteasome function, and induced apoptosis in MDA-MB-231 breast cancer cells. Proteasome inhibition was confirmed by decreased proteasome activity, increased levels of ubiquitinated proteins, and the accumulation of proteasome target protein p27 both in intact MDA-MB-231 breast cancer cells and in MDA-MB-231 xenografts treated with gold(III) dithiocarbamate compound. These data suggest proteasome as a target for gold(III) dithiocarbamates both in vitro and in vivo confirming that inhibition of the proteasomal activity is strongly associated with apoptotic death of cancer cells (Casini and Messori 2011).

Protein Kinase C Iota as a Potential Therapeutic Target in Gold(III) Compound–Mediated Cytotoxicity

Protein kinase C (PKC) is protein kinase involved in cellular proliferation, cell cycle control, differentiation, polarity, and survival. It was found that altered PKC activity, localization, and/or expression correlates with malignant phenotype suggesting several PKC isozymes, particularly protein kinase C iota (PKCi), as potential therapeutic target.

PKCi, structurally and functionally distinct from other PKCs, is a bonafide human oncogene required for the transformed growth of human cancer cells. PKCi-Rac1-Pak-Mek-Erk signaling axis plays an important role in the cell growth of nonsmall cell lung cancer (NSCLC) both in vitro and in vivo. Aurothioglucose blocks PKCi-dependent signaling to Rac1 and inhibits transformed growth of NSCLC cells, while aurothiomalate forms gold-cysteine adducts on target cellular proteins (in particular with Cys69), located within the active site of the PKCi. It could be concluded that the selective targeting of PKCi might explain the inhibition of NSCLC cancer cell growth by aurothioglucose and aurothiomalate (Nobili et al. 2010).

Apoptosis as a Major Mechanism for Gold(III) Porphyrin 1a Treatment of Human Nasopharyngeal Carcinoma

One of the most usually proposed mechanisms for gold(III) compound–induced cytotoxicity is the induction of apoptosis by mitochondrial death pathways related to ROS. This was clearly documented in studies that investigated cytotoxic effects of gold(III) porphyrin 1a in human nasopharyngeal carcinoma cell lines. It has been shown that, in HONE1 human nasopharyngeal carcinoma cells, gold(III) porphyrin 1a directly causes alteration of bcl-2 family proteins and translocation of the apoptosis inducing factor (AIF) to the nucleus, and induces release of cytochrome c, and activation of caspase-9 and caspase-3 subsequently producing cleavage of Poly (ADP-ribose) polymerase enzyme involved in DNA repair (Casini and Messori 2011).

A wide series of protein alterations were detected in SUNE1 human nasopharyngeal carcinoma cell line after treatment with gold(III) porphyrin 1a (Nobili et al. 2010). The main targets were stress-related chaperone proteins, proteins involved in ROS (e.g., stomatin-like 2, peroxiredoxin 1 and 6, and thioredoxin), enzyme proteins and translation factors (e.g., mitochondrial single-stranded DNA binding protein-mtSSB, splicing factor 17, peptidylprolyl isomerase F, and cyclophilin F-CypF), proteins that mediate cell proliferation or differentiation (e.g., cyclophilin A, porin isoform 1 voltage-dependent anion channel 1-VDAC1, calcyclin binding protein, Siah-interacting protein-CacyBP, and Ras-related nuclear protein), and proteins that belong to the internal degradation system (e.g., proteasome a type 3, proteasome b type 4, and proteasome a type 6).

Gold(III) porphyrin 1a treatment causes cell cycle arrest initially at G2-M phase, then at G0-G1, and upregulation of the proapoptotic protein p53 in SUNE1 cells. All the above discussed results strongly suggest gold(III) porphyrin 1a as a novel therapeutic agent in the treatment of human nasopharyngeal carcinoma.

Biological Effects of Gold(III) Complexes Tested In Vitro and In Vivo

Cytotoxic effects of gold(III) compounds were measured on different tumor cell lines in vitro and in vivo and the obtained results were promising.

Several new gold(III)-dithiocarbamate derivatives in vitro have even better cytotoxic properties than cisplatin toward a series of human tumor cell lines. It was clearly documented that some of the gold (III)-dithiocarbamate derivatives could be used as a novel therapeutic agent in the treatment of prostate cancer (Cattaruzza et al. 2011). It has been recently shown that gold(III)-dithiocarbamate derivatives [$AuCl_2$(DMDT)] and [$AuBr_2$(ESDT)] were highly active against the androgen-resistant prostate cancer cell lines PC3 and DU145 and managed to significantly inhibit tumor growth in vivo, causing minimal systemic toxicity (Cattaruzza et al. 2011). In addition, these complexes were more active than the reference drug cisplatin. The complex [$AuCl_2$(DMDT)] was cytotoxic against cisplatin-resistant R-PC3 cells ruling out the occurrence of cross-resistance phenomena and suggesting different mechanisms of action compared to cisplatin. Gold(III)-dithiocarbamate derivatives induced caspase-mediated apoptosis of prostate cancer cells: pro-apoptotic Bax was upregulated while the expression of anti-apoptotic Bcl-2 and phosphorylated form of the EGFR was reduced in [$AuCl_2$(DMDT)] treated PC3 cells. In addition, gold(III)-dithiocarbamate complexes altered mitochondrial functions by inducing mitochondrial membrane permeabilization and Cyt-c release, stimulating ROS generation, and strongly inhibiting the activity of the TrxR (Cattaruzza et al. 2011).

Several gold(III) complexes were active against leukemia cells in vitro. The water-soluble gold(III) complex [Au(III)(butyl-C^N)biguanide]Cl (butyl-HC^N = 2-(4-*n*-butylphenyl)pyridine, BCN) induced endoplasmic reticulum damage and S-phase cell cycle arrest in HeLa cells. Gold(III)-methylsarcosinedithiocarbamate derivatives [Au(MSDT)Cl_2] and [Au(MSDT)Br_2] were cytotoxic against human acute myeloid leukemia (AML) cells. In addition, both gold(III) complexes were significantly more active than cisplatin in inhibiting the clonogenic growth of AML cell lines. Gold(III)-ethylenediamine complex [Au(en)Cl_2] induced apoptosis in human chronic lymphocytic leukemia (CLL) cells (Milovanovic et al. 2010).

The same complex was highly active against 4T1 mouse breast cancer cells both in vitro and in vivo and managed to significantly reduce breast cancer tumor growth in mice (Volarevic et al. 2010). Although both [Au(en)Cl_2] and cisplatin showed similar cytotoxic effects against tested cancer cells, [Au(en)Cl_2] was tolerated much better than cisplatin.

Although in the presence of aurothioglucose, A549 cell tumors exhibited a marked reduction in growth kinetics compared with controls, several studies showed that gold(III) complexes have a cytostatic rather than cytotoxic effect on A549 tumors (Arsenijević et al. 2012).

Gold(III) tetraphenylporphyrin [Au(TPP)]+ downregulates the expression of genes playing roles in angiogenesis and inhibits microvessel formation of the epithelial cells. Also, [Au(TPP)]+ could inhibit migration and invasion of nasopharyngeal carcinoma cells. According to these data, [Au(TPP)]+ could be further explored for the treatment of cancer metastasis (Che and Sun 2011).

Gold(III) dithiocarbamate compound significantly inhibited tumor growth after subcutaneous injection in MDA-MB-231 tumor-bearing nude mice as a consequence of proteasomal inhibition and apoptosis induction. It is interesting to note that during the 29-day treatment with this compound at 1–2 mg/kg/day, no toxicity was observed, and mice did not display signs of weight loss, decreased activity, or anorexia (Nobili et al. 2010).

Conclusions and Future Directions

In the design of gold(III)-based drugs, there needs to be a more balanced consideration on the cytotoxic properties and aqueous solubility. Previous studies on different types of gold-based lipophilic cations revealed that introduction of lipophilic substitution(s) generally enhance cellular uptake and hence cytotoxic activities. However, aqueous solubility of these complexes would decrease, resulting in lowering the bioavailability in the biological system. Accordingly, in vivo data available for gold(III) compounds as anticancer agents are limited, probably due to the high redox potential and relatively poor kinetic stability of several of these compounds under physiological conditions. Some gold(III) complexes could be significantly stabilized, even at neutral pH, with the appropriate choice of the inert ligands, preserving its peculiar biological properties. The presence of at least two nitrogen donors directly coordinated to the gold(III) center leads to a significant decrease in the reduction potential of the complex. It is evident that more extensive in vivo examinations are recommended for gold compounds which have already demonstrated to display promising in vitro activities. To date, gold(III) compounds have not been formally investigated as anticancer drugs in the clinic and no published clinical data is available. In order to provide more clinically relevant information for clinical use, further in vivo examinations, such as biodistribution, biotransformation (metabolite(s) formation), pharmacokinetics, as well as safety pharmacological evaluation of selected gold (III) complexes, should be conducted.

Cross-References

- Biomineralization of Gold Nanoparticles from Gold Complexes in Cupriavidus Metallidurans CH34
- Gold and Nucleic Acids
- Gold Complexes as Prospective Metal-Based Anticancer Drugs
- Gold Nanomaterials as Prospective Metal-based Delivery Systems for Cancer Treatment
- Gold Nanoparticle Platform for Protein-Protein Interactions and Drug Discovery
- Gold Nanoparticles and Proteins, Interaction
- Gold(III), Cyclometalated Compound, Inhibition of Human DNA Topoisomerase IB
- Platinum-Resistant Cancer

References

Arsenijević M, Milovanovic M, Volarevic V et al (2012) Cytotoxicity of gold(III) complexes on A549 human lung carcinoma epithelial cell line. Med Chem 8:2–8

Casini A, Messori L (2011) Molecular mechanisms and proposed targets for selected anticancer gold compounds. Curr Top Med Chem 11:2647–2660

Cattaruzza L, Fregona D, Mongiat M et al (2011) Antitumor activity of gold(III)-dithiocarbamato derivatives on prostate cancer cells and xenografts. Int J Cancer 128:206–215

Che CM, Sun RW (2011) Therapeutic applications of gold complexes: lipophilic gold(III) cations and gold(I) complexes for anti-cancer treatment. Chem Commun 47:9554–9560

Djeković A, Petrović B, Bugarčić ZD et al (2012) Kinetics and mechanism of the reactions of Au(III) complexes with some biologically relevant molecules. Dalton Trans 41: 3633–3641

Milovanovic M, Djekovic A, Volarevic V et al (2010) Ligand substitution reactions and cytotoxic properties of [Au(L)Cl2](+) and [AuCl2(DMSO)2]+ complexes (L = ethylenediamine and S-methyl-l-cysteine). J Inorg Bioch 104:944–949

Nobili S, Mini E, Landini I (2010) Gold compounds as anticancer agents: chemistry, cellular pharmacology, and preclinical studies. Med Res Rev 30:550–580

Volarevic V, Milovanovic M, Djekovic A et al (2010) The cytotoxic effects of some selected gold(III) complexes on 4T1 cells and their role in the prevention of breast tumor growth in BALB/c mice. J BUON 15:768–773

Gold(III), Cyclometalated Compound, Inhibition of Human DNA Topoisomerase IB

Silvia Castelli[1], Oscar Vassallo[1], Prafulla Katkar[1], Chi-Ming Che[3], Raymond Wai-Yin Sun[3] and Alessandro Desideri[1,2]
[1]Department of Biology, University of Rome Tor Vergata, Rome, Italy
[2]Interuniversity Consortium, National Institute Biostructure and Biosystem (INBB), Rome, Italy
[3]Department of Chemistry, State Key Laboratory of Synthetic Chemistry and Open Laboratory of Chemical Biology of the Institute of Molecular Technology for Drug Discovery and Synthesis, The University of Hong Kong, Hong Kong, China

Synonyms

[Au(C^N^C)(IMe)]CF_3SO_3; Anticancer drug; Catalytic inhibitor; Inhibitory mechanism; Topoisomerase IB

Definition

The interaction of the gold (III) compound [Au(C^N^C)(IMe)]CF3SO_3 with topoisomerase IB consists in the inhibition of the cleavage reaction, and it is produced through a binding to the enzyme that does not permit anymore the binding of the DNA substrate.

Effect of a Gold Compound on Human Topoisomerase IB Activity

Square-planar d8 metal complexes exhibit promising anticancer activities, by covalent crosslink of d8 metal ions to DNA, by intercalation of the planar metal complexes between the DNA base pairs, and/or by inhibition of enzymes activity (Tiekink 2002; Che and Siu 2010). Different stable anticancer gold (III) complexes have been prepared, and some of these were identified as possible new anticancer drugs (Gold Complexes as Prospective Metal-Based

Anticancer Drugs). Gold(III) complexes containing dianionic $[C^N^C]^2$ and neutral auxiliary N-heterocyclic carbene $(NHC)^2$ ligands have been suggested to be promising candidates since they are more cytotoxic to cancerous cells than to normal cells and can be easily modified (▶ Gold(III) Complexes, Cytotoxic Effects).

Topoisomerases are key enzymes that control the topological state of DNA. These enzymes are involved in many vital cellular processes that influence DNA replication, transcription, recombination, integration, and chromosomal segregation (Champoux 2001). All the topoisomerases act introducing transient strand breaks in a DNA double-strand molecule. There are two classes of topoisomerases: type II enzymes which nick both DNA strands and whose activity is dependent on the presence of ATP and type I enzymes, which act by transiently nicking one of the two DNA strands. In particular, human topoisomerase IB forms a covalent bond with the 3′-phosphate end of the cleaved strand. During this state, the broken strand can rotate around the uncleaved strand leading to DNA relaxation. To restore the correct DNA double-strand structure, topoisomerase I catalyzes the religation of the 5′-hydroxyl termini (Stewart et al. 1998).

A number of antitumor agents have topoisomerases as their target, and they act through different mechanisms that are related to the different catalytic steps. In particular, the drugs inhibiting topoisomerase I can be divided into two classes: poisons and catalytic inhibitors. Topoisomerase I poisons, such as camptothecin (CPT) and its analogs, reversibly stabilize the topoI-DNA complex by inhibiting DNA religation (Liu et al. 2000). Collision of the stabilized complexes with advancing replication forks leads to the formation of irreversible strand breaks that are ultimately responsible for cell death (Hsiang et al. 1989). Conversely, catalytic inhibitors exert their cytotoxicity by preventing topoisomerase I binding to the DNA and/or the cleavage activity of the enzyme, resulting in the inhibition of DNA relaxation (Castelli et al. 2009).

The gold (III) compound [Au(C^N^C) (IMe)] CF_3SO_3 (here abbreviated Gold III) belongs to the class of the catalytic inhibitors since it inhibits the topoisomerase I cleavage reaction by not permitting the binding of the DNA substrate (Castelli et al. 2011).

The inhibition activity is demonstrated by a DNA relaxation assay. In this assay, the activity of topoisomerase IB is assayed in 20 μl of reaction volume containing 0.5 μg of negatively supercoiled pBlue-Script KSII(+) DNA and Reaction Buffer (20 mM Tris–HCl, 0.1 mM Na_2EDTA, 10 mM $MgCl_2$, 50 μg/ml acetylated BSA, and 150 mM KCl, pH 7.5). The effect of Gold III on enzyme activity is measured by adding different concentrations of the compound, at different times. Reactions are stopped with a final concentration of 0.5% SDS after 30 min or after each time-course point at 37°C. The samples are electrophoresed in a horizontal 1% agarose gel in 50 mM Tris, 45 mM boric acid, and 1 mM EDTA. The gel is stained with ethidium bromide (5 μg/ml), destained with water and photographed under UV illumination. Assays are usually performed at least three times. A typical gel is represented in Fig. 1 that shows that Gold III has an inhibitory effect on the human topoisomerase I relaxation activity. The assay detects the different electrophoretic mobility of the DNA supercoiled plasmid in comparison to the relaxed form produced by the enzyme and indicates that Gold III inhibits the human topoisomerase I in a dose-dependent manner (Fig. 1a, lanes 3–7), a full inhibition being achieved at around 15 μM. The compound is able to fully inhibit the enzyme relaxation activity even at a lower concentration if it is preincubated with the enzyme.

In detail, the enzyme fully relaxes the substrate in the absence of the compound (Fig. 1b, lanes 2–5) and partially relaxes the substrate in presence of 6.5 μM Gold III (Fig. 1b, lanes 6–9), but it is fully inhibited when it is preincubated with the same amount of Gold III (Fig. 1b, lanes 14–17).

A two- or fourfold dilution of the preincubated Gold III-enzyme mixture permits to verify if the inhibition is reversible or irreversible. This represents a useful and simple procedure, and in this case, it shows that a full inhibition remains also after dilution, indicating that the inhibitory effect of the compound is irreversible (Fig. 1c, lanes 6–9). The procedure also requires a measurement of the activity of the enzyme preincubated with DMSO in the absence of the inhibitor to verify that the enzyme maintains its activity at the same degree of dilution. In the case of the Gold III compound, this is really what is observed (Fig. 1c, lanes 2–5), confirming that the compound is an irreversible inhibitor.

The mechanism of inhibition can be better understood performing a so-called cleavage/religation equilibrium experiment. In this experiment, the oligonucleotide

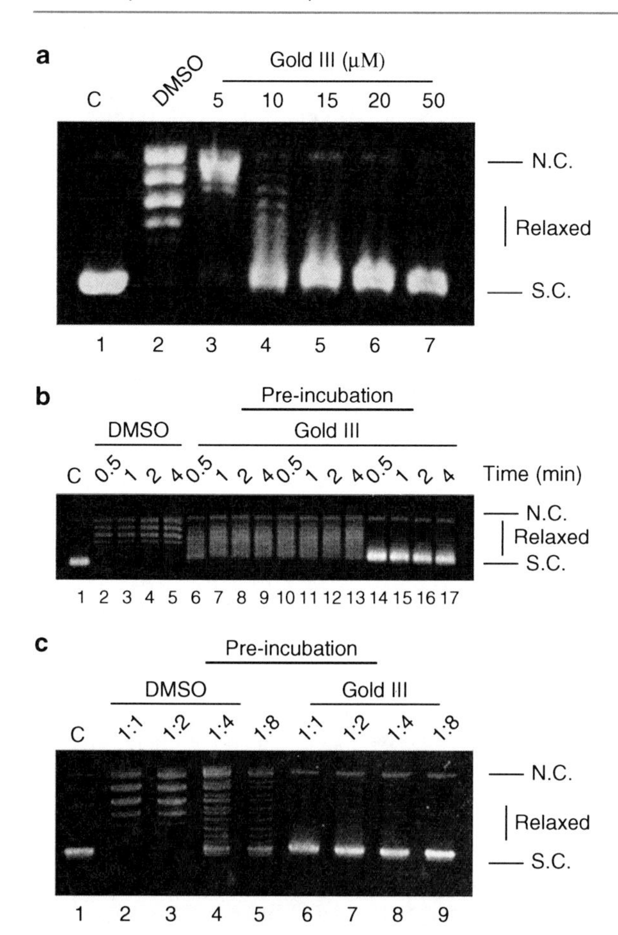

Gold(III), Cyclometalated Compound, Inhibition of Human DNA Topoisomerase IB, Fig. 1 (**a**) Relaxation of negative supercoiled plasmid DNA by topoisomerase IB in the presence of increasing concentrations of Gold III (lanes 3–7). The reaction products were resolved in an agarose gel and visualized with ethidium bromide. Lane 1, no protein added. Lane 2, control reaction with DNA and enzyme in the presence of only DMSO. *NC* nicked circular plasmid DNA; *SC* supercoiled plasmid DNA. (**b**) Relaxation activity assay of topoisomerase IB on negative supercoiled plasmid in presence of only DMSO (lanes 2–5), 6.5 μM Gold III (lanes 6–9), after 5-min 6.5 μM Gold III-substrate preincubation (lanes 10–13), and after 5-min preincubation of 6.5 μlM Gold III with the enzyme (lanes 14–17). Lane 1, no protein added. (**c**) Relaxation assay of topoisomerase IB in DMSO diluted two–four- or eightfold (lanes 2–5) or in the presence of Gold III (20 μM) diluted two–four- or eightfold (lanes 6–9) for 30 min at 37°C. Lane 1, no protein added

DNA substrate CL25 (5′-GAAAAAAGACTTA-GAAAAATTTTTA- 3′) is firstly radiolabelled with [γ-^{32}P]ATP at its 5′ end, while the CP25 complementary strand (5′-TAAAAATTTTTCTAAGTCTTTTTTC-3′) is phosphorylated at its 5′ end with unlabeled ATP. The two strands are annealed at a twofold molar excess of CP25 over CL25 as previously described (Chillemi et al. 2005). An excess of topoisomerase IB enzyme is then incubated at 37°C with the CL25/CP25 duplex (final concentration of 20 nM) in Reaction Buffer in presence or absence of the inhibitor. Dimethylsulfoxide (DMSO) is added in the controls not containing the drug to have the same quantity of DMSO in the solution containing or not containing the inhibitor. After different times, the reaction is stopped by adding 0.5% SDS and digested with trypsin after ethanol precipitation. Reaction products are resolved in 20% acrylamide-7 M urea gel (25 mA for 30 min). Also in this case, the experiments must be repeated at least three times.

The assay is performed as function of time for the enzyme alone or in the presence of 50 μM Gold III or 50 μM CPT or after preincubating the enzyme with 50 μM Gold III and then adding 50 μM CPT. The reactions is stopped with SDS, the samples digested with trypsin, and the products are analyzed by polyacrylamide–urea gel electrophoresis (Fig. 2). In presence of DMSO, the cleavage/religation equilibrium is shifted toward religation, showing only a tiny detectable band, corresponding to the trapped

a (CL25) 5′-p-GAAAAAAGACTTAGAAAAATTTTTA-3′
(CP25) 3′-CTT T TT TCTGAATC T TTTTAAAAAT-5′

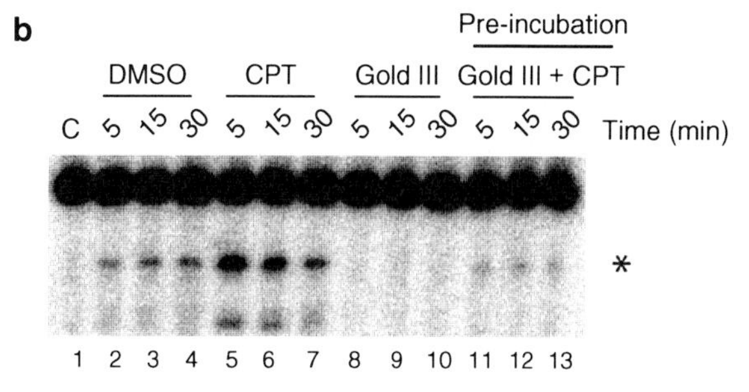

Gold(III), Cyclometalated Compound, Inhibition of Human DNA Topoisomerase IB, Fig. 2 Cleavage/religation equilibrium of the duplex substrate. (**a**) DNA substrate used. The substrate contains the 5′ end-labeled CL25 and the preferred topoisomerase IB binding site, indicated by an asterisk. (**b**) Gel electrophoresis of the products coming from the incubation of duplex substrate with the enzyme at 37°C for 5, 15, and 30 min in presence of DMSO (lanes 2–4) or in presence of 50 μM CPT (lanes 5–7), 50 μM Gold III (lanes 8–10), and after 5-min preincubation with 50 μM Gold III at 37°C and subsequent addition of 50 μM CPT (lanes 11–13). Lane 1, no protein added. The slowest migrating band corresponds to the uncleaved oligonucleotide. The *asterisk* indicates the band corresponding to the preferential cleavage site (CL1)

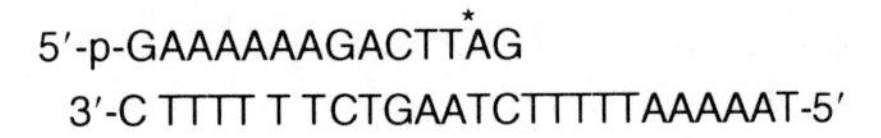

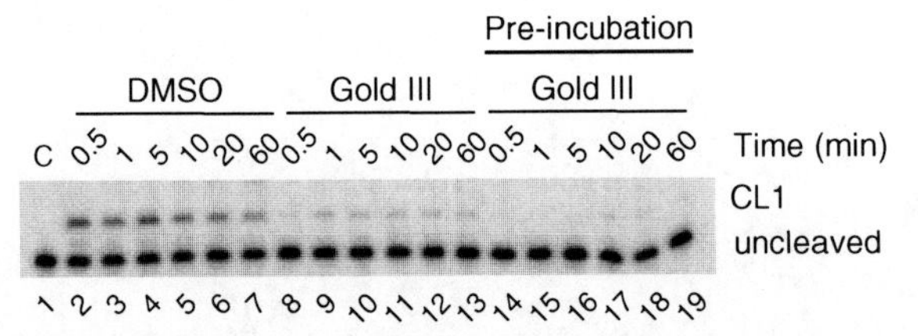

Gold(III), Cyclometalated Compound, Inhibition of Human DNA Topoisomerase IB, Fig. 3 Suicide cleavage kinetics. The CL14/CP25 substrate was incubated with topoisomerase IB alone (lanes 2–7), in presence of 50 μM Gold III (lanes 8–13), or after 5-min enzyme-Gold III pre-incubation (lanes 14–19). Lane 1, no protein added. CL1 represents the DNA fragment cleaved at the preferred enzyme site. CL14/CP25 substrate is shown in the *upper* part; the *asterisk* indicates the band corresponding to the preferential cleavage site

cleavable complex (Fig. 2, lanes 2–4, asterisk). In the presence of 50 μM Gold III, no detectable trapped cleavable complex can be detected (Fig. 2, lanes 8–10). When the enzyme is incubated with 50 μM CPT, the cleavage/religation equilibrium is shifted toward cleavage since the band corresponding to the substrate cleaved at the preferred CL1 site is clearly detectable (Fig. 2, lanes 5–7). This result is in line with the known CPT inhibition mechanism involving the reversible binding of drug to the DNA–enzyme covalent complex that is stabilized by CPT (Svejstrup et al. 1991). Preincubation of Gold III with the enzyme before the addition of the duplex oligonucleotide and CPT does not permit the stabilization of the cleavable complex (Fig. 2, lanes 11–13). These data show that Gold III either inhibits the cleavage or induces a faster religation so that the cleavable complex cannot be formed or cannot be stabilized by CPT.

To clarify whether Gold III affects the cleavage/religation equilibrium perturbing the religation or the cleavage reaction, the two processes are evaluated in separate experiments. To evaluate the cleavage process, a CL14/CP25 suicide substrate, shown in the upper part of Fig. 3, is produced annealing the two strands at a twofold molar excess of CP25 over CL14. The oligonucleotide substrate CL14 (5′-GAAAAAAGACTTAG-3′) is radiolabelled with [γ-^{32}P]ATP at its 5′ end. The CP25 complementary strand (5′-TAAAAATTTTTCTAAGTCTTTTTTC-3′) is phosphorylated at its 5′ end with unlabeled ATP. The suicide cleavage reactions are carried out, incubating 20 nM of this partial duplex substrate with an excess of enzyme in Reaction Buffer at 37°C and in presence of 50 μM Gold III. DMSO is added to no-drug control. Before the addition of the protein, 5-μl sample of the reaction mixture is removed and used as the zero time point (C). At different time points, 5-μl aliquots are removed and the reaction stopped with 0.5% SDS. After, ethanol precipitation samples are resuspended in 6 μl of 1 mg/ml trypsin and incubated at 37°C for 1 h. Samples are analyzed using denaturing urea/polyacrylamide gel electrophoresis. The experiment is replicated at least three times.

A time-course experiment of a suicide cleavage substrate incubated with the enzyme alone, with the enzyme in the presence of 50 μM Gold III, or with the enzyme preincubated with 50 μM Gold III is shown in Fig. 3, where the band corresponding to the cleaved DNA fragment is indicated as CL1. The cleavage reaction is fast in absence of the inhibitor since 90% of the cleavage product is produced within the first 30 s (Fig. 3, lanes 2–7) and in less than 5 min, the maximum quantity of cleaved substrate is obtained. In the presence of Gold III, the cleavage reaction is clearly inhibited since the band of the cleaved substrate corresponds to not more than 30% even after 1 h incubation (Fig. 3, lanes 8–13), and cleavage inhibition is almost complete when the enzyme is preincubated with 50 μM Gold III (Fig. 3, lanes 14–19).

The religation reaction is monitored incubating suicide CL14/CP25 substrate (20 nM) with an excess of topoisomerase IB enzyme for 30 min at 37°C in Reaction Buffer. A 5-μl sample of the reaction mixture is removed and used as the zero time point. Religation reactions are initiated by adding a 200-fold molar excess of R11 oligonucleotide (5′-AGAAAAATTTT-3′) over the duplex CL14/CP25 in the presence or absence of 50 μM Gold III, as shown in Fig. 4. DMSO is added to no-drug controls. At time-course points, 5 μl aliquots are removed and the reaction stopped with 0.5% SDS. After, ethanol precipitation samples are resuspended in 5 μl of 1 mg/ml trypsin and incubated at 37°C for 30 min. Samples are analyzed by denaturing urea/polyacrylamide gel electrophoresis. The experiment also in this case must be replicated three times.

A 200-fold molar excess of the complementary R11 oligonucleotide is added in the absence or

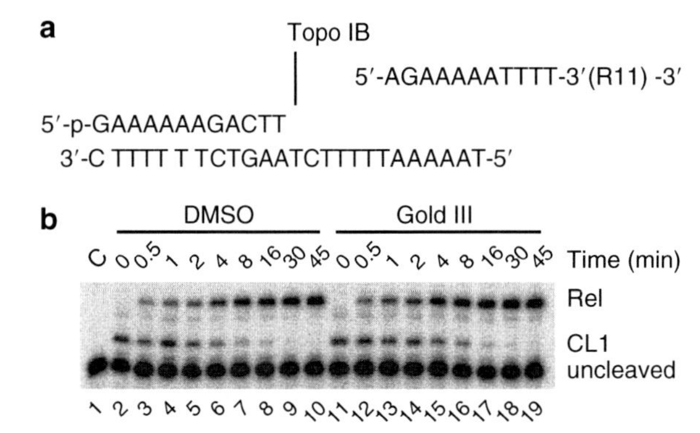

Gold(III), Cyclometalated Compound, Inhibition of Human DNA Topoisomerase IB, Fig. 4 Religation kinetics: (**a**) The CL14/CP25 suicide substrate and the R11 complementary oligonucleotide used to measure the religation kinetics of the enzyme. (**b**) Gel analysis of the religation kinetics for topoisomerase IB in absence (lanes 2–10) or presence of 50 μM Gold III (lanes 11–19). CL1 represents the DNA fragment cleaved at the preferred enzyme site. *Rel* represents the religation product

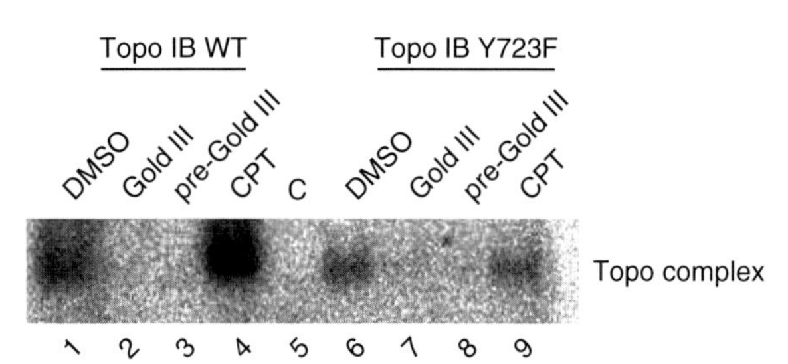

Gold(III), Cyclometalated Compound, Inhibition of Human DNA Topoisomerase IB, Fig. 5 Electrophoretic mobility-shift assay (EMSA) with the wt and inactive Y723F mutant enzyme (respectively, lanes 1 and 6), in presence of 50 μM Gold III (respectively, lanes 2 and 7), after 5- min enzyme pre-incubation with 50 μM Gold III (respectively, lanes 3 and 8) and 100 μM CPT (respectively, lanes 4 and 9). All the sample was analyzed after 30 min. Lane 1, the only radiolabelled CL25/CP25 DNA substrate

presence of 50 μM Gold III to the suicide cleavage substrate, incubated with an excess of native enzyme (Fig. 4a and b). The urea–polyacrylamide gel of different aliquots, analyzed as a function of time, indicates that an identical religation rate is observed (Fig. 4b). The same result was obtained when the suicide DNA complex was preincubated with 50 μM Gold III before addition of R11 (data not shown).

These experiments demonstrate that after preincubation, Gold III produces a full inhibition of the enzyme cleavage activity without affecting the religation rate, so they explain why no band is observed in the cleavage/religation equilibrium experiment, shown in Fig. 2, upon addition of CPT to the preincubated enzyme-Gold III mixture. However, these experiments cannot clarify whether this effect is due to the prevention of the DNA–enzyme binding or to inhibition of the catalytic reaction.

The binding assay is done using the 25-bp double-stranded oligonucleotide CL25/CP25, prepared as for the cleavage/equilibrium assay. The reactions are carried out using the catalytically inactive mutant Y723F (Fig. 5, lanes 6–9) as well as the wt enzyme as a control (Fig. 5, lanes 1–4). The enzymes are incubated in standard reaction conditions (20 mM Tris–HCl, pH 7.5, 0.1 mM Na_2EDTA, 10 mM $MgCl_2$, 50 μg/ml acetylated BSA, and 150 mM KCl) with the radiolabelled CL25/CP25 substrate in the presence of 1% (v/v) DMSO, 50 μM Gold III, or 100 μM CPT at 37°C for 30 min in a final volume of 20 μl. In the case of preincubation, the enzymes are incubated with Gold III for 5 min before adding the DNA substrate and letting the binding reactions proceed for 30 min at 37°C. Reactions are stopped by the addition of 5 μl of dye (0.125% Bromophenol Blue and 40% (v/v) glycerol). Samples are loaded onto 6% (v/v) native polyacrylamide gels and electrophoresed at 40 V in TBE3 (12 mM Tris, 11.4 mM boric acid, and 0.2 mM EDTA) at 4°C for 4 h. Products are visualized by PhosphorImager. The inactive Y723F human topoisomerase IB mutant, incubated with the radiolabelled CL25/CP25 DNA substrate in presence of DMSO, shows a tiny slowly migrating band, corresponding to a non-covalent DNA-topoisomerase IB complex (Fig. 5, lane 6). The band has a similar intensity in presence of CPT (Fig. 5, lane 9), while it is not observed when Gold III is present (Fig. 5, lane 7) or is preincubated with the enzyme (Fig. 5, lane 8). As a control, the same assay was performed with the wt enzyme (Fig. 5, lanes 1–4). The enzyme is able to bind the substrate in presence of only DMSO (Fig. 5, lane 1) through a non-covalent and covalent bond that is stabilized in presence of CPT, as shown in Fig. 5, lane 4. The wt enzyme, as well as the mutant, cannot bind the substrate in the presence of Gold III (Fig. 5, lanes 2 and 3), indicating that the compound acts by preventing the non-covalent topoisomerase IB–DNA complex formation.

Cross-References

▶ Gold Complexes as Prospective Metal-Based Anticancer Drugs
▶ Gold(III) Complexes, Cytotoxic Effects

References

Castelli S, Campagna A, Vassallo O et al (2009) Conjugated eicosapentaenoic acid inhibits human topoisomerase IB with a mechanism different from camptothecin. Arch Biochem Biophys 486:103–110

Castelli S, Vassallo O, Katkar P et al (2011) Inhibition of human DNA topoisomerase IB by a cyclometalated gold III compound: analysis on the different steps of the enzyme catalytic cycle. Arch Biochem Biophys 516:108–112

Champoux J (2001) DNA topoisomerases: structure, function, and mechanism. Annu Rev Biochem 70:369–413

Che CM, Siu FM (2010) Metal complexes in medicine with a focus on enzyme inhibition. Curr Opin Chem Biol 14:255–261

Chillemi G, Fiorani P, Castelli S et al (2005) Effect on DNA relaxation of the single Thr718Ala mutation in human topoisomerase I: a functional and molecular dynamics study. Nucleic Acids Res 33:3339–3350

Hsiang YH, Lihou MG, Liu LF (1989) Arrest of replication forks by drug-stabilized topoisomerase I-DNA cleavable complexes as a mechanism of cell killing by camptothecin. Cancer Res 49:5077–5082

Liu LF, Desai SD, Li TK, Mao Y, Sun M, Sim SP (2000) Mechanism of action of camptothecin. Ann N Y Acad Sci 922:1–10

Stewart L, Redinbo MR, Qiu X, Hol WG, Champoux JJ (1998) A model for the mechanism of human topoisomerase I. Science 279:1534–1541

Svejstrup JQ, Christiansen K, Gromovat II et al (1991) New technique for uncoupling the cleavage and religation reactions of eukaryotic topoisomerase I. the mode of action of camptothecin at a specific recognition site. J Mol Biol 222:669–678

Tiekink ER (2002) Gold derivatives for the treatment of cancer. Crit Rev Oncol Hematol 42:225–248

Gold, Physical and Chemical Properties

Fathi Habashi
Department of Mining, Metallurgical, and Materials Engineering, Laval University, Quebec City, Canada

Gold is the first element recognized by man as a metal. Ancient Egypt was the principal gold producing country and maintained that status until ca. 1500 BC. Gold has been valued since ancient times because of its color, its occurrence in the native state, and its non-tarnishing properties. It was established as the standard for currency by Newton in 1717 when he was Master of the Mint in London. Gold Rushes were responsible for creating large cities like San Francisco, Johannesburg, Sydney, and others.

Physical Properties

Atomic number	79
Atomic weight	196.97
Naturally occurring isotope	197
Relative abundance in Earth's crust, %	1×10^{-7}
Melting point, °C	1064.43
Boiling point, °C	2,808
Unit cell	Face-centered cubic
Lattice constant (a_0), nm	0.40781
Density, g/cm^3	
At 20°C	19.32
At 900°C	18.32
At 1,000°C	18.32
At 1,065°C	17.32
At 1,200°C	17.12
At 1,300°C	17.00
Vapor pressure, Pa	
At 1,064°C	0.002
At 1,319°C	0.1
At 1,616°C	10
At 1,810°C	100
At 2,360°C	10,000
Atomic volume at 20°C, cm^3/mol	10.21
Electrical resistivity at 0°C, W cm	2.06×10^{-6}
Thermal conductivity at 0°C, W cm^{-1} K^{-1}	3.14
Specific heat, J g^{-1} K^{-1}	0.138
Enthalpy of fusion, kJ/mol	12.77
Enthalpy of vaporization, kJ/mol	324.4
Tensile strength, N/mm^2	127.5
Hardness on the Mohs' scale	2.5
Brinell hardness	18
Standard potential, V	
Au/Au^{3+}	+1.498
Au/Au^{+}	+1.68
Au^{+}/Au^{3+}	+1.29

Gold is one of the metals that can be prepared readily to the highest purity 99.99^{+}% and is the most ductile of all metals. It can be cold drawn to give wires of less than 10 mm diameter, and beaten into gold foil with a thickness of 0.2 mm. Because of its softness, gold can be highly polished. Very thin gold foil is translucent; transmitted light appears blue-green.

Chemical Properties

Gold in solution exists as Au^{+} and Au^{3+}. Gold docs not react with water, dry or humid air, oxygen (even at high temperature), ozone, nitrogen, hydrogen, fluorine, iodine, sulfur, and hydrogen sulfide under normal conditions. Sulfuric, hydrochloric, hydrofluoric, phosphoric, nitric, and practically all organic acids have no effect on gold, either in concentrated or dilute solutions and at temperatures up to the boiling point. A mixture of hydrochloric and nitric acids known as *aqua regia* or *royal water* dissolves gold due to the formation of chlorine and nitrosyl chloride:

$$HNO_3 + 3HCl \rightarrow Cl_2 + NOCl + 2H_2O$$

Gold can also be dissolved in chlorine water (the Plattner process). Aqueous solutions of alkali metal hydroxides and alkali metal sulfides do not attack gold. However, gold dissolves in solutions of alkali metal cyanides in the presence of oxygen:

$$\begin{aligned} &2Au + 4CN^{-} + O_2 + 2H_2O \\ &\quad \rightarrow 2\left[Au(CN)_2\right]^{-} + H_2O_2 + 2OH^{-} \end{aligned}$$

Gold can be precipitated from cyanide solution by zinc powder:

$$2Na\left[Au(CN)_2\right] + Zn \rightarrow Na_2\left[Zn(CN)_4\right] + 2Au$$

Gold can be recovered from solution by electrolytic deposition or by chemical reduction. If the tetrachloroaurate(III) complex is present, then iron (II) salts, tin(II) salts, sulfur dioxide, hydrazine, or oxalic acid can be used as reducing agents. Fused caustic alkalis do not attack gold, provided air and other oxidizing agents are excluded. Gold reacts vigorously with alkali metal peroxides to form aurates. It is inert to sodium borate which can therefore be used as slagging agents for removing metallic impurities from gold. Gold reacts readily with dry chlorine. The maximum reactivity occurs at 250°C.

Colloidal gold forms hydrosols of an intense red or violet color. Gold alloys readily with mercury at room temperature to form an amalgam. The mercury can be distilled out by heating. This property is utilized in the amalgamation process and in fire gilding.

References

Habashi F (2009) Gold. History, metallurgy, culture. Métallurgie Extractive Québec, Québec City. Distributed by Laval University Bookstore "Zone", www.zone.ul.ca

Renner H, Johns MW (1997) Gold. In: Habashi F (ed) Handbook of extractive metallurgy. Weinheim, WILEY-VCH, pp 1183–1214

G

Gold, Ultrasmall Nanoclusters and Proteins, Interaction

Lennart Treuel[1,2] and Gerd Ulrich Nienhaus[1,3]
[1]Institute of Applied Physics and Center for Functional Nanostructures (CFN), Karlsruhe Institute of Technology (KIT), Karlsruhe, Germany
[2]Institute of Physical Chemistry, University of Duisburg-Essen, Essen, Germany
[3]Department of Physics, University of Illinois at Urbana-Champaign, Urbana, IL, USA

Synonyms

Protein adsorption onto ultrasmall gold nanoclusters; Protein corona formation around ultrasmall gold nanoclusters

Definition

Ultrasmall gold nanoclusters (Au NCs) consist of a few to roughly one hundred gold atoms. They are promising candidates for the development of novel fluorescence markers in biological applications because of attractive features including small size (diameter < 2 nm), low toxicity, and good photophysical properties. In biological applications, Au NCs and other nanoparticles become exposed to biological fluids and, typically, a protein adsorption layer forms around the particle surface.

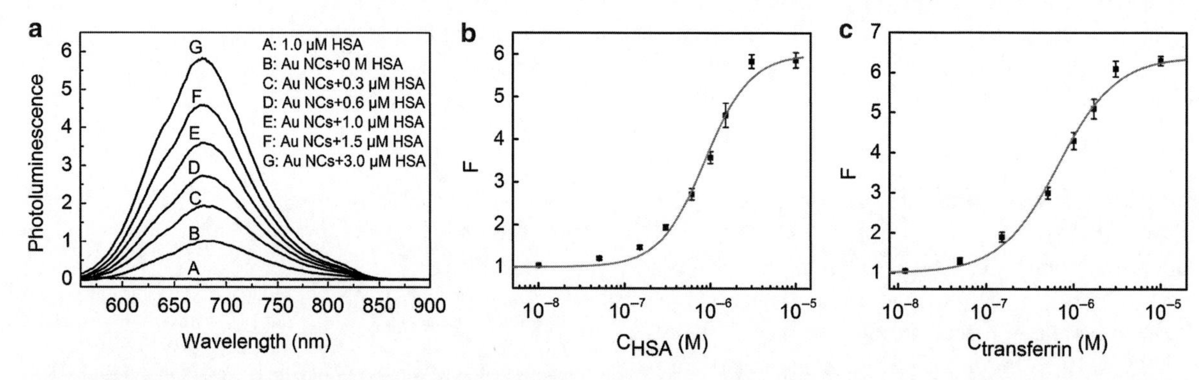

Gold, Ultrasmall Nanoclusters and Proteins, Interaction, Fig. 1 (**a**) Fluorescence emission spectra (excitation: 550 nm) of pure HSA (curve A) and Au NCs exposed to buffer solutions with varying concentrations of HSA (curves B–G). Fluorescence intensities of Au NCs, plotted as a function of (**b**) HSA and (**c**) transferrin protein concentration in the solution. The *grey lines* represent fits to the data points using the adapted Hill equation (Reproduced with kind permission from Shang et al. 2012)

Au NCs in Biological Systems

Upon nanoparticle (NP) exposure to an organism, proteins from body fluids bind to NP surfaces (Röcker et al. 2009; Maffre et al. 2011). Consequently, living systems interact with protein-coated rather than bare NPs. This so-called protein corona largely defines the biological identity of the NP, and the efficiency of this interaction can be a decisive factor of the biological response of an organism to NP exposure. Nel and coauthors (2009) have presented an in-depth discussion of the basic physical interactions occurring at the nano-bio interface. Biophysical mechanisms associated with nanoparticle-protein interactions on a molecular scale have been reviewed recently (Treuel and Nienhaus 2012).

Protein adsorption onto NP surfaces can cause structural changes of the protein (Gebauer et al. 2012; Treuel et al. 2010). Altered protein conformations can result in a loss of the biological activity and exposure of novel "cryptic" peptide epitopes that can promote autoimmune diseases (Nel et al. 2009).

A variety of synthesis strategies have been described for ultrasmall Au NCs, peptides, and proteins have also been employed successfully as templates for synthesizing highly fluorescent metal NCs (Shang et al., 2011).

Quantitative Aspects of Protein Adsorption

In ultrasmall Au NCs, the particle size is comparable to the Fermi wavelength of electrons resulting in discrete electronic states giving rise to size-dependent fluorescence properties (Lin et al. 2009). Furthermore, the strong, size-dependent surface plasmon absorption characteristic of larger Au NPs is completely absent.

Fluorescence emission intensity was used as a readout in a quantitative study (Shang et al. 2012) of protein adsorption onto ultrasmall Au NCs (1.4 ± 0.3 nm diameter) coated with dihydrolipoic acid (DHLA). Protein binding resulted in a substantial increase in the fluorescence intensity of the Au NCs (Fig. 1). By using this effect as a quantitative experimental readout, apparent binding affinities of four different proteins, human serum albumin (HSA), transferrin, lysozyme, and apolipoprotein E4 (ApoE4) to these Au NCs were determined. They were all in the micromolar concentration regime (K_D [HSA] = 0.9 μM, K_D [transferrin] = 0.7 μM, K_D [lysozyme] = 3.0 μM, K_D [ApoE4] = 2.7 μM). However, the increase in fluorescence intensity may not scale with the fraction of proteins bound to the NCs, so the data in Fig. 1b, c should not be viewed as proper binding isotherms.

Au NCs in Sensing Applications

The effect of protein adsorption onto Au NCs on their fluorescence properties can be put to good use in biosensor application. Selective receptor molecules (e.g., antibodies) have been attached to the surfaces of fluorescent Au NCs for the development of fluorescence-based protein biosensors.

As an early application, Leblanc and coworkers (Triulzi et al. 2006) reported an immunoassay based on electrostatic conjugation of polyclonal anti-IgG antibodies with dendrimer-encapsulated Au NCs. Linear fluorescence quenching was observed in response to changes in the surface electrostatic properties of the Au NCs caused by formation of an antigen-antibody immunocomplex. Further examples can be found in a recent review (Shang et al. 2011).

How Protein Adsorption Affects Fluorescence Properties of Au NCs

To explore the underlying mechanisms of the observed fluorescence increase upon protein adsorption, the effect on fluorescence lifetime was studied (Shang et al. 2012), revealing an average lifetime of 450 ± 5 ns for bare Au NCs. HSA adsorption onto these Au NCs slowed the luminescence decay, resulting in an average lifetime increase to 980 ± 40 ns. A detailed analysis of the luminescence decay kinetics revealed that the main contribution to this change is associated with the presence of an Au(I)-thiol complex on the particle surface.

Moreover, this study showed a shift in the electron binding energy of the Au core levels upon protein adsorption. They are sensitive to the local chemical or physical environment around the surface gold atoms. In the Au 4f XPS spectra of the Au NCs, the binding energies of the Au $4f_{5/2}$ and Au $4f_{7/2}$ levels in the absence of HSA, 88.3 ± 0.1 and 84.6 ± 0.1 eV, respectively, shifted to 88.0 ± 0.1 and 84.3 ± 0.1 eV, upon HSA adsorption onto the NC surface. Overall, these results indicated that photophysical properties of the Au NCs are directly altered by protein binding to the NC surface.

Cross-References

- ▶ Aluminum, Physical and Chemical Properties
- ▶ Biomineralization of Gold Nanoparticles from Gold Complexes in Cupriavidus Metallidurans CH34
- ▶ Colloidal Silver Nanoparticles and Bovine Serum Albumin
- ▶ Gold and Nucleic Acids
- ▶ Gold Nanomaterials as Prospective Metal-based Delivery Systems for Cancer Treatment
- ▶ Gold Nanoparticle Platform for Protein-Protein Interactions and Drug Discovery
- ▶ Gold Nanoparticles and Proteins, Interaction

References

Gebauer JS, Malissek M, Simon S, Knauer SK, Maskos M, Stauber RH, Peukert W, Treuel L (2012) Impact of the nanoparticle-protein corona on colloidal stability and protein structure. Langmuir 28:9181–9906

Lin CAJ, Lee CH, Hsieh JT, Wang HH, Li JK, Shen JL, Chan WH, Yeh HI, Chang WH (2009) Synthesis of fluorescent metallic nanoclusters toward biomedical application: recent progress and present challenges. J Med Biol Eng 29:276–283

Maffre P, Nienhaus K, Amin F, Parak WJ, Nienhaus GU (2011) Characterization of protein adsorption onto FePt nanoparticles using dual-focus fluorescence correlation spectroscopy. Beilstein J Nanotechnol 2:374–383

Nel AE, Mädler L, Velegol D, Xia T, Hoek EM, Somasundaran P, Klaessig F, Castranova V, Thompson M (2009) Understanding biophysicochemical interactions at the nano-bio interface. Nat Mater 8:543–557

Röcker C, Pötzl M, Zhang F, Parak WJ, Nienhaus GU (2009) A quantitative fluorescence study of protein monolayer formation on colloidal nanoparticles. Nat Nanotechnol 4:577–580

Shang L, Dong S, Nienhaus GU (2011) Ultra-small fluorescent metal nanoclusters: synthesis and biological applications. Nano Today 6:401–418

Shang L, Brandholt S, Stockmar F, Trouillet V, Bruns M, Nienhaus GU (2012) Effect of protein adsorption on the fluorescence of ultrasmall gold nanoclusters. Small 8:661–665

Treuel L, Nienhaus GU (2012) Toward a molecular understanding of nanoparticle–protein interactions. Biophys Rev 4:137–147

Treuel L, Malissek M, Gebauer JS, Zellner R (2010) The influence of surface composition of nanoparticles on their interactions with serum albumin. Chem Phys Chem 11:3093–3099

Triulzi RC, Micic M, Giordani S, Serry M, Chiou WA, Leblanc RM (2006) Immunoasssay based on the antibody-conjugated PAMAM-dendrimer-gold quantum dot complex. Chem Commun 48:5068–5070

Gold-Based Nanomedicine

▶ Gold Nanomaterials as Prospective Metal-based Delivery Systems for Cancer Treatment

G-Quartet

▶ Strontium and DNA Aptamer Folding

Gray or Grey Arsenic

▶ Arsenic in Pathological Conditions

Green Remediation

► Lead and Phytoremediation

G-Rich

► Selenoprotein K

Group 13 Elements NMR in Biology

► NMR spectroscopy of Gallium in Biological Systems

Growth Factor Regulated Channel (GRC)

► Transient Receptor Potential Channels TRPV1, TRPV2, and TRPV3

Growth Inhibitory Factor (GIF)

► Zinc Metallothionein-3 (Neuronal Growth Inhibitory Factor)

G-Tetrad

► Strontium and DNA Aptamer Folding

GTF

► Chromium and Glucose Tolerance Factor

Hafnium, Physical and Chemical Properties

Fathi Habashi
Department of Mining, Metallurgical, and Materials Engineering, Laval University, Quebec City, Canada

Hafnium is a refractory metal with a high melting point. It occurs mainly in association of zirconium in the mineral is zircon, $ZrSiO_4$, in a concentration of about 1%. Both zirconium and hafnium are transition metals, i.e., they are less reactive than the typical metals but more reactive than the less typical metals. The transition metals are characterized by having the outermost electron shell containing two electrons and the next inner shell an increasing number of electrons. Although hafnium [and zirconium] has two electrons in the outermost shell and it would have been expected that to have valency of 2, yet its valency is 4.

Physical Properties

Hafnium is a heavy, hard, and ductile metal. Although chemically similar to zirconium with which it occurs in nature in a concentration of about 1%, it has twice the density of zirconium, a higher phase transition temperature, and a higher melting point. Hafnium has a high-thermal neutron absorption coefficient whereas that for zirconium is very low (1.8×10^{-29} m^2).

Property		
Atomic number	72	
Relative abundance, ppm	2.8–4.5	
Atomic weight	178.49	
Naturally occurring isotopes		%
	^{174}Hf	0.16
	^{176}Hf	5.2
	^{177}Hf	18.6
	^{178}Hf	27.1
	^{179}Hf	13.7
	^{180}Hf	35.2
Melting point, °C	2,227	
Boiling point, °C	4,602	
Density, g/cm^3	13.31	
Thermal conductivity, W m^{-1} K^{-1}	23.0	
Coefficient of linear expansion (25–1,000°C), K^{-1}	5.9×10^{-6}	
Specific heat (25°C), J kg^{-1} K^{-1}	117	
Electrical resistivity (25°C), $\Omega \cdot$ m	3.51×10^{-7}	
Thermal neutron absorption cross-section, m^2	1.04×10^{-26}	
Crystal structure	Hexagonal close-packed ⇆ Body-centered cubic	
Temperature of transformation, 1,760°C		

V.N. Uversky et al. (eds.), *Encyclopedia of Metalloproteins*, DOI 10.1007/978-1-4614-1533-6,

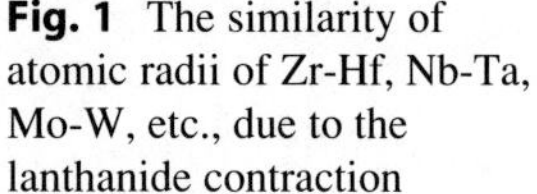

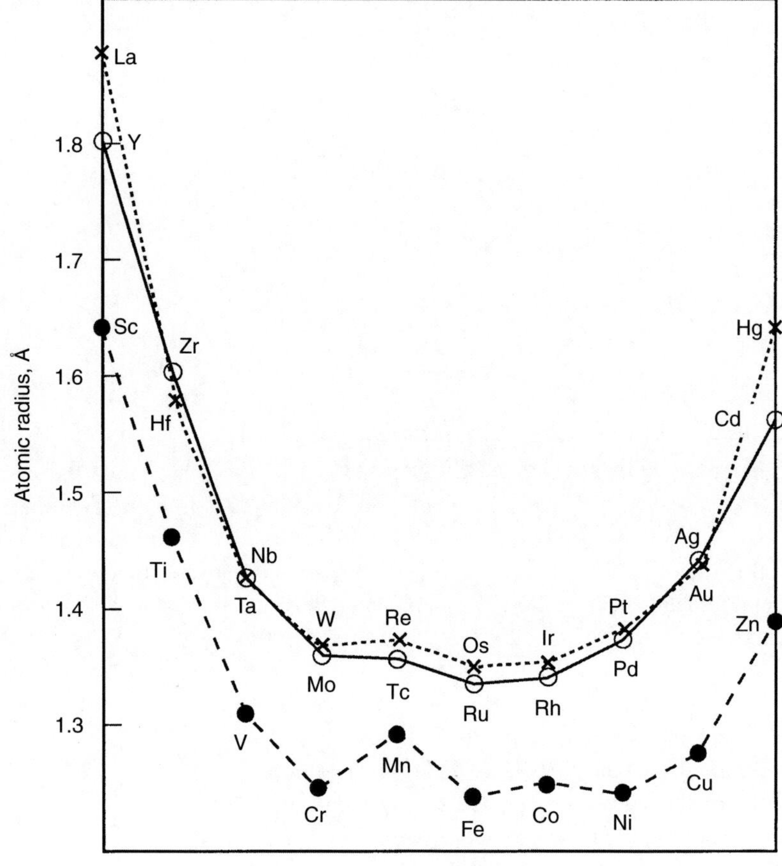

Hafnium, Physical and Chemical Properties, Fig. 1 The similarity of atomic radii of Zr-Hf, Nb-Ta, Mo-W, etc., due to the lanthanide contraction

Chemical Properties

The ionic radii of hafnium and zirconium are almost identical because of the lanthanide contraction (Fig. 1). Both elements exhibit a valence of four. Therefore, the chemistry of hafnium is similar to that of zirconium.

Elemental hafnium reacts with hydrogen (>250°C), carbon (>500°C), and nitrogen (>900°C) to form brittle, nonstoichiometric interstitial compounds with metal-like conductivity. In molten salts, hafnium is normally quadrivalent, but in anhydrous molten halide salts, it can be reduced to hafnium(III) and hafnium(II). In aqueous solution, hafnium is always quadrivalent, with a high coordination number (6, 7, or 8). In dilute acid, hafnium slowly hydrolyzes and polymerizes. Hafnium hydrous oxide precipitates from aqueous solution at ca. pH 2. The only inorganic compounds with significant solubility in neutral or slightly basic aqueous solution are the ammonium–hafnium carbonate and potassium–hafnium carbonate complexes.

Both metals are used in nuclear reactors: zirconium as a container for uranium fuel element because of its low-neutron cross-section, and hafnium as control rods because of its high-neutron cross-section.

References

Habashi F (2003) Metals from ores. An introduction to extractive metallurgy. Métallurgie extractive Québec, Quebec City, Canada. Distributed by Laval University Bookstore, www.zone.ul.ca

Nielsen RH (1997) Hafnium and hafnium compounds. In: Habashi F (ed) Handbook of extractive metallurgy. Wiley-VCH, Weinheim, pp 1459–1470

Heat Shock Proteins

▶ Arsenic-Induced Stress Proteins

Heavy Metal Efflux

► CusCFBA Copper/Silver Efflux System

Heavy Stone

► Tungsten Cofactors, Binding Proteins, and Transporters in Biological Systems

Helix-Loop-Helix Ca^{2+}-Binding Protein

► EF-hand Proteins and Magnesium

Helix-Loop-Helix Calcium-Binding Poteins

► EF-Hand Proteins

Heme

► Heme Proteins, Heme Peroxidases

Heme Proteins, Cytochrome c Oxidase

Oliver-M. H. Richter and Bernd Ludwig
Institute of Biochemistry, Goethe University, Frankfurt, Germany

Synonyms

Cellular respiration; Energy transduction; Terminal oxidases

Definition

Cytochrome *c* oxidases are integral membrane proteins in energy-transducing systems, acting as molecular machines: They couple the free energy of electron transport to oxygen in this segment of the respiratory chain to the uphill translocation of protons against a concentration gradient, eventually powering ATP synthesis. Such redox-driven proton pumps are widespread in nature, from the classical enzyme in mitochondria to numerous oxidase variants found in bacteria, collectively termed heme/copper oxidases.

Oxidases as Components of Respiratory Chains

Cytochrome *c* oxidase (complex-IV; COX) is the terminal member of the respiratory chain localized in the mitochondrial inner membrane, or the cytoplasmic membrane of many bacteria (Richter and Ludwig 2009; Kaila et al. 2010; Renger and Ludwig 2011; Sousa et al. 2012; Yoshikawa et al. 2011). On oxidizing reduced cofactors such as NADH, electrons are transferred via complex-I (NADH:ubiquinone oxidoreductase) and complex-III (ubiquinol:cytochrome *c* oxidoreductase) to cytochrome *c*. This small carrier protein is a single-electron donor to COX where in four subsequent, highly concerted transfer steps, dioxygen is eventually reduced to water, see (1). All of the above redox complexes are membrane-embedded, multi-subunit protein assemblies acting as redox-coupled proton translocation units ("proton pumps") utilizing the free energy of their partial reactions to establish a proton gradient across the membrane which is subsequently used by ATP synthase (complex-V) to convert ADP and phosphate into ATP, the general energy "currency" of any cell. COX typically operates at an outward-directed pump stoichiometry of $1H^+$/electron transferred (see below), leading to the uptake of a total of eight protons from the organellar matrix (the bacterial cytoplasm, resp.); four of these protons are destined to reach the binuclear center of oxidase (BNC, see below) required in water formation.

$$4\text{cyt}c^{2+} + 8\text{H}^+_{\text{in}} + \text{O}_2 \rightarrow 2\text{H}_2\text{O} + 4\text{cyt}c^{3+} + 4\text{H}^+_{\text{out}} \quad (1)$$

Subunits, Genetic Origin, and the Oxidase Superfamily

Depending on the organism, mitochondrial COX (Yoshikawa et al. 2011) consists of up to 13 different protein subunits (SU) encoded by both the mitochondrial genome (SU I–III) as well as the nuclear genome (up to 10 further SU), calling for a tight nuclear-organellar cross talk in expression regulation, and an elaborate system for protein import and for assembly in (to) the organelle for all cytoplasmically translated SU. Bacterial COX representatives (Sharpe and Ferguson-Miller 2008; Richter and Ludwig 2009; see Fig. 1 and below), on the other hand, exhibit a much simpler SU composition, typically consisting of three (+/− 1) SU, all related in sequence and 3-D structure to the three largest, mitochondrially encoded SU, a fact explained by the endosymbiotic origin of mitochondria. In functional terms, bacterial COX appear virtually identical in most aspects to their mitochondrial counterparts, thus making them popular model systems for genetic manipulations in structure/function analyses. Along with a large number of related bacterial COX species differing in their particular electron donor (cytochrome *c*, or a quinol moiety), their heme type composition, and their specific expression pattern under different growth conditions, they nevertheless are all grouped into the large (super-)family of heme/copper terminal oxidases (HCO; see below), named for the two metal cofactors in the binuclear center(BNC) of SU I (see Sousa et al. 2012).

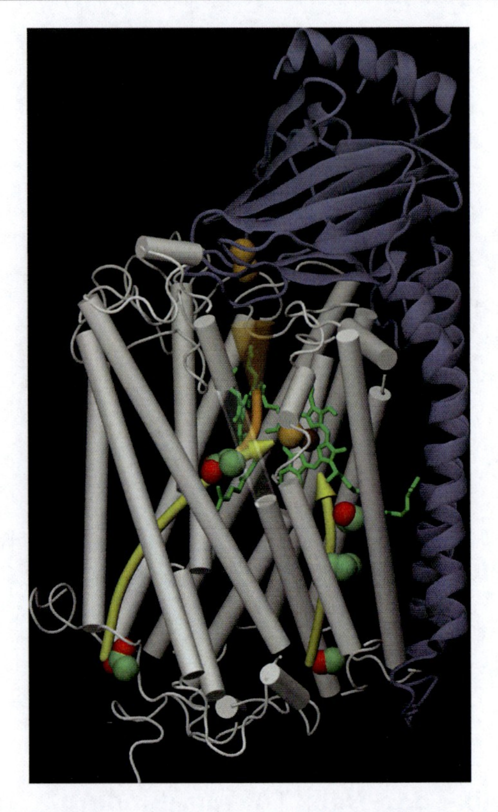

Heme Proteins, Cytochrome c Oxidase, Fig. 1 Three-dimensional structure of the two core subunits of the cytochrome *c* oxidase complex, as exemplified for the bacterial enzyme isolated from *Paracoccus denitrificans* (pdb 1AR1), containing all redox-active cofactors. SU I, depicted with its 12 transmembrane helices as *gray rods*, houses the low-spin heme *a* (porphyrin body in *green* with its central iron atom in *brown*); to its right the binuclear center responsible for oxygen reduction, consisting of the high-spin heme a_3 and the neighboring, electronically coupled copper ion (Cu_B) depicted in *gold*. The two canonical proton pathways are shown by *bent yellow arrows* trailing through the TMH scaffold, with some characteristic amino acid residues highlighted: *right*, the K pathway delivering protons from the cytoplasmic (matrix, resp.) side exclusively to the BNC (Ser295/Lys354/Thr351); *left*: the D pathway starting at Asp124 extending to Glu278. This latter side chain may act as a crucial gating device to direct protons either to the BNC, or all the way across the membrane ("pumped protons") to the periplasmic (intermembrane) space, as symbolized by the upward-facing conical element (helix VI presented partially translucent for better visualization). In *blue*, secondary structure elements of SU II with its two TMH and an extensive β-sheet hydrophilic domain extending into the periplasm/intermembrane space that accommodates the two copper ions forming the Cu_A site where electrons donated from cytochrome *c* enter the complex; for further details, see text

Core Subunits and Arrangement of Their Redox-Active Cofactors

Subunit I is the central component of all oxidases, consisting of a core structure of 12 helices spanning the lipid membrane, see Fig. 1. Three of the four redox centers of oxidase are firmly embedded in this framework, to about one third of the membrane depth: The two A-type hemes (*a* and a_3) are aligned perpendicular to the membrane plane, with an interplanar angle of 108°; the low-spin iron of heme *a* is liganded by two histidine residues, the high-spin iron of heme a_3 has a single histidine as the proximal ligand. Within a distance of approx. 5 Å, a copper ion (termed Cu_B) is found, being liganded by three histidine residues, one of which (His276 in *P. denitrificans*) is covalently cross-linked to Tyr280. Heme a_3 • Cu_B together form the BNC of SU I, the site of oxygen binding and reduction, representing most likely also the coupling device between electron transport (ET) and proton translocation, see below.

Subunit II houses the homo-binuclear copper center termed Cu_A, the first ET acceptor in COX, in a hydrophilic domain anchored to the membrane by two N-terminal TMHs, see Fig. 1. The two copper ions are 2.6 Å apart, close enough to allow electronic coupling.

SU III, not depicted in Fig. 1, is a highly hydrophobic component with a 7 TMH structure devoid of redox cofactors; even though found in many bacterial complexes, it is not strictly required for COX activity. In mitochondrial COX, this SU is encoded by the organellar genome as well.

Up to 10 additional SU make up the complexity of eukaryotic enzyme complexes; their structural genes (even with isoforms observed in some cases) are expressed from the nuclear genome in a tissue- and development-specific manner, imported into and assembled in the organelle, and have generally been implicated in the general regulation of cellular energy metabolism (see Pierron et al. 2012).

Electron Pathways in COX

Reduced cytochrome *c* interacts with the COX complex mainly electrostatically and delivers one electron at a time to Cu_A in SU II, a homo-binuclear, mixed valence copper center (see Fig. 1) some 5 Å below the presumed docking surface. A tryptophan residue protruding from the docking site (Trp121 in *P. denitrificans*) has been assigned a pivotal role in mediating this ET between COX and its substrate. From Cu_A, electrons reach the redox centers in SU I: first the heme *a*, followed by transfer to the BNC in close proximity, made up of heme a_3 and Cu_B both closely spaced and electronically coupled. This BNC is buried by about one third into the depth of the hydrophobic core of the membrane, making any charge movements energetically highly unfavorable without a compensating countercharge, as governed by the electroneutrality principle (Rich et al. 1996). All internal ET distances, including the Cu_A to heme *a* edge-to-edge distance of around 16 Å, fall well within the limits of biologically relevant electron tunneling rates (Moser et al. 2006),and are compatible with macroscopically determined continuous turnover rates for isolated COX preparations of around 10^3electrons per sec.

Oxygen Reaction Cycle

Complete reduction of molecular oxygen at the BNC requires the delivery of four electrons and four protons in a highly controlled manner, such as to avoid generation and potential leakage of damaging reactive oxygen species (ROS) resulting from partial reduction. Figure 2 gives a simplified overview of redox intermediates during a full reduction cycle, focusing on the BNC and its immediate surroundings, and assuming the stepwise supply of electrons from the entry site, the isopotential Cu_A/heme *a* couple (not shown). For detailed spectral and kinetic distinctions of these major intermediates identified over the past 50 years, see, e.g., Kaila et al. 2010; Yoshikawa et al. 2011.

In the "O" state, both BNC metal ions are oxidized (Fe^{3+}in heme $a_3$and Cu_B^{2+}), to be reduced by the first (yielding the "E" state) and second electron to enter the BNC, arriving at "R." Only with this stage reached, an oxygen molecule, entering from a nearby cavity, is able to bind and gets trapped between the two metal centers to form the short-lived adduct "A" intermediate. In a concerted reaction sequence, four electrons are transferred to the oxygen molecule to split the O–O bond and reduce both oxygen atoms to the formal oxidation state of water. This "P_M" ("mixed valence," historically assumed to represent the peroxi form of dioxygen) state is reached within less than 100 μs, and has been identified as an oxoferryl (Fe(IV) = O) state of the heme. While this reaction sequence has been acknowledged as a crucial kinetic safeguard to avoid formation of ROS, the source of the fourth electron required for this step remained unclear until experimental evidence was provided for a radical state intermediate in the redox cycle (Kaila et al. 2010; Renger and Ludwig 2011), transiently providing the fourth electron, next

H

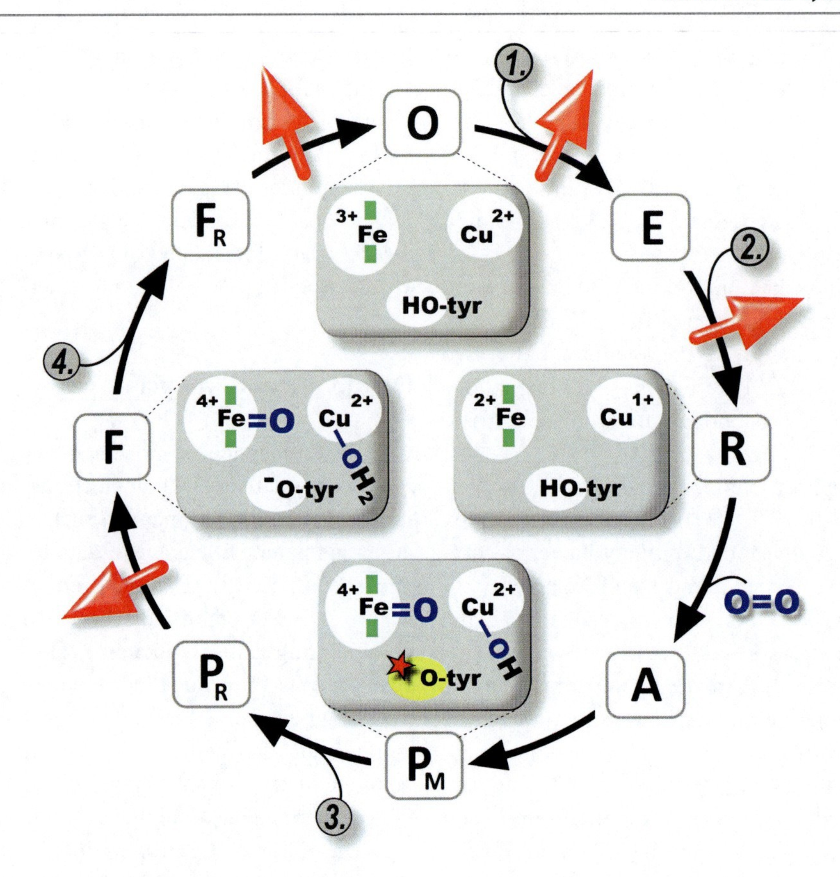

Heme Proteins, Cytochrome c Oxidase, Fig. 2 Dioxygen reduction cycle catalyzed by COX. *Inner circle*: *gray* background icons illustrate BNC redox states for heme a_3, Cu_B and a crucial cross-linked tyrosine residue as well as oxygen species present during key intermediate redox states of COX (*boxed* letters O, R, P_M, and F). *Outer circle*, stepwise entry of electrons (numbers outlined in *black*) originating from cytochrome *c*, to reduce the oxidized ("O") state of COX to the "E" and then the "R" state, now ready to bind oxygen (shown in *blue*). Note that in the follow-up intermediate of the oxidative half-cycle, P_M, the dioxygen bond is already broken, and both oxygen atoms have formally acquired the redox state of water by the internal transfer of four electrons (two from the heme iron, one from Cu_B, and one from the tyrosine side chain to form an intermediate radical state, *red star symbol*). Later electron entry steps 3 and 4 are used to abandon the radical state of the tyrosine (leading to "F") and recover the Fe^{3+}state of the heme iron in "O." *Large red arrows* pointing outward: proton pumping steps and their presumed timing subsequent to each electron entry step. No reference is given to other oxygenous ligands such as water or hydroxyl ions in the BNC, nor does the scheme specify the entry of protons for water formation or the use of particular protonic pathways as shown in Fig. 1; for further details, see text (Figure adopted from Renger and Ludwig 2011)

to the two electrons supplied from the heme a_3 iron (II–IV oxidation state) and one from Cu_B (see Fig. 2).

Present focus for this radical side chain contribution is on a tyrosine (Fig. 2; Tyr280 in *P. denitrificans*) conspicuous for the fact that it is cross-linked to a histidine liganding the Cu_B ion, thus presumably yielding unique chemical and redox properties to influence both the proton and the electron affinity in the BNC (Kaila et al. 2010). Kinetic evidence, e.g., by rapid freeze quenching studies, has also identified nearby aromatic residues which may be involved in a radical migration mechanism.

The third and fourth input electrons in the reaction cycle of Fig. 2 are destined to abrogate the radical, reaching state "F," and subsequently reduce the iron ferryl state to conclude the cycle at "O."

Pump Steps and Proton Pathways in COX

Concepts of coupling ET to proton translocation in COX have been a matter of considerable controversy in the past, and a distinct molecular mechanism is still at large today. It is generally agreed that (1) any charge translocations, be it protons or electrons, into and from a highly insulated region such as the membrane-embedded BNC or a heme *a* alone should immediately be counterbalanced by opposing charge movements, providing the overall driving force for proton pumping, and (2) the stoichiometry of coupling, at least in most cases studied in detail, is unity, meaning that for every electron passing through the COX redox centers, one proton is translocated across the membrane. What still is less accepted presently is the precise timing of each of the four pumping steps, the proton pathways utilized in any given COX complex, and the nature of the gating device to avoid short-cutting of an imposed proton gradient.

Figure 2 illustrates the notion, recently confirmed experimentally in different systems, that every electron input step (1–4) is more or less immediately followed by a transmembrane pump event, i.e., chemical reactions involving oxygen species bound in the BNC drive protonic movements (red outward-pointing arrows). While this concept may appear most convincing in the oxidative part of the cycle following O_2 binding, evidence for proton translocation steps following electron input steps 1 and 2 has been provided more recently. Note, however, that extra oxygenous ligands have been omitted for clarity from the BNC sketch in the reductive part in Fig. 2, and that during continued turnover, intermediate "O" may represent a high-energy state to allow for proton pumping with the first electron entering the BNC (Kaila et al. 2010).

Even before solving the 3-D structure(s) of COX, extensive mutagenesis studies provided first evidence for two different pathways for protons operating within SU I, interpreted at that time such that protons to be translocated would use a distinct path different from those destined for the water formation reaction at the BNC. Careful time-resolved electrometric approaches later revealed that out of the eight protons being taken up from the inside (matrix or bacterial cytoplasm, respectively) during one O_2 cycle, six protons travel via the D pathway, named for a highly conserved aspartate residue at its entrance (Asp124 in *P. denitrificans*; see Fig. 1), while the two other protons would use the K pathway (named for a conserved lysine residue). This dual-purpose D channel usage posed a serious subsequent problem still largely unsolved, invoking a switching device to "instruct" or guide four of the six protons for transmembrane pumping, and the remaining two required for oxygen reduction at the BNC. A highly conserved glutamate residue, with alternative side chain orientations documented experimentally, Glu278 in *P. denitrificans* (see Fig. 1), may take this role as a control element for protons.

Recently, a further protonic conduit has been suggested for the pumped protons in the mitochondrial COX, and experimental evidence given for this H channel (Yoshikawa et al. 2011), which, however, has no counterpart in the bacterial COX complexes analyzed.

The exit routes of pumped protons past the BNC level in SU I, and that of the two water molecules produced, are less clear, but both paths may coincide and pass the (redox-inactive) Mg/Mn site being located at the intermembrane space- (periplasmatically) oriented interface between SU I and II.

Many current aspects of COX could not be addressed here in any detail, ranging from distinct deviations in ET and proton translocation mechanisms in some of the deviant bacterial members of the HCO superfamily (Sousa et al. 2012), to questions of biogenesis and the role of chaperones (e.g., Richter and Ludwig 2009) required for metal cofactor insertion both in pro- and eukaryotes; nor has the intricate relationship and cross talk between the nuclear and organellar genome (see Pierron et al. 2012) in gene expression and regulation of the mitochondrial COX, or the contribution of computational and modeling approaches (see, e.g., Kim and Hummer 2012) to our understanding of oxidase function been elucidated here.

Cross-References

- ▶ Biological Copper Transport
- ▶ Copper, Biological Functions
- ▶ Copper-Binding Proteins
- ▶ Cytochrome c Oxidase, CuA Center

References

Kaila VR, Verkhovsky MI, Wikström M (2010) Proton-coupled electron transfer in cytochrome oxidase. Chem Rev 110(12):7062–7081

Kim YC, Hummer G (2012) Proton-pumping mechanism of cytochrome *c* oxidase: a kinetic master-equation approach. Biochim Biophys Acta 1817(4):526–536

Moser CC, Page CC, Dutton PL (2006) Darwin at the molecular scale: selection and variance in electron tunneling proteins including cytochrome *c* oxidase. Phil Trans R Soc B 361:1295–1305

Pierron D, Wildman DE, Huttemann M, Chand G, Markondapatnaikuni AS, Grossman LI (2012) Cytochrome *c* oxidase: evolution of control via nuclear subunit addition. Biochim Biophys Acta 1817(4):590–597

Renger G, Ludwig B (2011) Mechanism of photosynthetic production and respiratory reduction of molecular dioxygen: a biophysical and biochemical comparison. In: Peschek G et al (eds) Bioenergetic processes of cyanobacteria. Springer, Dordrecht, pp 337–394

Rich PR, Meunier B, Mitchell R, Moody AJ (1996) Coupling of charge and proton movement in cytochrome *c* oxidase. Biochim Biophys Acta 1275:91–95

Richter OMH, Ludwig B (2009) Electron transfer and energy transduction in the terminal part of the respiratory chain – lessons from bacterial model systems. Biochim Biophys Acta 1787:625–633

Sharpe MA, Ferguson-Miller S (2008) A chemically explicit model for the mechanism of proton pumping in heme-copper oxidases. J Bioenerg Biomembr 40(5):541–549

Sousa FL, Alves RJ, Ribeiro MA, Pereira-Leal JB, Teixeira M, Pereira MM (2012) The superfamily of heme-copper oxygen reductases: types and evolutionary considerations. Biochim Biophys Acta 1817(4):629–37

Yoshikawa S, Muramoto K, Shinzawa-Itoh K (2011) Proton-pumping mechanism of cytochrome *c* oxidase. Annu Rev Biophys 40:205–223

Heme Proteins, Heme Peroxidases

Thomas L. Poulos
Department of Biochemistry & Molecular Biology, Pharmaceutical Science, and Chemistry, University of California, Irivine, Irvine, CA, USA

Synonyms

Electron transfer; Enzyme mechanism; Heme; Peroxidases

Definition

Peroxidases are heme containing enzymes that utilize the oxidizing power of H_2O_2 to oxidize a variety of molecules.

Introduction to Peroxidases

Heme peroxidases occur throughout the biosphere and catalyze the oxidation of various substrates by peroxide. Peroxidases are best known for their participation in various aspects of plant metabolism but also are quite important in humans. During infection human neutrophil cells produce a variety of oxidants as a defense against the invading pathogen. One of these is HOCl (common household bleach), which is produced by the oxidation of Cl^- by H_2O_2 to HOCl, a reaction catalyzed by the neutrophil enzyme, myeloperoxidase. However, given that peroxidases are more widespread in nonmammalian organisms, much of what is known about peroxidase structure and function derives from the nonmammalian enzymes. Nonmammalian peroxidases have been divided into three general classes based on sequence alignments and crystal structures. The three classes are intercellular (Class 1), extracellular fungal (Class II), and extracellular plant (Class III). Initially Class III peroxidases received most of the attention since these were fairly easy to purify from plant tissues. Much of the early enzyme kinetics and mechanistic understanding of peroxidases derived from work with horseradish peroxidase (HRP). The heart of peroxidases is the heme prosthetic group (Fig. 1) which provides the brownish-red color to the purified enzyme. The ring part of the heme, called the porphyrin, and the iron directly participate in peroxidase redox chemistry. The following reaction scheme summarizes the overall catalytic mechanism of HRP which applies to most heme peroxidases.

1. $Fe^{3+}\,P + H_2O_2 \rightarrow Fe^{4+} = OP^{\bullet+} + H_2O$
   Compound I
2. $Fe^{4+} = OP^{\bullet+} + S \rightarrow Fe^{4+} = OP + S^{\bullet}$
   Compound II
3. $Fe^{4+} = OP + S \rightarrow Fe^{3+}\,P + H_2O + S^{\bullet}$

During the catalytic cycle the heme undergoes oxidation reduction reactions which make it quite easy to follow formation and decay of distinct enzyme intermediates since each intermediate exhibits a different color (Fig. 1). In step 1, peroxide removes two electrons from the enzyme to give what is called Compound I. Compound I is green in color compared to the starting brownish/red enzyme. The peroxide binds to the heme iron which is followed by cleavage

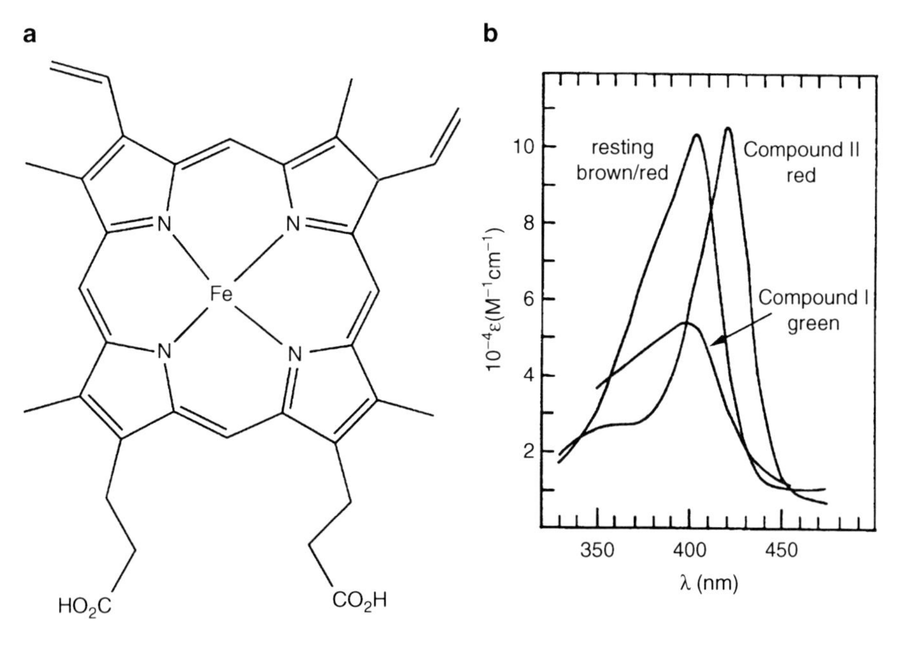

Heme Proteins, Heme Peroxidases, Fig. 1 (**a**) The heme group found in many heme proteins including peroxidases. The iron is coordinated by the 4 N atoms of the porphyrin macrocycle. This leaves two additional axial coordination positions (above and below the plane of the heme) for coordinating two additional ligands to the iron. One is provided by the protein which is His in peroxidases. This leaves the remaining axial coordination position available for coordinating the substrate, H_2O_2. (**b**) The UV/Vis spectrum of resting HRP, Compound I, and Compound II (Adapted Dunford (1982) Advances in Inorganic Biochemistry 4: 41–68.) The stability and clearly defined spectral differences between these species makes it fairly easy to follow the kinetics of peroxidase reactions

of the peroxide O–O bond and the departure of water. This leaves behind an oxygen atom bound to the iron. Since the O atom has only six valence electron, it is a potent oxidizing agent and thus removes two electrons from the heme. One electron is removed from the iron to give the oxy-ferryl ($Fe^{4+} = O$) species and a second electron is removed from the porphyrin (P in the above scheme) to give a porphyrin π cation radical. In some peroxidases, an amino acid side chain is oxidized to a radical rather than the porphyrin. In step 2, the porphyrin radical is reduced by a substrate molecule giving a substrate radical, S˙, and Compound II. Owing to porphyrin reduction, Compound II is no longer green but red. Finally in step 3, Compound II is reduced by a second substrate molecule. In plant peroxidases, the substrates normally are small aromatic molecules and once substrate radicals form, they dimerize or disproportionate in nonenzymatic reactions.

Although HRP was the early favorite in working out the peroxidase mechanism, yeast cytochrome c peroxidase (CCP) provided much of the initial insights into structure-function relationships since CCP exhibits some distinct experimental advantages. Most importantly CCP can be easily purified from common Baker's yeast and readily crystallizes by simply dialyzing the enzyme against water. With a bit more care high quality crystals can be obtained that diffract to high resolution while the many isozymes of HRP did not yield diffraction quality crystals, probably due to the extensive glycosylation of HRP while CCP has no sites of glycosylation. In addition, unlike HRP, CCP has no disulfide bonds which made it much easier to clone and express CCP in *Escherichia coli* thus opening the way for protein engineering studies. Proper formation of S–S bonds in *E. coli* expression systems often proves difficult and with HRP a complicated refolding protocol of the unfolded enzyme is required to enable the S–S bonds to correctly form and generate soluble active enzyme. Finally CCP oxidizes a more interesting substrate, cytc. For some time CCP and cytc were the only

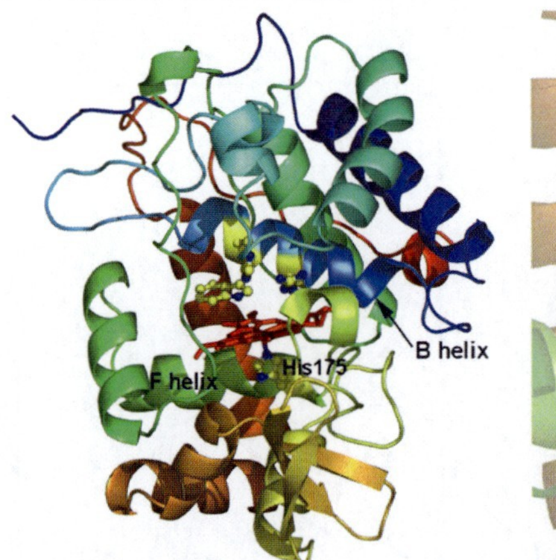

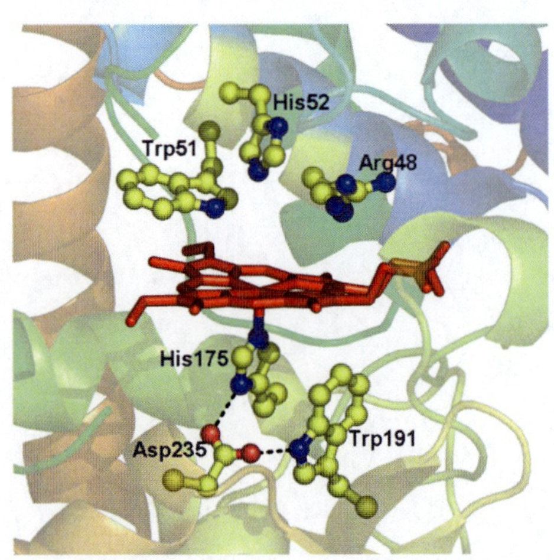

Heme Proteins, Heme Peroxidases, Fig. 2 The crystal structure of yeast cytochrome c peroxidase (CCP). All three classes of nonmammalian heme peroxidases exhibit a similar 10 helical fold. The heme is held in place between the B and F helices. At the C-terminal end of the F helix His175 coordinates the heme iron. His175 donates an H-bond to Asp235 which increases the imidazolate (anionic) character of His175. This in turn helps to stabilize the Fe^{3+} relative to Fe^{2+}. The B helix provides key groups (Arg48, Trp51, and His52) that form the distal binding pocket where H_2O_2 binds

redox partners where both crystal structures were known. This together with the ease of preparing significant amounts of each protein has enabled the CCP-cytc system to serve as a central paradigm in the area of inter-protein electron transfer.

Crystal Structure

Yeast CCP was the first peroxidase and first heme enzyme structure to be solved back in the early 1980s. Since then several other peroxidase structures from all three classes have been solved. The various peroxidases all consist of a similar 10 helical core, two of which, the B and F helices, sandwich the heme in place (Fig. 2). The various peroxidases differ in structure primarily on the surface. For example, Class III peroxidases have two additional helices on the surface (F′ and F″) where CCP has beta structure. The conserved His175 serves as one iron ligand while in the resting Fe^{3+} state water is situated above the iron on the opposite (distal) side of the heme but is too far for direct bonding to the iron. The His175 ligand donates an H-bond to the conserved Asp235. This interaction is thought to impart a partial negative charge to the His ligand which can stabilize the Fe^{3+} state relative to Fe^{2+}. Porphyrin alone (heme without the iron) has two of the 4 N atoms protonated. For iron to bind, these two protons are stripped off giving a net −2 charge to the porphyrin core. The coordination of Fe^{3+} to the porphyrin core thus gives a net charge of +1. Providing a nearby negative charge in the form of partially anionic His175 therefore stabilizes the Fe^{3+} redox state relative to Fe^{2+} and lowers the redox potential. Peroxidases are thus good reducing agents which means the equilibrium between Fe^{3+} and Fe^{2+} favors Fe^{3+}. This is one reason why peroxidases exhibit a lower redox potential in the Fe^{3+}/Fe^{2+} couple than the globins even though both the globins (i.e., hemoglobin) and peroxidases both use a His ligand. The His ligand in the globins has no nearby negative charge to stabilize Fe^{3+} relative to Fe^{2+}. Directly adjacent to His175 is Trp191 which also H-bonds with Asp235. In most Class II and III peroxidases Trp191 is replaced by a Phe. This is an important difference since, as will be discussed further on, Trp191 plays an essential catalytic role in CCP. On the distal side of the heme, Arg48, Trp51, and His52 form a pocket above the iron where H_2O_2 will bind and coordinate the heme iron. These residues are conserved in peroxidases except Trp51 which is Phe in Class II and III peroxidases.

Mechanism of Compound I Formation

The rate of Compound I formation is fast, on the order of 10^7–$10^8 M^{-1}s^{-1}$. The reaction between peroxidases and H_2O_2 behaves as a second order reaction that is stoichiometric and irreversible. If Compound I formation required formation of an enzyme-peroxide complex first, then one should observe a maximum rate at high peroxide concentration levels. That is, the rate would flatten out at high peroxide concentrations. However, this is not observed and plots of peroxide concentration versus rate are linear and extrapolate back to zero peroxide concentration. This type of kinetic behavior is consistent with a second order reaction but not formation of pre-reaction enzyme-peroxide complex. This raises some interesting questions on the meaning of Km obtained from steady state kinetics. However, it is possible to obtain kinetics that are consistent with formation of a pre-reaction enzyme-peroxide complex when the kinetics are measured at substantially reduced temperatures. Assuming that the mechanism does not change under conditions required to follow the reaction at reduced temperatures, it would appear that peroxidases do form an enzyme-peroxide complex even at room temperature but the reaction is simply too fast to achieve saturation behavior consistent with a pre-reaction complex kinetics.

With respect to the detailed mechanism of Compound I formation, much was known about the chemistry of Compound I formation at the time the CCP structure was solved. For example, it was known that the peroxide O–O bond cleaves heterolytically versus homolytically (Fig. 3). In order for this to occur efficiently it was generally thought that acid-base catalysis must be involved. That is, the peroxide O2 O atom (Fig. 3) must be protonated to ensure heterolytic rupture of the O–O bond. It also was known that the iron is in the Fe^{4+} oxidation state. The CCP structure provided the structural underpinning of these fundamental early studies. There has been some fine-tuning of the mechanism with the most important being the possible involvement of water. HIs52 is really too far from the peroxide O1 atom to accept a proton so it has been suggested that water may mediate the process of shuttling a proton from the peroxide O1 to His52 (Fig. 3).

Compound I Structure

Given that Compound I is stable in many peroxidases, it has been possible to utilize a variety of biophysical methods, including crystallography, to study the structure of Compound I. However, the intense synchrotron x-ray sources often used these days generate hydrated electrons which are potent reductants of metal centers. This is especially troublesome for metals that exhibit high redox potentials such as the Fe^{3+}/Fe^{4+} couple with a redox potential in the 900 mV range so the Compound I Fe^{4+} center is a powerful oxidant and is readily reduced in the x-ray beam. This has proven especially troublesome since there has been a long-standing discrepancy on the nature of the Fe^{4+}-O bond between crystallography and other techniques like Extended X-ray Atomic Fine Structure (EXAFS) and resonance Raman. A majority of crystal structures show a long Fe^{4+}-O bond in the range of 1.8 Å suggesting a Fe^{4+}-OH single bond while the other methods give a shorter bond on the order of 1.6–1.7 Å indicating $Fe^{4+} = O$ double bond. This is not a trivial difference since the reactivity of each species is expected to be quite different. To resolve these discrepancies has required careful x-ray data collection techniques coupled with single crystal spectroscopy. It now is possible to obtain the UV/Vis absorption spectrum of a single crystal before, during, and after data collection. The spectrum provides an estimate on the extent of reduction of Fe^{4+} as a function of time in the x-ray beam enabling one to develop a composite data collection strategy ensuring that the final combined x-ray data set is $\approx 90\%$ or greater Compound I. This requires that each crystal be exposed for a very short period of time (on the order of seconds) and the merging together of a large number of data sets collected from many crystals. While still difficult, this type of composite data collection is more feasible with the advent of cryogenic robotics for mounting and dismounting crystals. With this approach recent structures of both HRP and CCP Compound I give Fe^{4+}-O distances of 1.70 Å and 1.72 Å, respectively, which are much closer to the EXAFS and Raman data. This means that Compound I is best described as $Fe^{4+} = O$ and that previous estimates of a longer Fe–O bond were due to reduction of $Fe^{4+} = O$ in the x-ray beam to give Fe^{3+}-OH.

Electronic Structure of Compound I

The basic job of peroxidases is to store the oxidizing power of H_2O_2 in the enzyme and then utilize this oxidizing power to oxidize specific substrates. This is

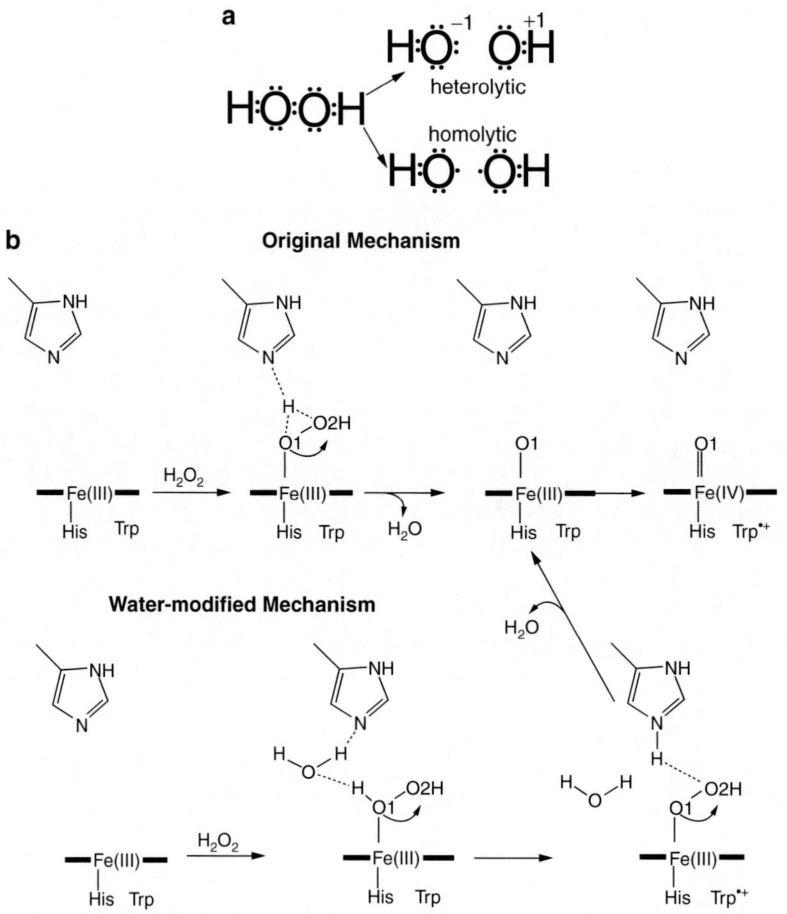

Heme Proteins, Heme Peroxidases, Fig. 3 (**a**) Homolytic cleavage of the peroxide O–O bond gives two hydroxyl radicals. Heterolytic cleavage gives a hydroxyl anion and a hydroxyl cation. The cation has only six valence electrons and thus is a potent oxidizing agent. (**b**) The original peroxidase mechanism is shown at the top. The distal His serves as an acid-base catalyst and assists in the transfer of the peroxide O1 proton to the peroxide O2 atom. The doubly protonated O2 is a good leaving group which ensures heterolytic cleavage of the peroxide O–O bond. O2 now has only six valence electrons and thus is a powerful oxidizing agent since the most stable form of an O atom is to have eight valence electrons. One electron is removed from the iron to give Fe^{4+} and one from Trp191 to give a cationic radical. One problem with this mechanism is that the His is too far from O1 to accept an H-bond from the peroxide so a water molecule may mediate the transfer of the proton from O1 to the His. This is consistent with crystal structures since the structure of Compound I shows a water molecule H-bonded with the O1 atom of Compound I

one way Nature converts a potent yet nonspecific oxidant into an oxidant with high selectivity. As outlined in the previous section, a good deal is known about the basic catalytic process for oxidizing peroxidases to Compound I and where the peroxide oxidizing equivalents are stored. Another intriguing question that has been more difficult to answer is how the protein stabilizes the potent oxidants in Compound I. In most peroxidases Compound I consists of two powerful oxidants: Fe^{4+} and the porphyrin radical. In CCP, however, the radical is stored on Trp191 (Fig. 3). Initially this was easily explained based on sequence information. Most peroxidases have a Phe at this position and also form porphyrin radicals. A Phe side chain is more difficult to oxidize than a Trp side chain so it was reasoned that oxidation of Trp191 in CCP is thermodynamically more favorable. This argument was strengthened by an experiment where Trp191 was converted to Phe and the resulting mutant formed a short-lived porphyrin radical. However, as more sequences and structures became available this simple view came into question. The first hint that something might not be quite right came from the structure and characterization of ascorbate peroxidase (APX), the second Class I peroxidase to be well characterized. APX has a Trp located in exactly the same place as Trp191 in CCP yet APX does not form a Trp radical but instead forms the more traditional porphyrin radical. The reason for this difference was attributed to differential electrostatic stabilization of the Trp cationic radical. In APX there is K^+ ion about 8Å away from the Trp. A positive charge so close to the Trp should make it much more difficult to oxidize the Trp to a cationic radical. This explanation gained merit when the APX K^+ site was engineered into CCP and the mutant CCP was found to form a dramatically less stable Trp191 radical. Moreover, the Trp-to-Phe mutant exhibits very little activity. Further engineering studies showed other residues near Trp191 also contribute to electrostatic stabilization of the Trp191 cationic radical. The electrostatic stabilization idea recently has gained further support by the characterization of a peroxidase from *Leishmania major* (LmP for *L. major* peroxidase). LmP has the same Trp as CCP and APX, has the APX K^+ binding site near the active site Trp, yet LmP also forms a stable Trp radical. The reason why LmP still forms a Trp radical yet has the APX K^+ site nearby is due to a unique Cys residue near the LmP Trp which provides additional electrostatic stabilization to the Trp cationic radical positive charge.

An important observation that has been derived from these studies is that in the various mutants of CCP that have a destabilized Trp191 radical, the initial reaction with H_2O_2 still forms a Trp191 but the radical rapidly migrates to other sites, usually Tyr residues. In wild-type CCP the Trp191 radical is stable for many minutes and is easy to trap, while in the mutants the stability is down in the seconds or less range. This is a large change but in terms of free energy difference, the change is not so great. For example, changing the stability of thc radical from 30 min to 10 ms is a factor 180,000 which translates to a free energy change of about 7 kcal/mole. Simple Coulomb's law, with the assumption that the internal dielectric constant of protein is low (in the range of 8), shows that this 7 kcal/mole can be entirely accounted for by a −1.0 charge being within about 4 Å of the Trp191 cationic radical positive charge. In other words, it does not take much in terms of local electrostatics to dramatically change the kinetic stability of cationic radicals. In the peroxidase active site the stabilization is achieved by not just a single close by negative charge, but a series of side chains and peptide backbones that each provide partial negative charges that add up to give large kinetic stability.

CCP-Cytc Complex

Unlike most other peroxidases that catalyze the oxidation of small organic molecules. CCP catalyzes the oxidation of cytc. Cytc operates as a soluble electron transfer protein that shuttles electrons between membrane bound components of the mitochondrial respiratory chain. Like cytc CCP is located between the inner and outer mitochondrial membranes and thus has ready access to cytc and, like cytc oxidase, CCP can serve as a cytc oxidant. Early on CCP and cytc served as ideal proteins for studying electron transfer since both proteins could be readily purified and CCP is a much simpler protein than the far more complex and membrane bound cytc oxidase. For a time CCP and cytc were the only electron transfer components where both crystal structures were known which further spurred interest in the CCP-cytc system. The CCP-cytc crystal structure is known (Fig. 4) and has been thoroughly tested using primarily site directed mutagenesis. Prior to the structural work it was known the CCP-cytc complex weakens as the ionic strength increases which indicates the importance of electrostatic interactions. It also was known that Lys and Arg residues surrounding the exposed heme edge of cytc are important for interactions with redox partners, so it was anticipated that CCP would provide Glu and Asp residues that ion pair with the Lys and Arg residues in cytc. As shown in Fig. 4 CCP and cytc have complementary charged surfaces, with CCP providing the negative charges and cytc the positive charges. It thus was expected that Lys/Arg residues in cytc would form specific contacts with Asp/Glu residues in CCP. It therefore was a surprise to find from the crystal structure of the CCP-cytc complex that there are no specific ionic interactions in the range of 3–4 Å. Instead ordered water molecules are situated between CCP and cytc to form water bridges between polar groups. This structure has required a modification on the way redox complexes and electrostatic interactions are thought about. There is no doubt that the electronegative surface of CCP interacts with the electropositive surface of cytc but the structure shows that

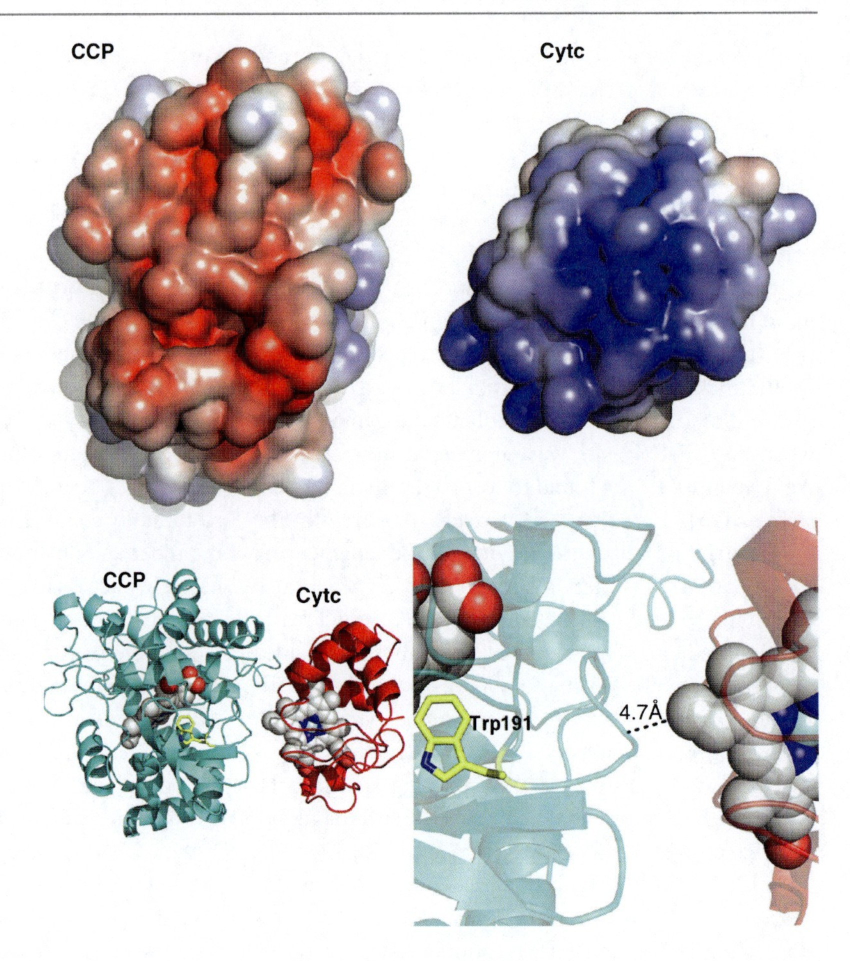

Heme Proteins, Heme Peroxidases, Fig. 4 The CCP-cytc complex. The top panels show the electrostatic (*red*=negative electrostatic potential and *blue*=positive electrostatic potential) surfaces of CCP and cytc that interact with one another. The crystal structure of the complex shows that the heme of cytc is only ≈4.7Å from CCP. As a result, there is only one "through space" jump of the electron from cytc to the CCP polypeptide directly connected to Trp191

specific ionic contacts are not involved. Here it is important to note that electrostatic interactions can be strong between two opposite charges that are not within 3–4 Å of direct contact as long as the dielectric milieu between charges is small. Thus one should think about the CCP-cytc interaction as two oppositely charged surfaces (see Fig. 4) that have an optimum orientation for maximum electrostatic stability but specific ionic interactions are not required. This also implies that the stability difference between a nonproductive and productive electron transfer complex is small and that the initial encounter between CCP and cytc need not be the productive complex. However, within the nonproductive encounter complex, cytc samples the surface of CCP and eventually ends up in the low energy productive complex observed in the crystal structure. This view of inter-protein electron transfer provides a balance to the requirement of high turnover rates and specificity. If the complex were too tight and too specific, then cytc dissociation would greatly limit turnover. To achieve high turnover cytc must bind, deliver its electron, and then leave. This would not be possible if the complex were tight with specific salt bridges at the interfaces.

Another complexity of the CCP-ctyc system is that two electrons are delivered to CCP Compound I, one to the Trp191 radical and the second to $Fe^{4+} = O$. The structure shows that the cytc heme is only 4.7Å (Fig. 4) from Ala193 which means there is only one "through space" jump from the cytc heme to CCP followed by a short "through bond" transfer to the Trp191 radical. The overall electron transfer mechanism is outlined in the following scheme.

1. Fe^{4+} $Trp^{\bullet+}$ + cyt*c* (Fe^{2+}) → Fe^{4+} Trp + cyt*c* (Fe^{3+})
2. Fe^{4+} Trp → Fe^{3+} $Trp^{\bullet+}$
3. Fe^{3+} $Trp^{\bullet+}$ + cyt*c* (Fe^{2+}) → Fe^{3+} Trp + cyt*c* (Fe^{3+})

In step 1 an electron from cytc reduces the Trp radical. In step two, there is an internal electron transfer from Trp191 to the Fe^{4+} iron to give Fe^{3+} $Trp^{\bullet+}$. Finally, the second cytc electron reduces the Trp radical thus returning CCP to the Fe^{3+} Trp resting state.

Since the Trp191 radical is the site of electron entry for both electron transfer steps, the same electron transfer complex is used for both steps.

Cross-References

- ▶ Heme Proteins, Cytochrome c Oxidase
- ▶ Heme Proteins, the Globins
- ▶ Peroxidases

References

Dunford HB (2010) Peroxidases and catalases: biochemistry, biophysics, biotechnology, and physiology. 2nd edn. Wiley, Hoboken

Erman JE, Vitello LB (2002) Yeast cytochrome c peroxidase: mechanistic studies via protein engineering. Biochem Biophys Acta 1597:193–200

Poulos TL (2010) Thirty years of heme peroxidase structural biology. Arch Biochem Biophys 500:3–12

Volkov AN, Nicholls P, Worrall JA (2011) The complex of cytochrome c and cytochrome c peroxidase: the end of the road? Biochim Biophys Acta 1807:1482–1503

Zamocky M, Furtmuller PG, Obinger C (2010) Arch Biochem Biophys 500:45–57

Heme Proteins, the Globins

Andrea Bellelli and Maurizio Brunori
Department of Biochemical Sciences, Sapienza University of Rome, Rome, Italy

Synonyms

Heme-containing oxygen carriers; Nitrogen oxide(s) scavengers; Respiratory proteins

Definitions

Oxygen carriers: proteins capable of reversible combination with oxygen. Nitrogen oxides: a family of highly reactive compounds that may possess unpaired electrons, produced by the cell metabolism. Nitrogen oxide scavenger: usually an oxidoreductase capable of converting these compounds to the biologically inactive ions nitrite or nitrate.

Introduction

Globin is the name given to the protein moiety of a class of hemoproteins that includes hemoglobins and myoglobins from different phyla. Human hemoglobin (Hb) and sperm whale myoglobin (Mb) are the best characterized representatives of the class (and possibly of all proteins), and for almost a century, the physiological function of these proteins, primarily related to oxygen (O_2) transport, has been assumed as typical of the entire class. However, over the last two decades, it emerged that: (1) globin-like proteins are present in the genomes of eubacteria, green algae, Archaea, protozoans, and plants; and (2) they carry out physiological functions over and above O_2 transport, including several reactions related to the detoxification of O_2 reactive species and nitrogen oxides. Other important families of proteins containing protoheme are vital to the cell physiology. The crucial electron transfer hemeproteins, forming the large family of the cytochromes, essential components of the respiratory chain, are reviewed by B. Ludwig (this volume). Moreover numerous peroxidases, primarily involved in the metabolism of hydrogen peroxide, have been extensively studied and are presented in this volume by T. Poulous.

The Globin Fold

All globins share a conserved *protein fold* essentially similar to that of sperm whale Mb originally unveiled by J. C. Kendrew and coworkers, and depicted in Fig. 1. Most globins are composed of a polypeptide chain of 130–160 amino acid residues, folded into 7 or 8 α-helices named A through H; vertebrate Mb and the β subunits of Hb have 8 helices, whereas the α subunits of Hb lack the D helix.

The prosthetic group of globin is the heme (or iron protoporphyrin IX), containing one iron atom which binds the external ligands (primarily O_2). Heme is bound in a crevice of the protein located in between helices E and F (Fig. 1). In the canonical globin fold, helices A, B/C, and D (where present) are on the same side as helix E (*the distal side*), whereas helices F, G, and H lie on the *proximal side*; the two bundles are connected via the EF turn. This overall 3D topology is generally referred to as the *3/3 helical structure*.

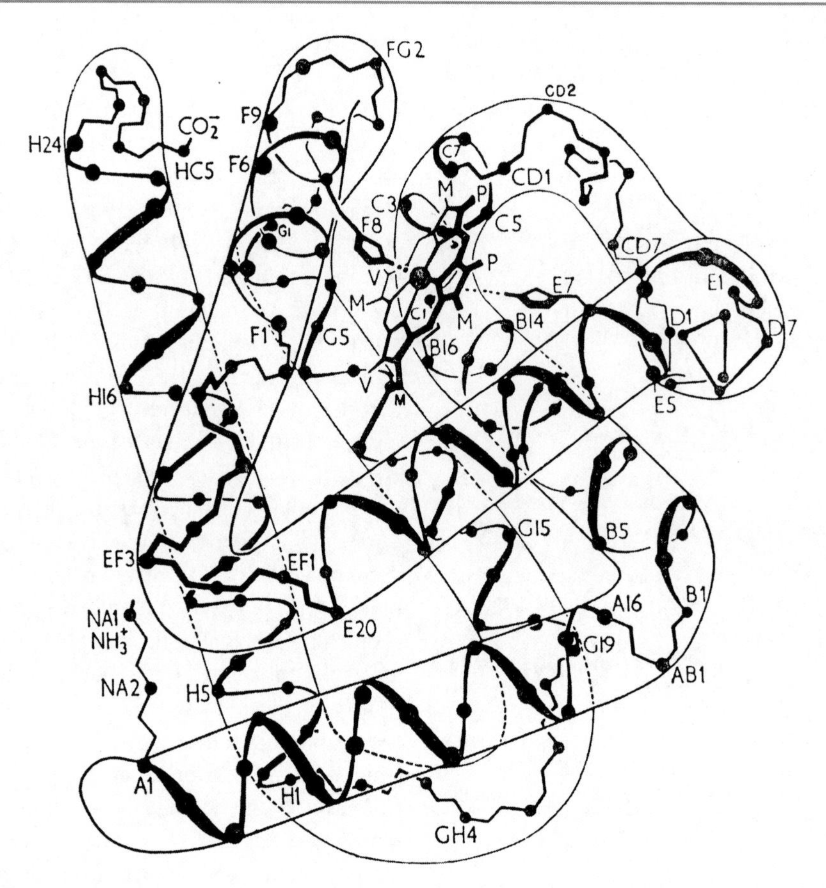

Heme Proteins, the Globins, Fig. 1 The *globin fold*. Schematic representation of the 3D structure of sperm whale Mb, a protein of 153 amino acids. Only the position of the Cα-atoms (*solid circles*) and the overall polypeptide, as a continuous line, are shown. The typical fold of vertebrate globins includes two to three unstructured residues at the amino terminus (NA), before helix A; helix B, which follows either immediately or with the interposition of a single residue, continues into helix C with a sharp turn where the α-carbonyl of residue C1 accepts a H-bond from the α-amino group of the residue at B13. A coil segment called the CD corner continues with helix D, and thereafter helix E and the EF corner. The F helix precedes the last two helices G and H; two to four unstructured residues follow the H helix at the C terminus. The heme is shown inserted in the crevice in between helices E and F; the ligand-binding pocket is on the *right*, toward His(E7)

In bacteria, shorter or "truncated" globins were discovered over the last 20 years. These lack completely the N terminal A helix and have the F helix reduced to a loop; thus, the helical bundles on the *distal side* of the heme are reduced to B/C and E while G and H are on the *proximal side*. This variant of the classical Mb fold has been called *2/2 helical structure* typical of truncated globins (Wittenberg et al. 2002) (Fig. 2).

Globins bind the heme iron via one axial ligand on the *proximal side* provided by a His residue which – in typical 3/3 globins – occupies the eight position of the F helix (F8) (see Figs. 1, 2). In the 2/2 truncated globins, the proximal His has the same topological position as in the 3/3 globins, even though the F helix is reduced to a loop.

In all globins, the heme is surrounded by the side chains of the amino acid residues coating the heme pocket and controlling heme reactivity, variable as it may be. In particular, residues on the *distal side* belonging to helices E and B may come close to the sixth coordination position of the metal and stabilize the external ligand, as in Mb and Hb where the *distal* His(E7) at the seventh position of helix E establishes a H-bond with bound O_2. In other globins, such as neuroglobin (Ngb) and cytoglobin, His(E7) directly coordinates the heme iron, thereby controlling O_2 binding.

In eukaryotes, the genes that code for globins contain two introns (three in the case of plant legHbs); the three exons roughly correspond to helices A-B, B-G,

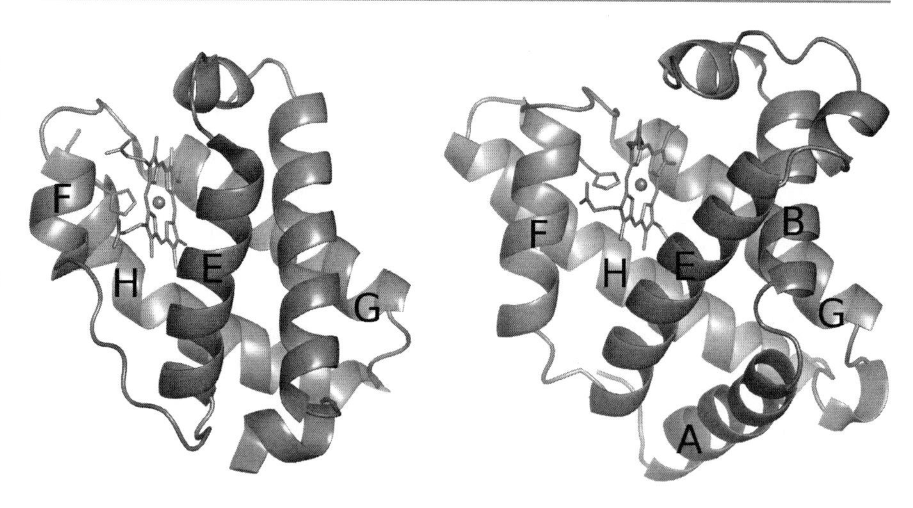

Heme Proteins, the Globins, Fig. 2 Comparison of 2/2 and 3/3 globin structures. *Left*: shows the 3D structure of *Paramecium caudatum* 2/2 Hb, with the heme and the proximal His residue highlighted (PDB: ldlw). Helices are labeled according to the conventional globin fold nomenclature (A through H). *Right*: the 3D structure of sperm whale myoglobin in the same orientation as *above* (PDB: 1a6n1)

and G-H respectively (Eaton 1980; Go 1981). It has been observed that the region coded by the central exon corresponds to the heme-binding domain. Selective proteolysis has been used to convert Mb into a heme-binding polypeptide which mimics the core of the second exon; this construct, called miniMb, binds the heme and is capable of reversible ligand binding (De Sanctis et al. 1986). The regions coded by the three exons have been considered sufficiently stable to correspond to spontaneously folding units ("foldons"), although the junctions occur within helices B and G. The possible role of exons in evolution is still a matter of debate.

Molecular Weights

Vertebrate O_2-binding globins, the best studied members of the class, occur in solution either as monomers (e.g., Mb and Ngb) or as tetramers. The former have molecular weights of approximately 17,000 Da, while the latter have MW of approximately 65,000 Da. In invertebrates, other types of Hbs characterized by MWs different from the canonical ones have been reported (Antonini and Brunori 1971).

Tetrameric Hbs of vertebrates are constituted by two couples of identical subunits, named α and β, assembled with a tetrahedral geometry such that each subunit makes contact with all other subunits (with the exception of the two β chains, see below). The four subunits (identified as α_1, α_2, β_1, and β_2) are stabilized by contacts along six interfaces ($\alpha_1\beta_1$ and symmetrical $\alpha_2\beta_2$; $\alpha_1\beta_2$ and symmetrical $\alpha_2\beta_1$; $\alpha_1\alpha_2$ and $\beta_1\beta_2$).

The surface area buried in the $\alpha_1\beta_1$ interface amounts to 1870 A^2 in deoxy Hb and 2050 Å in HbO_2. The $\alpha_1\beta_2$ interface is less extended and the contacts change upon binding of O_2 to deoxygenated Hb, the $\alpha_1\beta_2$ buried surface area corresponding to 1470 $Å^2$ in deoxy Hb and 1160 $Å^2$ in HbO_2. The relatively minor $\alpha_1\alpha_2$ interface is contributed by the N- and C- terminal residues and by the last residues of the H helix of both α chains. Finally, the $\beta_1\beta_2$ interface is described as "loosely packed" (Dickerson and Geis 1983).

Upon dilution of oxygenated Hb down to the μM range, the tetramer dissociates reversibly by cleavage along the $\alpha_1\beta_2$ interface, forming two $\alpha_1\beta_1$ dimers; thus, Hb is conveniently described as a symmetric dimer of $\alpha_1\beta_1$ dimers, that can eventually dissociate into α and β chains:

$$2(\alpha\beta)_2 \leftrightarrow 4\alpha\beta \leftrightarrow 4\alpha + \beta_4 \qquad (1)$$

Deoxygenated Hb is a considerably more stable tetramer, and dissociation of $\alpha_1\beta_1$ dimers only occurs at nM concentrations or below. This different stability is consistent with the different surface area buried by the inter-subunit interfaces in liganded and unliganded Hb.

Evolution of Globins

Evolutionary studies based on amino acid sequence alignments indicate that there are three major lineages of globins, namely: (1) the *3/3 globin family* that includes animal Hbs and Mbs, bacterial flavoHbs,

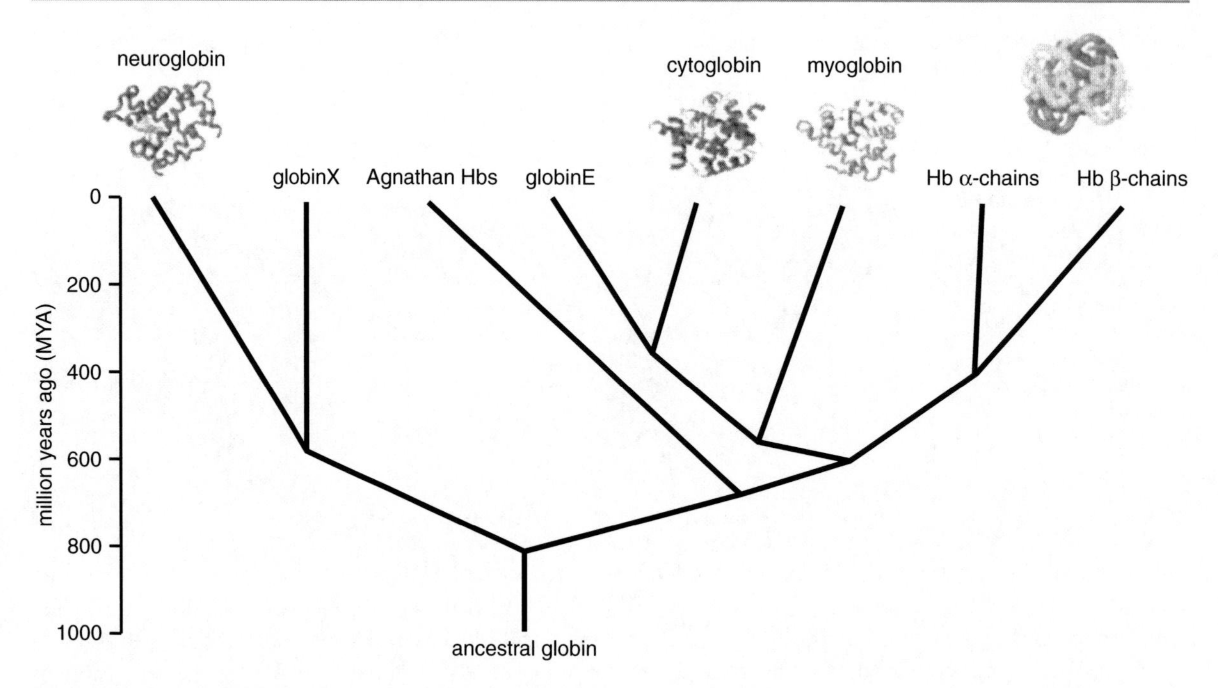

Heme Proteins, the Globins, Fig. 3 A simplified phylogenetic tree of vertebrate globins (From Burmester et al. 2004, modified)

and related proteins; (2) the *3/3 globin family* that includes globin-coupled sensors and protoglobins, found in prokaryotes and Archaea, but not in eukaryotes; and (3) the *2/2 globin family* found in bacteria, green algae, Archaea, protozoans, and plants (Vinogradov et al. 2007, 2008). As to the distribution of these three lineages, it has been observed that bacteria consistently possess all three, whereas archaeans lack representatives of the first lineage and eukaryotes lack representatives of the second. Thus, it has been suggested that the globin superfamily emerged in bacteria probably starting with a 3/3 ancestor. Since bacteria do not need O_2 transport and storage, evolutionary data suggest the primordial function of globins to be catalytic, likely related to the radical chemistry.

In spite of such a wide structural and functional divergence, the members of this protein superfamily have in common some crucial structural features: the heme-binding site in between helices E and F, the helical 3D structure with the 2/2 or 3/3 topologies, and the invariant position of the proximal His to coordinate the heme iron (Figs. 1 and 2).

Sequence alignments consistently indicate that globin-coupled sensors and protoglobins, the 3/3 globins that only occur in prokaryotes and Archaea, are distantly related to the 3/3 globin lineage that includes flavoHbs and other eukaryotic Hbs; indeed both lineages of 3/3 globins are more closely related to the 2/2 globins than with each other. Although decidedly insufficient, the available information is compatible with the hypothesis that an ancestral 3/3 single domain globin gene did appear very early in prokaryotes, and evolved to yield the three main globin classes, some being laterally transferred to Archaea, and much later to eukaryotes.

The evolutionary trees based on the distribution of globins and the comparative analysis of their sequence suggest that these proteins are of very remote origin (~ 3,500 mya), having probably arisen in prokaryotes before the appearance of photosynthetic organisms, thus in a quasi O_2-free environment. Several factors contribute to make uncertain the early stages of the evolution of globins: the frequent occurrence of lateral gene transfer in prokaryotes, the functional plasticity of these proteins, and the presumably strong selective pressure on their functions. By contrast, the recent evolution of globins in vertebrates is relatively clear and can be described by a consistent phylogenetic tree with fairly reliable datings (Fig. 3). The relatively recent discovery of neuroglobin (Burmester et al. 2004) moved backward in time the earliest separation from the presumed ancestral vertebrate globin.

Sources and Distribution of Globins

Prokaryotes. A survey of globin distribution and evolution (Vinogradov et al. 2007, 2008) highlights that approximately 65% of the known bacterial genomes do contain one or more putative globin gene(s). Prokaryotes are noteworthy because their genome includes representatives of all three major globin lineages; indeed they express 2/2 truncated Hbs with enzymatic functions, most notably NO metabolism. Prokaryotes also produce 3/3 flavoHbs, which are involved in the scavenging of various N-oxides and in the reduction of organic peroxides (essentially damaged membrane lipids). Finally, prokaryotes produce 3/3 globin-coupled sensors.

Archaea produce 2/2 truncated globins and globin-coupled sensors, while they seem to lack Hbs and flavoHbs. According to Vinogradov et al. (2007) only 25% of the screened archaean genomes contain one or more globin genes.

Globin genes have been found in over 90% of the eukaryotic genomes; however, these are conveniently described with reference to the classical groups: protozoans, fungi, plants, and animals.

Protozoans. Several single cell protozoans produce Hbs such as the flavoHb isolated form *Giardia intestinalis*, a human parasite. Globins are conspicuously absent in the genomes of *apicomplexans Entamoeba and Plasmodium*, and in that of *Trypanosoma*; this observation is noteworthy because all these protozoans are human (and animal) pathogenic parasites, capable of evading oxidative attack of their host(s).

Fungi produce two types of Hbs, the 2/2 truncated Hbs and the 3/3 flavoHbs; neither is well characterized. Putative genes coding for flavoHb have been indentified in *Aspergilli*.

Plants may produce 3/3 and 2/2 truncated Hbs. The first plant Hbs to be discovered, and the best characterized, are the so-called legHbs, produced by the plant cells in the root nodules of legumes. These specialized cells host in their Hb-rich cytoplasm symbiotic bacteroids capable of nitrogen fixation. Nitrogen fixation depends on a fully functional nitrogenase, which is poisoned by O_2. LegHbs allow the plant cell to meet the two contrasting requirements of keeping a low intracellular O_2 concentration while maintaining a high reserve of the gas for aerobic energy production. The role of legHbs is essential as demonstrated by the fact that suppression of their biosynthesis by RNA interference also suppresses nitrogen fixation. Somewhat confusingly, legHbs are often called symbiotic plant Hbs, in spite of the fact that they are produced by the plant cells and not by the symbiotic bacteroids.

Invertebrates. A variety of Hbs are found in invertebrates ranging from the monomeric Mb-like hemoprotein from *Chironomus thummi thummi* larvae to the giant erythrocruorins from *Lumbricus terrestris* worms. Polychaete worms may also produce a strictly related hemeprotein called chlorocruorin, whose pigment, the chloroheme, is a variant of protoporphyrin IX in which vinyl at position 3 is replaced by a formyl. The accepted physiological function of these proteins is O_2 transport (Antonini and Brunori 1971).

Vertebrates produce at least one (often several) tetrameric Hb contained in the red blood cells, and at least one monomeric Mb in the striated muscle cells. In addition, several tissues produce specialized intracellular, usually monomeric Hbs collectively called cytoglobins, such as the well-characterized neuroglobin (in neural cells).

Myoglobin (Mb)

Myoglobin (Mb) is a monomer found in the red muscles and the heart at an intracellular concentration of 0.1–0.3 mM (in heme), which binds O_2 reversibly ($p_{1/2}$ = 0.5–0.7 mmHg at 20°C, see below) via coordination to the ferrous heme iron. The O_2 adduct entails a partial electron transfer from the metal to the gas, to form a ferric iron-superoxide complex, whose dissociation (a rare event) leads to the liberation of superoxide and oxidation of the iron to the physiologically inactive ferric state (Antonini and Brunori 1971). Other Lewis bases coordinate the ferrous heme iron, the best characterized being CO, NO, isocyanides, and nitrosoaromatic compounds; of these, NO and perhaps CO play a physiological role.

Coordination of ligands has geometrical effects on the structure of the heme. The ionic radius of high-spin unliganded ferrous iron is too large to fit among the porphyrin's nitrogens; as a consequence, the heme of deoxy Mb is domed and the iron is out of the porphyrin plane. Upon O_2 binding, the iron becomes low spin and the decrease of the ionic radius allows the metal to fit among the nitrogen atoms; the heme flattens and pulls the proximal His(F8) residue, the only amino acid covalently bound to the metal.

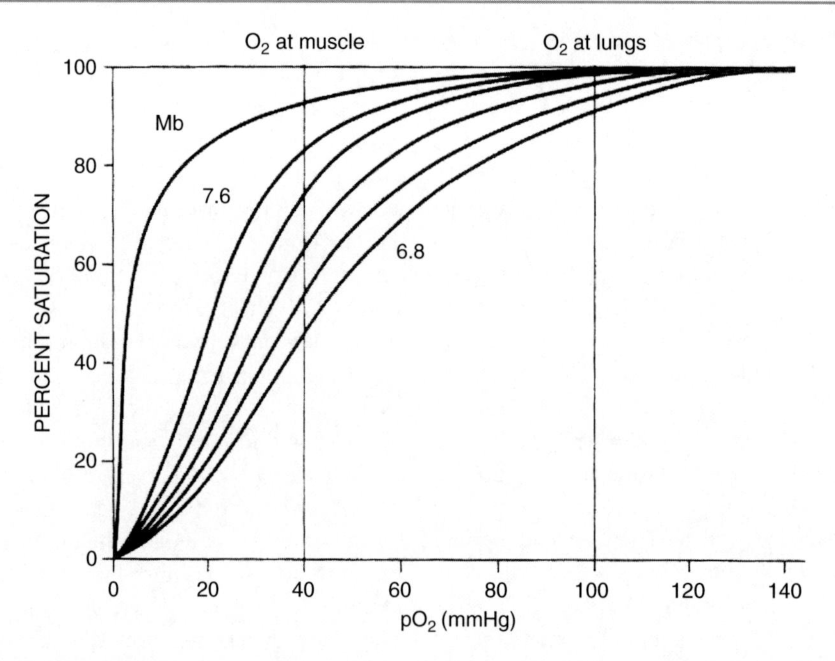

Heme Proteins, the Globins, Fig. 4 Oxygen-binding curves for Mb and Hb. The binding curve for Mb, depicted to the *left*, is a hyperbola with high O_2 affinity ($p_{½}$ ~ 0.5–0.7 mmHg at 20°C) (Antonini and Brunori 1971). On the *right*, human HbA-binding curves under various conditions. From *left* to *right* at five different pH values: 7.6; 7.4; 7.2; 7.0; and 6.8. The dependence of O_2 affinity with pH is called the Bohr effect. The O_2 equilibrium curve of fresh blood from healthy individuals (male and female) yields: $p_{½} = 27.4$ mmHg and $n_{max} = 3$ under physiological conditions (i.e., pH = 7.4; $pCO_2 = 40$ mmHg; [BPG] = 5 mM; HbCO = 1% T = 37°C) (Data reported by Imai 1982)

Reversible O_2 coordination to the heme iron in Mb (and also in monomeric Hbs) obeys the simple mass law:

$$Mb + O_2 \leftrightarrow MbO_2; \quad K = [MbO_2]/[Mb]pO_2 \quad (2)$$

The fraction of oxygenated Mb (usually indicated by Y) is a hyperbolic function of pO_2 (Fig. 4); the pO_2 required to achieve 50% oxygenation (p_{50} or $p_{½}$) equals the reciprocal of the equilibrium affinity constant $p_{50} = 1/K$. Indeed from eq. 2 we derive:

$$Y = KpO_2/(1 + KpO_2) \quad (3)$$

Classically Mb's physiological role is reported to be O_2 storage for tissue respiration (Brunori 2010). This is surely important in diving mammals whose muscles contain Mb at a concentration ~ 10-fold greater than in humans; however, in man and the other mammals, it was calculated that O_2 bound to intracellular Mb would be sufficient to sustain aerobic metabolism for approximately a couple of seconds.

Another function of Mb is to facilitate O_2 diffusion from capillaries to mitochondria, given its high concentration and free translational diffusion in the cytoplasm. Mb's role in facilitating O_2 diffusion and in keeping O_2 concentration essentially constant throughout the cell is supported by in vitro and in vivo data, including careful experiments carried out on knockout mice that do not express Mb. These animals have a normal life but the density of capillaries in the heart is increased significantly to compensate for the lack of Mb, an additional evidence for its role in O_2 diffusion (see Brunori 2001).

One overlooked role for Mb is scavenging NO, thereby protecting the energy-producing machinery by preventing inhibition of the respiratory enzyme (Brunori 2001). The mechanism of quenching is based on the efficient reaction of oxygenated MbO_2 with NO, yielding nitrate and ferric Mb^{+3} (*see below*). In addition, it has been demonstrated that under anaerobic conditions, deoxy Mb in the heart catalyzes the production of NO from NO_2^- (that is known to be at ~ 10 μM in the tissues) (Gladwin et al. 2003). The cardioprotective effect that follows is so evident that clinical trials to assess the therapeutic role of nitrite administration on the heart have been undertaken. Thus, Mb effectively controls free NO concentration in the heart.

The Architecture of Hemoglobin

Hemoglobin (Hb), contained in the red blood cells, reversibly binds O_2 at the ferrous heme. Hb's concentration in normal human erythrocytes is extremely high (340 g/dm^3), corresponding to 5.2 mM (for a molecular mass of 65,000 g/mol). Its primary role is the transport of O_2 from the lungs (or the gills) to the peripheral tissues, and the transport of CO_2 from the tissues to the lungs. Moreover, Hb significantly contributes to the buffer capacity of blood (Bunn and Forget 1986).

The four polypeptide chains of tetrameric Hb are identical in pairs, its "formula" being $\alpha_2\beta_2$. Each α chain is a polypeptide of 141 amino acid residues, and each β chain has 146 amino acid residues. The α and β subunits have different amino acids sequences but both acquire the characteristic globin fold (see Fig. 1).

Besides $\alpha_2\beta_2$ (also called adult Hb or HbA or HbA_0), human erythrocytes contain other Hb components. In HbA_2 and HbF (or fetal Hb), normal α chains are combined with subunits different from the adult β chains, and called respectively δ chains (in HbA_2 or $\alpha_2\delta_2$) and γ chains (in HbF or $\alpha_2\gamma_2$). Humans have also embryonic Hbs: Hb Gower-1 of composition $\zeta_2\varepsilon_2$, Hb Gower-2 or $\alpha_2\varepsilon_2$, and Hb Portland or $\zeta_2\gamma_2$.

Posttranslational modifications of Hb are frequent due to reaction with small molecules, forming covalent adducts. Of significance because of their clinical implications in diabetes are the glycated Hbs which are modified by reaction with glucose (or glucose derivatives) at the N-terminus of the β chains; HbA_{1c} is the most abundant among the minor Hb components, reaching as much as 4% of the total or more in diabetics.

The Reaction with Heme Ligands

Ferrous Hb with no ligand bound to the heme iron is called deoxy or unligated Hb. It may be prepared by removing (chemically or physically) the heme ligands such as O_2, CO, NO, and others (Antonini and Brunori 1971). OxyHb (or HbO_2) is subject to slow spontaneous oxidation of the heme iron producing ferric or metHb, which cannot combine with O_2 but binds water and several anionic ligands (OH^-, F^-, OCN^-, SCN^-, N_3^-, or CN^-) (see also above under myoglobin).

The binding of O_2 or other ligands of the ferrous heme is associated to many changes in the physicochemical properties of Hb, including spectral and magnetochemical parameters. The heme spectral changes in the visible region (400 to 700 nm) are very large and thus have been used to follow ligand binding.

Similarly to Mb (see above), the reversible combination of Hb with O_2 is illustrated by the ligand-binding curve, wherein the fraction of oxygenated hemes (Y) is plotted against the partial pressure of O_2 (Fig. 4). The sigmoidal shape of the saturation curve implies that O_2 binding is "cooperative" and reflects heme-heme interactions within tetrameric Hb (Antonini and Brunori 1971; Imai 1982). A quantitative rigorous treatment of the cooperative O_2 binding to Hb requires some kind of assumption (see below, allostery); however, a simplified empirical treatment is possible, representing the overall reaction by the following scheme:

$$Hb + nO_2 \leftrightarrow Hb(O_2)_n \quad (4)$$

The fraction of hemes oxygenated is given by:

$$Y = \frac{[Hb(O_2)_n]}{[Hb] + [Hb(O_2)_n]} \quad (5)$$

Thereby the association equilibrium constant for Hb can be written as follows:

$$K = \frac{[Hb(O_2)_n]}{[Hb]p^n} \quad (6)$$

where O_2 activity is expressed in terms of its partial pressure p. It follows that:

$$\frac{Y}{1-Y} = Kp^n \quad \text{or} \quad \log\frac{Y}{1-Y} = \log K + n\log(p) \quad (7)$$

According to the logarithmic Hill plot (Eq. 7), the binding curve is defined numerically by two parameters: $p_{1/2}$ which is a measure of the *overall* affinity of Hb

Heme Proteins, the Globins, Table 1 Functional properties of human hemoglobin in 0.05 M Tris, pH 7,4; $[Cl^-] = 0.10$ M, and 20°C (From Imai 1982, modified)

Oxygen equilibria in the presence and absence of BPG						
Conditions	$p_{1/2}$ (torr)	n	K_T (μM^{-1})	K_R (μM^{-1})	L_0	$\Delta G°$ (kJ/mol)
Stripped	5.3	2.8	0.018	3.9	7.3×10^5	13.0
+2 mM BPG	14.0	3.1	0.008	3.0	3.0×10^6	16.3

for O_2, and n the slope of the Hill plot which is related to the shape of the ligand-binding curve and therefore is a measure of cooperativity. A hyperbolic binding curve, which is characteristic of Mb (Fig. 4), yields a value of $n = 1$. In the case of normal human Hb, the slope of the sigmoidal O_2-binding curve changes in going from very low ($Y < 0.03$) to very high ($Y > 0.97$) saturations; the data between 10 and 90% saturation can be fitted well with $p_{1/2} = 10$ mmHg at 20°c (Fig. 4) and the Hill parameter $n = 3$. This value implies strong heme-heme interactions since it is quite close to the limiting value corresponding to the total number of hemes per hydrodynamic unit (clearly 4 for tetrameric Hb).

Regulation of O_2 Affinity

The cooperative O_2-binding curve is a general property of all tetrameric Hbs, with some variations in the values of the overall affinity ($p_{1/2}$) and cooperativity (since n may vary from ~ 3 to almost 1 or even less in different species). Two representative conditions are illustrated in Table 1. Moreover, cooperativity is not unique to O_2 but it has been demonstrated for all ligands of ferrous Hb, and most carefully studied for CO and NO.

Cooperativity is direct evidence for "homotropic" interactions, i.e., functional coupling between the active sites binding the same type of ligand, the four hemes. Moreover, ligand affinity and cooperativity are regulated by the concentration of several small molecules which bind to sites on the protein physically removed from the hemes; this type of phenomenon, which has been called "heterotropic," has very important physiological implications in controlling the amount of O_2 delivered to tissues (Antonini and Brunori 1971; Dickerson and Geis 1983).

The classical "heterotropic" ligands are protons and carbon dioxide. When the concentration of CO_2 is increased, the overall O_2 affinity of human blood is decreased, the binding curve being shifted to the right (with an increase in $p_{1/2}$) (Fig. 4). This phenomenon, which was correlated to the acidification of the medium induced by hydration of CO_2, is called the Bohr effect from the name of the Danish physiologist Christian Bohr (father of the famous physicist Niels Bohr).

The effect of pH and CO_2 concentration on the ligand-binding properties of Hb has been investigated in great detail since its discovery in 1904. It implies that binding to the protein of these heterotropic ligands is different for oxy and deoxyHb: protons are released in the medium upon O_2 binding because some ionizable groups of the protein decrease their pKs due to a conformational change coupled to ligand binding at the heme. The identification of the groups involved has been a matter of long scientific debates, but by-and-large the interpretation proposed by M.F. Perutz (1970, 1998) based on a comparison of the 3D structure of oxy- and deoxyHb is generally accepted.

Other effective heterotropic ligands (such as chloride and 2,3 glycerate-bis-phosphate BPG) have been discovered and characterized. The effect of BPG is particularly important in physiology, the molecule being a crucial regulator of O_2 delivery (Table 1). This polyphosphate binds in a cavity of the deoxyHb tetramer in between the two β chains, via specific interactions with several positively changed amino acid side chains. Upon O_2 (or CO) binding to the heme, the conformational change of the protein is associated to expulsion of the BPG from the binding pocket.

In deionized solution, the Bohr protons that are released upon O_2 uptake originate largely from HisHC3(146)β, a residue that in the T structure forms a H-bond with AspFG1(94)β. In the presence of Cl^- and/or BPG, other residues (e.g., ValNA1(1)α and LysEF6(82)β) contribute additional Bohr protons.

Cooperativity and Allostery

Cooperativity implies that binding of a ligand at one heme affects binding of a second ligand at another heme *via* a conformational change; likewise, binding of O_2 at the heme is affected by the association of other ligands (e.g., H^+ or BPG) at different non-heme sites on the protein. These features promoted Hb to the rank of the prototype for an allosteric protein. O_2 binding data on human Hb were analyzed in the famous theoretical paper published by Monod et al. (1965). The expression "allosteric inhibition" however had been coined by J. Monod and F. Jacob in 1961, in commenting the data obtained by J-P Changeux on L-threonine deaminase. Changeux has published the story of the discovery of the nonoverlapping mechanism of control linked to binding of a small molecule at a site removed from the active site of enzymes (Changeux 2011).

The sigmoidal O_2-binding curve and the Bohr effect, which had been quantitatively characterized and extensively analyzed, were therefore the best examples of allosteric behavior; moreover, the 3D structure of Hb solved by M. F. Perutz and coworkers showed the hemes to be separated by large distances (approximately 30 Å) and the structure to change upon ligand binding, with tertiary and quaternary conformational changes (Perutz 1970; Perutz et al. 1998). In 1951, Wyman and Allen had already proposed that the Bohr effect implied a ligand-linked conformational change of the protein; Perutz's direct structural data proved this hypothesis.

The basic idea of allostery is that the intrinsic O_2 affinity at the heme is controlled by the overall 3D structure of the protein, rather than by the number of ligand molecules already bound to the tetramer (see Cui and Karplus 2008 for recent review). This model assumes that there are only two different quaternary structures (Fig. 5): one with fewer and weaker interactions among the subunits (fully ligated Hb or relaxed: R), and the other with more numerous and stronger bonds between the subunits (deoxy Hb, or tense: T) (Monod et al. 1965; Eaton et al. 1999; Cui and Karplus 2008). The overstabilization of the T state has been attributed by Perutz (1970, 1987) to eight salt bridges (either in between or within the subunits) which are broken upon O_2 binding at the heme. In the R state, Hb binds O_2 with high affinity, while in the T state, the O_2 affinity is reduced and the protein binds more strongly to heterotropic ligands (H^+, Cl^-, BPG, CO_2). Oxygenation of the subunits within each quaternary state is associated with smaller tertiary structural changes. Therefore, although binding to the four hemes in each of the two quaternary states is noncooperative, the shift of the T ↔ R equilibrium gives rise to a sigmoid O_2-binding curve. The MWC model allows one to describe the ligand-binding curve of Hb with only three independent parameters, i.e., L_0, the constant that governs the equilibrium between the two quaternary conformations of Hb in the absence of heme ligands, and the intrinsic affinities of the R and T state (K_R and K_T respectively).

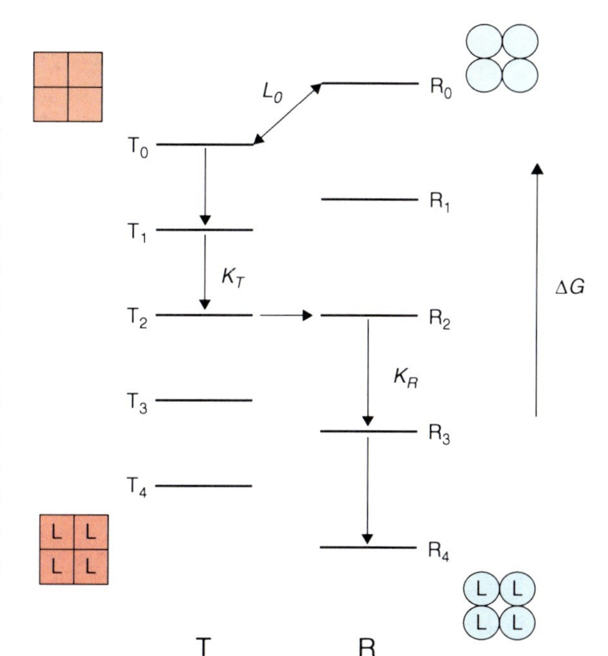

Heme Proteins, the Globins, Fig. 5 The two-state MWC allosteric model. Energy level diagram indicating the 10 states (T_0 to T_4 and R_0 to R_4) and the fundamental equilibrium constants of the MWC model, namely, K_T and K_R, the O_2 dissociation constants of the two allosteric states T (tense) and R (relaxed), and L_0, the population ratio of the two states in the fully deoxygenated tetrameric Hb. Drawings depict the two different quaternary states; the assumption implicit is the fully concerted quaternary transition which (in the specific case of a symmetric binding curve) occurs at the level T_2–R_2 (as shown) (switch-over point). Typical values of the parameters for human Hb at neutral pH and 20°C are approximately: $L_0 = [T_0]/[R_0] = 10^5$, and $c = K_T/K_R = 0.01$

The effect of heterotropic ligands (H^+, Cl^-, BPG, CO_2) can also be understood in terms of the MWC allosteric model given that addition, e.g., of BPG (which binds to deoxyHb in a 1:1 ratio) shifts the equilibrium in favor of the T structure; the same effect is obtained by lowering pH from 9 to 6. Other heterotropic ligands also lower the O_2 affinity by forming additional bonds that specifically stabilize the T structure (Cui and Karplus 2008).

Perutz's Stereochemical Mechanism

The conformational changes responsible for cooperative O_2 binding in Hb involve a large quaternary structural transition, with a rearrangement of the contacts between the two αβ dimers, as well as motions within each monomer which, though more subtle, are crucial. Upon ligand binding the two αβ dimers slide by 1 Å and rotate by 15° with respect to each other along the $\alpha_1\beta_2$ interface, which is called the "cooperative" interface (Fig. 6). Analysis of the molecular contacts reveals a machinery devised to perform two fundamental tasks: (1) the amplification of a signal coming from the heme iron, and (2) the all-or-none switch from the low to the high affinity conformation, and vice versa.

The main leverage is provided by the rigid F helix which contributes both to the heme pocket and to the $\alpha_1\beta_2$ interface (through the FG corner). As already described for Mb, indeoxy Hb the bond between the proximal His(F8) and the metal, the large radius of the high-spin ferrous iron, and its pentacoordinate state are associated to a displacement of the metal out of the heme plane by approximately 0.5 Å and a slight doming of the porphyrin skeleton (Fig. 7). Upon binding of O_2, six-coordination of the iron and reduction of its spin state and ionic radius are associated to collapse into the heme plane and pulling on His(F8); moreover (1) the porphyrin tends to flatten and (2) the heme plane tilts slightly. Though the motion of the metal is small, the movement of the proximal His forces a larger motion (1 Å or more) of the F helix and the FG corner in both the α and β chains (Fig. 7) (Perutz et al. 1987). An important additional contribution to the movement of the FG corner comes from the tilting of the liganded heme, which causes a displacement of the invariant Val(FG5) and perturbs its H-bond with the equally invariant penultimate Tyr(HC2).

Since these tertiary structural changes are mostly confined to the E and F helices and the FG corner, the relative positions of the C helix and FG corner of each subunit change, imposing a strain on the $\alpha_1\beta_2$ interface that triggers the quaternary allosteric transition. This "cooperative" interface comprises the two contact regions between αC-βFG and αFG-βC, the side chain interactions of the latter being maintained during the allosteric transition, whereas those in the αC-βFG region experiencing a gross change in the packing of its side chains: (1) His(FG4)97β logs in between Pro (CD2)44α and Thr(C6)41α in the T state but moves in between Thr(C6)41α and Thr(C3)38α in the R state; and (2) H-bonding at the beginning of the G helix involves Tyr(C7)42α to Asp8G1)99β in the T state, and Asp(G1)94α to Asn(G4)102β in the R state (Perutz 1970; Perutz et al. 1987).

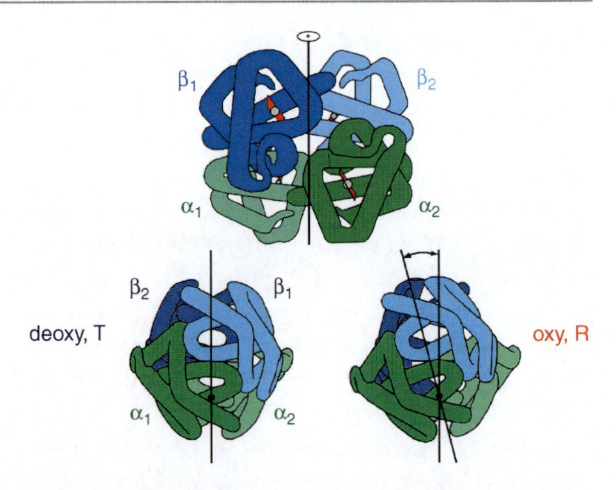

Heme Proteins, the Globins, Fig. 6 The quaternary conformational change of Hb. At the *top*, schematic representation of the Hb tetramer, showing the relative topology of the α and β chains ($\alpha_1\alpha_2\beta_1\beta_2$); below the two views looking down the pseudo twofold axis of symmetry. The conformational change consists of a rotation by ~15° of the symmetrically related αβ dimers relative to each other, and a translation of ~0.1 nm along the rotation axis (Modified after Eaton et al. 1999)

An important feature of Perutz's stereochemical model is that it recognizes both tertiary effectors, i.e., molecules or ions whose binding is O_2 linked but not strictly coupled to the allosteric transition, and quaternary effectors, whose binding biases the allosteric equilibrium. An example of a tertiary effector is provided by H^+ which is released in the course of the tertiary relaxation, and thus follows oxygenation rather than the allosteric transition (Antonini and Brunori 1971; Perutz et al. 1987); the opposite holds for BPG, which is released upon the T to R switch. The effect of

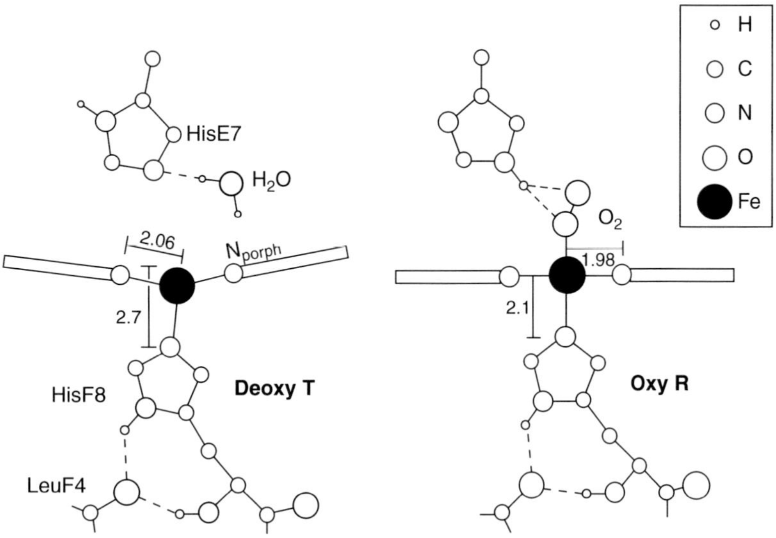

Heme Proteins, the Globins, Fig. 7 Perutz's stereochemical mechanism. The bond between His(F8) and the metal is a major element in the transfer of information from the ligand bound at the heme iron to the protein, as demonstrated by the shortening of the bond length (from 2.7 to 2.1 Å) in going from T-deoxy to R-oxy Hb. Loss of this coordination bond upon mutation of His (F8) → Gly of both chains leads to modifications in structure and ligand affinity consistent with Perutz's stereochemical mechanism

Cl^- is more complex since this anion binds to several O_2-linked sites, involved either in the tertiary or in the quaternary relaxation. It is essential to realize, however, that a tertiary effector will also bias the allosteric transition, and vice versa a quaternary effector imposes additional constraints on the tertiary structure of subunits.

Reactions Involving NO

Globins are capable of several reactions that involve NO and other nitrogen oxides as substrates or products, reactions which were investigated in eukaryotes and prokaryotes, where they were shown to play an essential role (Poole 2008; Stamler et al. 2008; Gladwin et al. 2003; Wittenberg et al. 2002).

NO is unusual in so far as it binds reversibly to both the ferrous and the ferric states of the heme iron; in the latter case, however, NO may act as a reductant and be released as the nitrosonium cation:

$$\mathrm{HbFe(II) + NO \leftrightarrow HbFe(II)NO} \quad (8)$$

$$\mathrm{HbFe(III) + NO \leftrightarrow HbFe(III)NO \leftrightarrow HbFe(II) + NO^+} \quad (9)$$

The complex of the ferrous heme with NO (8) is extremely stable and in the case of human Hb a dissociation constant of approximately 20–40 pM has been estimated, the extremely high affinity being largely assigned to a very low dissociation rate constant. HbFe(II)NO may also react with O_2 yielding ferric Hb and nitrate, or be oxidized yielding ferric Hb and the nitroxyl anion (NO^-). There is evidence that in mammalian Hbs, NO may be transferred from the ferrous heme iron to a conserved Cys residue at position 93 of the β subunits:

$$\mathrm{SH - CysHbFe(II)NO \rightarrow SNO - CysHbFe(II) + H_3O^+ + e^-} \quad (10)$$

This reaction requires oxidation of NO to NO^+ but the physiological oxidant is unuclear; this reversible reaction would effectively buffer the blood concentration of this important vasoactive messenger (Stamler et al. 2008).

Under physiological conditions, dissociation of NO from the S-nitroso Cys (S-NO) occurs via transfer to other thiols (e.g., glutathione) rather than by the inverse of reaction (10), which in addition has been questioned (Gladwin et al. 2003).

Oxygenated hemoproteins catalyze very efficiently the scavenging of NO according to:

$$\mathrm{HbFe(II)O_2 + NO \rightarrow HbFe(III)ONO_2^- \rightarrow HbFe(III) + NO_3^-} \quad (11)$$

This reaction has been demonstrated to occur both in vitro and in vivo. In the cell, it is coupled to the re-reduction of the ferric heme to its ferrous state by several reducing cellular systems. Reaction (11) is fast and irreversible, the second order rate constant ($3 \div 5 \times 10^7\ M^{-1}\ s^{-1}$) approaching the diffusion limit. The short lived intermediate, an iron-peroxynitrite complex, has been detected at high pH.

Oxidation of NO is an important physiological function of all cytoglobins, including Mb and Ngb. Moreover, this is the main function attributed to bacterial 2/2 truncated Hbs; e.g., it has been demonstrated that the 2/2 HbN protects the pathogen *M. bovis* from NO produced by the granulocytes of the infected host. It is relevant that *Mycobacterium tuberculosis*, the closely related bacterium which causes tuberculosis in humans, has a very similar truncated Hb.

Under anaerobic conditions, Hb and Mb can produce NO from nitrite according to:

$$\mathrm{HbFe(II)} + \mathrm{NO_2}^- + \mathrm{H_2O} \leftrightarrow \mathrm{HbFe(III)} + \mathrm{NO} + 2\mathrm{OH}^- \quad (12)$$

This reaction is supposed to be an important protective mechanism in the heart since it produces NO for vasodilation under anaerobiosis; the inhibition of cytochrome c oxidase by NO which follows is a mechanism to avoid O_2 radical burst upon reoxygenation of the tissue (Gladwin et al. 2003).

In bacteria and fungi, Hbs and flavo Hbs have been suggested to play a role in the process of denitrification (reduction of nitrite to nitrogen). Notably, it has been demonstrated that in *Alcaligenes eutrophus*, flavoHb is required for the production of nitrous oxide (N_2O); however, the details of the putative reaction mechanism have not been characterized.

Other Functions

All globins can scavenge hydrogen peroxide, in a peroxidase-like reaction that yields an oxo-ferryl derivative of the heme. This reaction, however, is hardly considered physiologically relevant since much more efficient peroxidases are present in the cell; moreover, reaction of H_2O_2 with the ferrous heme is often followed by oxidative damage of the porphyrin and may lead to irreversible inactivation. On the other hand, bacterial flavoHbs may be efficient peroxidases, capable of detoxifying hydrogen peroxide and/or organic peroxides, as demonstrated in the cases of *Escherichia coli* and *Thermobifida fusca* flavoHb, the reaction being as follows:

$$\mathrm{R-O-OH} + \mathrm{NAD(P)H} + \mathrm{H}^+ \rightarrow \mathrm{R-OH} + \mathrm{H_2O} + \mathrm{NAD(P)}^+. \quad (13)$$

A peculiar property of at least some Hbs in the ferric state is the high affinity for hydrogen sulfide. It has been suggested that this property may be physiologically relevant in view of the production of this substance during the metabolism of cysteine.

Protoglobins are 3/3 globins found in Prokaryotes and Archaea, whose functions are still a conundrum. It has been suggested that protoglobins are the evolutionary precursors of globin-coupled sensors and perhaps of all Hbs. They are assumed to be coupled in vivo to other proteins and to function as sensors for the concentration of diatomic gases and other compounds capable of reversible coordination to the heme iron.

At least two protoglobins have been characterized: those from the aerobic hyperthermophile *Aeropyrum pernix*, and from the anaerobic methanogen *Methanosarcina acetivorans*, both of which bind the classical diatomic gases. The structural and functional properties of the protoglobin from *M. acetivorans* are peculiar in so far as it is a homodimer characterized by anticooperative ligand binding (i.e., the first ligand molecule bound decreases the affinity of the second site); moreover, because of the stereochemistry of the heme pocket and the distortion of the protoporphyrin ring, it binds O_2 with an affinity higher than CO.

Globin-coupled sensors are two-domain proteins in which one domain has the 3/3 globin structure and contains a heme group (the sensor), whereas the other domain has a signal-transduction activity (Gilles-Gonzalez and Gonzalez 2005). Typical examples are globin-coupled kinases and globin-coupled nucleotide cyclases. The physiological role of the globin moiety is reversible combination with ligands, most often diatomic gases similar to those discussed above. The best characterized function of globin-coupled sensors is aerotactic: these proteins are used by at least some bacteria to recognize and follow O_2 gradients. An aerophylic response mediated by a globin sensor has been identified in *Bacillus subtilis* and an aerophobic one in the Archeon *Halobacterium salinarium*.

Cross-References

- ▶ Cytochromes, a class of heme-containing respiratory proteins not directly related to globins, are explained by B. Ludwig (this volume)
- ▶ Heme-Containing Peroxidases, the enzymes involved in the metabolism of peroxides, are explained by T. Poulos (this volume)

References

Antonini E, Brunori M (1971) Hemoglobin and myoglobin in their reactions with ligands. North Holland, Amsterdam

Brunori M (2001) Nitric oxide moves myoglobin centre stage. Trends Biochem Sci 26:209–210

Brunori M (2010) Myoglobin strikes back. Protein Sci 19:195–201

Bunn HF, Forget B (1986) Hemoglobin: molecular, genetic, and clinical aspects. WB Saunders, Philadelphia

Burmester T, Haberkamp M, Mitz SA, Roesner A, Schmidt M, Ebner B, Gerlach F, Fuchs C, Hankeln T (2004) Neuroglobin and cytoglobin: genes, proteins and evolution. IUBMB Life 56:703–707

Changeux J-P (2011) 50th anniversary of the word "allosteric". Protein Sci 20:1119–1124

Cui Q, Karplus M (2008) Allostery and cooperativity revisited. Protein Sci 17:1295–1307

De Sanctis G, Falcioni G, Giardina B, Ascoli F, Brunori M (1986) Mini-myoglobin: preparation and reaction with oxygen and carbon monoxide. J Mol Biol 188:73–76

Dickerson RE, Geis I (1983) Hemoglobin: structure, function, evolution, and pathology. The Benjamin/Cummings, Menlo Park

Eaton WA (1980) The relationship between coding sequences and function in haemoglobin. Nature 284:183–185

Eaton WA, Henry ER, Hofrichter J, Mozzarelli A (1999) Is cooperative binding by hemoglobin really understood? Nat Struct Biol 6:351–358

Gilles-Gonzalez MA, Gonzalez G (2005) Heme-based sensors: defining characteristics, recent developments, and regulatory hypotheses. J Inorg Biochem 99:1–22

Gladwin MT, Lancaster JR, Freeman BA, Schechter AN (2003) Nitric oxide's reactions with hemoglobin: a view through the SNO-storm. Nat Med 9:496–500

Go M (1981) Correlation of DNA exonic regions with protein structural units in haemoglobin. Nature 291:90–92

Imai K (1982) Allosteric effects in hemoglobin. Cambridge University Press, Cambridge, UK

Monod J, Wyman J, Changeux J-P (1965) On the nature of allosteric transitions: a plausible model. J Mol Biol 12:88–118

Perutz MF (1970) Stereochemistry of cooperative effects in haemoglobin. Nature 228:726–734

Perutz MF, Fermi G, Luisi B, Shaanan B, Liddington RC (1987) Stereochemistry of cooperative mechanisms in hemoglobin. Acc Chem Res 20:309–321

Perutz MF, Wilkinson AJ, Paoli M, Dodson GG (1998) The stereochemical mechanism of the cooperative effects in hemoglobin revisited. Annu Rev Biophys Biomol Struct 27:1–34

Poole RK (2008) Globins and other nitric oxide-reactive proteins. Methods Enzymol Part A 436:XXIII–XXIV

Stamler JS, Singel DJ, Piantadosi CA (2008) SNO-hemoglobin and hypoxic vasodilation. Nat Med 14:1008–1009

Vinogradov SN, Hoogewijs D, Bailly X, Mizuguchi K, Dewilde S, Moens L, Vanfleteren JR (2007) A model of globin evolution. Gene 398:132–142

Vinogradov SN, Moens L (2008) Diversity of globin function: enzymatic, transport, storage and sensing. J Biol Chem 283:8773–8777

Wittenberg JB, Bolognesi M, Wittenberg BA, Guertin M (2002) Truncated hemoglobins: a new family of hemoglobins widely distributed in bacteria, unicellular eukaryotes, and plants. J Biol Chem 277:871–874

Wyman J, Allen DW (1951) The problem of the heme interactions in hemoglobin and the basis of the Bohr effect. J Polymer Sci VII:499–518

Heme-Containing Oxygen Carriers

▶ Heme Proteins, the Globins

Hemochromatosis

▶ Iron Homeostasis in Health and Disease

Hemocuprein

▶ Zinc in Superoxide Dismutase

Hepatolenticular Degeneration

▶ Zinc and Treatment of Wilson's Disease

Heptaplatin

▶ Platinum Anticancer Drugs

hERG (KCNH2) Potassium Channel, Function, Structure and Implications for Health and Disease

Henry J. Duff
Libin Cardiovascular Institute of Alberta, Calgary, AB, Canada

Synonyms

Channelopathies; Ion channels structure–functions relations; K^+ dependent proteins

hERG Channel: Implications to Health and Disease

The hERG potassium channel was first identified by Trudeau et al. (1995). Mutations in fruit fly genes manifest physiologically as exaggerated flight patterns after exposure to ether – thus, the nomenclature – ether-a-go-go-related gene. Seminal work by Keating et al. reported that mutations in hERG underlie the long QT2 syndrome in humans (Splawski et al. 1997). The hERG channel encodes the pore-forming subunit of a multiprotein potassium channel complex that includes beta subunits – KCNE1 and KCNE2. The gene is alternatively processed resulting in various isoforms with distinct structure and function. Mutations in the hERG channel that are associated with disease are cause by either loss of function or gain of function. Loss of function mutations generally result in dominant negative or haploinsufficiency long QT2 phenotypes. These loss-of-function mutations can result from mutations within the alpha subunits or the beta subunits. Generally, the loss of function relates to unfolding of the channel's tertiary structure, retention in the endoplasmic reticulum and golgi, and ubiquitination followed by transport to the proteosome. The gain-of-function mutations in hERG result in the short QT syndrome. The short QT syndrome is a genetic clinical condition in which a missense mutation in hERG potassium channel disrupts or shifts the normal C-type inactivation to more depolarized potentials. This mechanism increases the time-dependent current at the expense of decreasing the tail current. The net result is a shortened repolarization time on the surface ECG of patients (short QT). This is associated with propensity to ventricular fibrillation. The mechanism of the susceptibility to ventricular fibrillation is likely exaggerated transmural dispersion of repolarization.

The hERG potassium channel not only has roles in the adult heart but also is widely distributed in the nervous system where it has roles in setting resting potential and firing frequency in some neurons. It is extensively expressed in chromaffin tissues, autonomic ganglia, chemo-sensing ganglia, hippocampus, and midbrain tissues. Changes in autonomic nervous system activity may be modulated by hERG-related currents (I_{hERG}) and thus have an indirect impact of cardiac repolarization and propensity to arrhythmias. In embryos, hERG plays pivotal roles in blastula formation and cardiogenesis. Mutations in hERG or pharmacologic block of hERG can result in embryo-lethality and cardiovascular defects such as tetralogy of Fallot (TOF) and coarctation of the aorta. Mutations in hERG can also lead to one form of the Sudden Infant Death Syndrome.

The hERG channel also plays an important role in regulation of cell growth and apoptosis. The hERG channel appears to be naturally overexpressed in some cancers. Pharmacologic block of hERG appears to manifest antiproliferative activity in some tumors and indeed can induce apoptosis. Accordingly, the hERG channel may be a target for development of new oncolytic drugs. The major hurdle to this strategy is that hERG is extensively expressed in human ventricle and drugs that block hERG cause the main acquired form of the human long QT syndrome (Fernandez et al. 2005).

The biological mechanisms underlying decreases in I_{Kr} activity expression in the plasmalemma is dependent on the balance between synthesis/degradation. There is little doubt that many missense mutations of hERG cause trafficking defects due to conformations that do not foster exportation of the channel from the endoplasmic reticulum to the plasmalemma. Other defects in structure cause endocytosis of the channel or a decrease in its stability within the plasmalemma. More rarely, hERG channels containing a mutation are adequately trafficked, but they are nonfunctional because of disruption of the selectivity filter (G628S). Most expression of activity and function is regulated posttranscriptionally.

The Structure of hERG Channel

In previous chapters, the common structures of potassium channels have been reviewed. Similar to other voltage-gated K-channels, the hERG channel is a homo-tetramer. While a number of ion channels have been crystallized, a crystal for the S1–S6 portion hERG has yet to be developed. Accordingly, homology models based on similarity to crystallized ion channels have been developed predominantly to well-conserved pore domain (see Fig. 1). Some insights have been achieved using these homology models. These insights related to interacting elements between helical segments and the topology of drug-induced I_{hERG} block. Another challenge in this area of research is developing models of the channels in different states: closed, open, and inactivated.

Soluble domains of the hERG potassium channel have been crystallized or their structure has been established by NMR, and the structure-function of these elements has been more thoroughly examined (Torres et al. 2003). Similar to other K_v channels, the full- functional assembled form of hERG1 consists of the pore and voltage-sensing domains and large intracellular C- and N-termini. The C-terminus domain has a structural fold homologous to the cyclic-nucleotide-binding domain (cNBD) found in cyclic-nucleotide-gated channels (CNG). The exact structural role of this domain is unclear. The N-terminus is a very elongated, 400+ residues long with residues 1–135 forming the EAG domain. The crystal structure for residues 26–135 of the EAG domain is available from MacKinnon lab and was shown to fold into a Per-Arnt-Sim (PAS) domain (Cabral et al. 1998). The PAS domain interfaces with elements of the S1–S6 domain to dominantly regulate deactivation kinetics of I_{Kr}. Recent data indicated that PAS domain containing N-terminal elements are sufficient to regulate expression and function of heteromultimeric hERG1a/herg1b and homoteterameric hERG1b channels, but not homotetrameric hERG1A channels. Scanning mutagenesis and molecular dynamic modeling of the cyclic-nucleotide-binding domain suggest that negatively charged patches on its cytoplasmic surface form an interface with the N-terminal 1–26 domain of hERG1a. In this modeling, the NT1-26 obstructs gating motions of the cyclic-nucleotide-binding domain to allosterically stabilize the open conformation of the pore. Recent studies show the presence of an amphipathic alpha helix from residues 13–22. The first 25 residues form a flexible unstructured domain (2–9) and an alpha helical segment (13–22). Residues within the initial flexible segment and the alpha–helix are required, but neither alone is sufficient to produce the slow deactivation kinetics typical of the wild-type channel. The residues R5 and G6 in the initial segment were important slow deactivation (Wang et al. 2011). Other studies provide evidence for interaction of the PAS N-terminal domain with the cyclic-nucleotide-binding domain of hERG. Although, the PAS domain of hERG1 has been crystallized, and its presence alters the deactivation of hERG1, but how it molecularly interfaces with the voltage-sensing/activating domains of hERG1A has not been examined (Morais Cabral et al. 1998).

Drug Blockade to hERG1 Channel: Cardiovascular Drug Development and Drug Safety

The best-known feature of the hERG1 channel is its unique promiscuity in binding of diverse organic molecules. A broad panel of organic compounds used both in common cardiac and noncardiac medications (i.e., antibiotics, antihistamines, and antibacterial) are thought to cause a reduction in the repolarizing current I_{Kr} by blocking of the central cavity leading to ventricular arrhythmia. Several Food and Drug Administration (FDA) approved drugs (i.e., terfenadine, cisapride, astemizole, and grepafloxin) have been withdrawn from the market, while use of others such as thioridazine, haloperidol, sertindole, and pimozide was restricted. The drug-related arrhythmias led to an implementation of mandatory drug screen for hERG1 blockade by both the FDA and the European Medicines Agency (EMEA). The recommended drug-screening methods include traditional patch clamp techniques, radiolabelled-drug-binding assays, 86RB-flux assays, and high-throughput cell-based fluorescent dyes and stably transfected hERG ion channels from Chinese hamster ovary (CHO) cells. The ability of nonclinical tests to determine whether drugs have a real risk for QT prolongation in humans is fairly good; however, the

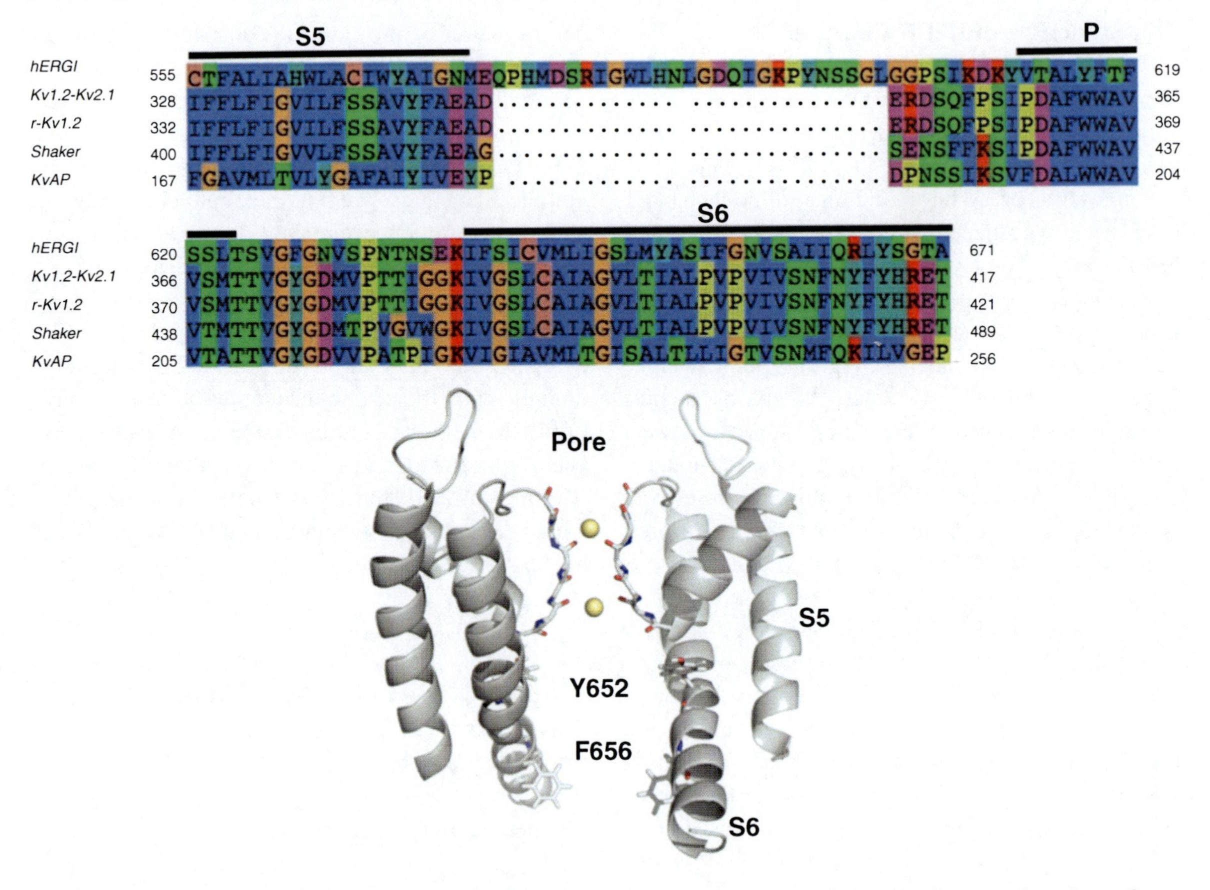

hERG (KCNH2) Potassium Channel, Function, Structure and Implications for Health and Disease, Fig. 1 Sequence alignment for pore domain of hERG1 (S5-pore-S6 helices) and channels with available crystal structures. A homology model built on this alignment. Locations of crystallographic ions in the selectivity filter are shown; Y652 and F656 are shown in sticks

ability of these techniques to identify drugs carrying no risk is not well established. The combination of ligand-based QSAR and receptor-based molecular-docking methods to a homology model has lead to the development of a universal pharmacophore model, which appears to provide rapid and relatively robust assessment of ability to chemicals to block the hERG1 channel. The pivotal binding site for the hERG blockers resides in the pore region. The most important amino acid determinants of pharmacologic block of hERG are F656 and Y562 in the distal S6 (shown in Fig. 1). Binding of most drugs requires access to the inner pore guarded by the inner activation gate. For some drugs such as dofetilide, inactivation of the channel appears to be necessary for high-affinity binding (see Fig. 2 for examples of blockers and their proposed binding mode).

hERG Channel Activators

While the best feature characterizing hERG1-drug interactions is a promiscuous intracavitary blockade by a variety of drugs, a rapidly emerging strategy focuses on channel activation by small molecules – channel activators aka openers (Stoehr et al. 2007). An implemented screen of novel drug candidates for ability to attenuate hERG1 function has led to an identification of compounds capable of hERG1 current enhancement (Xu et al. 2008). The mechanism by which activators increase hERG tail currents is not homogeneous. Some activators shift the voltage dependence of activation to hyperpolarized potentials (termed facilitators) (Perry et al. 2010), while some activators affect deactivation or attenuate inactivation (see Fig. 3 for examples of well-studied activators).

Dofetilide (Neutral and ionized)

Cisapride

d-sotalol

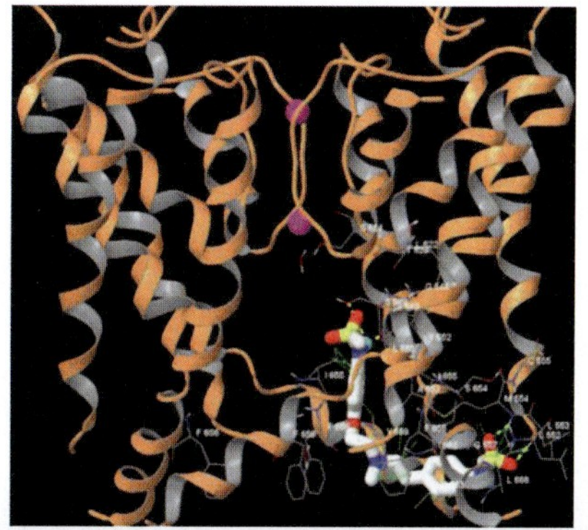

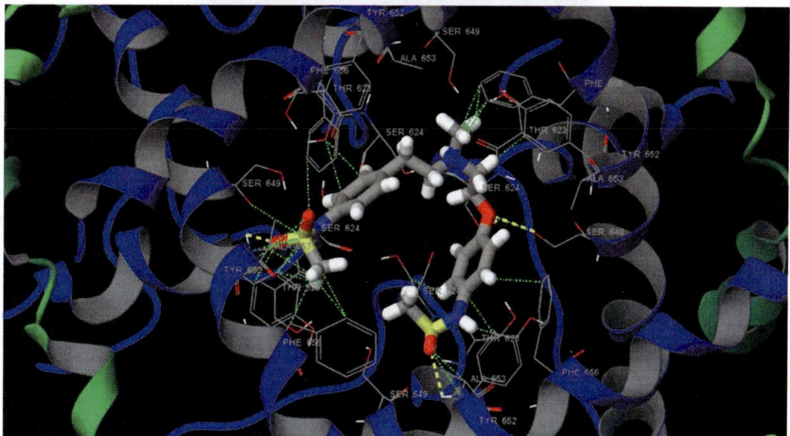

hERG (KCNH2) Potassium Channel, Function, Structure and Implications for Health and Disease, Fig. 2 *Top Panel*: List of common blocker structures: state-dependent high-affinity blocker dofetilide, low-affinity blocker cisapride, and medium-range blocker D sotalol. Middle Panel: Example of the binding pose for dofetilide at the hERG1 cavity. *Bottom Panel*: Zoom-out to the dofetilide binding pose in the open-activated state of hERG1 cavity. The residues involved into formation of the binding pocket are labeled

Type 1 activators such as RPR260243 attenuate inactivation and slow deactivation, whereas Type 2 activators, such as PD118057, are considered to attenuate inactivation. NS1643 is considered a Type 2 activator. The potential binding site for the Type 1 activator RPR260243 was mapped by Perry et al. (Perry et al. 2010) They proposed the binding site of the activator at the intersection of the cytosolic ends of S5 (L553, F557) and adjacent region of S6 (N658, V659) segments of a single hERG1 subunit shown in Fig. 3. RPR260243 binding also enhances current by attenuating inactivation and slowing the rate of the channel deactivation. In contrast, the Type 2 hERG activator PD118057 binding is thought to rely on the pocket formed by hydrophobic residues from two channel subunits; one with L646 in the S6 domain of one subunit and other in the L622 and F619 in the pore helix of the adjacent subunit. Interactions of

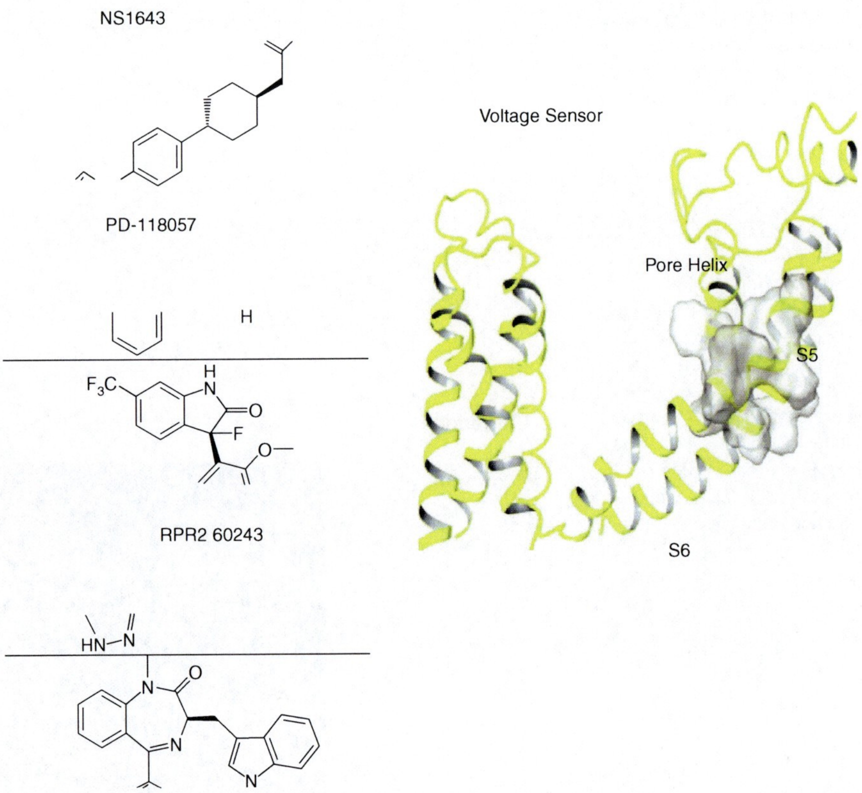

hERG (KCNH2) Potassium Channel, Function, Structure and Implications for Health and Disease, Fig. 3 List of known hERG channel activators molecules and mapping of essential amino acids found by Grunnet et al. (2011) and hERG S1S6 TM (only one monomer shown) model developed independently by Subbotina et al. (2010)

PD118057 with these sites may decrease channel's inactivation and possibly increase open probability. Many hERG1 activators also behave as blockers of the hERG1 channel by binding at an intracavitary site in the distal S6 domain at higher concentrations. For example, an opener PD118057 blocks hERG1 at 30 μM. It was shown that the F656 residue in the distal S6 was an important determinant of PD118057 binding to the blocking site.

Mutations that disrupt inactivation or alter K^+ selectivity suppress the ability of the activator NS1643 to increase hERG1 currents (Xu et al. 2008). These mutations included T613C during MTS treatment and S631C during with H_2O_2 treatment. It was shown that cells with a loss of inward rectification had diminished responsiveness to NS1643. Recently, Grunnet et al. (2011) examined the impact of a wide range of amino acid mutations in hERG1 on response to NS1643. Their study proposes a binding site in the vicinity of the extracellular end of the S5/S6 segments of two adjacent hERG subunits. In addition, F619A, I567A, Y652, and C643A enhanced response to NS1643. The proposed binding site for an activator is shown in Fig. 3. To supplement electrophysiological recording, NS1643 was also docked to a homology model of hERG based on the crystal structure of the mammalian Shaker Kv1.2 potassium channel. The following amino acids were in close proximity to the drug (I560, L622, L646, L650, and M651). We have recently reported a range of amino acids that appear to be important determinants of NS1643 interface with hERG. Two domains, one near the selectivity folder N629 and one on the S4-S5 linker appear important (Durdagi et al. 2012).

In review, hERG is a potassium channel that plays pivotal biologic roles in growth and development and physiologic function of heart, brain, autonomic nervous system, and chromaffin cells. Abnormalities in structure or function of the hERG potassium channel predisposes to death and disability. Moreover, the biologic mechanisms underlying defects in protein trafficking and dysfunction have general applicability to a number of other ion-channel diseases.

Cross-References

- ▶ Cellular Electrolyte Metabolism
- ▶ Potassium Channel Diversity, Regulation of Potassium Flux across Pores
- ▶ Potassium Channels, Structure and Function
- ▶ Potassium Dependent Sodium/Calcium Exchangers

References

Cabral JHM, Lee A, Cohen SL, Chait BT, Li M, Mackinnon R (1998) Crystal structure and functional analysis of the HERG potassium channel N terminus: a eukaryotic PAS domain. Cell 95:649–655

Durdagi S, Guo J, Lees-Miller JP, Noskoy SY, Duff HJ (2012) Structure-guided topographic mapping to elucidate binding sites for the human ether-a-go-go-related potassium channel (KCNH2) activator NS1643. J Pharmacol exp ther 441–452

Fernandez D, Piper D, Tristani-Firouzi M, Sanguinetti MC (2005) HERG channel gating and block. J Mol Cell Cardiol 39:168–169

Grunnet M, Abbruzzese J, Sachse FB, Sanguinetti MC (2011) Molecular determinants of human ether-a-go-go-related gene 1 (hERG1) K^+ channel activation by NS1643. Mol Pharmacol 79:1–9

Morais Cabral JH, Lee A, Cohen SL, Chait BT, Li M, Mackinnon R (1998) Crystal structure and functional analysis of the HERG potassium channel N terminus: a eukaryotic PAS domain. Cell 95:649–655

Perry M, Sanguinetti M, Mitcheson J (2010) Revealing the structural basis of action of hERG potassium channel activators and blockers. J Physiol-London 588:3157–3167

Splawski I, TristaniFirouzi M, Lehmann MH, Sanguinetti MC, Keating MT (1997) Mutations in the *hminK* gene cause long QT syndrome and suppress I-Ks function. Nat Genet 17:338–340

Stoehr SJ, Dickson JA, Castle NA, Gerlach AC (2007) Identification of a potent hERG channel activator with a novel mechanism of action. FASEB J 21:A539

Subbotina J, Yarov-Yarovoy V, Lees-Miller J, Durdagi S, Guo J, Duff HJ, Noskov SY (2010) Structural refinement of the hERG1 pore and voltage-sensing domains with ROSETTA-membrane and molecular dynamics simulations. Proteins 78:2922–2934

Torres AM, Bansal PS, Sunde M, Clarke CE, Bursill JA, Smith DJ, Bauskin A, Breit SN, Campbell TJ, Alewood PF, Kuchel PW, Vandenberg JI (2003) Structure of the HERG K^+ channel S5P extracellular linker – role of an amphipathic alpha-helix in C-type inactivation. J Biol Chem 278:42136–42148

Trudeau MC, Warmke JW, Ganetzky B, Robertson GA (1995) Herg, a human inward rectifier in the voltage-gated potassium channel family. Science 269:92–95

Wang DT, Hill AP, Mann SA, Tan PS, Vandenberg JI (2011) Mapping the sequence of conformational changes underlying selectivity filter gating in the K(v)11.1 potassium channel. Nat Struct Mol Biol 18:35–42

Xu X, Recanatini M, Roberti M, Tseng GN (2008) Probing the binding sites and mechanisms of action of two human ether-a-go-go-related gene channel activators, 1,3-bis-(2-hydroxy-5-trifluoromethyl-phenyl)-urea (NS1643) and 2-[2-(3,4-dichloro-phenyl)-2,3-dihydro-1H-isoindol-5-ylamino]-nicotinic acid (PD307243). Mol Pharmacol 73:1709–1721

Hexavalent Chromium

▶ Chromium Toxicity, High-Valent Chromium

Hexavalent Chromium and Cancer

Yana Chervona and Max Costa
Department of Environmental Medicine, New York University Medical School, New York, NY, USA

Synonyms

Chromium [Cr (VI)] compounds

Definition

Chromium (VI) is one of the two stable species of chromium (Cr) that can be found in biological systems. Cr(VI) compounds can both be soluble or insoluble and exist in a *+6* oxidative state. At neutral pH, hexavalent chromium exists as a mixture of chromate (CrO_4^{2-}) or hydrochromate ($HCrO_4^-$) anions, with an approximate ratio of 3:1. Due to their structural resemblance to phosphate (CrO_4^{2-}vs. PO_4^{2-}), the chromate anions are actively transported into all cells in lieu of phosphate, and are extremely toxic. Upon entering the cell, Cr(VI) is reduced to a number of unstable, highly reactive intermediates such as Cr(V) and Cr(IV), and finally to the stable Cr(III) species, which are produced at high levels inside the cell and react with cellular ligands. The acute reactivity of its intermediate species, Cr(V) and Cr(IV), and the high ligand-binding affinity of its stable counterpart, Cr(III), therefore, allow hexavalent chromium to have a multifaceted approach in causing cellular damage and carcinogenesis.

Introduction

Hexavalent chromium [Cr (VI)] is a well-recognized human carcinogen that has gained notoriety both in the environmental and occupational settings. It has been implicated in the occurrence of a number of malignancies, including bone, brain, prostate, Hodgkin's lymphomas, leukemia, stomach, genital, renal, and bladder cancers (Costa 1997). Cr(VI) compounds are widely utilized in manufacturing, chrome plating, ferrochrome production, and stainless steel welding. Occupational chromium exposure affects roughly half a million industrial workers in the USA and several million worldwide. Chromium contamination also has profound effects on the environment, polluting the water supply of millions of people (Salnikow and Zhitkovich 2008).

Basic Characteristics

The valence state of chromium ions plays a very important role in their chemistry because it determines the ions' chemical reactivity and uptake. Cr(III) and Cr(VI) have very distinctly different bioavailabilities and biological half-lives, with the latter having about two order-of-magnitudes higher bioavailability (6.9% vs. 0.13%) and upon absorption, a one order-of-magnitude higher half-life than the former (39 h vs.10 h) (Kerger et al. 1996). Upon entering the body, hexavalent chromium is readily absorbed by a number of different tissues. The readiness of hexavalent chromium's absorption is due to the fact that the chromium anion structurally resembles a phosphate ion (PO_4^{2-}) and is mistakenly transported across the cellular membrane through the chloride phosphate channels. It has been called a Trojan horse. Therefore, to enter the cell, Cr(VI) passes through the cell membrane via a divalent anion concentration gradient in a chloride phosphate anionic channel and invades the cell as a tetrahedral divalent chromate (CrO_4^{2-}) anion (Chiu et al. 2010).

Once inside the cell, Cr(VI) is first reduced to two unstable and highly reactive intermediates, Cr(V) and Cr(IV), and is then finally reduced to Cr(III), a stable species that is ultimately formed at high levels inside the cell and has a high binding affinity to cellular ligands. This is the scenario when glutathione is the reductant; however, with ascorbate, Cr(VI) is reduced by two electron reductions to Cr(III). With respect to other human cocarcinogens, hexavalent chromium is a bit of an anomaly because in mammalian cells its metabolism does not require any enzymes and solely relies on the direct electron transfer from ascorbate and nonprotein thiols, such as glutathione and cysteine. While ascorbate is the leading biological reducer of Cr(VI) in vivo, accounting for nearly 90% of chromium metabolism in cells, in cell cultures where levels of ascorbate are low (50–60 μM in vitro vs. 1.3 mM in vivo), human and nonhepatic cells rely on thiols for reducing chromium (Salnikow and Zhitkovich 2008).

The reduction of hexavalent chromium within the cell is extremely important. If reducing agents are not present, Cr(VI) cannot and does not react with DNA. When it is reduced, however, its respective reactive intermediates are produced via Fenton-like and/or Haber-Weiss reactions, to yield free radicals and cause substantial DNA damage. The first Cr(VI) intermediate, Cr(V), for example, has a half-life of 37 min and is quite reactive. When generated, it can produce hydroxyl radicals (OH), which can induce NF-kappa B activation, DNA strand breaks, and 2-deoxyguanosine (dG) hydroxylation. The second intermediate, Cr(IV), which can be produced via the reduction of Cr(V) by cellular reducing agents, such as glutathione (GSH), is also capable of producing hydroxyl radicals via a Fenton-like reaction and in turn can generate DNA damage (Chiu et al. 2010). The third and final reduced species that is produced in the cells is the stable trivalent chromium species, which is produced at high levels inside the cell, has a strong binding affinity to many cellular ligands, and is kinetically inert to ligand substitution reactions (Costa 1997). With respect to its ligands, Cr (III) assumes an octahedral geometry and usually binds six ligands; however, depending on the nature of the ligands, it may continue to bind other molecules. Ligand exchange will cease once the trivalent chromium has reacted with ligands that are not easily displaced, such as two molecules of cysteine or glutathione (Costa and Klein 2006).

Mechanisms of Carcinogenesis

It has been demonstrated that hexavalent chromium is capable of inducing double- and single-stranded break(s), DNA protein, DNA intrastrand cross-links, Cr adducts, and oxidized nucleotide bases

(Salnikow and Zhitkovich 2008). Furthermore, chromate exposure has been associated with centrosome abnormalities, aneuploidy, microsatellite instability, as well as repression of DNA mismatch repair genes (i.e., hMHL1) (Hirose et al. 2002; Takahashi et al. 2005; Holmes et al. 2006). Hexavalent chromium is particularly active in inducing apoptosis in cells with normal p53 activity. The extensive damage caused by chromium-induced DNA breaks and adducts likely serves as a potent signal for p53 activation in the cell cycle checkpoint, and if the detected damage cannot be repaired, the cell undergoes an apoptotic death. Moreover, the reduction of Cr(VI) to Cr(III) also generates a large number of oxygen radicals, which in turn can activate several other antiapoptotic signaling processes (i.e., PI3K and AKT), allowing the cell to evade apoptosis and transform into a malignant cell (Costa and Klein 2006).

DNA Breaks

Hexavalent chromium is a well-known DNA damaging agent. A vast body of evidence suggests that Cr(VI) exposure causes DNA strand breaks. Single-strand breaks (SSBs) are the most common type of strand break occurring in vitro and in vivo after hexavalent chromium exposure (Salnikow and Zhitkovich 2008). Double-stranded breaks (DSBs), however, also occur and have been detected post-Cr(VI) exposure by a number of different assays, including the Comet assay and positive visualization of the activated DNA damage sensor ATM as well as phosphorylated nuclear gamma-H2AX by immunofluorescence microscopy and Western blots. The production of these strand breaks is most likely produced by oxidative species generated during the metabolism of Cr(VI). Specifically, the DNA strand breaks occur during the conversion of chromium(VI) to chromium(III) by glutathione (Chiu et al. 2010).

Cr-DNA Adducts

In mammalian cells, small Cr-DNA adducts are the most common type of DNA lesion. These small lesions are thought to be responsible for all the mutagenic damage produced during the reduction of hexavalent chromium by cysteine and ascorbate. The majority of the adducts formed in vitro are binary and are not very mutagenic due to the fact that their existence is compromised by the presence of numerous Cr(III) species binding to small molecules. Ternary Cr-DNA adducts are, however, larger and are the principal Cr-DNA complexes in the cell. They are formed during the assault of DNA by preformed ligand-Cr(III) complexes. Ternary adducts are comprised of four major DNA cross-links: glutathione-Cr-DNA, cysteine-Cr-DNA, histidine-Cr-DNA, and ascorbate-Cr-DNA complexes. The N_7 of dG as well as the phosphate group on the DNA backbone serves as the primary site of attachment for these chromium adducts, and both types of adducts are subject to nucleotide excision repair (NER). The ternary adducts are much more mutagenic than the binary adducts, and therefore are the likely culprits responsible for altering the DNA code during the cell transformation process (Costa and Klein 2006).

DNA-Protein and DNA-Intrastrand Links

Cr(VI)-induced DNA-protein cross-links (DPCs) are well established in studies conducted using both in vivo and in vitro systems. Their estimated overall yield is fairly low, about 1% of all the Cr-DNA adducts but can be higher in vitro. The biological significance of the Cr(VI)-induced DPCs remains unclear, however, a substantial amount and size of these lesions could very likely hinder the replication and transcription processes. Unlike DPCs, interstrand DNA cross-links have only been detected in vitro and are most likely produced by Cr(III) oligomers. While their existence may be a purely in vitro phenomenon that arise when high Cr(III) and low ligand concentrations are present, if formed, the interstrand cross-link would be a powerful obstacle for cellular DNA replication (Salnikow and Zhitkovich 2008).

DNA Base Pair Damage

DNA base pair damage can appear in two forms: either as a loss of bases (i.e., production of abasic sites) or as chemical modifications with the maintenance of the altered bases in the DNA double helix. The most likely culprit involved in Cr-induced DNA base pair damage is the reduction of hexavalent chromium. In vitro studies of Cr(VI) reductions with ascorbate and glutathione demonstrated that the generation of abasic sites closely reflected the yield of SSB and involved the same reactive species. The reactive oxygen species, such as OH, produced during Fenton-like reactions in the course of chromium reduction readily interact with guanine residues at several positions to produce a range of lesions. The ROS preference for this base is due to its high oxidation potential relative to cytosine, thymine, and adenine. The most abundant and studied chromium-induced DNA lesions are 8-hydroxydeoxyguanosine, 8-hydroxyguanosine, and 7,8-dihydro-8-oxoguanine (Chiu et al. 2010).

Epigenetic Effects of Hexavalent Chromium

Until recently, DNA damage, genetic instability, and changes in gene expression were considered to be the underlying mechanisms for Cr(VI)-induced carcinogenesis. However, an emerging body of evidence is suggesting that epigenetic effects of Cr(VI) may play an important role in its genotoxicity and carcinogenicity. It has been demonstrated that potassium dichromate could induce DNA methylation and silence the transgene expression of a cell line expressing a bacterial *gpt* reporter gene. Moreover, the same compound was shown to increase genome-wide cytosine hypermethylation in the CCGG-DNA sequence of the *Brassica napus* L. plants. An increase in DNA methylation was also observed in the promoter region of the tumor suppressor gene p16 and the DNA mismatch repair (hMLH1) gene in chromate-induced human lung cancers. In addition, it was demonstrated that Cr(VI) could cross-link the histone deacetylase 1-DNA methyltransferase 1 complexes to the chromatin of the *Cyp1a1* promoter and inhibit histone marks induced by AHR-mediated gene transactivation, including phosphorylation of histone H3 Ser-10, trimethylation of H3 Lys-4, and several acetylation marks in histones H3 and H4. Most recently, it was observed that exposure of human lung carcinoma A549 cells to potassium chromate was able to generate global changes in number of histone tail modifications. Furthermore, chromate exposure induced an increase in H3K9 dimethylation in the promoter of the MLH1 gene, causing a decrease in its mRNA expression. Taking all this evidence into consideration, one can compellingly postulate that other than previously acknowledged genotoxicity effects, hexavalent chromium's ability to alter epigenetic homeostasis may further contribute to its carcinogenesis (Arita and Costa 2009).

History of Cr(VI)-Induced Carcinogenesis

Hexavalent chromium has been known to be a health hazard for nearly a century. The first epidemiological study that linked respiratory cancer to chromate exposure was performed by Machle and Gregorius in 1948 and found that 21.8% of the mortality rates of the chromate workers were due to respiratory cancers compared to only about 1.4% of control population (Holmes et al. 2008). Furthermore, an increased risk of lung cancer was also identified by Thomas F. Mancuso and Wilhelm C. Hueper, who studied the mortality rates of chromium-exposed workers who were employed at the Painesville, Ohio facility between 1931 and 1937. They first published their results in 1951, and later provided updates on the cohorts in 1975 and 1997, consistently observing an excess risk of lung cancer among the exposed workers (Michaels et al. 2006). Today, the association between hexavalent chromium exposure in the occupational setting and lung cancer has been documented in numerous epidemiological studies worldwide. Chromate, chromate pigment, and chromate plating workers in the USA, Federal Republic of Germany, France, the Netherlands, Norway, and the UK have consistently demonstrated excess risks for lung cancer (IARC 1997). During chromate productions, trivalent chromite ore is combined with soda ash and is heated to oxidize chromite to soluble sodium chromate. If and when lime was added to the production process, insoluble calcium chromate is also generated (Holmes et al. 2008). The production of chromate pigment entails reacting sodium chromate with zinc or lead, respectively. While zinc and lead chromate compounds are the predominant compounds used in the chromate pigment industries, insoluble barium and strontium chromate are also utilized (Holmes et al. 2008)

The exposure to both insoluble and soluble hexavalent chromium compounds in the raw materials used in their production is associated with an increase risk of cancer. The excess risk of lung cancer is particularly well established in zinc chromate facilities. Workers in the chromium plating industries (i.e., die casting and plating) also demonstrated increased risks of lung cancer. This risk was particularly elevated among those whose works were employed at the chrome baths for at least 10 years. In addition to lung cancers, cases of sinonasal cancer were also described in epidemiological studies of primary chromate production workers in Japan, the UK and the USA, of chromate pigment production workers in Norway, and of chromium platers in the UK, suggesting a strong pattern of association between Cr(VI) exposure and increased risk of these rare tumors (IARC 1997).

Cr(VI) Compounds

Given the extensive and undisputable body of epidemiological evidence, the carcinogenicity of various hexavalent chromium compounds was also examined using both in vivo an in vitro models. The following is a brief description of the findings by the International Agency for Research on Cancer (IARC):

Calcium chromate was tested in the murine and hamster models either via inhalation, intratracheal, intrabronchial, intrapleural, subcutaneous, and intramuscular modes of administration. Calcium chromate inhalation resulted in a borderline increase in the incidence of lung adenomas in mice, while intratracheal intrabronchial administrations resulted in the induction of lung tumors in rats. Furthermore, intrapleural and intramuscular administrations also produced local tumors in rats and mice, respectively.

Chromium trioxide (chromic acid) has been tested as a mist by inhalation in mice as a solid by intrabronchial implantation in rats. In mice, a low incidence of lung adenocarcinomas was reported at the higher dose of exposure and of nasal papillomas at the lower dose, while perforation of the nasal septum was observed at both dose levels. Intrabronchial administration of chromic acid in rats produced a few lung tumors.

Sodium dichromate has been primarily tested in rats via inhalation, intratracheal, intrabronchial, intrapleural, and intramuscular administration. Benign and malignant lung tumors were observed in the studies by inhalation and by intratracheal administration. However, there was no apparent increase in tumorigenesis after intrabronchial, intrapleural, or intramuscular administration.

Barium chromate has also been primarily tested in rats by intrabronchial, intrapleural, and intramuscular implantation. However, intrapleural and intramuscular implantation studies were inadequate in their evaluation of the carcinogenicity of the compound, while one of the intrabronchial implantation did not produce an increase in the tumor incidence.

Lead chromate and derived pigments have been tested in rats by subcutaneous and intramuscular injection, causing malignant tumors at the site of injection and, in one study, renal carcinomas, while intrabronchial implantation in rats did not yield tumors. Furthermore, no increase in tumor incidence was observed when lead chromate was administered intramuscularly to mice, and a single subcutaneous injection of basic lead chromate produced a high incidence of local sarcomas in rats.

Zinc chromates have been tested in rats by intrabronchial implantation, causing bronchial carcinomas; by intrapleural administration, yielding local tumors; and by subcutaneous and intramuscular injection, generating local sarcomas.

Strontium chromate was tested in rats by intrabronchial implantation, producing an elevated incidence of bronchial carcinomas, while the intrapleural and intramuscular injections of strontium chromate produced local sarcomas.

In addition to the compounds listed above, the carcinogenic potential of various other hexavalent chromium compounds was also examined. The literature has shown that potassium dichromate, sodium dichromate, ammonium dichromate, potassium chromate, sodium chromate, ammonium chromate, chromium trioxide, calcium chromate, strontium chromate, and zinc yellow induced a cornucopia of cellular injury effects, including DNA damage, gene mutation, sister chromatid exchange, chromosomal aberrations, cell transformation, and dominant lethal mutation both in vitro and in vivo. Moreover, potassium chromate was able to induce aneuploidy in insects, while potassium dichromate produced recombination, gene mutation, and aneuploidy in fungi (IARC 1997).

Characteristics of Cr(VI)-Induced Tumors

The majority of lung cancers associated with hexavalent chromium exposure are squamous cell carcinomas. Human pathology studies have shown that Cr(VI) deposits and persists at the bronchial bifurcations and that is consistent with particulate chromate exposure. Furthermore, it has been reported that the malignancy of the tumors is significantly associated with the Cr(VI) accumulation in the lung (Holmes et al. 2008). With respect to the molecular mode of Cr(VI)-induced carcinogenesis, several studies have reported that lung tumors from chromate workers exhibited chromosomal instability (CIN) and microsatellite instability (MIN). Specifically, CIN was found to be significantly increased in chromate-induced tumors, and MIN was reported to increase with longer time of exposure and was strongly correlated with decreased expression of hMHL1, a DNA mismatch repair gene (Hirose et al. 2002; Takahashi et al. 2005; Holmes et al. 2008). In addition to chromosomal and microsatellite instabilities, changes in the gene expression of cyclin D1, p16, and surfactant protein B(SP-B) were also reported in Cr(VI)-induced tumors. Cyclin D1 expression was significantly increased in the chromate-induced tumors but was not elevated in the adjacent normal epithelia; p16 expression was decreased in nearly 87% of chromate-induced lung tumors and was associated with promoter methylation; and deletion or insertion variations in the SP-B genes were significantly increased in workers with chromate-induced lung tumor (Holmes et al. 2008).

Summary

Due to its compelling and multifaceted genotoxicity, hexavalent chromium is a potent human carcinogen. IARC classifies Cr(VI) as a Group 1 carcinogen, acknowledging that there is ample evidence of its carcinogenicity in humans (IARC 1997). As previously noted, occupational exposure to Cr(VI) compounds, in particular, is a well-documented cause of respiratory cancers. It has been demonstrated that hexavalent chromium is capable of inducing double- and single-stranded break(s), DNA protein, DNA intrastrand cross-links, Cr adducts, and oxidized nucleotide bases (Salnikow and Zhitkovich 2008). Furthermore, chromate exposure has been associated with centrosome abnormalities, aneuploidy, microsatellite instability, changes in gene expression of cyclin D1, p16, and surfactant protein B, as well as repression of DNA mismatch repair genes (i.e., hMHL1) (Hirose et al. 2002; Takahashi et al. 2005; Holmes et al. 2006, 2008). Several recent epidemiological and risk-assessment studies have found that the lifetime risk of dying of lung cancer under the current permissible exposure limit, 5 $\mu g/m^3$, is as much as 25%. The most common form of Cr(VI)-induced occupational lung cancers is squamous cell carcinomas, the majority of which are located in the central part of the lung (Holmes et al. 2008; Salnikow and Zhitkovich 2008).

Although, all Cr(VI) compounds are deemed to be carcinogenic, there is some uncertainty as to whether their carcinogenic potency is influenced by their water solubility. Less-soluble chromates have been implicated in the occurrence of respiratory cancers based upon epidemiological studies. However, there is also ample evidence suggesting that human exposure to soluble Cr(VI) compounds significantly increases the risk of lung cancer (Mancuso 1997). Cr(VI) compounds of high water solubility are likely to penetrate throughout the body and induce a wide variety of cancers, including nonrespiratory malignancies. Hexavalent chromium exposure has also been shown to elevate the incidence of bone, brain, prostate, Hodgkin's lymphomas, leukemia, stomach, genital, renal, and bladder cancers in workers with a higher exposure to Cr(VI) compounds (Costa 1997). Furthermore, it was also shown that chromium workers exhibited an increase in mental, psychoneurotic, and personality disorders and experienced an increase in suicides (Costa and Klein 2006). This is not surprising since hexavalent Cr compounds are readily absorbed by inhalation and ingestion. Furthermore, soluble Cr(VI) is widely distributed in vivo and even enters the brain because, as noted earlier, at a physiological pH, it is structurally analogous to phosphate and is rapidly transported into cells. Thus, it can affect many different organs, causing the wide variety of cancers.

References

Arita A, Costa M (2009) Epigenetics in metal carcinogenesis: nickel, arsenic, chromium and cadmium. Metallomics 1:222–228

Chiu A, Shi XL, Lee WK, Hill R, Wakeman TP, Katz A, Xu B, Dalal NS, Robertson JD, Chen C, Chiu N, Donehower L (2010) Review of chromium (VI) apoptosis, cell-cycle-arrest, and carcinogenesis. J Environ Sci Health C Environ Carcinog Ecotoxicol Rev 28:188–230

Costa M (1997) Toxicity and carcinogenicity of Cr(VI) in animal models and humans. Crit Rev Toxicol 27:431–442

Costa M, Klein CB (2006) Toxicity and carcinogenicity of chromium compounds in humans. Crit Rev Toxicol 36:155–163

Hirose T, Kondo K, Takahashi Y, Ishikura H, Fujino H, Tsuyuguchi M, Hashimoto M, Yokose T, Mukai K, Kodama T, Monden Y (2002) Frequent microsatellite instability in lung cancer from chromate-exposed workers. Mol Carcinog 33:172–180

Holmes A, Wise S, Sandwick S, Lingle W, Negron V, Thompson W, Wise JS (2006) Chronic exposure to lead chromate causes centrosome abnormalities and aneuploidy in human lung cells. Cancer Res 66:4041–4048

Holmes A, Wise S, Wise JP (2008) Carcinogenicity of hexavalent chromium. Indian J Med Res 128(4):353

International Agency on Cancer Research (IARC) (1997) Chromium, nickel and welding. Summary of data reported and evaluation, World Health Organization in Leon, France

Kerger BD, Paustenbach DJ, Corbett GE, Finley BL (1996) Absorption and elimination of trivalent and hexavalent chromium in humans following ingestion of a bolus dose in drinking water. Toxicol Appl Pharmacol 141:145–158

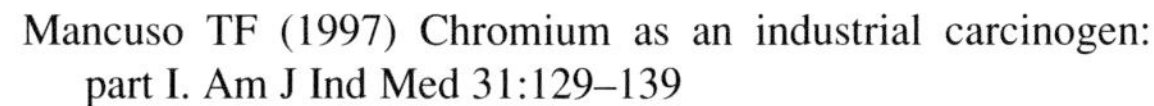

Mancuso TF (1997) Chromium as an industrial carcinogen: part I. Am J Ind Med 31:129–139
Michaels D, Monforton C, Lurie P (2006) Selected sciecne: an industry campaign to undermine an OSHA hexavalent chromium standard. Environ Health 5:5
Salnikow K, Zhitkovich A (2008) Genetic and epigenetic mechanisms in metal carcinogenesis and cocarcinogenesis: nickel, arsenic, and chromium. Chem Res Toxicol 21:28–44
Takahashi Y, Kondo K, Hirose T, Nakagawa H, Tsuyuguchi M, Hashimoto M, Sano T, Ochiai A, Monden Y (2005) Microsatellite instability and protein expression of the DNA mismatch repair gene, hMLH1, of lung cancer in chromate-exposed workers. Mol Carcinog 42:150–158

Hexavalent Chromium and DNA, Biological Implications of Interaction

John Pierce Wise Sr. and Qin Qin
Wise Laboratory of Environmental and Genetic Toxicology, Maine Center for Toxicology and Environmental Health, Department of Applied Medical Sciences, University of Southern Maine, Portland, ME, USA

Synonyms

Biological implications: Cell death; Biological implications: Chromosomal aberrations; Biological implications: Genomic instability; Biological implications: Mutagenesis; Chromium: Chromate; Chromium: Chromium(VI); Chromium: Cr(VI); Chromium: Cr^{6+}; DNA Lesions: Adducts; DNA Lesions: Cross-links; DNA Lesions: DNA double strand breaks; DNA Lesions: DNA single strand breaks

Definition

Chromium: Chromium is a metallic element with oxidation states ranging from +2 to +6. Hexavalent chromium rarely occurs naturally, but is produced from anthropogenic sources. Chromium in the hexavalent state occurs naturally only in the rare mineral crocoite ($PbCrO_4$).
DNA lesions: Damage to DNA includes alterations to and breaks in the DNA backbone or bases.
Biological implications: Potential negative biological outcomes resulting from cellular exposure to Cr(VI) and DNA exposure to the products of the intracellular reduction of Cr(VI) include chromosomal aberrations, cell death, mutagenesis, genomic instability, and carcinogenesis.

Main Text

Cr(VI) itself does not interact with DNA directly. However, the chemical products of the intracellular reduction of Cr(VI) interact with DNA in a variety of ways to form adducts, cross-links, DNA single-strand breaks, and DNA double-strand breaks leading to a number of negative biological outcomes including chromosomal aberrations, cell death, mutagenesis, genomic instability, and carcinogenesis, among others (De Flora and Wetterhahn 1989; Holmes et al. 2008). These aspects are discussed in further detail below.

Physico-chemistry

The biological implications of the relationship of Cr(VI) and DNA begin with its physicochemical mechanism. Epidemiology, animal, and cell culture transformation studies all indicate that the water-insoluble or particulate Cr(VI) compounds, such as lead chromate and zinc chromate, are the carcinogenic forms, while soluble Cr(VI) compounds are genotoxic but do not appear to be as potent as the particulate compounds (International Agency for Cancer Research IARC 1990). The explanation for this difference in potency remains uncertain as cell culture studies indicate that the genotoxic activity of the particulate compounds is due to the Cr(VI) anion, though the cations, such as zinc and lead, may have some activities, such as inhibiting DNA repair (Wise et al. 2008). There are data that suggest these cations can indeed inhibit repair (Beyersmann and Hartwig 2008), but so far this effect has not been tested with Cr(VI).

Another possibility is that the potency is due to differential delivery mechanisms with the Cr(VI) particles impacting and persisting at bronchial bifurcation sites in the lungs, while the soluble forms are readily cleared from the lungs. This possibility is consistent with human pathology studies that show that both high tissue chromium levels and Cr(VI)-tumors occur at lung bifurcation sites where Cr(VI) particles would be expected to impact and persist (Holmes et al. 2008). Once inside the cell, Cr(VI) is rapidly reduced to trivalent chromium [Cr(III)]. This metabolism is carried out by intracellular molecules such as ascorbate, glutathione, amino acids, and cytochrome P450 (De Flora and Wetterhahn 1989). In lung tissue, ascorbate levels are quite low and it is unlikely that it contributes to Cr(VI) reduction in lung tissue (Holmes et al. 2008). The reduction process produces reactive oxygen species (ROS) and highly reactive chromium intermediates, such as pentavalent chromium [Cr(V)] and tetravalent chromium [Cr(IV)] (De Flora and Wetterhahn 1989). Cr(III) or one of these reactive intermediates is the ultimate carcinogenic species.

DNA Lesions

The reactions of Cr(VI) with DNA occur through its reductive intermediates. Cr(VI), itself, does not interact with DNA, but all of its reduction products do (De Flora and Wetterhahn 1989). ROS can cause oxidative damage and cross-links. Cr(III), Cr(IV), and Cr(V) can all cause primary DNA adducts as well as more complex adducts such as DNA–Cr–amino acid cross-links, DNA–Cr–DNA cross-links and DNA–Cr–protein cross-links (De Flora and Wetterhahn 1989; Wise et al. 2008).

The spectrum of Cr(VI)-induced oxidative DNA damage includes base and sugar lesions, strand breaks, DNA-protein cross-links, and abasic sites affecting both purines and pyrimidines (Wise et al. 2008). If they are not repaired they can lead to detrimental biological outcomes like cell death, mutations, and malignant transformation. The most commonly reported lesion is 7,8-dihydro-8-oxoguanine (8-oxoG). However, that lesion is oxidized to spiroiminodihydantoin (Sp), which may prove to be more toxicologically significant than the 8-oxoG lesion.

Oxidative lesions are typically repaired by base excision repair (BER). Cells treated with Cr(VI) have been shown to activate BER pathways (Wise et al. 2008). Studies in Chinese hamster cells, human cancer cell lines, and yeast show that BER repair genes, such as apn1, ntg1, ntg2, OGG1, NEIL1, and XRCC1, respond to Cr(VI) exposure. In some of the studies, loss of the BER genes led to mutagenesis or increased DNA damage compared to the BER-proficient cells, suggesting that for negative biological outcomes to occur, the BER repair pathway must become inactivated or overwhelmed.

Cr(VI)-induced ternary adducts also lead to detrimental biological outcomes, such as DNA polymerase arrest, changes in gene expression, cell death, mutagenesis, and DNA strand breaks including single-strand breaks (SSBs) and double-strand breaks (DSBs) (Wise et al. 2008). One long-standing challenge continues to be isolating and detecting specific Cr-DNA adducts. Over the past few decades, numerous laboratories have attempted and failed to isolate a Cr-DNA adduct. Current techniques can only measure Cr associated with DNA. These outcomes suggest that Cr is not forming a true adduct and instead is forming an electrostatic bond with the DNA backbone or base rather than a covalent bond. Regardless of how it interacts with DNA, the consequence of this association is a stalled replication fork leading to DNA strand breaks.

Nucleotide excision repair (NER) can repair relatively simple Cr-DNA adducts. Studies in hamster, human, and yeast cells show that NER does occur after Cr(VI) exposure with genes such as XPA, XPC, XPD, and XPF/ERCC1 responding to Cr(VI) exposure in cultured cells (Wise et al. 2008). In some of the studies, loss of the NER gene led to decreased mutagenesis and DNA damage indicating that negative biological outcomes can come from the NER-mediated repair of these lesions.

Mismatch repair (MMR) corrects mispaired bases and may address some of these Cr(VI)-induced lesions. Studies in human and mouse cells show that MMR occurs after soluble Cr(VI) exposure and genes, such as MLH1 and MSH2, respond to Cr(VI) exposure in cultured cells (Wise et al. 2008). Cr(VI)-induced human tumors appear to have acquired MMR deficiency due to the prevalence of microsatellite instability (MIN) in these tumors. However, as observed for NER, proficient MMR in Cr(VI)-treated cells is more mutagenic than MMR-deficient repair. Thus, the MMR deficiency observed in the tumors was probably acquired late in the overall carcinogenic process after the changes related to DNA damage have occurred.

DNA cross-link repair can also repair some of these lesions. Studies in cultured hamster and human cells show that Cr(VI) does induce DNA cross-link repair and genes such as FANCA, FANCD2, and FANCG respond to Cr(VI) exposure in cultured cells

(Wise et al. 2008). However, the normal process of this repair pathway induces a DNA double-strand break to remove the cross-link indicating that negative biological outcomes can come from the repair of these lesions.

DNA Strand Breaks and Chromosomal Aberrations

Cr(VI) does not appear to cause direct DNA strand breaks (Wise et al. 2008). The SSB appears to occur secondary to oxidative DNA damage and the DSB occurs secondary to a stalled replication fork or DNA cross-link repair. Indeed, the DSBs only form in G2 phase of the cell cycle. The SSBs are repaired rapidly by BER pathways and the DSBs are repaired by homologous recombination repair (HRR). Studies in hamster, mouse, and human cells show that double-strand break repair genes involved in HRR repair, such as H2A.X, Mre11, NBS1, ATM, and RAD51, respond to Cr(VI) exposure in cultured cells. These breaks can have negative biological outcomes, such as cell death, mutagenesis, chromosomal aberrations, and carcinogenesis.

Mutagenesis

Mutations are at the foundation of the multistage carcinogenesis paradigm that fits for some cancers, such as colorectal tumors. The concept is that carcinogenesis is a multistep process involving progressive accumulation of mutations in key genes resulting in a series of cellular and molecular changes. For this paradigm to apply to a chemical, it must induce significant mutagenesis to affect key genes. Cr(VI) can induce mutations (Holmes et al. 2008). Animal data show Cr(VI)-induced mutations were G:C targeted base substitutions with few deletions. Cell culture data also show G:C base substitution mutations but usually after exposure to very highly cytotoxic levels of Cr(VI). Moreover, cell culture studies suggest that for the mutations to occur, ascorbate levels typically need to be at levels greater than a millimolar (mM). By contrast, data from human lung tumors induced by Cr(VI) appear to contradict these studies as mutations are infrequent in Cr(VI)-induced tumors.

The explanation for this discrepancy is probably that Cr(VI) is only weakly mutagenic (Holmes et al. 2008). Cr(VI)-induced mutations can only be found in experimental studies, because they can employ very high Cr(VI) doses that do not occur in the workplace. Moreover, experimental studies can add millimolar levels of ascorbate to help induce these mutations, while in human lung tissue, ascorbate levels are actually 5.5 times lower. Thus, the reason mutation levels are low in Cr(VI)-induced human lung tumors may be due to the fact that both Cr(VI) level and ascorbate levels are too low.

Because Cr(VI) is a weak mutagen, occupational Cr(VI) exposures are not likely to induce a high enough mutation frequency to induce mutations in multiple key genes. For example, when cells are not overloaded with high ascorbate, Cr(VI) induced a two- to fourfold increase in mutation frequency over the spontaneous mutation rate (Holmes et al. 2008). Estimates indicate that in the normal lifespan of a cell, given about 70,000 genes in the human genome and a spontaneous mutation rate of 1.4×10^{-10} mutations per base pair per cell generation, only one mutation would arise spontaneously. Thus, by increasing mutation frequency by two- to fourfold, Cr(VI) would only produce two to four mutations in the lifespan of a cell. The likelihood that two to four random mutations would successfully mutate a key tumor suppressor gene or oncogene seems low. In addition, inactivating tumor suppressor genes often requires that both alleles be mutated, which also seems unlikely given the low mutation rate of Cr(VI). Thus, while Cr(VI) can be induced experimentally to cause mutations, Cr(VI)-induced mutations seem unlikely to be involved in the biological outcome of Cr(VI) exposure.

Genomic Instability

Acquiring an unstable genome is at the foundation of the genomic instability paradigm that fits for some cancers, such as lung tumors (Holmes et al. 2008). The concept is that disruption of the control of genomic stability results in a cascade of changes in the whole genome. There are two types of genomic instability: microsatellite instability (MIN) and chromosome instability (CIN). Some cancers, such as lung tumors, can have both MIN and CIN others, such as colorectal tumors, have MIN or CIN.

MIN is characterized by changes in the lengths of microsatellites which are series of repetitive noncoding DNA sequences that are abundant in the human genome. MIN has been observed in Cr(VI)-induced human tumors, but since MIN is a product of failed MMR, as discussed above, this outcome may be occurring late in the carcinogenesis pathway for Cr(VI) (Holmes et al. 2008).

CIN is a change in the frequency of the normal number (diploid) or structure of chromosomes, typically analyzed in metaphase chromosomes. Numerical CIN involves the loss and gain of whole chromosomes, while structural CIN involves chromosome translocations, fusions, and breaks. Cr(VI)-induced CIN has not been investigated in vivo or in chromate-induced tumors. However, human lung cell culture studies show Cr(VI) induces both forms of CIN (Holmes et al. 2008; Wise and Wise 2010).

A number of mechanisms underlie numerical CIN such as centrosome amplification, spindle assembly checkpoint bypass, malfunctions in sister chromatid cohesion, and abnormalities in kinetochore structure or function. Human lung cell culture studies show particulate and soluble Cr(VI) induce centrosome amplification and spindle assembly checkpoint bypass (Holmes and Wise 2010; Wise and Wise 2010). Structural CIN is most likely caused by Cr(VI)-induced DNA double-strand breaks (discussed above). Studies in hamster cells suggest that HRR protects against structural CIN, while nonhomologous end-joining (NHEJ) does not (Wise et al. 2008).

Considering the body of the Cr(VI) literature, it shows that genomic instability is the most likely mechanism that leads to Cr(VI)-induced carcinogenesis (Holmes et al. 2008). The reduction of Cr(VI) creates metabolic intermediates that adduct DNA and lead to ternary cross-links. These cross-links cause DNA double-strand breaks from stalled and collapsed replication forks and as a step in their repair. Chronic production of these breaks leads to structural CIN, centrosome amplification, and spindle assembly checkpoint bypass, which lead to numerical CIN. Combined, the two types of CIN result in genomic instability and ultimately lung cancer.

Other Biological Implications

Mutagenesis, genomic instability, and DNA strand breaks have well-established and profound implications for teratogenesis, reproductive success, and the development of offspring (Handel and Schimenti 2010). Increased occurrence of these events can lead to spontaneous abortions and failed implantation, decreasing reproductive success. They can lead to developmental abnormalities, both structural and behavioral.

More recent data show that DNA strand breaks can lead to neurological and immunological deficits (McKinnon and Caldecott 2007). These deficits occur because of loss of gene expression. It is likely that similar outcomes will be found in other organ systems as studies progress. Thus, the potential biological implications of the interaction of Cr(VI) with DNA are likely to be quite extensive.

Cross-References

- ► Chromium Binding to DNA
- ► Chromium Toxicity, High-Valent Chromium
- ► Chromium(III) and Immune System
- ► Chromium(VI), Oxidative Cell Damage
- ► Chromium, Physical and Chemical Properties
- ► Hexavalent Chromium and Cancer

References

Beyersmann D, Hartwig A (2008) Carcinogenic metal compounds: recent insight into molecular and cellular mechanisms. Arch Toxicol 82(8):493–512

De Flora S, Wetterhahn K (1989) Mechanisms of chromium metabolism and genotoxicity. Life Chem Rep 7:169–244

Handel MA, Schimenti JC (2010) Genetics of mammalian meiosis: regulation, dynamics and impact on fertility. Nat Rev Genet 11(2):124–136

Holmes AL, Wise JP Sr (2010) Mechanisms of metal-induced centrosome amplification. Biochem Soc Trans 38(6): 1687–1690

Holmes AL, Wise SS, Wise JP Sr (2008) Carcinogenicity of hexavalent chromium. Indian J Med Res 128:353–372

International Agency for Cancer Research (IARC) (1990) Monographs on the evaluation of carcinogenic risks to Humans: chromium, nickel and welding, vol 49. IARC, France

McKinnon PJ, Caldecott KW (2007) DNA strand break repair and human genetic disease. Annu Rev Genomics Hum Genet 8:37–55

Wise SS, Wise JP Sr (2010) Aneuploidy as an early mechanistic event in metal carcinogenesis. Biochem Soc Trans 38(6): 1650–1654

Wise SS, Holmes AL, Wise JP Sr (2008) Hexavalent chromium-induced DNA damage and repair mechanisms. Rev Environ Health 23(1):39–57

Hg

- ► Mercury and Immune Function
- ► Mercury Neurotoxicity

Hg in Plants

► Mercury in Plants

Hg^{2+} Excretion/Efflux

► Mercury Transporters

Hg^{2+} Resistance

► Mercury Transporters

Hg^{2+} Uptake

► Mercury Transporters

High Molecular Mass Aluminum Species in Serum

► Aluminum Speciation in Human Serum

Histological Diagnosis

► Silver Impregnation Methods in Diagnostics

Holmium

Takashiro Akitsu
Department of Chemistry, Tokyo University of Science, Shinjuku-ku, Tokyo, Japan

Definition

A lanthanoid element, the tenth element of the f-elements block (yttrium group), with the symbol Ho, atomic number 67, and atomic weight 164.93032. Electron configuration $[Xe]4f^{11}6s^2$. Holmium is composed of stable (^{165}Ho, 100%) and four synthetic radioactive (^{163}Ho; ^{164}Ho; ^{166}Ho; ^{167}Ho) isotopes. Discovered by P. T. Cleve in 1879. Holmium exhibits oxidation state III; atomic radii: 176 pm, covalent radii 193 pm; redox potential (acidic solution) Ho^{3+}/Ho −2.319 V; Ho^{3+}/Ho^{2+} −2.9 V; electronegativity (Pauling) 1.23. Ground electronic state of Ho^{3+} is $^5I_4$ with $S = 2$, $L = 6$, $J = 8$ with $\lambda = -520\ cm^{-1}$. Most stable technogenic radionuclide ^{163}Ho (half-life 4570 years). The most common compounds: Ho_2O_3, HoF_3, and $Ho(OH)_3$. Biologically, holmium is of low to moderate toxicity, can inhibit activity of reticuloendothelial system after intravenous infusion (Atkins et al. 2006; Cotton et al. 1999; Huheey et al. 1997; Oki et al. 1998; Rayner-Canham and Overton 2006).

Cross-References

► Lanthanide Ions as Luminescent Probes
► Lanthanide Metalloproteins
► Lanthanides and Cancer
► Lanthanides in Biological Labeling, Imaging, and Therapy
► Lanthanides in Nucleic Acid Analysis
► Lanthanides, Physical and Chemical Characteristics

References

Atkins P, Overton T, Rourke J, Weller M, Armstrong F (2006) Shriver and Atkins inorganic chemistry, 4th edn. Oxford University Press, Oxford/New York
Cotton FA, Wilkinson G, Murillo CA, Bochmann M (1999) Advanced inorganic chemistry, 6th edn. Wiley-Interscience, New York
Huheey JE, Keiter EA, Keiter RL (1997) Inorganic chemistry: principles of structure and reactivity, 4th edn. Prentice Hall, New York
Oki M, Osawa T, Tanaka M, Chihara H (1998) Encyclopedic dictionary of chemistry. Tokyo Kagaku Dojin, Tokyo
Rayner-Canham G, Overton T (2006) Descriptive inorganic chemistry, 4th edn. W. H. Freeman, New York

Holo α-LA Apo α-LA

► α-Lactalbumin

HSPs

► Arsenic-Induced Stress Proteins

Human Health Effects of Arsenic Exposure

► Arsenic

Hybrid Catalysts

► Palladium-catalysed Allylic Nucleophilic Substitution Reactions, Artificial Metalloenzymes

Hydrargyrum

► Mercury Neurotoxicity

Hydration

► Nitrile Hydratase and Related Enzyme

Hydrogen-Peroxide Oxidoreductase Enzymes

► Manganese and Catalases

Hydrolases

► Cobalt-containing Enzymes

Hydroxyl Radicals

► Arsenic, Free Radical and Oxidative Stress

Hyphenated Techniques

► Aluminum Speciation in Human Serum

I

Immune Responses

▶ Chromium(III) and Immune System

Immunity

▶ Magnesium and Inflammation

Immunolabeling

▶ Palladium, Colloidal Nanoparticles in Electron Microscopy

In Silico Prediction

▶ Zinc-Binding Proteins, Abundance

Incorporation of Selenium into Protein

▶ Artificial Selenoproteins

Indium, Physical and Chemical Properties

Fathi Habashi
Department of Mining, Metallurgical, and Materials Engineering, Laval University, Quebec City, Canada

Indium is a silvery-white low-melting-point metal belonging to the less typical metals group, i.e., when it loses its outermost electrons, it will not have the electronic structure of inert gases; hence, it will not be reactive as the typical metals. It is softer than lead, ductile, malleable, and retains its plastic properties at cryogenic temperatures. It becomes superconducting at 3.37 K. It does not form its own minerals but is found in trace amounts in many minerals, e.g., sphalerite, ZnS. Low-melting alloys represent a major use of indium. Indium has the highest solubility in mercury [57%].

Physical Properties

Atomic number	49
Atomic weight	114.82
Relative abundance in the Earth's crust, %	1×10^{-5}
Melting point, °C	156.6
Boiling point, °C	2,072
Density g/cm^3	
At 20°C	7.31
At melting point	7.02

(*continued*)

V.N. Uversky et al. (eds.), *Encyclopedia of Metalloproteins*, DOI 10.1007/978-1-4614-1533-6,

Heat of fusion, kJ·mol^{-1}	3.281
Heat of evaporation, kJ·mol^{-1}	231.8
Specific heat at 25°C, kJ mol^{-1} K^{-1}	26.70
Entropy at 25°C, kJ mol^{-1} K^{-1}	57.7
Entropy of fusion, kJ mol^{-1} K^{-1}	7.58
Entropy of vaporization, kJ mol^{-1} K^{-1}	98.56
Coefficient of linear expansion (0–100°C), K^{-1}	3×10^{-5}
Vapor pressure p, kPa, at temperature, K [between melting and boiling points]	$\mathrm{Log}\, p = 9.835 - 12{,}860/T - 0.7\log T$
Atomic radius (coordination number 12), nm	0.162
Ionic radius (In^{3+}, coordination number 6), nm	0.081
Atomic volume, m^3/mol	15.73×10^{-6}
Crystal structure Parameters, nm	Tetragonal A6 $a_0 = 0.458$ $c_0 = 0.494$
Thermal conductivity (0–100°C), W m^{-1} K^{-1}	71.1
Surface tension at temperature T, N/m	$0.602 - 10^{-4}\, T$
Electrical resistivity at 3.38 K At 273.15 K, Ω m	Superconducting 8.4×10^{-8}
Temperature coefficient (0–100°C), K^{-1}	4.9×10^{-3}
Brinell hardness	0.9
Tensile strength, MPa	2.645
Modulus of elasticity, GPa	10.8
Thermal neutron cross section at 2,200 m/s Absorption, m^2/atom Scattering	 $(194 \pm 2) \times 10^{-28}$ $(2.2 \pm 0.5) \times 10^{-28}$
Standard electrode potentials, V $In^{3+} + 3e^- \leftrightarrows In$ $In^{3+} + 2e^- \leftrightarrows In^+$ $In^{2+} + e^- \leftrightarrows In^+$ $In^{3+} + e^- \leftrightarrows In^{2+}$	 −0.338 −0.40 −0.40 −0.49

Chemical Properties

The most common valence of indium is three. Monovalent and bivalent compounds of indium with oxygen, sulfur, and halogens are also known. The trivalent indium compounds are the most stable. The lower valence compounds tend to disproportionate to give the corresponding trivalent compounds and indium metal. The semiconducting compounds *indium antimonide*, InSb, *indium arsenide*, InAs, and *indium phosphide*, InP, are prepared by direct combination of the high-purity elements at high temperature. Indium phosphide is also obtained by thermal decomposition of a mixture of a trialkyl indium compound and phosphine (PH_3).

References

Felix N (1997) Indium. In: Habashi F (ed) Handbook of extractive metallurgy. Wiley, Weinheim, pp 1531–1542

Habashi F (2001) Cadmium, indium, and thallium production. In: Encyclopedia of materials: science & technology, pp 879–880

Inflammation

▶ Chromium(III), Cytokines, and Hormones

▶ Selenium and Glutathione Peroxidases

▶ Zinc Leukotriene A_4 Hydrolase/Aminopeptidase Dual Activity

Inhibit: Hinder

▶ Monovalent Cations in Tryptophan Synthase Catalysis and Substrate Channeling Regulation

Inhibition of DNA Repair

▶ Antimony, Impaired Nucleotide Excision Repair

Inhibitory Mechanism

▶ Gold(III), Cyclometalated Compound, Inhibition of Human DNA Topoisomerase IB

Inner Coordination Sphere: Coordination Bond

▶ Monovalent Cations in Tryptophan Synthase Catalysis and Substrate Channeling Regulation

Inner-Sphere Cr-DNA Complexation

▶ Chromium Binding to DNA

Inorganic Chromium

▶ Chromium and Nutritional Supplement

Inorganic Mercury

▶ Mercury Nephrotoxicity
▶ Mercury Toxicity

Inositol Triphosphate (IP_3)

▶ Calcium Sparklets and Waves

Inositol Triphosphate Receptors (IP_3Rs)

▶ Calcium Sparklets and Waves

Insulin Signal Transduction

▶ Chromium and Insulin Signaling

Insulin Signaling

▶ Chromium and Insulin Signaling

Insulin-Signaling Pathway

▶ Chromium and Insulin Signaling

Insusceptibility

▶ Cadmium Exposure, Cellular and Molecular Adaptations

Integrins

▶ Nanosilver, Next-Generation Antithrombotic Agent

Interaction of Cadmium and Metallothionein

▶ Cadmium and Metallothionein

Interactions of *Cupriavidus Metallidurans* CH34 with Mobile, Ionic Gold

▶ Gold Biomineralization in Bacterium Cupriavidus metallidurans

Interactions of Mercury with Low Molecular Mass Substances

▶ Mercury and Low Molecular Mass Substances

Interdependence Between Manganese and Other Metal Ions

▶ Manganese, Interrelation with Other Metal Ions in Health and Disease

Intermediate: Metabolite

▶ Monovalent Cations in Tryptophan Synthase Catalysis and Substrate Channeling Regulation

Intermolecular Cr-DNA Cross-Link

► Chromium Binding to DNA

Intracellular Signaling

► C2 Domain Proteins

Intramolecular G-Quadruplex Structure

► Strontium and DNA Aptamer Folding

Iodothyronine Deiodinase, DIO

► Selenoproteins and Thyroid Gland

Ion Channel

► Potassium Channel Diversity, Regulation of Potassium Flux across Pores
► Potassium Channels, Structure and Function

Ion Channel Structure

► Potassium Channels and Toxins, Interactions

Ion Channels

► Magnesium and Cell Cycle

Ion Channels Structure–Functions Relations

► hERG (KCNH2) Potassium Channel, Function, Structure and Implications for Health and Disease

Iridium, Physical and Chemical Properties

Fathi Habashi
Department of Mining, Metallurgical, and Materials Engineering, Laval University, Quebec City, Canada

Iridium is a member of the group of six platinum metals which occur together in nature and is located in the center of the Periodic Table.

Fe	Co	Ni
Ru	Rh	Pd
Os	*Iridium*	Pt

Physical Properties

Iridium is a transition metal like the other members of the group. They are less reactive than the typical metals but more reactive than the less typical metals. The transition metals are characterized by having the outermost electron shell containing two electrons and the next inner shell an increasing number of electrons. Because of the small energy differences between the valence shells, a number of oxidation states occur.

Atomic number	77
Atomic weight	192.22
Relative abundance	1×10^{-7}
Density at 20°C, g/cm^3	22.63
Major natural isotopes	191 (37.3%)
	193 (62.7%)
Crystal structure	Face centered cubic
Atomic radius, nm	0.136
Melting point, °C	2,410
Boiling point, °C	4,550
Specific heat at 25°C, C_p, J g^{-1} K^{-1}	0.13
Thermal conductivity λ, W m^{-1} K^{-1}	59
Brinell hardness	172
Tensile strength σ_B, N/mm^2	490.5
Specific electrical resistance at 0°C, μΩ cm	4.71
Temperature coefficient of electrical resistance (0–100°C), K^{-1}	0.0043
Thermoelectric voltage versus Pt at 100°C E, V	+0.66

Chemical Properties

Because of the small energy differences between the valence shells, a number of oxidation states occur: −1, 0, +1, +2, +3, +4, +5, +6. The tetravalent state is the most common. The most important iridium compound is ammonium hexachloroiridate, $(NH_4)_2[IrCl_6]$, which is precipitated by NH_3 from aqua regia leach solution as a black powder. The powder is reduced by hydrogen at 800°C to produce iridium metal.

References

Habashi F (2003) Metals from ores. An introduction to extractive metallurgy. Métallurgie extractive Québec, Quebec City, Canada. Distributed by Laval University Bookstore, www.zone.ul.ca

Renner H (1997) Platinum group metals. In: Habashi F (ed) Handbook of extractive metallurgy. Wiley-VCH, Weinheim, pp 1269–1360

Iron Absorption

► Iron Homeostasis in Health and Disease

Iron Binding Protein

► Iron Proteins, Transferrins and iron transport

Iron Bionanomineral Encapsulated in Protein

► Iron Proteins, Ferritin

Iron Deficiency

► Iron Homeostasis in Health and Disease

Iron Homeostasis in Health and Disease

Gregory J. Anderson
Iron Metabolism Laboratory, Queensland Institute of Medical Research, PO Royal Brisbane Hospital, Brisbane, QLD, Australia

Synonyms

Anemia; Hemochromatosis; Iron absorption; Iron deficiency; Iron metabolism; Iron overload; Regulation of iron traffic

Definitions

The trace element iron is essential for the function of many proteins, but can also be toxic when present in excess. Normal physiology requires cellular and body iron levels to be kept within strict limits and thus these processes are tightly controlled. Too little iron can lead to anemia and suboptimal cellular function, while excess iron (iron overload; hemochromatosis) can damage cells and tissues. Either too little or too much iron can have significant clinical consequences.

Introduction

Iron is essential for a wide range of cellular functions. Its ability to readily accept and lose electrons makes it a critical component of the active site of a large number of enzymes, and in the oxygen-carrying proteins hemoglobin and myoglobin. Iron can also be toxic when present in excess as it can catalyze the production of reactive oxygen species through the Fenton and Haber-Weiss reactions. Because of this dual nature, essential but toxic, homeostatic processes maintain cellular and body iron levels within strict physiological limits (Crichton 2009). This entry focuses on mammalian, and specifically human, iron homeostasis, but almost all organisms have an absolute requirement for iron and must regulate its intake, utilization, and export.

The body of the average adult human male contains 3.5–4.5 g of iron (50 mg/kg). Of this, approximately 65% (2,600 mg) is found in circulating hemoglobin, approximately 20% (1,000 mg) is storage iron (which is in excess of immediate body needs), and the remaining 15% is distributed throughout various body tissues. A small, but vital component is the 3 mg of iron bound to transferrin in the circulating blood. This essential plasma protein binds iron with extremely high affinity (10^{23} M^{-1}) and is the major form by which iron is delivered to cells to meet their metabolic needs. Women have a smaller amount of body iron (approximately 3–4 g; 40 mg/kg) and a disproportionately lower pool of storage iron (10–15%; 300 mg).

A summary of the major proteins involved in maintaining iron homeostasis in cells and tissues is provided in Table 1 and many of these will be considered in more detail in the following text.

Normal Iron Homeostasis

Body iron levels are tightly conserved. Humans do not possess a mechanism for the active excretion of iron, so under normal circumstances, the small amounts of iron that are lost from the body come from the sloughing of intestinal epithelial cells, skin cells, and in the urine. This equates to losses of approximately 1 mg per day. Both men and women have similar basal losses of iron, but women incur additional losses through menstrual bleeding and during pregnancy. The loss of iron is balanced by a similar amount of iron absorption from the diet (1–2 mg/day) such that there is no net loss or gain of iron. Although body iron intake represents only a few milligrams per day, the cells of the body have daily metabolic requirements of approximately 30 mg, most of which (24 mg) is needed by the erythroid marrow for hemoglobin synthesis. These requirements are met predominantly through the recycling of iron from senescent red cells. At the end of their life, erythrocytes are phagocytosed by macrophages and the hemoglobin is broken down and iron released to plasma transferrin. Thus, it is the combination of internal iron recycling and dietary iron absorption that maintains body iron homeostasis.

Intestinal iron absorption involves the transport of iron across the mature villus enterocytes of the proximal small intestine (Fig. 1) (Anderson and Vulpe 2009; Anderson and McLaren 2012). The gut can utilize iron from a wide range of dietary sources, but ultimately these are considered as either heme or nonheme iron. Heme iron is particularly rich is dietary components such as red meat. It is less influenced by the composition of the diet than nonheme iron and is generally more efficiently absorbed, but how heme iron traverses the intestinal epithelium is poorly understood. The form of nonheme iron can be quite diverse and includes elemental or ionic forms of iron, as well as protein-bound iron, such as that found in the iron storage protein ferritin (Anderson and McLaren 2012). Solubilized ionic iron is transported across the enterocyte brush border (apical) membrane through a multispanning membrane protein known as divalent metal-ion transporter 1 (DMT1). Dietary nonheme iron is principally in the oxidized or ferric (Fe^{3+}) form and must be reduced before it can be utilized by DMT1 which is a ferrous (Fe^{2+}) iron transporter. Duodenal cytochrome b (Dcytb) has been identified as one brush border ferric reductase, but others are likely involved. Ferritin appears to enter the intestinal epithelial cells via an endocytic pathway. The fate of iron within the enterocyte is unclear, but as in other cells it is probably bound by low molecular weight organic acids or loosely to proteins. If the body has an immediate need for iron, the metal will move quickly across the basolateral membrane and into the circulation. Iron that does not immediately pass into the circulation is sequestered within ferritin within the enterocyte. Although this iron can subsequently be absorbed if required, most of it is lost when the enterocytes are shed into the intestinal lumen at the end of their 3–4 day life span. Iron that is needed is transported across into the circulation through the iron exporter ferroportin 1 (FPN1). For this process to occur efficiently, FPN1 must act in conjunction with the iron oxidase hephaestin. This basolateral transport of iron is rate limiting for iron absorption and represents a major point of control of body iron intake.

Iron is transported around the body bound to the abundant plasma protein transferrin (TF) (Anderson and Vulpe 2009; Anderson and McLaren 2012). TF accepts iron from the small intestine and iron storage sites and delivers it to cells where it is required. TF can bind two atoms of iron and it is this diferric TF that is most efficiently delivered to cells via its interaction with transferrin receptor 1 (TFR1) on the plasma membrane. Cells with particularly high iron requirements, such as developing erythroid cells (which require iron for hemoglobin synthesis) and rapidly dividing cells (e.g., the enterocytes of the intestinal crypts), express

Iron Homeostasis in Health and Disease, Table 1 Some mammalian proteins of iron homeostasis and its regulation

Protein	Protein abbreviation	Gene symbol[a]	Function/role in iron metabolism
Iron transport and uptake			
Transferrin	TF	*TF*	Plasma iron transport
Transferrin receptor 1	TFR1	*TFRC*	Internalization of diferric transferrin
Divalent metal-ion transporter 1	DMT1	*SLC11A2*	Ferrous iron importer
Natural resistance-associated macrophage protein 1	NRAMP1	*SLC11A1*	Phagosomal iron transport in macrophages
Lipocalin-2	LCN2	*LCN2*	Binds bacterial siderophores and delivers them to mammalian cells
Six transmembrane epithelial antigen of prostate protein 3	STEAP3	*STEAP3*	Iron reductase of erythroid cells
Exocyst complex component 6	SEC15L1	*EXOC6*	Trafficking of TFR1-containing vesicles
Duodenal cytochrome B	DCYTB	*CYBRD1*	Enterocyte brush border reductase
Heme carrier protein 1/Proton-coupled folate transporter	HCP1/PCFT	*SLC46A1*	Low affinity heme importer and high-affinity folate transporter
Iron scavenging and recycling			
Hemopexin	HPX	*HPX*	Scavenges and clears heme
Haptoglobin	HP	*HP*	Scavenges and clears hemoglobin
Heme oxygenase 1	HO-1	*HMOX1*	Recycling of hemoglobin iron in macrophages
Iron storage			
Ferritin H chain	HFT	*FTH1*	Intracellular iron storage (in combination with LFT). Exhibits ferroxidase activity
Ferritin L chain	LFT	*FTL*	Intracellular iron storage (in combination with HFT)
Iron export			
Ferroportin 1	FPN1	*SLC40A1*	Ferrous iron exporter
Ceruloplasmin	CP	*CP*	Systemic (secreted) ferrous iron oxidase
Hephaestin	HEPH	*HEPH*	Ferrous iron oxidase (gut and CNS)
Amyloid beta A4 precursor protein	APP	*APP*	Ferrous iron oxidase (CNS)
Feline leukemia virus, type C, receptor	FLVCR	*FLVCR1*	Heme export protein
Mitochondrial iron metabolism			
Mitoferrin	MFRN	*SLC25A37*	Mitochondrial iron importer
Mitochondrial ferritin	MtF	*FTMT*	Stores iron in mitochondria
ABC transporter type B7	ABCB7	*ABCB7*	Mitochondrial Fe-S exporter
Regulation of cellular iron metabolism			
Iron regulatory protein 1	IRP1	*ACO1*	Iron-dependent RNA-binding protein
Iron regulatory protein 2	IRP2	*IREB2*	Iron-dependent RNA-binding protein
Regulation of systemic iron metabolism			
Hepcidin	HEPC	*HAMP*	Regulator of iron release into plasma
Hemochromatosis protein	HFE	*HFE*	Positive regulator of hepcidin
Transferrin receptor 2	TFR2	*TFR2*	Positive regulator of hepcidin
Hemojuvelin	HJV	*HFE2*	Positive regulator of hepcidin
Transmembrane protease, serine 6 (Matriptase-2)	TMPRSS6	*TMPRSS6*	Negative regulator of hepcidin

[a]Human genome organization approved symbol

particularly high levels of TFR1 (Fig. 1). TF-bound iron can also be taken up by certain cells via TFR1-independent pathways that remain poorly characterized. Under normal physiological conditions, circulating TF is only approximately 30% saturated with iron, so there is a considerable reserve capacity to sequester any large amounts of potentially toxic iron that enter the plasma. In certain pathological

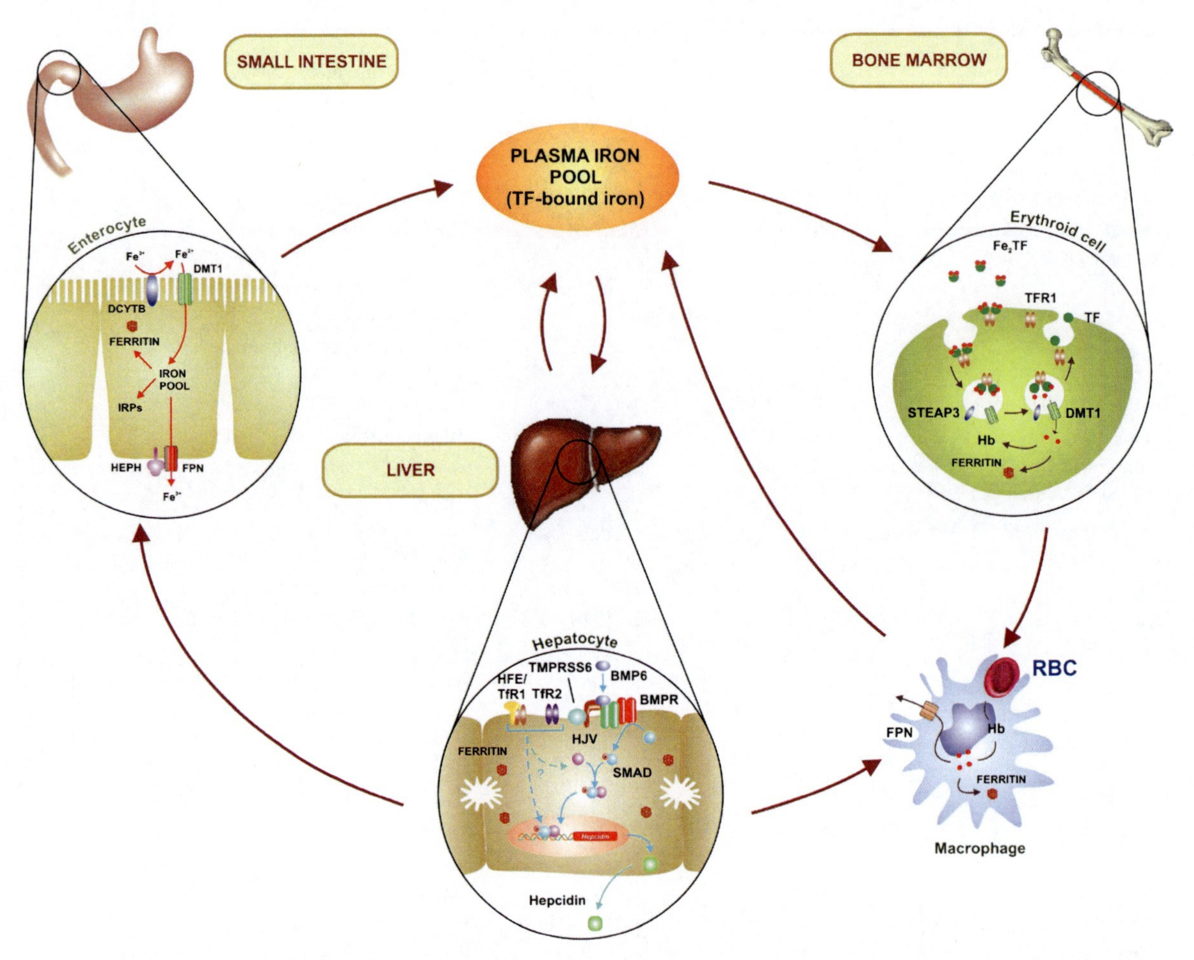

Iron Homeostasis in Health and Disease, Fig. 1 Overview of body iron homeostasis. Dietary iron traverses the enterocytes of the proximal small intestine, crossing the apical membrane via DMT1 and the basolateral membrane via FPN1. This iron binds to TF in the plasma pool and is distributed to cells throughout the body. Most iron is directed to immature erythroid cells in the bone marrow which take up TF-bound iron very efficiently and use it for hemoglobin synthesis. At the end of their life, red blood cells are phagocytosed by macrophages and the hemoglobin-derived iron is recycled to the plasma pool. A range of signals, including plasma iron levels, act on hepatocytes through several key signal transduction pathways to modulate the expression of the iron regulatory hormone hepcidin. Hepcidin in turn acts on macrophages, enterocytes, and body iron cells to limit their capacity to export iron to the plasma

situations, the capacity of TF to bind iron is exceeded and non-transferrin-bound iron may appear in the circulation. This form of iron can very rapidly enter cells, at least partially via DMT1, and is potentially highly cytotoxic. Small amounts of ferritin iron can also be found in the circulation and these can be delivered to certain cell types via receptors including TIM2 and Scara5 (Anderson and Vulpe 2009).

After TF binds to TFR1 the complex undergoes receptor-mediated endocytosis. The endosomes are acidified, iron is released from TF and transported into the cytoplasm (Fig. 1) (Anderson and Vulpe 2009). The release of iron requires reduction and this is carried out by a member of the STEAP family of oxidoreductases. The resulting ferrous iron is transported across the endosomal membrane via DMT1, the same protein found on the brush border of intestinal enterocytes. This iron can have a range of fates, depending on the needs of the cell. It can be used for metabolic purposes, (e.g., via incorporation into mitochondria for the synthesis of heme or iron-sulfur proteins), to regulate cellular iron homeostasis (e.g., through modulating iron regulatory proteins 1 and 2), or it can be stored within ferritin and hemosiderin. Some cells (e.g., hepatocytes, macrophages) are particularly well adapted for iron storage. This stored iron can be used at a later time either by the cell itself or by other cells. In the latter case, the iron must be exported from the storage cell and this is achieved via FPN1. The plasma ferroxidase ceruloplasmin is required for efficient iron release from FPN1.

The Regulation of Iron Metabolism

Both cellular and body iron trafficking are tightly regulated to ensure an adequate supply of iron for metabolic needs while preventing the accumulation of potentially toxic excess iron (Hentze et al. 2010; Ganz 2011; Knutson 2010). Body iron homeostasis is regulated by a small peptide known as hepcidin. Hepcidin is secreted by the liver and its main function is to control the amount of iron that is released into the circulation (Fig. 1). It does this by binding to the iron exporter FPN1 and facilitating its internalization and degradation, thereby limiting the capacity cells to export iron. When body iron requirements are high (e.g., under iron deficient conditions or when erythropoiesis is elevated), hepcidin levels are reduced, cell surface FPN1 expression is preserved and enterocytes, macrophages, and other cells are able to donate iron to the plasma. In contrast, when iron demands are reduced, high levels of hepcidin lead to a reduction in cellular iron export. Thus, there is a close inverse relationship between hepcidin levels and cellular iron release.

While the basic facets of the mode of action of hepcidin are now understood (Hentze et al. 2010; Ganz 2011; Knutson 2010), how hepcidin itself responds to changes in body iron demand is less well defined and this remains an area of intense investigation. Nevertheless, a number of the key signaling pathways in hepatocytes that modulate transcription of the *HAMP* gene (which encodes hepcidin) have been identified (Fig. 1). The most important of these is BMP/SMAD signaling. Bone morphogenetic proteins (BMPs) are members of the TGFβ superfamily of proteins and are able to stimulate *HAMP* expression. They bind to a complex of Type I and Type II BMP receptors (BMPRs) on the plasma membrane and this complex in turn phosphorylates the signaling components SMADs 1, 5, and 8. These subsequently combine with a co-SMAD (SMAD4) and the complex translocates to the nucleus where it binds to SMAD response elements in the *HAMP* promoter. Although a number of BMPs have been shown to enhance hepcidin expression, BMP6 appears to be the most important of these in vivo. Signaling through the BMPRs is enhanced by the co-receptor hemojuvelin (HJV). Patients with mutations in HJV have severely reduced hepcidin expression and develop early-onset iron-loading disease (see below), confirming the physiological importance of HJV in hepcidin regulation. Interestingly, while membrane-bound HJV stimulates BMP-SMAD signaling, a soluble form of HJV can inhibit the process. The source of sHJV in vivo is not known. The activity of HJV can also be reduced by the membrane-bound serine protease matriptase-2 (or TMPRSS6). Patients with mutations in the TMPRSS6 gene have enhanced hepcidin production and chronic iron deficiency anemia, indicating that the matriptase-2 plays an important role in maintaining normal hepcidin expression.

The other well-characterized signaling pathway modulating *HAMP* expression is its response to inflammatory stimuli (Ganz 2011). Inflammation is a strong inducer of hepcidin and the hypoferremia that accompanies inflammation can be explained, at least in part, by increased plasma hepcidin levels. Pro-inflammatory cytokines such as interleukin-6 (IL-6) enhance hepcidin expression by signaling through the JAK-STAT pathway. However, there is cross talk between the JAK-STAT and BMP-SMAD signaling pathways as a functional SMAD-binding site in the *HAMP* promoter is necessary for the stimulation of *HAMP* expression.

How a number of the other known modulators of hepcidin exert their effects is poorly understood. Increased body iron stores will enhance hepcidin expression and at least two regulatory pathways have been proposed (Ganz 2011; Knutson 2010). One is through direct stimulation of BMP-SMAD signaling by BMP6 as the levels of this protein have been shown to be enhanced when body iron stores are increased. Body iron levels may also be communicated via the hepatocyte membrane proteins HFE and TFR2. Mutations in each of these proteins lead to reduced hepcidin expression and body iron loading, so they act as upstream regulators of *HAMP* expression. The link between body iron demand and these proteins may be provided by diferric TF. Diferric TF binding stabilizes TFR2 while both TF and HFE can compete for binding to TFR1. With increased body iron load, diferric TF levels rise and this may favor signaling through TFR2 and HFE (perhaps as a complex). There is some evidence that these proteins can enhance BMP-SMAD signaling, but whether they act solely through this pathway or have some independent functions has yet to be resolved.

Since the major sink for iron in the body is the erythroid marrow, it is no surprise that increased erythropoiesis reduces hepcidin expression, thus increasing iron supply by promoting iron release from macrophages and other storage sites, and increasing iron absorption (Ganz 2011). There are several potential mechanisms by which erythropoiesis could reduce *HAMP* expression, and all may operate together. Firstly, by creating a high demand for iron and thereby reducing systemic iron levels, hepcidin levels may be reduced through the iron-dependent mechanisms described above. Secondly, increased erythropoiesis could be associated with systemic hypoxia. The *HAMP* promoter contains two hypoxia response elements and it has been demonstrated that HIF-1α signaling will reduce *HAMP* transcription. Finally, there is evidence that erythroid-specific factors may reduce hepcidin regulation. Two marrow-derived proteins, GDF15 and TWSG1, have been shown to act on hepatic cells to reduce hepcidin expression, likely through their effects on BMP signaling, but whether they are the physiologically important molecules or whether others are involved is still under investigation.

Disorders of Iron Metabolism

Either too little or too much iron can have significant clinical consequences and disorders of iron homeostasis represent a significant class of human diseases (Anderson and McLaren 2012; Wessling-Resnick 2010; Camaschella and Poggiali 2011; Porter 2009; Rivella 2009). The main cause of low iron levels in the body is inadequate dietary intake and iron deficiency is the world's most significant nutritional deficiency. In contrast, body iron overload usually results from mutations in genes encoding proteins regulating body iron intake. Some of these conditions are covered in more detail below and a summary is provided in Table 2.

The diagnosis of disturbed iron homeostasis is usually made on the basis of an abnormal blood test and it is usual for a number of different markers of iron status to be taken into account. The parameters usually measured and their normal ranges are shown in Table 3. Direct measurement of serum iron tends to be indicative of iron status but is rarely used on its own. Transferrin saturation tends to be more widely used, but again it is usually not employed in isolation. In the absence of inflammation, the serum ferritin concentration usually provides an accurate representation of the amount of storage iron, so it is a particularly useful parameter for assessing excess iron in the body. A low serum ferritin ($<$12–15 ng/mL) is indicative of iron depletion, but it is not a good indicator of the degree of depletion. However, after storage iron dissipates and body iron levels continue to decrease, soluble transferrin receptor (sTFR) increases in proportion to the degree of iron depletion. The ratio of sTFR to serum ferritin has proven a useful index of body iron levels over quite a wide range. Various red cell parameters (hemoglobin, MCV, RDW, reticulocyte hemoglobin) are useful in the diagnosis of iron deficiency anemia. More invasive procedures such as bone marrow biopsy and liver biopsy can be used in the diagnosis of iron repletion or iron overload respectively, but these are not used as frequently today as they were in the past.

Iron Deficiency

Iron deficiency is normally considered in terms of stages (Anderson and McLaren 2012). A normal adult human has a significant reserve of stored iron that can be drawn upon in times of need. As body iron levels decline, these stores are eroded. Iron depletion is the situation where little stored iron remains but hematopoiesis remains unaffected and plasma iron indices are only slightly reduced. With further declines in iron content, the condition of iron deficient erythropoiesis is reached where iron supply to the marrow is impaired. This stage is associated with significant declines in plasma iron levels and impaired red cell production, but red cell morphology is usually normal. As body iron levels decline further, erythropoiesis becomes more severely impaired and iron deficiency anemia results. At this stage frank decreases in the hemoglobin level and a reduction in red cell size are evident. Although it is changes in the erythroid compartment that are most apparent, iron supply to all body cells becomes limiting. The consequences of anemia and tissue iron deficiency include lethargy, reduced work and exercise performance, defective immune

Iron Homeostasis in Health and Disease, Table 2 Major disorders of iron homeostasis

Disorder or cause	Gene	Clinical features
Iron deficiency and anemia		
Due to inadequate iron intake, increased body iron requirements, blood loss, gastrointestinal pathology	–	Lethargy; reduced work and exercise performance; defective immune function; impaired motor and intellectual development in infants; adverse outcomes in pregnancy; anemia
Iron-refractory iron deficiency anemia (IRIDA)	*TMPRSS6*	Reduced iron absorption secondary to increased hepcidin; features as above; limited response to iron therapy
Chronic inflammation and disease	–	Anemia and the other general features of iron deficiency
Primary iron overload		
HFE-related hemochromatosis (Type 1)	*HFE*	Parenchymal iron overload, liver disease
TFR2-related hemochromatosis (Type 3)	*TFR2*	Parenchymal iron overload, liver disease
FPN-related hemochromatosis (Type 4)	*FPN1*	Transport defective form – parenchymal and reticuloendothelial iron overload in the liver; Hepcidin resistance form – parenchymal iron overload
Juvenile hemochromatosis (HJV-related) (Type 2A)	*HJV*	Severe parenchymal iron overload, cardiac disease, liver cirrhosis, endocrine failure, diabetes, arthropathy
Juvenile hemochromatosis (hepcidin-related) (Type 2B)	*HAMP*	Severe parenchymal iron overload, cardiac disease, liver cirrhosis, endocrine failure, diabetes, arthropathy
Secondary iron overload		
Ineffective erythropoiesis (e.g. β-thalassemia, sickle-cell anemia, X linked sideroblastic anemia)	e.g., *HBB*, *ALAS2*	Parenchymal and reticuloendothelial iron overload; anemia; reticulocytosis; increased absorption secondary to enhanced erythropoiesis; compounded by transfusion
Chronic hemolytic anemias (e.g., pyruvate kinase deficiency, hereditary spherocytosis)	e.g., *ALAS2*, *PKLR*	Parenchymal iron load, anemia, jaundice, and splenomegaly
Hypoplastic anemias (e.g., Fanconi anemia, Diamond-Blackfan anemia, chronic renal failure)	e.g., *FANC and other genes*	Iron loading and anemia; a range of other sequelae such as pancytopenia, myelodysplasia, red cell aplasia, predisposition to leukemia
Some other iron-loading disorders		
Friedreich Ataxia	*FXN*	Mitochondrial iron overload, neurological and heart disease
Hereditary atransferrinemia	*TF*	Severe hemosiderosis of the heart and liver
Hereditary aceruloplasminemia	*CP*	Iron loading, degeneration of retina and basal ganglia, DM

function, impaired motor and intellectual development in infants, and adverse outcomes in pregnancy.

The main cause of iron deficiency globally is inadequate iron intake (Anderson and McLaren 2012). Throughout much of the world, particularly in developing countries, this reflects general malnutrition, but even in developed countries iron deficiency can be common in certain groups. Young infants and children, adolescent girls, and pregnant women tend to be at increased risk of iron deficiency due to rapid growth, the onset of menses combined with rapid growth, and the iron requirements of the fetus, placenta, and expanded maternal red cell mass, respectively. In general, women are more prone to iron deficiency than men as they have lower basal iron stores and regularly lose blood through menstruation. The nature of the diet can have a significant influence on iron status. People consuming diets rich in protein-bound iron (heme and ferritin iron), and in substances that promote the absorption of elemental iron (e.g., ascorbic acid) are less prone to iron deficiency than those with minimal protein-bound iron intake consuming diets rich in inhibitors of elemental iron absorption. Phytates, fiber, polyphenols, and tannins (rich in plant-based diets) as well as calcium can greatly inhibit the absorption of elemental iron but appear to have limited effect on the absorption of protein-bound iron.

Iron deficiency can also result from a range of pathological conditions (Anderson and McLaren 2012). Most iron deficiency in adults in developed countries is the result of blood loss. This could reflect bleeding through gastrointestinal malignancies, ulcers, or from the prolonged consumption of aspirin or nonsteroidal anti-inflammatory drugs, or intravascular

Iron Homeostasis in Health and Disease, Table 3 Selected markers for the evaluation of body iron status

Test	Normal range	Comments
Serum iron	10–30 μmol/L	Low in iron deficiency and high in iron overload; variable; requires fasting; specific in uncomplicated iron deficiency but reduced by inflammation.
Transferrin saturation	20–50%	Low in iron deficiency and high in iron overload; variable; requires fasting; specific in uncomplicated iron deficiency but reduced by inflammation.
Total iron-binding capacity (TIBC)	47–70 μmol/L	High in iron deficiency and low in iron overload; increased in pregnancy and by oral contraceptives; not used in isolation.
Serum ferritin	15–300 μg/L	Low in iron deficiency and high in iron overload; correlates well with iron stores in uncomplicated iron loading; increased by inflammation.
Soluble transferrin receptor (sTFR)	Varies according to method	Increased in iron deficiency and in response to enhanced erythropoiesis; not influenced by inflammation.
sTFR/ferritin	Varies according to sTFR method	High in iron deficiency and low in iron overload; reflects iron stores over a broad range; can be influenced by inflammation.
Hemoglobin	Women >120 g/L Men >130 g/L	Reduced in advanced iron deficiency/iron deficiency anemia; influenced by smoking and altitude.
Reticulocyte hemoglobin content		Early indicator of iron deficient erythropoiesis; not readily available; requires fresh sample.
Red cell zinc protoporphyrin	<80 μmol/mol Hb	Indicator of iron deficient erythropoiesis; increased by lead poisoning.
Hepatic iron concentration	3–33 μg/g dry weight	Precise measurement of iron overload; can be sampling error; invasive.
Bone marrow iron	Staining grade	Reliable indicator of general iron status; useful to separate iron deficiency anemia from the anemia of inflammation; invasive.

hemolysis. In developing countries, parasitic infections, particularly hookworm, can make a significant contribution to gastrointestinal blood loss. Conditions that limit iron absorption, either through a reduction in the absorptive surface area of the small intestine or through impaired gastric acid secretion, can also lead to iron deficiency. These include celiac disease, autoimmune gastritis, and *Helicobacter pylori* infection. Inherited disorders leading to reduced iron intake are rare, but it has been recognized recently that many cases of iron deficiency anemia that are refractory to iron therapy result from mutations of the *TMPRSS6* gene, which encodes the upstream regulator of hepcidin matriptase-2. In affected individuals, hepcidin levels are unusually high, and this leads to reduced iron absorption.

The Anemia of Inflammation

The anemia associated with a range of chronic inflammatory conditions, known as the anemia of chronic disease (ACD) or the anemia of inflammation, is the most common form of anemia in developed countries after iron deficiency anemia (Anderson and McLaren 2012; Wessling-Resnick 2010). It is not caused by overall body iron deficiency, but rather a sequestration of iron within the tissues such that plasma iron levels are decreased and the amount of iron available for erythropoiesis becomes limiting. This form of anemia is typically associated with conditions such as rheumatoid arthritis, systemic lupus erythematosus, inflammatory bowel disease, various types of cancers, and infection. The pathogenic mechanisms underlying the ACD are complex. The expression of hepcidin can be strongly induced by inflammatory stimuli, and elevated hepcidin will favor the sequestration of iron within macrophages and other cells, explaining the inability to mobilize body iron stores. However, there also appear to be direct effects of various pro-inflammatory cytokines on macrophages that lead to iron retention. Serum ferritin levels are also increased by inflammatory stimuli and this parameter is not a good indicator of iron stores (particularly depleted iron stores) in the presence of inflammation.

Iron Overload

Although less common than iron deficiency on a global scale, iron overload syndromes can be relatively common in certain populations, particularly where iron intake is adequate (Anderson and McLaren 2012;

Camaschella and Poggiali 2011). Iron-loading disorders characterized by defects in proteins directly involved in body iron acquisition are known as primary iron-loading diseases. In these conditions, intestinal iron absorption is dysregulated such that the body continues to accumulate iron in the presence of adequate or even enhanced iron stores. Secondary iron loading usually results from blood transfusions. Patients who receive transfusions to treat hemoglobinopathies, bone marrow failure, or hemolytic syndromes accumulate iron in their tissues after the iron derived from the transfused red cells is released in macrophages following erythrophagocytosis. Hematological conditions where erythropoiesis is markedly ineffective may also be associated with increased intestinal iron absorption and this compounds their iron loading. A summary of the various forms of iron loading is provided in Table 2.

Primary Iron-Loading Disorders

Primary iron loading usually results from increased intestinal iron absorption and this in turn reflects reduced hepcidin expression. The newly absorbed iron is deposited in a range of tissues, but notably the liver, pancreas, and heart. If iron accumulates to such an extent that the capacity of the cells to effectively sequester it in a nontoxic form is exceeded, then cellular and tissue damage can result. Hepatic fibrosis and cirrhosis, cardiomyopathy, and diabetes frequently accompany advanced iron loading (Camaschella and Poggiali 2011).

The most common primary iron overload disease is HFE-related (Type 1) hemochromatosis. This autosomal recessive disorder is usually caused by a homozygous mutation in the *HFE* gene, leading to the C282Y substitution in the HFE protein, although other mutations have also been described. It is essentially a Caucasian disease and in populations of northern European origin homozygosity for the C282Y mutation can be 1 in 200 or higher. Although most genetically predisposed individuals will accumulate excess iron in their body, few women and only approximately 30% of men will progress to significant disease. This is likely to reflect mainly physiological and environmental factors (e.g., menstrual blood loss, childbirth, dietary composition, dietary iron content), but genetic influences may also be involved. HFE-related hemochromatosis is a relatively mild form of iron loading in that signs of iron excess usually do not become apparent until the 4th or 5th decade of life, even though increased iron absorption is apparent from an early age. Iron initially deposits in the hepatocytes surrounding the portal tracts of the liver, but in advanced cases iron deposition may be panlobular in the liver and may extend to other organs such as the heart, pancreas, and pituitary. The resident macrophages of the liver (Kupffer cells) are characteristically low in iron among hemochromatosis patients in the early stages, but may contain iron in advanced cases. The diagnosis of HFE-related hemochromatosis begins with blood iron tests which indicate a raised serum ferritin level and increased transferrin saturation. A gene test for *HFE* mutations is then usually prescribed and a positive test confirms the diagnosis. In some cases, tissue iron levels can be assessed by noninvasive MRI methodologies, and in advanced cases, a liver biopsy may be performed to assess tissue damage. The treatment for hemochromatosis is repeated phlebotomy. The removal of blood stimulates erythropoiesis and this in turn means that iron is mobilized from the organs for new hemoglobin synthesis. Aggressive phlebotomy (once each week or even more frequently) may be required to reduce a heavy iron load, but once stores have been depleted they can be maintained at a low level by less frequent venesections.

A very rare form of adult-onset primary iron loading is associated with mutations in the transferrin receptor 2 (*TFR2*) gene (Type 3 hemochromatosis). TFR2 is a homologue of TFR1, but rather than playing a major role in iron uptake, it acts as an upstream regulator of hepcidin. Patients with TFR2 mutations show a similar histological pattern of iron deposition to that seen in HFE-related disease and the consequences of the iron loading are essentially the same. Type 3 hemochromatosis appears to be slightly more severe than HFE-related iron loading.

Severe, early-onset iron loading is known as juvenile (or Type 2) hemochromatosis. In this rare condition, symptoms usually manifest themselves in affected patients in the second decade of life and by the third decade organ damage may occur. Cardiac dysfunction is far more prevalent in juvenile hemochromatosis than in patients with mutations in *HFE* or *TFR2*. Overall, however, the pattern of iron deposition is similar to that seen in HFE-related hemochromatosis and is consistent with greatly elevated dietary iron absorption. Most cases of juvenile hemochromatosis result from mutations in the *HJV* gene (Type 2a), but in some cases the *HAMP* gene itself is mutated (Type 2b). In contrast to Types 1 and 3 hemochromatosis, where hepcidin levels are moderately reduced, in juvenile hemochromatosis hepcidin levels are negligible or undetectable. This explains the uninhibited iron absorption in these conditions and the rapid iron deposition in the tissues.

Type 4 hemochromatosis results from mutations in the gene encoding FPN1 and is often called ferroportin disease. This autosomal dominant form of iron loading shows a relatively complex pathogenesis. In some cases, plasma iron levels are increased and iron accumulates predominantly in hepatic parenchymal cells, consistent with increased intestinal iron absorption, but in most cases reticuloendothelial iron deposition is more prominent and plasma iron levels are low. These differences reflect the underlying mutation. If the mutation reduces the iron transport capacity of FPN1, then total body iron load is not increased, but the release of iron from macrophages and other cells is impeded. Consequently, iron accumulates in the tissues but plasma iron levels are low. On the other hand, mutations in FPN1 that affect its ability to bind hepcidin lead to an iron-loading phenotype which is similar to that of patients with Types 1–3 hemochromatosis. Ferroportin disease is the most frequently reported form of non-HFE primary iron overload.

Secondary Iron-Loading Disorders

Iron overload can also occur when the pathways regulating body iron homeostasis, including iron absorption, are operating normally. In such cases, the extra iron in the body usually comes from repeated blood transfusions (transfusional siderosis) or from increased iron absorption that is secondary to various pathologies (Porter 2009; Rivella 2009). Diseases such as β-thalassemia or sickle-cell anemia are treated by blood transfusions as their red cells do not function effectively and have a reduced life span. For each liter of blood transfused, approximately 0.5 g of iron is delivered to the body in the form of hemoglobin. Since humans do not possess a mechanism for the active excretion of iron, this hemoglobin-derived iron accumulates in the body and can rapidly reach pathological levels. The consequences of the excess iron are similar to those associated with primary iron-overload disorders, but cardiac manifestations are more prominent. While the precise reasons for enhanced cardiotoxicity are not clear, the route of iron delivery appears to be important and a smaller proportion of the iron is sequestered in the liver than in conditions where iron absorption is elevated. Patients with transfusional iron overload are subsequently treated with iron chelators to reduce their iron load. Desferrioxamine, which is administered through prolonged subcutaneous infusion, has been used for this purpose for many years, but more recently oral iron chelators, such as deferiprone and deferasirox, have been increasingly employed for de-ironing.

Increased iron absorption can also occur secondary to some other disease conditions and this can also lead to iron loading. Disorders of the red cell, including those associated with ineffective erythropoiesis (e.g., β-thalassemia, sideroblastic anemia), chronic hemolytic anemias (e.g., pyruvate kinase deficiency, sickle-cell anemia), or hypoplastic anemias (e.g., chronic renal failure, aplastic anemias), are good examples of such conditions. The enhanced erythropoiesis in these conditions reduces hepcidin expression and this leads to enhanced iron absorption. In addition to increased iron absorption, the clearance of defective red cells by reticuloendothelial macrophages can also contribute to iron accumulation in various tissues.

Other Forms of Iron Loading

Iron loading can also result from a range of other conditions, most of which are quite rare (Anderson and McLaren 2012). (1) *Atransferrinemia.* Although TF plays a critical role in the regulated delivery of iron to most body cells, patients with congenital TF deficiency develop severe anemia and massive tissue iron loading. The anemia stimulates iron absorption and the newly absorbed non-transferrin-bound iron is taken up very efficiently by non-erythroid cells. Patients can be effectively treated by infusions of normal plasma or transferrin. (2) *Aceruloplasminemia.* Another prominent plasma protein, ceruloplasmin, plays an important role in iron homeostasis by acting in conjunction with FPN1 to facilitate tissue iron efflux. In the absence of ceruloplasmin, iron accumulates in macrophages, hepatocytes, and other cells, although plasma iron levels are reduced. (3) *H-ferritin-associated iron overload.* This is a very rare form of autosomal dominant iron overload where affected patients carry mutations in the iron responsive element (IRE) in the 5′-untranslated region of the H-ferritin mRNA. As a result, H-ferritin synthesis is reduced and there is a concomitant increase in the L-ferritin subunit. How this leads to iron loading is unclear. Interestingly, mutations in the L-ferritin IRE lead to an increase in serum ferritin and bilateral cataracts, but no iron loading. (4) *Neonatal hemochromatosis.* A rare form of fetal hepatic iron loading of unknown cause, although an immune etiology is suspected. Perinatal lethality is common. (5) *African iron overload.* A form of iron overload that was formerly common (less so today) in sub-Saharan Africa. Excessive iron intake through locally brewed beer was long suspected as the major cause, but more recent studies have shown that mutations in *FPN1* are also a contributing factor. (6) *Medicinal iron loading.* Acute iron poisoning can occur when large doses of oral iron supplements are ingested. This is rarely a problem in adults but can be seen in children who ingest the supplements.

Iron and Other Disorders

Since iron is a strong pro-oxidant, excess iron has the potential to increase the severity of a wide range of disorders that would not be considered primary diseases of iron metabolism (Anderson and McLaren 2012). For example, excess iron has been implicated in the pathogenesis of a number of common neurodegenerative disorders including Alzheimer's disease, Parkinson's disease, Huntington's disease, and amyotrophic lateral sclerosis, as well as in the neuromuscular condition Friedreich ataxia. However, iron is also critical for the function of the central nervous system and an inadequate supply of iron has been linked to another CNS disorder, restless legs syndrome. Iron excess has been shown to be an important comorbidity in a range of chronic liver diseases, and the pathogenesis of fatty liver disease, alcoholic liver disease, and hepatitis C infection is worse in the presence of excess iron. Strong links have also been drawn between increased iron and cancer, with iron acting either as a mutagen or promoter of carcinogenesis (e.g., hepatocellular carcinoma in Type 1 hemochromatosis) or as a growth factor for rapidly growing tumor cells. The need for iron for the growth of microorganisms also can explain why some infections are more severe in the presence of excess iron. The recognition that iron can play a role in such diverse pathologies underscores the importance of delineating basic mechanisms of iron homeostasis and the potential therapeutic benefits of systemic or selective iron depletion.

Acknowledgments GJA is supported by a Senior Research Fellowship from the National Health and Medical Research Council of Australia.

Cross-References

▶ Iron Proteins, Ferritin
▶ Iron Proteins, Plant Iron Transporters
▶ Iron Proteins, Transferrins and Iron Transport
▶ Iron, Physical and Chemical Properties

References

Anderson GJ, McLaren GD (eds) (2012) Iron physiology and pathophysiology in humans. Humana Press, New York
Anderson GJ, Vulpe CD (2009) Mammalian iron transport. Cell Mol Life Sci 66:3241–3261
Camaschella C, Poggiali E (2011) Inherited disorders of iron metabolism. Curr Opin Pediatr 23:14–20
Crichton RR (2009) Iron metabolism: from molecular mechanisms to clinical consequences, 3rd edn. Wiley, Chichester
Ganz T (2011) Hepcidin and iron regulation, 10 years later. Blood 117:4425–4433
Hentze MW, Muckenthaler MU, Galy B, Camaschella C (2010) Two to tango: regulation of mammalian iron metabolism. Cell 142:24–38
Knutson MD (2010) Iron-sensing proteins that regulate hepcidin and enteric iron absorption. Annu Rev Nutr 30:149–171
Porter JB (2009) Pathophysiology of transfusional iron overload: contrasting patterns in thalassemia major and sickle cell disease. Hemoglobin 33(Suppl 1):S37–S45
Rivella S (2009) Ineffective erythropoiesis and thalassemias. Curr Opin Hematol 16:187–194
Wessling-Resnick M (2010) Iron homeostasis and the inflammatory response. Annu Rev Nutr 30:105–122

Iron Metabolism

▶ Iron Homeostasis in Health and Disease

Iron Overload

▶ Iron Homeostasis in Health and Disease

Iron Proteins, Ferritin

Elizabeth C. Theil
Children's Hospital Oakland Research Institute, Oakland, CA, USA
Department of Molecular and Structural Biochemistry, North Carolina State University, Raleigh, NC, USA

Synonyms

Ferritin-(1) Iron bionanomineral encapsulated in protein or (2) Reversibly mineralizing protein nanocage

Definitions

Oxidoreductase (ferroxidase) sites: The di-iron binding sites in ferritin proteins cages where catalysis occurs or alternatively, the sites of ferritin catalytic activity, where two Fe^{2+} + O_2 or H_2O_2 react to form biomineral precursors (Fe^{2+}-O(H)-Fe^{3+}).

Ferritin subunits: H- *H*igh catalytic activity; also called H′, or M or H-1, H-2, etc. (older definitions: *H*eart, *H*eavy, *H*igher in SDS gels); L (animals only) – *L*acks catalytic activity (older definitions: *L*iver, *L*ight, *L*ower (SDS gels)).

Ferritin ion channels: Subdomains in ferritin proteins cages that connect the external environment (cytoplasm) to the inside of the protein cage are the ion channels. Fe^{2+} entry to active sites and Fe^{2+} exit from the mineral is controlled by the ion channels. The channels, 1/3 subunits, are identified by structural (protein crystallography) and functional (Fe^{2+}/O_2 catalysis or Fe^{2+} exit/chelation) effects of amino acid substitutions at conserved ion channel amino acids. Fe^{2+} entry to active sites and Fe^{2+} exit from the mineral are controlled by the ion channels.

Ferritin nucleation channels: Subdomains are found in eukaryotic ferritins that connect the oxido- reductase sites to the mineralization cavity at the center of the protein cage. The channels, 1/oxidoreductase site, are identified by effects of Fe^{3+}O mineral precursors on the NMR properties of nearby residues and by functional effects of amino acid substitutions at conserved, nucleation channel amino acids.

Introduction: The General Properties of Ferritins

Iron in ferritin is very different from the one or two iron atoms/protein in other iron proteins such as globins, cytochromes or iron-sulfur proteins. Ferritins can contain thousands of iron atoms! Some of the iron atoms bind to the ferritin protein, as in other iron –proteins, after passing through the ion entry channels. At the ferritin di-iron binding sites Fe^{2+} reacts with oxidant (O_2 or H_2O). After oxidation, Fe^{3+} atoms-linked by oxygen atoms to other Fe^{3+} atoms accumulate inside the protein cage as a solid, iron- oxy mineral, or nanorock, $Fe_2O_3 \cdot H_2O$. Ferritin mineral growth occurs in the large, central, protein cavity, which is $\sim$30% of the protein cage (Fig. 1). Only a few, if any of the iron atoms in ferritin minerals are actually in contact with the protein. Ferritins are the only proteins that reversibly control the formation/dissolution of a biomineral (Theil 2011; Theil et al. 2012). Another major biomineral, $CaPO_4$ in bone and teeth, by contrast, form outside cells in the body and requires several specialized cell types and matrices to form and dissolve. Ferritins are found in all organisms, whether anaerobes or aerobes, single-celled or multicellular, animal or plant (Andrews 2010; Briat et al. 2010; Chiancone and Ceci 2010; Deng et al. 2010; Le Brun et al. 2010; Theil 2011). Understanding how ferritin protein cages function has growing applications in nanotechnology; a recent illustration is the use of radioheating of intracellular ferritin nanominerals to control, in mice, the synthesis of insulin and glucose metabolism (Stanley et al. 2012).

Why do ferritins have so much iron? The amount of iron that living cells need to make proteins with iron cofactors (heme, FeS clusters or di-iron centers in oxygenases), can be so high at certain times during the life of cells that transport of iron, from outside through the cell membrane, is too slow. Examples of high cellular iron need are cell division (doubling amounts of other iron proteins) or mitochondrial doubling (increasing cytochromes, the heme electron transfer proteins). A special case in animals occurs in red blood cell formation. Here the hemoglobin content of each cell increases from zero (erythroid stem cell) to 90% of the cell protein (erythrocyte). After hormone (erthropoitin) signaling large amounts of iron accumulate inside the developing red cells inside ferritin at the same time transferrin receptors in the cell membrane

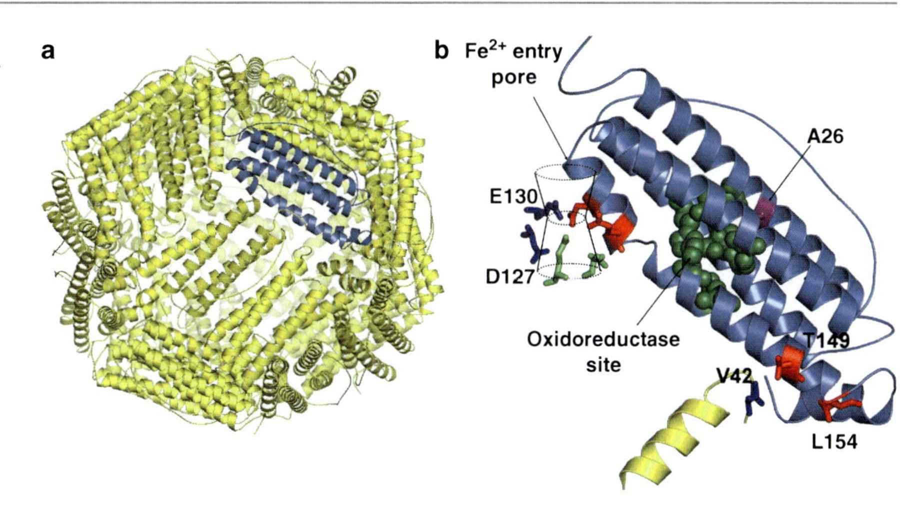

Iron Proteins, Ferritin, Fig. 1 *Ferritin protein cage structure*. (**a**) External view. (**b**) A single subunit (From Haldar et al. 2011)

increase iron transport into the cells, in preparation for increased heme synthesis, the cofactor, and globin protein synthesis. Iron also accumulates in ferritin of cells of certain organs in preparation for the periods of low external food intake immediately after birth in newborn mammals, or during metamorphosis in amphibia. Other examples of iron accumalations in ferritin occur before the rapid replacement of embryonic by fetal/adult red blood cells in vertebrate animals or before the synthesis of nitrogenase (32 Fe/molecule) and leghemoglobin during nitrogen-fixation in plant root nodules. Changes in the cellular ferritin content of single celled organisms relate to cyclical culture properties such as stationery phase (Andrews 2010), or to nitrogen-fixation, or responses to oxidative stress (Andrews 2010; Chiancone and Ceci 2010; Le Brun et al. 2010). In animals small amounts of ferritin, usually containing very little iron, occur in serum. Phenomenological analyses have shown that immunological measurements of the serum ferritin protein concentration ("serum ferritin") have clinical uses as a measure of iron deficiency, distinguishing among different types of anemia, and as an inflammatory marker. The preponderance of ferritin, however, is inside cells.

The role of ferritins in oxidant responses. Ferritins are antioxidant proteins important during cell stress and inflammation in humans and other animals and in plants (Briat et al. 2010; Theil et al. 2012). When iron proteins are damaged, ferritin "saves" the released iron for cell recovery. The antioxidant properties of ferritins depend on consumption of the substrates, Fe^{2+} and O_2 or H_2O_2 during synthesis of the ferritin mineral (1). [O] can be contributed by dioxygen, as in eukaryotes or from H_2O_2 in microbial mini-ferritins.

Overall ferritin reaction

$$2\,Fe^{2+}{\bullet}H_2O + 2[O] + H_2O \Leftrightarrow [Fe_2O_3{\bullet}H_2O]_{solid} + 2H^+. \quad (1)$$

O_2/H_2O_2 in living cells forms toxic ROS (reactive oxygen species), which participate in chain reactions (Fenton chemistry). Normally, ferritin protein cages are only partly filled with ferritin iron mineral (1,000–2,000 Fe atoms/protein cage), so they can "buffer" iron concentrations inside living cells (Arosio et al. 2009). When a host is infected by a microbial pathogen a battle occurs using ferritins as weapons: (1) The host macrophages retain iron in ferritin, making the host anemic (anemia of chronic disease), to keep the iron out of the serum and away from the pathogen. (2) The host cells also secrete H_2O_2 to kill the pathogen, but many pathogens respond by making mini-ferritins. The mini-ferritins consume H_2O_2 + Fe^{2+} to resist the host H_2O_2 defenses; they were first identified as *D*NA *p*rotection during *s*tress proteins (Chiancone and Ceci 2010); hence the alternative name for mini-ferritins is Dps proteins.

The ferritin family is ancient and large (10–12 nm). Ferritin members share a conserved quaternary structure (the hollow protein cage) (Fig. 1), conserved secondary structure (12 or 24 polypeptides folded

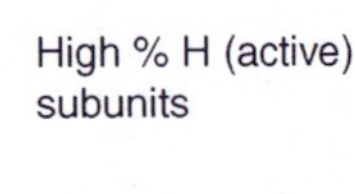

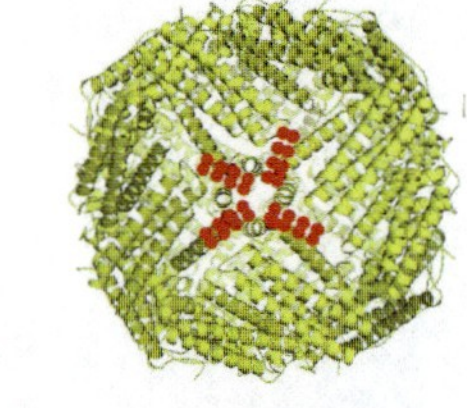
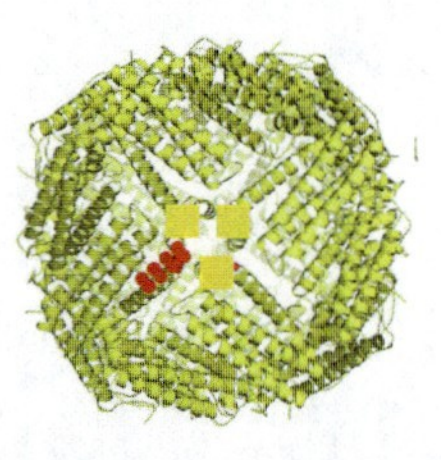

Iron Proteins, Ferritin, Fig. 2 *Protein-based control of ferritin mineral crystallinity in eukaryotic ferritins.* Fe^{3+} O mineral nuclei are guided/extruded through nucleation channel deep in the protein cage, in eukaryotic ferritins (Using data from Turano et al. (2010))

into four α-helix bundles) (Fig. 1), conserved function (oxidoreductase activity) and conserved synthesis of hydrated ferric oxide biominerals. The symmetry of self-assembling ferritin protein cages is amazingly high and relates to the coordinated distribution of Fe^{2+} to the active sites (three-fold symmetry) and apparent build up of ordered minerals (four-fold symmetry). The threefold symmetry axes of the cages, at the N-terminal end of each subunit, penetrate ion channels that deliver Fe^{2+} to active sites in as many as three subunits. The four-fold symmetry axes, at the C-terminal end of each subunit (Figs. 1 and 2), are exits from channels where mineral nuclei exit the protein cage into the mineral growth cavity; understanding of the four-fold symmetry axes of ferrtin protein cages as control centers of ordered mineral growth in eukaryotic ferritins is beginning to emerge (Turano et al. 2010; Theil 2011). The 24 subunit maxi-ferritins occur in all organisms, while to date, the 12 subunit mini-ferritins have only been found in single-celled microbes. In spite of the conservation of ferritin secondary and quaternary structure with function throughout biology, the amino acid sequences vary as much as 80% between bacteria and animals. In plants and animals the amount and subunit composition of ferritin in differentiated cell types varies. Ferritin subunits with different sequences, encoded in different genes, can co-assemble. The crystallinity of ferritin minerals also varies in different tissues (St Pierre et al. 1991) but coincides with differences in ferritin protein cage subunit mixtures.

Catalytic (Fe/O oxidoreductase) sites. The catalytic sites in ferritin proteins cages are present in each of the subunits of ferritin protein cages except in animal ferritins. Thus, ferritins have 24 active sites (12 in mini-ferritin) except in animals. Here, ferritin protein cages are mixtures of two gene products: catalytically active (H/H′/M) ferritin subunits and inactive ferritin subunits (L). The H and L designations for ferritin subunits as Heavy and Light masses is a misconception based on early studies of ferritins in rats and humans. (See the section "Definitions.") When multiple H-type genes and subunits occur in animals or plants, each type of tissue has a different combination of the H, H′, M, H-1, H-2, etc., or L subunits; H type subunits with amino acid sequences that differ as little as 15% are found in different amounts in different tissues and coincide with different activity. The kinetics of oxidoreductase are influenced by the different subunits (Bou-Abdallah 2010; Theil et al. 2012). Each microbial ferritin cage assembles from a single type of subunit (gene product) in contrast to animal or plant ferritins. Examples include the maxi-ferritins FTNA and heme containing BFR, or the mini-ferritin, Dps (Andrews 2010; Le Brun et al. 2010). In bacteria, rather than the tissue type determining the ferritin genes used as in plants and animals, the environmental conditions determine which bacterial ferritin gene is used.

Biological dependence on ferritins is illustrated by the lethality of deleting a ferritin gene in mice and the oxidant sensitivity of deleting ferritin genes in bacteria (Chiancone and Ceci 2010). In bacteria, the multiple ferritin genes are expressed under different environmental conditions or culture phase (Andrews 2010). The synthesis of such a large and important proteins as ferritin, which requires such large amounts of cell energy (ATP and GTP) very tightly regulated. In animals, ferritin synthesis uses both DNA (transcription) and mRNA (translation) control mechanisms (Nandal et al. 2010; Theil 2011).

There are four types of iron in ferritin: (1) Fe^{2+} in transit into and out of the ion channels that connect the interior of the ferritin protein cage (Fig. 1)

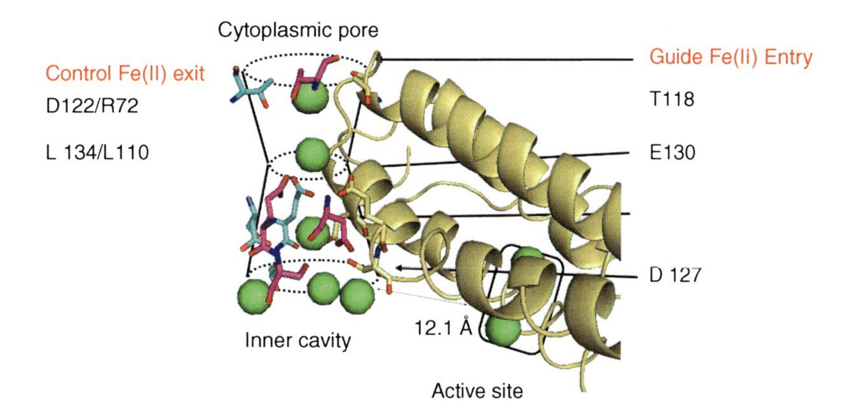

Iron Proteins, Ferritin, Fig. 3 *Ferritin Fe^{2+} ion entry/exit channels*. The two groups of residues specifically controlling channel structure, Fe^{2+} entry and Fe^{2+} exit are indicated. Note that different carboxylate groups are important in Fe^{2+} entry or Fe^{3+} exit (Modified from Tosha et al. (2010))

to the cytoplasm. (2) Fe^{2+} at the catalytic coupling sites within in the protein cage. (3) Ferritin iron oxidation products and protein based mineral nucleation in eukaryotic ferritins. (4) Hydrated $Fe_2O_3 \bullet H_2O$ minerals. Most of the iron in ferritin is encased by the proteins cage and in the inorganic, solid mineral, and is not bound to the protein.

Ferritin Fe^{2+} Ion Entry/Exit Channels

Iron enters ferritin protein cages as Fe^{2+} ions and exits as Fe^{2+} ions through the ion channels that connect the cytoplasm and the inside of the cage (Fig. 3-ion channels). The line of metal ions in Fig. 3 (green spheres) appears much the same as those observed in potassium ion channels (Gouaux and Mackinnon 2005). Ferritins are the first known, water soluble arrays of ion channels. Most ion channels are embedded in membranes and connect the external environment of living cells to the cytoplasm. All ion channels, including those in ferritins share the property of controlling ion transit across a relatively hydrophobic matrix (cell or organelle membrane or ferritin protein cage) between different internal and external environments. Effects of ferritin protein structure on Fe^{2+} entry and exit, "gating" are measured spectrophotometrically (Fig. 4).

Ferritin Fe^{2+} ion channels form in the cage during self-assembly of the polypeptide subunits. Short helix-loop helix segments (from helices 3 and 4) of each 4-helic subunits align around the threefold symmetry axes of the cage to form eight iron entry/exit channels in 24 subunit ferritins and four iron entry exit channels in mini-ferritins. The highly conserved amino acids that are required for correct iron passage through the channel are different for Fe^{2+} entry and Fe^{2+}exit in 24 subunit ferritins, where both negatively charged and hydrophobic amino acids play a role (Theil 2011). In mini-ferritins, by contrast, the same residues affect both Fe^{2+} entry and Fe^{2+}exit in the four Fe^{2+} entry/exit channel (Chiancone and Ceci 2010). Even in living cells, the surface-limited dissolution of Fe^{2+} from ferritin minerals is changed when the protein channels are altered in variant ferritins; when ferritin pores are unfolded, cultured human cells have more chelatable iron.

The natural biological regulators of ferritin ion channel entry and exit remain unidentified. Potential models for such regulators are two small, tight-binding, peptides, which were selected, from a combinatorial library of 10^9, for tight binding to ferritin channels. The selected peptides altered Fe^{2+} exit rates; see references in (Theil 2011). Interestingly, physiological concentrations of urea (1–10 mM) also regulate ferritin pore/channel folding in solution; see references in (Theil 2011). Whether or not changes in cellular urea concentration regulate ferritin pore function in vivo is unknown. For ferritins, still unknown are the routes for Fe^{2+} ions between the ion channels and the active sites, or from the dissolving mineral surface to the outside, for oxidant (O_2 or H_2O_2) to the active sites, or for electrons to the mineral surface(s), or for protons to and from the mineral surface.

Ferritin Fe^{2+} at the Catalytic Coupling Sites in Ferritin Protein Cages

Two Fe^{2+} ions bind at two different iron sites in the catalytic centers of maxi-ferritins (Fig. 5). The results are based on a variety of spectroscopic studies,

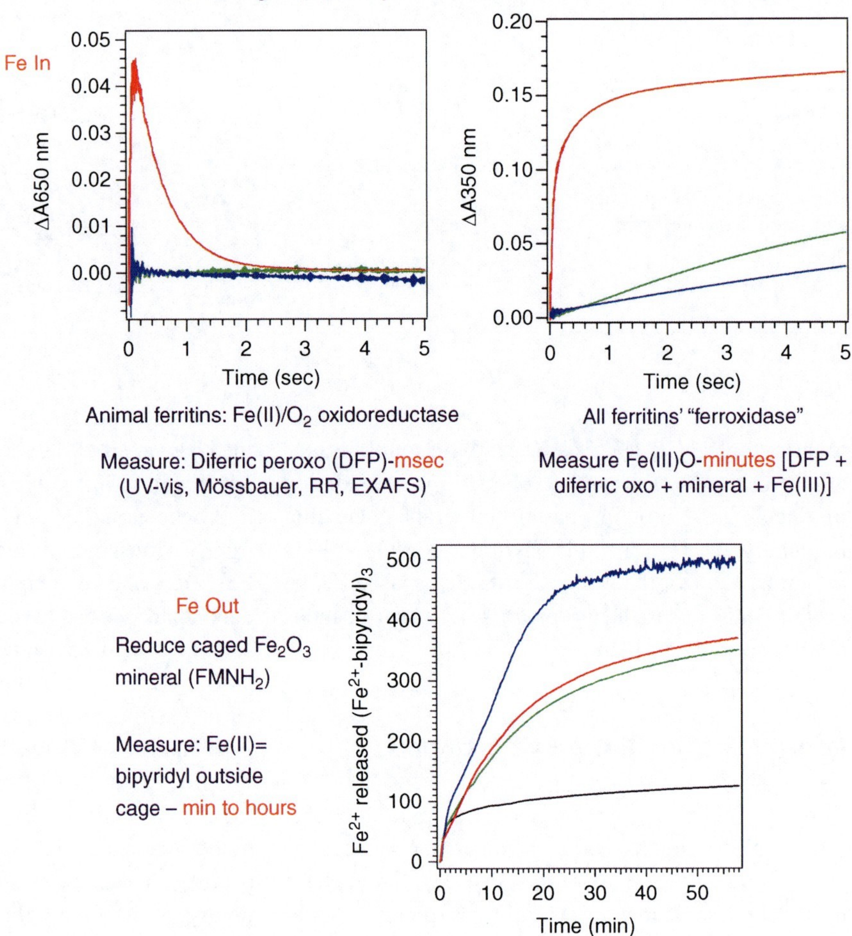

Iron Proteins, Ferritin, Fig. 4 *Spectrophotmetric analyses of ferritin activity. Top*: Oxidoreductase activity. *Bottom*: NADH/FMN-triggered mineral dissolution and Fe^{2+} exit from the protein cage with bipyridyl chelation

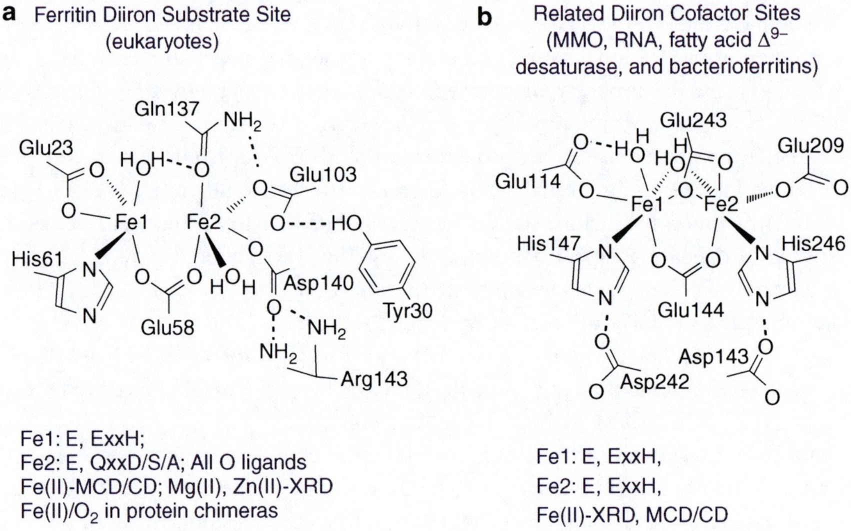

a Ferritin Diiron Substrate Site (eukaryotes)

Fe1: E, ExxH;
Fe2: E, QxxD/S/A; All O ligands
Fe(II)-MCD/CD; Mg(II), Zn(II)-XRD
$Fe(II)/O_2$ in protein chimeras

b Related Diiron Cofactor Sites (MMO, RNA, fatty acid Δ^{9-} desaturase, and bacterioferritins)

Fe1: E, ExxH,
Fe2: E, ExxH,
Fe(II)-XRD, MCD/CD

Iron Proteins, Ferritin, Fig. 5 *Oxidoreductase sites.* (**a**) A eukaryotic ferritin site where Fe^{2+} is a substrate. (**b**) An iron cofactor site from iron-methane monooxygenase as representative of iron ribonucleotide reductase, fatty acid desaturase, and bacterioferritin, the heme containing ferritin in some bacteria reviewed in (Bou-Abdallah 2010; Theil 2011). Each five-coordinate Fe^{2+} site contains one coordinated water and a vacant site for dioxygen, but differ in coordination geometry. Insertional and deletional mutagenesis correctly predicted the active site residues but could not predict the differences in site geometry and coordination number. Studies of protein crystals have not provided information on Fe^{2+} binding, since there are no anaerobic studies of co-crystals of ferritin with Fe^{2+}. Fe^{3+} sites in ferritin protein-iron cocrystals relate to the catalytic products, which, based on NMR studies using Fe^{3+} paramagnetic effects (Turano et al. 2010), move away from the active sites to nearby residues such as conserved A26. All the residues at eukaryotic maxi-ferritins active sites are conserved (Fig. 5), except one residue which can be aspartate, serine or alanine. When alanine, serine or aspartate are compared as variants in the overall ferritin backbone sequence differences in oxidoreductase kinetics are observed. The combination of the different kinetics and the synthesis of the different ferritin subunits in different tissues indicates biological importance, but understanding is far from complete.

The mechanism by which the diferric dimer product is released from ferritin catalytic sites is not yet known. Whether or not the two iron sites release both Fe^{3+} in a concerted reaction or one at a time remains to be determined. Ferritin active sites share similarities with di-iron cofactor sites in enzyme of oxidative chemistry (Fig. 5) such as ribonucleotide reductase, methane monooxygenase, stearoyl-acyl carrier protein desaturase, and myo-inositol oxygenase (Krebs et al. 2011). Many of the enzymes form the same diferric peroxo intermediate as the transition intermediate in eukaryotic ferritins, Fe^{3+}-O-O-Fe^{3+}, which has been characterized extensively by Mössbauer, resonance Raman and EXAFS spectroscopies (Theil 2011).

The oxidoreductase mechanisms in ferritins from bacteria can differ from those in eukaryotic ferritins. The diferric peroxo intermediate cannot be trapped, or does not form; Fe^{2+} binding can be monitored by quenching of tryptophan fluorescence. In *E. coli* bacterioferritins, stoichiometries of 2 Fe/active site were observed (Le Brun et al. 2010). The two iron atoms are bound very tightly by two sets of E, ExxH ligands. In fact the di-iron center in *E. coli* BFr, and some other bacterial ferritins, appear to functions as cofactors (Le Brun et al. 2010). In mini-ferritins only one 1 Fe^{2+}/site is bound in the absence of oxidant. The second Fe^{2+} is predicted to bind after binding of the oxidant (Chiancone and Ceci 2010). In mini-ferritins the predicted ligands for the active sites are weaker than in maxi-ferritins, and generally are located at the interface between two subunits.

When the specific diferric peroxo intermediate (λmax 650 nm) cannot be detected, such as in bacterial ferritins or mini-ferritins, ferritin oxidoreductase activity is monitored as the nonspecific $Fe^{3+}O$ absorbance, in the range from 310 to 450 nm. There are several limitations to the data collected this way. First, there is no UV-vis spectroscopic distinction, to date, among the diferric catalytic product, mineral nuclei and mineral, so there is no decay component in the catalytic progress curve. Second, since the absorbance band is broad, it is difficult to separate contributions of overlapping, absorbing species. This is a particular problem at shorter wavelengths at low iron: protein stiochiometries where the changing contributions of aromatic amino acids to the absorbance need to be considered. However, where the diferric peroxo intermediate cannot be trapped, or does not form, the information from the absorbance of $Fe^{3+}O$ has provided enormous amount of very important information. (For examples see the reviews Chiancone and Ceci (2010) and Le Brun et al. (2010)).

Characterization of ferric reaction intermediates is still an actively developing subject in eukaryotic ferritins (Bou-Abdallah 2010; Turano et al. 2010; Theil 2011) and in microbial ferritins (Le Brun et al. 2010). Variations in ferritin active site mechanisms and structure coincide with large differences in ferritin primary structure and remind us that evolution has preserved/selected only the overall features of ferritin hollow protein cages with the capacity to reversibly synthesize encapsulated iron oxy biominerals. Only time will tell if the variations among ferritin active sites are the consequences of convergent evolution, or of specialization matched to environmental niches, or of physiological matching between ferritin protein cage and biomineral structure for different biological needs.

Ferritin Iron Oxidation Products and Mineral Nucleation

Iron biomineral growth in the large (5–8 nm) central ferritin protein cavity is envisioned as an inorganic process in many current models. In such models the protein cage contributes nothing beyond ion entry/exit channels and oxidoreductase sites in mini-ferritins (Chiancone and Ceci 2010) or in bacterioferritins (Le Brun et al. 2010). The physical distance between the active sites and the mineral growth cavity is relatively short in such protein-based control of mineral nucleation and nanomaterial order, based on recent studies (Turano et al. 2010; Theil 2011).

Fe^{3+}-O products of ferritin catalysis, in eukaryotic ferritins, are guided through an internal channel (~20 Å), from the catalytic sites to the mineralization cavity (Turano et al. 2010; Haldar et al. 2011). The $Fe^{3+}O$ channels were recently identified using an iron "titration" where Fe^{3+} paramagnetism of the oxidoreductase products "erased" proximal (<5 Å) amino acid signatures, of site-specifically assigned in ^{13}C-^{13}C solution NOESY spectra; note that only postoxoxidation sites, and not the active sites residues, could be examined in this study. For each catalytic reaction cycle (turnover), the ferritin active sites were titrated with saturating amounts of Fe^{2+} in the presence of O_2. $Fe^{3+}O$ reached the mineral growth cavity only after three to four active site turnovers. During passage or extrusion of $Fe^{3+}O$ through the channels Fe^{3+}-O-Fe^{3+} catalytic products appeared to be reacting with each other to make $Fe^{3+}O$ multimers (mineral nuclei) (Turano et al. 2010). The nucleation/extrusion channels from each active site end at openings around the four-fold symmetry axes of the ferritin protein cage, near the channels openings in three other subunits (Fig. 5). The combined trajectories of iron moving through ferritin protein cages from the Fe^{2+} ion entry channel through the active sites to the $Fe^{3+}O$ nucleation/extrusion channel exits is ~40–50 Å (Fig. 1) (Turano et al. 2010; Theil et al. 2012). The extraordinarily high symmetry of ferritin protein cages in eukaryotes, at least in animal ferritins where they have been studied, has functional significance related to ferritin mineral crystallinity (Figs. 2 and 3).

Iron biominerals synthesized in ferritin vary in size, crystallinity, and phosphate content (Theil 2011). Experimentally, large amounts of phosphate decrease ferritin mineral order. In Nature, ferritin biominerals high in phosphate (Fe:P ~1:1) are synthesized in plants and bacteria; minerals of animal ferritin have a phosphate content that is lower (Fe:P ~8:1). The phosphate content of microbial cytoplasm is higher than eukaryotes, which explains the higher phosphate content of microbial ferritins and plants ferritins as well, because in plants ferritin mineralization occurs in plastids, which are derived from endosymbiosis of single-celled microbes.

Natural ferritin minerals from animal tissues display variability in crystallinity that coincides with the differences in the protein cage. The differences in ferritin mineral crystallinity of animals can be explained by the contributions of eukaryotic ferritin oxidoreducatse sites with the nucleation/extrusion channels (Fig. 2). For example, heart ferritin mineral has high crystallinity and liver ferritin mineral has relatively low crystallinity (St Pierre et al. 1991). The relative number of H and L subunits varies in heart and liver, as well as in other tissues. Heart ferritin has a high H:L ferritin subunit ratio (~2:1) and liver ferritin has a low H:L subunit ratio (~1:4). With a large fraction of the subunits as H, the mineral nuclei will be extruded from the nucleation channels of each subunit near other H subunits around the four-fold symmetry axes, which facilitates ordered structure in the growing mineral. When the fraction of L subunits in a ferritin cage is large, as in animal liver (Arosio et al. 2009), the probability decreases that an $Fe^{3+}O$ mineral precursor will be extruded from an H-subunit near another H subunit $Fe^{3+}O$ mineral precursor (Fig. 2). No only do L subunits lack active site residues they lack other conserved residues such as A26 at the extrusion channel entrance near the active sites (Theil et al. 2012). In ferritins, with few H subunits, many mineral nuclei extruded from H subunits will diffuse within the internal ferritin cavity before contact with other $Fe^{3+}O$ mineral nuclei and will not be oriented by the protein channels. Thus, tissue specific H:L ferritin subunit expression contributes to properties of the ferritin mineral in each tissue. The physiological significance of animal-specific, L ferritin subunits, is not clear, but manipulation with L ferritin subunit RNAi alters cell growth rates, possibly because changes in the "hardness"/crystallinity of ferritin minerals slows iron release and iron protein synthesis. The high L ferritin subunit content of liver ferritin protein cages, leading to relatively disordered and easier to dissolve iron mineral, may reflect the role of liver iron in the body. Biologically, liver ferritin iron is used in the rapid response to sudden losses of body iron (hemorrhage) when repairs (new red cell and hemoglobin synthesis) are needed.

Among ferritins, only quaternary cage structure, biomineral cavity and the overall reaction of ferrous and oxidant (1) are conserved throughout evolution.

Differences in protein cage structure (size, amino acid sequence, catalytic mechanisms, and mineral crystallinity) appear matched to specific intracellular, and for higher organisms, whole body stability.

Control of Ferritin Protein Biosynthesis

Ferritin synthesis is tightly regulated because of the energy costs (ATP, GTP), and the protection ferritin provides from oxidant damage and iron reactivity. The human body needs to manage 30 mg of iron/day from old, red blood cells. Given the low solubility of ferric iron under physiological conditions (10^{-18} M), 10 trillion liters of water or 37 gal of citric acid (rich orange juice) would need to be consumed each day to prevent excreted iron from rusting, if the iron were not effectively retained and recycled using macrophage ferritin. The synthesis of ferritin in the cells that recycle the iron (spleen macrophages) and store the iron (liver), and in other cells as well, is controlled by the amount of ferritin mRNA (transcription) and also by utilization of ferritin mRNA (translation/ribosome binding) (Goss and Theil 2011); in plants and bacteria only ferritin mRNA (transcription) is regulated. Regulating ferritin mRNA translation appeared very early, in animals as primitive as sponges (Piccinelli and Samuelsson 2007); IRE riboregulators have spread to other mRNAs related to iron homeostasis (Wallander et al. 2006).

Why translational regulation of ferritin mRNA is absent in plants is unclear. Adding the animal riboregulator to plant ferritin mRNA actually decreases the riboregulator effectiveness. Plant cells need to coordinate ferritin synthesis with the expression of the extra genome in the plastids may override the need for a ferritin mRNA riboregulator in contemporary plants. Other possible explanations for the difference in regulation of plant and animal ferritins include the relatively shorter life expectancy of individual plant cells compared to animal cells, the fact that plants do not readily exchange genetic information because they are relatively stationary, or chance.

The signals that control ferritin synthesis in plants, animals and bacteria are iron and oxidants. In *E. coli*, for example, maxi-ferritin FTNA protein synthesis is increased by iron (increased DNA transcription/ferritin mRNA synthesis) (Nandal et al. 2010), while. *E. coli* mini-ferritin Dps synthesis is increased by H_2O_2 (and other types of stress). Plant ferritin synthesis is regulated by oxidants and by such large amounts of iron that distinction from oxidant stress is unclear (Briat et al. 2010). By contrast, animals use the dual controls of ferritin synthesis transcription *plus* translation. The DNA sequence that regulates ferritin gene transcription in animals is called ARE (*A*ntioxidant *R*esponse *E*lement), which binds two different protein repressors. One of the ARE-DNA binding repressor is Bach 1 with heme as a signal (Theil 2011) and the other is ATF-1 with a serine phosphorylation signal (Hailemariam et al. 2011). A number of genes, in addition to ferritin, have ARE elements that coordinate repair from (oxidative) stress damage. Thus a single stress inducer will increase the transcription and synthesis of all the proteins encoded in ARE genes such as heme oxygenase and ferritin H and L, proteins important in recovering iron from oxidant-damaged iron-proteins. Two other ARE genes encode important in repairing oxidative damage are NADPH quinone reductase 1 and thioredoxin reductase 1. ARE genes in animals, including ferritin H and L, all respond to the same environmental oxidative signals.

The physiological signal that regulates ferritin mRNA (translation) is Fe^{2+} (Goss and Theil 2011; Ma et al. 2012). A noncoding, three-dimensional, flexible, stem loop RNA structure (Walden et al. 2006) embedded in the mRNA is a noncoding riboregulator of ferritin mRNA activity in protein synthesis. Named IRE (*I*ron *R*esponsive *E*lement), the ferritin mRNA regulator, IRE-RNA, like the ferritin gene regulator, DNA- ARE, occurs in a number of animal mRNAs to coordinate synthesis of proteins homeostasis. IRE-RNAs are the classical examples of mRNA regulation. [Descriptions of multiple IREs in different mRNAs are easily confused with another set of mRNA regulatory structures named IRES (Internal Ribosome Entry Sites); to be clear, IRE-RNA is used here.] IRE-RNAs bind repressor proteins, IRPs that block protein synthesis (translation) by inhibiting ribosome binding. IRE-RNAs also bind an initiation factor (eIF4F) protein to enhance ribosome binding (Ma et al. 2012). Earlier ideas suggested that cellular iron caused degradation of the IRP repressor proteins and insertion of an iron-sulfur cluster into unbound IRP, but these are down stream effects. Fe^{2+} binds to IRE-RNA, changing the IRE-RNA conformation and the binding of IRP and eIF4F; thus, the atomic/molecular form of the initial biological iron signal is Fe^{2+} which is

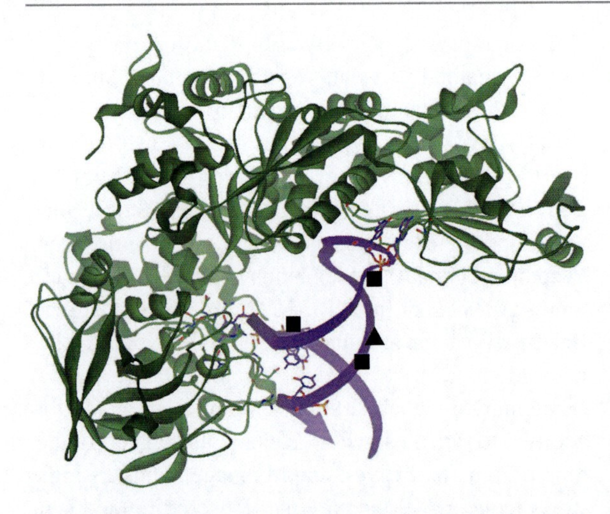

Iron Proteins, Ferritin, Fig. 6 *Structure of the ferritin mRNA riboregulator (IRE), which controls ferritin protein synthesis (translation) bound to the protein repressor (IRP).* Note the exposed sites on the IRP-complexed IRE-riboregulator that invite other interactions such as Fe^{2+} binding and/or protein translation activators (Drawn using data in PDB 1IPY, see (Goss and Theil 2011) and references therein)

transduced by IRE-RNA binding to IRE-RNA conformational change (Goss and Theil 2011; Ma et al. 2012). EIF4F is a large protein complex that binds ribosomes, in addition to IRE-RNA. The first clue that metal-RNA interactions could be involved in iron signaling was the observation that when ferritin IRE-RNA is bound to the IRP repressor in crystals, an RNA surface known to bind metal ions is exposed (Fig. 6). In the reducing environment of animal cells, where dioxygen is compartmentalized and free iron is Fe^{2+}, increasing the Fe^{2+} concentrations will release IRP and bind eIF4F to IRE-mRNAs.

RNA structures such as the IRE-RNA, make handsome targets for small molecule regulation of specific proteins. Such riboregulatory mRNA structures display selective, three dimensional binding surfaces that, unlike DNA, are specifically synthesized in every cell. In addition, mRNAs, unlike proteins, are present in low concentrations, making a single mRNA species easier than protein to manipulate with small doses of a therapeutic small molecule. In the case of ferritin, increased availability of free ferritin mRNA and ferritin protein biosynthesis for example, could find use in diseases treated by therapeutic hypertransfusion (Sickle Cell disease and Thalassemia). In such conditions, increases in ferritin synthesis lag behind increases in tissue iron and the buffering capacity of ferritin protein becomes saturated. In such conditions, toxic hemosiderin (damaged ferritin) forms and oxidative damage occurs. Specific manipulation of ferritin synthesis through IRE-RNA targeting not only is important therapeutically for iron-related disease, but is "proof of principle" for mRNA targeting in diseases with newly discovered mRNA riboregulators.

Perspective

Ferritins, the protein cages that reversibly synthesize internal ferric oxide minerals nanometers in diameter, are unique. Members of the ancient ferritin family are found everywhere from anaerobic archaea to humans. The consumption of iron and oxygen to make caged ferritin minerals confers antioxidant activity on ferritins, in addition to the role of ferritins in providing metabolic iron concentrates. A role for ferritins in the terrestrial transition from anaerobic to aerobic life is easy to imagine. Plants use ferritin in seeds to nurture the developing embryo; the seed ferritin iron is also bioavailable to humans and animals. Ferritin accumulates in animal tissues at key stages of development, an apparent analogy to seed ferritins. In single cell organisms ferritins prepare cultures for rapid growth when the environment becomes favorable, and protect them against environmental or host oxidants.

The self-assembling, quaternary cage structure of ferritin, built out of multiple subunits with the four α-helix bundle motif, is extraordinarily symmetric. Each type of ferritin subunit is encoded in a different gene. In animals and plants, multiple copies of each type of subunit can coassemble to form a single cage. In animals, both DNA and three-dimensional mRNA noncoding structures (riboregulators) combine with specific regulatory proteins to control ferritin protein synthesis rates; Fe^{2+} binding to IRE-RNA is the metabolic signal. All ferritins have two and three fold symmetry. In 24 subunit maxi-ferritins, the cages have, in addition, fourfold symmetry. Ion entry/exit channels connect the cytoplasmic surface of ferritins with the central mineral growth cavity and form around the three-fold symmetry axes. The ion channels around the threefold axes distribute entering Fe^{2+} ions to as many as three catalytic sites each. Recent studies indicate an unpredicted role for the four-fold symmetry axes of 24 subunits ferritins in animals: influencing mineral order during growth of the internal $Fe_2O_3 \cdot H_2O$. Mineral order in ferritin minerals of plants (in plastids) and bacteria, by contrast, appears to depend more on environmental phosphate concentration but direct experimental investigations are needed.

Puzzles still needing solutions for understanding the complex structural and functional properties of the ferritins include: (1) The ferritin protein cage assembly code (amino sequences diverge as much as 80%). (2) The mechanisms of $Fe^{3+}O$ product release from oxidoreductase sites. (3) Mechanisms of mineral nucleation/extrusion channel function. (4) The trajectories of H^+ and ε^- entering the cage to dissolve ferric oxide minerals and of Fe^{2+} released from the mineral surface. (5) The mechanism of selection among the multiple active sites for Fe^{2+} substrate exiting the entry channels. (6) The identity and function of cytoplasmic molecules (ferritin iron chaperones) involved in ferritin pore folding/unfolding. (7) The identity of intestinal endocytic receptors for absorption of ferritin. (8) The mechanisms and RNA ligands for Fe-IRE-RNA interactions and ribosome binding. Nature has been tweaking ferritins for so very long that human solutions to the difficult problems of ferritin understanding will take only the "blink of an eye" relatively speaking. However, obtaining the answers to the remaining questions about ferritins is fundamental to knowing how biology works with hazardous materials such as iron and oxygen, and how to use ferritins more effectively in medicine, nutrition and nanotechnology.

Acknowledgement The author would like to thank all the members of the Theil Group and our many Collaborators for their enthusiastic contributions to the work of the author cited here. That work has been supported by the NIH (DK20251) and the CHORI Foundation.

Cross-References

▶ Heme Proteins, Cytochrome c Oxidase
▶ Iron Homeostasis in Health and Disease
▶ Iron Proteins, Transferrins and Iron Transport
▶ Iron-Sulfur Cluster Proteins, Nitrogenases
▶ Ribonucleotide Reductase

References

Andrews SC (2010) The ferritin-like superfamily: evolution of the biological iron storeman from a rubrerythrin-like ancestor. Biochim Biophys Acta 1800:691–705

Arosio P, Ingrassia R, Cavadini P (2009) Ferritins: a family of molecules for iron storage, antioxidation and more. Biochim Biophys Acta 1790:589–599

Bou-Abdallah F (2010) The iron redox and hydrolysis chemistry of the ferritins. Biochim Biophys Acta 1800:719–731

Briat JF, Ravet K, Arnaud N, Duc C, Boucherez J, Touraine B, Cellier F, Gaymard F (2010) New insights into ferritin synthesis and function highlight a link between iron homeostasis and oxidative stress in plants. Ann Bot 105:811–822

Chiancone E, Ceci P (2010) The multifaceted capacity of Dps proteins to combat bacterial stress conditions: detoxification of iron and hydrogen peroxide and DNA binding. Biochim Biophys Acta 1800:798–805

Deng J, Liao X, Yang H, Zhang X, Hua Z, Masuda T, Goto F, Yoshihara T, Zhao G (2010) Role of H-1 and H-2 subunits of soybean seed ferritin in oxidative deposition of iron in protein. J Biol Chem 285:32075–32086

Goss DJ, Theil EC (2011) Iron responsive mRNAs: a family of Fe^{2+} sensitive riboregulators. Acc Chem Res 44:1320–1328

Gouaux E, Mackinnon R (2005) Principles of selective ion transport in channels and pumps. Science 310:1461–1465

Hailemariam K, Iwasaki K, Huang BW, Sakamoto K, Tsuji Y (2011) Transcriptional regulation of ferritin and antioxidant genes by HIPK2 under genotoxic stress. J Cell Sci 123:3863–3871

Haldar S, Bevers LE, Tosha T, Theil EC (2011) Moving Iron through ferritin protein nanocages depends on residues throughout each four {alpha}-Helix bundle subunit. J Biol Chem 286:25620–25627

Krebs C, Bollinger JM Jr, Booker SJ (2011) Cyanobacterial alkane biosynthesis further expands the catalytic repertoire of the ferritin-like "di-iron-carboxylate" proteins. Curr Opin Chem Biol 15:291–303

Le Brun NE, Crow A, Murphy ME, Mauk AG, Moore GR (2010) Iron core mineralisation in prokaryotic ferritins. Biochim Biophys Acta 1800:732–744

Ma J, Haldar S, Khan MA, Sharma SD, Merrick WC, Theil EC, Goss DJ (2012) Fe^{2+} binds iron responsive element-RNA, selectively changing protein-binding affinities and regulating mRNA repression and activation. Proc Natl Acad Sci USA 109:8417–8422

Nandal A, Huggins CC, Woodhall MR, McHugh J, Rodriguez-Quinones F, Quail MA, Guest JR, Andrews SC (2010) Induction of the ferritin gene (ftnA) of *Escherichia coli* by Fe(2+)-Fur is mediated by reversal of H-NS silencing and is RyhB independent. Mol Microbiol 75:637–657

Piccinelli P, Samuelsson T (2007) Evolution of the iron-responsive element. RNA 13:952–966

St Pierre T, Tran KC, Webb J, Macey DJ, Heywood BR, Sparks NH, Wade VJ, Mann S, Pootrakul P (1991) Organ-specific crystalline structures of ferritin cores in beta-thalassemia/hemoglobin E. Biol Met 4:162–165

Stanley SA, Gagner JE, Damanpour S, Yoshida M, Dordick JS, Friedman JM (2012) Radio-wave heating of iron oxide nanoparticles can regulate plasma glucose in mice. Science 336:604–608

Theil EC (2011) Ferritin protein nanocages use ion channels, catalytic sites, and nucleation channels to manage iron/oxygen chemistry. Curr Opin Chem Biol 15:304–311

Theil EC, Behera RK, Tosha T (2012) Ferritins for chemistry and for life. Coord Chem Rev Coor. Chem. II, article 00302, Elsevier, Oxford (in press)
Tosha T, Ng H-L, Bhatassali O, Alber T, Theil EC (2010) Moving metal ions through ferritin protein nanocages from the three-fold pores to catalytic sites. J Am Chem Soc 132:14562–14569
Turano P, Lalli D, Felli IC, Theil EC, Bertini I (2010) NMR reveals pathway for ferric mineral precursors to the central cavity of ferritin. Proc Natl Acad Sci USA 107:545–550
Walden WE, Selezneva AI, Dupuy J, Volbeda A, Fontecilla-Camps JC, Theil EC, Volz K (2006) Structure of dual function iron regulatory protein 1 complexed with ferritin IRE-RNA. Science 314:1903–1908
Wallander ML, Leibold EA, Eisenstein RS (2006) Molecular control of vertebrate iron homeostasis by iron regulatory proteins. Biochim Biophys Acta 1763:668–689

Iron Proteins, Mononuclear (non-heme) Iron Oxygenases

Michael M. Mbughuni and John D. Lipscomb
Department of Biochemistry, Molecular Biology, and Biophysics, University of Minnesota, Minneapolis, MN, USA

Synonyms

Dioxygenase; Mixed-function-oxidase; Monooxygenase; Oxygenase

Definitions

Mononuclear nonheme iron oxygenases are a large group of iron-containing enzymes known to catalyze the transfer of one or both oxygen atoms from dioxygen (O_2) into organic substrates.

Introduction

Mononuclear nonheme iron oxygenases (Costas et al. 2004; Bruijnincx et al. 2008; Kovaleva and Lipscomb 2008) are distinct from the more familiar heme-dependent oxygenases primarily due to the set of ligands used to coordinate the catalytic iron atom. Ligands to the nonheme sites are protein-derived amino acids (most commonly His, Tyr, Asp, and Glu) along with solvent molecules, while heme sites utilize a porphyrin macrocyclic ring to coordinate the metal. Consequently, the nonheme iron sites exhibit a much wider diversity of metal site ligand compositions and coordination geometries (Solomon et al. 2003). One ramification of this structural diversity is that nonheme sites are known to stabilize iron at redox potentials spanning more than 1 V. In some cases, this allows the iron to be stabilized in one oxidation state throughout the O_2 activation reaction. In contrast, heme oxygenases always invoke a redox active metal center. Another significant difference between heme and some types of nonheme iron sites is that the latter can utilize multiple iron ligand sites to bind more than one reaction substrate at the same time. The importance of this for catalysis is seen in the examples given here. Oxygenase chemistry within the broad nonheme iron family proceeds by a significantly more diverse set of mechanistic strategies than found in the heme oxygenase family. However, at the heart of each of these mechanisms, is the strategy for activation of molecular oxygen, O_2.

O_2 is a unique reagent for biology because it combines availability and high potential reactivity with inertness at ambient temperature (Klinman 2007). The latter is true because the molecular structure of O_2 would cause a violation of a quantum mechanical rule governing the conservation of spin angular momentum if it were to react with most biological molecules. The reaction is thus termed "spin forbidden." Nature uses oxygenases to unleash this inherent free energy stored in the O–O bond specifically when and where it is required. Consequently, an important aspect of oxygenase mechanisms is the manner in which the enzymes regulate the O_2 activation reaction to assure specificity and avoid the release of activated oxygen species which would randomly attack biological molecules.

In broad terms, there are two large subfamilies represented within the mononuclear nonheme iron oxygenase class, namely, dioxygenases and monooxygenases. Dioxygenases transfer both atoms of oxygen from O_2 into one or more of the organic substrates or cofactors (Kovaleva and Lipscomb 2008; Vaillancourt et al. 2006; Bugg 2011), whereas monooxygenases transfer one atom of oxygen from O_2 into the organic substrate with reduction of the second atom to water (Hausinger 2004; Krebs et al. 2007). A few subfamilies can catalyze both types of oxygenase reaction using a shared reaction scheme

(Boyd et al. 2003). However, most of the time, monooxygenases and dioxygenases use significantly different mechanistic strategies.

Mononuclear nonheme iron oxygenases usually function by using a reductive strategy for activation of O_2, but the source of the required electrons varies significantly. Common sources include oxidation of the substrate itself, oxidation of a specific cofactor, and use of the energy stored in universal metabolic cofactors such as NAD(P)H. The reduction of O_2 in and of itself overcomes the spin forbidden nature of the reaction with common biological molecules. However, it may also be a precursor to O–O bond cleavage to yield highly reactive oxidizing species containing only a single oxygen atom.

The iron plays many different roles in the process of O_2 activation through its ability to bind O_2 and/or organic substrates, assume multiple oxidation and spin states, and promote electron spin exchange via the unpaired electrons in its *d*-orbitals (Kovaleva and Lipscomb 2008; Pau et al. 2007). Here the strategies for the most prevalent subfamilies of the mononuclear nonheme iron oxygenases are illustrated. The discovery over 55 years ago of the first nonheme oxygenase enzymes fostered recognition of both the diversity of this family the essential roles it plays in the synthesis and degradation of biological molecules.

Mononuclear Nonheme Iron Oxygenase Subfamilies

Aromatic Ring Cleaving Dioxygenases

Plants convert a significant portion of the global CO_2 reservoir into stable aromatic polymers which compose structural scaffolds in woody tissue. This carbon is returned to the global carbon cycle by the action of microorganisms that first breakdown the aromatic polymer and then metabolize the individual aromatic molecules for growth (Bugg et al. 2011). A key step in this process is the opening of the aromatic ring which is catalyzed in the aerobic biosphere by one of two types of dioxygenases that bind Fe(II) or Fe(III), respectively, in their active site (Lipscomb and Orville 1992; Vaillancourt et al. 2006). Both types result in the complete incorporation of oxygen from O_2 into the ring-open product. The Fe(II) class is divided into enzymes that cleave catechols (1,2-dihydroxybenzene adducts) (Fig. 1a) and those that cleave gentisates (2,5-dihydroxybenzoate adducts). The catechol-cleaving Fe(II) enzymes open the ring adjacent to the vicinal hydroxyl functions to yield yellow muconic semialdehyde products. If the aromatic ring has another substituent, then this so-called extradiol dioxygenase class can be further divided into proximal and distal cleaving depending upon the relation of the cleavage site to the position of the substituent. Gentisate dioxygenases cleave between the carboxylate and 2-OH substituents to yield maleylpyruvic acid. Other types of substrates such as hydroquinones and salicylates are ring opened by similar Fe(II) dioxygenases. The Fe(III) class is largely composed of enzymes that catalyze ring cleavage between the vicinal hydroxyls of catechol adducts to yield muconate derivatives and thus are called intradiol dioxygenases (Fig. 1b).

The Fe(II) atom of extradiol dioxygenases is bound by 2 His and 1 Glu/Asp, all oriented on one side of the metal coordination sphere (Fig. 1a). The opposite side is bound by displaceable solvents, and thus, can be used to bind substrates upon displacement of the solvents. This so-called 2-His-1-carboxylate facial triad is the most common iron binding motif within the mononuclear nonheme Fe(II) oxygenase family, and it was first recognized in the extradiol subfamily (Koehntop et al. 2005; Lipscomb 2008; Han et al. 1995). This ligand set and the fact that the iron is ferrous results in a nearly colorless enzyme. The redox potential of the iron is high so that the Fe(II) state is stabilized and maintained throughout the catalytic cycle.

The Fe(III) of intradiol dioxygenases is bound by 2 Tyr and 2 His amino acid ligands as well as a deprotonated solvent molecule (OH^-) (Fig. 1b) (Lipscomb and Orville 1992). The Tyr ligands transfer charge to (but do not reduce) the Fe(III) site yielding a burgundy chromophore. In this case, the ligand set causes the redox potential to be very low so that the Fe (III) state is constant during catalysis (Pau et al. 2007).

Due to the different metal binding environments of Fe (II) and Fe(III) dioxygenases, the O_2 activation and ring cleaving strategies are fundamentally different. Homoprotocatechuate 2,3-dioxygenase (HPCD) will be used to illustrate the extradiol dioxygenase mechanism (Fig. 2). The reaction cycle begins with the catecholic substrate (HPCA) binding to the Fe(II) via both of its OH groups (a chelate), only one of which deprotonates. This directly displaces two solvent ligands and indirectly displaces (or weakens) the third, allowing O_2 to bind in the newly available ligand site adjacent to the HPCA.

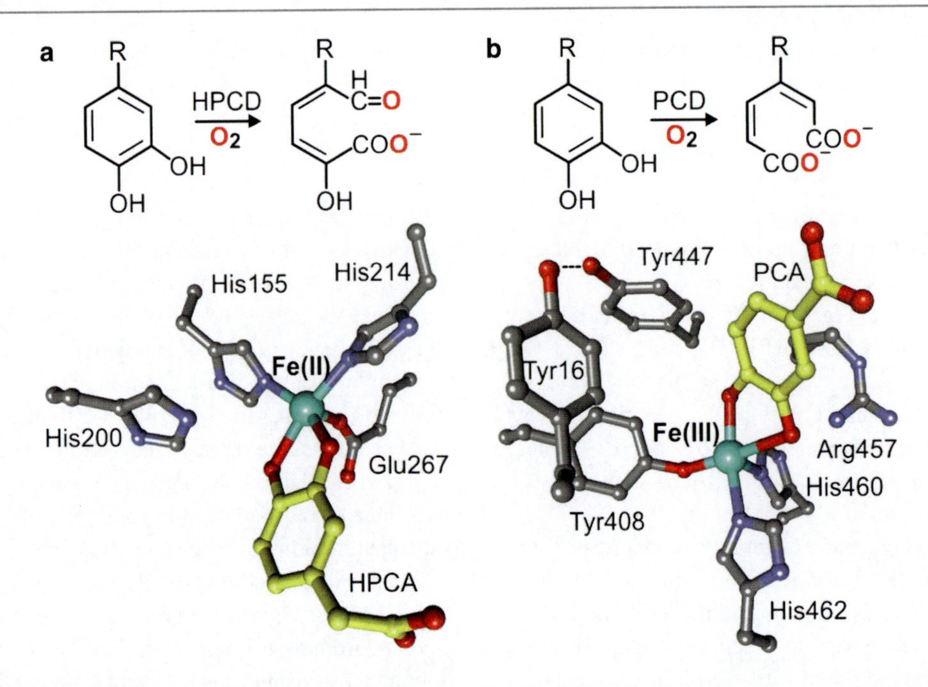

Iron Proteins, Mononuclear (non-heme) Iron Oxygenases, Fig. 1 Typical reactions and metal coordination environments of aromatic ring cleaving dioxygenases. (**a**) Reaction of HPCD and the X-ray crystal structure of the Fe(II) 2-His-1-carboxylate ligand environment with the substrate (*yellow*) bound (PDBID: 1Q0C). The loss of solvent ligands from the iron as substrate binds creates a vacant site for O_2 to bind. (**b**) Reaction of PCD and the X-ray crystal structure of the Fe(III) ligand environment of the substrate (*yellow*) complex containing one Tyr and 2 His ligands (PDBID: 3PCA). An open site is created on the Fe(III), but O_2 cannot bind without simultaneously reacting with the substrate

The Fe(II) serves two functions. First, it organizes the substrates to direct the upcoming reaction. Second, it serves as a conduit to allow an electron to move from the substrate to the O_2. In this way, both the substrate and the O_2 take on radical character (Lipscomb and Orville 1992). Recombination of the radicals produces an alkylperoxo intermediate. This intermediate is thought to breakdown in what is termed a Criegee rearrangement in which the alkylperoxo O–O bond breaks heterolytically and one oxygen is inserted into the aromatic ring to form a lactone adduct. Finally, the lactone is hydrolyzed by the second oxygen from O_2 to yield the reaction product that can dissociate from the enzyme. In the case of the extradiol dioxygenases, the Criegee rearrangement is catalyzed by an active site His residue (His200 for HPCD) that acts as an acid catalyst probably using the proton removed from the second OH of the HPCA (Fig. 1a). It is important to recognize that the reaction between the iron-bound substrate and oxygen radicals is not "forbidden" and thus readily occurs. The enzyme uses iron to facilitate the generation of the radicals. The activation only occurs when both the HPCA and the O_2 are present and properly oriented, which guarantees a specific reaction. Many aspects of this mechanism have been experimentally tested, including studies that have reported electronic and three-dimensional structures of proposed reaction intermediates (Kovaleva and Lipscomb 2007; Mbughuni et al. 2010; 2011).

Gentisate and salicylate dioxygenases also form a substrate-Fe(II) chelate, in this case via the carboxylate and 2-OH substituents (Lipscomb and Orville 1992). Structural studies have shown that there is an open Fe (II) coordination site adjacent to the substrate binding site, implying that O_2 binding and activation occurs in a similar manner to that just described for the extradiol dioxygenases (Chen et al. 2008). However, the iron coordination geometry contains 3 His residues rather than the usual 2-His-1-carboxylate motif suggesting that there may be subtle differences.

Protocatechuate 3,4-Dioxygenase (PCD) will be used to illustrate the O_2 activation strategy of Fe(III) dioxygenases (Fig. 3). Fe(III) sites cannot bind O_2 directly and the redox potential is much too low to be reduced by catecholic substrates, so the O_2 activation strategy must take a very different course than just

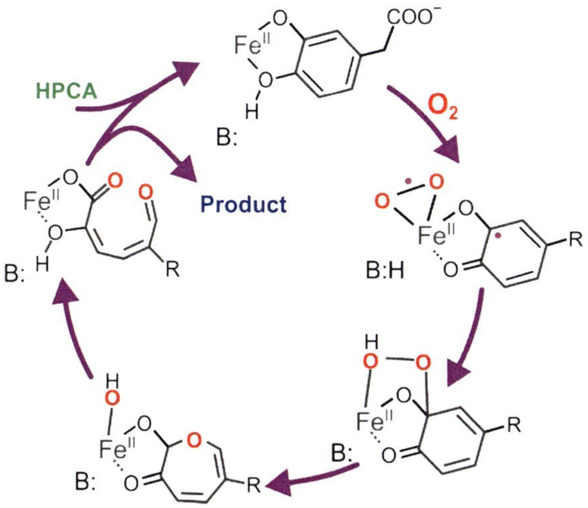

Iron Proteins, Mononuclear (non-heme) Iron Oxygenases, Fig. 2 Proposed reaction cycle for Fe(II) extradiol dioxygenases. The reaction cycle is illustrated for homoprotocatechuate 2,3-dioxygenase

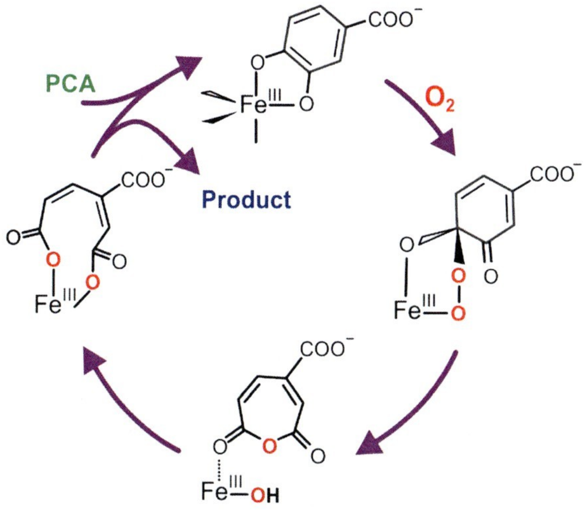

Iron Proteins, Mononuclear (non-heme) Iron Oxygenases, Fig. 3 Proposed reaction cycle for Fe(III) intradiol dioxygenases. The reaction cycle is illustrated for protocatechuate 3,4-dioxygenase

described for the Fe(II) class (Lipscomb and Orville 1992). The reaction begins by the substrate catechol (PCA) binding in a multistep process to the iron as a fully deprotonated chelate. During the binding process, the solvent OH^- and one Tyr (Tyr447 for PCD) are dissociated and the iron coordination geometry changes (Fig. 1b). The original trigonal bipyramidal geometry (no open coordination sites) shifts to square pyramidal (one open site). This allows O_2 to bind in a unique concerted process in which bonds to the substrate and the iron are formed simultaneously to form an alkylperoxo intermediate (Pau et al. 2007). A Criegee rearrangement similar to that described above ensues, but this time an anhydride intermediate is formed which can be hydrolyzed by the second oxygen from O_2 to yield a muconic acid adduct as the product.

Rieske Mono- and Dioxygenases

Aromatic compounds are often encountered in the environment not as catechols or gentisates but rather as unactivated single or multi-ring species such as benzene or naphthalene. These may appear as even less reactive compounds when halo, nitro, and other deactivating substituents are present. To deal with these compounds, microorganisms often utilize Rieske mono- or dioxygenases that can introduce one or two O_2-derived hydroxyl substituents (Gibson and Parales 2000; Ferraro et al. 2005). The Rieske dioxygenases introduce both atoms of oxygen from one O_2 into the same face of the aromatic ring at adjacent carbons to form a nonaromatic *cis*-dihydrodiol product. This product is rearomatized by the next enzyme in the pathway to yield a catechol that can be ring cleaved by the dioxygenases discussed above.

The Rieske mono- and dioxygenases have very similar structures and O_2 activation mechanisms, so the naphthalene 1,2-dioxygenase system (NDOS) will be used to illustrate the common aspects of catalysis (Fig. 4). NDOS is called a system because it consists of three protein components. A reductase (NDR) containing an FAD flavin and a Fe_2S_2 iron sulfur cluster is reduced by NAD(P)H. The reduced NDR then reduces a Fe_2S_2 Rieske ferredoxin component (NDF) which transfers the electrons to the oxygenase component (NDO). NDO contains a Rieske type Fe_2S_2 cluster and a mononuclear nonheme Fe(II) bound in a 2-His-1-carboxylate facial triad. Rieske clusters differ from the usual Fe_2S_2 clusters in that the two Cys residues that normally bind one of the irons are replaced by His residues. The mononuclear iron of Rieske dioxygenases differs from that in ring cleaving extradiol dioxygenases in that it is redox active and can be stabilized in either the Fe(III) or the Fe(II) state. The NDO is a trimer of dimers in which the β subunits appear to be primarily structural and the α subunits hold both of the metal centers. The α subunit structure

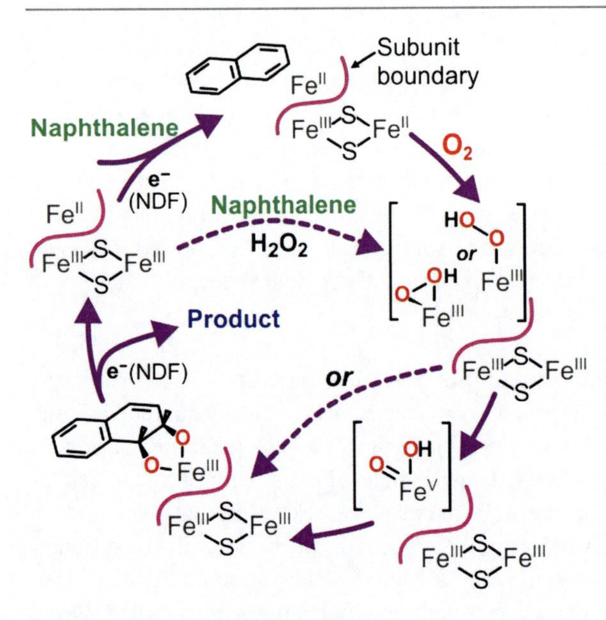

Iron Proteins, Mononuclear (non-heme) Iron Oxygenases, Fig. 4 Proposed reaction cycle for redox-cycling Rieske oxygenases. The reaction cycle is illustrated for naphthalene 1,2-dioxygenase

is such that the metal centers are far apart. However, when the enzyme assembles, the three α subunits align head to tail so that the Rieske cluster of one α subunit is only 12 Å from the mononuclear site of the adjacent α subunit. This cross-boundary pair of metals centers appears to be the active unit (Ferraro et al. 2005).

Catalysis is highly regulated in NDOS. The cycle begins by reduction of the enzyme by two electrons so that the mononuclear iron is in the Fe(II) state and the Rieske cluster holds one electron. Reduction of the mononuclear iron causes a conformational change that allows the substrate naphthalene to bind. Once the substrate and both electrons are present, O_2 is allowed to bind to begin the activation process. If the Rieske cluster is not reduced, O_2 will not bind even when the substrate is present and the mononuclear iron is reduced. This was explained by MCD and ENDOR studies of NDO and structural studies in the Rieske monooxygenase 2-Oxoquinoline 8-Monooxygenase (OMO) (Kovaleva and Lipscomb 2008). Substrate binding near (but not to) the mononuclear Fe(II) causes the loss of one solvent ligand from the iron, opening a potential site for O_2, but the site is sterically blocked. Reduction of the Rieske cluster initiates a series of cross-boundary structural changes initiated by protonation of one of the His ligands of the cluster. This results in the mononuclear iron being drawn away from the substrate, providing space for O_2 to bind.

The mechanism of Rieske dioxygenases remains controversial. However, two recent experimental approaches suggest that a two electron reduced form of oxygen is involved (Kovaleva and Lipscomb 2008). First, two-electron reduced NDO was shown to react with substrate and O_2 in a single turnover reaction to form product in near stoichiometric yield and leave the mononuclear iron in the Fe(III) state. Second, both NDO and a fully oxidized form of benzoate dioxygenase (BZDO) were shown to utilize H_2O_2 as a source of two oxygens and two electrons to catalyze formation of product in a single turnover. In the normal turnover reaction, it is proposed that O_2 accepts one electron from the mononuclear Fe(II) and one transferred electron from the Rieske cluster as it binds. This would result in an Fe(III)-(hydro)peroxo adduct. Accordingly, a crystal structure of an intermediate formed after adding O_2 to a single crystal with naphthalene bound shows a side-on bound Fe-(hydro) peroxo in which the O–O bond is aligned with the substrate ring C–C bond that will be dihydroxylated (Karlsson et al. 2003). Moreover, in the H_2O_2 shunt reaction with BZDO, an intermediate can be trapped from solution that has spectroscopic characteristics of an Fe(III)-(hydro)peroxo species (Kovaleva and Lipscomb 2008).

Once the Fe(III)-(hydro)peroxo species is formed, the reaction with the aromatic substrate may occur directly or the species may undergo O–O bond cleavage to yield a HO–Fe(V)=O reactive intermediate. Density functional theory (DFT) computations suggest that NDO reacts via the Fe(III)-hydroperoxo species to yield a substrate epoxide that can open up to form a cation and then the *cis*-dihydrodiol product after reaction with the second Fe-bound oxygen (Bassan et al. 2004). However, when NDO catalyzes monooxygenase reactions with radical clock substrates, there is clear formation of a long-lived radical intermediate which is inconsistent with attack by a peroxo species, but is the expected reaction for the HO–Fe(V)=O intermediate (Kovaleva and Lipscomb 2008). DFT calculations show that the OH–Fe(V)=O species is the most likely reactive species in the reactions of small molecule chelate complexes that catalyze *cis*-dihydroxylation and there is considerable experimental support for this proposal. These complexes differ in that the iron is low spin in the precursor

Fe(III)hydroperoxo species in contrast to the high spin Fe(III)(hydro)peroxo species of the enzymes.

Rieske monooxygenases such as OMO or oxygenating *O*-demethylases, such as 4-Methoxybenzoate Monooxygenase or Dicamba Monooxygenase, are likely to activate O_2 in the same manner as the dioxygenases (Bruijnincx et al. 2008). The true oxygenating species is unknown, although exclusive monooxygenation of the aromatic ring of 2-oxoquinoline and the above mentioned radical intermediate in the monooxygenation of radical clocks favors the HO–Fe(V)=O species.

2-Oxo-Acid-Dependent Mononuclear NonHeme Fe(II) Dioxygenases

A large and diverse family of mononuclear nonheme iron oxygenases exists that utilizes 2-oxo acids as cosubstrates in reactions that result in hydroxylation of another substrate molecule (Hausinger 2004). Many of these use α-ketoglutarate (αKG) as the cosubstrate and bind Fe(II) in a 2-His-1-carboxylate binding motif in the active site. In the canonical reaction, the αKG chelates the iron via its carboxylate and oxo groups, resulting in release of the solvent ligands and opening an adjacent site for O_2 binding. However, the O_2 does not bind until the substrate to be hydroxylated binds in a position near the iron, showing once again that the O_2 activation process is highly regulated.

The proposed mechanism for 2-oxo-acid-dependent dioxygenases is illustrated in Fig. 5 for the reaction of Taurine Dioxygenase (TauD) (Hausinger 2004; Bollinger et al. 2005). Oxygen binding to the Fe(II) begins the activation process. However, in contrast to the Fe(II) ring cleaving dioxygenase substrates, αKG lacks the extended double bond character to transfer an electron to the iron, so an Fe(III)–$O_2^{\bullet-}$ rather than the Fe(II)–$O_2^{\bullet-}$ is likely to form. It is proposed that the bound $O_2^{\bullet-}$ attacks the 2-oxo carbon to generate an alkylperoxo intermediate. The role of many 2-oxo-acid dioxygenases is to hydroxylate a stable substrate, so a relatively powerful activated oxygen species must ultimately be generated. It is proposed that the O–O bond of the alkylperoxo species breaks heterolytically to yield an Fe(IV)=O reactive species and form succinate and CO_2 from the αKG. In this case, the 2 electrons required to promote O–O bond cleavage are derived from breaking the C–C bond of αKG rather than an external source such as NADH. The Fe(IV)=O species is proposed to attack the substrate by hydrogen

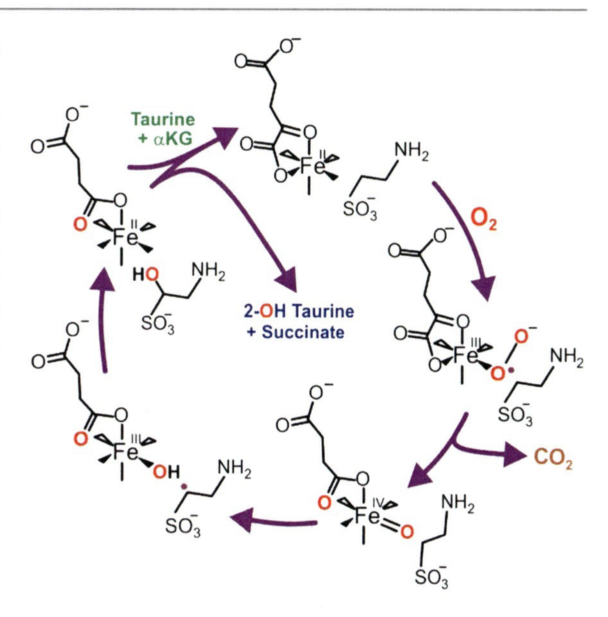

Iron Proteins, Mononuclear (non-heme) Iron Oxygenases, Fig. 5 Proposed reaction cycle for redox-cycling 2-oxo-acid-dependent dioxygenases. The reaction cycle is illustrated for taurine dioxygenase

atom abstraction. Then rebound of $OH^{\bullet}$ and substrate radical yield the hydroxylated product and the Fe(II) resting state of the enzyme. In this enzyme subfamily, one oxygen atom from O_2 is found in the hydroxylated product and the other in succinate, preserving the dioxygenase stoichiometry.

Support for the proposed mechanism has come through transient kinetic studies that allowed intermediates in the reaction cycle to be detected and in some cases trapped and spectroscopically characterized (Krebs et al. 2007). The number and general kinetic properties of the intermediates are in accord with the proposed mechanism. Perhaps the most diagnostic intermediate, an Fe(IV)=O species termed J, was trapped by rapid freeze quench and characterized by Mössbauer spectroscopy (Price et al. 2003). Intermediate J was shown to exhibit a large KIE in its reaction with substrate, as expected for the proposed hydrogen atom abstraction mechanism.

The 2-oxo-acid linked dioxygenases catalyze a remarkable number of reactions in organisms from bacteria to mammals. The types of reactions they catalyze include hydroxylation, epoxidation, epimerization, desaturation, ring closure and expansion, and substituent migration (Hausinger 2004). These reactions are characteristic of oxygenases that employ high valent iron intermediates in support of the proposed mechanism. While most of the enzymes use αKG or a similar separate 2-oxo-acid cosubstrate, one important class incorporates the 2-oxo acid into the substrate itself. An example is 4-Hydroxyphenylpyruvate Dioxygenase (HPPD) where the 2-oxo acid of the aromatic substrate side chain is key to the reaction. O_2 activation in this enzyme is thought to follow a similar course to that shown in Fig. 5 except that the Fe(IV)=O intermediate may directly attack the aromatic ring to yield an arenium cation (Bruijnincx et al. 2008). This would allow a so-called NIH shift to occur resulting in migration of the ring substituents en route to formation of the observed homogentisate product.

Another recently described class of mononuclear nonheme Fe(II) dioxygenases has some similarities to the 2-oxo-acid dioxygenase family in that in the reactive tautomer of the substrate one keto- and one anionic oxygen from the substrate bind to adjacent ligand sites on the Fe(II). Also, the two oxygens from O_2 are found in two different products resulting from substrate cleavage. No additional cofactor is required. The only well-studied example is diketone cleaving enzyme (DEK1) which catalyzes the cleavage of acetylacetone to form methyl glyoxal and acetate (Diebold et al. 2011). The mechanistic theory for this type of enzyme is still evolving, but it is proposed to involve attack of an initially formed Fe(III)–$O_2^{\bullet-}$ on the central carbon of the substrate β-dicarbonyl. This is similar to the first step of both the 2-oxo acid- and the ring cleaving dioxygenase families. However, in the case of DEK1, the oxygen bound to the iron is proposed to migrate to one of the dicarbonyl oxygens to form a dioxetane which breaks down to form the products. DEK1 is unusual in that it employs 3 His ligands to bind Fe(II) rather than the 2-His-1-carboxylate facial triad. It has been proposed that the difference in net charge for these two binding motifs has a significant effect on the direction taken in the reaction chemistry (Diebold et al. 2011).

Tetrahydropterin-Dependent Mononuclear Nonheme Fe(II) Oxygenases

The tetrahydropterin-dependent mononuclear nonheme Fe(II) oxygenases are responsible for the biosynthesis of molecules such as DOPA, serotonin, and tyrosine in humans (Fitzpatrick 1999). Similar to the Rieske dioxygenases, these enzymes harbor both an 2-His-1-carboxylate facial triad Fe(II) site and a nearby reduced cofactor. However, in this case, the tetrahydropterin cofactor can supply two rather than one election, and it can bind O_2. Structural studies show that the tetrahydropterin and the Fe(II) are aligned so that O_2 can bind to both, forming a peroxo bridge (Bruijnincx et al. 2008; Kovaleva and Lipscomb 2008).

Phenylalanine hydroxylase is used to illustrate the reaction mechanism of this subfamily in Fig. 6. It is proposed that the tetrahydropterin provides two electrons to promote heterolytic O–O bond cleavage to yield 4a-hydroxypterin and an Fe(IV)=O intermediate. The latter species has recently been directly observed and trapped for spectroscopic studies (Eser et al. 2007). The x-ray crystal structures show that the aromatic substrate also binds near the Fe(II) site. Thus, it is proposed that the Fe(IV)=O species reacts with the aromatic ring in a one or two step reaction to yield an Fe(II)–O–substrate arenium cation (Fitzpatrick 1999; Bassan et al. 2003). This intermediate would convert to the hydroxylated product via an NIH shift and return the iron in the enzyme to the resting Fe(II) state.

The ability of the tetrahydropterin to provide two electrons suggests that the reactive state of this oxygenase class is one electron more reduced than the intermediate formed in the Rieske dioxygenases, making it formally a less potent reagent. Nevertheless, it is evident that the Fe(IV)=O species is often selected by nature as the reagent of choice to catalyze many types of difficult hydroxylation reactions (Krebs et al. 2007). In fact, the 2-oxo acid and tetrahydropterin perform similar functions in their respective oxygenases by directly participating in the formation of an initial peroxo species and then supplying the two electrons required to ultimately form the common Fe (IV)=O reactive intermediate.

Related Enzymes

The mononuclear nonheme iron oxygenase subclasses discussed here form the majority of the general class. However, there are numerous other enzymes that either employ iron to carry out oxygenase reactions using different strategies or activate O_2 for oxidase or related reactions using one of the mechanistic strategies just introduced. An example of the former class is the lipoxygenase family (Klinman 2007). The enzyme catalyzes incorporation of O_2 into polyunsaturated fatty acids in plants to generate chemicals involved in development and related processes, and into

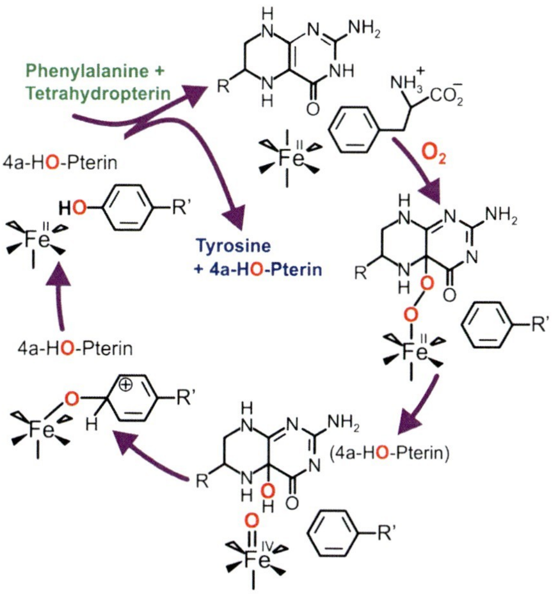

Iron Proteins, Mononuclear (non-heme) Iron Oxygenases, Fig. 6 Proposed reaction cycle for redox-cycling tetrahydropterin-dependent oxygenases. The reaction cycle is illustrated for phenylalanine hydroxylase

arachidonic acid in mammals to form various eicosanoids. In the resting state, these enzymes bind a nonheme Fe(II) via three His ligands, a carboxylate oxygen from a C-terminal isoleucine, a weakly associated Asn (plants) or His (mammals), and a solvent water. In contrast to the oxygenases described this far, the O–O bond is not cleaved in the reaction. The enzyme is activated by oxidation of the resting Fe(II) to the Fe(III) state. This causes the solvent to become OH^- which then serves to abstract a hydrogen atom from the 1,4-pentadiene moiety found in the substrates to yield a substrate radical and an Fe(II)–OH_2 complex. O_2 then directly attacks the substrate radical to form an hydroperoxy radical intermediate, which is then reduced by reclaiming the hydrogen atom from the Fe(II)–OH_2 complex to regenerate the Fe(III)–OH^- state ready for another turnover.

Several nonheme Fe(II) containing oxygenases have been identified that share the atypical 3-His Fe (II) ligation described above for DEK1 and the gentisate dioxygenases. For example, dioxygenases with this ligation have been shown to catalyze oxygen insertion reactions using cysteine, cysteamine, 4-mercaptopropionate, acireductone, and quercetin (de Visser and Straganz 2009; Schaab et al. 2006). The mechanism proposed for each of these enzymes is generally similar to that of one of the major classes described above. It is clear from the initial studies that the mechanistic diversity of the 2-His-1-carboxylate class also applies to the new 3-His class of oxygenases.

Numerous biosynthetic or biodegradative oxidase enzyme are structurally and mechanistically more related to the oxygenase enzymes discussed here than to more familiar oxidases involved in bioenergetics as terminal electron acceptors. Some well-studied examples of enzymes in this category are Isopenicillin *N*-synthase, Aminocyclopropane-1-carboxylate oxidase, (S)-2-Hydroxypropylphosphonic Acid Epoxidase, and CytC3 Halogenase (Clifton et al. 2006; Krebs et al. 2007; Bruijnincx et al. 2008).

Perspective

Research in the mononuclear nonheme iron oxygenase field continues to expand rapidly due in part to the ability to recognize these enzymes in the ever-growing data base of genome sequences. Although they were initially recognized in bacteria and soon thereafter in mammals, the true extent of their pervasive and diverse roles in the metabolism of all aerobic organisms has only recently been fully appreciated. Future directions in this field will likely involve both identification of more members of this large family and in depth investigations of the molecular mechanisms of O_2 activation and insertion. The identification of new members will include new members of known subfamilies that perform unrecognized metabolic roles such as the recently described demethylation reactions of DNA (Yi et al. 2010) as well as new metal site environments. The search for new approaches to detect, capture, and chemically and structurally characterize intermediates is likely to allow rapid advances in mechanistic theory. Moreover, the characterization of intermediates will advance the application of correlated computational techniques to allow an understanding of the activation of molecular oxygen at a level significantly beyond what is now possible.

Cross-References

▶ Acireductone Dioxygenase
▶ Iron, Physical and Chemical Properties

References

Bassan A, Blomberg MRA, Siegbahn PEM (2003) Mechanism of aromatic hydroxylation by an activated FeIV=O core in tetrahydrobiopterin-dependent hydroxylases. Chem Eur J 9:4055–4067

Bassan A, Blomberg MRA, Siegbahn PEM (2004) A theoretical study of the *cis*-dihydroxylation mechanism in naphthalene 1,2-dioxygenase. J Biol Inorg Chem 9:439–452

Bollinger JM Jr, Price JC, Hoffart LM, Barr EW, Krebs C (2005) Mechanism of taurine: alpha-ketoglutarate dioxygenase (TauD) from *Escherichia coli*. Eur J Inorg Chem 2005:4245–4254

Boyd DR, Sharma ND, Bowers NI, Boyle R, Harrison JS, Lee K, Bugg TDH, Gibson DT (2003) Stereochemical and mechanistic aspects of dioxygenase-catalyzed benzylic hydroxylation of indene and chroman substrates. Org Biomol Chem 1:1298–1307

Bruijnincx PCA, van Koten G, Klein Gebbink RJM (2008) Mononuclear non-heme iron enzymes with the 2-His-1-carboxylate facial triad: recent developments in enzymology and modeling studies. Chem Soc Rev 37:2716–2744

Bugg TDH (2011) Non-heme iron-dependent dioxygenases: mechanism and structure. In: de Visser SP, Kumar D (eds) Iron-containing enzymes: versatile catalysts of hydroxylation reactions in nature. The Royal Society of Chemistry, Cambridge, pp 42–66

Bugg TDH, Ahmad M, Hardiman EM, Singh R (2011) The emerging role for bacteria in lignin degradation and bio-product formation. Curr Opin Biotechnol 22:394–400

Chen J, Li W, Wang M, Zhu G, Liu D, Sun F, Hao N, Li X, Rao Z, Zhang XC (2008) Crystal structure and mutagenic analysis of GDOsp, a gentisate 1,2-dioxygenase from *Silicibacter pomeroyi*. Protein Sci 17:1362–1373

Clifton IJ, McDonough MA, Ehrismann D, Kershaw NJ, Granatino N, Schofield CJ (2006) Structural studies on 2-oxoglutarate oxygenases and related double-stranded beta - helix fold proteins. J Inorg Biochem 100:644–669

Costas M, Mehn MP, Jensen MP, Que L Jr (2004) Dioxygen activation at mononuclear nonheme iron active sites: enzymes, models, and intermediates. Chem Rev 104:939–986

de Visser SP, Straganz GD (2009) Why do cysteine dioxygenase enzymes contain a 3-His ligand motif rather than a 2His/1Asp motif like most nonheme dioxygenases? J Phys Chem A 113:1835–1846

Diebold AR, Straganz GD, Solomon EI (2011) Spectroscopic and computational studies of α-keto acid binding to Dke1: understanding the role of the facial triad and the reactivity of β-diketones. J Am Chem Soc 133:15979–15991

Eser BE, Barr EW, Frantom PA, Saleh L, Bollinger JM Jr, Krebs C, Fitzpatrick PF (2007) Direct spectroscopic evidence for a high-spin Fe(IV) intermediate in tyrosine hydroxylase. J Am Chem Soc 129:11334–11335

Ferraro DJ, Gakhar L, Ramaswamy S (2005) Rieske business: structure-function of Rieske non-heme oxygenases. Biochem Biophys Res Commun 338:175–190

Fitzpatrick PF (1999) Tetrahydropterin-dependent amino acid hydroxylases. Annu Rev Biochem 68:355–381

Gibson DT, Parales RE (2000) Aromatic hydrocarbon dioxygenases in environmental biotechnology. Curr Opin Biotechnol 11:236–243

Han S, Eltis LD, Timmis KN, Muchmore SW, Bolin JT (1995) Crystal structure of the biphenyl-cleaving extradiol dioxygenase from a PCB-degrading pseudomonad. Science 270:976–980

Hausinger RP (2004) Fe(II)/alpha -ketoglutarate-dependent hydroxylases and related enzymes. Crit Rev Biochem Mol Biol 39:21–68

Karlsson A, Parales JV, Parales RE, Gibson DT, Eklund H, Ramaswamy S (2003) Crystal structure of naphthalene dioxygenase: side-on binding of dioxygen to iron. Science 299:1039–1042

Klinman JP (2007) How do enzymes activate oxygen without inactivating themselves? Acc Chem Res 40:325–333

Koehntop KD, Emerson JP, Que L Jr (2005) The 2-His-1-carboxylate facial triad: a versatile platform for dioxygen activation by mononuclear non-heme iron(II) enzymes. J Biol Inorg Chem 10:87–93

Kovaleva EG, Lipscomb JD (2007) Crystal structures of Fe^{2+} dioxygenase superoxo, alkylperoxo, and bound product intermediates. Science 316:453–457

Kovaleva EG, Lipscomb JD (2008) Versatility of biological non-heme Fe(II) centers in oxygen activation reactions. Nat Chem Biol 4:186–193

Krebs C, Fujimori DG, Walsh CT, Bollinger JM Jr (2007) Non-heme Fe(IV)-oxo intermediates. Acc Chem Res 40:484–492

Lipscomb JD (2008) Mechanism of extradiol aromatic ring-cleaving dioxygenases. Curr Opin Struct Biol 18:644–649

Lipscomb JD, Orville AM (1992) Mechanistic aspects of dihydroxybenzoate dioxygenases. Metal Ions Biol Syst 28:243–298

Mbughuni MM, Chakrabarti M, Hayden JA, Bominaar EL, Hendrich MP, Münck E, Lipscomb JD (2010) Trapping and spectroscopic characterization of an FeIII-superoxo intermediate from a nonheme mononuclear iron-containing enzyme. Proc Natl Acad Sci USA 107:16788–16793

Mbughuni MM, Chakrabarti M, Hayden JA, Meier KK, Dalluge JJ, Hendrich MP, Münck E, Lipscomb JD (2011) Oxy-intermediates of homoprotocatechuate 2,3-dioxygenase: facile electron transfer between substrates. Biochemistry 50:10262–10274

Pau MYM, Lipscomb JD, Solomon EI (2007) Substrate activation for O_2 reactions by oxidized metal centers in biology. Proc Natl Acad Sci USA 104:18355–18362

Price JC, Barr EW, Tirupati B, Bollinger JM Jr, Krebs C (2003) The first direct characterization of a high-valent iron intermediate in the reaction of an α-ketoglutarate-dependent dioxygenase: a high-spin Fe(IV) complex in taurine alpha-ketoglutarate dioxygenase (TauD) from *E. coli*. Biochemistry 42:7497–7508

Schaab MR, Barney BM, Francisco WA (2006) Kinetic and spectroscopic studies on the quercetin 2,3-dioxygenase from *Bacillus subtilis*. Biochemistry 45:1009–1016

Solomon EI, Decker A, Lehnert N (2003) Non-heme iron enzymes: contrasts to heme catalysis. Proc Natl Acad Sci USA 100:3589–3594

Vaillancourt FH, Bolin JT, Eltis LD (2006) The ins and outs of ring-cleaving dioxygenases. Crit Rev Biochem Mol Biol 41:241–267

Yi C, Jia G, Hou G, Dai Q, Zhang W, Zheng G, Jian X, Yang C-G, Cui Q, He C (2010) Iron-catalysed oxidation intermediates captured in a DNA repair dioxygenase. Nature 468:330–333

Iron Proteins, Plant Iron Transporters

Khurram Bashir[1] and Naoko K. Nishizawa[1,2]
[1]Graduate School of Agricultural and Life Sciences, The University of Tokyo, Tokyo, Japan
[2]Research Institute for Bioresources and Biotechnology, Ishikawa Prefectural University, Ishikawa, Japan

Synonyms

MAs: Mugineic acid family phytosiderophores; *MATE:* Multidrug and toxic compound extrusion (MATE); *NA:* Nicotianamine; *NRAMP:* Natural resistance-associated macrophage protein; *PEZ:* Phenolics efflux zero; *ROS:* Reactive oxygen species; *TOM:* Transporter of MAs; *YS1:* Yellow stripe 1; *YSL:* Yellow stripe 1-like

Definition

Iron (Fe) is an essential micronutrient for all plant species and its deficiency severely impairs plant growth and development. Despite its abundance in mineral soils, it is often not readily available to plants, especially under neutral to alkaline conditions. Excess Fe, however, is toxic to cells through reactive oxygen species evolution, and thus Fe uptake and translocation must be strictly regulated. Plants have developed sophisticated and complex systems to take up Fe from soil. Understanding the components of plant Fe uptake from soil and translocation to the shoots, and ultimately the seeds, is of the utmost importance for mitigating the problem of widespread Fe deficiency. During recent years, several genes involved in plant Fe transport have been characterized in detail. In this report, the role of these Fe transporters throughout plant development has been summarized.

Introduction

Plants, like other organisms, require iron (Fe) as an essential micronutrient. It is needed for numerous functions such as mitochondrial electron transport, heme biosynthesis and function, Fe-sulfur (Fe-S) cluster synthesis, chlorophyll biosynthesis, and photosynthetic electron transport. Although Fe is abundant in soils, it is largely present as insoluble Fe(III) compounds and as a result, plants are unable to readily utilize it. This problem is particularly severe in calcareous soils, which are estimated to cover approximately 800 million hectares worldwide. Under Fe-deficient conditions, plants fail to develop and stabilize adequate chlorophyll and consequently turn yellow. Indeed, the low chlorophyll content (chlorosis) of young leaves and poor plant growth are the most obvious visible symptoms of Fe deficiency, and yet Fe deficiency also affects roots morphogenesis, flower development, and flower color. Also, since Fe is a transition metal and readily accepts and donates electrons, it is involved in the production of reactive oxygen species (ROS). Plants, therefore, have to strictly control the movement of Fe, and as such have developed sophisticated mechanisms to absorb Fe from soil and transport it from roots to vegetative organs and seeds. Plants are classified into two broad categories based upon the strategy they adopt to acquire soil Fe (Marschner 1995). All plants, with the exception of graminaceous grasses {e.g., arabidopsis (*Arabidopsis thaliana*), tomato (*Lycopersicum esculentum*), and soybean (*Glycine max*)}, utilize strategy I, i.e., release protons to solubilize Fe(III). In most cases, strategy I plants also release chelating/reducing compounds mainly phenolics. After solubilization, Fe(III) is reduced to Fe^{2+} by a membrane-bound Fe(III) reductase oxidase and Fe^{2+} is taken up through an iron-regulated transporter (IRT1). On the other hand, graminaceous plants {e.g., barley (*Hordeum vulgare*), maize (*Zea mays*), rice (*Oryza sativa*), and wheat (*Triticum aestivum*)} secrete mugineic acid (MA) family phytosiderophores through transporter of MAs 1 (TOM1) to chelate Fe(III) (Nozoye et al. 2011). The resulting Fe(III)-MA complex is taken up by the yellow stripe family transporters. Rice (*Oryza sativa* L.) is unique in the sense that it utilizes partial strategy I. Although rice is not reported to possess a functional Fe(III) chelate reductase oxidase, and that H^+-ATPase in rice roots is not induced by Fe deficiency, rice takes up Fe^{2+} (which is abundant under submerged paddy conditions) through OsIRT1. As free Fe catalyzes the production of highly toxic ROS and is harmful to cells, Fe is bound to its target metalloproteins or chelated to organic ligands. In plants, the important Fe chelators are mugineic acid family phytosiderophores (MAs), nicotianamine (NA), phenolics, and citrate. NA, phenolics, and citrate are present in all higher plants, whereas MAs are specifically synthesized in graminaceous plants. NA and DMA share the biosynthetic pathway from L-methionine (Met). NA synthase (NAS) catalyzes the trimerization of *S*-adenosyl Met to NA. In graminaceous plants, NA is then converted to a 3″-keto intermediate through the transfer of an amino group using NA aminotransferase (NAAT). The subsequent reduction of the 3″-keto group by deoxymugineic acid synthase (DMAS) produces 2′-deoxymugineic acid (DMA), which is the first species of MAs synthesized in this conserved pathway. In some graminaceous plants, DMA is further converted to other kinds of MAs (reviewed by Bashir et al. 2010). The production and secretion of MAs significantly increases in response to Fe deficiency, and tolerance to Fe deficiency in graminaceous plants is strongly correlated with the MAs secreted. For example, rice secretes only DMA in relatively low amounts and is more sensitive to low Fe availability. In contrast, barley secretes large amounts of different MAs, including mugineic acid, 3-epihydroxy-2′-deoxymugineic acid, and 3-epihydroxymugineic acid, and is therefore more tolerant to low Fe availability. The genes of the MAs biosynthetic pathway have been cloned and characterized from various graminaceous plants including barley, rice, and maize. The expression of these genes is significantly upregulated by Fe deficiency through transcriptional regulation mediated by various combinations of *cis*-acting elements and *trans*-acting factors, and consequently the synthesis and secretion of MAs increases with limiting Fe availability (reviewed by Kobayashi and Nishizawa 2012). The expression of genes for MAs biosynthesis changes in diurnal rhythm and MAs are also secreted in diurnal fashion with a peak in the morning to avoid bacterial degradation. Phenolics and citrate also form complexes with Fe. Phenolics such as protocatechuic acid (PCA) and caffeic acid (CA) have the ability to chelate Fe(III), and solubilize and reduce it to Fe^{2+} in vitro. Citrate form complex with Fe and the citrate-Fe complex is especially significant in xylem Fe transport.

In this entry, members of the complex Fe-transport networks in plants and the role of individual transporter proteins at each developmental stage will be discussed. Due to the space limitation, discussion will be focused on Fe transport mainly in model plants such as rice and *Arabidopsis*.

Plant Fe Transporters Belong to Different Families

Fe transport is a complex process involving many genes belonging to different families. Among the transporters involved in Fe homeostasis, the yellow stripe-like (YSL) family (named due to phenotype of mutants from which the ZmYS1 was cloned), iron-regulated transporters (IRT), and citrate transporters AtFRD3 and OsFRDL1 have been characterized in detail. Recently, the MAs efflux transporter TOM1, NA efflux transporter ENA1 and phenolics efflux transporters PEZ1 and PEZ2 have also been characterized. Moreover, the transporters for Fe within the cell have also been characterized.

IRT transporters belong to the zinc-regulated transporter/iron-regulated transporter (ZRT/IRT1)-related protein (ZIP) family. Members of this family are involved in Cd, Fe, Mn, and Zn transport. IRT1 is a key Fe^{2+} transporter in plants with representatives in *Arabidopsis* (IRT1) and rice (OsIRT1) localized to the plasma membrane and functioning to transport Fe^{2+} from the rhizosphere (apoplasm) to root cells. The expression of *IRT1* in rice and *Arabidopsis* is regulated by Fe availability. The *Arabidopsis* IRT2 localizes to vesicles (reviewed by Conte and Walker 2011) within root epidermal cells, while OsIRT2 localizes to the plasma membrane. Like IRT1, IRT2 also transports Fe^{2+} and its expression is also regulated by Fe deficiency both in rice and *Arabidopsis*.

Maize *yellow stripe 1* (*YS1*), which encodes an Fe(III)–MA transporter, belongs to oligopeptide family of transporters that function as proton-coupled symporters for various DMA-bound metals (reviewed by Walker and Waters 2011). The rice genome contains 18 putative YSL family genes, among which *OsYSL2*,

OsYSL15, and *OsYSL18* have been characterized in detail. The protein encoded by *OsYSL15* transports Fe(III)–DMA from the rhizosphere to roots and to developing seeds. *OsYSL18* encodes a functional Fe(III)–DMA transporter involved in Fe distribution in reproductive organs. In comparison, OsYSL2 is a phloem Fe–NA and Mn-NA transporter. The YSL transporters from barley (HvYS1 and HvYS5) have also been characterized and eight putative members have been reported in *Arabidopsis*. As Strategy I plants do not synthesize and utilize MAs, YSL genes in these plants are thought to play a role in Fe transport within the plant body. The most characterized *Arabidopsis* YSL members, AtYSL1, AtYSL2, and AtYSL3, transport the Fe(II)–NA complex. In rice the expression of OsYSL2 and OsYSL15 is strongly upregulated by Fe deficiency, however in *Arabidopsis* the expression of YSLs is not upregulated by Fe deficiency (reviewed by Conte and Walker 2011).

Members of natural resistance-associated macrophage proteins (NRAMPs), a large family of integral membrane proteins, have been identified in several plant species. The NRAMP family is involved in Fe transport, as well as the transport of Mn and Cd. In *Arabidopsis*, NRAMP genes induced by Fe deficiency in the roots include *AtNRAMP1*, which controls Fe homeostasis, and AtNRAMP3 and AtNRAMP4, which mobilize vacuolar Fe (reviewed by Jeong and Guerinot 2009). Likewise, OsNRAMP1 in rice is a plasma membrane protein involved in Fe transport, while OsNRAMP5 transports Fe, Mn, and Cd. The *Arabidopsis* FRD3 and rice FRDL1 (OsFRDL1), members of the multidrug and toxic compound extrusion (MATE) transporter family, are citrate transporters localized to pericycle cells and are important for efficient Fe translocation from roots to shoots. Other members of the MATE family include the phenolic efflux transporters PEZ1 and PEZ2 (phenolic efflux zero 1 and 2; Pez1 and Pez2; Ishimaru et al. 2011; Bashir et al. 2011b). MA efflux transporters belong to the major facilitator family of proteins (MFS), a large family of membrane transporters found ubiquitously in a wide variety of organisms including eukaryotes, and the DMA efflux transporters, transporter of mugineic acids (TOM) have been characterized in rice (TOM1) and barley (HvTOM1). The role of different transporters in plant Fe transport is summarized in Table 1.

Fe Homeostasis During Germination

The first source of Fe for the germinating seed is stored in the seed itself. In *Arabidopsis*, transporters involved in both vacuolar Fe influx (VIT1) and efflux (AtNRAMP3 and AtNRAMP4) have been shown to be essential for germination and seedling development. VIT1 deposits Fe into the vacuole during seed development, while AtNRAMP3 and AtNRAMP4 are thought to export vacuolar Fe to the cytoplasm during germination. When seeds are grown under Fe-limiting conditions, *AtNRAMP3* and *AtNRAMP4* are expressed during the early stages of germination (days 1 and 2), while *IRT1* expression is not observed until the 3rd day, when its expression is strong, indicating its importance during this process. The citrate efflux transporter FRD3 also plays a major role during germination. *frd3* mutants exhibit slow root growth and chlorotic cotyledons during the first 3 days of germination, specifically under Fe-deficient conditions. Fe supplementation as well as germination in the presence of citrate rescues the growth of *frd3* seedlings, suggesting that limited Fe availability and decreased citrate excretion causes improper Fe nutrition of *frd3* embryos. During early germination, FRD3 is primarily expressed in the cotyledons and to some extent in hypocotyls; later expression is observed in the cell layers, the embryo, and areas of the seed coat surrounding the developing embryo. These results clearly suggest a role for FRD3 in citrate efflux and Fe mobilization during germination.

In rice, the OsFRDL1 is neither strongly expressed nor does its expression change significantly during germination. Besides Fe, rice seeds also contain NA and DMA, with the amount of the latter exceeding the former significantly. During germination, Fe is mobilized along with NA or DMA, as revealed by an analysis of rice seeds by X-ray fluorescence imaging. During germination, Fe is detectable in the dorsal vascular bundle, aleurone layer, and the embryo and the localization of Fe changes during germination. Upon germination, *OsIRT1* expression is induced in the dorsal vascular bundle, and specifically in the vascular bundle of the scutellum and the leaf primordium of the embryo. During germination, the expression of various MA biosynthetic genes increases sharply. NA seems to be produced in the endosperm, while DMA is produced in the embryo, especially in the coleorhizae and bud scales, with NA and DMA

Iron Proteins, Plant Iron Transporters, Table 1 Plant Fe transporters

Transporter	Substrate	Strategy	Functions
IRT1 (Arabidopsis), OsIRT1, OsIRT2,	Fe^{2+}	I	Fe uptake from soil
IRT2 (Arabidopsis)	Fe^{2+}	N/A	Fe transport to vesicles
OsNRAMP1	Fe^{2+}	N/A	Fe transport
OsNRAMP5	Fe^{2+}	N/A	Fe uptake from soil
AtFPN1 (IREG1)	Fe^{2+}	N/A	Fe efflux to xylem
VIT1(Arabidopsis)	Fe^{2+}	N/A	Fe efflux to vacuoles
AtNRAMP3, AtNRAMP4	Fe^{2+}	N/A	Fe transport from vacuoles
OsYSL2	NA-Fe(II)	N/A	Fe transport to phloem
AtYSL1, AtYSL2, AtYSL3	NA-Fe(II)	N/A	Fe uptake from xylem, Fe transport to phloem, Fe transport to developing seeds
OsYSL18, OsYSL16	DMA-Fe(III)	II	Fe transport to phloem, Fe transport to developing seeds
OsYSL15	DMA-Fe(III)	II	Fe uptake from soil, Fe transport to developing seeds
TOM1 (rice), HvTOM1	DMA	II	Fe acquisition from soil, Fe transport to reproductive organs
FRD3 (Arabidopsis), OsFRDL1	Citrate	I & II	Citrate efflux to xylem
PEZ1, PEZ2 (Rice)	PCA, CA	I & II	Phenolics efflux to apoplasm
MIT (Rice)	Fe	N/A	Fe transport to mitochondria
PIC1 (Arabidopsis)	Fe	N/A	Involved in chloroplastic Fe homeostasis

At: *Arabidopsis thaliana* (Arabidopsis), Hv: *Hordeum vulgare* (Barley).
*N/A. Strategy I and strategy II classification does not apply to these genes.

production overlapping in the embryo. Hence, NA and DMA are synthesized during germination and are involved in Fe translocation during germination. Furthermore, the expression of *OsIRT1* indicates that rice can utilize both Fe^{2+} and Fe(III) during germination. In addition, the expression of Fe transporters *OsYSL2* and *OsYSL18* are observed to increase during this time. *OsYSL2* is expressed mainly in the epithelium, vascular bundle of the scutellum, and leaf primordium, suggesting that OsYSL2 is also important for Fe translocation from seeds. In *OsYSL2* knockdown RNAi plants, reduced translocation of Fe to the roots and shoots of seedlings during germination result in growth defects. In addition, *OsYSL15* expression is gradually activated during seed germination and is seen in the scutellum in the first 3 days after sowing, with a peak in the 2nd day. Its expression is also observed in the vascular bundle of the scutellum, shoot apex, ventral scale, and bud scale in 2- and 3-day-old seedlings. The growth of *OsYSL15* knockdown seedlings is arrested at the early stages of growth including germination, and the mutants can be rescued by adding high levels of Fe, indicating that *OsYSL15* plays a crucial role in Fe uptake and homeostasis during early growth. The expression of TOM1 is also observed in the epithelium of germinating seeds 3 days after germination. The germination ratio in *TOM1*-inhibited RNAi plants significantly decreased, indicating that it has a role in Fe acquisition during germination. The expression of *mitochondrial Fe transporter* (*MIT*) is also observed during rice seed germination. MIT transports Fe to the mitochondria, but which form the Fe is in during transport is still not clear (Bashir et al. 2011a). During germination, *MIT* expression is specific to the embryo and was seen on the 1st day of germination in the leaf primordia and coleorhizae, after which its expression increased so that by the 3rd day it was observed in the whole embryo. Thus, various families of transporters participate in mobilizing Fe during seed germination.

Fe Uptake from Soil

Fe is abundant in soils, but it is largely present as insoluble Fe(III), which is unavailable to plants. For this reason, plants have developed sophisticated mechanisms to solubilize and utilize Fe from the soil, often classified as Strategy I and II. Strategy I plants (non-graminaceous plants) release protons from the root plasma membrane via H^+-ATPases to lower the pH, secrete phenolics to solubilize precipitated Fe(III), reduce Fe(III) by membrane-bound ferric reductase oxidase2 (FRO2), and eventually take up Fe^{2+} through the IRT1 transporter. The Strategy I responses are also characterized by morphological changes in root architecture, such as the formation of transfer cells and extra root hairs. Although the significance of phenolic secretion is well established in Fe uptake from the rhizosphere, phenolic efflux transporter has not been identified in Strategy I plants. Recently, phenolic efflux transporters (PEZ1 and PEZ2) were identified in rice, and along with a possible role in caffeic acid (CA) transport, have been involved in the efflux of protocatechuic acid (PCA) to the apoplasm. The orthologs of PEZs are suggested to be involved in phenolics efflux in Strategy I plants.

In contrast, graminaceous plants (Strategy II plants) rely on an Fe(III) chelation system that involves the secretion of MAs. Graminaceous plants secrete MAs into rhizosphere through TOM1. TOM1 and HvTOM1 transport DMA but not NA, indicating that these proteins are the MAs efflux transporters. Under Fe-deficient conditions, *TOM1* and *HvTOM1* are expressed at high levels in rice and barley roots, respectively. The efficiency of DMA secretion increases by overexpression of *TOM1* and decreases by its repression. Further, rice lines overexpressing *TOM1* exhibit increased tolerance to Fe deficiency. The secreted MA binds to Fe(III) and the resulting Fe(III)–MA complexes are absorbed through the YSL family of transporters. YS1 was first identified from maize and since then, Fe(III)–MAs transporters have been characterized from rice (OsYSL15) and barley (HvYS1).

Rice is unique in the sense that it utilizes partial Strategy I system in addition to graminaceous-specific Strategy II. Besides secreting DMA, it also absorbs Fe^{2+}. Although two homologs of ferric-chelate reductase are present in rice, the expression of these genes is not observed under Fe-deficient or -sufficient conditions, and the level of Fe(III) chelate reductase activity is very low compared to that in other plants. As rice plants are adapted to grow under submerged conditions, where Fe^{2+} is abundant, they may have lost the ability to reduce Fe(III) through the development of a functional Fe(II)-regulated transporter, instead of a ferric-chelate reductase. *OsIRT1* and *OsIRT2* expression is observed only in Fe-deficient roots. Phenolics such as PCA are released to the apoplasm and are reported to chelate Fe(III), and solubilizing and reducing it to Fe^{2+} in vitro. Although the expression of PEZ2 is strong in vascular tissue, it could also be observed in epidermis indicating that PEZ2 may be involved in root Fe acquisition besides it role in solubilizing apoplasmic Fe in xylem. The existence of partial Strategy I and complete Strategy II in rice allows it to utilize both Fe^{2+} and Fe(III), depending on their availability and the environmental conditions. The Fe uptake in rice is summarized in Fig. 1.

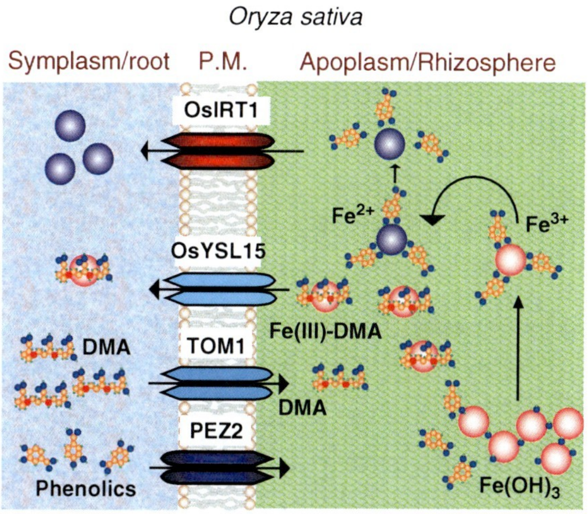

Iron Proteins, Plant Iron Transporters, Fig. 1 Rice (*Oryza sativa* L.) genes involved in Iron uptake from soil. PEZ2 (phenolics efflux zero 2) is suggested to secrete phenolics to rhizosphere. TOM1 (transporter of MAs) secretes DMA to rhizosphere and OsIRT1 (Iron-regulated transporter 1) and OsYSL15 (*yellow stripe*-like 15) transports Fe^{2+} and Fe(III)-DMA, respectively. *P.M.* plasma membrane

Fe Transport During Vegetative and Reproductive Stages

Fe transport from the roots to the shoots is essential for normal plant growth. Many similarities exist between Strategy I and Strategy II plants in terms of root to shoot Fe transport. In xylem, the major form of Fe is in a complex with citrate, as indicated by the strong positive correlation in the levels of citrate and Fe in xylem sap. Moreover, several factors suggest that a significant portion of Fe in xylem is bound to be citrate, such as the pH of xylem, existence of Fe–citrate complexes in the xylem and the fact that the level of organic acids (citrate, malate, and succinate) in the xylem of Fe-deficient plants increases. As discussed above, the *Arabidopsis* FRD3 loads citrate into the xylem; and in xylem sap of *frd3* mutants, the concentration of citrate and Fe is significantly lower compared to those in wild-type (WT) plants. The *FRD3* gene is expressed only in the pericycle and the cells surrounding the vascular tissue. In the *frd3* mutants, Fe accumulates adjacent to the xylem in the cells where *FRD3* is normally expressed. All three Strategy I Fe-deficiency responses, i.e., secretion of protons and phenolics, activity of FRO and expression of IRT1 in *frd3* mutants are activated despite the fact that leaf Fe content is almost double compared to WT plants. The knockout of rice citrate transporter *osfrdl1* results in chlorotic plants with low levels of Fe in the leaves and Fe precipitation in the root stele. *osfrdl1* shows a reduced concentration of citrate and Fe(III) in the xylem sap compared to WT rice. The expression of *OsFRDL1* is not regulated by Fe availability and could be observed in cells involved in long-distance transport as well as in reproductive organs. These results suggest a role for *OsFRDL1* in xylem Fe homeostasis. PEZ1 and PEZ2 localize to the plasma membrane and transports PCA when expressed in *Xenopus laevis* oocytes. *PEZ1* localizes mainly in the stele of roots. Fe concentrations increases in the roots of *pez1* plants, whereas it decreases in the xylem sap, suggesting that the defect in PCA secretion increases precipitated apoplasmic Fe in the stele and reduces soluble Fe in the xylem. The growth of *PEZ1* overexpression lines was severely restricted, and these lines accumulated more Fe as a result of the high solubilization of precipitated apoplasmic Fe in the stele. The results indicate that PEZ1 secretes phenolics into the apoplasm; it is responsible for an increase in PCA concentration in the xylem sap and is essential for the utilization of apoplasmic precipitated Fe in the stele. Besides citrate and phenolics, Fe must also be exported to the xylem. *Arabidopsis* Ferroportin 1 (FPN1/IREG1) is suggested to be involved in Fe loading to the xylem. FPN1 localizes to the plasma membrane and is expressed in the stele of the root, at the root–shoot junction, and in the leaf veins. *fpn1* mutants exhibit symptoms of Fe deficiency indicating a role in Fe efflux into the apoplast and/or the xylem. The levels of Fe in *fpn1* mutants are not significantly different compared to those in WT plants, indicating that other mechanisms also play a major role in xylem Fe loading.

In graminaceous plants, Fe can be transported in various forms through the xylem and phloem, including Fe–citrate, Fe(III)–DMA, Fe(II)–NA, and Fe-phenolics. Although DMA is involved in Fe acquisition from soil, its role in internal Fe homeostasis cannot be overlooked. In rice and barley, DMA has been detected in shoots under Fe-sufficient conditions, and the amount of DMA increases under Fe deficiency. The amount of DMA is higher in rice leaves compared to barley leaves under Fe-sufficient and -deficient conditions, although barley secretes larger amounts of MA. DMA detected in Fe-sufficient rice leaves is suggested to be translocated from roots in a complex with Fe. In contrast, DMA, at least partially synthesized in Fe-deficient shoots, is thought to be involved in Fe homeostasis in shoots and flowers.

Among the 18 putative YSL family genes in rice, Fe deficiency induces the expression of *OsYSL2* and *OsYSL15*, and possibly also *OsYSL9* and *OsYSL16*. Although the expression of the other *OsYSL* genes is not significantly regulated by Fe, at least in root tissues, the involvement of these genes in Fe homeostasis cannot be ruled out. Phylogenically, the YSL family in plants can be grouped into four subgroups, and among these *OsYSL1*, *3-4*, *7-8*, and *17-18* form a graminaceae-specific group. In this group, only *YSL18* has been characterized as an Fe(III)-DMA transporter, and the graminaceae-specific nature of this subgroup indicates that other genes may also transport DMA-metal complexes. Among these genes, *OsYSL2*, *OsYSL15*, *OsYSL16*, and *OsYSL18* have been characterized in detail. As noted above, *OsYSL15* transports Fe(III)–DMA from the rhizosphere to the roots and is involved in internal Fe homeostasis. *OsYSL15* promoter-driven GUS expression was not only observed in leaf tissue but also at the flowering stage. These results indicate that *OsYSL15* is involved in Fe transport to rice grains. Furthermore, *OsYSL18* encodes a functional Fe(III)–DMA transporter involved in DMA-mediated Fe distribution in reproductive organs, lamina joints, and phloem cells at the base of the leaf sheath.

NA, besides being an MA precursor, also serves as a chelator of divalent metals. In rice, *OsYSL2* has been identified as an Fe–NA transporter. *OsYSL2* is expressed at the reproductive stage, and it seems to contribute to Fe accumulation in seeds.

Furthermore, *OsYSL2* RNAi and 35S-*OsYSL2* lines have been characterized. At the vegetative stage, Fe and Mn concentrations decrease in the shoots of *OsYSL2*-inhibited RNAi plants, while the Fe concentration increases in the roots. At the reproductive stage, Fe translocation to the shoots and seeds is suppressed in *OsYSL2*-inhibited RNAi plants. The Fe and Mn concentrations also decrease in *OsYSL2* RNAi seeds, especially in the endosperm. The Fe concentration in *OsYSL2* overexpressing plants is lower in seeds and shoots but higher in roots compared to WT plants, indicating the role of *OsYSL2* in Fe homeostasis.

The efficient transfer from the xylem to phloem is extremely important for Fe translocation to leaves and seeds. Both citrate and phenolics (e.g., PCA and CA) seem to play important roles in xylem Fe transport, but the transfer of Fe from the xylem to phloem is relatively poorly understood. The YSL family of transporters plays a very significant role in phloem transport of Fe and other metals, both in Strategy I and Strategy II plants. The OsYSL15 transports Fe(III)–DMA complexes mainly from the rhizosphere to the root, but its expression is also observed during seed development, indicating a role for it in Fe transport to the developing seed. OsYSL18 is involved in DMA-mediated Fe distribution in reproductive organs. In comparison, the OsYSL2 has been reported as a phloem Fe–NA transporter. AtYSL1, AtYSL2, andAtYSL3 are *Arabidopsis* Fe(II)–NA transporters expressed most strongly in leaf vascular parenchyma cells closely associated with the xylem, indicating a role in xylem to phloem transport of Fe(II)–NA.

Subcellular Fe Trafficking

For the proper functioning of cells, Fe must be delivered to different subcellular compartments like chloroplasts and mitochondria, while the excess Fe must be stored precisely to avoid the generation of ROS; however, plant subcellular Fe transport is still not fully understood. In plants, up to 80–90% of the total cellular Fe can be accumulated in the chloroplast due to its high demand for Fe in the photosynthetic apparatus. As discussed above, Fe is required for the development and stabilization of chlorophyll, and Fe deficiency severely affects the functioning of electron transfer within the chloroplast. *Arabidopsis* permease in chloroplast (PIC1) localizes to the inner chloroplast membrane, complementing the yeast metal uptake defects, and is proposed to play a role in transport Fe into the chloroplast. *PIC1* knockout mutants exhibit a dwarf chlorotic phenotype. In *Arabidopsis*, the reduction of Fe before its transport into the chloroplast has also been suggested. Among the FRO family proteins in *Arabidopsis*, FRO7 localizes to the chloroplast. The Fe accumulation in the chloroplast of *fro7* mutants significantly decreases, resulting in impaired photosynthesis and ultimately, severe chlorosis in alkaline conditions, suggesting that FRO7 is required for Fe import into the chloroplast.

After the chloroplast, the mitochondria utilize most of the remaining Fe. Recently, the mitochondrial Fe transporter, MIT, has been identified by screening a rice T-DNA library for symptoms of Fe deficiency while grown in the presence of Fe. MIT is homologous to known mitochondrial Fe transporters such as yeast *Mrs3*/*Mrs4* and animal *Mitoferrin* genes. MIT localizes to the mitochondria and complements the growth of Δ*mrs3*Δ*mrs4* yeast defective in mitochondrial Fe transport. The homozygous *MIT* knockout mutant is lethal, while *MIT* knockdown lines exhibit impaired growth in spite of high Fe accumulation in the shoots. In MIT knock down mutant, the activity of the mitochondrial and cytosolic Fe-S enzyme, aconitase, decreases indicating that Fe-S cluster synthesis is affected in *mit-2* plants. MIT is expressed during all developmental stages in rice, indicating its importance for plant growth and development.

Unlike animal cells, plants do not possess cytoplasmic ferritin, so the excess Fe could either be stored in chloroplastic, or in the vacuole. The vacuole generally functions as a metal pool to help the plant avoid metal toxicity. Vacuolar Fe transporter 1 (VIT1) is an efflux Fe transporter depositing Fe from cytoplasm into the vacuole. *vit1* mutants show aberrant Fe localization in seeds and poor germination in alkaline soil. The *Arabidopsis* AtNRAMP3 and AtNRAMP4 are metal influx transporters involved in retrieving Fe from the vacuolar globoids.

Fe Transport to Seeds

Fe deficiency is also a very serious nutritional problem in humans affecting more than a billion human beings. Comprehensively understanding the mechanisms of Fe

I

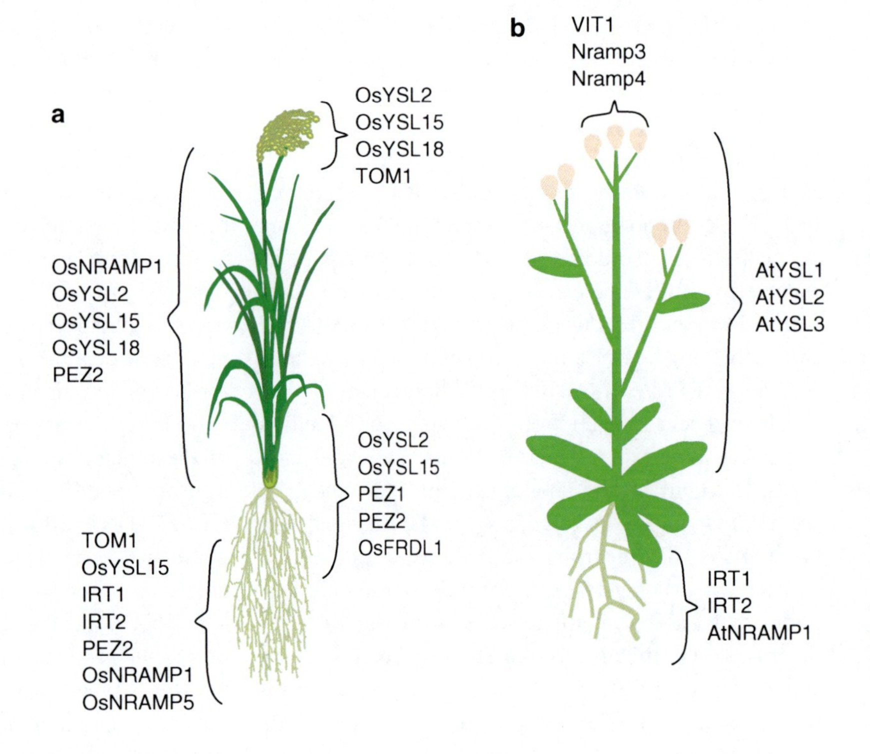

Iron Proteins, Plant Iron Transporters, Fig. 2 Genes involved in Fe uptake in rice and arabidopsis. Genes expressed in different tissue are mentioned in (**a**) rice and (**b**) arabidopsis

transport to seeds is important in developing effective Fe biofortification strategies to mitigate human Fe deficiency. In plants, the major players in Fe transport to the seeds are YSLs transporters. In rice, the expression pattern of genes involved in NA and MA synthesis indicate that NA and DMA are being synthesized and that DMA–Fe is being transported to seeds. The expression of OsYSL2, OsYSL15, OsYSL16 and OsYSL18 also suggests that these genes are involved in Fe transport to seeds. It has been demonstrated that the transport to the seed could be effectively increased (up to 4.4 times) by regulating the expression of OsYSL2 through the sucrose transporter promoter.

The subcellular Fe transporters like MIT, VIT1, AtNRAMP3, and AtNRAMP4 also play a very critical role during seed development. The expression of *MIT*, *AtNRAMP3*, and *AtNRAMP4* observed during seed development, which is significantly affected in the knockout or knockdown mutants of respective genes. X-Ray fluorescence microtomography of *Arabidopsis* seeds has revealed that Fe is predominantly localized in the provascular strands of the embryonic hypocotyl, radicle, and cotyledon of WT plants. Fe localization in *vit-1* mutants significantly changes and could be observed at the abaxial sides of the cotyledons, where normally Mn is localized. Since Fe in vacuoles is detected in *vit-1* mutants, the possibility that other transporters, besides VIT1, are involved in Fe transport to vacuoles cannot be ruled out.

Summary

Fe is not readily available to plants, and thus higher plants have developed sophisticated mechanisms (classified as Strategy I and Strategy II) to acquire soil Fe. All plants, with the exception of graminaceous plants, depend on Strategy I. The first step in a plant's life is to mobilize seed Fe during germination followed by its uptake from the soil. Fe transport from the soil and inside the plant body involves complex processes with many genes including those synthesizing Fe chelators (e.g., NA and DMA) and various sets of influx and efflux transporters (Fig. 2), the expression of which is strictly regulated to avoid Fe toxicity. Subcellular Fe homeostasis is also extremely important, and failure to transport enough Fe to subcellular compartments (e.g., mitochondria) may result in a lethal phenotype. Despite the increasing knowledge about Fe transport in plants, many questions remain, such as in which form is Fe transported to different subcellular organs? Understanding Fe uptake mechanisms, modifying Fe uptake, and efficiently controlling the Fe flow to seeds are all strategies that may increase seed Fe content, helping to overcome widespread Fe deficiency in humans.

Acknowledgments We are grateful to all the researchers working to elucidate Fe transport in plants for providing material for this report. The references could not be added to this essay due to style restrictions. We are particularly thankful to Dr. Takanori Kobayashi (Ishikawa Prefectural University) for critically reading the manuscript. This work was supported by a grant from the Ministry of Agriculture, Forestry, and Fisheries of Japan (Green Technology Project IP-5003).

Cross-References

- ▶ Iron Homeostasis in Health and Disease
- ▶ Iron Proteins, Ferritin
- ▶ Iron Proteins, Transferrins and Iron Transport

References

Bashir K, Ishimaru Y, Nishizawa N (2010) Iron uptake and loading into rice grains. Rice 3:122–130
Bashir K, Ishimaru Y, Shimo H, Nagasaka S, Fujimoto M, Takanashi H, Tsutsumi N, An G, Nakanishi H, Nishizawa NK (2011a) The rice mitochondrial iron transporter is essential for plant growth. Nat Commun 2:322
Bashir K, Ishimaru Y, Shimo H, Kakei Y, Senoura T, Takahashi R, Sato Y, Sato Y, Uozumi N, Nakanishi H, Nishizawa NK (2011b) Rice phenolics efflux transporter 2 (PEZ2) plays an important role in solubilizing apoplasmic iron. Soil Sci Plant Nutr 57:803–812
Conte SS, Walker EL (2011) Transporters contributing to iron trafficking in plants. Mol Plant 4:464–476
Ishimaru Y, Kakei Y, Shimo H, Bashir K, Sato Y, Sato Y, Uozumi N, Nakanishi H, Nishizawa NK (2011) A rice phenolic efflux transporter is essential for solubilizing precipitated apoplasmic iron in the plant stele. J Biol Chem 286:24649–24655
Jeong J, Guerinot ML (2009) Homing in on iron homeostasis in plants. Trends Plant Sci 14:280–285
Kobayashi T, Nishizawa NK (2012) Iron uptake, translocation, and regulation in higher plants. Annu Rev Plant Biol 63:131–152
Marschner H (1995) Mineral nutrition of higher plants. Academic, London
Nozoye T, Nagasaka S, Kobayashi T, Takahashi M, Sato Y, Sato Y, Uozumi N, Nakanishi H, Nishizawa NK (2011) Phytosiderophore efflux transporters are crucial for iron acquisition in graminaceous plants. J Biol Chem 286:5446–5454
Walker EL, Waters BM (2011) The role of transition metal homeostasis in plant seed development. Curr Opin Plant Biol 14:318–324

Iron Proteins, Transferrins and Iron Transport

Anne B. Mason[1] and Brian E. Eckenroth[2]
[1]Department of Biochemistry, University of Vermont, Burlington, VT, USA
[2]Department of Microbiology and Molecular Genetics, University of Vermont, Burlington, VT, USA

Synonyms

Iron binding protein; Iron transport protein

Definition

Human serum transferrin (hTF) is a bilobal glycoprotein able to tightly but reversibly bind ferric iron (Fe^{3+}) in each lobe for delivery to proliferating cells. Iron is delivered following binding to a specific transmembrane cellular transferrin receptor (TFR); diferric hTF bound to the TFR enters the cell by endocytosis. Iron release within the endosome involves a combination of lower pH, salt, an unidentified chelator, and the TFR. In the process, Fe^{3+} is reduced to ferrous iron (Fe^{2+}) for transport by a divalent metal transporter. Because of its natural abundance (~35 μM in serum, 2.8 mg/mL), hTF has been extensively studied since its discovery in 1946.

Introduction

Iron is the most abundant element on Earth with a mass percentage of ~32%, similar to oxygen at 30%, with carbon some 45-fold less abundant. However, in the ocean, iron is barely a trace element while oxygen is at 85%. Even without a consideration of biology, the reason for this discrepancy is obvious. Specifically, oxygen favors the highly insoluble ferric (Fe^{3+}) over the reduced and reactive ferrous (Fe^{2+}) form. Hence, the human body which is largely comprised of water and high concentrations of oxygen is faced with the dilemma of how to utilize iron. During evolution, the introduction of oxygen into the atmosphere of the Earth offered challenges that required new proteins and processes to harness the useful redox properties of iron while minimizing its potentially detrimental effects.

Fortunately, this disadvantage is compensated for by the more robust energy generation provided by aerobic respiration (Sheftel et al. 2012). Clearly the same chemical properties that make iron reactive with water and oxygen make it useful to carry out biological reactions. However, the reactivity requires a constant chaperone for travel throughout the aqueous milieu of the body and within cells.

Key Concepts of Iron Metabolism

In humans, iron is needed to support the daily synthesis of the 200 billion red blood cells derived from reticulocytes, which synthesize massive quantities of hemoglobin. Since no effective excretion mechanism for iron (or any metal) exists within the body, control is completely at the level of uptake. Maintenance of iron metabolism requires the uptake of only about 1–2 mg of iron from the diet each day; the remaining ~25 mg of iron needed daily is captured and reused following the destruction of senescent red blood cells by macrophages. The intricate iron sensing system regulating iron uptake is beyond the scope of this short entry (Gkouvatos et al. 2012). Suffice it to say that iron acquisition is tightly controlled and that diseases such as anemia and hemochromatosis are illustrative of the detrimental effects of either negative or positive iron imbalances (Andrews and Schmidt 2007). In the search of the underlying causes of these diseases many proteins and peptides involved in the regulation of iron have been identified. Naturally occurring mutations and knockout mice have proved particularly helpful in this endeavor.

The majority of the ~4–5 g of iron in a normal human is found in hemoglobin and the iron storage and antioxidant response protein ferritin. Other proteins utilizing iron to perform important functions include: myoglobin which uses iron bound to heme, a variety of iron-sulfur cluster proteins involved in mitochondrial electron transport, the citric acid cycle enzyme aconitase, the radical generating *S*-adenosylmethionine enzymes, ribonucleotide reductase, and catalase, to name just a few. These examples illustrate how the redox properties of iron have been exploited to carry out metabolic and enzymatic functions critical to human survival. It is important to note that ribonucleotide reductase is *the* rate limiting enzyme for DNA synthesis.

In mammals, iron absorption from the diet occurs by different pathways depending on the iron source, i.e., heme, ferritin or elemental. Following absorption by the intestinal enterocytes (by a non-TF-dependent process), most Fe^{3+} is chaperoned throughout the body by TF making it is a central player in iron metabolism. Non-transferrin iron delivery occurs in certain tissues such as the intestine, as well as in some disease states usually characterized by iron overload (Andrews and Schmidt 2007; Gkouvatos et al. 2012).

Evolution of Transferrin

TFs are members of a family with substantial structural homology which span the animal kingdom, including invertebrates (Lambert 2012). As the result of a probable gene duplication event, hTF is comprised of two homologous lobes termed the N-and C-lobes. This bilobal form is the predominant form of TF for higher ordered animals, although both mono-lobed and tri-lobed TF homologues have been identified in urochordates and crustaceans, respectively (Lambert 2012). The first member of the family to be discovered was ovotransferrin (oTF), which makes up ~12% of egg white protein. Another important member is lactoferrin (LTF) found in human milk and many bodily secretions. In these milieus oTF and LTF serve a bactericidal function by binding iron with such high affinity that bacteria are deprived of this essential metal and therefore unable to replicate. Notably, TF has also been found in the serum of fish, at least one marsupial, and in the hemolymph of insects. Other structurally related family members include melanotransferrin, saxiphilin, and the inhibitor of carbonic anhydrase (ICA). These more recently identified members of the family transport or sequester molecules other than iron. For example, saxiphilin binds with high affinity to the neurotoxin, saxitoxin. ICA, originally isolated from pig plasma, by its tight binding to carbonic anhydrase (CA) immobilized on an affinity column, is a strong inhibitor of CA. Although unable to bind iron due to lack of conservation of some of the iron binding residues in each lobe, ICA shares 63% identity with pig TF. As indicated above, proteins with homology to TF have been identified in a green alga, a crab, and in sea urchins (Lambert 2012).

Tissue-Specific Synthesis of Transferrin

In all mammals, TF is synthesized by hepatocytes as a precursor protein with a cleavable 19 amino acid signal peptide that targets it for delivery to the ER, transport to the Golgi for glycosylation and eventual delivery into the serum. Due to its adverse effect on the liver, excess alcohol has been shown to result in carbohydrate-deficient hTF. Specifically, the four terminal sialic acid residues normally present on the two biantennary glycans may be truncated or absent in persons who abuse alcohol since the liver enzymes responsible for carbohydrate synthesis are especially sensitive to alcohol consumption. Since these changes in the glycosylation do not appear to significantly affect the production or function of hTF and because of its natural abundance in the serum, hTF is an attractive target for clinical screening (asialo or disialo-isoforms of hTF predominate in heavy drinkers). The Scandinavian countries have been particularly active in this effort.

As a result of the presence of a number of barriers to the free flow of blood in the body, smaller amounts of TF are expressed in other tissues and cells besides hepatocytes. For example, although controversial, the so-called blood brain barrier is proposed to prevent TF from crossing into the brain. In culture, neurons, oligodendrocytes, and possibly astrocytes synthesize TF. TF is also expressed by cells of the choroid plexus and is the main constituent of the β-globulin fraction of the central nervous system. Additionally, granulosa cells in the ovary, Sertoli cells of the testis, and placental syncytiotrophoblasts are probably capable of making TF. In fowl, the tubular gland cells of the oviduct produce TF (formerly called conalbumin due to its tendency to co-purify with albumin) and now referred to as ovotransferrin. Since avians have a single gene for TF, the same protein serves dual functions of iron transport in serum and as an antimicrobial in egg white (Mason and Everse 2008). Iron status, steroids, and an assortment of cis acting elements have been reported to both positively and negatively influence the synthesis of TF in a tissue-specific manner (Gkouvatos et al. 2012).

Function of Transferrin

The only proven function of hTF is the transport of iron from the circulation into the cell by receptor-mediated endocytosis (Sheftel et al. 2012). It is notable that hTF bound to its specific receptor (TFR1) is *the* model system for this mechanism. The challenge is to explain how the iron which is held in each lobe of hTF with such high affinity ($K = \sim 10^{22}\ M^{-1}$ at pH 7.4) is extracted so rapidly within a cell. Within an endosome a combination of lower pH (supplied by a v-ATPase pump), an unknown chelator, salt, and the TFR itself results in the efficient release of iron from each lobe. To exit the endosome through the divalent metal transporter, Fe^{3+} must be reduced to Fe^{2+} (see Fig. 1). The discovery of the Steap family of reductases provides a source of electrons to reduce the ferric iron. Alternatively it has been proposed that the binding of hTF to the TFR may raise the reduction potential to promote reduction in situ. In either case, after delivering its payload of iron in an estimated 2–6 min time frame, hTF devoid of iron (apohTF) remains bound to the TFR and is returned to the cell surface (in a time frame estimated to be as short as 5 min up to 30 min or more). At neutral pH, the apohTF either dissociates from the TFR or is displaced by the tighter binding Fe_2hTF. ApohTF can then acquire more iron and repeat the cycle as many as 100 times.

Beginning with its discovery, a large number of studies have been carried out to try to understand the process by which iron is delivered to cells. A substantial disagreement in the literature existed; briefly there was a question of whether iron delivery occurred at the surface of a cell or whether endocytosis was required. The main approach used to resolve this issue involved protocols in which hTF was labeled with the radioisotopes ^{125}I and ^{59}Fe. The ^{125}I-labled hTF provided a sensitive and convenient way to measure binding as a function of time or hTF concentration, and the ^{59}Fe label was used to follow iron uptake. Addition of increasing amounts of ^{125}I-hTF to a constant number of cells under conditions in which iron uptake was inhibited (4°C or NH_4Cl) allowed determination of affinity constants and the number of TFR molecules per cell. Kinetic studies provide rate constants to define the uptake and release processes (see below). Early work mainly utilized isolated reticulocytes. Later studies used cell lines of both erythroid and non-erythroid origin which may differ in some details. A large number of studies over several decades under different conditions (pH, salt, chelators, and temperature) have varied so greatly that comparisons are difficult.

Transferrin has been identified as an essential component of tissue culture medium to support the growth

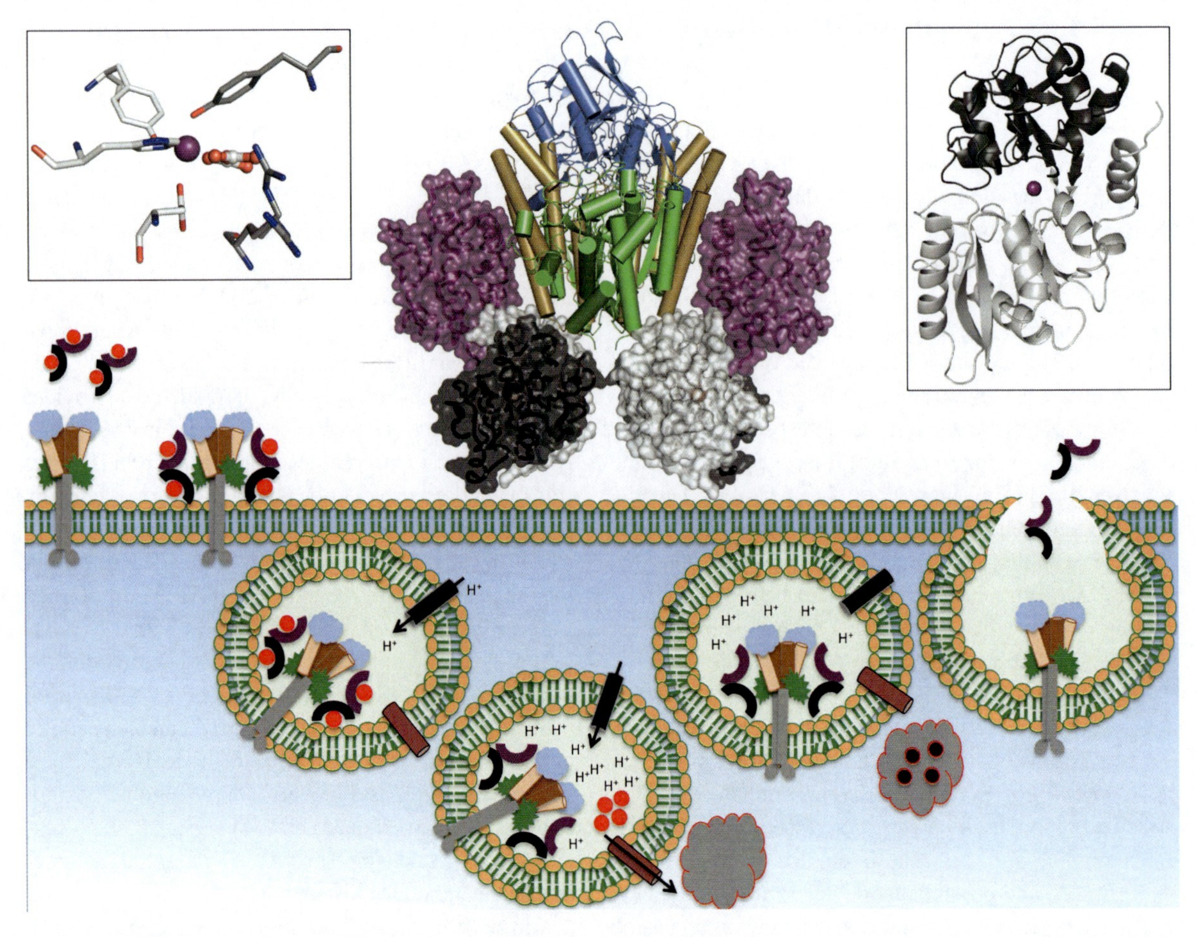

Iron Proteins, Transferrins and Iron Transport, Fig. 1 Structures and schematic of the transport of iron into a cell via the hTF/TFR complex. Coordination of the Fe^{3+} in the N-Lobe of hTF (*Upper left*); *Ribbon* diagram of the N-Lobe of hTF with bound Fe^{3+} (*Upper right*). Note that the two lobes of hTF have homologous structures and iron binding ligands; The structure of the complex between hTF with iron in the N-lobe (N-lobe in *black* (N2 subdomain and *gray* (N1 subdomain), C1 subdomain in *purple*) and the ectodomain of TFR (protease-like domain in *green*, apical domain in *blue*, helical domain in tan) is shown (*Upper middle*). Simplified cartoon representation of the endocytic cycle for delivery of Fe^{3+} to the cell (*Bottom*)

and proliferation of cells by providing the iron that is essential to these processes. With a few exceptions, a TF of a given species binds best to its own homologous TF receptor. It is important to appreciate that bovine transferrin binds poorly to the human TFR, such that cultured human derived cells are starved for iron and respond by expressing large numbers of TF receptors. Thus, for example, a typical HeLa cell passaged in fetal calf serum has one to two million TF receptors *per* cell.

It is increasingly clear that connections exist between iron metabolism and the immune system (Johnson and Wessling-Resnick 2012). Likewise, there are numerous connections with the nervous system (Ponka 2004; Andrews and Schmidt 2007). It is often difficult to ascertain whether the effects are directly related to the vital role of iron in proliferation or of a more general nature. Additionally, the high affinity of the interaction of hTF with TFR and the subsequent internalization make it an attractive candidate for delivery of therapeutics via a Trojan Horse design, including to the highly proliferative cells characteristic of various cancers.

The natural abundance of hTF and the relative ease of isolation and purification led to extensive study. The complication of studying the two lobes with distinctly different properties encouraged development of a number of strategies to dissect and understand the origin of the differences. In fact, the unique spectral properties of hTF, its bilobal nature and the substantial conformational change in each lobe resulting from iron

Iron Proteins, Transferrins and Iron Transport, Table 1 X-ray crystal structures of human TF and closely related pig and rabbit TF

Transferrin	Source	Resolution	pH	N-lobe	C-lobe	PBD ID
N-lobe (Fe)	Human	1.6 Å	5.75	Closed	None	1A8E
N-lobe (apo)	Human	2.2 Å	5.3	63°	None	1BP5
TF	Human	2.7 Å	6.5	59.4°	49.5°	2HAU
TF	Pig	2.15 Å	8.5	Closed	Closed	1H76
TF	Rabbit	2.6 Å	5.4	Closed	Closed	1JNF
TF with Fe in N-lobe / sTFR complex	Human	3.22 Å	7.5	Closed	Missing density for C2 subdomain	3S9L 3S9M 3S9N

binding or release has served as a model system for development and optimization of an impressive number of biophysical techniques (Mason and Everse 2008). Included are: extended X-ray absorption fine structure, resonance Raman spectroscopy, chemical relaxation studies, differential scanning calorimetry, titration calorimetry, high-resolution mass spectroscopy, electron paramagnetic resonance spectroscopy, nuclear magnetic resonance spectroscopy, and X-ray scattering, as well as modeling studies. A serendipitous discovery in 1976 was that 6 M urea gels could be used to simultaneously visualize the four possible forms of hTF including iron free (apohTF), monoferric hTF with iron in either the N-lobe or the C-lobe, and Fe_2hTF (Sheftel et al. 2012). In particular, structures provided by X-ray crystallography, some of which are given in (Table 1), have provided a wealth of useful details that have helped to elucidate how hTF is able to bind and release iron and to interact with the TFR with such high avidity and specificity (Mason and Everse 2008).

Recombinant expression of hTF and of the soluble portion of the TFR (sTFR), allows mutation of individual amino acids to determine the role of a particular residue in the function of each. It has become clear that functional hTF cannot be produced in a conventional manner using a recombinant bacterial expression system such as *E. coli*. Human TF has 38 cysteine residues (16 in the N-lobe and 22 in the C-lobe), all of which are engaged in disulfide bonds making the correct folding of newly synthesized hTF extremely challenging and precludes proper folding in the prokaryotic environment. Successful production has been achieved in eukaryotic systems including: insect cells infected with baculovirus, yeast, and baby hamster kidney (BHK) cells. Most systems utilize plasmids that retain the signal peptide that results in secretion of newly synthesized hTF from hepatocytes into the serum. Inclusion of this signal peptide allows secretion of the recombinant hTF (and mutants thereof) as well as the sTFR from BHK cells into the tissue culture medium. Addition of a hexa histidine tag as well as a factor Xa cleavage site at the amino terminus simplifies downstream purification and provides a means to remove the His tag if desired. More recently, commercial production of hTF for use in tissue culture medium has been reported in yeast (*S. cerevisiae*) and rice (*Oryza sativa*) (Sheftel et al. 2012). The soluble portion of the TFR has been produced in Chinese hamster ovary cells, in an insect cell baculovirus system and in BHK cells using similar protocols to those used for production of recombinant hTF.

Structural and Spectral Properties of Transferrin

Human serum TF is comprised of 679 amino acids, and has two N-linked biantennary glycans attached to C-lobe residues Asn413 and Asn611. The molecular mass of hTF is 75,143 kDa with an additional 4,412 kDa due to the carbohydrate, for a total of 79,555 kDa. Glycans do not appear to play a role in receptor binding, iron release or any other aspect of hTF chemistry that has been experimentally evaluated. A reasonable, although untested, possibility is involvement of the carbohydrate in whole body clearance of hTF; specifically the loss of the terminal sialic acid residues over time could result in binding of asialo-hTF to hepatic asialo-receptors for degradation.

The bilobal structure of TF resembles two oysters connected together at their hinge region with an iron atom, in place of a pearl in each. Just like the oyster, extracting the pearl/iron is difficult requiring a prying open of the cleft. Human TF tightly, but reversibly binds iron thereby preventing Fe^{3+} hydrolysis at physiological pH. In the circulation hTF is only about 30%

iron saturated providing a buffer against iron excess, as well as the potential to transport other metals. The homologous N- and C-lobes fold into separate globular domains joined by a short bridge of 7 amino acids; each lobe is further divided into two subdomains comprising a deep cleft that forms the iron-binding site. These subdomains are designated as N1 (residues 1–93 and 247–331), C1 (residues 339–425 and 573–679), and N2 (94–246) and C2 (426–572). Iron coordination in each lobe involves the same four amino acid ligands (two tyrosines, 95 and 188 in the N-lobe and 426 and 517 in the C-lobe, a histidine, 249 in the N-lobe and 585 in the C-lobe, and an aspartic acid, 63 in the N-lobe and 392 in the C-lobe) with two additional ligands provided by a synergistic anion which is carbonate. The carbonate is stabilized by binding to a highly conserved arginine residue (Arg124 in the N-lobe and Arg456 in the C-lobe). A unique feature of the hTF structure is the large distance separating the iron binding ligands in the primary sequence.

The effect of simple anions on the behavior of hTF was noted early on in the history of hTF and has been extensively studied. In this regard, a distinction has been made between synergistic and non-synergistic anions. Nearly all synergistic anions contain a carboxylate group that binds to a conserved arginine residue with an electron donor group one or two carbon atoms away binding to the iron. In contrast, non-synergistic anions bind to hTF but, by definition, are unable to participate in or to support iron binding. It is thought that these anions affect iron release from hTF by binding to sites and possibly inducing conformational changes which ultimately are required for, and lead to, iron release (Mason and Everse 2008). Emphasizing the allosteric effect on iron release from hTF, some non-synergistic anion binding sites have been referred to as kinetically significant anion binding or KISAB sites. The only documented KISAB site, Arg143 was identified in the N-lobe by mutagenesis studies (Steere et al. 2012).

The binding of non-synergistic anions can be observed by spectroscopic techniques including EPR, NMR, nuclear magnetic resonance, and UV difference spectra. Thus, titration of apohTF with a variety of monovalent or divalent anions results in a negative electronic absorbance spectrum that resembles the reciprocal of the positive spectrum when iron or other metals are added to hTF. Using these techniques, the binding strength of many different anions (both synergistic and non-synergistic) has been measured.

Each subdomain of hTF is composed of similar α/ß folds in which a number of helices are packed against a central six stranded mixed ß-sheet (Fig. 1, upper right). The domains are connected by a hinge composed of two antiparallel β-strands that allow the subdomains to open and close by ~50° during the binding or release of iron. This large conformational change is accompanied by substantial differences in the resistance of hTF to temperature and to proteolysis; iron saturated hTF features an impressive imperviousness to degradation in comparison to the iron free apo-conformation. The geometry of each cleft appears to be optimal for the binding of Fe^{3+} and the synergistic carbonate anion, although it remains flexible enough to accommodate other metals and anions. The ability of hTF to bind at least 30 other metals is well documented and the physiological significance of transporting other metals has been reviewed (Sun et al. 1999). Additionally, although the four amino acids which bind iron within each lobe of hTF are identical, the kinetic and thermodynamic properties of the two lobes differ considerably (see below). Equilibrium dialysis studies showed that at pH 7.4, the C-lobe binds iron with an affinity constant that is sixfold stronger than the N-lobe (Sheftel et al. 2012). The difference between the two lobes is further illustrated by the 9 nm difference in absorbance maximum for the two monoferric species (Table 2).

Although the primary ligands in both lobes of nearly all TF family members are the same, there are substantial differences in the binding and release kinetics of lobes within a given TF and also between TF family members. These differences have been attributed in part to the amino acid residues immediately surrounding the primary ligands. These "second shell" residues are defined as amino acids bound to or close to the liganding residues that affect the properties of a given lobe through the extensive hydrogen bonding network within the cleft. Specific examples include a pH-sensitive dilysine pair in the N-lobe comprised of two lysine residues (Lys206 and Lys296), one in each subdomain, which share a hydrogen bond in the iron containing N-lobe, but are 9 Å apart in the apoN-lobe. These lysine residues reside in a hydrophobic "box" likely to change the pK_a of one or both the lysine residues to promote sharing of a single proton at neutral pH. A triad of residues in the C-lobe (Lys534, Arg632, and Asp634) seems to provide a similar, although less responsive, pH sensitive motif.

Iron Proteins, Transferrins and Iron Transport, Table 2 Millimolar absorption coefficients (ε) at 280 nm for full length hTF. All samples were assayed in 100 mM HEPES, pH 7.4 (iron-bound) or 100 mM acetate, pH 4.0 containing 4 mM EDTA (Apo)

Protein	Calculated ε_{280} (Apo)[a]	Experimental ε_{280} (Apo)[b]	Experimental ε_{280} (Iron)[b]	% Increase due to iron	Visible A_{max}	A_{280}/A_{max}
Fe_2 hTF	85.1	84.0 ± 0.2	103.9 ± 0.2	23.7	466	21.3
Fe_N hTF[c]	82.1	81.4 ± 0.3	92.5 ± 0.3	13.6	461	42.2
Fe_C hTF[d]	82.1	81.5 ± 0.2	92.1 ± 0.2	13.0	470	41.4
ApohTF[e]	79.2	80.1 ± 0.4	–	–	None	

[a]$\varepsilon(280)(M^{-1}cm^{-1})$ = (# Trp)(5500) + (# Tyr) (1490) + (# cystine) (125)
[b]Values are means ± STD three or more determinations
[c]Recombinant His-tagged hTF with Y426F and Y517F mutations to prevent iron binding in the C-lobe
[d]Recombinant His-tagged hTF with Y95F and Y188F mutations to prevent iron binding in the N-lobe
[e]Recombinant His-tagged apohTF (Y95F/Y188F/Y426F/Y517F) unable to bind iron in either lobe

The importance of each of these residues in the two lobes has been well documented by extensive mutagenesis studies (Mason and Everse 2008). Unique to the crystal structure of the N-lobe of hTF is the presence of Arg124 and the synergistic carbonate anion in two positions partially occupying both mono- and bidentate positions for iron binding. This synergistic carbonate/bicarbonate may provide access to the iron site and could serve as target to promote iron release although in the complex with the TFR this idea seems less likely. As elaborated below, it is now clear that residue His349 in the C-lobe of hTF is *the* residue that dictates iron release form the C-lobe of hTF when bound to the sTFR. Thus, His349 acts as a pH-inducible switch that accelerates iron release from the C-lobe of hTF (Steere et al. 2012).

A hallmark of hTF biology is its unique spectral signature. Critically important to the spectral properties of hTF is that binding of iron to the two tyrosine ligands in each lobe (in the form of tyrosinates) results in formation of a ligand to metal charge transfer band in the visible spectrum (466 nm) giving rise to the characteristic salmon pink color of Fe_2hTF. Furthermore, iron coordination disrupts the π to π^* transition of the tyrosine ligands, which increases the UV absorbance and overlaps the intrinsic tryptophan fluorescence signal. The disappearance of the visible absorbance signal can be monitored to directly determine rate constants for iron release from hTF under conditions that promote such release (various combinations of low pH, salt, and/or an appropriate chelator to capture the iron). Additionally, the presence of iron in each lobe strongly quenches the fluorescent signal (produced in large part by three Trp residues in the N-lobe and five Trp residues in the C-lobe). Thus iron release can also be monitored by recording the increase in the fluorescent signal (Steere et al. 2012). These characteristics have been used extensively to investigate TF binding and release of Fe^{3+}.

Kinetics of Iron Release

In vitro, it is often experimentally simpler to measure the release of iron than to determine iron uptake by hTF in solution. Because binding protocols must use a chelator to keep the Fe^{3+} in solution, the hTF competes with the chelator to acquire the iron. Thus, at least some hTF mutants with compromised binding are not able to effectively compete with the chelator which would displace it from the iron binding cleft. Many early iron release studies with isolated hTF were carried out at neutral pH. This is clearly appropriate to identify effective chelators for whole body clearance, but is not relevant to the endocytic release of Fe^{3+} from hTF. Likewise, although it is a necessary control, the kinetic behavior of hTF in solution in the absence of the TFR differs considerably from the behavior of hTF when bound to the TFR (Steere et al. 2012). As first suggested by the Aisen laboratory, the TFR plays a critical role in balance that allows efficient release of iron from both lobes during the transit of hTF through the endosome. Rate constants for each step in the transition from Fe_2hTF to apohTF have been determined by using mutants that are either unable to bind or to release Fe^{3+} from one or both lobes (Steere et al. 2012). Significantly, under the specific conditions of pH (5.6), temperature (25°C), and chelator (4 mM EDTA) binding of hTF to the TFR decreases the rate of iron release from the N-lobe (6–15 fold, relative to hTF alone) and increases the rate of iron release from the C-lobe (7–11 fold). This balancing of the rates is important for the efficient release of iron from each lobe during a given cycle of hTF through the cell. Without knowledge of the precise endosomal milieu, application of in vitro findings to describe the mechanism of iron release from cells remains speculative. An important observation from our cumulative studies is that the effect of the TFR on iron release surpasses the effect of anion binding and dominates the kinetic profile.

The Structure of hTF with Iron in the N-Lobe Bound to the Soluble Portion of the TFR

As indicated above, the interaction between TF and TFR is required for entry into the cell and subsequent iron delivery. Since the kinetic studies reveal that this interaction dramatically impacts the iron release from hTF, it is crucial to understand how this occurs at a molecular level. The recently reported crystal structure of the complex between the TFR ectodomain and monoferric hTF containing N-lobe bound iron (Fig. 1) provides insights into the interactions between the two binding partners, as well as new structural details for the sTFR (Eckenroth et al. 2011). As previously established, the homodimeric TFR ectodomain is comprised of three distinct domains per monomer: a protease-like domain (domain I, 121–188 and 384–606), an apical domain (domain II, 189–383), and a helical domain (domain III, 607–760). A 7.5 Å cryoEM model provided the first overview of the disposition of the two lobes of hTF in relation to the TFR. The crystal structure provides important details with regard to three previously modeled motifs in hTF which contact the TFR with obvious implications for the mechanism of iron release during the endocytic cycle. Two of the three motifs are contributed by the N-Lobe of hTF and represent 40% of the total contact surface area. Motif 1 occurs between the N1 subdomain of hTF and the helical domain of the TFR and is comprised of a single salt bridge (Arg50 and Glu664 in sTFR), and an extensive hydrogen bonding network that would be relatively insensitive to changes in pH. The second motif occurs between the hTF N2 subdomain and the protease-like domain of the TFR and is comprised primarily of van der Waals forces involving a pair of proline residues (142 and 145) in a loop that contains the KISAB site, Arg143 in hTF. This motif, while also not pH sensitive, is substantially weaker than the N1subdomain. Thus, we suggest that the repulsion force provided by the dilysine trigger in the N-lobe at low pH in solution would cause displacement of this N2 motif from the TFR to allow opening of the cleft. The need to disrupt the N2 motif explains the slower iron release rate for the hTF/sTFR complex in comparison to hTF in solution.

Motif 3 is extensive, contributing 60% of the total binding surface area within the complex. It is comprised of the C1 subdomain of hTF and the helical domain of the sTFR. Among the many interactions is a strong salt bridge between C-lobe residue Asp356 and Arg651 of the sTFR (2.7 Å). Motif 3 is juxtaposed to a region on the TFR that experiences significant conformational change upon binding of either hTF (or the hemochromatosis protein HFE). Remarkably, the conformational change in the TFR involves the apical/protease domains of the one TFR monomer interacting with the helical domain of the other monomer within the complex. (Note that the apical domain is the only domain in the TFR which does not directly bind to hTF). Binding of hTF results in a 15 Å movement of a TFR loop (residues 305–325) that brings TFR residue His318 to the interface of motif 3, close to TFR residue Trp641 in the other monomer and within 5 Å of the critical His349 in hTF. Protonation of His349 at endosomal pH converts it to a strong cation-π interaction possibly with Phe760 (the C-terminal amino acid of the sTFR) or allows it to make a salt bridge with Asp757. In either case a conformational change results in iron release from the C-lobe. Additionally, the sTFR loop contains one of three TFR glycosylation sites (Asn317) also shown to have an impact on complex stability. The TFR is further stabilized by a Ca^{2+} ion in octahedral coordination with four residues in the interface between the apical and protease-like domains.

Besides the change in orientation of the apical domain, the helical domain of the sTFR also undergoes a shift in orientation upon binding hTF. Of significance, motif 1 and motif 3 (which appear to remain bound throughout the endocytic cycle), are connected to opposite ends of a long α-helix (αIII-3, residues 640–662) in the helical domain of the TFR. It is possible that this fortuitous arrangement could allow communication of the iron status of the two lobes of hTF to occur through the TFR. The αIII-3 is oriented perpendicularly at the N-terminal end to α-helix I of the C1 subdomain of hTF while the N1 subdomain of hTF interacts with a seven residue loop that connects the C-terminal end of αIII-3 to αIII-4. The connectivity of αIII-3 to other elements in the TFR also provides an explanation for the experimentally observed pH dependent control of iron release dictated by the TFR. The C-terminus of the TFR, which has been shown to significantly contribute to the TFR mediation of iron release, interacts with αIII via a number of residues including Arg655 and Asp659. The connection of αIII-3 to αIII-4 also allows communication between TFR monomers because this helix provides two of four histidine residues at the dimerization interface which also undergo significant conformational changes upon binding of hTF. While the details are yet to be elucidated, it is apparent that the interaction between hTF and the TFR has been optimized through evolution to allow small structural changes within one partner (the TFR) to communicate large changes to the other (hTF).

An enduring question has been why hTF is comprised of two lobes? Depending on the experimental method used to evaluate binding, the isolated N-lobe of hTF is completely incapable of binding to the TFR and the isolated C-lobe binds weakly, or not at all. In contrast, the two monoferric hTF species bind with nearly equal affinity (~10-fold lower than Fe_2hTF) as measured by a number of methods. The suggestion was made many years ago that kidney retention evolutionarily favored the bilobal structure. Other ideas include the lobe-lobe interactions (cooperativity) that might provide some mechanistic advantage. Given the tremendous effect of the TFR on the kinetics of iron release, we suggest that binding of Fe_2hTF may maximally potentiate the interaction and cause appropriate conformational changes in each to elicit iron release.

Perspective

In summary, hTF and other family members have been studied extensively for many years. Collectively, the TF family is fascinating with distinctive spectral properties that have long intrigued researchers. For hTF the critical role of the TFR in affecting iron release has recently been substantiated and given a more detailed structural basis. In spite of all the progress and work, many questions remain with regard to both mechanisms and regulation of iron trafficking in the body (Sheftel et al. 2012; Gkouvatos et al. 2012). For example, it is unknown precisely how iron released from hTF within the early endosome is delivered to the terminal enzyme in heme synthesis for insertion into heme, ferrochelatase, which resides in the mitochondrion. A 2012 special issue of Biochem Biophys Acta (Volume 1820) contains 24 articles on the history of TF with far more detailed descriptions of its role in health and in disease and many useful references to the primary literature.

Acknowledgments This work was funded by USPHS R01 DK21739 (ABM).

Cross-References

- ▶ Gallium Uptake and Transport by Transferrin
- ▶ Heme Proteins, the Globins
- ▶ Iron Homeostasis in Health and Disease
- ▶ Iron Proteins, Ferritin
- ▶ Iron Proteins, Mononuclear (non-heme) Iron Oxygenases
- ▶ Iron Proteins, Plant Iron Transporters
- ▶ Iron, Physical and Chemical Properties
- ▶ Iron-Sulfur Cluster Proteins, Fe/S-S-adenosylmethionine Enzymes and Hydrogenases
- ▶ Iron-Sulfur Cluster Proteins, Ferredoxins
- ▶ Lanthanide Substitution of Iron-Containing Transferrin, Lactoferrin, and Ferritin, Development of Luminescent Reporter Proteins
- ▶ Ribonucleotide Reductase

References

Andrews NC, Schmidt PJ (2007) Iron homeostasis. Ann Review Physiol 69:69–85

Eckenroth BE, Steere AN, Chasteen ND, Everse SJ, Mason AB (2011) How the binding of human transferrin primes the transferrin receptor potentiating iron release at endosomal pH. Proc Natl Acad Sci USA 108:13089–13094

Gkouvatos K, Papanikolaou G, Pantopoulos K (2012) Regulation of iron transport and the role of transferrin. Biochim Biophys Acta 1820:188–202

Johnson EE, Wessling-Resnick M (2012) Iron metabolism and the innate immune response to infection. Microbes Infect 14:207–216

Lambert LA (2012) Molecular evolution of the transferrin family and associated receptors. Biochim Biophys Acta 1820:244–255

Mason AB, Everse SJ (2008) Iron transport by transferrin. In: Fuchs H (ed) Iron metabolism and disease. Research Signpost, Kerala

Ponka P (2004) Hereditary causes of disturbed iron homeostasis in the central nervous system. Ann N Y Acad Sci 1012:267–281

Sheftel AD, Mason AB, Ponka P (2012) The long history of iron in the Universe and in health and disease. Biochim Biophys Acta 1820:161–1187

Steere AN, Byrne SL, Chasteen ND, Mason AB (2012) Kinetics of iron release from transferrin bound to the transferrin receptor at endosomal pH. Biochim Biophys Acta 1820:326–333

Sun H, Li H, Sadler PJ (1999) Transferrin as a metal ion mediator. Chem Rev 99:2817–2842

Iron Transport Protein

▶ Iron Proteins, Transferrins and iron transport

Iron, Biological Properties, Introduction to Iron Entries

Elizabeth C. Theil
Children's Hospital Oakland Research Institute, Oakland, CA, USA
Department of Molecular and Structural Biochemistry, North Carolina State University, Raleigh, NC, USA

Iron in biology is unique among transition metals because of its widespread use in biological reactions. Iron is also very abundant in the terrestrial environment, compared to other redox active metals in biology. Whether it is the special chemical properties of iron or the abundance, or the combination that led to the dependence of life on iron, remains a subject for debate.

Iron complexes with specific shapes such as iron-sulfur clusters or iron-porphyrin, common in metalloproteins today, were present in prebiotic times. Some scientists have even proposed that the geometric shapes of iron-sulfur clusters influenced the selection of shapes and chirality in the evolution of life.

Proteins with iron cofactors or substrates participate in the key reactions of life: cell division (synthesis of deoxyribose from ribose by ribonucleotide reductase), respiration (mitochondrial heme electron transfer proteins, such as cytochrome oxidase), photosynthesis (ferredoxins), nitrogen fixation (nitrogenase), hydrogen splitting (hydrogenases), oxidant protection (ferritin, peroxidases), and O_2 transport (globins). More specific reactions catalyzed by iron proteins occur in steroid hormone and vitamin synthesis, fatty acid desaturation, and the activation of O_2 (nonheme iron oxygenases) in the metabolism of many, small, organic substrates. Iron ions are also emerging as regulators for genes in a feedback loop for proteins of iron metabolism (homeostasis).

Oxygen-containing molecules (dioxygen, hydrogen peroxide, lipids) react with redox active metal ions to generate oxygen radical species (ROS – reactive oxygen species). There are two properties of iron that make the biological iron reactions particularly difficult for living organisms to manage. First, the abundance of iron in the environment and second, the insolubility oxidized iron (Fe^{3+}) under physiological conditions. To avoid biological rusting, specialized iron proteins such as transferrin and other iron transporters bring iron inside the cells of living organisms from the environment and transport it within the cells. While atmospheric oxygen complicates the problem of managing biological iron, even anaerobic organisms generate oxygen-containing molecules and use some of the same strategies as air-living organisms to protect themselves. In the iron volume of the *Encyclopedia of Metalloproteins*, most of the entries introduce a specific family of iron proteins. In the first entry, the role of iron in health and disease describes iron biology from the perspective of human biology.

Cross-References

▶ Heme Proteins, Cytochrome c Oxidase
▶ Heme Proteins, Heme Peroxidases
▶ Heme Proteins, the Globins
▶ Iron Homeostasis in Health and Disease
▶ Iron Proteins, Ferritin
▶ Iron Proteins, Mononuclear (non-heme) Iron Oxygenases
▶ Iron Proteins, Plant Iron Transporters
▶ Iron Proteins, Transferrins and Iron Transport
▶ Iron-Sulfur Cluster Proteins, Fe/S-S-adenosylmethionine Enzymes and Hydrogenases
▶ Iron-Sulfur Cluster Proteins, Ferredoxins
▶ Iron-Sulfur Cluster Proteins, Nitrogenases
▶ Ribonucleotide Reductase

Iron, Physical and Chemical Properties

Fathi Habashi
Department of Mining, Metallurgical, and Materials Engineering, Laval University, Quebec City, Canada

Iron is an ancient metal that became the basis of our civilization. It is usually used in the form of steel which contains 0.5% and 1% carbon. World steel production in 1 year equals the production of all other metals combined in 10 years. It is a transition metal, that is, it is less reactive than the typical metals but more reactive than the less-typical metals. It occurs as an alloy with nickel in meteorites.

Physical Properties

Iron exists in several allotropic forms:

- α-Iron: Magnetic and stable to 768°C, crystallizes in a body-centered cubic. It dissolves very little carbon (0.025% at 721°C).
- β-Iron: It is a form stable between 768°C and 910°C. It is alpha iron that has lost its magnetism. It does not dissolve carbon.
- γ-Iron: This form is stable between 910°C and 1,390°C, crystallizes in face-centered cubic. It is nonmagnetic and dissolves 2% carbon at 1,102°C. Carbon in solution forms a hard carbide.
- δ-Iron: It is nonmagnetic and stable between 1,391°C and the melting point crystallizes in body-centered cubic.

The most important characteristic of steel is that it may be hardened if heated to about 900°C, and suddenly cooled. This is based on the fact that iron-carbon alloys with carbon content below 1.7% are formed by heating. When this is quenched it does not have time to precipitate the carbon and the carbide phase remains in a supersaturated solution.

Atomic number	36
Atomic weight	55.85
Relative abundance in Earth's crust, %	5
Density, g/cm^3	
At 20°C	7.874
At melting point	6.98
Atomic radius, pm	126
Thermal expansion coefficient at 20°C	11.7×10^{-6}
Crystal structure at room temperature	Body-centered cubic
Lattice constant, cm	2.861×10^{-8}
Temperature of α, γ transformation on heating, °C	910
Temperature of γ, δ transformation on cooling, °C	1,400
Melting point, °C	1,539
Boiling point, °C	~3,200
Heat of fusion, kJ·mol^{-1}	13.81
Heat of vaporization, kJ·mol^{-1}	340
Molar heat capacity, J·mol^{-1}·K^{-1}	25.10
Resistivity at 20°C, Ω cm	9.7×10^{-6}
For commercial iron	11×10^{-6}
Temperature coefficient of resistance	0.0065
Compressibility, cm^2/kg	0.60×10^{-6}
Specific heat, cal g^{-1} K^{-1}	0.105
Heat of fusion, cal/g	64.9
Heat of α, γ transformation, cal/g	3.86
Heat of γ, δ transformation, cal/g	1.7
Linear correction at α, γ on heating	0.0026
Linear correction at γ, δ on heating	0.001–0.003
Modulus of elasticity, dynes/cm^2	21×10^{11}
Young's modulus, GPa	211
Shear modulus, GPa	82
Bulk modulus, GPa	170
Poisson ratio	0.29
Mohs hardness	4
Vickers hardness, MPa	608
Brinell hardness, MPa	490

Chemical Properties

The pure metal is not often encountered in commerce, but is usually alloyed with carbon or other metals. Chemically pure iron can be prepared either by the reduction of pure iron oxide (best obtained for this purpose by heating the oxalate in air) with hydrogen, or by the electrolysis of aqueous solutions of iron(II) salts, for example, of iron(II) ammonium oxalate. On the technical scale, pure iron is prepared chiefly by the thermal decomposition of iron pentacarbonyl, $Fe(CO)_5$. The so-called carbonyl iron prepared in this way initially contains some carbon and oxygen in solid solution. These impurities can be removed by suitable after treatment. Iron carbonyl is highly toxic.

Iron has a great affinity for oxygen. It rusts in moist air, that is, the surface gradually becomes converted into iron oxide hydrate. When heated in air it forms the intermediate oxide, Fe_3O_4. In a finely divided state, iron is pyrophoric. Iron dissolves in dilute acids with the evolution of hydrogen and the formation of iron(II) salts. If dipped into a copper sulfate solution, it becomes covered with metallic copper: $Fe + Cu^{2+} \rightarrow Fe^{2+} + Cu$.

Iron forms compounds mainly in the +2 and +3 oxidation states. Iron also occurs in higher oxidation states, for example, potassium ferrate, K_2FeO_4, which contains iron in its +6 oxidation state. Iron(IV) is a common intermediate in many in biochemical oxidation reactions. The color of blood is due to the hemoglobin, an iron-containing protein. Numerous organometallic compounds contain oxidation states of +1, 0, −1, or −2. There are also many mixed valence compounds that contain both iron(II) and iron(III) centers, such as magnetite and Prussian blue, $Fe_4[Fe(CN)_6]_3$. Iron(II) compounds tend to be oxidized to iron(III) in air. However, iron(II) ammonium sulfate, $FeSO_4(NH_4)SO_4{\cdot}6H_2O$, known as Mohr's salt, is soluble in water, easily purified, and more stable to oxidation than iron(II) sulfate heptahydrate. Iron does not form amalgams with mercury. As a result, mercury is stored in flasks made of iron.

References

Habashi F et al (1997) Iron. In: Habashi F (ed) Handbook of extractive metallurgy. Wiley, Weinheim, pp 29–268

Iron-Sulfur Cluster Proteins, Fe/S-S-adenosylmethionine Enzymes and Hydrogenases

Adam V. Crain, Kaitlin S. Duschene, John W. Peters and Joan B. Broderick
Department of Chemistry and Biochemistry, Montana State University, Bozeman, MT, USA

Synonyms

Enzyme activation; Metal cluster biosynthesis; Metalloenzymes; Radical reactions

Definition

The radical SAM superfamily of enzymes contains a site-differentiated [4Fe-4S] cluster that is chelated by *S*-adenosylmethionine (SAM), a cofactor or co-substrate required for the production of a 5′-deoxyadenosyl radical and subsequent hydrogen atom abstraction. Radical SAM enzymes carry out diverse reactions such as sulfur insertion, DNA repair, and posttranslational modifications to name a few.

Iron-Sulfur Clusters in Biology

Iron-sulfur clusters are ubiquitous in biology and take on many forms in nature (Beinert 2000; Johnson et al. 2005). The simplest clusters in proteins are the [2Fe-2S], [3Fe-4S], and [4Fe-4S] clusters; these are most often coordinated to the protein by cysteinal ligation of all irons of the cluster; however, arginine, histidine, glutamic acid, and aspartic acid have all been observed coordinating to biological iron-sulfur clusters. Iron-sulfur clusters span a remarkable range of reduction potentials from less than −400 to greater than +400 mV, giving them unusual versatility in functions in biology (Beinert 2000; Johnson et al. 2005). Biological iron-sulfur clusters exhibit intriguing similarities to abundant iron-sulfur minerals; this similarity has led to the hypothesis that iron-sulfur minerals played a central role in the emergence of life by catalyzing the formation of the basic building blocks of biological molecules, and that these mineral clusters were incorporated into early proteins as essential components of catalysis (Wachterhauser 1992; McGlynn et al. 2009). In modern biology, iron-sulfur clusters have such diverse and essential functions as electron transfer in photosynthesis and respiration, catalysis in central metabolic pathways, gene regulation in response to environmental signals, and protein structure stabilization (Beinert 2000; Johnson et al. 2005). Further, these clusters exhibit structural diversity; in addition to the simple [2Fe-2S], [3Fe-4S], and [4Fe-4S] clusters mentioned above, more complex clusters containing heterometals, nonprotein ligands, and alternate structures are found in a number of enzymes, including those involved in the metabolism of one carbon compounds, hydrogen, and nitrogen (Beinert 2000; Drennan and Peters 2003).

Radical SAM Enzymes: Historical Perspective

Radical SAM enzymes are a superfamily of enzymes that use a [4Fe-4S] cluster and *S*-adenosylmethionine (SAM) to initiate diverse radical reactions (Fig. 1). Although the superfamily was first identified in 2001, the earliest studies on radical SAM enzymes preceded this date by approximately 30 years, originating in the laboratories of Knappe and Barker in the late 1960s and early 1970s. The Knappe laboratory identified a novel anaerobic enzymatic reaction requiring SAM, Fe^{2+}, flavodoxin, and two protein components to convert pyruvate to formate and acetyl-CoA (Knappe et al. 1965; Knappe and Blaschkowski 1975). The first protein component, pyruvate formate-lyase (PFL), was later found to contain a catalytically essential glycyl radical in its active state; the second protein component (pyruvate formate-lyase activating enzyme, PFL-AE) was found to catalyze the formation of this glycyl radical (Knappe et al. 1984). PFL-AE was ultimately shown to require an iron-sulfur cluster, rather than simply Fe^{2+} (Broderick et al. 1997). A similar system involved in anaerobic ribonucleotide reduction that also utilized an iron-sulfur cluster and SAM to generate a catalytically essential glycyl radical was also described prior to the identification of the superfamily (Eliasson et al. 1990; Mulliez et al. 1993). The activating enzymes for PFL and the anaerobic ribonucleotide reductase are part of a subclass of radical SAM enzymes referred to as glycyl radical enzyme-activating enzymes (GRE-AE, Fig. 1). Lysine 2,3-aminomutase (LAM) was also discovered in the late 1960s, in *Clostridium subterminale* during the anaerobic fermentation of lysine to acetate, butyrate, and ammonia (Chirpich et al. 1970). LAM enzyme activity was initially reported to be dependent on Fe^{2+}, SAM, PLP, and reducing equivalents; however, later work showed the requirement for iron-sulfur clusters rather than Fe^{2+} (Petrovich et al. 1991). The final two early iron-sulfur cluster–SAM enzymes were lipoate synthase and biotin synthase, both of which were shown to catalyze the insertion of sulfur into C-H bonds (Sanyal et al. 1994; Busby et al. 1999).

Therefore, prior to the identification of the radical SAM superfamily, five different enzyme systems had been identified in which iron-sulfur clusters and *S*-adenosylmethionine were required for catalysis. The bioinformatics study of Heidi Sofia published in 2001 showed that these five enzymes were part of a much larger superfamily of enzymes that spanned the phylogenetic kingdom (Sofia et al. 2001). Although the superfamily members showed little overall sequence homology, most contained a conserved cysteine motif, CX_3CX_2C, to coordinate an iron-sulfur cluster. Other limited sequence conservation appears to be associated with the binding of SAM. The discovery of the radical SAM superfamily and the identification of hundreds of additional members generated an increase in research on radical SAM enzymes, with numerous discoveries of new enzymatic reactions catalyzed through the use of iron-sulfur clusters and SAM (Frey et al. 2008; Shepard and Broderick 2010). It is now clear that radical SAM enzymes are capable of mediating numerous and diverse reactions, and the superfamily may in fact represent the most common method in biology for initiating radical reactions (Challand et al. 2011).

Mechanism and Structure in Radical SAM Enzymes

The diverse reactions catalyzed by radical SAM enzymes are thought to utilize a common mechanism, some details of which have been deconvoluted based primarily on the studies of PFL-AE and LAM (Frey et al. 2008; Shepard and Broderick 2010). A novel aspect of this chemistry is the direct coordination of SAM, via the amino and carboxylate moieties, to the unique site of a site-differentiated [4Fe-4S] cluster (Fig. 2). The coordination of SAM to the unique iron of the site-differentiated [4Fe-4S] cluster is a novel feature of radical SAM enzymes which was first identified by spectroscopic studies of PFL-AE. Mössbauer studies provided the first direct evidence for a unique iron site (Krebs et al. 2002), while electron-nuclear double resonance (ENDOR) spectroscopy confirmed direct coordination of the cluster by amino and carboxylate moieties of SAM (Walsby et al. 2002b, 2005). Orbital overlap of the iron-sulfur cluster and the sulfonium sulfur of SAM was also demonstrated in ENDOR studies, suggesting an inner-sphere electron transfer between these moieties (Walsby et al. 2002a, 2005).

The catalytically active state of the iron-sulfur cluster is the $[4Fe\text{-}4S]^+$ state (Henshaw et al. 2000), which transfers an electron to the bound

Iron-Sulfur Cluster Proteins, Fe/S-S-adenosylmethionine Enzymes and Hydrogenases, Fig. 1 Representative radical SAM reactions. Shown are the reactions catalyzed by the glycyl radical enzyme-activating enzymes (GRE-AE), lysine 2,3-aminomutase (LAM), biotin synthase (BioB), a hydrogenase maturation protein (HydG), an enzyme involved in the synthesis of the molybdopterin cofactor (MoaA), and the spore photoproduct lyase (SPL)

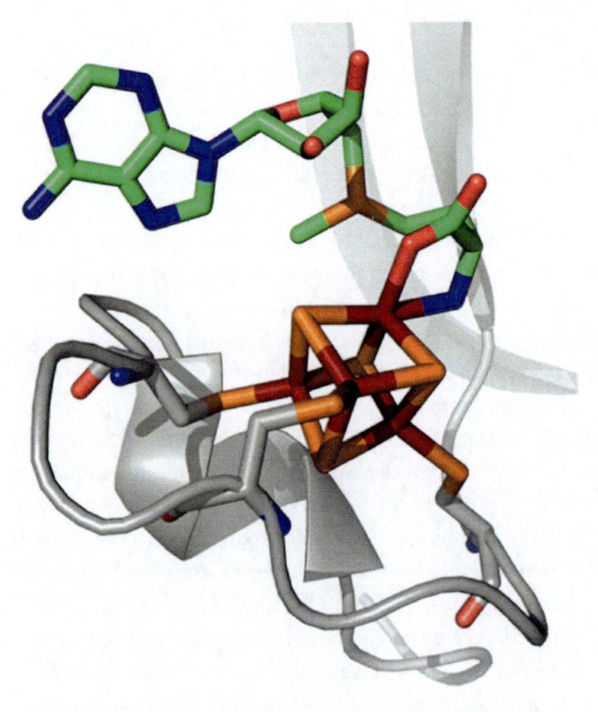

Iron-Sulfur Cluster Proteins, Fe/S-S-adenosylmethionine Enzymes and Hydrogenases, Fig. 2 Site-differentiated [4Fe-4S] cluster chelated at the unique iron site by amino and carboxylate moieties of *S*-adenosylmethionine in PFL-AE (3CB8)

S-adenosylmethionine via inner-sphere electron transfer to the sulfonium sulfur (Fig. 3). The reduction of SAM results in homolytic cleavage of the S-C5′ bond to produce a 5′-deoxyadenosyl radical intermediate and methionine. Although the 5′-deoxyadenosyl radical intermediate has never been directly observed, its involvement has been inferred from label transfer studies, as well as from observation of the stabilized allylic radical formed when 3′,4′-anhydro-SAM is used in place of SAM in reactions of LAM (Magnusson et al. 1999; Magnusson et al. 2001). The highly reactive 5′-deoxyadenosyl radical then abstracts a hydrogen atom from the substrate to initiate catalysis. In most of the radical SAM enzymes characterized to date, SAM is a co-substrate, and the SAM cleavage products (methionine and 5′-deoxyadenosine) must dissociate from the enzyme and be replaced by another SAM molecule before a second round of catalysis can occur. Some radical SAM enzymes, including LAM and spore photoproduct lyase, use SAM as a catalytic cofactor that reversibly generates the 5′-deoxyadenosyl radical in a manner analogous to coenzyme B12 (Magnusson et al. 1999; Cheek and Broderick 2002). Further, there is emerging evidence that a subcategory of radical SAM enzymes

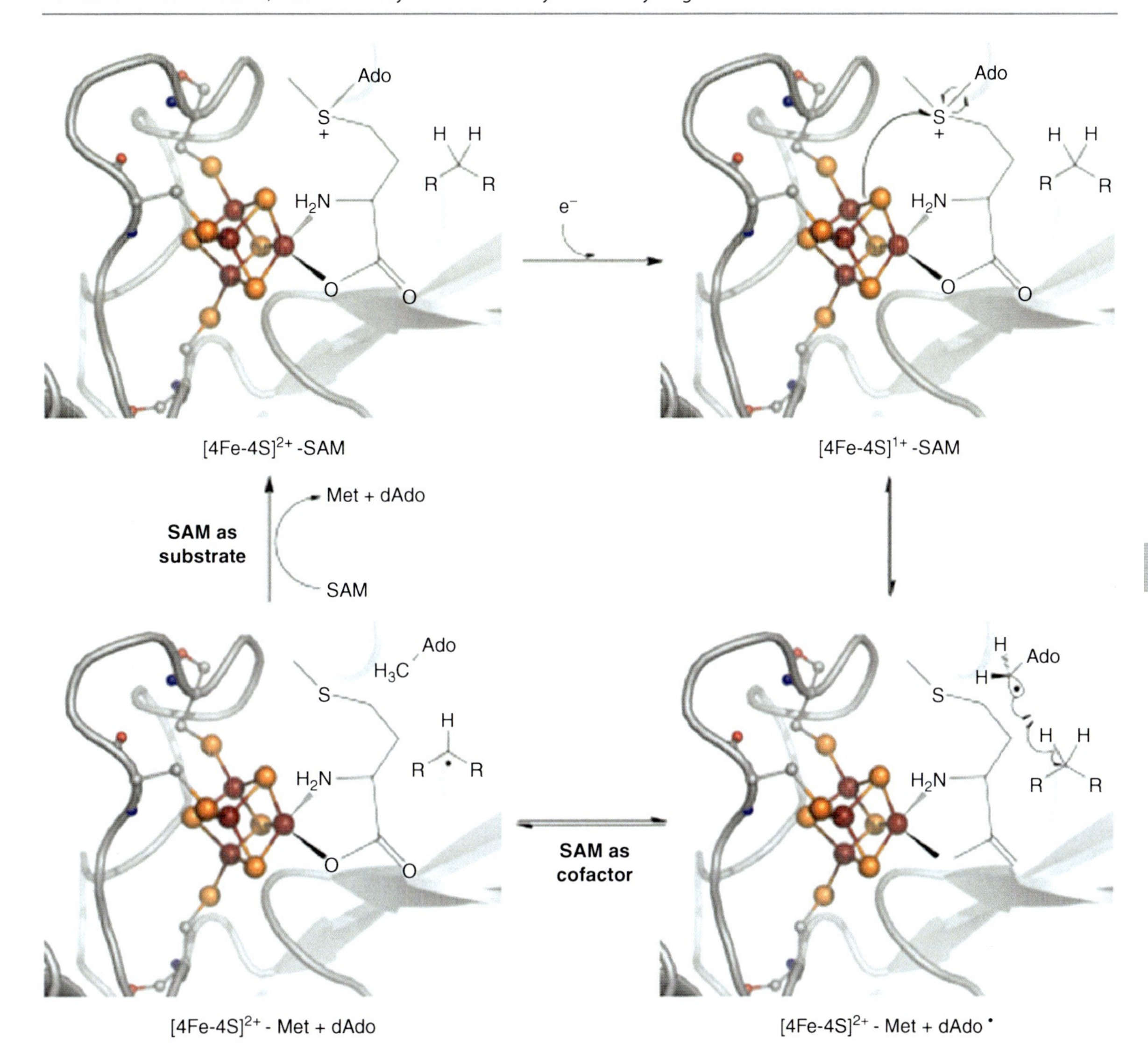

Iron-Sulfur Cluster Proteins, Fe/S-S-adenosylmethionine Enzymes and Hydrogenases, Fig. 3 Proposed mechanism for radical SAM enzymes. SAM is shown bound to the unique iron of the oxidized $[4Fe\text{-}4S]^{2+}$ cluster (*upper left*). Reduction of this cluster generates the catalytically active $[4Fe\text{-}4S]^{+}$ state, which undergoes inner-sphere electron transfer to the sulfonium of SAM, thereby promoting the cleavage of the S-C(5′) bond of SAM (*upper right*). The homolytic cleavage of SAM produces the 5′-deoxyadenosyl radical intermediate, which directly abstracts a hydrogen atom from substrate (*lower right*). The substrate radical generally undergoes further radical-mediated chemical transformations. In most radical SAM enzymes characterized to date, the methionine and 5′-deoxyadenosine (*lower left*) are released as products of the reaction; in these cases, SAM is acting as a substrate of the reaction. In a few examples of radical SAM enzymes, however, a rearranged product radical reabstracts a hydrogen atom from 5′-deoxyadenosine to regenerate the 5′-deoxyadenosyl radical intermediate, which recombines with the methionine to regenerate SAM; in these cases, SAM acts as a cofactor rather than a substrate

may catalyze reductive cleavage of SAM in which an alternative S-C bond is cleaved, and thus, an alternate radical intermediate (other than the 5′-deoxyadenosyl radical) is generated (Zhang et al. 2010; Demick and Lanzilotta 2011). Once the 5′-dAdo (or alternative) radical is generated, the reaction diverges depending on the substrate involved leading a variety of chemically distinct reactions (Frey et al. 2008; Shepard and Broderick 2010; Challand et al. 2011).

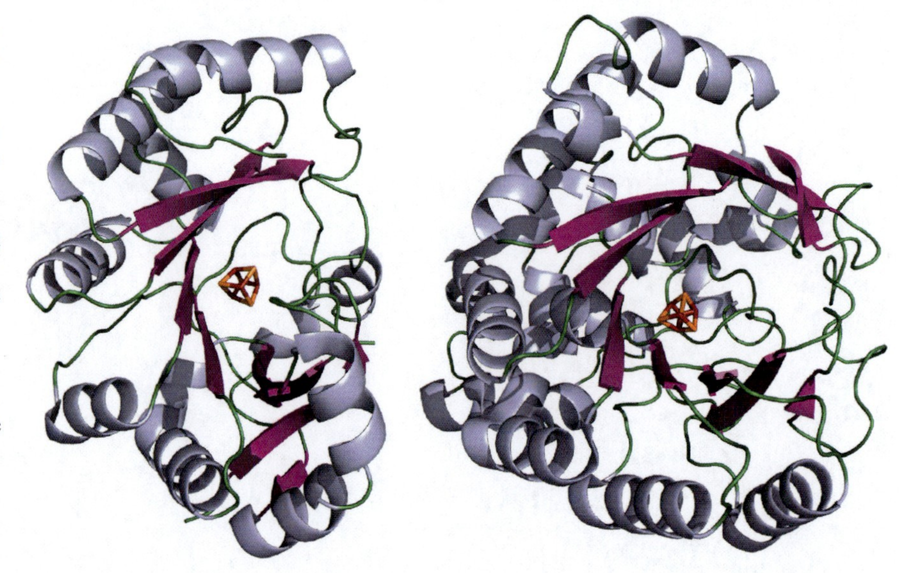

Iron-Sulfur Cluster Proteins, Fe/S-S-adenosylmethionine Enzymes and Hydrogenases, Fig. 4 Structures of the radical SAM enzymes PFL-AE (3CB8, *left*) and LAM (2A5H, *right*). The completeness of the TIM barrel structure correlates with the size of the substrate: PFL-AE acts on a very large protein substrate and has an incomplete and thus quite open barrel, while LAM acts on a small-molecule substrate and its barrel is more complete

Significant insights into the energetics of radical SAM reactions have been revealed by studies of LAM (Wang and Frey 2007). The reduction potentials of the [4Fe-4S] clusters in radical SAM enzymes are in the range of ~ − 400 to − 550 mV; however, best estimates for the reduction potential of SAM are ~ −1.7 V; this > 1 V gap helps to explain the prohibitive nature of electron transfer from the catalytic cluster to an isolated SAM to promote reductive cleavage. However, the potential of the cluster in LAM in the presence of SAM, PLP, and lysine decreases to − 600 mV, while that of SAM increases to − 990 mV, decreasing the energy barrier from ~ 32 kcal/mole to 9 kcal/mole, making reductive SAM cleavage more feasible (Hinckley and Frey 2006; Wang and Frey 2007). It has also been proposed that electron transfer from the pentacoordinate $[4Fe\text{-}4S]^{+}$ cluster (with SAM coordinated as a bidentate ligand to the unique site) to the sulfonium in SAM results in a more favorable hexacoordinate $[4Fe\text{-}4S]^{2+}$ (with methionine coordinated in a tridentate fashion to the unique iron) and may facilitate inner-sphere electron transfer (Wang and Frey 2007). LAM remains one of the best characterized examples of a radical SAM enzyme, and putative characteristics of other members of the superfamily have been extrapolated using LAM as a model system.

Radical SAM enzymes also exhibit common structural features, as elucidated by spectroscopic and structural studies of a number of superfamily members (Nicolet and Drennan 2004; Vey and Drennan 2011; Dowling et al. 2012). All of the radical SAM crystal structures reported to date show the presence of a partial or complete TIM (triose phosphate isomerase) barrel, with the degree of completeness of the barrel correlating with the size of the substrate. For example, when the substrate is a large macromolecule, as in the case of PFL-AE, whose substrate is the 170 kDa PFL, the barrel is less complete, leaving a wide opening for binding the large substrate (Fig. 4); radical SAM enzymes that act on smaller substrates have more complete barrels, thereby limiting active site accessibility to small molecules (Nicolet and Drennan 2004). Regardless of the completeness of the TIM barrel, the [4Fe-4S] cluster binding site is at one end of the barrel. Three of the four irons of the [4Fe-4S] cluster are coordinated by cysteine residues from a conserved motif (usually CX_3CX_2C but variations exist in the superfamily), while the fourth iron is chelated by SAM. Substrate binds in close proximity to SAM, thereby allowing for direct H-atom abstraction from substrate by the 5′-deoxyadenosyl radical intermediate.

Diverse Reactions Catalyzed by Radical SAM Enzymes

Radical SAM reactions are involved in diverse reactions in all kingdoms of life, including anaerobic glycolysis, nucleotide metabolism, DNA repair, vitamin and cofactor biosynthesis, and synthesis of complex metal clusters, as well as many more unmentioned and yet to be discovered reactions (Fig. 1). These reactions range from simple ones whereby H-atom abstraction is the initial and only step required to convert substrate to product, to reactions in which H-atom abstraction initiates complex and sometimes perplexing transformations. Several examples of radical SAM enzymes are described in the following paragraphs, while more complete discussions of superfamily members can be found in recent reviews (Shepard and Broderick 2010; Challand et al. 2011; Vey and Drennan 2011).

PFL-AE is a radical SAM enzyme that activates PFL, a central enzyme in anaerobic glycolysis, by generating a catalytically essential radical at G734 of PFL; PFL-AE is, therefore, a GRE-AE (Knappe et al. 1984). Active PFL catalyzes the reaction of pyruvate and CoA to formate and acetyl-CoA and provides the sole source of acetyl-CoA for the citric acid cycle under anaerobic conditions. The catalytically active $[4Fe\text{-}4S]^{+}$ cluster of PFL-AE is oxidized to the $[4Fe\text{-}4S]^{2+}$ in a 1:1 stoichiometric ratio with glycyl radical formation, indicating that the reduced cluster provides the electron necessary for the reductive cleavage of SAM to generate the deoxyadenosyl radical intermediate, which in turn abstracts a hydrogen atom from G734 of PFL (Henshaw et al. 2000). 5′-Deoxyadenosine and methionine are products of the PFL-AE reaction, and deuteron transfer from G734 to the 5′-dAdo product demonstrates a direct H-atom abstraction (Frey et al. 1994). Further biochemical experiments demonstrated that this H-atom abstraction was stereospecific for the pro-*S* hydrogen of G734 (Frey et al. 1994). Recent biophysical studies of PFL suggest that major structural rearrangement of PFL must occur in order for the normally buried G734 to bind in the active site of PFL-AE such that direct H-atom abstraction can occur (Peng et al. 2010).

Lysine 2,3-aminomutase (LAM) is one of a group of radical SAM enzymes that catalyze isomerization reactions, and is one of only two known to use SAM as a catalytic cofactor rather than a co-substrate(Frey et al. 2008). In the presence of reducing equivalents, SAM, PLP, and a [4Fe-4S] cluster, LAM catalyzes the conversion of L-α-lysine to L-β-lysine (Petrovich et al. 1991). Evidence for a 5′-deoxyadeosyl radical intermediate in the LAM mechanism was demonstrated using the SAM analog 3′, 4′-anhydro-*S*-adenosylmethionine (anSAM) in conjunction with electron paramagnetic resonance (EPR) measurements to characterize the allylic-stabilized 3′,4′-anhydroadenosyl radical intermediate (Magnusson et al. 1999). Subsequent ENDOR experiments demonstrated that anSAM, PLP, and the [4Fe-4S] cluster were within van der Waals distance of each other, making direct hydrogen abstraction feasible (Lees et al. 2006).

Biotin synthase (BioB) is an example of a radical SAM enzyme that catalyzes a sulfur insertion reaction. BioB contains a [4Fe-4S] cluster chelated at the unique iron by SAM and also contains a labile [2Fe-2S] cluster that is used as sulfur source for the synthesis of biotin from dethiobiotin (Benda et al. 2002). Two SAM molecules are required to carry out two sequential H-atom abstraction reactions from positions C6 and C8 of dethiobiotin; these react with a bridging sulfide of the [2Fe-2S] cluster to generate the thiophane ring of biotin (Taylor et al. 2008). The [2Fe-2S] cluster of biotin synthase is thus proposed to act as a substrate in the reaction, and must, therefore, be rebuilt after each turnover; this need to rebuild the substrate cluster is thought to be at least partly responsible for the limited amount of turnover observed for BioB in vitro.

Spore photoproduct (SP) lyase is a radical SAM enzyme that repairs the thymine dimer 5-thyminyl-5,6-dihydrothymine (SP), a major UV photoproduct in spore DNA. This repair reaction is initiated by direct H-atom abstraction from the C-6 position of SP by the 5′-dAdo radical intermediate (Cheek and Broderick 2002), with SAM serving as a catalytic cofactor (Buis et al. 2006). Enzymatic assays carried out with synthetic, stereochemically defined SP dinucleosides and dinucleotides have demonstrated that the repair reaction is stereospecific for the 5*R* diastereomer of SP (Chandra et al. 2009; Heil et al. 2011).

The molybdenum cofactor biosynthesis enzyme MoaA catalyzes the remarkably complex rearrangement of 5′-GTP to form precursor Z in molybdenum cofactor biosynthesis (Hänzelmann and Schindelin 2004). Subsequent enzymatic reactions generate the dithiolene group of molybdopterin, which coordinates molybdenum. The formation of precursor Z also requires the presence of an additional enzyme known as MoaC, although the precise roles of each enzyme remain to be elucidated. MoaA contains an N-terminal radical SAM [4Fe-4S] cluster and an additional C-terminal cluster that has been shown to bind GTP (Hänzelmann and Schindelin 2006; Lees et al. 2009).

Two radical SAM enzymes (HydE and HydG) have been identified as essential in maturation of the [FeFe]-hydrogenase. These two enzymes, together with the GTPase HydF, are in fact the only proteins required for heterologous expression of an active [FeFe]-hydrogenase (HydA) (Posewitz et al. 2004). The reactions catalyzed by these radical SAM enzymes, and their roles in [FeFe]-hydrogenase maturation, will be discussed below.

The [FeFe]-Hydrogenase and Its Maturation

Hydrogenases catalyze the reversible interconversion of protons and molecular hydrogen, $2H^+ + 2e^- \rightleftharpoons H_2$. Two types of hydrogenases catalyze this reaction: the [NiFe]-hydrogenase and the [FeFe]-hydrogenase. These enzymes are evolutionarily unrelated and yet contain some unusual active site features in common, including the presence of biologically unusual CO and CN^- ligands to the active site metals. The metal content is correlated with the catalytic bias, with hydrogen oxidation the favored direction of catalysis in the [NiFe]-enzymes (Vignais and Billoud 2007). In [FeFe]-hydrogenases, hydrogen production is greatly favored over hydrogen oxidation, making it an ideal catalyst for alternative energy especially when coupled to photosynthesis. Unfortunately several barriers exist to photosynthetic commercial hydrogen production such as competition between hydrogenase and NADPH-dependent carbon dioxide fixation and the strict requirement of anaerobic conditions (Yacoby et al. 2011).

[FeFe]-hydrogenases from *Clostridium pasteurianum* (CpI) and *Desulfovibrio desulfuricans* have been structurally characterized (Peters et al. 1998; Nicolet et al. 1999). These X-ray structures, together with detailed spectroscopic studies, have provided a picture of the remarkable active site H-cluster. The H-cluster is composed of a [4Fe-4S] cluster coordinated to the protein by four cysteinate ligands, with one of these cysteinates bridging to a 2Fe cluster whose remaining ligands consist of three CO and two CN^-, as well as a bridging dithiolate. The identity of the central atom in the dithiolate linker has been an area of controversy; however, recent spectroscopic studies favor N as the central atom, such that the bridging ligand is dithiomethylamine (Silakov et al. 2009). The distal iron of the H-cluster has been proposed to be the site of hydrogen oxidation and proton reduction. The structures of both of the above mentioned hydrogenase enzymes also show accessory clusters or F clusters made up of two ferredoxin like [4Fe-4S] clusters, one histidine ligated [4Fe-4S] cluster and one [2Fe-2S] cluster. The F clusters have been proposed to be involved in shuttling electrons to and from the H-cluster via donor and acceptor proteins which bind acidic and basic patches on either side of the domain (Peters et al. 1998).

Biosynthesis of the H-cluster and maturation of an active [FeFe]-hydrogenase requires three maturase proteins known as HydE, HydF, and HydG; HydE and HydG were proposed from sequence annotations to be radical SAM enzymes, while HydF was suggested to be a GTPase (Posewitz et al. 2004). Recent biochemical and spectroscopic studies have begun to reveal the specific roles for each of these accessory proteins in hydrogenase maturation (Challand et al. 2011; Shepard et al. 2011). It was shown that when HydE, HydF, and HydG were expressed together in *E. coli*, they generated an activating component associated with HydF, that could be transferred to the inactive [FeFe]-hydrogenase (HydA) to generate the active enzyme (McGlynn et al. 2007, 2008). EPR and Fourier transform infrared (FTIR) spectroscopic studies have provided evidence that this activating component is a 2Fe H-cluster precursor containing CO and CN- ligands, and that this activating component is generated by the actions of the radical SAM enzymes HydE and HydG (Shepard et al. 2010b). Consistent with these results, HydA expressed in *E. coli* in the absence of the accessory proteins contains not the entire H-cluster, but rather a simple [4Fe-4S] cluster that is essential for subsequent maturation (Mulder et al. 2009). This [4Fe-4S] cluster lies at the end of a cationic channel that presumably functions in transfer of the 2Fe H-cluster precursor into the active site of HydA (Mulder et al. 2010). Together, these results point to a model in which the radical SAM enzymes HydE and HydG build a 2Fe H-cluster precursor on HydF, and this precursor is then delivered to HydA to assemble the full H-cluster.

What are the specific functions of the radical SAM enzymes in H-cluster maturation? It seemed likely that these enzymes would be involved in synthesizing the unique ligands of the H-cluster; however, identifying specific functions first required that the substrates for these enzymes be identified. HydG has significant sequence similarity to ThiH, another radical SAM

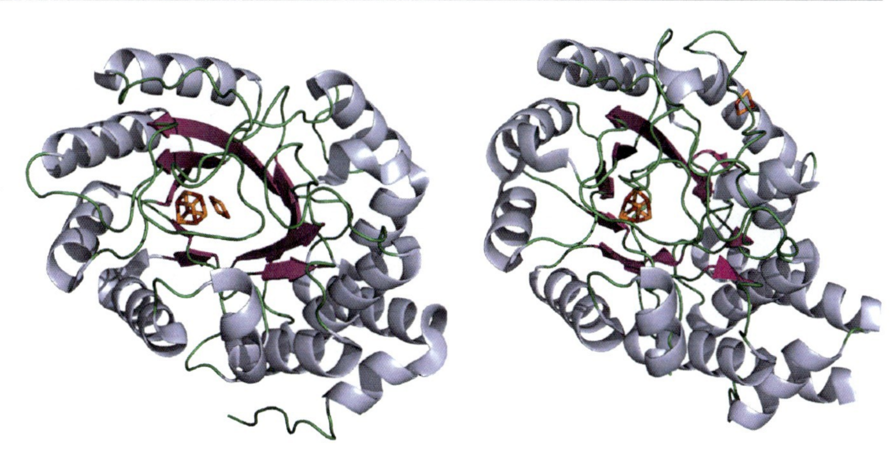

Iron-Sulfur Cluster Proteins, Fe/S-S-adenosylmethionine Enzymes and Hydrogenases, Fig. 5 Structural similarity between BioB (1R30, *left*), which catalyzes a radical SAM sulfur insertion reaction, and HydE (3CIX, *right*), whose specific reaction is currently unknown

enzyme that catalyzes the cleavage of tyrosine to form *p*-cresol and dehydroglycine, the latter being a key component in thiamine biosynthesis. This sequence homology inspired exploration of the possibility that tyrosine could be cleaved by HydG, and indeed HydG was shown to cleave tyrosine to produce *p*-cresol under anaerobic reducing conditions in the presence of SAM (Pilet et al. 2009). Subsequent studies designed to detect any CN^- and CO produced in this reaction demonstrated that the HydG-catalyzed cleavage of tyrosine was accompanied by the production of these two diatomic ligands. These experiments utilized isotopically labeled tyrosine in order to unequivocally demonstrate that the CO and CN^- were derived from tyrosine (Driesener et al. 2010; Shepard et al. 2010a). A recent study using cell-free extracts and isotopic labeling demonstrated that all five CO and CN^- diatomic ligands of the H-cluster originate from the amino and carboxylate moieties on tyrosine (Kuchenreuther et al. 2011). This data supports a model for H-cluster assembly in which HydG synthesizes the diatomic ligands and delivers them to HydF, where the 2Fe H-cluster precursor is assembled.

If HydG synthesizes the diatomic ligands to the H-cluster, then it is likely that HydE should synthesize the other unique feature of this cluster, the nonprotein dithiolate. Synthesis of such a ligand could proceed by radical-SAM mediated insertion of sulfur into a C-H bond in a manner analogous to BioB (Fig. 1). However, identification of a substrate for HydE and, thus, any further details of its function remain elusive. The crystal structure of HydE from *Thermotoga maritima* has been solved and *in silico* efforts have identified three anion-binding sites and a binding site for thiocyanate (Nicolet et al. 2008). HydE has a very similar fold to BioB, and both contain a [4Fe-4S] cluster that binds SAM and an additional labile [2Fe-2S] cluster (Fig. 5). It is possible that the [2Fe-2S] cluster in HydE could be used for sulfur insertion reactions similar to BioB, in synthesis of the dithiolate linker. In vitro studies have shown that hydrogenase activity is increased in the presence of SAM, tyrosine, and cysteine, indicating that cysteine may act as a substrate in the formation of the dithiolate bridge of the H-cluster (Kuchenreuther et al. 2009).

Overall, the model for biosynthesis of the H-cluster and maturation of the active [FeFe]-hydrogenase is one in which radical SAM chemistry plays a central role (Fig. 6). HydE is proposed to catalyze a BioB type reaction to insert the bridging sulfides from a [2Fe-2S] cluster on HydF into C-H bonds of an unknown substrate to generate the bridging dithiolate of the H-cluster. HydG then catalyzes the cleavage of tyrosine, generating one equivalent each of CO and CN^-, which are delivered to the modified 2Fe cluster on HydF. Two additional turnovers of tyrosine catalyzed by HydG would be required to generate the remaining CO ligands for the H-cluster. However, since the H-cluster has three CO ligands but only two CN- ligands, this scenario leaves one cyanide leftover after assembling the H-cluster precursor. The fate of this "extra" cyanide is unknown at this time.

Summary

Radical SAM enzymes catalyze an incredibly diverse array of reactions in biochemistry using a novel radical initiation mechanism and many additional features to control radical chemistry. The common intermediate

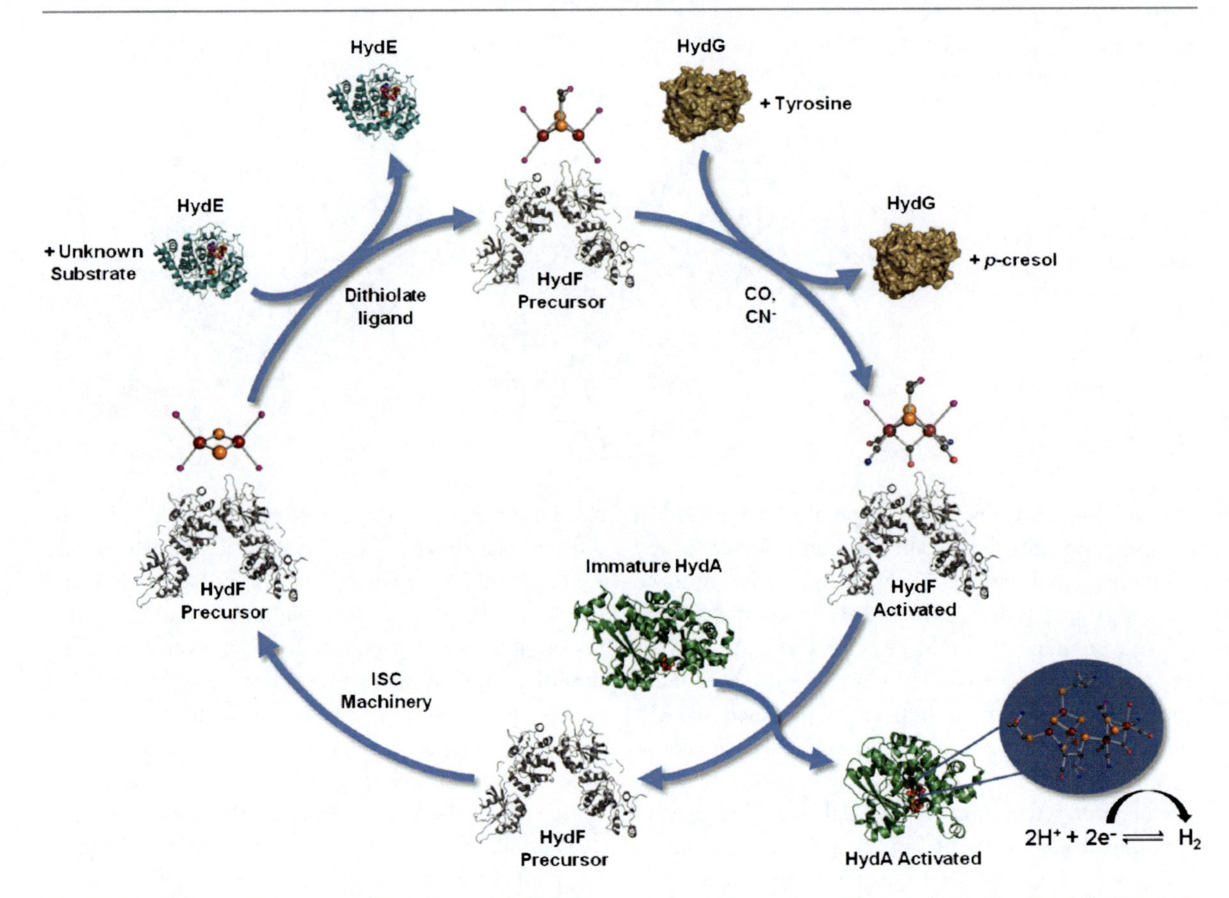

Iron-Sulfur Cluster Proteins, Fe/S-S-adenosylmethionine Enzymes and Hydrogenases, Fig. 6 Schematic representation of our current model for the maturation of the [FeFe]-hydrogenase. The HydF protein contains a [2Fe-2S] cluster (*left*) that serves as a substrate/precursor for synthesis of the 2Fe subcluster of the H-cluster. This cluster is acted on by HydE, utilizing an unknown substrate, to synthesize the bridging dithiolate of the H-cluster (*top*). HydG then catalyzes the decomposition of tyrosine to produce the diatomic ligands to the cluster. The resulting "activated" form of HydF (*right*) then transfers the 2Fe H-cluster precursor to HydA to generate the active hydrogenase. The individual proteins in the scheme are represented by their crystal structures where available (HydA, 3LX4; HydF, 3QQ5; HydE, 3IIZ). No crystal structure is available for HydG, and so it is represented using a surface rendering of the HydE structure

in radical SAM reactions, the SAM derived 5′-deoxyadenosyl radical intermediate, is an extremely reactive primary carbon radical that could in principle abstract a hydrogen atom from nearly any organic species; therefore, substrate binding in proximity to SAM and orientation of the substrate for abstraction of the *correct* H atom are presumably critical features of catalysis. Characterized radical SAM reactions include DNA and RNA modifications, vitamin synthesis, sulfur insertion reactions, cofactor synthesis, and complex substrate catalysis to name a few. These enzymes are also seminal in hydrogenase maturation in eucarya, archaea, and bacteria, underlining their important role in emerging biotechnology and alternative energy. Newly characterized motifs and mechanisms are emerging rapidly and will require many years of additional research before the full potential of radical SAM enzymes are realized.

Cross-References

- ▶ [NiFe]-Hydrogenases
- ▶ CO-dehydrogenase/Acetyl-CoA Synthase
- ▶ Iron Homeostasis in Health and Disease
- ▶ Iron Proteins, Ferritin
- ▶ Iron, Physical and Chemical Properties
- ▶ Iron-Sulfur Cluster Proteins, Ferredoxins
- ▶ Iron-Sulfur Cluster Proteins, Nitrogenases
- ▶ Ribonucleotide Reductase

References

Beinert H (2000) Iron-sulfur proteins: ancient structures, still full of surprises. J Biol Inorg Chem 5:2–15

Benda R, Bui BTS et al (2002) Iron-sulfur clusters of biotin synthase in vivo: a Mössbauer study. Biochemistry 41:15000–15006

Broderick JB, Duderstadt RE et al (1997) Pyruvate formate-lyase activating enzyme is an iron-sulfur protein. J Am Chem Soc 119:7396–7397

Buis JM, Cheek J et al (2006) Characterization of an active spore photoproduct lyase, a DNA repair enzyme in the radical S-adenosylmethionine superfamily. J Biol Chem 281(36):25994–26003

Busby RW, Schelvis JPM et al (1999) Lipoic Acid Biosynthesis: LipA is an iron-sulfur protein. J Am Chem Soc 121:4706–4707

Challand MR, Driesener RC et al (2011) Radical S-adenosylmethionine enzymes: mechanism, control and function. Nat Prod Rep 28:1696–1721

Chandra T, Silver SC et al (2009) Spore photoproduct lyase catalyzes specific repair of the 5R but not the 5 S spore photoproduct. J Am Chem Soc 131:2420–2421

Cheek J, Broderick JB (2002) Direct H atom abstraction from spore photoproduct C-6 initiates DNA repair in the reaction catalyzed by spore photoproduct lyase: evidence for a reversibly generated adenosyl radical intermediate. J Am Chem Soc 124:2860–2861

Chirpich TP, Zappia V et al (1970) Lysine 2,3-aminomutase. J Biol Chem 245:1778–1789

Demick J, Lanzilotta WN (2011) Activation of the B12-independent glycerol dehydratase results in formation of 5′-deoxy-5′-(methylthio)adenosine and not 5′-deoxyadenosine. Biochemistry 50:440–442

Dowling DP, Vey JL et al (2012) Structural diversity in the AdoMet radical enzyme superfamily. Biochim Biophys Acta

Drennan CL, Peters JW (2003) Surprising cofactors in metalloenzymes. Curr Opin Struct Biol 13(2):220–226

Driesener RC, Challand MR et al (2010) [FeFe]-hydrogenase cyanide ligands derived from S-adenosylmethionine-dependent cleavage of tyrosine. Angew Chem Int Ed Engl 49:1687–1690

Eliasson R, Fontecave M et al (1990) The anaerobic ribonucleoside triphosphate reductase from *Escherichia coli* requires S-adenosylmethionine as a cofactor. Proc Natl Acad Sci USA 87:3314–3318

Frey M, Rothe M et al (1994) Adenosylmethionine-dependent synthesis of the glycyl radical in pyruvate formate-lyase by abstraction of the glycine C-2 pro-S hydrogen atom. J Biol Chem 269(17):12432–12437

Frey PA, Hegeman AD et al (2008) The radical SAM superfamily. Crit Rev Biochem Mol Biol 43(1):63–88

Hänzelmann P, Schindelin H (2004) Crystal structure of the S-adenosylmethionine-dependent enzyme MoaA and its implications for molybdenum cofactor deficiency in humans. Proc Natl Acad Sci USA 101(35):12870–12875

Hänzelmann P, Schindelin H (2006) Binding of 5′-GTP to the C-terminal FeS cluster of the radical S-adenosylmethionine enzyme MoaA provides insights into its mechanism. Proc Natl Acad Sci USA 103(18):6829–6834

Heil K, Kneuttinger AC et al (2011) Crystal structures and repair studies reveal the identity and the base-pairing properties of the UV-induced spore photoproduct DNA lesion. Chem Eur J 17(35):9651–9657

Henshaw TF, Cheek J et al (2000) The $[4Fe\text{-}4S]^{+}$ cluster of pyruvate formate-lyase activating enzyme generates the glycyl radical on pyruvate formate-lyase: EPR-detected single turnover. J Am Chem Soc 122:8331–8332

Hinckley GT, Frey PA (2006) Cofactor dependence of reduction potentials for [4Fe-4S]2+/1+ in lysine 2,3-aminomutase. Biochemistry 45:3219–3225

Johnson DC, Dean DR et al (2005) Structure, function, and formation of biological iron-sulfur clusters. Annu Rev Biochem 74:247–281

Knappe J, Blaschkowski HP (1975) Pyruvate formate-lyase from *Escherichia coli* and its activation system. Methods Enzymol 41:508–518

Knappe J, Bohnert E et al (1965) S-adenosyl -L-methionine, a component of the clastic dissimilation of pyruvate in *Escherichia coli*. Biochim Biophys Acta 107:603–605

Knappe J, Neugebauer FA et al (1984) Post-translational activation introduces a free radical into pyruvate formate-lyase. Proc Natl Acad Sci USA 81:1332–1335

Krebs C, Broderick WE et al (2002) Coordination of adenosylmethionine to a unique iron site of the [4Fe-4S] of pyruvate formate-lyase activating enzyme: a Mössbauer spectroscopic study. J Am Chem Soc 124(6):912–913

Kuchenreuther JM, Stapleton JA et al (2009) Tyrosine, cysteine, and S-adenosylmethionine stimulate invitro [FeFe] hydrogenase activation. PLoS One 4(10):e7565

Kuchenreuther JM, George SJ et al (2011) Cell-free H-cluster synthesis and [FeFe] hydrogenase activation: all five CO and CN- ligands derive from tyrosine. PLoS One 6(5):e20346

Lees NS, Chen D et al (2006) How an enzyme tames reactive intermediates: positioning of the active site components of lysine 2,3-aminomutase during enzymatic turnover as determined by ENDOR spectroscopy. J Am Chem Soc 128:10145–10154

Lees NS, Hänzelmann P et al (2009) ENDOR spectroscopy shows that guanine N1 binds to [4Fe-4S] cluster II of the S-adenosylmethionine-dependent enzyme MoaA: mechanistic Implications. J Am Chem Soc 131:9184–9185

Magnusson OT, Reed GH et al (1999) Spectroscopic evidence for the participation of an allylic analogue of the 5′-deoxyadenosyl radical in the reaction of lysine 2,3-aminomutase. J Am Chem Soc 121:9764–9765

Magnusson OT, Reed GH et al (2001) Characterization of an allylic analogue of the 5′-deoxyadenosyl radical: an intermediate in the reaction of lysine 2,3-aminomutase. Biochemistry 40:7773–7782

McGlynn SE, Ruebush SS et al (2007) In vitro activation of [FeFe] hydrogenase: new insights into hydrogenase maturation. J Biol Inorg Chem 12(4):443–447

McGlynn SE, Shepard EM et al (2008) HydF as a scaffold protein in [FeFe] hydrogenase H-cluster biosynthesis. FEBS Lett 582(15):2183–2187

McGlynn SE, Mulder DW et al (2009) Hydrogenase cluster biosynthesis: organometallic chemistry nature's way. Dalton Trans 39(22):4274–4285

Mulder DW, Ortillo DO et al (2009) Activation of HydAΔEFG requires a preformed [4Fe-4S] cluster. Biochemistry 48(26):6240–6248
Mulder DW, Boyd ES et al (2010) Stepwise [FeFe]-hydrogenase H-cluster assembly revealed in the structure of HydAΔEFG. Nature 465:248–251
Mulliez E, Fontecave M et al (1993) An iron-sulfur center and a free radical in the active anaerobic ribonucleotide reductase of *Escherichia coli*. J Biol Chem 268:2296–2299
Nicolet Y, Drennan CL (2004) AdoMet radical proteins–from structure to evolution–alignment of divergent protein sequences reveals strong secondary structure element conservation. Nucleic Acids Res 32(13):4015–4025
Nicolet Y, Piras C et al (1999) Desulfovibrio desulfuricans iron hydrogenase: the structure shows unusual coordination to an active site Fe binuclear center. Structure 7:13–23
Nicolet Y, Rubach JK et al (2008) X-ray structure of the [FeFe]-hydrogenase maturase HydE from *Thermotoga maritima*. J Biol Chem 283:18861–18872
Peng Y, Veneziano SE et al (2010) Pyruvate formate-lyase, evidence for an open conformation favored in the presence of its activating enzyme. J Biol Chem 285(35):27224–27231
Peters JW, Lanzilotta WN et al (1998) X-ray crystal structure of the Fe-only hydrogenase (CpI) from Clostridium pasteurianum. Science 282:1853–1858
Petrovich RM, Ruzicka FJ et al (1991) Metal cofactors of lysine-2,3-aminomutase. J Biol Chem 266:7656–7660
Pilet E, Nicolet Y et al (2009) The role of the maturase HydG in [FeFe]-hydrogenase active site synthesis and assembly. FEBS Lett 583:506–511
Posewitz MC, King PW et al (2004) Discovery of two novel radical S-adenosylmethionine proteins required for the assembly of an active [Fe] hydrogenase. J Biol Chem 279:25711–25720
Sanyal I, Cohen G et al (1994) Biotin synthase: purification, characterization as a [2Fe-2S] cluster protein, and in vitro activity of the *Escherichia coli* BioB gene product. Biochemistry 33:3625–3631
Shepard EM, Broderick JB (2010) S-Adenosylmethionine and iron-sulfur clusters in biological radical reactions: the radical SAM superfamily. In: Mander LN, Liu HW (eds) Comprehensive natural products II: chemistry and biochemistry, vol 8. Elsevier Press, Oxford, pp 625–662
Shepard EM, Duffus BR et al (2010a) [FeFe]-hydrogenase maturation: HydG-catalyzed synthesis of carbon monoxide. J Am Chem Soc 132:9247–9249
Shepard EM, McGlynn SE et al (2010b) Synthesis of the 2Fe subcluster of the [FeFe]-hydrogenase H cluster on the HydF scaffold. Proc Natl Acad Sci USA 107:10448–10453
Shepard EM, Boyd ES et al (2011) Biosynthesis of complex iron-sulfur enzymes. Curr Opin Chem Biol 15(2):319–327
Silakov A, Wenk B et al (2009) ^{14}N HYSCORE investigation of the H-cluster of [FeFe] hydrogenase: evidence for a nitrogen in the dithiol bridge. Phys Chem Chem Phys 11(31):6592–6599
Sofia HJ, Chen G et al (2001) Radical SAM, a novel protein superfamily linking unresolved steps in familiar biosynthetic pathways with radical mechanisms: functional characterization using new analysis and information visualization methods. Nucleic Acids Res 29(5):1097–1106
Taylor AM, Farrar CE et al (2008) 9-mercaptodethiobiotin is formed as a competent catalytic intermediate by *Escherichia coli* biotin synthase. Biochemistry 47:9309–9317
Vey JL, Drennan CL (2011) Structural insights into radical generation by the radical SAM superfamily. Chem Rev 111:2487–2506
Vignais PM, Billoud B (2007) Occurrence, classification, and biological function of hydrogenases: an overview. Chem Rev 107:4206–4272
Wachterhauser G (1992) Groundworks for an evolutionary biochemistry: the iron-sulphur world. Prog Biophys Mol Biol 58(2):85–201
Walsby CJ, Hong W et al (2002a) Electron-nuclear double resonance spectroscopic evidence that S-adenosylmethionine binds in contact with the catalytically active $[4Fe\text{-}4S]^+$ cluster of pyruvate formate-lyase activating enzyme. J Am Chem Soc 124(12):3143–3151
Walsby CJ, Ortillo D et al (2002b) An anchoring role for FeS clusters: chelation of the amino acid moiety of S-adenosylmethionine to the unique iron site of the [4Fe-4S] cluster of pyruvate formate-lyase activating enzyme. J Am Chem Soc 124(38):11270–11271
Walsby CJ, Ortillo D et al (2005) Spectroscopic approaches to elucidating novel iron-sulfur chemistry in the "radical SAM" protein superfamily. Inorg Chem 44(4):727–741
Wang SC, Frey PA (2007) Binding energy in the one-electron reductive cleavage of S-adenosylmethionine in lysine 2,3-aminomutase, a radical SAM enzyme. Biochemistry 46(45):12889–12895
Yacoby I, Pochekailov S et al (2011) Photosynthetic and electron partitioning between [FeFe]-hydrogenase and ferredoxin: NADP + -oxidoreductase (FNR) enzymes in vitro. Proc Natl Acad Sci USA 108(23):9396–9401
Zhang Y, Zhu X et al (2010) Diphthamide biosynthesis requires an organic radical generated by an iron-sulphur enzyme. Nature 465:891–896

Iron-Sulfur Cluster Proteins, Ferredoxins

Jean-Marc Moulis
Institut de Recherches en Sciences et Technologies du Vivant, Laboratoire Chimie et Biologie des Métaux (IRTSV/LCBM), CEA-Grenoble, Grenoble, France
CNRS, UMR5249, Grenoble, France
Université Joseph Fourier–Grenoble I, UMR5249, Grenoble, France

Synonyms

Cluster dynamics; Electron transfer shuttles; Functional genomics; Origin of life; Reduction potential; Spectroscopic properties of bioinorganic moieties

Definition

- Reorganization energy (λ) is the energy required for all structural adjustments (in the reactants and in the surrounding solvent molecules) which are needed in order that the electron donor and acceptor assume the configuration required for the transfer of the electron (IUPAC Compendium of Chemical Terminology).
- Prosthetic group: A molecule different from the polypeptide chain but needed for the activity of a protein or an enzyme. This molecule may not bear the activity itself, but it induces it when bound to the polypeptide.
- Apoprotein: A protein without its prosthetic group (s), thus without activity.
- Holoprotein: A protein with its prosthetic group(s).
- Prebiotic: Relative to the terrestrial conditions before the advent of life

Introduction

In an enlightening example of the strength of a multidisciplinary approach in Science, the spreading use of physical methods, including spectroscopies, to study biological systems in the middle of the twentieth century led to the discovery of iron atoms in proteins and enzymes that could not be accounted for by the already known hemes (Beinert 1994). Chemical analysis of the iron species in acidic conditions resulted in the release of hydrogen sulfide. Soon after, major progress in protein purification techniques at the beginning of the 1960s afforded the first examples of samples in which iron and sulfur were associated at the active site of electron shuttles (Mortenson et al. 1962; Tagawa and Arnon 1962). These pioneering studies identified proteins in which iron and sulfur assembled in two different types of clusters based on the nuclearity of the prosthetic groups: they were used to define the "plant-type" and "bacterial-type" iron-sulfur proteins. This classification still underlies many contemporary studies, but the advent of large-scale DNA sequencing methods now allow a more detailed sorting according to phylogeny. Indeed, analysis of all domains of life (Andreini et al. 2007) has now identified thousands of gene sequences encoding protein products that are genuine *ferredoxins*. However, cataloguing ferredoxins (Fdx) from large-scale sequencing is not straightforward, and automatic annotation of crude genomes based on sequencing patterns with closely spaced cysteines does not provide unambiguous output.

The original definition of Fdx was an electron transferring protein containing iron, which was rapidly shown to be combined with "inorganic (or labile) sulfur" in iron-sulfur clusters. The absence of labile sulfur in small nonheme iron proteins purified in the early 1960s does not change the ability to exchange electrons, but such proteins received another name: *rubredoxins*. The initially identified soluble Fdx interact with several partners (see below), but other iron-sulfur proteins were later shown to exchange electrons exclusively within specialized protein complexes. The association/dissociation dynamics of soluble complexes with Fdx is generally poorly characterized. In contrast, Fdx in membrane-embedded complexes are actual intrinsic subunits. As an example of the difficulty of defining a Fdx, the "iron-protein" (product of the *nifH* gene) of the nitrogenase complex selectively transfers low potential electrons to the actual catalytic dimeric subunit (the "molybdenum (or vanadium) protein") with the help of ATP hydrolysis. This nitrogenase iron protein (NifH) meets the criteria of the Fdx definition, yet it is not usual practice to call it Fdx. The present entry is restricted to iron-sulfur proteins of a relatively small size (arbitrarily set below 15 kDa), the role of which, when known, is to transfer electrons to other proteins.

Electron Transfer

Electron transfer is considered here as a reaction in which electrons (strictly speaking one at a time, since simultaneous exchange of pairs is hardly relevant in Biology) travel through long distances relative to the length separating atoms, without being used to cleave or build chemical bonds. "Long distances" at this scale range between ca. 5 Å (1 Å = 10^{-10} m) and several nanometers (10^{-9} m). Biological membranes fall in this range and electrons can cross them via protein complexes to build energy-rich electrochemical gradients. In order to readily exchange electrons, partner molecules must accept or donate them from their highest occupied molecular orbitals at a reasonable energetic cost. Proteins are made of amino acid residues connected via peptide bonds which are not efficiently permeable to electric charges, that is, proteins are dielectric materials. This is why most electron transfer proteins need bound prosthetic groups or cofactors to increase electrical conduction through them. In the case of Fdx, these prosthetic groups are the iron-sulfur clusters.

Iron-Sulfur Clusters: The Core of the Matter

The small size of Fdx and the high purification yields for some of them, preferably isolated in the absence of oxygen, made Fdx 3D structures among the first to be solved by X-ray crystallography. Fdx are also amenable to NMR, but the spectra, although they can afford unique information, are complicated by the presence of the clusters. In terms of ease of interpretation and resolution, the NMR structural method for Fdx does not compete with diffraction data obtained from good crystals.

Solving the structure of Fdx has been instrumental in identifying the main classes of iron-sulfur clusters which have then been observed in scores of other proteins far beyond the mere Fdx (Meyer 2008). The basic patterns of the arrangement of iron and sulfur atoms are represented in Fig. 1. They are ranked as a function of the nuclearity of the cluster, with 1Fe (in rubredoxins), 2Fe, 3Fe, and 4Fe. The iron-sulfur bonds (2.2–2.35 Å long) can be separated into two categories: those linking the cluster irons to the peptide chain through cysteine residues, and those between iron and the unique kind of sulfur found in such clusters: "inorganic" sulfur. The latter is derived from the amino acid cysteine by the action of enzymes called desulfurases which bind the product as a persulfide. This sulfur product (sulfane, formally S^0) is then mobilized by a reducing acceptor molecule as sulfide (S^{2-}) in biosynthetic reactions incorporating sulfur, such as tRNA thiolation, molybdopterin synthesis, and building of iron-sulfur clusters.

Although the basic structures in Fig. 1 look different, they share similar features. Iron is bound by four sulfur ligands in the tetrahedral geometry found without inorganic sulfur in rubredoxins. Thus, the S-Fe-S angles distribute around the ideal value of 109.5°. Cysteines bind iron the usual way by the thiolate function. In a few cases, a minority of these thiolates are replaced by carboxylates from glutamate or aspartate, as in *Pyrococcus furiosus* Fdx, or imidazolium from histidine, as in the Rieske Fdx. Inorganic sulfur is coordinated in two different ways, bridging two (μ^2) or three (μ^3) iron atoms, the former in 2Fe-Fdx, the latter in 4Fe-Fdx. Both sorts of bridging sulfide occur in 3Fe clusters.

Before the availability of atomic resolution structures of Fdx, skilled chemists provided deep insight into the detailed structure and other properties of many iron-sulfur clusters, including biologically relevant ones (Rao and Holm 2004). Under proper experimental conditions, which are not too remote from those supposed to have prevailed in the prebiotic world (see below), a surprisingly simple mixture of iron salt, sulfide, and thiols afforded strikingly close analogs of [4Fe-4S]Fdx. This chemical work showed that the [4Fe-4S] structure is thermodynamically more stable than the [2Fe-2S] cluster in these experiments, since constraints through bidentate thiol ligands had to be introduced to force assembly of [2Fe-2S] analogs. Further development of ligand (thiol)-driven synthesis of iron-sulfur clusters produced [3Fe-4S] compounds. In addition to providing excellent structural mimics of biological clusters, these chemical approaches unraveled the synthetic pathways followed by the reactants and intermediates, such as mononuclear iron thiolates (analogs of rubredoxins) and iron-sulfur adamantane structures. These investigations led to the concept of self-assembly (from simple precursors/reactants) for metal clusters in proteins, which has now to be integrated with recently obtained knowledge about biosynthesis of iron-sulfur clusters, involving multiple carriers and scaffold proteins.

In contrast to the limited set of different iron-sulfur clusters found in Fdx (Fig. 1), a large variety of folding patterns has been evidenced for the protein around them (Meyer 2008). The clusters are located in protein cavities surrounded by main chain carbonyl and amide groups which contribute to the properties of the clusters. These cavities may be buried deep inside protein shells, as in the high-potential [4Fe-4S]Fdx, or closer to the surface, as in [2Fe-2S]Fdx of plants. Whereas the accessibility of solvent water tune electronic properties of these clusters, the actual location of the active site seems relatively neutral for the electron transfer properties of Fdx, probably because the small size and the likely optimized folding of these proteins do not significantly impair conductance.

Within the strict definition of Fdx given herein, these proteins are not prone to associate with other cofactors than iron-sulfur clusters. Still the presence

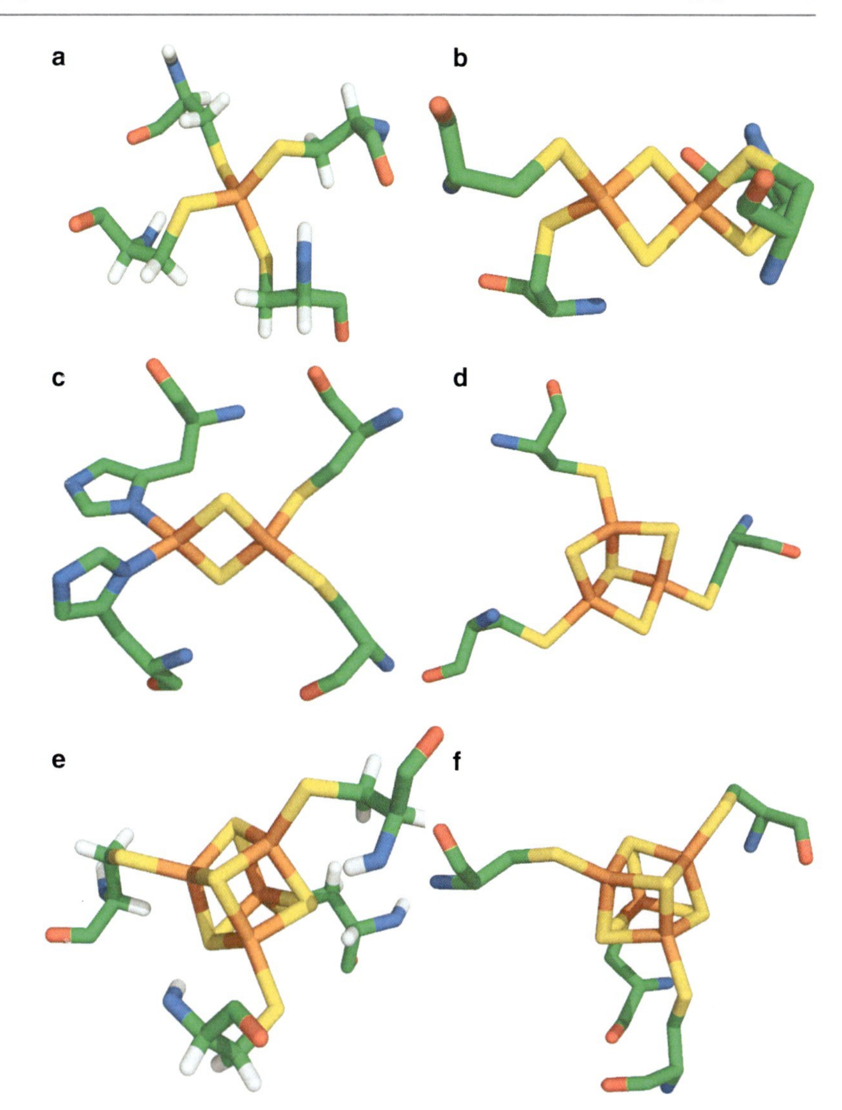

Iron-Sulfur Cluster Proteins, Ferredoxins, Fig. 1 Structures of iron-sulfur clusters in ferredoxins. The following models were retrieved from the Protein Data Bank (http://www.rcsb.org/): (**a**) *Clostridium pasteurianum* rubredoxin (1IRO); (**b**) *Anabaena* PCC7119 [2Fe-2S]Fdx (1CZP); (**c**) Rieske Protein from *Thermus thermophilus* (1NYK); (**d**) *Pyrococcus furiosus* [3Fe-4S]Fdx (1SJ1); (**e**) *C. acidurici* 2[4Fe-4S]Fdx (2FDN); (**f**) *Thermochromatium tepidum* high-potential [4Fe-4S]Fdx (2AMS). The color code is: iron atoms in *orange*; sulfur, *yellow*; nitrogen, *blue*; carbon, *green*; oxygen, *orange*. Clusters were drawn with the Pymol software (http://www.pymol.org; DeLano Scientific LLC)

of a zinc atom has been found in some of them from archaeal origin: zinc appears to play a structural role in the folding of a particular domain in these Fdx.

The nuclearity of their clusters (i.e., [2Fe-2S]Fdx) gives information about the properties of Fdx. However, some Fdx can accommodate different clusters/metal centers than the natural ones, mostly after in vitro manipulations. For instance, the four cysteines of rubredoxins natively bind a single iron atom, but they can also bind a [4Fe-4S] cluster, and removal of one of these cysteine triggers assembly of a [2Fe-2S] cluster. In addition, heterometal replacements for iron in Fdx have been described and many [3Fe-4S] clusters arise from the loss of one iron from a [4Fe-4S] cluster. The physiological meaning of such changes is generally unclear, but iron-sulfur cluster conversion is central to the function of the transcription factor FNR involved in the aerobic/anaerobic transition of bacteria. The structural dynamics of the active site of Fdx, and iron-sulfur proteins overall, help explain the biological need for chaperones and specific biosynthetic pathways for this kind of metalloproteins.

Iron-Sulfur Clusters Confer Unique Properties to Fdx

The chemical nature of the clusters implies they are intrinsically sensitive to oxidative conditions. Iron may be formally described as ferrous or ferric in iron-sulfur clusters, but the oxidation numbers of the sulfur components are low, meaning that these structures are fairly electron-rich. This property and the distribution of the plentiful electrons participate in the electron exchange ability of Fdx, but they also sensitize the clusters to reactions with a variety of molecules, such as the superoxide radical anion, nitrogen monoxide, or even protons, hydroxide anions, and dioxygen. Whereas some of these reactions are of physiological importance for the function of specific iron-sulfur proteins, exposure to such compounds in the case of Fdx leads to cluster degradation (i.e., [4Fe-4S] to [3Fe-4S] conversion) and quite often complete destruction. Consequently, the iron-sulfur clusters of Fdx must be chemically stabilized and a good deal of the Fdx peptide chain is used to shield the clusters from noxious molecules that may be present in the cellular environment. This protein contribution modulates the intrinsic stabilities of the clusters determined with chemical analogs: the less-easy-to-synthesize [2Fe-2S] clusters often occur in more stable, and also larger, Fdx than [4Fe-4S] centers. Accordingly, [2Fe-2S]Fdx are more frequent in aerobic organisms than [4Fe-4S]Fdx.

A key property of Fdx is their reduction potential (often indicated as E° and referenced, in volts, to the value of a known system/electrode), since this value determines the free energy of reaction with their electron transfer partners. The more positive is the E° value, the stronger the molecule accepts electrons. For iron-sulfur clusters in Fdx, E° values are generally low, reflecting the electron-rich character of these structures, but the range of reduction potentials for Fdx is large; it spans roughly 75% (ca. −700 to +450 mV with respect to the Normal Hydrogen Electrode, NHE) of the biological reduction potentials range. The latter is between ca. -700 mV for the electrons produced by Photosystem I and ca. +800 mV for the O_2/H_2O couple used by most aerobic respiring organisms (Fig. 2). To span the more than 1 V E° range, given the modest number of iron-sulfur basic structures in Fdx (Fig. 1), the reduction potential in Fdx proteins reflects adjustments controlled by the protein environment of the cluster(s). And, since the reaction of interest is electron acceptance or release, the modulating forces are of electrostatic origin. Factors of importance in this respect include charge-charge and charge-dipole interactions between the cluster on the one hand and the protein chain and the solvent on the other hand. Thus, depending on charged residues in the vicinity of the cluster, its solvent accessibility, or the distribution of polar groups (peptide carbonyl, polar amino acid residues, hydrogen bonds donors, etc) around it, the reduction potential of a cluster may shift by dozens of millivolts in different proteins. Of particular importance among these contributions are hydrogen bonds involving peptide amide as donor and clusters' sulfurs as acceptors. Their distribution around the structures of Fig. 1 is asymmetric and they participate significantly in the arrangement of partial charges over the clusters. Another way of changing the

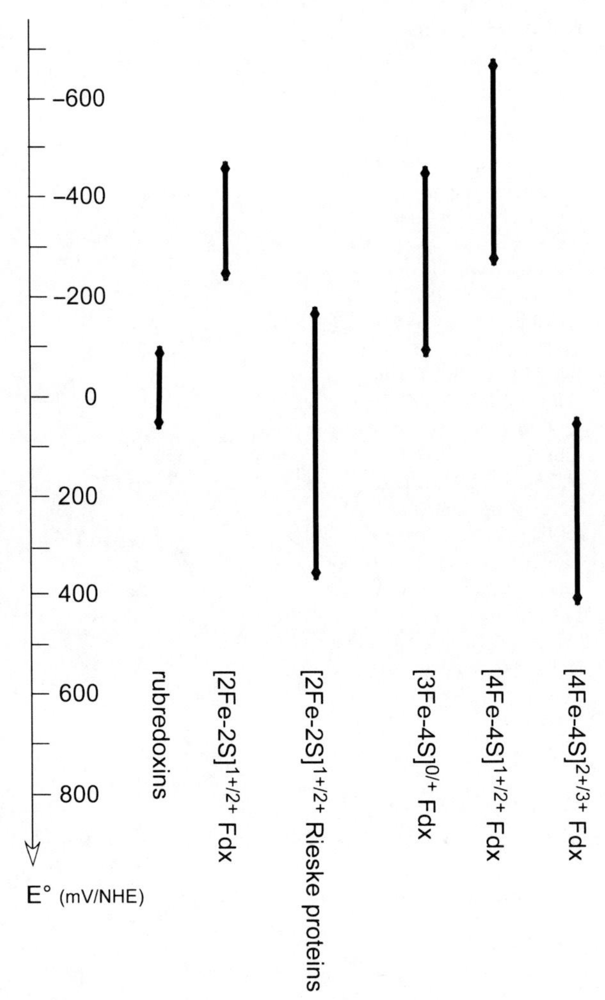

Iron-Sulfur Cluster Proteins, Ferredoxins, Fig. 2 Reduction potentials of ferredoxins. The range of reduction potentials displayed by each type of Fdx is represented along the energy axis given by standard values referenced to the normal hydrogen electrode

reduction potential of FeS custers in Fdx proteins is by replacing the iron ligands: the Rieske Fdx with two histidines binding one iron have E° values more than ca. 250 mV above those of [2Fe-2S]Fdx with full cysteine coordination. Furthermore, the presence of groups, such as the imidazolium of histidine, with neutral pKa in the vicinity of the iron-sulfur cluster may couple electron transfer with proton exchange.

In iron-containing electron transfer proteins, the metal cycles between ferrous (2+) and ferric (3+) ions, as in rubredoxins that display E° close to the reference NHE value (within 100 mV). For the iron-sulfur clusters of higher nuclearity, it might be expected that several electrons can be exchanged, one for each iron. This is not the case, and each cluster of Fdx has only two accessible redox levels. They are $[2Fe\text{-}2S]^{1+/2+}$, $[3Fe\text{-}4S]^{0/1+}$, and $[4Fe\text{-}4S]^{1+/2+}$, or $[4Fe\text{-}4S]^{2+/3+}$ in high-potential Fdx. This notation represents the formal charge of the cluster excluding the ligands, and it should not overshadow the anionic character of Fdx active sites with full cysteine coordination which adds four negative charges. Thus, Fdx are one-electron exchangers, in contrast to flavoproteins for instance. This is because other redox couples than the ones listed here are not accessible in Fdx under physiological conditions.

The presence of metals in proteins opens the way to a range of biophysical methods detecting and characterizing these metalloproteins. Some, such as Electron Paramagnetic Resonance, have been instrumental in the discovery of iron-sulfur proteins (Beinert 1994), and many, because they rely on the interaction of radiation (of different kinds depending on the method) and matter, provide detailed information on the intimate properties of iron-sulfur clusters in proteins. As in the case of X-ray crystallography, Fdx paved the way to detailed knowledge of biological iron-sulfur clusters by use of nuclear and electronic magnetic resonances and associated multiple spectroscopies, Mössbauer spectroscopy, resonance Raman, X-ray absorption spectroscopies (XANES and EXAFS), and many others. Yet, these sophisticated methods are ill adapted to the study of the most biologically relevant property of Fdx, namely, electron transfer. Indeed, they usually probe stable and static states and they do not have access to the dynamics of the electronic properties. One exception is NMR, which depends on relaxation times in the kinetic range of electron transfer in some complexes involving Fdx, and other electron transfer proteins (Bertini and Luchinat 1996). Consequently, specifically designed NMR experiments have measured the kinetic constants of some electron transfer reactions in Fdx. The main parameters for these reactions were derived from these experiments in the widely accepted theoretical framework linking rate constants to electronic coupling between donor and acceptor, their free energy difference, and the reorganization energy of the system upon electron transfer (Marcus and Sutin 1985). It appeared that the reorganization energy (λ) in the case of $[4Fe\text{-}4S]^{1+/2+}$ Fdx was fairly low, in agreement with a minimal hurdle to electron exchange (Kümmerle et al. 2001). These kinetic data are supported by the electronic delocalization over Fe-S clusters (i.e., high charge densities are not trapped in one spot of the structure and the orbitals involved in electron transfer are far reaching), and by the few high-resolution structural data obtained on Fdx at two different redox levels. In general terms, reduction increases (slightly) the distances within the clusters and decreases (i.e., strengthens) those of the hydrogen bonds involving their sulfurs. But these changes are very minor and only measured with very high quality data. The peptide moiety may also relax differently around the clusters, even at long distance, as a function of the redox level, but these changes can be predicted to be more important for recognition with the Fdx partners than for the efficiency of electron transfer. Overall, the structures of Fdx appear to be optimized for minimal reorganization (i.e., small λ) upon electron transfer.

Ferredoxins: Where Do They Come from and How Are They Made?

The initial identification of iron-sulfur clusters in proteins soon led scientists to wonder about the evolutionary rationale for their occurrence. Such questions led to the theory of a chemo-autotrophic origin of life (Wächtershäuser 2007), in which iron-sulfur minerals (pyrite) provided the main energy source for transforming prebiotic molecules, and acted as surface catalysts, with contributions from nickel, based on the known [Ni-Fe] hydrogenases, or other metals, to synthesize organic molecules as metabolic precursors. Additional observations, such as the ability of iron sulfide materials to form closed structures potentially confining reactants, backed this theory and suggested a succession of events to build a protocell that predates living cells as they are known today. While debate over the theory continues, it seems clear that iron ions and sulfide were abundant in the dioxygen-depleted Earth of Hadean ages, and that Fdx are the remnants of important actors involved in the origin of life.

Assembly of the active site of most Fe-S proteins was initially thought to occur spontaneously within cells, when the apoprotein was provided with the necessary reactants: iron ions and sulfide. Such reactions are relatively easy to carry out in vitro with pure apo-Fdx. This initial, chemically oriented, view now has to be integrated with the more recently identified biological pathways dedicated to the cellular synthesis of iron-sulfur clusters (Lill 2009). Still unclear is the role of the Fe-S biosynthetic systems in the coordinated assembly, matching, and effective insertion of a particular type of iron-sulfur cluster into the appropriate apoproteins. They may be required to avoid damage to the cluster or to the cell, or to simply provide proper iron and sulfur chemical forms in the conditions prevailing in contemporary cells, which are quite different from those of prebiotic times. The interwoven origins of present-day complex biological systems is exemplified by the mandatory involvement of a mitochondrial [2Fe-2S]Fdx in the de novo biogenesis of iron-sulfur proteins in eukaryotes.

The subcellular location of the mature holoprotein is a further complication in the assembly of iron-sulfur clusters in eukaryotes. Since iron-sulfur cluster(s) are important protein folding elements and translocation of fully folded proteins is an inefficient process, the Fe-S moiety must be assembled on the spot, where the active protein is to be used. This implies that either clusters or precursors have to be properly transported from the site of reactants availability (e.g., mitochondria in animal cells) to that of assembly into apoproteins, or high Fe-S requiring organelles must have their own Fe-S biosynthetic machinery. A good example of the latter is the chloroplast, which contains components of different biosynthetic systems identified in bacteria: these components are among the less sensitive to oxygen and they may be the most effective mediators of iron-sulfur biogenesis in the highly oxygenated environment of the chloroplasts.

Occurrence, Classification, and Binding Motifs

Despite the likely ancient origin of Fdx, these proteins are found in the most evolved forms of life, indicating their irreplaceable biological functions. This statement may sound surprising since electron transfer is one of the simplest reactions which can be imagined. Yet, Biology is not only a question of sophisticated chemical reactions, but also of specificity and regulation. Fully sequenced genomes show the wide distribution of Fdx in *Archeae* and most *Bacteria*, and in eukaryotes as well. In the eukaryotic green alga *Chlamydomonas reinhardtii*, for instance, six Fdx genes are predicted to encode chloroplast proteins, most with noncommutable roles. Genomic sequencing also indicates that Fdx containing [4Fe-4S] or [3Fe-4S] centers are far more frequent in prokaryotes than in eukaryotes, whereas $[2Fe\text{-}2S]^{1+/2+}$Fdx are found in all phylogenic branches.

Although other criteria may be proposed, nuclearity of the Fe-S cluster is the simplest way to classify Fdx proteins. Thus, the two main Fdx classes are proteins with [2Fe-2S] clusters and those with [3/4Fe-4S] clusters. The rationale to put [3Fe-4S] and [4Fe-4S]Fdx in the same group comes from the likely common origin of these proteins and their similar fold around the clusters (Meyer 2008). Classifying Fdx by their clusters requires the careful characterization of these prosthetic groups in the cellular context in which the proteins occur; Fdx manipulations in vitro that may alter the nuclearity of their clusters, as occurs often in recombinant, sometimes modified (e.g., His-tagged), proteins has to be avoided. Subdivisions of the [3Fe-4S] and [4Fe-4S]Fdx group may be based on the number (generally one or two) of clusters/protein, the accessible redox levels, the presence of additional posttranscriptional modifications (zinc atom, disulfide bridge) or of specific peptide domains, but they do not add significant improvement. By contrast, [2Fe-2S] Fdx can be conveniently separated into : (a) "conventional" Fdx, represented by the first plant [2Fe-2S]Fdx that was characterized; (b) [2Fe-2S]Fdx involved in biosynthesis of iron-sulfur clusters; (c) Rieske-type proteins in which two ligands are histidines; and (d) homodimeric [2Fe-2S]Fdx, which are exclusively found in bacteria and display a thioredoxin folding pattern.

Iron-Sulfur Cluster Proteins, Ferredoxins, Table 1 Ligands bearing sequences of Fdx

Fdx	Binding sequence	comments
Rubredoxin	CxxC + CxxC	
[2Fe-2S]Fdx	CxxxxCxxC + C	
[2Fe-2S]Fdx (Isc)	CxxxxxC + CxxC	Homologs of Fdx encoded by the bacterial Isc operon
[2Fe-2S]Fdx thioredoxin fold	$C(x)_zC$ + CxxxC	$z \in [10–12]$
[2Fe-2S] Rieske proteins	CxH + CxxH	
[3Fe-4S]Fdx	$C(x)_yC(x)_zCP$	$y = 5$. $z > 3$. The distant C is often followed by a P. Fdx in which $z = 3$ likely contain two clusters
[4Fe-4S]Fdx	$CxxC(x)_yC(x)_zCP$	$y \in [2–8]$. Same remark as above for z
High-potential [4Fe-4S]Fdx	$CxxC(x)_yC(x)_zC$	$y \in [8–16]$. $z \in [12–15]$

The coordinating amino acids for iron confer Fdx characteristic sequence patterns. Cysteines are (by far) the main iron-binding residues and cysteine-rich fragments are hallmarks of Fdx sequences (Table 1). However, other small cysteine-rich proteins are not Fdx, and they should not be confused with them. For instance, the multiple cysteines of thioneins and defensins do not readily bind iron-sulfur clusters, and these proteins do not participate in long-range electron transfer reactions. The requirement of binding iron-sulfur clusters surrounded by stabilizing hydrogen-bonds networks (see above) restricts the distribution of cysteines along the Fdx sequences. $[4Fe\text{-}4S]^{1+/2+}$ clusters in Fdx are linked by $CxxCxxC(x)_zC$ peptides, in which x is any amino acid, and the last cysteine is often followed by a proline. The distance of this last ligand (z amino acids) may be any number but not too small, even though z appears to be 3 in many sequences: indeed when $z = 3$, the forth cysteine binds another cluster (or does not bind any cluster) than the one bound by the first three, despite the fact that in vitro work has forced $[4Fe\text{-}4S]^{1+/2+}$ assembly in minimal peptides in which $z = 3$. Also, the gap between the second and third cysteine of the CxxCxxC sequence may be longer than two amino acids up to eight: this situation is a sign for the likelihood of a $[3Fe\text{-}4S]^{0/1+}$ cluster, as also occurs when the second cysteine is replaced by another amino acid. The binding pattern of $[4Fe\text{-}4S]^{2+/3+}$ clusters is more loose with another CxxC motif associated with two other remote cysteines. For conventional [2Fe-2S]Fdx, there is a $C(x)_4CxxC$ fragment and another distant cysteine, whereas Rieske-type proteins have two different cluster-binding fragments with close C and H residues. For the thioredoxin-like [2Fe-2S]Fdx, the $C(x)_{12}C$ and CxxxC peptides bearing the ligands still bind the cluster after significant rearrangements of the chelating groups in vitro, but the physiological significance of this feature is not yet known.

Biological Functions of Fdx

Two basic properties of Fdx used in biology are the ability of exchanging electrons, reflected by the E° value, and the efficiency of long-distance electron transfer. The latter is central to bioenergetics by using the energy of these electrons to reduce substrates or to generate proton motive forces across biological membranes. Fdx are thus heavily involved in using the reducing power generated by the metabolism for biosynthetic purposes. Key examples are the [2Fe-2S]Fdx that accept electrons from Photosystem I and donate them to reduce $NADP^+$ or thioredoxin in photosynthetic organisms. In anaerobic heterotrophic bacteria, 2[4Fe-4S]Fdx couple pyruvate oxidation to CO_2 fixation, NADPH (re)generation, sulfite reduction, hydrogen evolution, and, sometimes, nitrogen fixation (Mortenson 1963), all reactions of current ecological or biotechnological importance. These same functions of Fdx, as electron carriers coupling catabolic and anabolic pathways, can also be fulfilled by unrelated electron transfer proteins, flavodoxins, when Fdx are experimentally deleted or when iron is limiting in the environment. Apparently, flavodoxins are a backup system for bare survival, but it is not clear whether electron partitioning among the many different redox partners is the same with Fdx and flavodoxins, or between different Fdx that may coexist and may interact with the same enzymes in the same cell/compartment.

The mechanisms orienting the electron flow in Fdx are only partly understood. A large proportion of Fdx have E° values close to the lower end of the biological range (Fig. 2) and the reduction must be coupled to a strongly exergonic reaction, such as decarboxylation of organic compounds. However, association of Fdx in complex with redox partners may shift the Fdx E° value, and thus ease the thermodynamic hurdle required for Fdx reduction. Other questions are linked to the specificity of the interactions with redox partners belonging to different (branched) metabolic pathways in which Fdx function as *electron carriers* (i.e., shuttles) (Kümmerle et al. 2000). In contrast, the involvement of Fdx in linear electron transfer chains, in other words as *electron mediators*, is easier to understand. In such cases, Fdx are mandatory links between specific donors and acceptors. Examples include the Rieske proteins which are subunits of the membrane cytochromes bc_1 and b_6f complexes, as well as electron donors to dioxygenases involved in the degradation of bacterial substrates such as aromatic compounds. Other "conventional" [2Fe-2S]Fdx intervene in P450 monooxygenases hydroxylating cholesterol for steroid hormone biosynthesis or for bacterial terpenoid degradation, and high-potential [4Fe-4S]Fdx (re)reduce the tetraheme cytochrome of bacterial photosynthetic reaction centers or participate in bacterial aerobic respiratory chains.

A particular [2Fe-2S]Fdx has recently become famous because it is required for de novo iron-sulfur biosynthesis in mammalian cell mitochondria and in yeast. The corresponding bacterial protein belongs to the ISC (Iron-Sulfur Cluster) operon. Besides the surprising concept that this protein should be involved in the biosynthesis of its own active site, it, like other Fdx, conveys electrons from NADPH via a reductase to a yet unidentified substrate, possibly persulfide bound to cysteine desulfurases. In contrast to the above example of Fdx/flavodoxin substitution, the Fdx participating to the biosynthesis of iron-sulfur clusters cannot substitute for the Fdx paralog involved in steroid biosynthesis and vice versa. However, such specific examples of Fdx indispensability are limited because identifying a metabolic pathway in which a Fdx takes an essential part is a difficult task, including because gene deletion of such a Fdx is likely to be lethal.

Many turns in the expanding field of identifying biological roles for Fdx have been unexpected. For instance, Fdx importance in the virulence of parasitic protozoans, and in the secretion of cytotoxic proteins from pathogenic bacteria invading endothelial cells of the blood-brain barrier, has been recently documented. Furthermore, activation of pro-drugs such as metronidazole relies on the reducing power of Fdx in some anaerobic microbes targeted by this drug. A recently identified function of Fdx in bacteria is to assist in the recovery of iron from the ferritin mineral of the heme protein bacterioferritin, which illustrates the theme of the Iron-Sulfur World (see above) where metabolic proto-reactions occur in a confined space (the ferritin nanocage) with the energy provided by iron-sulfur clusters (from Fdx).

Conclusions

Despite the non-catalytic roles of Fdx, the number of biological functions in which they are involved is very large. The precise metabolic pathways in which the earliest studied representatives participate have been known for half a century. Nevertheless, the biological functions have been clearly identified for only a small fraction of Fdx. Scores of Fdx have been thoroughly characterized by physicochemical methods, with hardly any secrets remaining hidden in these tiny proteins and their particular prosthetic groups shown in Fig. 1. Yet, most of them still await a physiological placement. Indeed, most predicted Fdx Open Reading Frames, principally in *Bacteria* and *Archeae*, are not associated with any biological pathway. From some recently characterized examples of unexpected functions mentioned above, it may be anticipated that further surprises will arise in the years to come as to the biological functions of these "plain vanilla" electron transfer proteins.

Summary

Ferredoxins (in the sense defined in this entry) are small proteins organized around iron-sulfur clusters containing between two and four iron ions. They occur in the three domains of life and they are involved in long-range electron transfer reactions. Most of these proteins provide high-energy electrons to a variety of cellular processes, hence fueling many biosynthetic pathways. Some contribute to the catabolism of chemically inert bacterial substrates. Despite the presence of several potentially redox-active iron ions, ferredoxins exchange only one electron. At the active site, iron is mainly surrounded by sulfur atoms and the peptide binding the Fe-S clusters are cysteine-rich; however, parts of the cluster-binding sequences can also be found in other proteins than ferredoxins. Ferredoxins are often considered as remnants of some prebiotic chemistry having contributed to the origin of life, but they now require sophisticated biochemical systems to become active in contemporary cells. This suggests irreplaceable roles which have been kept throughout evolution, and most of the many ferredoxins predicted by whole genome sequencing still have not been included in biological pathways. Very recent data indicate that the latter may largely go beyond the bioenergetic ones identified for the better characterized ferredoxins.

Cross-References

- ▶ [NiFe]-Hydrogenases
- ▶ Iron-Sulfur Cluster Proteins, Fe/S-S-adenosylmethionine Enzymes and Hydrogenases
- ▶ Iron-Sulfur Cluster Proteins, Nitrogenases
- ▶ Nitrite Reductase

References

Andreini C, Banci L, Bertini I, Elmi S, Rosato A (2007) Non-heme iron through the three domains of life. Proteins 67:317–324

Beinert H (1994) Looking at enzymes in action in the 1950s. Protein Sci 3:1605–1612

Bertini I, Luchinat C (1996) NMR of paramagnetic substances. Coord Chem Rev 150:1–296

Kümmerle R, Kyritsis P, Gaillard J, Moulis J-M (2000) Electron transfer properties of iron-sulfur proteins. J Inorg Biochem 79:83–91

Kümmerle R, Gaillard J, Kyritsis P, Moulis J-M (2001) Intramolecular electron transfer in [4Fe-4S] proteins: estimates of the reorganization energy and electronic coupling in *Chromatium vinosum* ferredoxin. J Biol Inorg Chem 6:446–451

Lill R (2009) Function and biogenesis of iron-sulphur proteins. Nature 460:831–838

Marcus RA, Sutin N (1985) Electron transfers in chemistry and biology. Biochim Biophys Acta (BBA) 811:265–322

Meyer J (2008) Iron-sulfur protein folds, iron-sulfur chemistry, and evolution. J Biol Inorg Chem 13:157–170

Mortenson LE (1963) Nitrogen fixation: role of ferredoxin in anaerobic metabolism. Annu Rev Microbiol 17:115–138

Mortenson LE, Valentine RC, Carnahan JE (1962) An electron transport factor from *Clostridium pasteurianum*. Biochem Biophys Res Commun 7:448–452

Rao PV, Holm RH (2004) Synthetic analogues of the active sites of iron-sulfur proteins. Chem Rev 104:527–559

Tagawa K, Arnon DI (1962) Ferredoxins as electron carriers in photosynthesis and in the biological production and consumption of hydrogen gas. Nature 195:537–543

Wächtershäuser G (2007) On the chemistry and evolution of the pioneer organism. Chem Biodivers 4:584–602

Iron-Sulfur Cluster Proteins, Nitrogenases

Lance C. Seefeldt[1] and Dennis R. Dean[2]
[1]Department of Chemistry and Biochemistry, Utah State University, Logan, UT, USA
[2]Department of Biochemistry, Virginia Tech University, Blacksburg, VA, USA

Synonyms

Catalysis; Fe metabolism; Fe-S clusters; Nitrogen fixation

Definition

Biological nitrogen fixation, the reduction and hydrogenation of nitrogen (N_2) to yield ammonia (NH_3), is catalyzed by a complex metalloenzyme called nitrogenase. The best-studied member of this class of oxygen (O_2)-sensitive enzymes is comprised of two interacting protein partners called iron (Fe) protein and molybdenum-iron (MoFe) protein. Nitrogenase couples nucleotide hydrolysis to inter- and intra-electron transfer and N_2 activation and reduction. The overall stoichiometry of the reaction is represented by: $N_2 + 8e^- + 8\ H^+ + 16ATP \rightarrow 2NH_3 + H_2 + 16\ ADP + 16P_i$.

Principles of Nitrogen Fixation

Life depends on processes that utilize iron (Fe)-containing cofactors, often referred to as iron-sulfur (Fe-S) clusters, that couple electron transfer and catalysis (Beinert et al. 1997). Among these processes are photosynthesis, respiration, and nitrogen fixation. Indeed, life on Earth is likely to have emerged through reactions that occurred on metal-sulfur surfaces primarily involving Fe. Nitrogen fixation can occur either through a biological process, involving nitrogenases, or through the industrial Haber-Bosch process (Smil 2004). A fundamental aspect of both processes is the involvement of Fe. The agronomic and economic importance of industrial nitrogen fixation is evidenced by the enormous expansion in the human population since the advent of the Haber-Bosch process. For example, it is estimated that nearly 40% of the current human population would not exist without the availability of industrial nitrogen fixation (Smil 2004). However, there are a number of adverse consequences associated with industrial nitrogen fixation, including energy costs related to the consumption of nonrenewable resources, transportation costs, and pollution related to the application of nitrogenous fertilizers. A more effective use of biological nitrogen fixation or development of alternative catalysts informed by the biological reaction could potentially ameliorate some of these problems. For these reasons, there has been intensive research over the past five decades aimed at understanding the mechanistic features and regulation of expression of nitrogenase, as well as understanding the processes involved in the assembly of its associated cofactors (Burgess and Lowe 1996). These studies have revealed the fundamental importance of Fe in the catalytic mechanism, which involves both simple and complex Fe-S clusters.

Nitrogenase Types

The term "nitrogenase" defines a class of structurally related but genetically distinct enzymes that catalyze biological nitrogen fixation. There are three different types of nitrogenases, which are distinguished by the metal content of the corresponding cofactors that make up the N_2-binding site (Eady 1996). These cofactors contain an organic constituent, *R*-homocitrate, together with 7 Fe, 9S, and a variable metal that is either molybdenum (Mo), vanadium (V), or iron (Fe) (Fig. 1). The Mo-containing enzyme appears to be the most effective catalyst, with the other forms being produced and utilized only when Mo is unavailable. The Mo-containing enzyme is also the most widely distributed nitrogenase in nature. In spite of differences in the metal composition of their active site cofactors, all three types of nitrogenases are unified by common mechanistic features, as described below for the Mo-containing enzyme.

Nitrogenase Mechanism

The conversion of N_2 to 2 NH_3 is thermodynamically favorable, having a negative standard Gibb's free energy change, yet, the reaction is among the most difficult to achieve in nature because of the high activation energy required to break the N≡N triple bond (Smil 2004). A second challenge is that multiple electrons and protons are necessary to accomplish N_2 reduction (Burgess and Lowe 1996). For nitrogenase, a key to overcoming these challenges is the involvement of three different metal cofactors distributed between two protein partners. In the case of the Mo-containing enzyme, these protein partners are, respectively, designated as the Fe protein and the MoFe protein (Fig. 1). The Fe protein is a homodimer that contains two nucleotide-binding sites, one in each subunit, and a single [4Fe-4S] cluster bridging the subunits. The MoFe protein is an $\alpha_2\beta_2$heterotetramer with each $\alpha\beta$−pair acting as a catalytic unit (Seefeldt et al. 2009). Each $\alpha\beta$-pair contains two different cofactors called the P cluster ([8Fe-7S]) and the FeMo-cofactor ([7Fe-9S-1Mo-X-*R*-homocitrate]). An atom, X, has been detected in the middle of the six Fe atoms of the FeMo-cofactor. The identity of X has yet to be established, however, its electron density is consistent with carbon, nitrogen, or oxygen (Howard and Rees 2006).

A key challenge in understanding the nitrogenase mechanism is unraveling how the associated Fe-containing cofactors and nucleotide hydrolysis are linked to electron/proton transfer and N_2 binding and reduction. The cascade of events is initiated by interaction of the ATP-bound form of the Fe protein, having its [4Fe-4S] cluster in the reduced state, with the MoFe protein (Seefeldt et al. 2009). This interaction occurs such that the [4Fe-4S] cluster of the Fe protein is placed within close proximity to the P cluster, which is located near the surface of an $\alpha\beta$-unit interface of the MoFe protein. The protein-protein docking event is proposed to result in stimulation of electron delivery from the P cluster to the FeMo-cofactor, which is buried within the α-subunit of the MoFe protein. These events are controlled by large protein conformational changes and end in a transiently oxidized

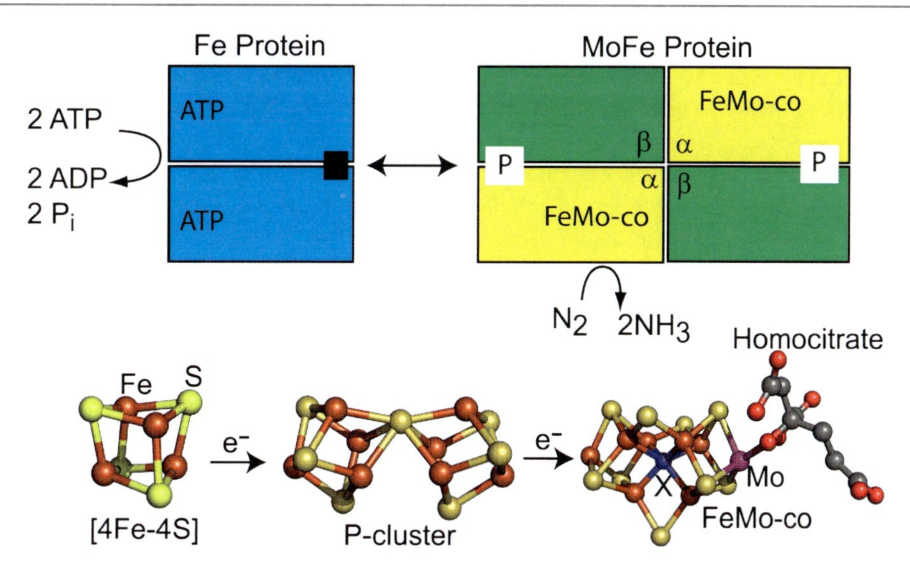

Iron-Sulfur Cluster Proteins, Nitrogenases, Fig. 1 Mo-containing nitrogenase and its associated cofactors. Mo-containing nitrogenase is comprised of two component proteins, called the Fe protein (*left*) and the MoFe protein (*right*). The Fe protein is a homodimer (represented by *blue boxes*) with two ATP-binding sites and a single [4Fe-4S] cluster (represented as a *black box* and in ball-and-stick form below). The MoFe protein is a $\alpha_2\beta_2$-tetramer (α subunit represented as a *yellow box* and ß subunit represented as a *green box*). The MoFe protein contains two P clusters (represented as a *white box* and in ball-and-stick form below), one in each αß pair, and two FeMo-cofactors, one in each α-subunit. For the ball-and-stick structures (*below*), Fe is shown in rust and sulfur in *yellow*. FeMo-cofactor also contains Mo (*magenta*), R-homocitrate (O in *red* and C in *gray*), and the atom X of unknown identity located in the middle of the central portion (*blue*)

P cluster and an FeMo-cofactor reduced by one electron relative to its resting state. This intramolecular electron transfer event is relatively slow compared to the re-reduction of the oxidized P cluster, which occurs through rapid intermolecular electron delivery from the Fe protein. In aggregate, these events ultimately result in the accumulation of one electron within the FeMo-cofactor and dissociation of the oxidized Fe protein. The dissociated, oxidized Fe protein can be readied for subsequent electron delivery by re-reduction of its [4Fe-4S] cluster by ferredoxin, flavodoxin, or a small molecule reductant (e.g., dithionite, in vitro) and exchange of ADP by ATP.

The above sequence of events only describes a mechanism for the delivery of a single electron to the FeMo-cofactor. However, nitrogenase catalysis involves accumulation of multiple electrons and protons. A current model suggests that such accumulation occurs in the following way (Hoffman et al. 2009). The delivery of a second electron to the FeMo-cofactor, using the same mechanism as described above, is coupled to formation of a hydride that is bridged between two Fe atoms located within the central portion of the FeMo-cofactor. This event returns the FeMo-cofactor to its resting oxidation state, readying it to receive additional electrons and protons. A proton, possibly bound to a bridging sulfur atom within the FeMo-cofactor, must also be available to balance electron accumulation. After the system has accumulated four electrons as two hydrides, it is now primed for the binding of N_2, which displaces one hydride and one proton to release H_2. Current evidence supports binding of N_2, or several nonphysiological substrates (e.g., acetylene), to one or more Fe atoms in the FeMo-cofactor. Subsequent reduction steps progressively reduce the bound N_2 through intermediate states, including those that are equivalent to diazene and hydrazine, ultimately resulting in the release of two NH_3 molecules. The overall reaction involves eight rounds of Fe protein and MoFe protein association and dissociation, resulting in the hydrolysis of a minimum of 16 ATP, the evolution of one H_2, and the formation of two NH_3.

Regulation of Nitrogenase Expression and Activity

In addition to the catalytic mechanism, there are a number of other aspects associated with the physiological capacity to fix N_2. One of these involves regulation of the expression and activity of nitrogenase. Cellular regulation of the formation and activity of nitrogenase is important because the reaction is energy intensive, it is extremely sensitive to inactivation by O_2, and it is not necessary when a supply of fixed nitrogen, such as NH_3, is already available. Different N_2-fixing microorganisms have evolved a wide variety of mechanisms to modulate the expression or activity of nitrogenase depending on the particular ecological niche they occupy (Dixon and Kahn 2004). For example, some microorganisms are able to protect nitrogenase from O_2 by producing the enzyme only when growing under anoxic conditions. Symbiotic nitrogen fixing bacteria, such as *rhizobia*, protect nitrogenase from O_2 by living in specialized structures, called nodules, produced by their plant partners (e.g., soybeans). Some oxygen evolving photosynthetic bacteria, such as *cyanobacteria*, protect nitrogenase by only producing the enzyme in oxygen impermeable cells called heterocysts.

In addition to protection of nitrogenase from inactivation by O_2 through exploitation of a particular physiology, cellular differentiation, or compartmentalization, N_2-fixing microorganisms have also evolved highly complex networks of regulatory circuits that control nitrogenase expression at the gene transcription level (Dixon and Kahn 2004). Such regulation depends on cellular energy status, the availability of fixed nitrogen, and the availability of transition metals that are necessary for formation of the metal-containing cofactors.

Nitrogenase Metal Cluster Assembly

The electron transfer and substrate binding properties of the nitrogenases depends on the three Fe-S cofactors contained within the protein partners. One of these cofactor types, the [4Fe-4S] cluster found in the Fe protein, is also present in a wide variety of electron carrier proteins and enzymes involved in other cellular processes, such as respiration, photosynthesis, and carbon metabolism. Although [4Fe-4S] clusters can readily form spontaneously, this does not occur in living cells because free Fe and S are highly toxic. Instead, [4Fe-4S] clusters are formed through the coordinated function of a protein that traffics activated S in the form of persulfide, derived from L-cysteine, to another Fe-binding protein upon which Fe-S clusters are assembled (Dos Santos and Dean 2011). Preassembled Fe-S clusters are then available for delivery to other proteins that require the cluster for electron transfer or catalysis. Although the same mechanism is essentially involved in the activation of the nitrogenase Fe protein as is found for other [4Fe-4S] cluster-containing proteins, specialized S delivery and Fe-S cluster assembly proteins are used for activation of the Fe protein. The need for specialized proteins is probably related to the fact that, because of their inefficient catalytic activities, nitrogenase proteins are expressed at very high levels in N_2 fixing cells. As a result, N_2-fixing cells require a boost in the capacity for [4Fe-4S] cluster formation, which is provided by these specialized assembly components.

In contrast to the formation of [4Fe-4S] clusters, which are commonly found in many proteins not associated with N_2 fixation, both the P cluster and the FeMo-cofactor are unique to the process of N_2 fixation. As a result, there are unique biosynthetic pathways for P cluster and FeMo-cofactor formation. A common feature of these cofactors is that they are topologically quite similar to each other and can be considered as being constructed from [4Fe-4S] cluster building blocks. However, the two cofactors are distinguished from one another in that current evidence indicates that assembly of the P cluster is completed within the MoFe protein and results from the fusion of [4Fe-4S] units, whereas the FeMo-cofactor is separately assembled intact and then inserted into the MoFe protein (Rubio and Ludden 2008; Hu et al. 2008).

Although a number of details of FeMo-cofactor assembly remain elusive, certain principles have emerged. Assembly is initiated through the same type of S mobilization and Fe acquisition mechanisms as found for all other Fe-S cluster-containing proteins. These basic building blocks are then processed through a series of steps that increase the complexity of the Fe-S core, insert the central atom X located within that core, insert a distal metal atom (Mo, V or Fe), and form and attach the organic constituent, *R*-homocitrate. These events also involve a series of carrier proteins, which provide platforms for the sequential formation of the FeMo-cofactor and its ultimate delivery for activation of the MoFe protein.

Summary

Biological nitrogen fixation is a complex process that requires the cooperation of a large suite of proteins working together to deliver electrons, assemble Fe-containing cofactors, and couple ATP hydrolysis so as to achieve the demanding reduction of N_2 to ammonia, a process upon which life depends.

Cross-References

- ▶ Iron, Physical and Chemical Properties
- ▶ Iron-Sulfur Cluster Proteins, Ferredoxins
- ▶ Molybdenum and Ions in Living Systems
- ▶ Molybdenum and Ions in Prokaryotes
- ▶ Molybdenum Cofactor, Biosynthesis and Distribution
- ▶ Molybdenum in Biological Systems

References

Beinert H, Holm RH, Münck E (1997) Iron-sulfur clusters: nature's modular, multipurpose structures. Science 277:653–659

Burgess BK, Lowe DJ (1996) Mechanism of molybdenum nitrogenase. Chem Rev 96:2983–3012

Dixon R, Kahn D (2004) Genetic regulation of biological nitrogen fixation. Nat Rev Microbiol 2:621–631

Dos Santos PC, Dean DR (2011) Co-ordination and fine-tuning of nitrogen fixation in *Azotobactervinelandii*. Mol Microbiol 79:1132–1135

Eady RR (1996) Structure – function relationships of alternative nitrogenases. Chem Rev 96:3013–3030

Hoffman BM, Dean DR, Seefeldt LC (2009) Climbing nitrogenase: toward a mechanism of enzymatic nitrogen fixation. Acc Chem Res 42:609–619

Howard JB, Rees DC (2006) How many metals does it take to fix N_2? A mechanistic overview of biological nitrogen fixation. Proc Natl Acad Sci USA 103:17088–17093

Hu Y, Fay AW, Lee CC, Yoshizawa J, Ribbe MW (2008) Assembly of nitrogenaseMoFe protein. Biochemistry 47:3973–3981

Rubio LM, Ludden PW (2008) Biosynthesis of the iron-molybdenum cofactor of nitrogenase. Annu Rev Microbiol 62:93–111

Seefeldt LC, Hoffman BM, Dean DR (2009) Mechanism of Mo-dependent nitrogenase. Annu Rev Biochem 78:701–722

Smil V (2004) Enriching the earth: Fritz Haber, Carl Bosch, and the transformation of world food production. MIT Press, Cambridge

Isomerases

▶ Cobalt-containing Enzymes

ITC, Isothermal Titration Calorimetry

▶ Zinc Aminopeptidases, Aminopeptidase from Vibrio Proteolyticus (Aeromonas proteolytica) as Prototypical Enzyme

K

K, *Kalium* (Latin)

▶ Potassium, Physical and Chemical Properties

K^+ Dependent Proteins

▶ hERG (KCNH2) Potassium Channel, Function, Structure and Implications for Health and Disease
▶ Potassium in Biological Systems

K^+-Dependent Na^+/Ca^{2+} Exchanger

▶ Potassium Dependent Sodium/Calcium Exchangers

K^+-Dependent Proteins

▶ Potassium-Binding Site Types in Proteins

Kalium (K^+)

▶ Potassium in Health and Disease

KChIP

▶ Calcium, Neuronal Sensor Proteins

Keshan Disease

▶ Selenium and Muscle Function

Kininase II

▶ Angiotensin I-Converting Enzyme

V.N. Uversky et al. (eds.), *Encyclopedia of Metalloproteins*, DOI 10.1007/978-1-4614-1533-6,

L

Labeling, Human Mesenchymal Stromal Cells with Indium-111, SPECT Imaging

Rasmus Sejersten Ripa[1,2], Mandana Haack-Sørensen[1], Jens Kastrup[1,3] and Annette Ekblond[1]
[1]Cardiology Stem Cell Laboratory, Rigshospitalet University Hospital, Copenhagen, Denmark
[2]Cluster for Molecular Imaging and Department of Clinical Physiology, Nuclear Medicine and PET, Rigshospitalet University Hospital, Copenhagen, Denmark
[3]The Heart Centre, Cardiac Catheterization Laboratory, Rigshospitalet University Hospital, Copenhagen, Denmark

Synonyms

Cell labeling techniques; Cell tracking; Magnetic resonance cell imaging; Radionuclide cell imaging; Reporter gene cell tracking; Visualization method

Definition

Indium-111 labeling is a technique for determining the presence and location of cells after injection into human tissues.

Aim of Cell Labeling

Stem cells are theoretically ideal for tissue regeneration in ischemic heart disease. This field has been dominated by a very quick translation from bench to bed where most clinical trials of cell therapy have had a pragmatic design with intracoronary infusion of autologous bone marrow-derived mononuclear cells. The cell processing procedure and method of infusion often vary between research groups. Additionally, choice of endpoint assessment in these clinical trials are often determined by local accessibility and expertise rather than scientific rational which has led to a number of clinical trials with divergent results. This has opened for a reverse translation from bedside to bench to clarify some of the unknown factors such as optimal cell type and number, optimal time of therapy, optimal patient selection, usefulness of repeated or combined treatments, etc.

Cell imaging after treatment has a pivotal role in this translation. Imaging based cell tracking can potentially elucidate issues regarding cardiac homing, engraftment and growth of cells following transplantation. Furthermore, identification of cell redistribution to other organs where potential side effects may take place is important.

Methods for Clinical Imaging of Transplanted Cells

The ideal imaging technique should allow for serial tracking in humans for a prolonged period of time with high spatial resolution and with the capability of tracking a few cells without affecting the cells or the organ. It is important that the marker remains in the viable cell but is quickly cleared from the tissue upon cell death.

V.N. Uversky et al. (eds.), *Encyclopedia of Metalloproteins*, DOI 10.1007/978-1-4614-1533-6,

No single modality meets all of the demands, and various modalities can thus be used for cell imaging such as radionuclide, optical, magnetic resonance, and ultrasound. The most effective imaging strategy must be determined in each case considering the need for high spatial resolution, sensitivity, or time of follow-up.

Cell Imaging Using Single Photon Emission Tomography (SPECT)

SPECT is a tomographic imaging technique using gamma-emitting radioisotope (radionuclides) with a three-dimensional acquisition allowing for free reconstruction of imaging planes. The image is often fused with a computed tomography (CT) scan to identify anatomical structures (Fig. 1). The advantages of SPECT for cell tracking are high sensitivity (picomolar) allowing for trace amounts of radionuclide probes, high tissue penetration, reasonable acquisition time, high accessibility, and translational potential (can be used in both animal and human trials). The disadvantages are the radiation exposure that can affect both the labeled cells and the patients, and the limited spatial resolution (about 10 mm in clinical SPECT and a factor 10 less in preclinical microSPECT).

Radionuclides for SPECT

A number of radionuclides can theoretically be used for stem cell labeling, the most commonly used are ^{111}In and ^{99m}Tc. The radiolabeling techniques were originally developed for imaging and quantification of blood cells, but have recently been applied to stem cell labeling for in vivo imaging. The main difference between ^{111}In and ^{99m}Tc are the half-life and energy. ^{99m}Tc has a short half-life (6 h) and a photon energy (140 KeV) optimal for SPECT. The short half-life allows for higher dose which result in improved image quality, but at the expense of a very limited time window for cell tracking (up to approximately 24 h).

^{111}In is a radiotracer with a half-life of 2.8 days, and a photon energy of 171/245 keV. Imaging of leukocyte distribution by gamma camera imaging using ^{111}In-tropolone or oxine is a safe clinical routine procedure used to diagnose bleeding or inflammation. The main advantage of stem cell labeling with ^{111}In is the long half-life, allowing in vivo tracking of the cells for up to 1 or 2 weeks.

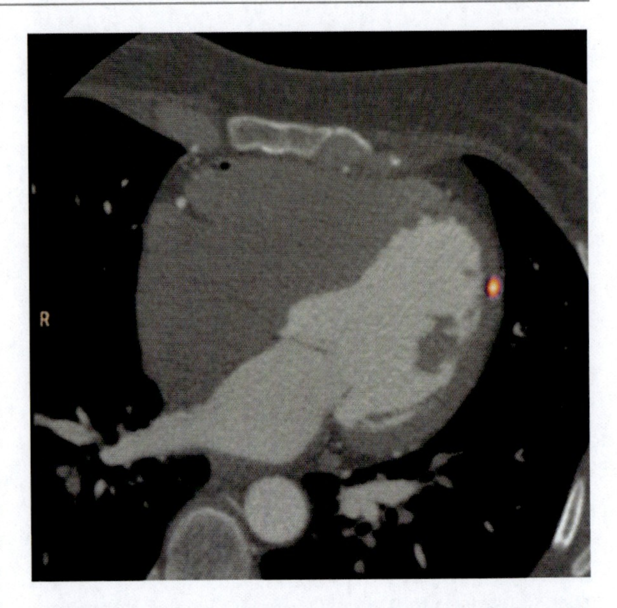

Labeling, Human Mesenchymal Stromal Cells with Indium-111, SPECT Imaging, Fig. 1 Animated fusion of CT image (*black* and *white*) with anatomical details of the heart and SPECT image (*color*) marking the radioactive tracer

Labeling of Mesenchymal Stromal Cells (MSC) for Clinical SPECT

A prerequisite for the use of ^{111}In labeling of MSC is that labeling does not interfere with cell viability and function. ^{111}In decays by a mechanism of electron capture and a two-peak gamma emission as well as an electron ejection process (Auger electrons). Auger electrons especially, pose considerable damage to labeled cells. Radiation damage on cells increases with the activity of the radionuclide and time, therefore the combination of radioactive labeling dose and incubation time has to be considered. Bindslev et al. (2006) recommend a dose of 30 Bq ^{111}In per cell and an incubation time of 20 min for primary human bone marrow–derived MSC labeling. Gildehaus et al. (2011) suggest only 10 Bq per cell for 20 min using an immortalized human MSC cell line. Activity uptake was 25% and 26%, respectively – but the labeling regime has to be addressed in each particular setup.

When performing ^{111}In labeling of MSC in vitro the aim is to reach a high level of uptake in each cell without cell damage. Also labeling has to be stabile, which means that it is bound inside the cell and does not leak to the extracellular environment over time. To reach a high level of ^{111}In uptake a chelator is used to enhance diffusion through the cell membrane. Traditional chelators are tropolone and oxine. Tropolone and oxine both create neutral and symmetrical 3:1 complexes with ^{111}In with a hydrophobic surface, which facilitates diffusion through the lipid bilayer of the cell membrane. Once inside the cell cytoplasm the complexes dissolve. ^{111}In is bound to macromolecules in the cell and tropolone or oxine diffuses out of the cell. The in vitro labeling environment affects ^{111}In uptake aided by tropolone or oxine as the presence of, e.g., metal ions or transferrin in serum or plasma competes with tropolone and oxine binding.

The MSC population itself has variable sensitivity to ^{111}In induced cell damage. A population of MSC is heterogeneous in spite of a homogeneous phenotype. Radiation damage is more pronounced in dividing cells than quiescent cells, and therefore differences in MSC origin and processing, culture passage and confluence, etc., should be considered when labeling cells with a radioisotope.

In short, the following parameters will affect both labeling efficiency and cell damage and have to be considered and optimized in each particular setup when labeling MSC with^{111}In: radionuclide activity, labeling time, choice of chelator, formulation of labeling medium, presence of confounding factors like serum and plasma and, not the least, the constitution of the MSC population. The detrimental consequences of not optimizing mentioned parameters is either inefficient labeling, acute cell death during labeling or accelerated senescence during in vitro culture/in vivo engraftment due to persistent Auger electron mediated DNA damage.

In order to ascertain that ^{111}In labeling does not hamper MSC regenerative capacity, a number of in vitro assays comparing with unlabeled cells, should be established. First, morphologic evaluation and a simple measure of MSC viability should be introduced. A variety of automated cell counting tools are available, but simple trypan blue exclusion is valid. The ability to grow/proliferate should be determined. Classical assays like BrdU (bromodeoxyuridine) incorporation measured by a colorimetric ELISA technique, indirect methods measuring mitochondrial activity with MTT (3-(4,5-Dimethylthiazol-2-yl)-2,5-diphenyltetrazolium bromide) tetrazolium salts or water soluble analogues, or the assessment of doubling time identified by incorporation of lipophilic fluorescent cell linker dyes and flow cytometry are a few among many options. Bindslev et al. (2006) found that previously mentioned labeling regime did not affect MSC population doubling time. To resolve the ability of MSC to home/migrate to a defect site, motility and a chemotaxis response should be assayed with a scratch assay or a Boyden chamber. Gildehaus et al. (2011) performed a scratch assay and found that motility was delayed in ^{111}In labeled MSC.

Working with MSC it is essential to ascertain that a fundamental stem cell character is maintained after ^{111}In labeling. MSC are characterized according to a particular set of criteria provided by *The International Society for Cellular Therapy* (ISCT) (Dominici et al. 2006). According to ISCT MSCs are plastic adherent, they express CD90 (Thy-1), CD73 (ecto 5 nucleotidase), CD105 (endoglin) surface markers, whereas they are negative for the hematopoietic and leukocyte markers CD34, CD45, CD14 or CD11b, CD79a or CD19, and HLA-DR. In addition, MSC have the capacity to differentiate to adipocytes, chondroblasts, and osteoblasts. Expression of the ISCT panel of markers after ^{111}In labeling can be verified by either immunocytochemistry or flow cytometry. Trilineage differentiation capacity can be demonstrated by culture in three special differentiation media followed by Oil Red O (adipocytes), Alizarin Red (osteoblasts), and Alcian Blue (chondroblasts) stainings. Bindslev et al. (2006) and Gildehaus et al. (2011) have addressed different aspects of MSC stem cell character, but both studies conclude that MSC stem cell character is intact after ^{111}In labeling.

Radionuclide retention should also be determined during culture to ascertain that cells are traceable for the desired period of time. Measuring gamma activity in the cell culture supernatant as a percentage of the activity in the cell monolayer illustrates leakage. Decay of ^{111}In within the cell monolayer should be accounted for.

Based on current in vitro data, efficient ^{111}In labeling of MSC without loss of vitality, proliferative capacity or stem cell character can be obtained. This has lead to the next step; verification in animal studies.

Experience with In Vivo ^{111}In Imaging of Stem- or Progenitor Cells

Several studies in pigs, dogs, and rats have used ^{111}In labeling of both MSC and endothelial progenitor cells and subsequent in vivo tracking of the radioactivity as a surrogate measure of cell engraftment and migration.

The Frankfurt group headed by Dr. Dimmeler performed two experiments with ^{111}In labeling of endothelial progenitor cells (Aicher et al. 2003) and hematopoietic cells (Brenner et al. 2004) in rats almost 10 years ago. The group found that the labeling procedure did not affect viability, proliferation, or migratory capacity of the endothelial progenitor cells whereas the less differentiated circulating hematopoietic progenitor cells were negatively affected resulting in complete impairment of proliferation and differentiation. In addition though, the group found evidence that radiolabeling with ^{111}In-oxine is a feasible in vivo method for monitoring the transplanted cells in a rat myocardial infarction model.

In a similar study Chin et al. (2003) labeled MSC with ^{111}In-oxine and infused the cells intravenously in two pigs after experimental myocardial infarction. The pigs were followed with SPECT imaging for 2 weeks. The authors found very high accumulation of ^{111}In in liver, spleen, bone marrow, and especially the lungs. It was concluded that radiolabeling of MSC is feasible and the in vivo SPECT imaging provided a noninvasive method for sequentially monitoring cell trafficking. The trial also revealed that intravenous infusion is not the optimal route of MSC administration since the vast majority of the cells were trapped in the lungs.

Mitchell et al. (2010) labeled endothelial progenitor cells with ^{111}In and injected the cells in a canine myocardial infarction model. Using this labeling method and SPECT follow-up of the animals it was demonstrated that cell retention following endocardial and epicardial injection were similar. This information has clinical impact since epicardial injections would require thoracotomy in the patients whereas endocardial injections can be performed via a minimal invasive left ventricular ejection catheter. Using the same cell type and canine model Sabondjian et al. (2012) recently demonstrated that hybrid SPECT/perfusion CT of ^{111}In labeled cells can localize the cells in relation to the infarcted myocardium after intramyocardial injection.

Of note, some of these animal studies indicated that there was a significant leak of radioactivity from the cells into surrounding tissue after transplantation. This would potentially confound the quantification and positive identification of the labeled cells.

In one human (Schächinger et al. 2008) trial ^{111}In labeled progenitor cells were infused into the coronary artery in patients after acute myocardial infarction ($N = 20$). One hour after infusion of progenitor cells, a mean of $6.9 \pm 4.7\%$ of total radioactivity was detected in the heart. Radioactivity remained in the heart after 3–4 days, indicating homing of progenitor cells to the myocardium.

Besides its use in cardiology, ^{111}In labeling of MSC has been used to determine homing after acute brain trauma in a rat model (Yoon et al. 2010) and homing to the liver after intravenous infusion in patients with advanced cirrhosis (Gholamrezanezhad et al. 2011).

Limitations with SPECT Imaging of In-Labeled Cells

It is imperative to acknowledge that SPECT only image the labeling compound (^{111}In). Therefore, in vivo tracking of the radioactivity is only a surrogate measure of cell engraftment and migration.

Lyngbaek et al. have in a recent trial (Lyngbaek et al. 2010) investigated whether the biodistribution and retention of ex vivo cultured MSC can be determined after direct percutaneously intramyocardial transplantation in a large animal model by ^{111}In-tropolone radiolabeling of human MSC. Labeled MSC were first transplanted into six pigs by trans-endocardial percutaneous injections. The ^{111}In activity in the heart was 35% (±11%) of the total activity in the pig 1 h after injection of viable ^{111}In labeled MSC, compared to only $6.9 \pm 4.7\%$ after intracoronary infusion and from $11.3 \pm 3\%$ to $20.7 \pm 2.3\%$ after trans-epicardial injection. SPECT imaging identified the ^{111}In within the myocardium corresponding to the locations of the intramyocardial injections. Whole body scintigraphy revealed focal indium accumulations in the cardiac region up to 6 days after injection (Fig. 2). The injected myocardium was analyzed by fluorescence in situ hybridization (FISH) and microscopy after euthanization of the animals. No human MSCs were identified with FISH, and microscopy identified widespread necrosis and acute inflammation. Two additional pigs were injected with ^{111}In-tropolone (without cells) to test the clearance of ^{111}In from the myocardium. A last pig was injected with dead ^{111}In labeled cells. The clearance of radioactivity of injected dead cells and of ^{111}In alone appeared

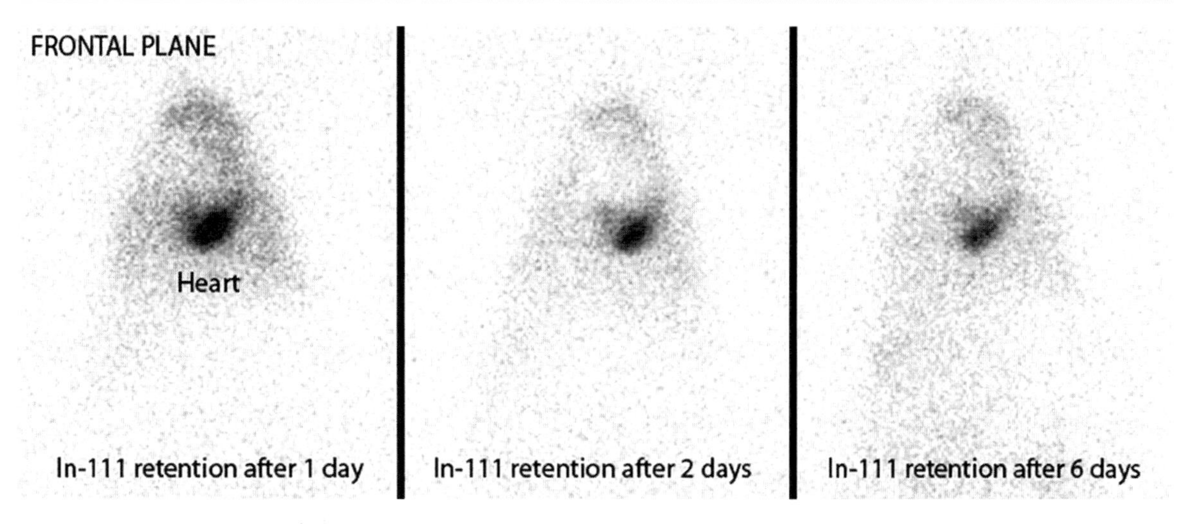

Labeling, Human Mesenchymal Stromal Cells with Indium-111, SPECT Imaging, Fig. 2 ^{111}In retention for up to 6 days after intramyocardial injection of labeled mesenchymal stromal cells in a pig

faster initially compared to that of viable cells, but retention after injection of viable cells, dead cells and ^{111}In followed a very similar pattern.

The results of this trial were potentially biased by the inclusion of a few animals in a prospective design. However, they did consistently in six animals observe intense radioactivity despite disappearance of the cells. A very high specificity (close to 100%) should be demanded of this labeling method for human trials, and the results are in conflict with this. In conclusion, Lyngbaek et al. found evidence that radioactivity from ^{111}In-labeled cells stays in the myocardium for a long time despite the disappearance of transplanted cells; clinical use of ^{111}In-labeled cells for monitoring of MSC in the *human* heart seems problematic unless viability can be determined by another method. This limitation is not a major issue in animal-trials where the presence and viability of the injected labeled cells can be determined by ex vivo methods after sacrifice of the animals.

Iron-oxide labeling for magnetic resonance imaging (MRI) tracking of injected cells has appeared as a suitable alternative to indium labeling, with the possibility of even longer follow-up. However, recent results very similar to Lyngbaek et al. using iron labeling have been reported. After 3–4 weeks no transplanted cells were detected. Instead a continued enhanced magnetic resonance signal was found from cardiac macrophages that engulfed the labeling particles suggesting that iron-oxide labeling is also an unreliable marker for monitoring cell survival and migration (Amsalem et al. 2007). This issue however, remains debatable among investigators.

Blackwood et al. has proposed a new quantitative in vivo method to assess survival of transplanted cells in the myocardium using SPECT and ^{111}In labeling (Blackwood et al. 2009). However, at present this method has not been validated by other.

Future Perspectives

Reporter gene imaging using clinical PET is a promising modality especially for long-term tracking. The reporter gene is expressed only in living cells and passes on to daughter cells during proliferation. The reporter gene is expected to be lost after cell death providing a very specific signal. This approach allows for repetitive imaging with PET/CT or PET/MRI over months to years and thus monitoring of cell trafficking, targeting, and replication. The biggest obstacle in reporter gene imaging is the need for genetic modification of the cells with potential issues of oncogenicity.

It is to be expected that the use of PET-based reporter gene imaging of stem cells will increase in the next years, although the translation of the reporter gene imaging technology into human clinical trials still needs more profound preclinical evaluation.

Cross-References

- ▶ Indium, Physical and Chemical Properties
- ▶ Technetium, Physical and Chemical Properties
- ▶ Thallium-201 Imaging

References

Aicher A, Brenner W, Zuhayra M et al (2003) Assessment of the tissue distribution of transplanted human endothelial progenitor cells by radioactive labeling. Circulation 107: 2134–2139

Amsalem Y, Mardor Y, Feinberg MS et al (2007) Iron-oxide labeling and outcome of transplanted mesenchymal stem cells in the infarcted myocardium. Circulation 116: I38–I45

Bindslev L, Haack-Sørensen M, Bisgaard K et al (2006) Labelling of human mesenchymal stem cells with indium-111 for SPECT imaging: effect on cell proliferation and differentiation. Eur J Nucl Med Mol Imaging 33:1171–1177

Blackwood KJ, Lewden B, Wells RG et al (2009) In vivo SPECT quantification of transplanted cell survival after engraftment using (111)In-tropolone in infarcted canine myocardium. J Nucl Med 50:927–935

Brenner W, Aicher A, Eckey T et al (2004) 111In-labeled CD34+ hematopoietic progenitor cells in a rat myocardial infarction model. J Nucl Med 45:512–518

Chin BB, Nakamoto Y, Bulte JW et al (2003) 111In oxine labelled mesenchymal stem cell SPECT after intravenous administration in myocardial infarction. Nucl Med Commun 24:1149–1154

Dominici M, Le BK, Mueller I et al (2006) Minimal criteria for defining multipotent mesenchymal stromal cells. The International Society for Cellular Therapy position statement. Cytotherapy 8:315–317

Gholamrezanezhad A, Mirpour S, Bagheri M et al (2011) In vivo tracking of 111In-oxine labeled mesenchymal stem cells following infusion in patients with advanced cirrhosis. Nucl Med Biol 38:961–967

Gildehaus FJ, Haasters F, Drosse I et al (2011) Impact of indium-111 oxine labelling on viability of human mesenchymal stem cells in vitro, and 3D cell-tracking using SPECT/CT in vivo. Mol Imaging Biol 13:1204–1214

Lyngbaek S, Ripa RS, Haack-Sorensen M et al (2010) Serial in vivo imaging of the porcine heart after percutaneous, intramyocardially injected 111In-labeled human mesenchymal stromal cells. Int J Cardiovasc Imaging 26: 273–284

Mitchell AJ, Sabondjian E, Sykes J et al (2010) Comparison of initial cell retention and clearance kinetics after subendocardial or subepicardial injections of endothelial progenitor cells in a canine myocardial infarction model. J Nucl Med 51:413–417

Sabondjian E, Mitchell AJ, Wisenberg G et al (2012) Hybrid SPECT/cardiac-gated first-pass perfusion CT: locating transplanted cells relative to infarcted myocardial targets. Contrast Media Mol Imaging 7:76–84

Schächinger V, Aicher A, Dobert N et al (2008) Pilot trial on determinants of progenitor cell recruitment to the infarcted human myocardium. Circulation 118:1425–1432

Yoon JK, Park BN, Shim WY et al (2010) In vivo tracking of 111In-labeled bone marrow mesenchymal stem cells in acute brain trauma model. Nucl Med Biol 37:381–388

Laccases

Thierry Tron
iSm2/BiosCiences UMR CNRS 7313, Case 342,
Aix-Marseille Université, Marseille, France

Synonyms

p-diphenol oxidase

Definition

Laccases (EC 1.10.3.2) are copper containing oxidases found in plants, fungi, and microorganisms. They catalyze the oxidation of diversely substituted aromatic substrates (e.g., phenols, anilines) with a concomitant reduction of dioxygen to water. A one-electron substrate oxidation occurs at a mononuclear copper site, while a triatomic copper cluster is responsible for the four-electron reduction of O_2. The overall outcome of the catalytic cycle is the reduction of one molecule of dioxygen into two molecules of water, coupled with the oxidation of four substrate molecules into four radicals that can form dimers, oligomers, and polymers (Scheme 1).

General Background

Since a description of the factor catalyzing the rapid hardening of latex from the Japanese lacquer trees (*Rhus sp.*) exposed to air at the end of the nineteenth century, laccases from different origins (plants, fungi, bacteria) have been continuously discovered and extensively studied. Hundreds of laccases encoding genes are now known, and crystal structures of laccases from 12 different organisms (one from bacterial and 11 from fungal origin) are currently available (www.rcsb.org/pdb/home/home.do).

$$4Sub \xrightarrow{} 4Sub^{\bullet} + 4H^+$$
$$O_2 + 4H^+ + 4e^- \xrightarrow{laccase} 2H_2O$$

Laccases, Scheme 1 Reaction catalyzed by laccases

Physiological roles of laccases encompass lignification (plant), delignification (fungi), oxidative stress management, morphogenesis, and virulence (Clauss 2004). Originally diverse, laccases are able to oxidize a disparate range of natural substrates: phenols, aromatic amines, and metal ions. Functionally diverse, robust, and environmentally friendly catalysts, laccases are largely studied for their potential uses in industrial processes (Gianfreda et al. 1999) and increasingly studied as catalysts in organic synthesis applications (Riva 2006). Molecular evolution techniques and overexpression of laccase variants in appropriate heterologous hosts (i.e. filamentous fungi) are used to produce tailored biocatalysts (Rodgers et al. 2010).

Structural, Spectroscopic, and Functional Features

Laccases are usually described as monomeric enzymes although oligomerization may have some functional relevance. The first structure of laccase has been obtained from a partially copper-depleted enzyme from the white-rot fungus *Coprinopsis cinerea* in 1998 (Ducros et al. 1998). Closely related to the plant ▶ ascorbate oxidase, the human ceruloplasmin and the yeast (Fet3) or bacterial (CueO) metal oxidases, laccases form with these enzymes the multicopper oxidases family.

Redox Sites

The surface located substrate oxidation center of laccases contains a blue ▶ copper-binding site (or T1) in which the copper ion is coordinated by the side chains of two histidines and one cysteine (in some cases, a methionine ligand occupies a fourth coordination position). This site exhibits an intense Cys(S) → Cu^{II} charge transfer (CT) band in its absorption spectrum ($\varepsilon \approx$ 5,000–6,000 M^{-1} cm^{-1} at about 600 nm) resulting in the intense blue color of the oxidized enzyme and a very small parallel hyperfine coupling constant ($A_{\parallel}$) in its electron spin resonance (ESR) spectrum. The source of these unique spectral features is the highly covalent Cu–S(Cys) π bond (Solomon et al. 1996). The redox potential of the blue copper site of laccases is comprised between 0.4–0.5 V (low-potential laccases) and 0.7–0.8 V (high-potential laccases) *versus* the normal hydrogen electrode (NHE). The dioxygen reduction center consists in a trinuclear copper site (TNC) structured around a pair of antiferromagnetically coupled copper ions coordinated to histidines (T3 copper site) and a histidines/H_2O coordinated copper ion (T2 copper site). The oxidized T3 copper site (similar to that present in ▶ Catechol Oxidase and Tyrosinase) is characterized by a bridging hydroxide to Cu^{II} CT transition at 330 nm ($\varepsilon \approx$ 5,000 M^{-1} cm^{-1}). The oxidized T2 copper site exhibits spectral features reminiscent of tetragonal cupric complexes (i.e. no detectable transition in the UV/VIS region and a "normal" $A_{\parallel}$ coupling constant in its ESR spectrum) (Solomon et al. 1996). Data on redox potential of the copper sites of the TNC are relatively rare, but both the T2 and the T3 coppers are apparently operating at potentials close to that of the T1 copper site. Substrate oxidation and oxygen reduction sites are located *ca.* 13 Å away from each other and are connected through a histidine-cysteine-histidine tripeptide (H-C-H) involved in the coordination of the metal ions (Fig. 1a).

Domains

The scaffold of laccases consists in a three times repeat of a homologous domain that shares distant homology to the single-domain ▶ monocopper blue proteins cupredoxins (Fig. 1b). The overall structure of cupredoxins (e.g., azurin, ▶ plastocyanin, rusticyanin) consists of β-strands arranged into two β-sheets forming a Greek key β-barrel structure. The evolutionary path from a single cupredoxin domain to a three-domain laccase (D1, D2, D3) is thought to involve a gene duplication and a domain recruitment (Nersissian and Shipp 2002). Evolving from an electron transfer protein to an oxidase, proto-laccase lost unnecessary blue copper sites in D1 and in D2 and acquired a T2–T3 trinuclear copper site mapping at the boundaries of D1 and D3 and substrates binding sites in neoformed clefts. Two cavities accessing the trinuclear copper cluster from the surface of the enzyme are thought to channel in and out O_2 and H_2O molecules.

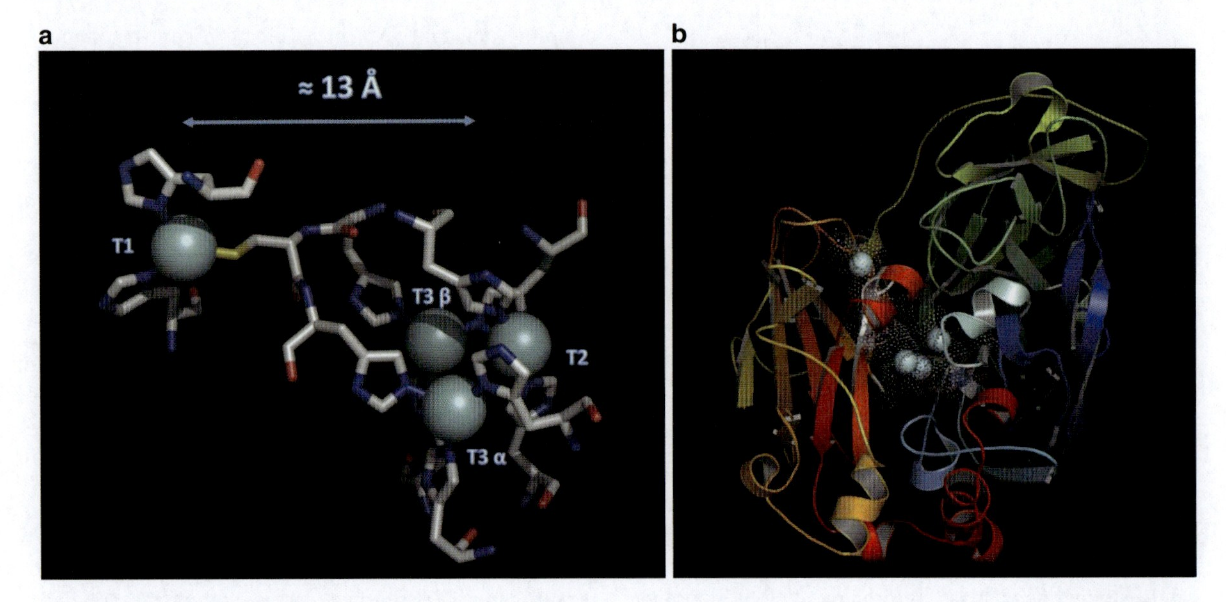

Laccases, Fig. 1 Structural aspects of laccases. (**a**) Structure of copper-binding sites with the tripeptide his-cys-his connecting the substrate oxidation center (T1 copper-binding site) to the dioxygen reduction center (TNC-binding site) and (**b**) overall structure of the enzyme distinguishing the three cupredoxin domains D1, D2, and D3 (counterclockwise), the surface located T1 copper (D3) and the buried TNC (at the interface between D1 and D3) (Views were generated using the coordinates from the Protein Data Bank entry 1HRG)

Substrate Oxidation

Laccases have low substrate specificity. The K_M values for the oxidation of phenols, aryl amines, or aminophenols are generally in the millimolar range. As confirmed by structural data, there is not really a binding pocket for substrate but rather a shallow depression at the bottom of which points the distal histidine coordinating the T1 copper ion. Substrate oxidation is strictly outer sphere and involves a nearby acidic residue for proton abstraction. The bimolecular reaction is fast ($\approx 10^5$–10^6 s^{-1} M^{-1}), and electron transfer (ET) rates can be even faster with nondeprotonable substrates. Subsequently, one electron is transferred from the reduced T1 copper ion (Cu^I) to the oxidized TNC (intramolecular electron transfer).

Dioxygen Reduction

Substantial progresses in understanding the mechanism of the four-electron reduction of O_2 in multicopper oxidases have been obtained since the 1970s, first with the use of various spectroscopic methods, including electronic absorption, ESR, circular dicroism, magnetic circular dicroism, X-ray absorption spectroscopy, resonance Raman spectroscopy (Solomon et al. 1996), and more recently with the obtention of crystallographic data. Mutagenesis techniques commonly available are used to generate laccase mutants (e.g., T1, T2, or T3 copper-depleted mutants) particularly useful for the fine description of the sequential steps leading from O_2 binding to the breakdown of the oxygen-oxygen bond, protonation, and water release. Globally, it is generally assumed that water formation processes through two sequential two-electron reduction steps (Solomon et al. 2008). First, dioxygen reacts with a reduced TNC generating a Cu^{II}–(O_2)–Cu^{II} peroxide species (Fig. 2a) that has been detected both spectroscopically (in T1 copper-depleted enzymes) and in several crystallized laccases. Second, a reductive cleavage of the O-O bond triggers movements of protonated oxygen atoms away from the cluster and neoformed water molecules enter the channels (Fig. 2b). From here, the enzyme can be fully reduced and can bind dioxygen again for a new catalytic cycle (Solomon et al. 2008).

Biotechnological Relevance

Laccases are largely studied for their potential uses in industrial processes. They generally work under mild conditions: room temperature, atmospheric pressure,

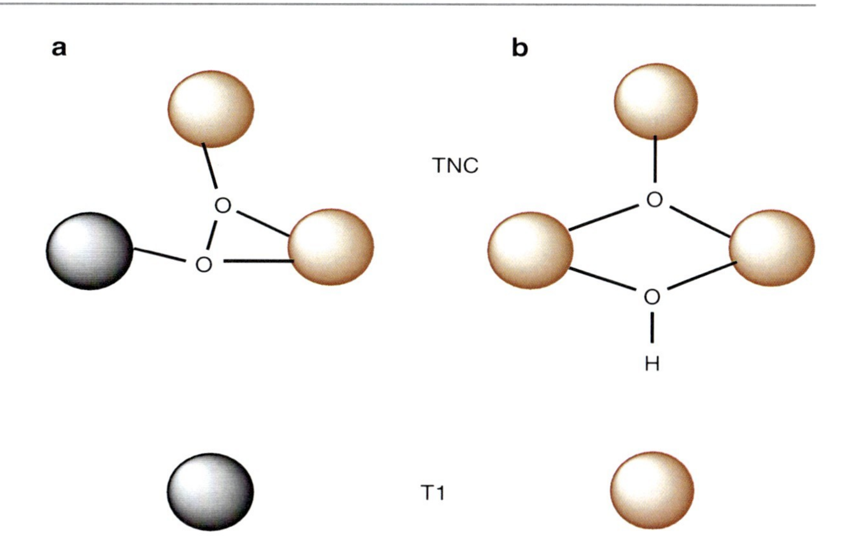

Laccases, Fig. 2 Steps in the dioxygen reduction mechanism by multicopper oxidases (see Solomon et al. 2008 for a review). (**a**) A Cu^{II}–(O_2)–Cu^{II} peroxo species resulting from the two-electron reduction of O_2 by copper ions from the TNC and (**b**) μ_2-hydroxo and μ_3-oxo species resulting from the two-electron reductive cleavage of the oxygen-oxygen bond. *Grey spheres*: reduced (Cu^I) copper atoms; *pale spheres*: oxidized (Cu^{II}) copper atoms

water as solvent, and dioxygen as electron acceptor. In the past decade, in addition to their use in technological and bioremediation processes, a significant number of reports focusing on applications of this ecofriendly enzyme in organic synthesis have been published (Riva 2006). For an industrial use, the current challenge is to obtain both enhanced expression levels and improved laccases with desirable physicochemical characters like a higher redox potential, an optimal activity at neutral or alkaline pH, and thermostability. Strategies to obtain such variants include natural biodiversity screenings and optimization of nature-derived scaffolds (Robert et al. 2011). Directed evolution is an effective strategy to engineer and optimize protein properties (Arnold 2001). Such strategies are particularly successful in achieving improvements in thermostability, altering substrate specificity and improving activity in organic solvents.

Prospect

Despite the fact that fungal laccases have been found to catalyze oxidation of aromatic alcohols to aldehydes, synthesis of substituted imidazoles, dimerization of penicillin or cephalosporin monomers, etc., they are not yet widely used as synthetic tools. This is mainly due to the fact that the phenoxy radical produced by the enzyme is persistent and of diffusible nature and can undergo C-C or C-O coupling in absence of any selectivity control by the enzyme. Regio- and stereoselective phenol coupling is, however, observed in nature, in plant secondary metabolism, in bacteria, in lichen, and in fungi. Introducing such selective properties into laccase would undoubtedly represent a major breakthrough in biocatalysis.

Cross-References

▶ Ascorbate Oxidase
▶ Catechol Oxidase and Tyrosinase
▶ Copper-Binding Proteins
▶ Monocopper Blue Proteins
▶ Plastocyanin

References

Arnold FH (2001) Combinatorial and computational challenges for biocatalyst design. Nature 409:253–257

Clauss H (2004) Laccases: structure, reactions, distribution. Micron 35:93–96

Ducros V, Brzozowski AM, Wilson KS, Brown SH, Ostergaard P, Schneider P, Yaver DS, Pedersen AH, Davies GJ (1998) Crystal structure of the type-2 Cu depleted laccase from *Coprinus cinereus* at 2.2 angstrom resolution. Nat Struct Biol 5:310–316

Gianfreda L, Xu F, Bollag J-M (1999) Laccases: a useful group of oxidoreductive enzymes. Bioremediation J 3:1–25

Nersissian AM, Shipp EL (2002) Blue copper-binding domains. Adv Protein Chem 60:271–340

Riva S (2006) Laccases: blue enzymes for green chemistry. Trends Biotechnol 24:219–226

Robert V, Mekmouche Y, Pailley PR, Tron T (2011) Engineering laccases: in search for novel catalysts. Curr Genomics 12:123–129
Rodgers CJ, Blanford CF, Giddens SR, Skamnioti P, Armstrong FA, Gurr SJ (2010) Designer laccases: a vogue for high-potential fungal enzymes? Trends Biotechnol 28:63–72
Solomon EI, Sundaram UM, Machonkin TE (1996) Multicopper oxidases and oxygenases. Chem Rev 96:2563–2605
Solomon EI, Augustine AJ, Yoon J (2008) O_2 reduction to H_2O by the multicopper oxidases. Dalton Trans 30:3921–3932

α-Lactalbumin

Eugene A. Permyakov
Institute for Biological Instrumentation, Russian Academy of Sciences, Pushchino, Moscow region, Russia

Synonyms

Calcium-binding constant; Holo α-LA apo α-LA

Definition

α-Lactalbumin (α-LA) is a small (Mr 14,200, 123-140 amino acid residues), acidic (pI 4–5), Ca^{2+}-binding protein, one of the major protein components of milk.

Molten globule is a compact state with partially conserved secondary structure and fluctuating tertiary structure.

Fibrillation is a formation of (protein) fibrils.

Whey is the watery part of milk that separates from the curds, as in the process of making cheese.

HAMLET *h*uman α-lactalbumin *m*ade *le*thal to *t*umor cells.

α-LAs belong to a super-gene family that also includes the genes for lysozymes *c* and calcium-binding lysozymes (reviewed by Permyakov 2005). Their genes are made up of four exons separated by three introns. The introns are located in homologous locations in all genes of this family. All but one intron in the six characterized α-LA genes are under 1,000 bases in length. In cow, genes for α-LA exist as a gene cluster, residing on bovine chromosome 5. In human, the gene is on chromosome 12. α-LA is expressed only in the mammary gland of mammals, and therefore only one tissue-specific promoter is required.

α-LA shares only 40% identity in amino acid sequence with lysozyme *c*, but it has a closer spatial structure and gene organization (reviewed by Permyakov 2005). Although structurally similar, functionally they are absolutely distinct. Analysis of phylogenetic trees constructed from amino acid sequences and metal-binding properties of various lysozymes *c* and α-lactalbumins showed that before the divergence of the lineages of birds and mammals, calcium-binding lysozyme diverged from non-calcium-binding lysozyme. α-LA evolved from the calcium-binding lysozyme along the mammalian lineage after the divergence of birds and mammals. Specific amino acid substitutions in α-LA result in the loss of the enzyme activity of lysozyme and the acquisition of the features necessary for its role in lactose synthesis. α-LA acts as a regulatory subunit of galactosyltransferase in lactose synthase, which catalyzes the synthesis of lactose from UDP-galactose and glucose.

Most α-LAs, including those from human, guinea pig, bovine, goat, camel, equine, and rabbit, consist of 123 amino acid residues, with a 19- or 20-amino acid residue signal peptide. Only rat α-LA contains 17 additional C-terminal residues. Usually, a small fraction of α-LA occurs in glycosylated form in fresh milk except for rat α-LA, which is mostly glycosylated and exists in multiple forms. Approximately, 1% of equine α-LA is N-glycosylated.

α-LA is synthesized only in the lactating mammary gland, where it combines with the Golgi-associated β-1,4-galactosyltransferase [EC 2.4.1.38] to form the lactose synthase [EC 2.4.1.22] complex (for review see McKenzie and White 1991).

Three-dimensional structures of several α-LAs have been determined (reviewed by Permyakov 2005). The molecule of α-LA consists of two domains: a large α-helical domain and a small, mostly β-sheet domain, which are connected by a calcium-binding loop (Fig. 1). Overall, the structure of α-LA is stabilized by four disulfide bridges: 6-120, 61-77, 73-91, and 28-111. The α-helical domain is composed of three major canonical α-helices (residues 5-11 (A), 23-34 (B), and 86-99 (C)) and two short 3_{10} helices (residues 17-21 and 115-119). While the canonical α-helix is characterized by 3.6 residues per turn and an interchain hydrogen bond loop containing 13 atoms (3.6_{13}-type helix), the 3_{10}-type helix has three residues per turn

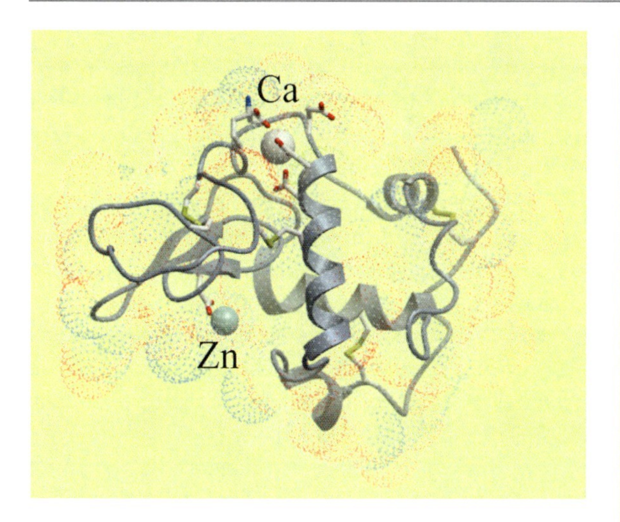

α-Lactalbumin, Fig. 1 Ribbon structure of human α-lactalbumin with bound calcium and zinc ions (PDB file 1HML). The α-helix domain is on the *right*; the β-strand domain is on the *left*

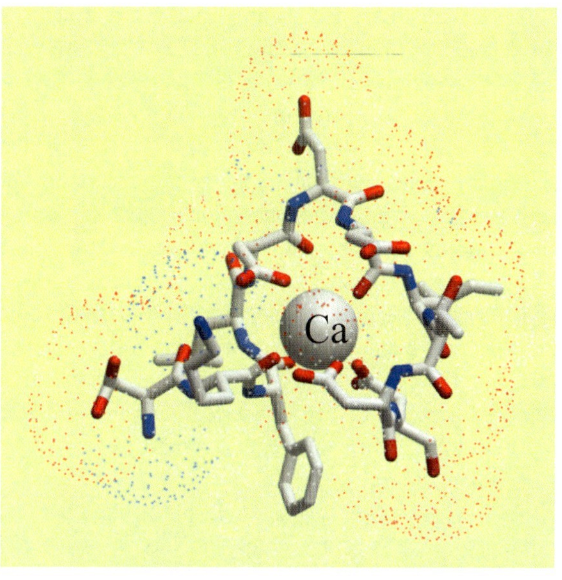

α-Lactalbumin, Fig. 2 Calcium-binding site in human α-lactalbumin (PDB file 1HML). The Ca^{2+} ion is coordinated by seven oxygen atoms in approximate pentagonal bipyramidal array

and ten atoms in the interchain hydrogen bond loop. The small β-sheet domain is composed of a series of loops, a small three stranded antiparallel β-pleated sheet (residues 40-43, 47-50, and 55-56), and a short 3_{10} helix (76-82). The two domains are separated by a deep cleft. At the same time, the two domains are held together by the cysteine bridge between residues 73 and 91, helping to form the Ca^{2+} -binding loop.

The protein possesses a single strong Ca^{2+}-binding site (apparent binding constant at room temperatures $\sim 10^8$ M^{-1}), which binds Mg^{2+}, Mn^{2+}, Na^+, and K^+ as well; it also contains several distinct Zn^{2+}-binding sites. The strong calcium-binding site is formed by oxygen ligands from carboxylic groups of three Asp residues (82, 87, and 88) and two carbonyl groups of the peptide backbone (Lys79 and Asp84) in a loop between two helices (Fig. 2). Overall the oxygen ligands form a distorted pentagonal bipyramidal structure. The binding of cations to the Ca^{2+} site increases protein stability against action of heat and various denaturing agents and proteases, while the binding of Zn^{2+} to the Ca^{2+} saturated protein decreases its stability and causes its aggregation. All the calcium, protein ligand (Ca–O) distances are between 2.2 Å and 2.6 Å. The region in the vicinity of the calcium-binding loop is the most rigid part of α-LA structure. Comparison of the calcium-binding loop in α-LA with the corresponding region of hen egg white lysozyme shows that the conformations of their main polypeptide chains are essentially the same, while the side chains are significantly different.

The well-known EF-hand Ca^{2+}-binding structural unit (reviewed by Permyakov 2009; Permyakov and Kretsinger 2011) is composed of two helices separated by a Ca^{2+}-binding loop, which is typically 12 amino acid residues long. The Ca^{2+} ion in the loop is roughly octahedrally coordinated with oxygen atoms. The Ca^{2+}-binding structure in α-LA, an "elbow," only superficially resembles the EF-hand and has no evolutionary relationship to it.

A secondary calcium-binding site was found by X-ray crystallography in human α-LA 7.9 Å away from the primary strong calcium-binding site. This secondary site is located near the surface of the α-LA molecule and has a coordination number smaller than the usual value of seven for Ca^{2+} ion. Four residues are involved in Ca^{2+} coordination at this site in a tetrahedral arrangement: side chain oxygens of Thr38, Gln39, Asp83 and the carbonyl oxygen of Leu81. No water molecules are present in the coordination sphere, although one internal water molecule is 3.8 Å from the secondary metal ion. The binding of the second Ca^{2+} ion does not produce any significant structural change in the protein.

Several distinct zinc-binding sites were found in α-LA, one of which is located in the "cleft" region, that is, the region which forms the active site of lysozyme. In the crystal structure of human α-LA the zinc is sandwiched between Glu49 and Glu116 of the symmetry-related subunit in the dimeric crystal unit cell. Studies on α-LA mutants showed that the strong Zn^{2+}-binding site in solution is not consistent with the site in the human α-LA crystal structure and, in fact, appears to be located near the N-terminus of the protein: site-directed mutagenesis of Glu1 to Met results in the disappearance of the strong Zn^{2+}-binding site in bovine α-LA. A proposed zinc-binding site involves Glu1, Glu7, Glu11, and Asp37 (reviewed by Permyakov 2005).

At low pH values (pH 2), α-LA is in the classical molten globule state (A-state) and has been studied extensively by many researchers (reviewed by Permyakov 2005). This A-state is defined as a compact state with fluctuating tertiary structure: it may have a high content of secondary structure, but many amino acid side chains are not entirely fixed. The radius of gyration of native Ca^{2+} loaded α-LA is 15.7 Å, but the acid molten globule has a gyration radius of 17.2 Å. The calcium depleted form of α-LA at neutral pH can be induced to adopt a partly folded state or molten globule upon moderate heating, by dissolving the protein in aqueous trifluoethanol, or by adding fatty acid. In the absence of calcium α-LA forms a molten globule intermediate even at neutral pH values at elevated temperatures above the thermal transition region. Molten globule α-LA still retains a globular shape, but is simply "swollen" from the native state.

When a Ca^{2+} ion is in a binding site, the negative charges of carbonyl and especially carboxyl oxygen atoms are partly compensated by its positive charge. The removal of the positive charge from the coordination sphere results in a sharp increase of the repulsive forces between the negatively charged oxygens. This must cause changes in protein conformation, at least near the calcium-binding sites. The changes might spread along the protein structure. If the binding constant is high enough, the energy of the interaction of cation with coordinating ligands contributes essentially to the total energy of stabilization of the protein. As a rule, the binding of calcium ions significantly increases the stability of the protein to the action of denaturing factors such as temperature, urea, guanidine hydrochloride, and others (reviewed by Permyakov and Kretsinger 2011). All of these results suggest that the structure of apoprotein should be less rigid and less stable in comparison with the structure of the Ca^{2+}-loaded protein. The binding of Ca^{2+} to α-LA at room and higher temperatures causes pronounced changes in structure, mostly in tertiary, but not secondary structure; these are clearly seen by many experimental methods (reviewed by Permyakov 2005). Both the Ca^{2+}-bound and Ca^{2+}-free forms of α-LA adopt practically the same folded conformation at low temperatures as evidenced by similarity in their circular dichroism and proton NMR spectra.

Chrysina et al. (2000) determined crystal structures of both apo- and holo-forms of bovine α-LA at 2.2 Å resolution. In both forms the protein is in a similar globular native conformation. Remarkably, the removal of calcium has only minor effects on the structure of the metal-binding site and the largest structural change is observed in the cleft on the opposite side of the molecule: Tyr103 is shifted toward the interior of the cleft, and water-mediated interactions with Gln54 and Asn56 replace the direct hydrogen bonds. The changes result in increased separation of the α and β domains, loss of a buried solvent molecule near the calcium-binding site, and the replacement of inter- and intra-lobe hydrogen bonds of Tyr103 by interactions with new immobilized water molecules. Apo-α-LA and holo-α-LA have respectively 6,512 $Å^2$ and 6,820 $Å^2$ accessible surface area. One hundred and sixty-seven bound water molecules were identified in apo-α-LA and 89 in holo-α-LA. A comparison of the crystallographic temperature (B)-factors in apo-α-LA and holo-α-LA indicates that the removal of calcium results in an increase in mobility in main chain and side chain atoms in the calcium-binding site and in the C-terminal region but a decrease in mobility in the β-lobe.

It should be noted, however, that in the crystal structure, apo-α-LA is probably stabilized in the native conformation by the high ionic strength of crystallization medium. The structural changes caused by the removal of calcium seem to be more pronounced in low ionic strength solution. Overall, the Ca^{2+}-binding loop in apo-α-LA has a slightly higher solvent accessibility and significantly increased temperature factors compared with the corresponding loop in holo-α-LA. In apo-α-LA, the side chain of Asp87 is reoriented but there is no other major structural change. Despite the minor structural changes in the Ca^{2+}-binding site of the apoprotein, it is reasonable to suggest that the larger change in the cleft region is propagated from the Ca^{2+}-binding loop. Charge repulsion between the carboxylates of the five aspartyl residues in this region is a probable source of this change; the reorientation of the side chain of Asp87 and a slight expansion of the loop appear to be in response to this, reducing the negative charge density.

Apparent binding constants of metal ions for bovine α-LA at pH 7–8 (Permyakov 2005)

Cation	Association constants (M^{-1})	
	37°C	20–25°C
Ca^{2+}	2×10^7	3×10^8
Mn^{2+}		3×10^4
Mn^{2+}		7.3×10^5
Mg^{2+}	211 ± 20; 46 ± 10	$2{,}000 \pm 100$; 200 ± 20
Cd^{2+}		4×10^5
Sr^{2+}		2.7×10^6
Ba^{2+}		9×10^4
Pb^{2+}	2×10^6	
Na^+	36 ± 10	100 ± 10
K^+	6 ± 3	8 ± 3

The values of the binding constants of α-LA for Mg^{2+}, Na^+, and K^+ ions are rather low but, considering their high concentrations in the cell, one might speculate that they could successfully compete with Ca^{2+} ions in vivo. Interestingly, it was found that equine and pigeon lysozymes bind calcium ions. The calcium-binding constants of equine and pigeon lysozymes were determined to be 2×10^6 and 1.6×10^7 M^{-1}, respectively, in 0.1 M KCl at pH 7.1 and 20°C.

The methods of protein engineering were applied for studies of α-LA. Surprisingly, wild-type recombinant α-LA has a lower thermostability and calcium affinity compared to the native protein (reviewed by Permyakov 2005). The only difference of the recombinant protein from the native one is an additional N-terminal methionine residue: the presence of the N-terminal N-formyl methionine in protein biosynthesis is ubiquitous in both prokaryotes and eukaryotes. Enzymatic removal of the N-terminal methionine leads to restoration of the native properties of the protein. Fluorescence, circular dichroism, and differential scanning calorimetry data show that recombinant wild-type α-LA in the absence of calcium ion is in a state, which is similar to the molten globule state. It is of interest that the ΔE1 mutant, in which the Glu1 residue of the native sequence is removed, leaving the N-terminal methionine in its place, shows almost one order of magnitude higher affinity for calcium and higher thermostability than the native protein. It means that the N-terminus of the protein dramatically affects both stability and function of α-lactalbumin as manifested in calcium affinity.

The equilibrium scheme of the binding of one metal ion (Me) to the protein molecule, taking into consideration equilibrium between native (P, PMe) and thermally changed (P^*, P^*Me) states of the protein, is:

$$
\begin{array}{ccc}
 & K_{Me} & \\
P + Me & \leftrightarrow & PMe \\
\gamma \downarrow\uparrow & & \downarrow\uparrow \beta \\
P^* + Me & \leftrightarrow & P^*Me \\
 & K_{Me}^* &
\end{array}
\qquad \text{Scheme A}
$$

in which K_{Me} and K_{Me}^* are intrinsic metal ion dissociation constants for the native and thermally denatured protein, respectively, and γ and β are equilibrium constants of the thermal denaturation of the protein in its apo and metal ion–bound forms, respectively.

$$K_{Me} = \exp[(\Delta H_{Me} - T \cdot \Delta S_{Me})/R \cdot T] \quad (1)$$

$$K_{Me}^* = \exp[(\Delta H_{Me}^* - T \cdot \Delta S_{Me}^*)/R \cdot T] \quad (2)$$

$$\gamma = \exp[-(\Delta H_\alpha - T \cdot \Delta S_\alpha)/R \cdot T] \quad (3)$$

$$\beta = \exp[-(\Delta H_\alpha - T \cdot \Delta S_\beta)/R \cdot T] \quad (4)$$

ΔH_{Me}, ΔH_{Me}^* and ΔS_{Me}, ΔS_{Me}^* are enthalpy and entropy changes for the metal ion binding to the native and thermally denatured protein. ΔH_α, ΔH_β and ΔS_α, ΔS_β are enthalpy and entropy changes for the thermal transitions in the apo and metal ion–bound protein. ΔH_α and ΔS_β can be determined using experiments studying the thermal denaturation of the apoprotein, while ΔH_β and ΔS_β can be determined from the thermal denaturation curve for the metal ion–bound protein.

The apparent metal ion dissociation constant determined experimentally is

$$1/K_{app} = \frac{[PMe] + [P^*Me]}{([P] + [P^*])[Me]} = 1/K_{Me}\frac{1+\beta}{1+\gamma} \quad (5)$$

$$K_{Me}^*/K_{Me} = \gamma/\beta \quad (6)$$

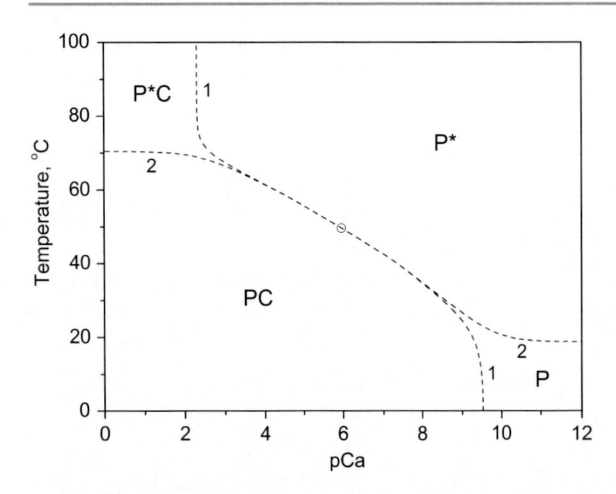

α-Lactalbumin, Fig. 3 Phase diagram of bovine α-LA in the free calcium concentration–temperature space, calculated according to the four-states Scheme A. Curves *1* and *2* correspond to half-transition for binding of calcium and thermal denaturation. *P, PMe,* P^*, and P^*Me are native and thermally changed states of the metal-free and metal-loaded protein

Knowledge of all the thermodynamic parameters for metal cation binding allows calculation of the thermal denaturation curve of the protein in the presence of any given concentrations of the metal ion. More important, this knowledge allows computation of a phase diagram in the free calcium, temperature coordinates. Figure 3 shows such phase diagrams for bovine α-LA (reviewed by Permyakov and Kretsinger 2011). The phase diagram depicts both regions of predominance of separate protein states and the areas of transitions between them, including lines of half transitions in temperature (1) and calcium (2) scales. Thus, knowing current temperature and pCa (-log[Ca^{2+}]) values, the actual protein state can be assessed. Moreover, limits of curves 1 and 2 obviously demonstrate mid-transition temperatures for apo- and calci-protein (curve 1, ordinate values) and calcium-binding affinities of native and denatured states of protein (curve 2, abscissa values).

It was shown that all four classes of surfactants (anionic, cationic, non-ionic, and zwitterionic) denature α-LA, making it an excellent model system to compare their denaturation mechanisms. The binding of metal cations significantly increases the stability of α-LA not only against temperature but also against the action of various denaturing agents such as urea or guanidine hydrochloride (reviewed by Permyakov 2005, 2009). In the presence of calcium, the unfolding of α-LA occurs at extremely high concentrations of the denaturants. Here, important features of the denaturation curves are distinct, intermediate molten globule-like states arising at intermediate denaturant concentrations. As a first approximation, the total unfolding can be expressed as a three-state mechanism involving the native (N), an intermediate (I), and the fully unfolded (U) states. Urea and alkali (pH higher than 10) induce unfolding transitions, which involve stable partially unfolded intermediates for all metal ion–bound forms of α-LA.

The equilibrium scheme accounting for the metal ion binding for urea or guanidine HCl induced unfolding process at low temperatures (below the temperature-induced transition) and neutral pH is

$$\begin{array}{ccc} P + Me & \leftrightarrow & PMe \\ \updownarrow & & \updownarrow \\ P^* + Me & \leftrightarrow & P^*Me \\ \updownarrow & & \updownarrow \\ P^{**} + Me & \leftrightarrow & P^{**}Me \end{array} \qquad \text{Scheme B}$$

where P and PMe are apo and metal ion–loaded native states, P* and P*Me are apo and metal ion–loaded intermediates, and P** and P**Me are apo and metal ion–loaded unfolded states of the protein. The scheme is not complete, since each state in it is in fact in equilibrium with the corresponding high temperature conformer.

Studies of the effects of pressure (up to 100 MPa) on the guanidine hydrochloride and temperature-induced unfolding processes of α-LA (reviewed by Permyakov 2005) indicate that the molar volume of α-LA only changes at the transition from the native to a molten globule state, and almost no volume change was found for the transition from a molten globule to the unfolded state. The binding of calcium stabilizes α-LA against pressure as monitored by a 200 MPa increase in the pressure where denaturation occurs. Interestingly, calcium binding increases the pressure stability of the calcium-binding loop to a greater extent than the pressure stability of the overall α-LA secondary structure.

It was found that at acidic pH values (pH 2) bovine α-LA in the classical molten globule conformation forms amyloid fibrils (reviewed by Permyakov 2005). S-(carboxymethyl)-α-LA, a disordered form of the protein with three out of four disulfide bridges reduced, is even more susceptible to fibrillation (faster formation of amyloid fibrills). The fibrillation is accompanied by a dramatic increase in the β-structure content monitored by Fourier transform infrared spectroscopy and a characteristic increase in the thioflavin T fluorescence intensity. Fibrillation of α-LA is extremely sensitive to the ionic strength of the solution: the lag time of fibrillation decreases from 48.9 h in the presence of 25 mM NaCl to 19.9 h in the presence of 150 mM NaCl.

The binding of Ca^{2+} and other physiologically significant ions to α-LA modulates all its properties including its ability to interact with various organic substances, peptides, and proteins. For example, α-LA binds UDP-galactose, the substrate of lactose synthase reaction, as well as UDP, and UTP (reviewed by Permyakov 2005). The binding parameters depend upon the metal-bound state of the protein, but the strongest binding constant for UDP-galactose falls in the range of 10^3–10^4 M^{-1}.

α-LA binds melittin, a short peptide from bee venom (reviewed by Permyakov 2005), which is frequently used as a model target protein for calmodulin and other calcium-binding proteins. Calmodulin and some other calcium-binding proteins bind melittin with higher affinity in the presence of Ca^{2+}. In contrast to such proteins as calmodulin or troponin C, α-LA binds melittin only in the absence of calcium ions. Apo-α-LA binds melittin with the binding constant 5×10^7 M^{-1}. The binding alters the melittin conformation from a random coil in solution to a helical structure in the binary complex with apo-α-LA. The usual scheme of intracellular calcium regulation suggests that the binding of Ca^{2+} to a set of Ca^{2+}-binding proteins changes their conformation and in this state they are able to interact with their target proteins. The interaction activates the proteins and triggers various intracellular processes. The finding that α-LA interacts with melittin only in the absence of Ca^{2+} shows that, in principle, it is possible that in the absence of calcium ions some regulatory Ca^{2+}-binding proteins interact with another target proteins and trigger alternative intracellular processes.

α-LA possesses several classes of fatty acid–binding sites (reviewed by Permyakov 2005). Bovine α-LA was shown by intrinsic protein fluorescence and electron spin resonance methods to interact with the spin-labeled fatty acid analog, 5-deoxylstearic acid, as well as stearic acid. In some cases, up to three classes of fatty acid–binding sites are visible with the dissociation constants from 10^{-6} to 10^{-4} M. The binding parameters of 5-deoxylstearic acid for apo- and Ca^{2+}-α-LA are an order of magnitude different from one another; the stronger one, apo-α-LA, exhibited a K_d of 35 μM.

The synthesis of lactose occurs in the lumen of the Golgi, a complicated membrane system; for this reason, it is important to study possible interactions of α-LA with lipid membranes. The self-incorporation of apo-α-LA into single unilamellar vesicles (SUV) of dimyristoylphosphatidylcholine (DMPC) and dipalmitoylphosphatidylcholine (DPPC) was demonstrated by column chromatographic analyses on Sephadex G-200 or Sepharose 4B and by intrinsic fluorescence emission of SUV-bound α-LA and scanning microcalorimetry (reviewed by Permyakov 2005). It was shown that apo-α-LA slowly incorporates into the DMPC vesicle bilayer after equilibrating different mixtures of protein and SUV for several hours. The addition of calcium causes a slow conversion from apo-α-LA to Ca^{2+}-α-LA by a mechanism consistent with passive diffusion of Ca^{2+} into the bilayer interior to the (buried) calcium-binding site. The interaction of α-LA with the liposomes affects the phase transition from gel to liquid-crystalline state in liposomes. The binding and the membrane-bound conformations of α-LA are highly sensitive to environmental factors, like calcium and proton concentrations, curvature, and charge of the lipid membrane. The interactions between the protein and the membrane result from a combination of hydrophobic and electrostatic interactions and the respective weights of these interactions depend on the physico-chemical conditions. As inferred by macroscopic as well as residue-level methods, the conformations of the membrane-bound protein range from native-like to molten globule-like states. However, the regions anchoring the protein to the membrane are similar and restricted to amphiphilic α-helices.

α-LA seems to fulfill several functions. It has long been known that α-LA is a component of lactose synthase (reviewed by Permyakov 2005), an enzyme system, which consists of galactosyltransferase and α-LA. α-LA complexes with galactosyltransferase only in the presence of substrates and modifies its affinity and specificity for glucose:

$$\text{UDP} - \text{Gal} + \text{glucose} \xrightarrow{\text{GT}/\alpha - \text{LA}} \text{lactose} + \text{UDP}$$

The reaction occurs in the Golgi lumen and presumably requires Mn^{2+} ions. Nevertheless, the role of metal cations in lactose synthase function is still far from clear. In the complex with galactosyltransferase α-LA holds and puts glucose right in the acceptor-binding site of galactosyltransferase, which then maximizes the interactions with glucose, thereby making it a preferred acceptor for the lactose synthase reaction. This interaction also stabilizes the sugar-nucleotide-enzyme complex, kinetically enhancing the sugar transfer. It was found that α-LA increases the effectiveness of galactosyltransferase in transfer of glucose from UDP-glucose to N-acetylglucosamine almost 30-fold. It also enhances the glucosyltransferase activity toward various N-acyl substituted glucosamine acceptors.

Another possible function of α-LA is connected with its bactericidal and antiviral activity. Proteolytic digestion of α-LA by trypsin and chymotrypsin yields three peptides with bactericidal properties. The polypeptides are mostly active against Gram-positive bacteria suggesting a possible antimicrobial function of α-LA after its partial digestion by endopeptidases.

It was found that some multimeric, yet not thoroughly characterized, human α-LA derivative is a potent Ca^{2+}-elevating and apoptosis-inducing agent with broad, yet selective, cytotoxic activity, killing all transformed, embrionic, and lymphoid cells tested. Similar results were obtained with HAMLET (*h*uman α-lactalbumin *m*ade *le*thal to *t*umor cells), which is native human α-LA converted in vitro to the apoptosis-inducing folding variant of the protein in complex with oleic acid (reviewed by Permyakov 2005, 2009). In tumor cells, HAMLET enters the cytoplasm, translocates to the perinuclear area, and enters the nuclei where it accumulates. In the cytoplasm, HAMLET targets ribosomes and activates caspases. Histones were identified as targets of HAMLET among nuclear constituents. HAMLET was found to bind histone H3 strongly and to a lesser extent histones H4 and H2B. In vivo in tumor cells, HAMLET co-localizes with histones and perturbs the chromatin structure; HAMLET associates with chromatin in an insoluble nuclear fraction resistant to salt extraction. In vitro, HAMLET binds strongly to histones and impairs their deposition on DNA. It was concluded that HAMLET interacts with histones and chromatin in tumor cell nuclei and suggested that this interaction locks the cells into the death pathway by irreversibly disrupting chromatin organization. It is of interest that monomeric α-LA, in the absence of fatty acids, is also able to bind efficiently to the primary target of HAMLET, histone HIII, regardless of Ca^{2+} content. Thus, the modification of α-LA by oleic acid is not required for binding to histones. It was suggested that interaction of negatively charged α-LA with the basic histone stabilizes apo-α-LA and destabilizes the Ca^{2+}-bound protein due to compensation for excess negative charge of α-LA's Ca^{2+}-binding loop by positively charged residues of the histone.

These results show that α-LA can alter its biological function depending on the conformational state. It is well known that α-LA is one of the most abundant proteins in human milk. Human milk has a high concentration of whey protein (70% of total protein). Of this, α-LA accounts for 41% of the whey and 28% of the total protein. Only 3% of the protein in bovine milk is α-LA. α-LA passes from the mammary gland through the gastrointestinal tract of a nursing baby. In the mammary gland, α-LA functions as a specifier protein in lactose synthesis. The low pH in the stomach of the breast-fed child results in release of calcium from α-LA and precipitation of casein. The acid lipase hydrolyses triglycerides, and fatty acids are released. These events seem to create conditions for formation of the HAMLET-like state of α-LA. Breast-feeding was proposed to protect both mother and child against cancer.

α-LA is a tryptophan-rich protein fraction in milk. A diet enriched with α-LA increases the ratio of tryptophan to the other large neutral amino acids, which may in turn increase brain serotonin content. In stress-vulnerable individuals, α-LA improved mood and attenuated the cortisol response after experimental stress.

Cross-References

▶ Bacterial Calcium Binding Proteins

References

Chrysina ED, Brew K, Acharya R (2000) Crystal structures of apo- and holo-bovine α-lactalbumin at 2.2-Å resolution reveal an effect of calcium on inter-lobe interactions. J Biol Chem 275(47):37021–37029

McKenzie HA, White FH Jr (1991) Lysozyme and α-lactalbumin: structure, function, and interrelationships. Adv Protein Chem 41:173–315

Permyakov EA (2005) α-Lactalbumin. Nova, New York

Permyakov EA (2009) Metalloproteomics. Wiley, Hoboken

Permyakov EA, Kretsinger RH (2011) Calcium binding proteins. Wiley, Hoboken

Lanthanide

▶ Lanthanides and Cancer

Lanthanide Analytical Probes

▶ Lanthanides in Nucleic Acid Analysis

Lanthanide Bioprobes

▶ Lanthanides in Biological Labeling, Imaging, and Therapy

Lanthanide Ions as Luminescent Probes

Juan C. Rodriguez-Ubis, Ernesto Brunet and Olga Juanes
Dept. Química Orgánica, Facultad de Ciencias, Universidad Autónoma de Madrid, Madrid, Spain

Synonyms

Lanthanide labels; Lanthanide luminescent labels; Lanthanide time-resolved fluorescence; Lanthanide-based luminescent assays

Definition

Luminescent lanthanide probes have highly unusual spectral characteristics that make them useful nonisotopic alternatives to organic fluorophores, particularly where there are problems of background autofluorescence. The applications range from luminescent labeling of biologically relevant molecules and the in vivo detection of cellular functions to the elucidation of the structure and function of enzymes and proteins. Despite the fact that the luminescence of aqueous solutions of simple salts of lanthanide ions, namely Tb(III) and Eu(III), is rather weak owing to their low absorption cross-section in the UV-visible region, the particular emission properties of these metal ion complexes have extensively been utilized in biological systems.

Personality and Role of Luminescent Lanthanide Probes

Lanthanide luminescence can be dramatically enhanced by chelating lanthanide ions with an appropriate organic ligand (antenna effect). The well-known emission in the visible region of certain lanthanide (III) salts has thus found a new dimension by the controlled assembly of supramolecular architectures in to efficiently enshroud and excite the metal. The emission properties of these systems find applications as luminescent probes in sensing schemes, interaction with biomolecules or light conversion systems. Various techniques have been employed, including enhancement or quenching of the lanthanide emission, resonance energy transfer, and delayed luminescence.

Differing from organic fluorescent dyes, emissive lanthanide complexes have several technical advantages, including very long decay time, large Stokes' shift, and sharp emission profile. These characteristics make them favorable for use as labeling reagents for millisecond time-resolved luminescence assays, because of the easy and efficient background discrimination in fluorimetric analyzis. Since the first immunoassay application was reported in 1983 by Finnish company Wallac, some commercial assay systems have been successfully developed, such as Wallac systems, DELFIA (dissociation lanthanide fluoroimmunoassay) and LANCE (Lanthanide Chelate Excite) multiple-label time-resolved fluoroimmunoassy (TR-FIA), CyberFluor (or FIAgen), CIS-Biosystems HTRF-TRACE (time-resolved amplified cryptate emission), Invitrogen LANTHASCREEN, or Perkin-Elmer AlphaLISA (Amplified luminescent proximity homogeneous assay).

Until recently, major reports about lanthanide luminescent probes focused on luminescent lanthanide complex-based labeling reagents and their applications to time-resolved luminescence bioassays. Now an increased number of lanthanide luminescent probes are being applied in other fields, as cell imaging, DNA and RNA analysis or their incorporation in nanoparticles for different bioanalytical applications.

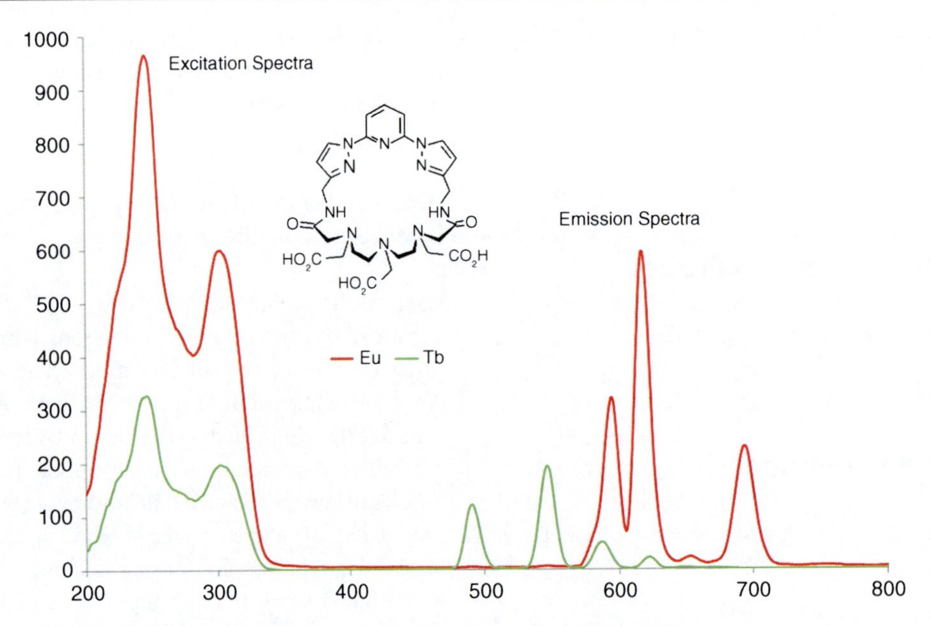

Lanthanide Ions as Luminescent Probes, Fig. 1 Excitation and emission spectra of characteristic Eu(III) and Tb(III) complexes

Photophysical Essentials of Sensitized Lanthanide Emission

The 15 elements that comprise the lanthanide series are present as tripositive ions [Ln(III)] in almost all their compounds, and their chemical properties are all very similar. Binding of ligands to the lanthanide ion occurs via electron transfer through f-orbitals, with highly electronegative donor atoms such as N and O preferred. The coordination of the ligand binding to the central lanthanide ion does not tend to be of fixed number and geometry as it is in transition metal ions; rather, it is somewhat malleable, with coordination numbers in the range of 9–12 observed. Where other more strongly binding ligands are not available, coordination sites are generally filled by water solvent molecules within the restrictions of inter-ligand repulsion.

Arising from their (inner-shell) 4f orbitals configurations, they differ markedly in their spectral properties. The complexes of several trivalent lanthanide ions, especially those with organic ligands that include aromatic or other π-bonded groups, exhibit a unique kind of luminescence. When irradiated with ultraviolet light, the complexes of the Eu(III), Tb(III), Sm(III), and Dy(III) ions emit visible light and, while the absorption and excitation spectra are broad and fairly similar for all complexes of the same ligand, the emission spectra consist of several very narrow bands that are characteristic of each Ln(III). As an example, Fig. 1 shows the excitation and emission spectra of Eu (III) and Tb(III) complexes of an organic macrocyclic chelating ligand that includes pyridine and pyrazole units.

The typical energy level diagrams for an organic fluorophore and a lanthanide chelate are compared and correlated in Fig. 2. For an organic fluorophore in solution, the various processes subsequent to light absorption from the ground state into the excited single state manifold generally can include internal conversion (IC; conversion of absorbed energy to heat), fluorescence (light emission during conversion of S_1 to S_0), intersystem crossing (ISC; nonradiative conversion from the singlet to the triplet state, or vice versa), phosphorescence (light emission during conversion of T_1 to S_0), and photodegradation. Figure 2 also illustrates the mechanism for the Eu(III) and Tb(III) luminescence in complexes of organic ligands.

Ligand-enhanced luminescence is a complicated energy transfer process involving the ligand energy levels, the ground and excited state energy levels of lanthanide ions, and environmental factors (e.g., solvent and temperature). The efficiency of lanthanide emission is rather sensitive to the vibration of neighboring X-H bonds. The vibronic coupling between the emissive levels of the metal and the O-H bonds of coordinated

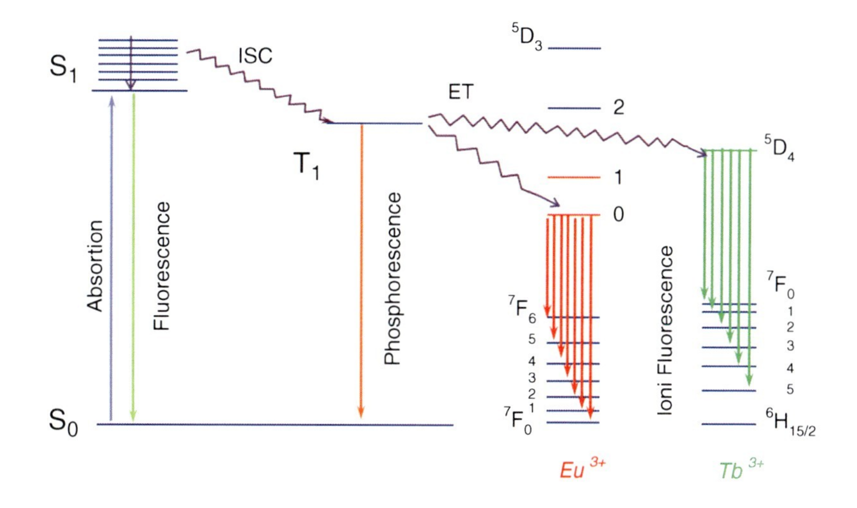

Lanthanide Ions as Luminescent Probes, Fig. 2 Photophysical pathway of Europium and Terbium emission by antenna excitation

water molecules is especially deleterious. Lanthanide ions themselves have very low absorption and emission probabilities as the transitions of interest are generally forbidden; thus in the absence of a strongly absorbing ligand, luminescence is comparatively weak. This emission consists of several discrete narrow (1–20 nm) bands, each of which may demonstrate some additional splitting due to ligand field effects.

When a suitable, strongly absorbing ligand or chelator with appropriately situated energy levels, as illustrated in Figs. 1 and 2, is bound to the lanthanide ion, absorption of light into the first excited singlet state of the ligand can be followed by intersystem crossing to the ligand triplet state and energy transfer to the lanthanide ion. Luminescence characteristic of the lanthanide ion will follow. The typical luminescence emission spectrum for a Eu(III) complex is illustrated in Fig. 1 (red line); in this case, the $L_1 \rightarrow L_0$ transition corresponds primarily to five transitions, ${}^5D_0 \rightarrow {}^7F_n$ (where n = 0–4), giving five distinct bands. The change in spin multiplicity during the transition (e.g., 5–7 for Eu(III)) results in a forbidden transition with a long luminescence lifetime. Because the luminescing states arise from excitation of f-electrons, which are relatively well shielded from most interactions with the environment, only the species in the first coordination sphere strongly affect the emission characteristics. Thus, in particular, if the chelate has a well-defined structure which does not include any weakly bound solvent or exchangeable ligand molecules in the first coordination sphere, it can be relatively insensitive to solvent and temperature effects. An antenna is necessary because of the inherently weak absorbance of the lanthanide ($1\ M^{-1}cm^{-1}$), compared with the conventional absorption of organic chromophores 10^4–$10^5 M^{-1}\ cm^{-1}$.

Although europium and terbium are by far the most useful lanthanides, dysprosium and samarium are the only other two lanthanides that emit in the visible region, but with much weaker intensity; their combined use enhances the multiplexing capabilities of bioanalytical assay.

The main disadvantages of most luminescent lanthanide chelates relative to organic fluorophores are their typically low luminescence in aqueous systems owing to quenching by coordinated water molecules, and the tendency to dissociation of the complex in some assay conditions. However, in most applications, these difficulties can be overcome and the advantages more than compensate.

Emission maxima is in the green (Tb(III), 545 nm) and red (Eu(III), 620 nm) regions. This large Stokes' shift and narrow emission bands of lanthanide chelates permit easy selective narrow-band wavelength filtering for the specific chelate signal, to remove scattered excitation light and more nonspecific backgrounds. Further, the very long luminescence lifetime permits time-gated detection on a micro- to millisecond timescale, to further select against typical short-lived nonspecific background signals, known as time-resolved "fluorescence" (TRF, Fig. 3).

Excitation of the antenna in the ultraviolet region, typically 300–350 nm is achieved with conventional flash Xenon lamps or nitrogen laser. Although longer excitation wavelengths are very convenient for cheaper excitation sources as tungsten or LED lamps,

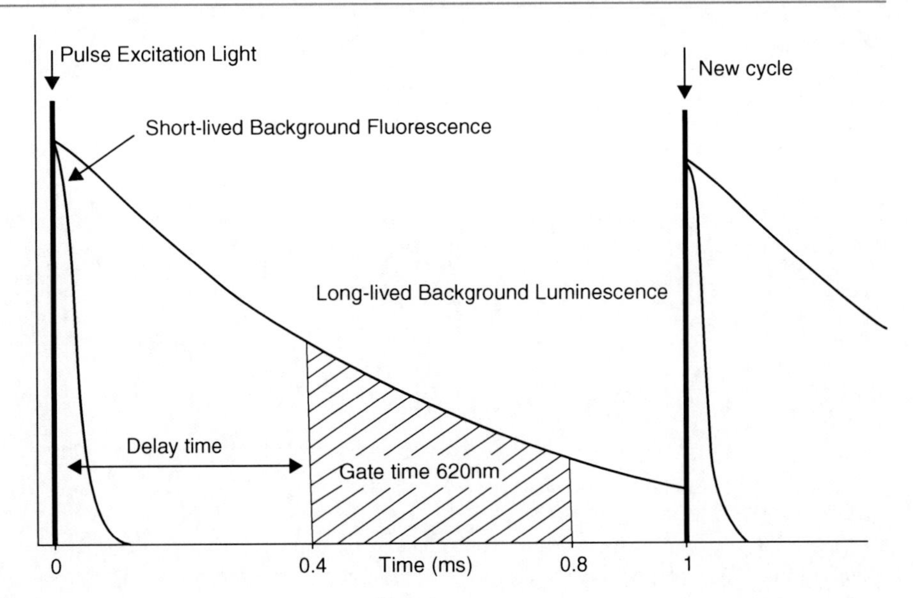

Lanthanide Ions as Luminescent Probes, Fig. 3 Measurement scheme of time-resolved luminescence

lower S_1 excitation reduce the transfer energy toward Eu(III) and Tb(III) emission (Latva et al. 1997).

The high-spin nature of the transitions is why emission is formally neither fluorescence (singlet-to-singlet transition) nor phosphorescence (triplet-to-singlet transition). Despite the unusual nature of the atomic states, emission primarily arises from electric dipole transitions. This is important because it is the same mechanism used by organic fluorophores. Hence, the electric field produced by a lanthanide donor and by an organic donor have the same distance dependence, decreasing as $1/R^3$ for distances lesser than working light wavelength. Ultimately, this leads to the same distance dependence, R^{-6}, for resonance energy transfer measurements (FRET), one of the applications of lanthanide luminescent probes.

Among the many described structures of luminescent lanthanide probes, Fig. 4 represents two schematic structures in which it is possible to integrate the number of luminescent lanthanide complexes described.

In type A structures, the chromophore antenna subunit is covalently linked to the coordinating ligand without being involved in coordination whereas in B type, the antenna is involved as a global or partial coordinating unit.

In order to be broadly useful as a label for bioassays and also for cell and tissue imaging, a lanthanide probe should meet several concurrent requirements. Additional biochemical necessities in dependence with the type of experiments performed (in vitro or in vivo) are named in the last three points:

1. High kinetic and thermodynamic stability of lanthanide complex to prevent the release of the lanthanide ion into the surrounding solution, or the exchange of lanthanide ions between two or more complexes
2. Absorbing a significant fraction of the exciting radiation
3. Efficiently converting the absorbed energy to intense light emission
4. Including a reactive functionality capable of covalent binding to a macromolecule
5. Being sufficiently soluble in aqueous systems to permit coupling and avoid precipitation of the labeled macromolecule
6. Ability of interacting specifically with the target without altering its bioaffinity properties
7. Absence of cytotoxicity
8. Easy excretion of the probe

Figure 5 represents eight prototypical structures related with some of the main systems developed for bioprobes applications. All contain an organic chromophore, which serves as a sensitizer or antenna, absorbing the excitation light and transferring the energy to the lanthanide ion. The complexes also contain chelating groups in order to tightly bind the lanthanide and shield it from the quenching effects of water.

In the structures 5c–h, the antenna is involved in binding the lanthanide; hence, logically, there is not a clear separation between chelate and antenna, whereas in the polyaminocarboxylate chelates such as DTPA-cs124 (5a) and DOTA Fig. (5b), the chelate

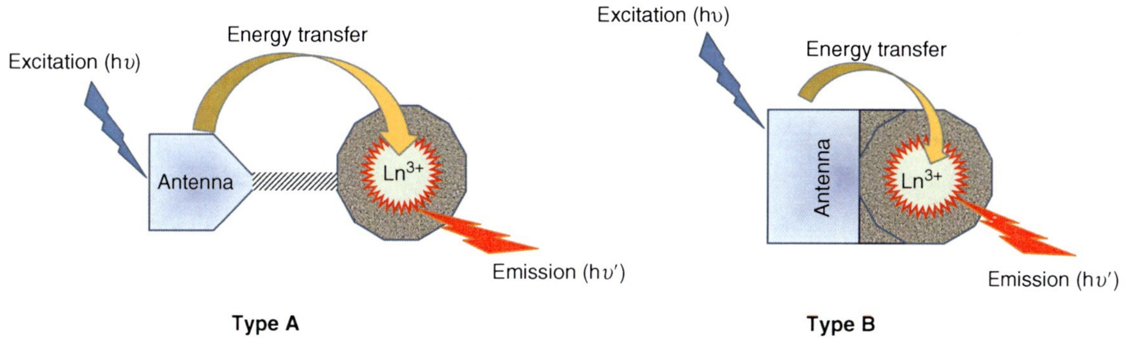

Lanthanide Ions as Luminescent Probes, Fig. 4 Schematic structure of representative chelates

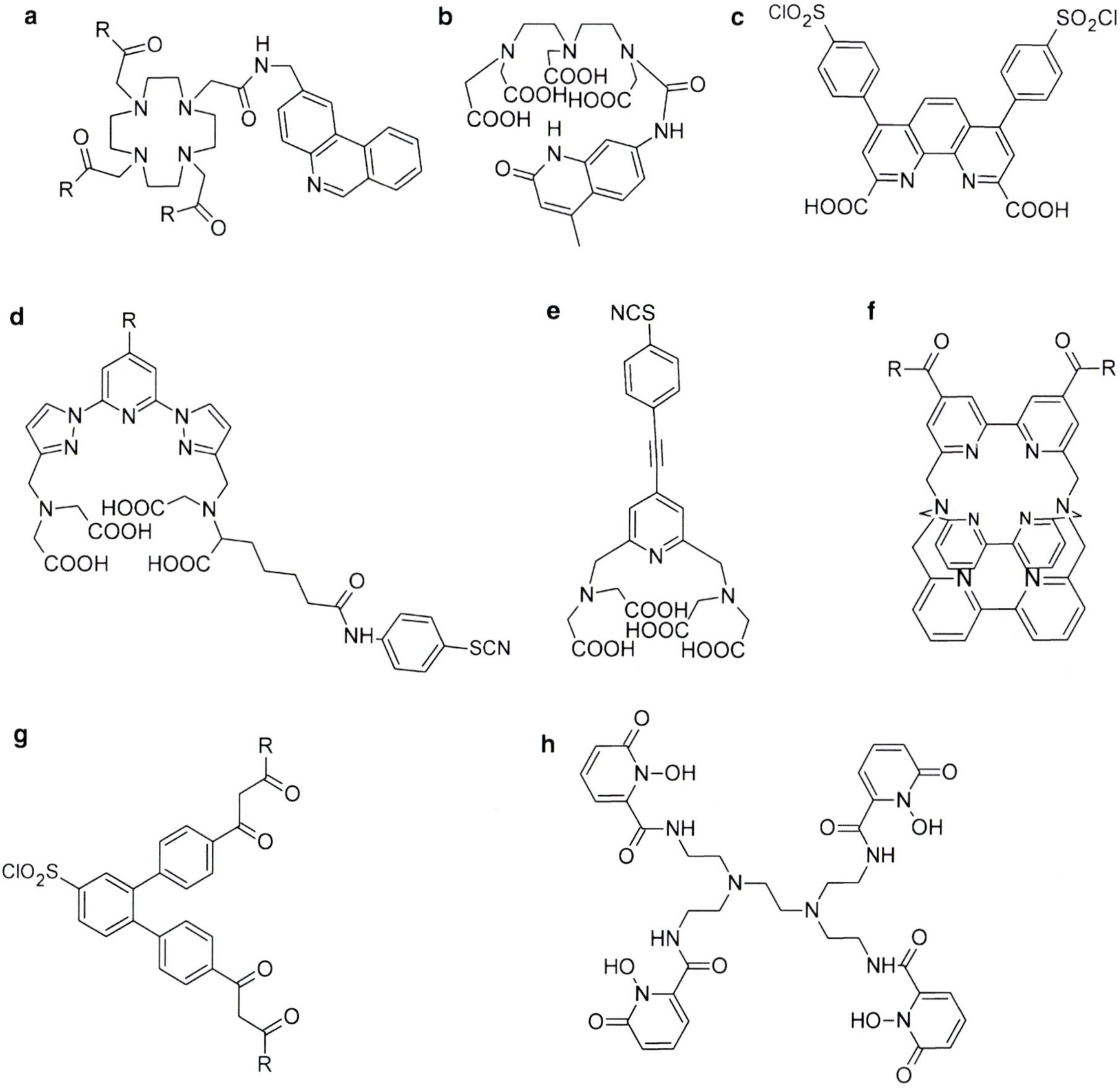

Lanthanide Ions as Luminescent Probes, Fig. 5 Structures of ligands commonly employed in luminescent lanthanide probes

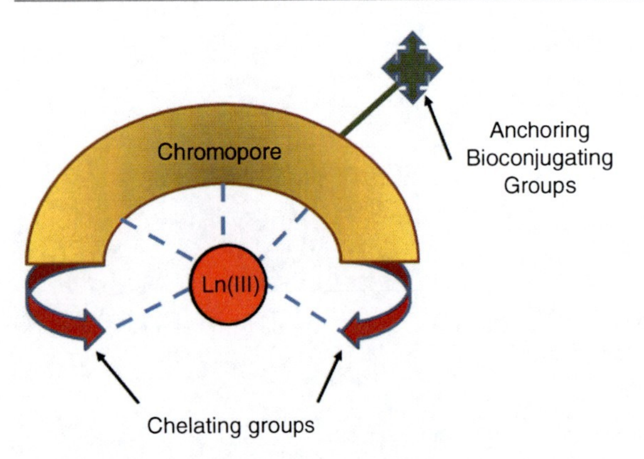

Lanthanide Ions as Luminescent Probes, Fig. 6 Schematic structure of a luminescent lanthanide probe

and chromophoric antenna are distinct entities. All eight sets of probes shown have been used as detection agents to replace either conventional fluorescent probes or radioactive probes where subpicomolar detection limits have been achieved. They have also been used in resonance energy transfer application.

The predominant application of the cryptates HTRF-TRACE (commercialized by CIS-Bio International) and chelates LANCE (commercialized by Perkin Elmer) has been in binding assays associated with high-throughput screening, whereas the primary application of the polyaminocarboxylate type A compounds has been in basic studies to measure conformational changes. However, all such chelates can be used in both applications.

Bioassays require specific targeting of the analyte, and therefore, the lanthanide luminescent probes have to be fitted with adequate functionalities able to couple with biological material, as it is shown in structures 5c–g, completing the structure of a luminescent lanthanide probe (Fig. 6).

Covalent coupling of a lanthanide luminescent chelate to bioactive molecules such as proteins, nucleic acids, or peptides relies on the presence of chemically reactive groups on these molecules. The most common ones are aliphatic amines. Thiol residues are other common reactive groups. The free thiol group (e.g., in cysteine) is more nucleophilic than amines and generally is the more reactive group in proteins. To couple with all these functionalities, the lanthanide chelate is first activated. In immunoassays, coupling between proteinic amines and lanthanide chelates fitted with isothiocyanato, chlorosulfonyl, or 2,4-dichloro-1,3,5-triazinyl groups proved to be the most successful. Another very convenient coupling group is *N*-hydroxysuccinimide.

A great number of chelates have been synthesized, but usually, only the most prominent structures were then converted to corresponding biomolecule labeling reactants. Whereas a large fraction of them are described in patent literature, the field remains quite active, with new chelators/sensitizer systems being proposed constantly (Hovinen and Guy 2009).

Lanthanide-Based Resonance Energy Transfer (LRET)

FRET (Fluorescence Resonance Energy Transfer) uses two fluorophores, a donor and an acceptor. Excitation of the donor by an energy source (e.g., flash lamp or fluorometer laser) triggers an energy transfer to the acceptor if they are within a given proximity to each other. The acceptor in turn emits light at its given wavelength.

Because of this energy transfer, molecular interactions between biomolecules can be assessed by coupling each partner with a fluorescent label and detecting the level of energy transfer. More importantly acceptor emissions, as a measure of energy transfer, can be detected without the need to separate bound from unbound assay components. This homogeneous assay format is extremely beneficial, reducing both assay time and costs.

FRET is governed by the physics of molecular proximity, which only allows energy transfer to occur when the distance between the donor and the acceptor is small enough. Despite lanthanide luminescence is not fluorescence, as it was mentioned above, it has the same distance dependence and in practice, lanthanide LRET systems are characterized by the Förster's radius (R0) distance at which FRET efficiency is 50%. This distance lies between 70 and 90 Å, depending on the acceptor used and the spatial arrangements of the fluorophores within the assay.

The limitations of traditional FRET chemistries are caused by background fluorescence from sample components such as buffers, proteins, chemical compounds, and cell lysate. Detected fluorescence intensities must be corrected for this autofluorescence which greatly handicaps assay sensitivity and complicates result interpretation. This type of background fluorescence is extremely transient (with a lifetime in the nanosecond range) and can therefore be eliminated using lanthanide-based time-resolved methodologies.

P.R. Selvin has described the biophysical applications of lanthanide-based resonance energy transfer (LRET), where donor and acceptor fluorophores have been conjugated to a variety of biomolecules creating functional assays such as protein-protein binding, antigen-antibody binding, ligand-receptor binding, DNA hybridization, and DNA-protein binding (Selvin 2002).

Based on this phenomenon, several commercial bioassays have been launched and will be shortly described in the following entry.

Bioassays with Lanthanide Luminescent Probes

Many lanthanide complexes have been synthesized in an effort to obtain species that would satisfy all the above requirements for general use. Especially difficult is the simultaneous achievement of the first three requirements: chemical stability and inertness, high absorbance, and intense luminescence. Because lanthanide complexes, unlike conventional fluorophores, do not undergo concentration quenching, the problem of their relatively low luminescence intensity can be overcome to some extent by employing very high levels of ligation to the targeting species.

One of the pioneer applications of lanthanides and most widely used has been the dissociative enhancement method (Soini and Lövgren 1987), which is suitable for heterogeneous competitive or noncompetitive formats.

In this method, the lanthanide chelate used for labeling, which has strong binding properties, is not itself luminescent in the assay medium owing to the absence of a suitable chromophore in the ligand; in that sense, it is not to be considered as a luminescent lanthanide probe. Rather, the lanthanide ion must be dissociated from the chelator and released into an enhancement solution in which it is luminescent and can be detected. The advantage of this detection system is that it does not require that the optimum binding and luminescence properties of the lanthanide chelate be combined into one label. The labeling chelate has very strong binding properties, is simple and easy to use for labeling, and is chemically and photochemically very stable. The enhancement solution has been chosen to optimize lanthanide luminescence, using a micellar environment containing the energy transfer reagent and various other luminescence enhancing reagents. In the original design, the labeling chelate was a derivative of diethylenetriaminopentaacetic acid (DTPA) bound with Eu(III). The enhancement solution consisted of 3-naphthoyltrifluoroacetone (β-NTA, the energy transfer chelator), trioctylphosphine oxide (TOPO, synergistic ligand), and Triton X-100 detergent in phthalate buffer, pH 3.2 (Figs. 7 and 8).

The low pH causes dissociation of the Eu(DTPA) chelate, followed by incorporation of the Eu(III) into a chelate consisting of 3-naphthoyltrifluoroacetone (β-NTA) and TOPO in the Triton X-100 micellar environment. This method has proven extremely sensitive and quite rugged and has been commercialized and automated. The detection limit for Eu(III), the lanthanide ion giving the highest sensitivity, is 5×10^{-14} mol l^{-1}.

The main disadvantages of the method are sensitivity to exogenous lanthanide ion contamination, and loss of signal localization which makes the method inappropriate for assay formats, such as most membrane blotting and in situ assays, which depend on spatial localization of the signal.

Improvements of these assay formats use a luminescent lanthanide chelate as a probe itself. No dissociation step to release the lanthanide ion is required. The luminescence signal from the lanthanide chelate, which directly labels the analyte, is detected from the surface of the solid phase used for immobilization (such as a plastic microwell, beads, or electrophoretic gel) or directly from the solution. Structures 5c–g (Fig. 5) are representative structures of the number of different probes developed. Most of the applications for directly luminescent lanthanide chelate labels have been in the area of conventional heterogeneous assay formats such as microwell- or membrane-based assays (Matsumoto and Yuan 2003). This methodology has been applied extensively to conventional microwell-based immunoassays for hormones, infectious agent antibodies and antigens, proteins, enzymes, toxins, peptides, antigens, metabolites, and therapeutic drugs (for a review detailing some methods and analytes for which immunoassays have been published or are commercially available, see (Gudgin Dickson et al. 1995)).

A number of lanthanide-based detection systems suitable for use in homogeneous assays have also

Lanthanide Ions as Luminescent Probes, Fig. 7 Assay principle of dissociation-enhanced lanthanide fluoroimmunoassay

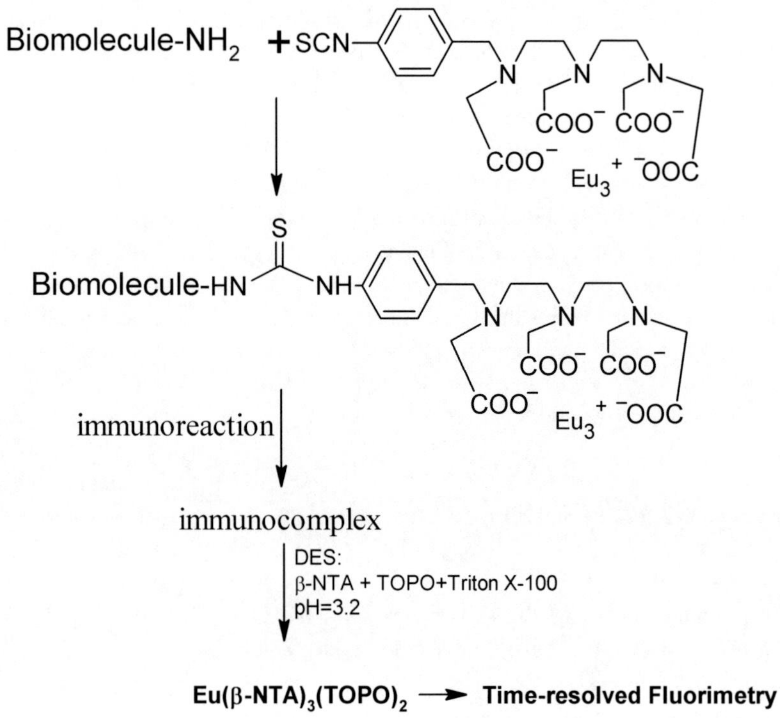

been developed. However, these assays require luminescent lanthanide probes with better sensitivities and selectivities that are much more difficult to obtain than in heterogeneous assay formats. All utilize the same fundamental principle: two or more reagents used in the assay, must be in proximity in order for the specific lanthanide luminescence signal to be either quenched or enhanced. Homogeneous assays rely on direct modulation of the probe luminescence during the biochemical reaction under inspection. The antigen of interest is coupled to two monoclonal antibodies, one draped with a lanthanide label playing as donor and the other one with an organic acceptor emitting at a wavelength distinct from the Ln(III) emission (Fig. 8). Once the immunoreactions take place, the sample is excited onto the excitation maxima of the lanthanide probe, and among the different photophysical processes that come about, measurement in time-resolved mode allows to isolate the wanted signal from the organic acceptor fed by the LRET process. In this way, removal of the unreacted conjugates is not necessary and the analysis time is reduced substantially. TR homogeneous assays have been commercialized under the trademarks: homogeneous time-resolved fluorescence (HTRF) and time-resolved amplification of cryptate emission (TRACE) from CisBio International, Lanthanide chelate energy transfer (LANCE) from Perkin-Elmer, and Lanthascreen from Invitrogen, among others. Figure 8 illustrate the fundamentals of the different systems.

DNA Analysis

During the past 15 years, there has been remarkable progress in the analysis and manipulation of DNA and its use in nanotechnology. DNA analysis is ubiquitous in molecular biology, medical diagnostics, and forensics. Much of the readout technology is based on fluorescence detection and lanthanide luminescence is taking a principal role in this field.

The DNA hybridization assay is one of the most widely used research tools for gene expression analysis and diagnosis of infectious, genetic, and neoplasmic diseases. In this field, luminescent lanthanide probes have been used for many homogeneous DNA assays, where immobilization, prehybridization, and washings are eliminated. Luminescent lanthanide chelates have also been successfully exploited in mixed-phase DNA hybridization assays, high-throughput helicase assays, as well as in real time PCR.

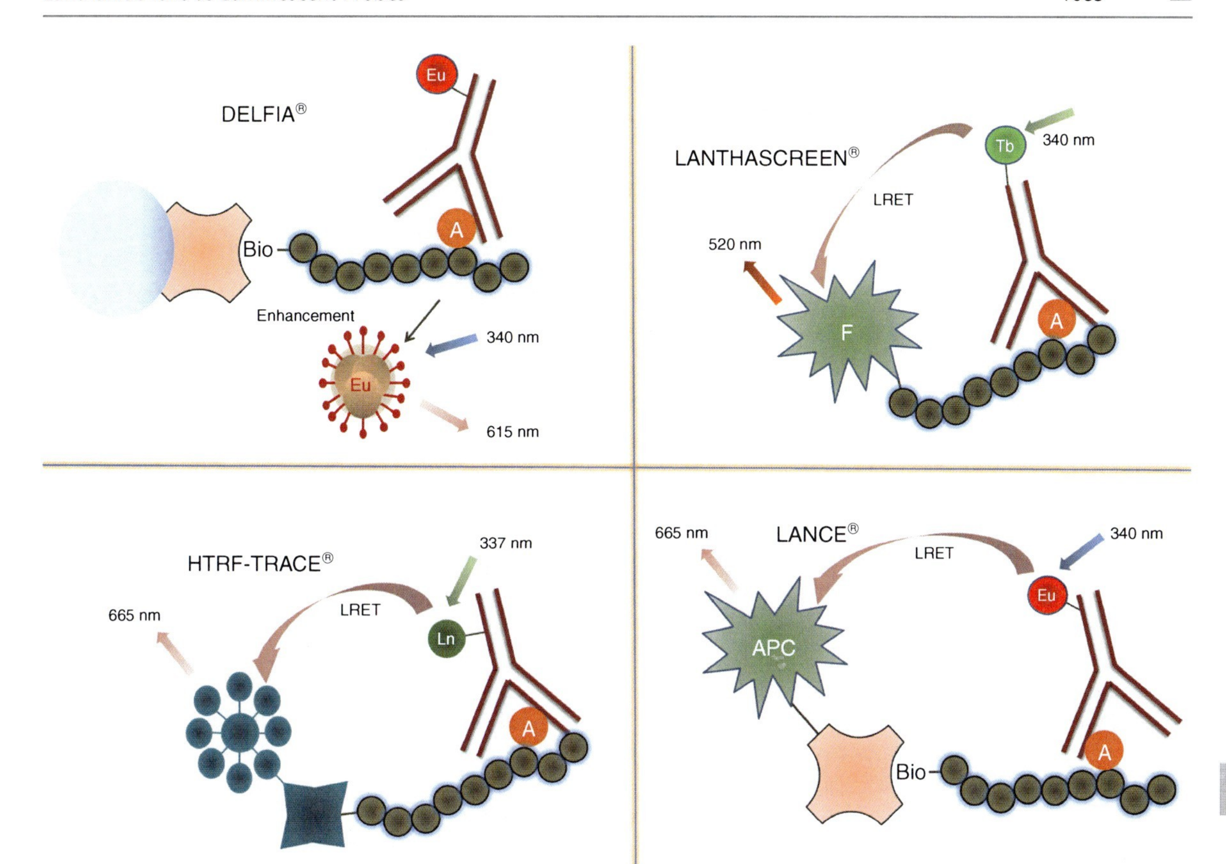

Lanthanide Ions as Luminescent Probes, Fig. 8 Different commercial methods for lanthanide-based heterogeneous and homogeneous assays

Several homogeneous time-resolved luminescence DNA hybridization assay methods based on LRET and luminescence quenching have been also developed (Selvin 2003).

Figure 9 shows a homogeneous DNA hybridization assay format based on the formation of a Eu(III) luminescent complex by approximation of an antenna ligand (AL) and an ion carrier chelate (ICC). The coordination vacants in ICC complements with the formation of a luminescent complex with coordinating AL, which reveals the existence of a target DNA sequence complementary to the sequence of AL plus ICC labeled oligonucleotides.

Lanthanide Luminescent Probes in Cell and Tissue Imaging Applications

Regardless of the success of luminescent lanthanide probes in bioassays shown until here, few studies on imaging took benefit until only very recent times. A key challenge in cell biology is gaining understanding about the structure and function of biological systems through methods that involve minimal perturbation of the system. The development of emissive optical probes is of paramount importance to the progression of the research. Lanthanide luminescent probes are also emerging in the study of cell and tissue imaging by microscopy, their sharp emission bands and long lifetimes offer considerable promise as intracellular optical probes. A very recent review of D. Parker gives account of the progress in the field (Montgomery et al. 2009).

Luminescent Lanthanide Probes in Nanoparticles

Nanoparticles composed of inorganic lanthanide compounds or doped with organic lanthanide complexes have attracted increasing interest for their potential application in fluoroimmunoassay, bioimaging or

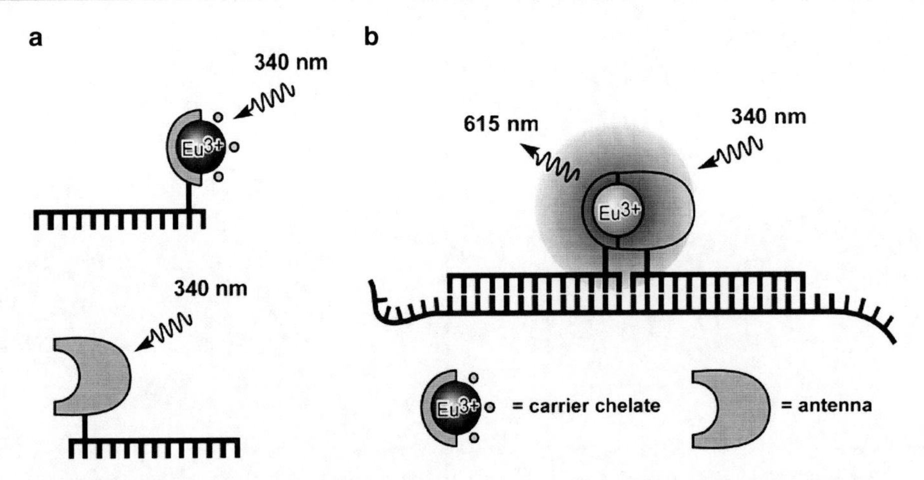

Lanthanide Ions as Luminescent Probes, Fig. 9 Schematic illustration of the oligonucleotide hybridization directed lanthanide chelate complementation assay. (**a**) Without target oligonucleotide, the probe A-EuIII-N1 and the probe B-antenna are nonluminescent. (**b**) Target oligonucleotide induces the formation of the highly luminescent mixed chelate complex, which is excited at 340 nm, and the long-lifetime luminescence emission is measured at 615 nm (Reprinted with permission from: Anal Chem, 2010, 82(2):751–754. Copyright 2010 American Chemical Society)

other techniques. In comparison with free Ln(III) complexes, the luminescent nanoparticles possess the additional benefits of remarkable signal amplification and enhanced photostability since each nanoparticle contains a lot of complex molecules shielded by a matrix.

An attractive approach to enhancing the signal intensity of lanthanide luminescence is to enclose a large amount of luminescent lanthanide chelates in one nanoparticle, which is then used as a probe in an affinity assay. Illustrative, commercially available (e.g., Seradyn Inc.) luminescent lanthanide nanoparticles are polystyrene-based latex particles (100–800 nm) containing a large amount of Eu(III) β-diketone, exhibiting similar spectroscopic properties to the free chelate complexes in solution (Ma and Wang 2010).

Another area of rapid development in lanthanide nanoparticles is up-conversion; this photophysical process shows interesting applications in biological labeling, imaging, and therapy. The term refers to the ability to convert two or more low-energy pump photons to a higher-energy output photon. After being recognized in the mid-1960s, up-conversion has attracted significant research interest as a new class of luminescent optical labels that have become promising alternatives to organic fluorophores and quantum dots for applications in biological assays and medical imaging.

Among the unique luminescent properties of lanthanide ions described until here, they present the ability by up-conversion to convert near infrared long-wavelength excitation radiation into shorter visible wavelengths. In recent years lanthanide-doped up-conversion nanocrystals have been developed for optical and magnetic applications (Wang et al. 2011). These nanocrystals offer low autofluorescence background, large anti-Stokes' shifts, sharp emission bandwidths, and high resistance to photobleaching, and also showing potential for improving the selectivity and sensitivity of conventional methods. Owing to their small physical dimensions and biocompatibility, up-conversion nanoparticles can be easily coupled to proteins or other biological macromolecular systems and used in a variety of assay formats ranging from bio-detection to cancer therapy. In addition, the intense visible emission generated from these nanoparticles under lower energy near-infrared excitation has several advantages, as it is less harmful to biological samples and has greater sample penetration depths than conventional ultraviolet excitation, enhancing their prospects as luminescent stains in bioimaging. Recent developments in optical biolabeling and bioimaging involving up-conversion nanoparticles simultaneously bring to the forefront the desirable characteristics, strengths, and weaknesses of these luminescent nanomaterials (Wang and Liu 2009).

Cross-References

- ► Lanthanide Substitution of Iron-Containing Transferrin, Lactoferrin, and Ferritin, Development of Luminescent Reporter Proteins
- ► Lanthanides in Biological Labeling, Imaging, and Therapy
- ► Lanthanides in Nucleic Acid Analysis
- ► Lanthanides, Physical and Chemical Characteristics
- ► Lanthanide/Actinide in Health and Disease
- ► Lanthanum, Physical and Chemical Properties
- ► Luminescent Tag
- ► NMR Structure Determination of Protein-Ligand Complexes using Lanthanides

References

Gudgin Dickson EF, Pollak A, Diamandis EP (1995) Ultrasensitive bioanalytical assays using time-resolved fluorescence detection. Pharmacol Therapeutics 66:207–235

Hovinen J, Guy PM (2009) Bioconjugation with stable luminescent lanthanide(III) chelates comprising pyridine subunits. Bioconjugate Chem 20:404–421

Latva M, Takalo H, Mukkala VM, Matachescu C, Rodriguez-Ubis JC, Kankare J (1997) Correlation between the lowest triplet state energy level of the ligand and lanthanide(III) luminescence quantum yield. J Lumin 75: 149–169

Ma Y, Wang Y (2010) Recent advances in the sensitized luminescence of organic europium complexes. Coord Chem Rev 254:972–990

Matsumoto K, Yuan J (2003) Lanthanide Chelates as Fluorescence Labels for Diagnostics and Biotechnology. In: Sigel A, Sigel H (eds) Metal ions in biological systems, vol 40. Marcel Dekker Inc, New York, 191–232

Montgomery GP, Murray BS, New EJ, Pal R, Parker D (2009) Cell-penetrating metal complex optical probes: targeted and responsive system based on lanthanide luminescence. Acc Chem Res 42:925–937

Selvin PR (2002) Principles and biophysical applications of lanthanide-based probes. Annu Rev Biophys Biomol Struct 31:275–302

Selvin PR (2003) Lanthanide-labeled DNA. In: Lakowicz J (ed) Topics in fluorescence spectroscopy, vol 7. Kluwer Academic / Plenum Publishers, New York, pp 177–212

Soini E, Lövgren T (1987) Time-resolved fluorescence of lanthanide probes and applications in biotechnology. CRC Crit Rev Anal Chem 18:105–154

Wang F, Liu X (2009) Recent advances in the chemistry of lanthanide-doped upconversion nanocrystals. Chem Soc Rev 38:976–989

Wang G, Peng Q, Li Y (2011) Lanthanide-doped nanocrystals: synthesis, optical-manetic properties and applications. Acc Chem Res 44:322–332

Lanthanide Labels

► Lanthanide Ions as Luminescent Probes

Lanthanide Luminescent Labels

► Lanthanide Ions as Luminescent Probes

Lanthanide Metalloproteins

► Lanthanide Substitution of Iron-Containing Transferrin, Lactoferrin, and Ferritin, Development of Luminescent Reporter Proteins

Lanthanide Substitution of Iron-Containing Transferrin, Lactoferrin, and Ferritin, Development of Luminescent Reporter Proteins

Satoshi Shinoda and Hiroshi Tsukube
JST, CREST, and Department of Chemistry, Graduate School of Science, Osaka City University, Sumiyoshi-ku, Osaka, Japan

Synonyms

Lanthanide ions as luminescent probes; Lanthanide metalloproteins

Definition

The physiological Fe^{3+} centers of transferrin, lactoferrin, and ferritin are substituted by lanthanide cations to develop luminescent proteins working in pH indication and protein recognition.

Metal Substitution Approach Toward Functionalization of Metalloproteins

Many chemical efforts have been devoted toward the understanding of biological protein functions, and have provided a useful basis for wide applications in

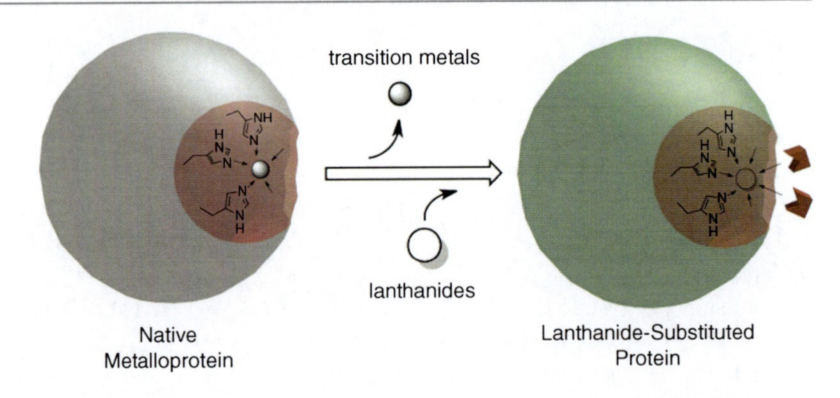

Lanthanide Substitution of Iron-Containing Transferrin, Lactoferrin, and Ferritin, Development of Luminescent Reporter Proteins, Fig. 1 Lanthanide substitution strategy for functionalization of metalloproteins

pharmacology, biosensing, and nano-biomaterials sciences. Chemical functionalization of metalloproteins is a typical example, though this is a more challenging problem than that of non-metalloproteins. The tuning of metal coordination structures, reorganization of substrate recognition modes, and addition of uncommon environmental domains successfully lead to functional conversion of a metalloprotein. Metal substitution can be considered as another promising approach, which tunes the structure and reactivity of the metalloprotein to generate nonnatural redox states, coordination arrangements, and varying functions. In addition to substitution with transition metals, lanthanide substitution of a metalloprotein can introduce a new class of protein-based functional materials such as MRI contrast agents, asymmetric catalysts, and luminescent devices (Fig. 1).

Trivalent lanthanide centers have large ionic radii (0.89–1.16 Å in octa-coordinated complexes) and high coordination numbers (8–12), which results in different coordination chemistry compared to common transition metal centers. Among them, Eu^{3+}, Tb^{3+}, Nd^{3+}, and Yb^{3+} cations are particularly useful in the development of luminescent labeling and sensing materials. Complexes of these metals have excited states insensitive to quenching by molecular oxygen, and exhibit emission spectra consisting of well-defined and narrow bands. The lanthanide luminescence has lifetimes in the microsecond range for Yb^{3+} and Nd^{3+} complexes, and in the millisecond range for Eu^{3+} and Tb^{3+} complexes. Since these lifetimes are much longer than those of most organic fluorophores (ca. 10 ns), the lanthanide complexes have potential as luminescent reagents applicable in fluorescence microscopy and bioassays, in which background emission and auto-fluorescence from the biological environment can then be excluded using time-gated measurements.

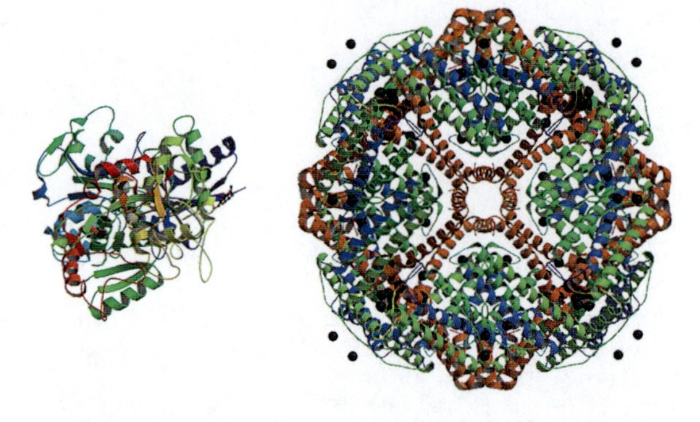

Lanthanide Substitution of Iron-Containing Transferrin, Lactoferrin, and Ferritin, Development of Luminescent Reporter Proteins, Fig. 2 Three-dimensional illustrations of biological transferrin and ferritin. PDB ID: transferrin, 1H76 (Hall et al. 2002); ferritin, 1FHA (Lawson et al. 1991)

Figure 2 schematically illustrates molecular structures of transferrin and ferritin proteins, which are common Fe^{3+} cation-carrying proteins. They function as supramolecular ligands to arrange several ligating amino acid residues for effective metal complexation. They also have mosaic-like charge distributions on their surfaces. Such ligand architecture cannot be easily accomplished by synthetic ligands. When the metal center of the metalloprotein is substituted by a luminescent lanthanide cation, the resulting complex is coordinatively unsaturated, and the remaining coordination sites are habitually occupied by counter-anions and/or labile solvent molecules to form a highly coordinated complex with an external guest. The formed ternary complex changes the coordination chemistry of the lanthanide center, and causes large changes in spectral shape and lifetime of the lanthanide emission. Therefore, lanthanide substitution of metalloproteins can be an effective tool for engineering precise luminescent sensory materials for many analytical applications.

Transferrin–Tb^{3+} Complex as a Luminescent pH Indicator

Transferrin and lactoferrin are dimeric glycoproteins that bind up to two Fe^{3+} cations for delivery in biological processes (Fig. 2, left). Each protein has two similar but not identical Fe^{3+} binding sites. The Fe^{3+} center is hexa-coordinated by four protein ligands (2 Tyr, 1 Asp, and 1 His) and a synergistic CO_3^{2-} anion in a bidentate fashion, which fits into a pocket between the metal center and *N*-terminus of an α-helix. (Sun et al. 1999; Hall et al. 2002). The two bound Fe^{3+} cations are released from the transferrin at pH 4.0–6.0, while the lactoferrin loses its Fe^{3+} cations at pH 2.5–4.0.

The transferrin protein has a sophisticated three-dimensional structure for forming luminescent lanthanide complexes (Pecoraro et al. 1981). When the Fe^{3+} cation was substituted by a lanthanide cation, an octa-coordinated complex was formed in which two water molecules directly coordinated to the lanthanide center with six other donor sites as found in the Fe^{3+} complexes. The Tb^{3+} cations were bound rapidly and reversibly to induce Tb^{3+} luminescence changes at pH 6.0–8.0 (Fig. 3) (Kataoka et al. 2009). Such pH-dependent luminescence changes were detectable both spectroscopically and visually. Lactoferrin formed a more stable Tb^{3+} complex than transferrin, and exhibited pH-dependent luminescence behavior at a lower pH. Since H^+ concentration plays a central modulating role in many metabolic and cellular events, a number of methods such as NMR, absorption, and fluorescence spectroscopies have been developed for pH measurements, in addition to using pH microelectrodes. Among them, fluorescence spectroscopy has significant advantages due to its nondestructive character, high sensitivity, excellent specificity, and wide availability. Two classes of fluorescent pH indicators are required for biological assays: one for cytosol at pH 6.8–7.4 and the other for lysosomes and endosomes at pH 4.5–6.0. The present type of luminescent

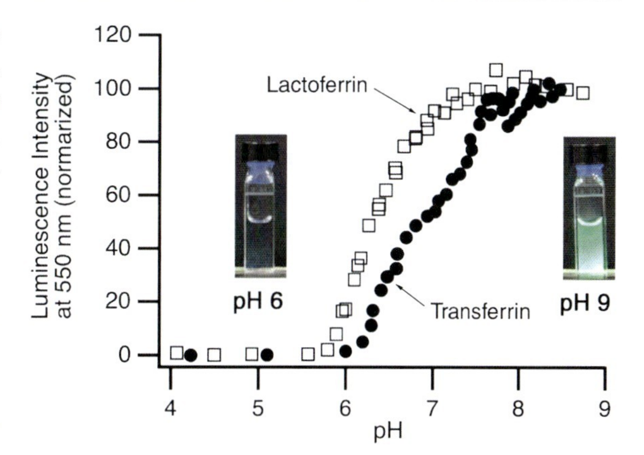

Lanthanide Substitution of Iron-Containing Transferrin, Lactoferrin, and Ferritin, Development of Luminescent Reporter Proteins, Fig. 3 pH-dependent emission profiles of transferrin–Tb^{3+} and lactoferrin–Tb^{3+} complexes

protein–lanthanide complexes functioned as tunable pH indicators for these purposes. Although transferrin has higher affinities for several transition metal cations (log K_1 = 21.4 for Fe^{3+} and 12.3 for Cu^{2+}) than Tb^{3+} cation (log K_1 = 9.0), non-chromophoric cyclen ligands effectively masked the transition metal cations in the luminescence detection process.

The transferrin–Tb^{3+} complex was successfully immobilized on monodispersed polystyrene beads (size: 1 μm) via biotin–avidin interaction. The immobilized transferrin–Tb^{3+} complexes were detected as green-luminescent particles at pH 7.0 by time-gated luminescence microscopy, while no luminescent particles were observed at pH 4.0. This fluorescence microscopic technique allows mapping of the spatial and temporal distribution of H^+ on the micrometer scale. Since the transferrin–Yb^{3+} complex was reported to permeate into the cell via a transferrin-receptor mechanism (Du et al. 2002), the present type of protein lanthanide complexes provides wide applications for pH determination.

Ferritin–Tb^{3+} Complex as a Luminescent Protein Probe

Ferritin is an iron-storage protein, which consists of 24 symmetrical subunits that form a spherical hollow shell of external diameter ca. 13 nm and internal diameter 7.5 nm (Fig. 2, right) (Ueno 2010). There are several different kinds of coordination sites, including

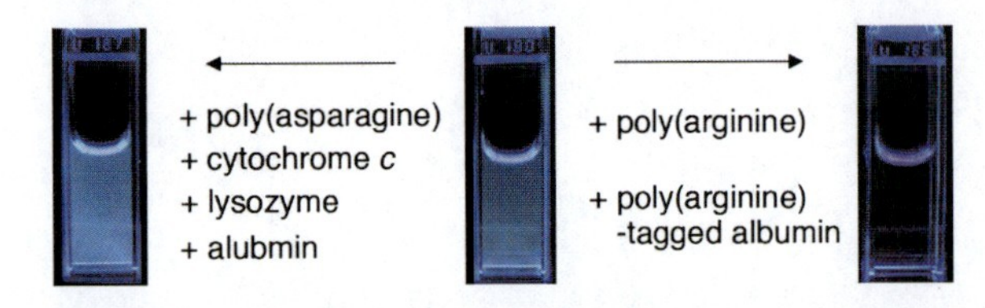

Lanthanide Substitution of Iron-Containing Transferrin, Lactoferrin, and Ferritin, Development of Luminescent Reporter Proteins, Fig. 4 Protein coprecipitation with ferritin–Tb^{3+} complex

Asp, Glu, Tyr, and His, on both interior and exterior surfaces. Although the lanthanide complexation behavior has not been detailed, ferritin accommodates Tb^{3+} cations in its protein domains to exhibit green luminescence in aqueous solution (Barnes et al. 2003). The ferritin–Tb^{3+} complex has several outstanding features that enable it to serve as an effective luminescence device for bio-analytical processes: (1) It maintains a spherical protein shell (diameter = ca. 13 nm) featuring anionic amino acid residues. (2) Excitation of the protein chromophore gives long-lived green luminescence in aqueous media. (3) Several functional groups and metals can be introduced on the surface of the protein shell. (4) Some derivatives were reported to be internalized into the cells.

Apoferritin and its Tb^{3+} complex exhibited unique self-assembling properties in aqueous solutions and formed coprecipitates with specific polycations (Fig. 4) (Tsukube et al. 2010). Since protein–protein interactions are central to many biological and biotechnological processes, the precipitation method is widely used both to characterize these interactions and for further concentration and fractionalization of the targeted proteins. Dynamic light scattering studies revealed that a solution of the apoferritin itself mainly contained monomeric, dimeric, and trimeric aggregates. Although Cd^{2+} and Fe^{3+} cations were reported to promote two-dimensional crystallization in solution, addition of Tb^{3+} cations gave higher order oligomers having an average diameter of 200 or 300 nm. The zeta potentials were measured as indications of surface electrical properties of the protein aggregates. Apoferritin and its Tb^{3+} complex exhibited small zeta potentials (–7.6 mV and –8.8 mV, respectively); therefore, they were not very stable under the employed conditions and favored higher aggregation.

When oligo(arginines) were added to an aqueous solution of ferritin–Tb^{3+} complex at neutral pH, precipitated materials appeared due to protein–protein interactions. The amounts of the precipitated materials significantly depended on the chain length of the oligo (arginines). Arginine, an amino acid, has a permanent positive charge on its sidearm, but rarely coprecipitates with ferritin–Tb^{3+} complex. In contrast, 20-mer and higher oligo(arginines) formed insoluble materials in quantitative yields, indicating that the polycationic sequence of the arginines was involved in the interaction with the ferritin complex. Since oligo(arginines) of the extended conformation were too large to penetrate the channel of the protein shell, they bound to the exterior surface of the ferritin. Various oligo (arginines) derivatives are known to play important roles in many biological protein recognition processes. A high number of arginines are often involved in protein–protein interactions. Attachment of the oligo (arginines) to the protein promotes internalization more effectively than lysine derivatives. The arginine-tag technology is also employed in the purification and detection of targeted proteins. In addition to the oligo(arginines) family, cytochrome c, lysozyme, albumin, oligo(arginines)-tagged albumin, and other protein substrates were examined in the coprecipitation experiments. Only oligo(arginines)-tagged albumin formed precipitates with the ferritin–Tb^{3+} complex. Since such specific precipitation phenomena were easily followed by monitoring the Tb^{3+} luminescence, the ferritin–Tb^{3+} complex was effective as a new luminescence sensing device specific for oligo (arginines)-tagged biomolecules.

Conclusion

Several kinds of lanthanide complexes were synthesized as new examples of luminescent materials that exhibited long-lived, guest-responsive luminescence. A general strategy for attaching the luminescent lanthanide complexes with the targeted proteins has been established, and can be successfully applied in cell biology and biochemical research. Lanthanide substitution of metalloproteins provides an alternative method for developing specific luminescent materials. Biological Fe^{3+}-carrying transferrin, lactoferrin, and ferritin were readily converted to luminescent reporter proteins by lanthanide substitution. The complexes were demonstrated to perform effectively as luminescent analytical tools in pH indication and protein recognition. The metal-substituted proteins typically showed identical mobilities in non-denaturing gel electrophoresis and circular dichroism spectra compared to the native ones, suggesting that they maintained biologically active structures. Further uses of the lanthanide-substituted proteins in combination with time-gated analysis, nanoparticle-functionalization, and micelle-immobilization will allow precise recognition and detection of various biochemical targets.

The authors gratefully acknowledge the financial support from the Japan Society for Promotion of Science (Nos. 20245014 and 22350065). They also thank American Science Publishers and Wiley-VCH for permission to use Figs. 3 and 4.

Cross-References

- ▶ Iron Proteins, Ferritin
- ▶ Iron Proteins, Transferrins and Iron Transport
- ▶ Lanthanide Ions as Luminescent Probes
- ▶ Lanthanide Metalloproteins
- ▶ Lanthanides in Biological Labeling, Imaging, and Therapy
- ▶ Lanthanides, Physical and Chemical Characteristics

References

Barnes CM, Petoud S, Cohen SM, Raymond KN (2003) Competition studies in horse spleen ferritin probed by a kinetically inert inhibitor, $[Cr(TREN)(H_2O)(OH)]^{2+}$, and a highly luminescent Tb(III) reagent. J Biol Inorg Chem 8:195–205

Du X-L, Zhang T-L, Yuan L, Zhao YY, Li R-C, Wang K, Yan SC, Zhang L, Sun H, Qian Z-M (2002) Complexation of ytterbium to human transferrin and its uptake by K562 cells. Eur J Biochem 269:6082–6090

Hall DR, Hadden JM, Leonard GA, Bailey S, Neu M, Winn M, Lindley PF (2002) The crystal and molecular structures of diferric porcine and rabbit serum transferrins at resolutions of 2.15 and 2.60A, respectively. Acta Crystallogr Sect D 58:70–80

Kataoka Y, Shinoda S, Tsukube H (2009) Transferrin–terbium complexes as luminescent pH sensing devices. J Nanosci Nanotechnol 9:655–657

Lawson DM, Artymiuk PJ, Yewdall SJ, Smith JMA, Livingstone JC, Treffry A, Luzzago A, Levi S, Arosio P, Cesareni G, Thomas CD, Shaw WV, Harrison PM (1991) Solving the structure of human H ferritin by genetically engineering intermolecular crystal contacts. Nature 349:541–544

Pecoraro VL, Harris WR, Carrano CJ, Raymond KN (1981) Siderophilin metal coordination. Difference ultraviolet spectroscopy of di-, tri-, and tetravalent metal ions with ethylenebis[(o-hydroxyphenyl)glycine]. Biochemistry 20:7033–7039

Sun H, Li H, Sadler P (1999) Transferrin as a metal ion mediator. Chem Rev 99:2817–2842

Tsukube H, Noda Y, Shinoda S (2010) Poly(arginine)-selective coprecipitation properties of self-assembling apoferritin and its Tb^{3+} complex: a new luminescent biotool for sensingof poly(arginine) and its protein conjugates. Chem Eur J 16:4273–4278

Ueno T (2010) An engineered metalloprotein as a functional and structural bioinorganic model system. Angew Chem Int Ed 49:3868–3869

Lanthanide Time-Resolved Fluorescence

▶ Lanthanide Ions as Luminescent Probes

Lanthanide/Actinide in Health and Disease

Karel Nesmerak
Department of Analytical Chemistry, Faculty of Science, Charles University in Prague, Prague, Czech Republic

Synonyms

Ln/An as chemotherapeutics; Ln/An as diagnostics

Definition

The lanthanides are a group of 15 elements with atomic numbers 57 through 71, from lanthanum to lutetium. The actinide series encompasses the elements with atomic numbers from 89 to 103, from actinium to lawrencium. None of these elements is essential for human or animals. The lanthanides have interesting biochemical effects, which are applicable in medicine. The effects of the actinides on organisms are dominated by their radioactivity; hence, a radiotherapy is the main medical application of the actinides.

Ln/An Biochemical Effects

Under physiological conditions, most of the lanthanides (abbreviated to Ln) are stable in the trivalent form, the only exceptions being Ce^{4+} and Eu^{2+}. Many of the biological properties of the lanthanides are a function of their ability to replace calcium in many biomolecules, without necessarily substituting for it functionally (Fricker 2006). The lanthanides have similar ionic radii to calcium, but due to a higher charge, they have a high affinity for Ca^{2+} sites on biological molecules.

The principal physiological effect of the Ln^{3+} is to block both voltage-operated and receptor-operated calcium channels. The Ln^{3+} can block the Na^+/Ca^{2+} synaptic plasma membrane exchange and inhibit skeletal, smooth, and cardiac muscle contraction by blocking the Ca^{2+}–ATPase in the sarcoplasmic reticulum of muscle. The lanthanides ions themselves are unable to cross cell membranes but act by blocking the exterior face of the calcium channel. Though the Ln^{3+} cannot gain access to intracellular organelles, they have been used as biochemical probes to study calcium transport in mitochondria and other organelles using techniques such as NMR, luminescence, or fluorescence spectroscopy (as the lanthanides spectroscopic properties resulted from their unusual electronic configuration). The potency of calcium channel blockade increases with the ionic radius of the lanthanide ion. Blockade of the T-type voltage-gated calcium channel by Ln^{3+} has been ranked in the order: $Ho^{3+} \approx Y^{3+} \approx Yb^{3+} \geq Er^{3+} > Gd^{3+} > Nd^{3+} > Ce^{3+} > La^{3+}$. Examples of receptor-operated calcium channel inhibition are a blockade of the vasopressin stimulated Ca^{2+} and Mn^{2+} influx across the hepatocyte membrane, and Ca^{2+}-dependent neurotransmitter release (e.g., epinephrine, serotonin, dopamine).

The lanthanides can also substitute for calcium (as well as other metals such as Mg^{2+}, Fe^{3+}, or Mn^{2+}) in proteins. Calcium-dependent enzymes can either be inhibited by lanthanides or, in some cases, be activated to a similar or greater extent than by calcium. Where calcium plays a catalytic role, the Ln^{3+} substitution leads to inactivation of the enzyme. For example, lanthanides interfere with calcium-dependent reactions involved in the blood-clotting cascade such as the activation of prothrombin and factor XIII. This results in the long-known anticoagulant effect of the lanthanides. If calcium plays a structural role, then Ln^{3+} substitution should lead to, at least, retention of activity (e.g., trypsin, acetylcholinesterase). The lanthanides can also inhibit calcium-mediated processes associated with immune cell function.

The effects of the actinides (abbreviated to An) on organisms predominantly result from their radioactivity. Thus, little is known on the actinides nonradioactive effects on biochemical reactions, but the analogy to the biochemical action of the lanthanides is assumed (Ansoborlo et al. 2006). Under physiological condition, plutonium and thorium are considered to be only present under An^{4+} form, americium and curium under An^{3+} form, neptunium can be present under both forms Np^{4+} and Np^{5+}, and uranium under U^{6+} form. The most actinides, when present in blood, are mainly bound to the iron transport protein transferrin. The apparent stabilities of the actinide–transferrin complexes are in the order: $Pa^{5+} > Pu^{4+} > Th^{4+} > Np^{4+} >> UO_2{}^{2+} > Cf^{3+} > Am^{3+} > Cm^{3+}$. The actinides affinity to proteins has been found also in liver, bone, and kidney (e.g., calmodulin, ferritin, sialoprotein). Thorium has been found to interact with Zn^{2+} in proteins. Uranyl ions alternate in some transport mechanisms across biological membranes (e.g., transport of phosphate or sodium), and they inhibit mitochondrial oxidative phosphorylation. Uranyl ions exchange with Ca^{2+} on the surface of bone mineral crystals although they do not participate in crystal formation or enter to existing crystals.

Ln/An Chemotherapeutic Applications

The earliest therapeutic application of a lanthanide was the use of cerium oxalate as an antiemetic. Formulations of cerium oxalate were available from the mid-nineteenth century to the mid-twentieth century. The mechanism of action has never been elucidated. In the early twentieth century, lanthanide salts were being used for the treatment of tuberculosis. Later, the lanthanides were found to have anticoagulant and antiatherosclerotic properties, but these applications were accompanied by severe side effects. The therapy of diabetes by uranium nitrate and treatment of flu by uranium acetate were the only chemotherapeutic applications of the actinides, but both are obsolete from a half of the twentieth century.

Antimicrobial Properties and the Treatment of Burn Wounds. Antibacterial effects of lanthanides salts were discovered at the end of the nineteenth century. Systematic studies later confirmed that cerium nitrate had broad-spectrum antibacterial activity against a range of bacteria including *Pseudomonas aeruginosa* and *Staphylococcus aureus*. Cerium nitrate has been shown to have beneficial effects on burn wounds (Fricker 2006). There is a high risk of mortality associated with major, full thickness burns, which is generally attributed to sepsis. Cerium nitrate is usually administered in combination with silver sulfadiazene, another metal-based agent with demonstrated efficacy in the treatment of burn wounds. The combination is manufactured commercially as Flammacerium® or Dermacerium®. The healing effect of cerium nitrate is based on its modulation of the burn-associated immune response. Burns of over 25% total body surface area disrupt immune defenses, and the immune regulation is endangered. This is manifested in part by immunosuppression, which makes the patient more susceptible to infection. Moreover, dysregulation of the immune response is a contributory factor to the systemic inflammatory immune response and subsequent multiorgan failure syndrome associated with mortality. The burn toxin responsible for the observed immune suppression was identified as a high-molecular-weight lipid protein complex (LPC) formed by heat-induced polymerization of six skin polypeptides. LPC induces a number of immune responses. Particularly, the LPC is a potent inhibitor of T-cell growth suggesting that, at least in part, the mechanism of immunosuppression in burn patients is due to LPC inhibition of T-cell activation. Though cerium nitrate can bind to and inactivate LPC, the immunomodulatory effect of cerium nitrate appears to be mediated by its effect on the burn eschar, the dry scab formed on the skin following a burn wound. Topical application of cerium nitrate leads to a firm, impermeable eschar, which is leather-like in appearance with a greenish discoloration and is firmly attached to the wound. The eschar deposits formed by cerium nitrate contain deposits of insoluble pyrophosphate, carbonate salts, and calcium. It has been proposed that cerium may bind pyrophosphate, thus removing inhibition to local calcium deposition. The resultant eschar acts as a biological dressing, forming an impermeable crust over the wound. This covering may prevent both ingress and egress of bacteria from the wound, thus preventing bacterial colonization, and egress of LPC into the systemic circulation, thus inhibiting activation of the sepsis cascade and multiorgan failure syndrome. This leaves the wound in a clean, healthy state, ready to accept a skin graft.

Management of Hyperphosphatemia. Hyperphosphatemia, increased serum phosphate levels, is one of the clinical consequences that accompany end-stage renal disease (ESRD). Normal adult serum phosphate levels range from 21 to 43 mg/L compared with the elevated levels of 62–93 mg/L found at ESRD patients. The average dietary phosphate intake is around 1,000–1,500 mg per day. At health subjects, phosphate is absorbed in the intestine and excreted via the kidney resulting in a net phosphate balance of zero. Phosphate metabolism is intimately linked with calcium metabolism and is regulated by parathyroid hormone (PTH) and vitamin D. PTH controls phosphate balance in the body by lowering tubular reabsorption of phosphate by the kidney. Consequently, during renal impairment, PTH secretion increases in an attempt to further decrease phosphate reabsorption and correct the hyperphosphatemia. Furthermore, vitamin D metabolism in the kidney is impaired in ESRD resulting in reduced calcium absorption and hypocalcemia. This decrease in calcium can in turn stimulate PTH secretion. In this way, the body attempts to correct for the high phosphate levels, but at the expense of increased parathyroid activity, a state described as secondary hyperparathyroidism. The pathological consequences of hyperphosphatemia are severe. These interacting events lead to several bone malformations in the joint, a disease known as renal osteodystrophy. Nearly half of the deaths of dialysis patients are due to cardiovascular complications prevalent in patients with ESRD. Cardiac and vascular tissue calcification is a major contributor to cardiovascular disease. Damage caused by calcification of cardiac tissue can lead to arrythmias, left ventricular dysfunction, damage to heart valves, and ultimately complete heart block. In addition, hyperphosphatemia and hypocalcemia are associated with atherosclerosis and a complication of calcification of soft tissues (e.g., lung, kidney). Unfortunately, hyperphosphatemia cannot be controlled by normal dialysis. Thus, the binding of dietary phosphate in the gut has been the favored option. There are several phosphate binders, such as aluminum hydroxide, calcium-based binders (e.g., calcium carbonate, acetate, ketoglutarate), or Renagel® (sevalamer hydrochloride), but every have had side effects. Recently, a new phosphate-binding drug based on lanthanum carbonate, Fosrenol®, has been introduced (Curran and Robinson 2009). It is based on the fact that the lanthanides formed precipitates with phosphate. Improved phosphate binding has been obtained with the lanthanum carbonate tetrahydrate, with optimal binding at pH 3–5 while retaining binding activity across the full pH range of 1–7. Thus, lanthanum carbonate can bind phosphate at the low pH of the stomach, as well as at the higher pH values found in the small intestine, duodenum, and jejunum, unlike calcium carbonate. In addition, lanthanum carbonate has little to no oral absorption and tissue accumulation and effectively completes elimination in the feces. Lanthanum carbonate is equally effective as aluminum hydroxide and more effective than either Renagel or calcium carbonate. It is well tolerated, poorly absorbed and therefore not accumulated in tissues, binds phosphate effectively across the physiological pH range of the upper gastrointestinal tract, and has no detrimental effect on calcium, vitamin D, or PTH metabolism.

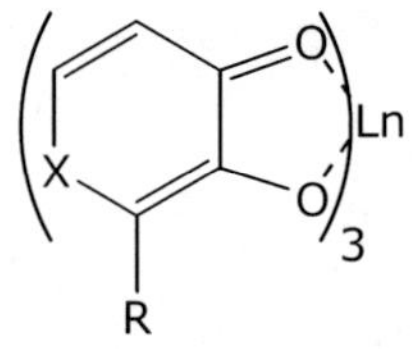

Therapeutic Care of Bone Disorders. The precipitation of the lanthanides with phosphate has been basis for experiments on the replacement or reinforcement of calcium in teeth by lanthanum. Prolonged treatment with La^{3+}-EDTA gives a lanthanum phosphate coating on the enamel, which is claimed to provide significant protection against dental caries. Another investigation of application of lanthanides in care of bone disorders has been focused on bone resorption disorders, including osteoporosis (Barta et al. 2007). Osteoporosis is characterized by low bone mineral density that leads to enhanced bone fragility and a consequent risk of low-impact bone fractures. The low bone mineral density is a result of an imbalance between bone resorption and bone formation. Normally, building and absorption of bone is a tightly regulated cycle wherein the bone matrix is manufactured by osteoblast cells and removed by osteoclast cells. Either increased activity of osteoclasts or decreased bone formation by osteoblasts leads to microarchitectural deterioration of bone tissue. Many contributing factors are known to influence the pathogenesis of the disease with the most prominent being inadequate calcium uptake. Lanthanum carbonate (Forsenol®, vide supra) has been proposed as a potential prophylactic agent for osteoporosis; however, gastrointestinal upset and the extremely low bioavailability are known negative side effects of this treatment. In an effort to find new orally active agents for these disorders, a series of Ln^{3+}-containing complexes incorporating small, nontoxic, bidentate pyrone and pyridinone ligands (Fig. 1) have been synthesized and tested. In vitro studies have been shown that the complexes investigated showed high hydroxyapatite binding (>98%) and did not appear to disturb the hydroxyapatite structure upon binding. These novel Ln^{3+} complexes thus have potential as orally available bone resorption inhibition drugs, with minimized side effects that may also serve as potential prophylactic agent for osteoporosis.

Lanthanide/Actinide in Health and Disease, Fig. 1 The structures of lanthanide complexes for osteoporosis treatment

Ln = La, Eu, Gd, Tb, Yb
X = O, NH, NCH_3
R = CH_3, CH_2CH_3

Future and Possible Chemotherapeutic Applications. The lanthanides have been reported to inhibit lymphocyte activation, neutrophil chemotaxis and aggregation, Kupffer cell activity, and histamine secretion from mast cells; reduce reactive oxygen species; reduce histamine- and serotonin-induced vascular permeability; and reduce carrageenin-induced inflammation. On the other hand, at low doses, they appear to enhance some aspects of the immune response such as antibody formation and lymphocyte activation (hormetic effect). The lanthanides have been investigated for the treatment of liver toxicity, artherosclerosis, and rheumatoid arthritis. A common theme linking these apparently disparate diseases is the interaction of the lanthanides with components of the immune system. The treatment of rheumatoid arthritis – an inflammatory disease characterized by a progressive erosion of the joints resulting in deformities, immobility, and a great deal of pain – is very promising field for lanthanide-based chemotherapeutics. Chemical mediators such as prostaglandins, leukotrienes, and cytokines drive this progressive inflammatory response. Tissue erosion is mediated by the release of degradative enzymes, such as collagenase, and reactive oxygen species. Chlorides of praseodymium, gadolinium, and ytterbium are able to reduce carrageenin-induced inflammation, and gadolinium chloride and cerium nitrate are able to modulate the levels of the inflammatory cytokines associated with toxicant-induced liver damage, suggesting that Ln^{3+} may be able to modulate the inflammatory process in rheumatoid arthritis. Additionally, it was found that Ln^{3+} could inhibit the activity of neutral metalloproteinases such as collagenase. Samarium ions proved the best inhibitor of gelatinase and caseinase, while lanthanum ions inhibited collagenase the most strongly. Furthermore, Ln^{3+} can reduce reactive oxygen species produced under inflammatory conditions. These observations led to the proposal that Ln^{3+} may have antiarthritic properties, but this application appears not to have advanced beyond biochemical studies up to the present time. Oral administration of lanthanum chloride in rabbit atherogenic diet resulted in inhibition of the development of atherosclerosis. Oxidation of low-density lipoprotein by reactive oxygen species is a key step in atherosclerotic plaque formation, so reduction of reactive oxygen species levels is one possible explanation for this observation. An alternative explanation is interference in the involvement of calcium in plaque deposition, as lanthanides can also inhibit platelet aggregation which is calcium dependent. However, the adverse effects of lanthanides on the cardiovascular system hindered their development as agents for the treatment of atherosclerosis. The hepatoprotective effect of gadolinium chloride towards toxicant-induced liver damage is an interesting potential application of lanthanide chemotherapeutics. Gadolinium chloride has been shown to protect against liver damage caused by a variety of toxicants including ethanol, tetrachloromethane, and cadmium. The Kupffer cells, which are resident macrophages in the liver, release cytokines and free radicals upon activation by chemicals such as alcohol that contribute to liver injury. The hepatoprotective effect of $GdCl_3$ is primarily due to the inactivation and destruction of the Kupffer cells. This results in a reduction in cytokine and reactive oxygen species production. Similarly, gadolinium chloride can also protect lungs from postischemic injury by lowering reactive oxygen species. Both findings suggest that there may be a therapeutic application for gadolinium chloride in the treatment of liver fibrosis. At least, the lanthanides can block stretch-activated channels in muscle, and this may have therapeutic applications for Duchenne muscular dystrophy, a degenerative muscle disease resulting in death due to respiratory muscle failure, in which stretch-induced muscle damage after exercise contributes to long-term muscle degeneration.

Ln/An in Diagnostics and Treatment of Cancer

Cancer is a class of diseases in which a group of cells display uncontrolled growth. Cancer is caused when genetic damage to the cells prevents them being responsive to normal tissue controls. Cancer is primarily an environmental disease (e.g., smoking, environmental pollutants, diet, infections, radiation), though a hereditary genetics influence the risk of some cancers. Cancer is one of the most frequent death causes (approximately about 15% of all human deaths worldwide); therefore, the timely diagnosis and efficient treatment are ways for suppression of cancer consequences. The lanthanides and their complexes have found a role in cancer treatment as anticancer agents, agents for photodynamic therapy, and magnetic resonance imaging agents for diagnostics. Ln/An radioisotopes have been used both for radioimaging and radiotherapy of tumors. The luminescent properties of the lanthanides also have been utilized in medical analysis (Bünzli 2010). A variety of luminescent bioassays and sensors have been developed that take advantage of the unique luminescent properties of these elements, such as a relatively long-lived emission.

Anticancer Potential of Lanthanides. Early clinical reports suggested that cerium iodide had activity against solid tumors. More recently, work has focused on lanthanide complexes. The lanthanides have been reported to inhibit reactive oxygen species (vide supra), which along with other factors disturb the cell proliferation, differentiation, and apoptosis. The inhibition is based on the fact that the lanthanides competitively suppress iron uptake to the cell. An iron overload is cause of iron-mediated reactive oxygen species generation. Lanthanide compounds have also been studied as agents for the prevention of cancer, based on their anticarcinogenic effects or inhibition of proliferation of tumor cells. For example, in vitro studies showed that Ln nitrate and Ln citrate inhibit the growth of human leukemia and human stomach cancer cells. Nevertheless, although a number of lanthanide compounds possess anticancer potency, rather high doses are needed, and the toxicity is thus significant. As the toxicity of Ln chelates is several orders of magnitude lower than inorganic compounds, the attention has been drawn to synthesis and application of lanthanide complexes (Kostova 2005). Preclinical studies of La^{3+}, Ce^{3+}, Nd^{3+} coumarin, and bis-coumarins complexes (Fig. 2) have indicated that the number of these compounds has antiproliferative activity on various cancer cell lines (e.g., P3HR1 Burkitt lymphoma cells, THP-1 myeloleukemia cells, Ehrlich ascites tumor cells). Lanthanide complexes have also been proved as agents for photodynamic therapy, a minimally invasive method for treatment of cancer. In this therapy, a drug, that preferentially localizes in rapidly growing cells and gets activated by the exposure of light in the presence of oxygen, generates very reactive cytotoxic species. The porphyrins show inherent tumor localizing properties coupled with the ability to generate reactive singlet oxygen when activated by light of particular wavelength. Texaphyrins, water-soluble tripyrrolic pentaaza-expanded porphyrins, are capable to form stable complexes with large metal cations including lanthanides (Fig. 3). Texaphyrins absorb and are activated by light with wavelengths greater than 700 nm. This allows for greater tissue penetration of the activating light, which facilitates their use as agents for photodynamic therapy. For example, gadolinium(III) texaphyrin complex (motexafin gadolinium) has been investigated as a radio- and chemosenzitizer for photodynamic treatment of brain metastases and primary brain tumors. Lutetium(III) texaphyrin has been applied for locally recurrent breast cancer, furthermore for the treatment of atherosclerotic plaque in coronary heart disease and for treatment of age-related macular degeneration.

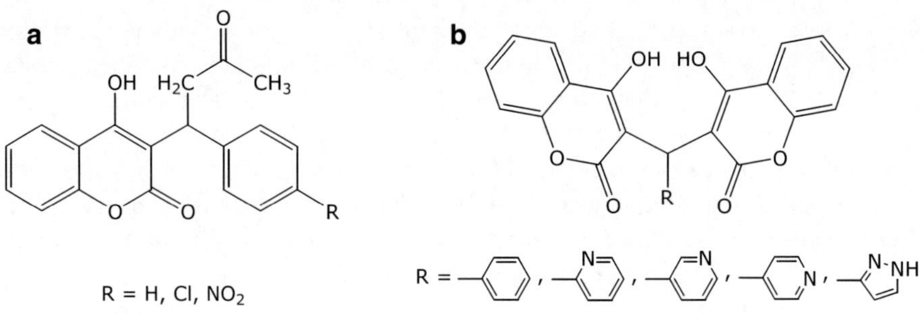

Lanthanide/Actinide in Health and Disease, Fig. 2 The structures of (**a**) coumarin and (**b**) bis-coumarin ligands for lanthanides complexes possess the anticancer effects

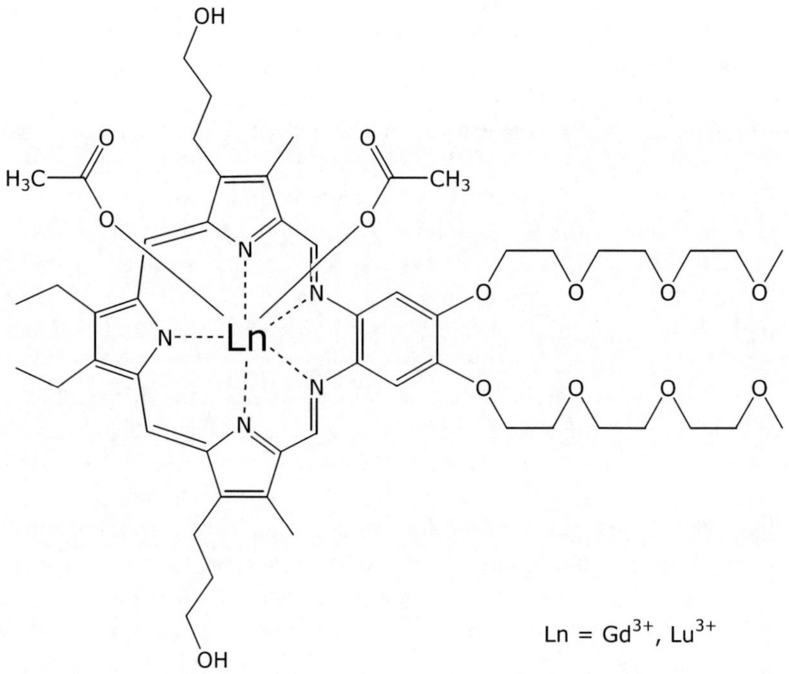

Lanthanide/Actinide in Health and Disease, Fig. 3 The structure of gadolinium and lutetium texaphyrin complexes

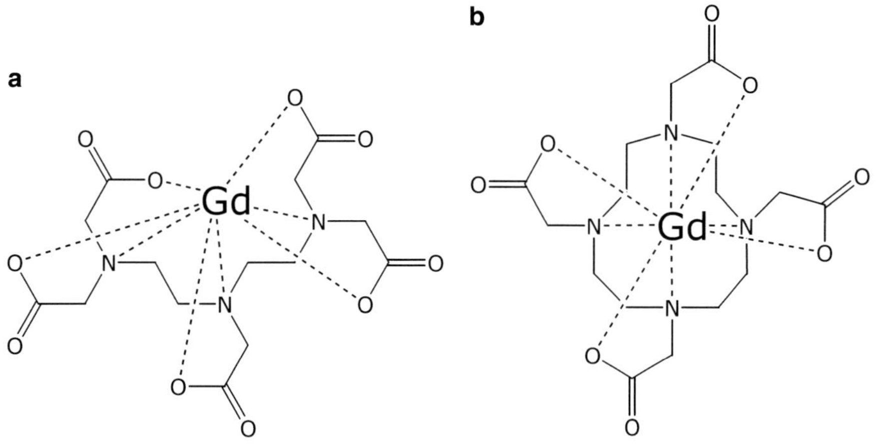

Lanthanide/Actinide in Health and Disease, Fig. 4 The structure of (**a**) Gd $(DTPA)^{2-}$ and (**b**) Gd $(DOTA)^{-}$ complexes used for magnetic resonance imaging

Ln in Magnetic Resonance Imaging. Magnetic resonance imaging is one of the most important and prominent noninvasive techniques in diagnostic clinical medicine (Laurent et al. 2009). Although some iron- and manganese-containing materials are commercially important, the predominant contrast agents in commercial use today, Magnevist® and Dotarem®, are comprised of a Gd^{3+} within chelating ligands based on a polyaminocarboxylate motif (linear DTPA and cyclic DOTA, respectively; Fig. 4). These agents are used for tumor detection, namely, those affecting parts of the central nervous system. Another example is the abovementioned gadolinium(III) texaphyrin complex, which has been proved to be selectively localized in tumor cells. In addition, pharmacokinetic studies have shown delayed clearance from tumors compared with rapid clearance from blood and normal tissues. Along with the continued dominance of gadolinium, in the last few years, there has been increasing interest in the utilization of other lanthanide ions (e.g., Eu^{2+}, Dy^{3+}) as contrast agents in magnetic resonance imaging (Bottrill et al. 2006). The research in this area is focused on development of site- or tissue-specific contrast agents.

Radioimaging and Radiotherapy. Many radioisotopes of the lanthanides and the actinides are available from nuclear reactors, generators, or cyclotrons. These radioisotopes show a variety of radiation characteristics that are suitable for applications ranging from diagnostics with positron-emitting tomography to radiotherapy (Table 1). Almost all radiopharmaceuticals are administered via intravenous injection. Therapeutic radiopharmaceuticals are molecules designed to deliver therapeutic doses of ionizing radiation to specific disease sites (most often cancerous tumors) (Volkert and Hoffman 1999). The design of each radiotherapeutic agent requires optimizing the balance between the specific in vivo targeting of the disease site and the clearance of radioactivity from nontarget radiosensitive tissues as well as the physical radioactive decay properties of the associated radionuclide. The various types of chelates are used for the design of radiotherapeutics. For example, aminophosphonate groups have been reported to have high affinity for hard cations such as Ca^{2+}. Metal complexes of polyaminophosphonates often localize in bone. Due to their high bone uptake, lanthanide complexes of polyaminophosphonates have been studied as therapeutic radiopharmaceuticals for bone pain palliation and for the treatment of bone cancer metastasis. The complex ^{153}Sm-ethylene diamine tetramethylene phosphonate (Quadramet®) is more widely used as an agent for the palliative treatment of bone metastases in terminally ill patients. Similar complex ^{166}Ho-1,4,7,10-cyclododecyl tetraamine tetramethylene phosphonate is applied for both bone pain palliation and the treatment of bone metastases. Ideally, therapeutic radiopharmaceuticals should localize in tumor site in sufficient concentration to deliver a cytotoxic radiation dose to tumor cells and at the same time clear rapidly from blood and noncancerous organs to minimize radiation damage to normal tissues (Liu 2008). A target-specific radiopharmaceutical is based on the receptor binding of a radiolabeled receptor ligand in the diseased tissue (radioimmunotherapy). In general, a target-specific radiopharmaceutical can be divided

Lanthanide/Actinide in Health and Disease, Table 1 Examples of lanthanide or actinide radionuclides used in radioimaging and radiotherapy

		Emissions	
Radionuclide	Half-life (days)	Particle	Energy (MeV)
^{149}Eu	93	γ	0.04
^{149}Gd	9.25	γ	0.04
^{149}Pm	2.2	β	1.07
^{149}Tb	0.17	α	4
^{153}Sm	1.9	β	0.8
^{166}Ho	1.1	β	1.6
^{177}Lu	6.7	β	0.50
^{225}Ac	10	α	6
^{255}Fm	0.84	α	7.02

into four parts: targeting biomolecule, pharmacokinetic modifying linker, bifunctional coupling or chelating agent, and radionuclide. The targeting biomolecule serves as a "carrier" for specific delivery of radionuclide to the diseased tissue, which is known to contain a substantial concentration of the targeted receptor. Many biomolecules, including monoclonal antibodies, small peptides, or nonpeptide receptor ligands, have been successfully used for target-specific delivery of radionuclides. The radiolabeled receptor ligand binds to receptors with high affinity and specificity, which results in selective uptake of the radiopharmaceutical. The chelating agents are similar to those used in magnetic resonance imaging, so they are based on a polyaminocarboxylate motif. There are several lanthanide isotopes to choose, but depending on the tumor size and location, the choice of the β-emitter may be different. For example, medium- or low-energy β-emitters such as ^{153}Sm and ^{177}Lu are better for smaller metastases while high-energy β-emitters such as ^{166}Ho are used for larger tumors. There has also been intensive investigation of α-emitting radionuclides (such as ^{225}Ac, ^{227}Th, ^{255}Fm) for clinical use, but they are still in the early stages of development and analysis. A diagnostic radiopharmaceutical is the molecule labeled with a gamma-emitting isotope for single photon emission computed tomography or a positron-emitting isotope for positron emission tomography. In general, diagnostic radiopharmaceuticals are used in very low concentrations (10^{-6}–10^{-8} mol/L) and are not intended to have any pharmacological effects.

Cross-References

- ▶ Actinides, Interactions with Proteins
- ▶ Actinides, Physical and Chemical Properties
- ▶ Lanthanide Ions as Luminescent Probes
- ▶ Lanthanide/Actinide Toxicity
- ▶ Lanthanides and Cancer
- ▶ Lanthanides in Biological Labeling, Imaging, and Therapy
- ▶ Lanthanides, Physical and Chemical Characteristics

References

Ansoborlo E, Prat O et al (2006) Actinide speciation in relation to biological processes. Biochimie 88:1605–1618

Barta CA, Sachs-Barrable K et al (2007) Lanthanide containing compounds for therapeutic care in bone resorption disorders. Dalton Trans 43:5019–5030

Bottrill M, Kwok L, Long NJ (2006) Lanthanides in magnetic resonance imaging. Chem Soc Rev 35:557–571

Bünzli JCG (2010) Lanthanide luminescence for biomedical analyses and imaging. Chem Rev 110:2729–2755

Curran MP, Robinson DM (2009) Lanthanum carbonate. A review of its use in lowering serum phosphate in patients with end-stage renal disease. Drugs 69:2329–2349

Fricker SF (2006) The therapeutic application of lanthanides. Chem Soc Rev 35:524–533

Kostova I (2005) Lanthanides as anticancer agents. Curr Med Chem Anti-Cancer Agents 5:591–602

Laurent S, Vander Elst L, Muller RN (2009) Lanthanide complexes for magnetic resonance and optical molecular imaging. Q J Nucl Med Mol Imaging 53:586–603

Liu S (2008) Bifunctional coupling agents for radiolabeling of biomolecules and target-specific delivery of metallic radionuclides. Adv Drug Deliv Rev 60:1347–1370

Volkert WA, Hoffman TJ (1999) Therapeutic radiopharmaceuticals. Chem Rev 99:2269–2292

Lanthanide/Actinide Toxicity

Karel Nesmerak
Department of Analytical Chemistry, Faculty of Science, Charles University in Prague, Prague, Czech Republic

Synonyms

Ln/An adverse effects

Definition

No lanthanide or actinide elements have known essential role in the normal biochemical reactions occurring in living organisms including plants, animals, and man. Consequently, their interactions with the various constituents of cells and tissues have the potential to induce both chemical and radiation toxicity. Lanthanides toxicity is predominantly a chemical one, based on competition with calcium in a number of calcium-mediated biological processes. Actinides show radiotoxicity in particular. The elements of both groups, like any exogenous element, should link with different biological ligands (proteins, amino acids, etc.) and take the place of natural biological elements by analogy (► Lanthanide/Actinide in Health and Disease).

Lanthanide Toxicity

The uses of lanthanide compounds in industry and technology are diverse, involving metallurgy, illumination, glass, ceramics, magnets, petroleum, electronics, medical imaging, and nuclear energy. Most of the industrial uses of the lanthanides require compounds (e.g., oxides) rather than pure elements. New and developing technologies increase the level of various lanthanides in the environment and, in turn, the exposure of man. Humans are also exposed to lanthanides directly as therapeutics and diagnostics. The lanthanide compounds are moderately to highly toxic. Generally, the toxicity depends on the nature of lanthanide compound (inorganic versus chelated compounds), on a route of administration, and on a dose.

Distribution in Environment and in Organisms. Although the lanthanides are sometimes termed "rare earth elements," they are widely distributed and are found in abundance in the Earth's crust over a relatively large range (cerium 60 mg/kg or lanthanum 30 mg/kg in comparison to thulium 0.5 mg/kg or lutetium 0.5 mg/kg). Of the 15 lanthanides, only promethium does not occur in nature; it is a man-made element. Levels of the lanthanides in healthy human tissues have been reported as follows (μg elements g^{-1} ash): liver, 0.005; kidney, 0.002; lung, 0.004; testes, 0.002; and bone, 0.2–1.0.

Toxicokinetics. According to animal models, the lanthanides are absorbed poorly (0.05% of the dose) in the gastrointestinal tract (Haley 1965). Following parenteral administration, the lanthanide ions form colloidal hydroxides or phosphates in blood. Toxicological studies of rats injected intravenously with chlorides of cerium, praseodymium, europium, dysprosium, ytterbium, and lutetium have revealed that these compounds accumulate in the liver (over 78% of the dose), bones, and spleen. Significant amounts have also been found to accumulate in the kidney and pancreas. Inhalation studies showed the highest accumulation of Ln in the lungs, gastrointestinal tract, liver, skeleton, and bronchial lymph nodes. In the lung, lanthanides accumulate in alveolar macrophages after inhalation or intratracheal instillation. Whole-body retention and tissue distribution of intravenous injected lanthanide compounds primarily depend on the stability of lanthanide in blood. Urinary excretion of cerium during 14 days was less than 1% of the dose following injection of $CeCl_3$ in mice, and a half-time of intravenous injected $CeCl_3$ was about 10 years in beagle dogs. After 1 day, about 80% of the dose of Ln chloride is distributed to the liver, bone, and spleen. Calcium accumulation has also been observed in liver, lung, kidney, and spleen after injection of the lanthanide chlorides. Although lanthanides disappear from the blood within 1 day, they are retained in organs as long as 45 days, especially the liver, bone, spleen, and lungs. On the other hand, chelated lanthanides seem to be excreted rapidly; a whole-body half-time of thulium citrate was about 2.5 h in rats, and approximately 50% of intravenous injected samarium ethylenediaminetetramethylene phosphonic acid was excreted in 8 h in humans. In general, chelated lanthanides, when injected intravenously, are excreted mainly in the urine after transient accumulation in kidney and have whole-body half-times of several hours. Based on this fact, an injection of Ca- or Zn-DTPA has been proven to be effective in removing Yb, Sc, and Ce from the body. In humans, occupational exposure may increase concentrations in specific organs. Although liver and kidney concentrations were equivalent, a twofold higher accumulation of lanthanides in lung was demonstrated in smelter workers compared with controls. The Ln concentration in these workers did not decline with time after exposure, indicating that these elements have a long biological half-life.

Mechanism of Toxicity. Lanthanides possess several biological properties, which include ability to replace calcium in many biomolecules, inhibition of lymphocyte activation and neutrophilic chemotaxis and aggregation, and inhibition of collagenase and stabilization of collagen fibrils (Palasz and Czekaj 2000). The neurotoxic components of such intoxication may be related to effects of lanthanide ions on neuronal ion flux (e.g., the GABA receptor-chloride channel complex). The rank order of efficacy on the flux is: Lu > Er > Tb > Eu > Nd > Ce > La, which is inversely correlated to the hydrated ionic radii of these elements. Lanthanum was more potent than the divalent cations Cd, Ni, or Co in blocking current through calcium channels in neuroblastoma cells. The lanthanides could compete for, or displace, Ca^{2+} from its binding site and act as a biological antagonist. Ln ions thus inhibited the electron transport processes in mitochondria, interfered with axonal activity, and inhibited contractions in both smooth and cardiac muscle. While most lanthanides promoted actin tubule formation, Er, Tm, Yb, Lu, and La interacted with actin but prevented tubule formation, which may be attributed to the high electrostatic charge of these specific lanthanide ions. Locally, direct eye or skin contact with the powder or liquid forms of the soluble lanthanide compounds can cause irritation. The irritation appears to be a result of exposure to the anion (e.g., nitrate) and not the lanthanide cation.

Acute Toxicity. Because human exposures have rarely reached toxic levels, only few instances of human toxicity have been observed. Therefore, the data that have been gathered are primarily from animal studies (Hirano and Suzuki 1996). Common symptoms of acute toxicity seen after these very high doses included writhing, ataxia, slightly labored and depressed respiration, arched back, stretching of limbs on walking, and lacrimation. Severe hepatotoxicity was observed after injection of lanthanide chlorides (Ce, Pr, Eu, Dy, Yb, Lu, and Y) each administered by single intravenous injection (9 or 10 mg/kg) to rats (Nakamura et al. 1997). In other studies, early biochemical changes seen in rats after acute intravenous injection of lanthanides included an increase in plasma free fatty acid levels, followed by an increase in fatty infiltration of the liver. It has been shown that intravenous injection of $CeCl_3$ caused fatty liver in female rats but not in male rats. Similarly, male rats are an order of magnitude less susceptible than female rats to oral as well as intraperitoneal administration of lanthanide compounds. After local injection or inhalation, the most pronounced effects of lanthanides are skin and lung granulomas.

Lanthanide/Actinide Toxicity, Table 1 An acute toxicity of lanthanide chlorides by various routes of administration (*i. v.* intravenous, *i. p.* intraperitoneal, *p. o.* per oral) to rats or mice (the sex is indicated by superscript F or M). Data from Lewis (2004)

		LD_{50} (mg/kg)		
Compound	Animal	i. v.	i. p.	p. o.
$LaCl_3$	Rat	4	106	4,184
	Mouse	18	213	
$CeCl_3$	Rat	5		2,111
	Mouse		353	5,277
$PrCl_3$	Rat	$4.3^F/49.6^M$		4,200
	Mouse		359	2,987
$NdCl_3$	Rat	$7.4^F/77.2^M$	245^F	
	Mouse		347	$5,250^M$
$SmCl_3$	Rat	$6.4^F/66.8^M$	270^F	$3,073^F$
	Mouse		585^F	$> 2,000^F$
$EuCl_3$	Mouse		550^M	3,527
$GdCl_3$	Mouse		550^M	
$TbCl_3$	Mouse		550^M	$5,100^M$
$DyCl_3$	Mouse		585^M	$7,650^M$
$HoCl_3$	Mouse		560^M	$7,200^M$
$ErCl_3$	Mouse		535^M	$6,200^M$
$TmCl_3$	Mouse		332	$6,250^M$
$YbCl_3$	Mouse		395^M	$6,700^M$
$LuCl_3$	Mouse		$290^F/315^M$	$7,100^M$

Transient pulmonary lesions indicative of acute chemical pneumonitis have been reported after intratracheal injection or prolonged inhalation of lanthanide mixtures in guinea pigs. Lanthanide salts injected into rats and guinea pigs precipitated at the injection site and initiated an inflammatory response. There is a delayed lethality with the death rate peaking between 48 and 96 h. If the animals survive for 30 days, there is generalized peritonitis, adhesions, and hemorrhagic ascitic fluid, and also true granulomatous peritonitis and focal hepatic necrosis. The effect of atomic weight on lethality is difficult to assess, but the transition elements (terbium group) appear to have a lesser toxicity than those above or below them in the periodic table (Table 1).

Chronic Toxicity. A number of studies have been completed in which the soluble lanthanide chlorides are given orally to animals over the course of months. Mice and rats exposed to the hydrated chloride forms of lanthanum, yttrium, and europium by oral gavage at doses of 0, 40, 200, or 1,000 mg/kg/day demonstrated a slight decrease in body weight due to decreased food intake at the 200 and 1,000 mg/kg/day doses. Mice fed varying levels of lanthanide oxides (La, Dy, Eu, Yb, Tb, and Sm) continually over three generations exhibited no differences in mortality, morbidity, morphological development, growth rate, reproductive capacity, or survival between the treated groups and controls. Repeated injection of La, Ce, Pr, or Nd chloride salts into rabbits produced hematic alterations, while no effects were observed after repeated oral administration. In humans, chronic exposure to the lanthanides occurs primarily via inhalation in occupational settings. Lanthanide-induced pneumoconiosis or progressive pulmonary fibrosis has been reported in occupationally exposed workers.

Genotoxicity and Reproductive Toxicity. The ability of Pr or Nd to induce chromosomal aberrations in vivo was evaluated in mouse bone marrow cells. A single exposure to Pr or Nd oxides (50–400 mg/kg, intraperitoneally) resulted in dose-related increases in chromatid and chromosome breaks at 6–12 h after treatment. La and Tb (0.1–1.0 mmol/L) are capable of promoting the neoplastic transformation of mouse epidermal cells. A decrease in the number of successful pregnancies and average litter size could be induced by a single injection of $LaCl_3$ (44 mg/kg, intraperitoneally) into pregnant mice.

Actinide Toxicity

All actinides are radioactive and release energy upon radioactive decay, their adverse effects to organisms are dominated by radiotoxicity. The actinides exist only as unstable, radioactive isotopes, which undergo radioactive decays to end up finally as stable isotopes of other elements (e.g., $^{232}Th \rightarrow {}^{208}Pb$ with half-live of 1.4×10^{10} years). Both the chemical toxicity and the radiotoxicity of the actinides may cause considerable hazards to living bodies. The actinides have some established daily-life applications, such as in smoke detectors (americium) and gas mantles (thorium), but they are mostly used as a fuel in nuclear reactors or in scientific research.

Distribution in Environment. Of the 15 actinides, only ^{227}Ac, ^{231}Pa, ^{232}Th, and ^{238}U (resp. ^{235}U, ^{234}U) occur in more than trace quantities in nature; the other actinides (so-called transuranium actinides) are synthetic elements, the result of single or multiple neutron capture by Th and U, followed by β-decay. For example, the Earth's crust contains about 2.4 mg/kg uranium, and seawater contains about 1–3 mg/kg. The burning of coal causes measurable anthropogenic emissions of natural actinides and their decay products. Due to their extremely low solubilities, soil in the vicinity of coal-fired power plants may contain increased concentrations of natural radionuclides. From the transuranium actinides, ^{239}Pu is globally found as a result of nuclear bomb testing.

Toxicokinetics. Actinide compounds are poorly absorbed from the lung (about 0.05–1.0% of the dose). The transfer of absorbed compounds to the blood from the gastrointestinal tract is strongly dependent on its chemical form (e.g., the values of the gastrointestinal uptake of uranium compounds range from less than 0.1% to about 6%) (Popplewell 1995). The behavior of actinides in blood is controlled largely by their strong tendency to hydrolyze and form complexes at physiological pH. Except for U^{6+} and Th^{4+}, with 70–90% associated with blood cells, the plasma is the predominant transport medium for An in blood (Taylor 1998). In plasma, Pu^{4+}, Th^{4+}, Pa^{5+}, U^{6+}, and Np^{5+} appear to bind strongly to the iron transport protein transferrin, although the binding characteristics may vary from element to element. The trivalent actinides, Ac^{3+}, Am^{3+}, Cm^{3+} and Cf^{3+}, apparently bind to transferrin but also to other proteins, and the association with transferrin appears to be much weaker than that of the other actinides. More than 70% of the actinides reaching the blood is assigned initially to bone surfaces and is subsequently transferred to bone marrow. The rest of the blood actinide content is distributed to liver, gonads, and other tissues. For thorium, the removal half-time from bone marrow to blood is assumed to be 0.25 years. On the other hand, the biological half-time for plutonium in skeleton has been estimated to 40 and 100 years, respectively.

Mechanism of Toxicity, Toxic Effects. The long-lived uranium and thorium isotopes (i.e., ^{235}U, ^{238}U, ^{232}Th) are both chemotoxic and radiotoxic (Bleise et al. 2003), whereas all other isotopes and decay products with their much shorter half-lives are dominated by their radiotoxicity (Table 2). Plutonium as well as all the other transuranium elements is radiotoxic. The dose-effect relationships for the chemical and the radiation toxicity of actinides are basically different. In general, chemical hazard displays a steep, dose-effect function with a threshold. This also holds for acute effects from high radiation doses from

Lanthanide/Actinide Toxicity, Table 2 Critical effects and occupational exposure limits for thorium, uranium, and decay products

Chemotoxicity				
Element	Target organ	Critical effect	Exposure limits (mg/m^3)	
Th		Cancer (liver)	0.2	
U	Kidneys	Renal failure, cancer	0.25	
Radiotoxicity				
			Exposure limits (Bq/year)	
Nuclide	Target organ	Critical effect	Inhalation	Ingestion
^{232}Th	Bone surface	Sarcoma	1.7×10^3	2.2×10^5
^{238}U	Bone surface	Sarcoma	3.5×10^3	2.6×10^6
^{226}Ra	Bone surface	Sarcoma	9.1×10^3	7.1×10^4
^{222}Rn	Lung	Lung cancer	600 Bq/m^3	
^{210}Pb	Bone surface	Cancer	1.8×10^4	2.9×10^4
^{210}Po	Whole body	Cancer	9.1×10^3	8.3×10^4
^{227}Ac	Bone surface	Cancer	3.0×10^2	1.8×10^4
^{231}Pa	Bone surface	Cancer	6.3×10^2	2.8×10^4
^{239}Pu	Bone surface	Cancer	2.4×10^3	2.2×10^6
^{241}Am	Bone surface	Cancer	7.4×10^2	1.0×10^5

therapeutic or accidental exposures, which cause widespread cell death and loss of tissue functions. However, ionizing radiation may cause additional changes, so-called stochastic effects, which are based on subtle changes in the genome of cells. Many experimental results and theoretical considerations suggest a linear, nonthreshold dose-effect relationship for a health detriment such as cancer from low-level α-irradiation. Due to the poor solubility of thorium and the low radiotoxicity of natural thorium, no acute effects on humans are reported. For humans and animals, uranium and its salts are highly chemotoxic. Dermatitis, renal damage, and acute arterial lesions may occur. Uranyl compounds readily complex with the phosphate-containing mineral matrix of bone. The renal toxicity of uranium in animals and man is caused by the precipitation of hexavalent uranium in the proximal kidney tubules in the process of clearance. The resulting tissue damage leads to kidney failure and the emergence of proteins, glucose, and creatinine in the urine. Acute intoxication may lead to irreversible damage and to death due to renal dysfunction. The LD_{50} indices have been established only for thorium and uranium compounds, as follows (Lewis 2004): Th $(NO_3)_4$ (i.v., mouse) = 45 mg/kg,$Th(NO_3)_4$ (p.o., mouse) = 1,760 mg/kg, $UO_2(NO_3)_2.6H_2O$ (i.p., rat) = 135 mg/kg. As for heavy metals, heavily charged actinide ions might theoretically show genotoxicity from their interaction with detoxification systems. Even for thorium and uranium, no genotoxicity effects are found in vivo. However, all substances emitting ionizing radiation must be considered mutagens and carcinogens. At higher doses, ionizing radiation is clearly teratogenic and may cause developmental defects leading to impaired brain functions (i.e., severe mental retardation). In humans, late effects from thorium are known. A considerable body of experimental animal data effects exists on the mutagenic and carcinogenic effects of transuranium actinides. For example, primary bone neoplasms developed in beagle dogs briefly exposed by inhalation to aerosols of $^{238}PuO_2$.

Cross-References

► Actinides, Interactions with Proteins
► Actinides, Physical and Chemical Properties
► Lanthanide/Actinide in Health and Disease
► Lanthanides, Physical and Chemical Characteristics
► Lanthanides, Toxicity

References

Bleise A, Danesi PR, Burkart W (2003) Properties, use and health effects of depleted uranium (DU) – a general overview. J Environ Radioact 64:93–112

Haley TJ (1965) Pharmacology and toxicology of the rare earth elements. J Pharm Sci 54:663–670

Hirano S, Suzuki KT (1996) Exposure, metabolism, and toxicity of rare earths and related compounds. Environ Health Perspect 104(Suppl 1):85–95
Lewis RJ (2004) Sax's dangerous properties of industrial materials, 11th edn. Wiley, Hoboken
Nakamura Y, Tsumura Y et al (1997) Differences in behavior among the chlorides of seven rare earth elements administered intravenously to rats. Fundam Appl Toxicol 37:106–116
Palasz A, Czekaj P (2000) Toxicological and cytophysiological aspects of lanthanides action. Acta Biochim Polonica 47:1107–1114
Popplewell DS (1995) Biokinetics and absorption of actinides in human volunteers. A review. Appl Radiat Isot 46:279–286
Taylor DM (1998) The bioinorganic chemistry of actinides in blood. J Alloys Compd 271–273:6–10

Lanthanide-Based Luminescent Assays

► Lanthanide Ions as Luminescent Probes

Lanthanide-Bioconjugate

► Lanthanides, Luminescent Complexes as Labels

Lanthanides

► Lanthanides, Toxicity

Lanthanides – Rare-Earths

► Actinide and Lanthanide Systems, High Pressure Behavior

Lanthanides and Cancer

Qiong Liu
College of Life Sciences, Shenzhen University, Shenzhen, P. R. China

Synonyms

Cancer; Lanthanide; Lanthanoid; Malignant neoplasm; Malignant tumor; Rare earth elements

Definition

Lanthanides

Lanthanides or lanthanoids (IUPAC nomenclature) comprise a group of 15 elements with atomic numbers from 57 (lanthanum, La) to 71 (lutetium, Lu), which fill gradually with the 4f electrons bexcept Lu, a d-block element. Thus, lanthanides have similar physicochemical properties. The 15 lanthanides, along with chemically similar scandium and yttrium, are often collectively known as the rare earth elements. The informal chemical symbol Ln is used in general discussions of lanthanide chemistry. All of the lanthanide elements can form trivalent cations, Ln^{3+}, whose chemistry is largely determined by the ionic radius, which decreases steadily from La to Lu.

Cancer

Cancer is a disease in which a group of cells display uncontrolled growth, invasion that intrudes upon and destroys adjacent tissues, and often metastasizes, where the tumor cells spread to other locations in the body via the lymphatic system or through the bloodstream. These three malignant properties of cancer differentiate malignant tumors from benign tumors, which do not grow uncontrollably, directly invade locally, or metastasize to regional lymph nodes and distant body sites like brain, bone, liver, or other organs.

Apoptosis

Apoptosis is the process of programmed cell death that may occur in multicellular organisms. Biochemical events lead to characteristic cell morphology changes and death. These changes include blebbing, cell shrinkage, nuclear fragmentation, chromatin condensation, and chromosomal DNA fragmentation. Unlike necrosis, apoptosis produces cell fragments called apoptotic bodies that phagocytic cells are able to engulf and quickly remove before the contents of the cell can spill out onto surrounding cells and cause damage.

Anoikis

Anoikis means "homelessness" in Greek. It is a form of programmed cell death, also known as suspension-induced apoptosis, which is caused by detaching from the surrounding extracellular matrix (ECM) in anchorage-dependent cells. Usually, cells stay close to the tissue to which they belong since the communication between proximal cells as well as between cells and ECM provides essential signals for growth or survival. When cells are detached from the ECM, i.e., there is a loss of normal cell–matrix interactions, they may undergo anoikis. This allows the body to rid itself of cells that are no longer needed and protects tissues from inappropriate colonization by nonadherent cells. However, metastatic tumor cells may escape from anoikis and invade other organs.

Drug Resistance

Drug resistance is the reduction in effectiveness of a drug such as an antimicrobial or an antineoplastic in curing a disease or condition. When the drug is not intended to kill or inhibit a pathogen, then the term is equivalent to *dosage failure* or drug tolerance. More commonly, the term is used in the context of resistance acquired by pathogens. When an organism is resistant to more than one drug, it is said to be multidrug resistant.

Radioimmunotherapy

Radioimmunotherapy (RIT) is a therapy for cancer treatment. It uses an antibody that recognizes tumor-associated antigens to carry a cytotoxic radionuclide to target and destroy cancer cells.

Photodynamic Therapy

Photodynamic therapy (PDT) uses photosensitizer to selectively and efficiently treat diseased cells. Photosensitizer, such as porphyrins and their derivatives, preferentially localizes in rapidly growing cells and gets activated by the exposure to light in the presence of oxygen to generate very reactive cytotoxic species. PDT has higher degree of selectivity and fewer side effects compared to chemotherapy and radiotherapy, which offers a promising treatment for cancers and a variety of other diseases.

Principles and Effects of Lanthanides on Cancer

Cancer is caused by environmental and hereditary factors. Cell reproduction is tightly regulated by several classes of genes, including oncogenes and tumor suppressor genes. Hereditary or acquired abnormalities in these regulatory genes can lead to the development of cancer. A small percentage of cancers, approximately 5–10%, are entirely hereditary. Though genetics influence the risk of some cancers, this disease is primarily environmentally dependent. Environmental factors cause or enhance abnormalities in the genetic material of cells, affecting cell propagation and forming tumors of varying degrees. Different therapies can be used, depending on how far cancer has spread. Chemotherapeutic agent is one of the therapies that kills rapidly dividing cells, thus slowing and stopping cancer from spreading.

Lanthanides (Ln) promote cell proliferation and induce apoptosis/anoikis depending on Ln species, concentrations, and cell types (Ni 2002). The antiproliferative activity of Ln appears at the concentration higher up to millimolar levels, while proliferative effect often occurs at micromolar concentration (Wang and Yang 2011). Ln complexes play an important role in cancer treatment. For example, Phase I, II, and III studies have shown that motexafin gadolinium (MGd), a redox modulator, is selectively taken up by tumor cells and may increase the therapeutic index of radiotherapy in the management of brain metastases and other brain tumors (Khuntia and Mehta 2004).

In addition, Lns also play important roles in the developing cancer therapies, including radioimmunotherapy (RIT) and photodynamic therapy (PDT), due to the radioactivity of some elements (^{166}Ho, ^{177}Lu, etc.) and their coordination capacity to form photosensitizing complex (e.g., gadolinium (III) texaphyrin, lutetium (III) texaphyrin). Traditional cancer therapies such as surgery, radiation therapy, and chemotherapy lack selectivity in removing or destroying diseased tissue and sparing normal healthy cells. They exert cytotoxic properties indiscriminately, which result in serious side effects. Thus, it is very important to develop new treatment protocols that display more judicious and effective discrimination of normal and diseased tissue.

Lanthanides Induce the Apoptosis of Cancer Cells

Ln nitrate, citrate, and chloride were reported to inhibit the growth of human leukemia K562 cells, stomach cancer PAMC82 cells, lung cancer PG cells, as well as the MNNG-induced cell transformation. The effect of Ln complexes on cell growth is bidirectional, depending on concentration and cells (Chen et al. 2000). Recently, some La(III) complexes have been synthesized with diverse ligands including chrysin, phenanthroline derivatives, and coumarin derivatives, whose IC_{50} values against various cancer cell lines are in the low micromolar range (see Table 1).

Lanthanides and Cancer, Table 1 The IC_{50} values of lanthanide complexes on cancer cells

Lanthanide[a]	Cell line[b]	IC_{50} (μmol/L)
$[LaCit_2]^{3-}$	HeLa	160
	SiHa	140
	MCF-7	180
	PC-3	1,550
	HepG2	2,710
$[YbCit_2]^{3-}$	HepG2	2,460
	HeLa	170
$[GdCit_2]^{3-}$	HeLa	240
La-1	HL-60	97.78
	BV-173	76.8
La-2	HL-60	168.44
	BV-173	160.72
La-3	HL-60	172.01
	BV-173	>200
Nd-1	HL-60	90.1
	HL-60/Dox cells	51.4
KP772	A459	1.5
	GLC-4	2.0
	Hep3B	2.2
	HL-60	1.2
	KB-3-1	2.5
	MDA-MB-231	1.3
	MCF7	3.7
	U373	1.2
	VL-6	5.1
	VL-8	1.5

[a]La-1: lanthanum complex of bis(4-hydroxy-2-oxo-2 H-chromen-3-yl)-piridin-2-yl-methane; La-2: lanthanum complex of bis(4-hydroxy-2-oxo-2 H-chromen-3-yl)-piridin-3-yl-methane; La-3: lanthanide complex of bis(4-hydroxy-2-oxo-2 H-chromen-3-yl)-piridin-4-yl-methane; Nd-1: neodymium with 3,3′-benzylidene-bis(4-hydroxy-2 H-1-benzopyran-2-one); KP772: [tris(1,10-phenanthroline)La(III)] trithiocyanate
[b]Human cell lines. HeLa, SiHa: cervical cancer; MCF-7: breast cancer; PC-3: prostate cancer; HepG2: hepatocellular carcinoma; HL-60, BV-173: leukemia; HL-60/Dox: the resistant subline of HL-60 cells; A459: lung adenocarcinoma; U373: astrocytoma; GLC-4: small cell lung carcinoma; Hep3B: hepatoma; KB-3-1: drug-sensitive (parental) cell line; MDA-MB-231: breast cancer; VL-6, VL-8: non-small cell lung cancer (NSCLC)

Effects of Ln citrate on the proliferation of different types of cancer cell lines were also investigated. Those cell lines included human cervical cancer HeLa and SiHa cells, breast cancer MCF-7 cells, prostate cancer PC-3 cells, and hepatocellular carcinoma HepG2 cells (Shen et al. 2009). Ln citrate, in the forms of $[LaCit_2]^{3-}$, $[YbCit_2]^{3-}$, and $[GdCit_2]^{3-}$, affected the proliferation of cancer cells dose-dependently and cell-type-dependently. The inhibition was significant at higher concentrations but not at lower concentration. The IC_{50} values of those Lns on various cancer cell lines were shown in Table 1, among which human cervical cancer cell lines SiHa and HeLa were more sensitive toward the treatment of $[LaCit_2]^{3-}$. HeLa cells treated with 0.15 mmol·L^{-1} $[LaCit_2]^{3-}$ for 24 h showed typical apoptotic changes at a rate of 61.65%, with a significant increase in sub-G1 phase and the arrest at G_0/G_1 cell cycle. The inhibition of $[LaCit_2]^{3-}$ on HeLa cells is related to the induction of cell apoptosis.

Lanthanides Induce the Anoikis of Cancer Cells

Cell–matrix interaction and cell–cell contact play important roles in regulating the proliferation, survival, and architecture of mammalian cells including normal cells and nontransformed tumor cells. If any of these two types of interaction becomes inadequate or inappropriate, a specific type of apoptosis, named anoikis, may be triggered. Transformed cells are often anchorage independent and lose contact inhibition during cell growth, which means they develop anoikis resistance during tumor genesis. This acquisition of anoikis resistance is considered to be a crucial step in the tumorigenic transformation of cells. For example, cells in primary tumors often lack contact with an organized basement membrane and thus must adapt to growth in matrix-poor or disorganized extracellular environments. Traversing the blood and lymph systems during metastasis also requires that cells survive in the absence of appropriate matrix contacts.

Lanthanum citrate was reported to be able to induce anoikis in HeLa cell at a concentration range of 0.001–0.1 mmol/L after 48-h treatment (Su et al. 2009). Before La treatment, HeLa cells were subjected to the anoikis-resistant selection to remove anoikis-sensitive cells and ensure the specificity of La-induced anoikis. In accordance with the anoikis induced by $[LaCit_2]^{3-}$ in HeLa cells, mitochondrial membrane potential was decreased, the cleavage of caspase-9 was detected, and Bax expression was increased. The intrinsic caspase pathway was proposed to be involved in the La-induced anoikis (as shown in Fig. 1). $[LaCit_2]^{3-}$ also caused the reorganization of actin cytoskeleton accompanied by an increase in

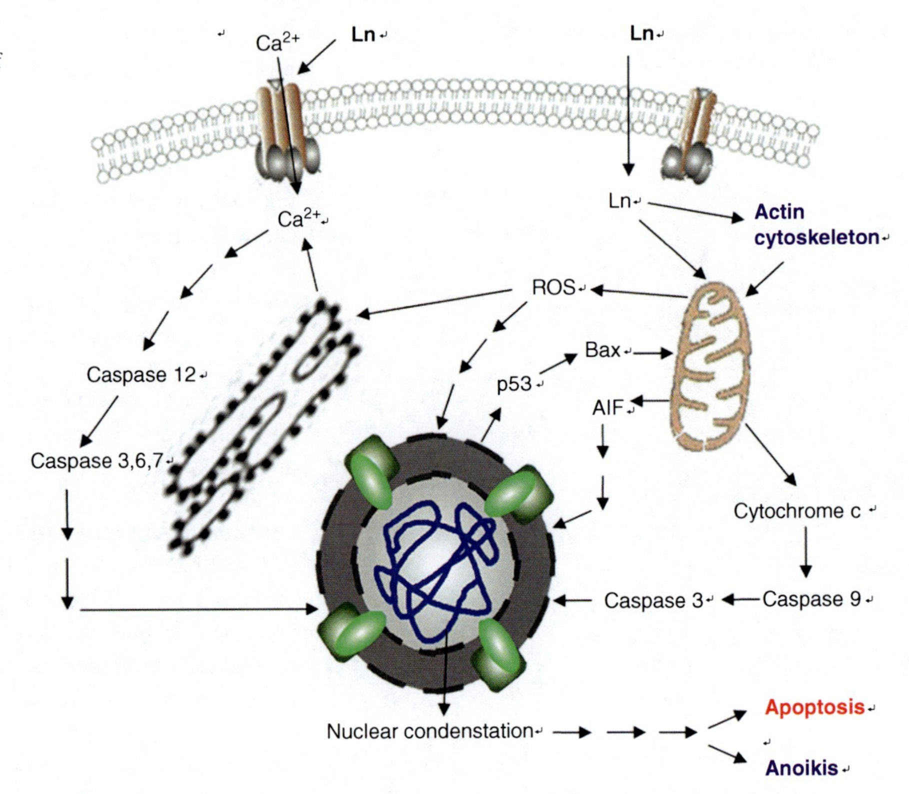

Lanthanides and Cancer, Fig. 1 Possible mechanism of lanthanide-induced apoptosis and anoikis

colocalization of F-actin with mitochondria, implying that both actin cytoskeleton and mitochondria may play important roles in La-induced anoikis.

Lanthanides Enhance the Anticancer Activity of Cisplatin

Platinum-based drugs, notably cisplatin and carboplatin, have dominated the treatment of various cancers by chemical agents. However, intrinsic drug resistance may develop in certain tumors, when treating patients with cisplatin. Development of cisplatin resistance is a major problem in this type of cancer treatment. To solve the problem, several compounds including Ln have been added together with cisplatin to the cancer cells. Terbium (Tb(III)) was found to bind the membrane of breast cancer cells and increased the cellular accumulation of cisplatin, thus enhanced the cytotoxicity of cisplatin (Kostova 2005). The combination of Tb(III) and cisplatin was later suggested in cancer chemotherapy, particularly for patients who have developed cisplatin resistance.

The ability of Tb(III) to modulate the cytotoxicity of cisplatin was detected in the cisplatin-sensitive MDA and cisplatin-resistant MDA/CH human breast cancer cells. MDA/CH cells were found to be approximately 3.3-fold more resistant to cisplatin than MDA cells. In both cell lines, the IC_{50} value for cisplatin was reduced twofold in the presence of 80 μM Tb(III), indicating the increased cytotoxicity of cisplatin with the addition of Tb(III). The enhancement of cisplatin cytotoxicity by Tb(III) is more effective in cisplatin-resistant MDA/CH cells than in cisplatin-sensitive MDA cells. Therefore, terbium is potentially useful in cisplatin combination therapy for breast cancer patients, especially for those patients who have developed resistance to the drug.

The synergistic action of Tb(III) and cisplatin was also confirmed by the investigation of FaDu and C13* human ovarian cancer cells. Binding of Tb(III) to a receptor was found to modulate the accumulation of cisplatin, leading to the synergic action. The cytotoxicity of cisplatin was approximately six times more potent than that of Tb(III). When cisplatin was combined with 80 μM Tb(III), the IC_{20} and IC_{50} values for cisplatin were reduced by 70% and 24%, respectively. The IC_{80} value, however, was increased by 124%. The results suggested that the cytotoxicity of cisplatin was enhanced by Tb(III) at low cisplatin concentrations.

It is notable that by the action of Tb(III), more cisplatin was accumulated in cisplatin-resistant MDA cells than in cisplatin-sensitive ones. A positive correlation between the membrane binding of Tb(III) and the cellular accumulation of cisplatin was also found in the MDA parent cell line and sublines ($r = 0.9$), where a specific Tb(III)-binding protein present in the plasma membrane was intimately associated with the accumulation of cisplatin in the breast tumor cells.

An alternative explanation for the synergistic effect of Ln on anticancer agents is to enhance permeability of cell membranes. $LaCl_3$ was shown to enhance the uptake of anticancer agents, such as hematoporphyrin derivatives, due to the Ln binding and the increase of cell membranes permeability for drugs (Kostova 2005). Thus, the mechanism of Ln enhancing cisplatin's anticancer activity remains to be clarified.

Lanthanide Complexes Synthesized for Anticancer Study

Metal ions and metal coordination compounds affect cells in not only natural processes, such as cell division and gene expression, but also nonnatural processes, such as toxicity, carcinogenicity, and antitumor chemistry. In chemotherapy, the key issue is killing the tumor cells, without causing too much harm to healthy cells. To study the Ln-based anticancer drugs, a number of Ln complexes have been synthesized and their cytotoxicity toward cancer cells was studied.

Coumarins have attracted significant attention as appropriate ligands for synthesis of new Ln coordination compounds. The complex of Ln and coumarins has demonstrated antiproliferative activity on various cancer cell lines. The coordination compounds of Ce(III), La(III), and Nd(III) with different coumarin derivatives, including Mendiaxon, Hymecromone, Umbelliferone, Warfarin, Coumachlor, and Niffcoumar, were assayed, and the complexes of Ce(III), La(III), and Nd(III) showed marginal cytotoxic activity against transformed leukemic cell lines P3HR1 and THP-1, as well as other investigated cancer cell lines, compared to the inorganic salts (Kostova 2005). The complex of Nd with 3,3′-benzylidene-bis(4-hydroxy-2 H-1-benzopyran-2-one) (named Nd-1 in Table 1) was the most active compound in inducing the cytotoxicity of both HL-60 cells and its resistant subline HL-60/Dox cells (IC_{50} values were 90.1 and 51.4 μM, respectively). Since the IC_{50} value of HL-60/Dox is almost twice smaller than that in the sensitive line HL-60, the complex is more sensitive to the resistant HL-60/Dox cells. Cytotoxic effects of the La(III) and bis-coumarins complexes were comparatively evaluated against the acute myeloid leukemia–derived HL-60 and the chronic myeloid leukemia (CML)-derived BV-173 cells. All of the complexes under investigation exhibited cytotoxic activity in micromolar concentrations; among them, the complex of La (III) with bis(4-hydroxy-2-oxo-2 H-chromen-3-yl)-piridin-2-yl-methane(La-1) (named La-1) proved to be superior in respect to relative potency and thus worthy for thorough pharmacological evaluation.

Complexes of La(III) with 1-aminocyclopentane, 1-aminocyclohexane, 1-aminocycloheptane, and 1-aminocyclo-4-ethylcyclohexanecarboxylic acids were also synthesized. Pharmacological study showed that all complexes manifested higher cytostatic and cytotoxic effects in comparison with $LaCl_3$. Much higher cytotoxic and cytostatic activity was found for the La complex with 1-aminocyclopentanecarboxylic acid. A hypothesis was proposed that La and plasma membrane interaction is the principal cause of the antitumor activity of La, either through structural changes of the cell membrane or through inhibition of the ATPase pumps.

[Tris(1,10-phenanthroline)La(III)] trithiocyanate (named KP772) was another compound that has promising anticancer properties in vivo and in vitro. Using ATP-binding cassette B1 (ABCB1)-overexpressing cells KBC-1 as multidrug resistance (MDR) model, KP772 hypersensitivity was demonstrated to be based on stronger apoptosis induction and/or cell cycle arrest at unaltered cellular drug accumulation. KP772 is hyperactive in MDR cells and might have chemosensitizing properties by blocking ABCB1 expression (Heffeter et al. 2008). Together with the disability of tumor cells to acquire KP772 resistance, KP772 should be especially active against notoriously drug-resistant tumor types and as second-line treatment after standard chemotherapy failure.

It should be noticed that drug resistance against Ln derivatives may develop easily. Despite this, Ln appears to be a bright and promising new development in the field of oncology (Kapoor 2009). The perspective of Ln is likely to be designed and developed to the dedicated drugs that are membrane transportable, intracellular survival, DNA binding, and, eventually, capable for excretion from the body with minimum side effects. Researches are ongoing in clinical application of Ln for the management of systemic malignancies and in disclosure of the anticarcinogenic activity and mechanism of Ln.

Anticarcinogenic Mechanism of Lanthanides

Although the anticarcinogenic mechanism of Ln remains to be unknown, some results have demonstrated its relevance with immune stimulation, DNA protection, enhancement in tumor-suppression gene expression, and induction of tumor cell apoptosis. Immune stimulation by Ln plays an important role in cancer inhibition (Chen et al. 2000). Lns stimulate immune function at low dose but inhibit it at high dose. A low dose of 0.04–0.4 mg/kg i.p. was found to exhibit an immune-stimulating action. T lymphocyte proliferation was increased by $La(NO_3)_3$, $Ce(NO_3)_3$, and LaCit at very low concentration. The function of macrophages and polymorphonuclear leukocytes was also enhanced by Ln nitrate and citrate at this Ln concentration but inhibited at high concentration. A daily oral dose of 0.2 or 2 mg/kg of Ln nitrate was recommended.

$LnCl_3$ was reported to effectively inhibit DNA damage. $LnCl_3$ at low concentration (0.5–1.5 mmol/L) inhibited the growth of PAMC82, and this was found to be associated with an enhanced expression of tumor suppressor genes, P53, P21, and P16. Ln^{3+} ions at 1 mmol/L inhibit the proliferation of B16 melanoma cells by arresting the cell transition from the G_0/G_1 to the S state, which is consistent with the inhibition of cell proliferation and DNA synthesis. Mechanism of the inhibition might also be traced along the signaling system. An appropriate concentration of La^{3+} cuts down the Ca^{2+} supply required for proliferation in human colon carcinoma cells HRT-18, leading to the apoptosis of cancer cells.

Recently, the strategies and techniques of proteomics and metabolomics have been applied to the mechanism study of Ln on cancer cells. Lns, in the forms of $[LaCit_2]^{3-}$, $[YbCit_2]^{3-}$, and $[GdCit_2]^{3-}$, were cultured, respectively, with HeLa, SiHa, and HepG2 cancer cells, and proteomic results revealed profound alteration in proteins classified into the following groups: redox reaction and oxidative stress, cell proliferation and apoptosis, transcription/translation and protein synthesis, signal transduction, and energy production (Fig. 2). Among those altered proteins, the group of redox reaction and oxidative stress accounts for the highest percentage, which includes superoxide dismutase (SOD1), peroxiredoxin (Prx1, 6), glutathione peroxidase (GPx1), ND2 protein (ND2), quinolinate phosphoribosyltransferase (QPRTase), prohibitin (PHB), etc. (Shen et al. 2009). Meanwhile, decreased mitochondrial transmembrane potential and increased generation of ROS were also detected in the Ln-treated cells, accompanied by the activation of caspase-9, specific proteolytic cleavage of PARP, increase in the proapoptotic protein bax, and decrease in the antiapoptotic protein Bcl-2. All those results suggested a mitochondrial pathway involved in the Ln-induced apoptosis of human cancer cells (see Fig. 1).

The metabolomic profiles of HeLa cells treated with gadolinium chloride ($GdCl_3$) were studied in time- and concentration- dependent manners (Long et al. 2011). A total of 48 metabolites released by HeLa cells are identified to be differentially expressed ($P < 0.05$) in different states. Metabolic pathway analyses reveal that the differential metabolites are mainly characterized by increased lipid and amino acid metabolisms and by decreased lipid, amino acid, and carbohydrate metabolisms for cells treated with $GdCl_3$ at lower and higher concentrations, respectively. Notably, in the higher-level $GdCl_3$ case, the down expressions of metabolites are predominantly in the glycolytic and the redox pathways. Those results disclosed that different cell signaling pathways are activated by $GdCl_3$ treatment with different concentrations, leading to inhibitory or promotional effect on HeLa cells.

Application of Lanthanides in Radioimmunotherapy

RIT uses an antibody that recognizes tumor-associated antigens to carry a cytotoxic radionuclide for targeting and destroying cancer cells. This therapy has the advantages of selective delivery of cytotoxic radiation following preferential uptake of radioantibody by tumor, high residence time in tumor, cross-fire effect by particle emissions, and minimal deleterious effect on normal tissues. The radioisotopes of the Ln have a variety of radiation characteristics that are suitable for the application in radiotherapy.

^{177}Lutetium is one of the radionuclides used in RIT. ^{177}Lu has a γ emission which permits external detection. This radiometal does not concentrate in

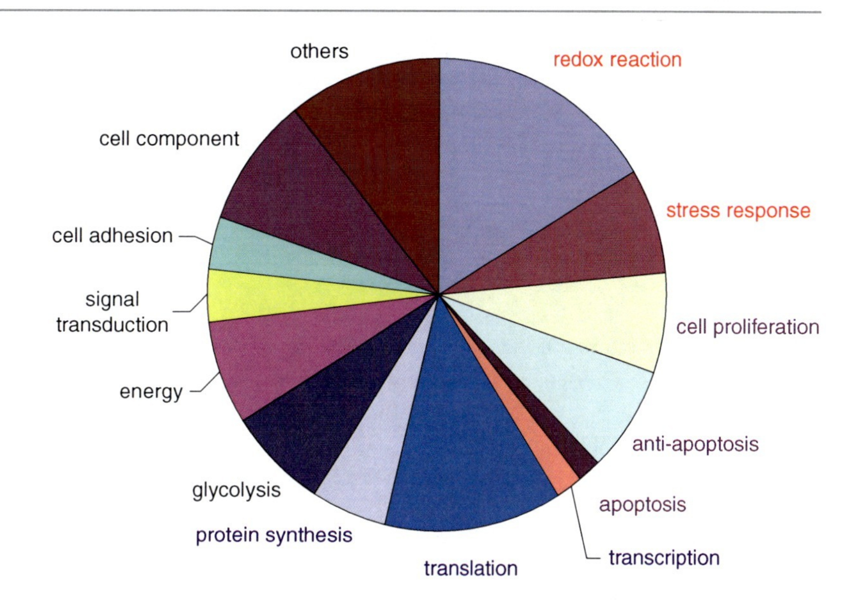

Lanthanides and Cancer, Fig. 2 Classification of the altered proteins induced by lanthanides in comparative proteomic investigation

bone. Compared with β-particle-emitting isotopes, α emitters can selectively kill individual cancer cells with far fewer decay events. Thus, α-particle-emitting radionuclides is good to be used in RIT. ^{149}Tb is one of the most promising α-particle-emitting isotope available for human clinical use. The selective cytotoxicity offered by ^{149}Tb constructs is due to the high linear energy transfer and short particle path length of ^{149}Tb. Preclinical and initial clinical use of ^{149}Tb conjugated to monoclonal antibodies has been carried out for the treatment of human neoplasia. However, the toxicity inherent in radioactivity necessarily means that the role of RIT will eventually have to be weighed against the benefits and side effects of a multitude of targeted molecular therapies.

Application of Lanthanides in Photodynamic Therapy

PDT is a very useful anticancer protocol. It uses photosensitizer to selectively and efficiently kill cancer cells. Photosensitizer is a drug that preferentially localizes in rapidly growing cancer cells and gets activated by the exposure to light in the presence of oxygen to generate very reactive cytotoxic species. Porphyrins and their expanded porphyrins, such as texaphyrins, have inherent tumor-localizing properties coupled with the ability to generate ROS when activated by light of particular wavelength. Those properties of porphyrins in turn results in cancer cytotoxicity, leading to the emergence of PDT as a therapeutic tool. PDT has higher degree of selectivity and fewer side effects compared to chemotherapy and radiotherapy, which offers a promising treatment for cancers and a variety of other diseases.

Porphyrins and expanded porphyrins are one class of molecules under intense investigation due to their photosensitizing ability for PDT application. Texaphyrins are water soluble tripyrrolic pentaaza-expanded porphyrin, which is capable of binding large metal cations including Ln^{3+}. Their ability to form stable complexes with Ln fetches them a unique position in the field of medicine. For example, gadolinium (III) texaphyrin is under Phase III trials for the treatment of brain metastases and Phase I trials for newly diagnosed primary brain tumors. Lutetium (III) texaphyrin is in Phase II clinical trials for locally recurrent breast cancer. Since any chromophore possible to induce phototoxicity upon illumination can selectively destroy diseased (malignant, premalignant, and benign) tissues, it stimulates researchers to incessantly design and develop many chromophores with optimal properties for treating a wider range of cancers.

Problems in Lanthanide Treatment

Although encouraging results have been increasingly reported on the cancer-preventive activity of Ln, it is still questioned by a series of problems, mainly related to the duality of these effects. The anticancer screening with animals performed in early stages of Ln study indicated that even though a number of complexes possess anticancer potency, rather high doses were needed. In addition, Ln^{3+} was found to induce malignant transformation of cells in vitro. For instance, La^{3+} and Tb^{3+} induced neoplastic transformation of TPA-sensitive and TPA-resistant preneoplastic mouse JB6 epidermal cells. It was also reported that pretreatment with $GdCl_3$ before surgery in mice caused increased tumor weight. Therefore, the risk in using Ln in anticancer treatment should also be sufficiently noted.

Cross-References

- ▶ Apoptosis
- ▶ Cisplatin
- ▶ Cytotoxicity
- ▶ Mitochondria

References

Chen X, Cheng Y, Ma K, et al (2000) Progress of rare earth science in China. In: Rare earths society. Metallurgy Press, Beijing (in Chinese)

Heffeter P, Jungwirth U, Jakupec M et al (2008) Resistance against novel anticancer metal compounds: differences and similarities. Drug Resist Updat 11:1–16

Kapoor S (2009) Lanthanum and its rapidly emerging role as an anti-carcinogenic agent. J Cell Biochem 106:193

Khuntia D, Mehta M (2004) Motexafin gadolinium: a clinical review of a novel radioenhancer for brain tumors. Expert Rev Anticancer Ther 6:981–989. doi:10.1586/14737140.4.6.981

Kostova I (2005) Lanthanides as anticancer agents. Curr Med Chem – Anti-Cancer Agents 5:591–602

Long XH, Yang PY, Liu Q, Yao J, Wang Y, He GH, Hong GY, Ni JZ (2011) Metabolomic profiles delineate potential roles for gadolinium chloride in the proliferation or inhibition of Hela cells. Biometals 24:663–677

Ni JZ (2002) Bioinorganic chemistry of rare earth elements. Science Press, Beijing (in Chinese)

Shen LM, Liu Q, Ni JZ et al (2009) A proteomic investigation into the human cervical cancer cell line HeLa treated with dicitratoytterbium (III) complex. Chem Biol Interact 181:455–462

Su X, Zheng X, Ni J (2009) Lanthanum citrate induces anoikis of Hela cells. Cancer Lett 285(2):200–209

Wang K, Yang XG (eds) (2011) Biological effect and pharmaceutical application of lanthanides, study in cellular inorganic chemistry. Peking University Medical Press, Beijing (in Chinese)

Lanthanides in Biological Labeling, Imaging, and Therapy

Jean-Claude G. Bünzli
Center for Next Generation Photovoltaic Systems, Korea University, Jochiwon–eup, Yeongi–gun, ChungNam–do, Republic of Korea
École Polytechnique Fédérale de Lausanne, Institute of Chemical Sciences and Engineering, Lausanne, Switzerland

Synonyms

Lanthanide bioprobes

Definition

A lanthanide bioprobe is a lanthanide chelate, or its bioconjugate, which exhibits characteristic magnetic or luminescent properties allowing its easy detection by spectroscopy or imaging techniques. Bioconjugates allow more specific detection to be carried out both in vivo and in vitro.

Lanthanide Bioprobes and Bioconjugates

Gaining greater understanding of the functional properties of living systems is a key challenge in biology and medicine. For instance, diagnosis and treatment of cancer are presently focusing on individualized approaches which, in turn, necessitate better and faster pathological analyses as well as highly contrasted, real-time bioimages. Noninvasive methodologies are required in order to perturb the investigated systems and organs as little as possible, and both magnetic resonance imaging and optical emissive probes are emerging as strong candidates for this purpose. A fortiori, a combination of both or a combination of optical probes with other existing imaging technologies will prove to be more informative; thus, bimodal agents are presently the subject of detailed and numerous investigations (Frullano and Meade 2007).

Trivalent lanthanide ions feature peculiar magnetic and photophysical properties due to their specific [Xe] $4f^n$ electronic configuration (see ▶ Lanthanum,

Physical and Chemical Properties) which makes them adequate for biosensing and bioimaging. In the mid-1970s, time-resolved luminescent assays (fluoroimmunoassays, FIA) were developed for quantifying immunoreactions with the purpose of replacing radio-immunoassays (RIA) which generate unwanted radioactive wastes. Chelates such as Eu^{III} and Tb^{III} polyaminocarboxylates and β-diketonates are particularly adequate for this goal, with Sm^{III} and Dy^{III} complexes being also useful. These analyses proved to be as sensitive as RIA and cheap since they only necessitate simple and inexpensive instrumentation. The 1980s have seen commercialization of these assays which are now ubiquitous in hospital, medical, and pharmaceutical laboratories. Almost in parallel, another lanthanide ion, trivalent Gd^{III}, was becoming commonplace in medical diagnostic. Indeed, the first experiments using a Gd^{III} chelate to enhance the contrast of brain tumor images collected by nuclear magnetic resonance were performed in 1984, and the first commercial contrast agent (CA), $[Gd(dtpa)(H_2O)]^{2-}$, was authorized in 1988, with a second one, $[Gd(dota)(H_2O)]^{-}$, following 1 year later. Since then, these contrast agents (CAs) and others developed later have been administered to millions of patients undergoing magnetic resonance imaging investigation (MRI). Other types of MRI contrast agents have been approved, for instance, iron oxide particles or manganese chelates, but Gd^{III} remains the dominant CA owing to its larger spin (7/2 as compared to 5/2 for Mn^{II}) and fast water exchange rate (10^7–10^9 s^{-1}). Basic imaging simply requires the contrast agent to be intravenously injected, and depending on the chelate, it localizes in the region of interest. However, more sophisticated experiments entail a more specific targeting. This is also the case of FIA.

Therefore, lanthanide chelates have to be fitted with adequate functionalities allowing bioconjugation with a specific antibody or peptide (Eliseeva and Bünzli 2010; Bünzli 2010). The coupling is achieved either directly, e.g, in immunoassays the lanthanide probe is linked to a monoclonal antibody mAb, or indirectly with the chelate being covalently bound to avidin (or biotin), and the resulting duplex then interacts with a biotinylated (or avidin derivatized) mAb via the strong avidin-biotin interaction ($\log K \approx 10^{15}$). Instead, avidin may be substituted by streptavidin or bovine serum albumin (BSA). Examples of indirect and direct couplings used for luminescent probes are depicted on

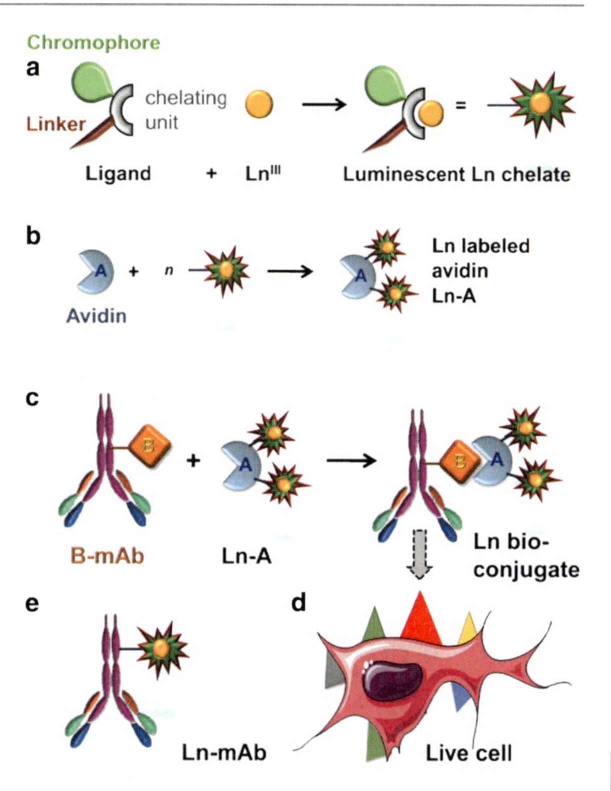

Lanthanides in Biological Labeling, Imaging, and Therapy, Fig. 1 Schematic representation of (**a**) the formation of a luminescent lanthanide chelate, (**b**) its linking to avidin (A), (**c**) subsequent conjugation to a biotinylated monoclonal antibody (B-mAb), (**d**) recognition of a biomarker expressed by a live cell, and (**e**) monoclonal antibody labeled with a luminescent chelate (Reproduced with permission from ref. (Bünzli 2010). © American Chemical Society 2010)

Fig. 1. The top part illustrates the formation of the luminescent chelate by reaction of a ligand decorated with chelating, chromophoric, and linking groups (a), followed by its covalent coupling to avidin (b). The bottom part describes the formation of the bioconjugate with a biotinylated monoclonal antibody (c) and the detection of a biomarker expressed by a live cell as an example of a specific immunocytochemical reaction (d); direct linking of the lanthanide label to a mAb is also illustrated (e).

Another methodology for achieving selectivity is to link the lanthanide chelate to a small peptide bearing a specific sequence. In turn, this sequence recognizes the targeted cells through the markers they express. Both the antibody and peptide methodologies apply equally to optical or magnetic probes. Regarding the latter, new contrast agents have been designed which can image a specific protein, for instance, fibrin which is present in all thrombi (arterial, venous, cardiac), opening the way to molecular imaging (Overoye-Chan et al. 2008).

Time-Resolved Luminescent Bioanalyses

Optical bioprobes are appreciated because when appropriate wavelengths are used, the penetration depth in biological tissues may be substantial, and light can reach regions of complex molecular edifices which are not accessible to other molecular probes. In addition, the emitted photons are easily detected by highly sensitive devices and techniques, including single-photon detection. When the lifetime of the excited emitting level is long enough, time-resolved detection (TRD) considerably enhances the signal-to-noise ratio. Organic luminophores are commonly fluorescent and highly emissive, but they are subject to photobleaching, and TRD necessitates sophisticated methodologies in view of the very short excited state lifetimes, a few ns. Trivalent lanthanide ions, Ln^{III}, present a valuable alternative to organic luminescent stains because they enable easy spectral and time discrimination of their emission bands which span both the visible and NIR ranges.

Time-resolved detection is illustrated on Fig. 2. Excitation of the sample is achieved by a pulsed light source. Sample autofluorescence and background emission develop immediately after the light pulse and decay on a fast time scale while population of the lanthanide excited level is much slower; in addition, Ln-centered luminescence decays slowly, on a microsecond to millisecond time scale. Therefore, if detection of the emitted light is delayed until all unwanted luminescence has disappeared, only the specific Ln^{III} emission is measured. In addition, the entire experiment lasts only a few milliseconds so that it can be repeated numerous times, typically 200–250 times per second. Since the signal-to-noise ratio is improved proportionally to the square root of the number of accumulated signals, a 100-fold improvement takes only a few minutes. This feature of lanthanide emitters is the key to all analyses taking advantage of such luminescent probes and is the reason why FIA analyses have met large success.

TRD is applied to immunoassays in view of the many undesirable substances coexisting in blood serum or in urine. The assays are divided into two

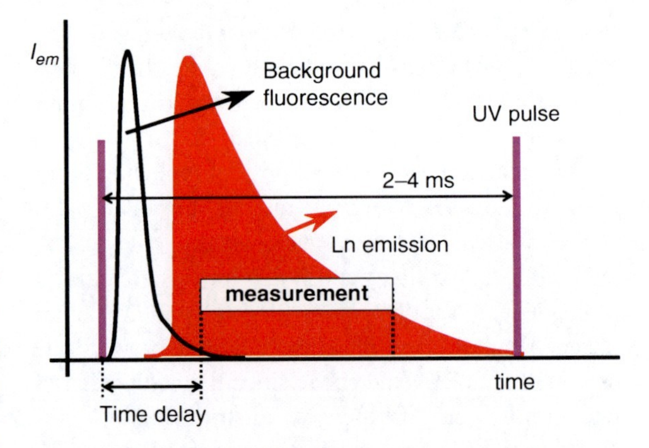

Lanthanides in Biological Labeling, Imaging, and Therapy, Fig. 2 Principle of time-resolved detection

categories, heterogeneous and homogeneous. The principle of the first ones is illustrated on Fig. 3 (top two panels). Two lanthanide chelates are used in the analysis format, which are commercialized under the name DELFIA® (Dissociation-Enhanced Lanthanide Fluorometric ImmunoAssay). The analyte is first linked to a specific binding agent immo?bilized on a solid support. Then, another specific immunoreaction couples a poorly luminescent lanthanide chelate with the analyte, and the unreacted reagents are removed. Further on, the chelate is dissociated in acidic medium and converted into another highly luminescent β-diketonate complex protected by a micelle. Subsequent time-resolved detection of the metal-centered luminescence yields the desired analytical signal. Antigens (e.g., hepatitis B surface antigen), steroids (e.g., testosterone or cortisol), as well as hormones (e.g., thyrotropin or luteotropin) are routinely assayed with this heterogeneous technique.

Homogeneous assays (HTRF®, homogeneous time-resolved fluorescence) rely on direct control of the bioprobe luminescence during the biochemical reaction under study. The analyte antigen is coupled to two mAbs, one decorated with a lanthanide label and the other one with an organic acceptor emitting at a wavelength distinct from the Ln^{III} emission. After completion of the immunoreactions, the sample is illuminated by UV light, and four types of luminescence develop: (1) two fast-decaying signals, background autofluorescence and fluorescence from the organic conjugate not bound to the antigen, and (2) two slow-decaying emissions, phosphorescence from the lanthanide conjugate not bound to the antigen (as well as

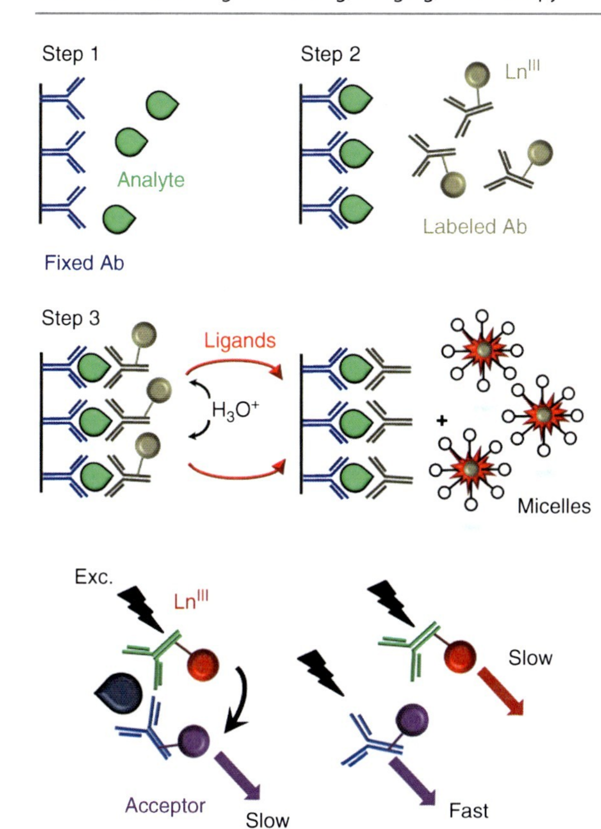

Lanthanides in Biological Labeling, Imaging, and Therapy, Fig. 3 Principle of a heterogeneous luminescent immunoassay (Ab = antibody) (*top and middle panels*). Principle of a homogeneous fluoroimmunoassay showing time discrimination (slow/fast) (*bottom panel*) (Reproduced with permission from ref. (Bünzli 2010). © American Chemical Society 2010)

residual luminescence from the bound one) and emission from the organic acceptor bound to the antigen and excited through Förster resonant energy transfer (FRET). Measurement in time-resolved mode allows one to eliminate the fast-decaying luminescence signals while spectral discrimination isolates the signal arising from the organic acceptor fed by FRET process (Fig. 3, bottom panel). In this way, removal of the unreacted conjugates is not necessary, and the analysis time is reduced substantially, making this technique ideally suited for high-throughput screening operations.

These new technologies have generated a broad interest and subsequent developments, such as dual assays (e.g., simultaneous detection of free and bound prostate specific antigen, PSA), optimization of bioconjugation methods for lanthanide luminescent chelates, and time-resolved luminescence microscopy (TRLM), resulted in applications of lanthanide probes in many fields of biology, biotechnology, and medicine, including analyte sensing, tissue and cell imaging, as well as monitoring drug delivery (Hemmilä 2008; Ghose et al. 2008).

Upconverting Nanoparticles (UCNPs)

Upconverting nanoparticles (UCNPs) are rare-earth-doped ceramic-type materials such as oxides, oxysulfides, fluorides, or oxyfluorides which convert red into visible light. They are usually synthesized as micro- or nanospheres and were introduced as probes for bioassays in the 1990s. Most of them contain Er^{III} ions as two-color (green, 540 nm; red, 654 nm) emitters and Yb^{III} ions as sensitizers, but other Ln^{III} pairs have also been proposed (e.g., Tm^{III}/Ho^{III}). UCNPs present several advantages over classical bioprobes, including high sensitivity, multiplexing ability if several different Ln^{III} ions are codoped, low sensitivity to photobleaching, and cheap laser diode excitation, in addition to the deep penetration of the excitation NIR light. Initially, they have been used in luminescent immunoassays, but presently, their applications are being extended to luminescence imaging of cancerous cells; novel imaging systems (e.g., NIR-NIR) based on UCNPs have been designed, and sensitivity down to single molecule detection within cells is foreseen (Cooper et al. 2007). Figure 4 gives an overview of the applications of UCNPs in biosciences, from bioanalyses to therapy.

Cell and Tissue Imaging with Luminescent Lanthanide Bioprobes (LLBs)

As the usefulness of LLBs unfolded its promising properties in bioanalyses, attempts to apply them for imaging purposes were a natural follow-up. Many lanthanide complexes are cell permeable and noncytotoxic and are able to signal the presence of important analytes (e.g., Ca^{II}, bicarbonate, ascorbate, urate), the fluctuations of which are tractable with high spatial resolution and yield information on cellular metabolism. Bioconjugated LLBs furthermore open the door to deciphering where analytes accumulate in live cells and what are their concentrations. Initial experiments dealing with cell imaging with the help of lanthanide chelates did not take advantage of TRD. However, it was soon realized that such detection

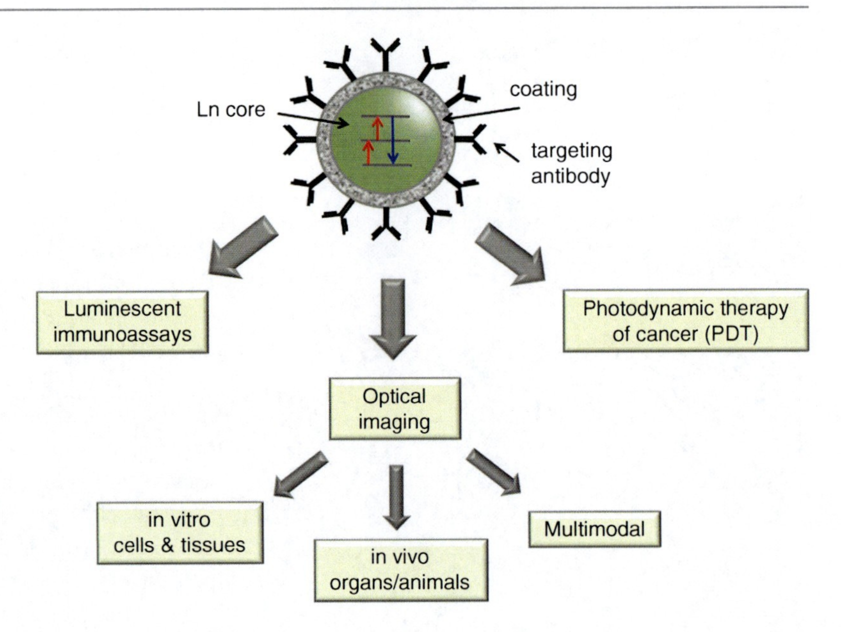

Lanthanides in Biological Labeling, Imaging, and Therapy, Fig. 4 Overview of bioapplications of UCNPs

method will considerably improve the quality of the images. The first report describing the design of a time-resolved microscope used in conjunction with a Eu^{III} probe dates back to 1990. Several designs were subsequently proposed, but given the millisecond lifetimes of the LLBs, the cheaper and simplest setups make use of a mechanical chopper. Multiparameter imaging is also feasible, for instance, a combination of up to four different metal luminophores in conjunction with an organic marker was used to image samples containing mixed populations of peripheral blood leukocytes; emission from the various luminescent stains was separated by time-resolved detection.

Several classes of chelating ligands yield adequate stains for cell imaging. We limit ourselves to two examples. The first one entails cyclen derivatives. In the course of an extensive investigation of over 60 different cyclen-based Eu^{III} and Tb^{III} LLBs fitted with various chromophores featuring tetraazatriphenylenes, acridones, azaxanthones, azathiaxanthones, or pyrazolyl azaxanthones, Parker's group has established that the nature of the chromophore and its attachment mode to the macrocycle primarily determines the cell uptake and localization and not the charge of the complex or its lipophilicity. Quantum yields are on the order of 10% and 40% for Eu^{III} and Tb^{III} chelates, respectively. Among the cell lines studied were mouse skin fibroblasts (NIH-T3 cells), Chinese hamster ovarian (CHO), or cervix carcinoma HeLa cells. Complexes staying in mitochondria have usually low IC_{50} values and cause cell apoptosis while those internalized in lysosomes (80% of the complexes) are nontoxic and therefore can act as responsive probes. The nature of the substituent on the sensitizer unit of the macrocyclic ligand is a key factor with respect to the sensitivity of the LLB; it also strongly affects protein affinity. The mechanism with which lanthanide cyclen chelates penetrate into live cells has also been investigated since it is critical to the development of imaging probes and therapeutic vectors. One of the major pathway is endocytosis involving a vesicular uptake of the incoming chelate followed by invagination of the cell membrane. Variants of this mechanism are macropinocytosis and clathrin/caveolae-independent cytosis. The dominant mechanism for cyclen Eu^{III} and Tb^{III} complexes seems to be macropinocytosis. An example of cell staining with a Eu^{III} cyclen derivative is presented in Fig. 5 (top panel).

Another class of cell-imaging LLBs are dinuclear helicates $[Ln_2L_3]$ which self-assemble in water at pH 7.4 and room temperature. The hexadentate ligands are derived from benzimidazolepyridine building blocks which act as adequate sensitizers of the luminescence of Ln^{III} ions. Polyoxyethylene pendants grafted either on the benzimidazole or pyridine aromatic moieties ensure water solubility and higher hydrophilicity of the helicates which are thermodynamically stable and kinetically inert. This framework sensitizes the luminescence of several lanthanides, primarily Eu^{III}

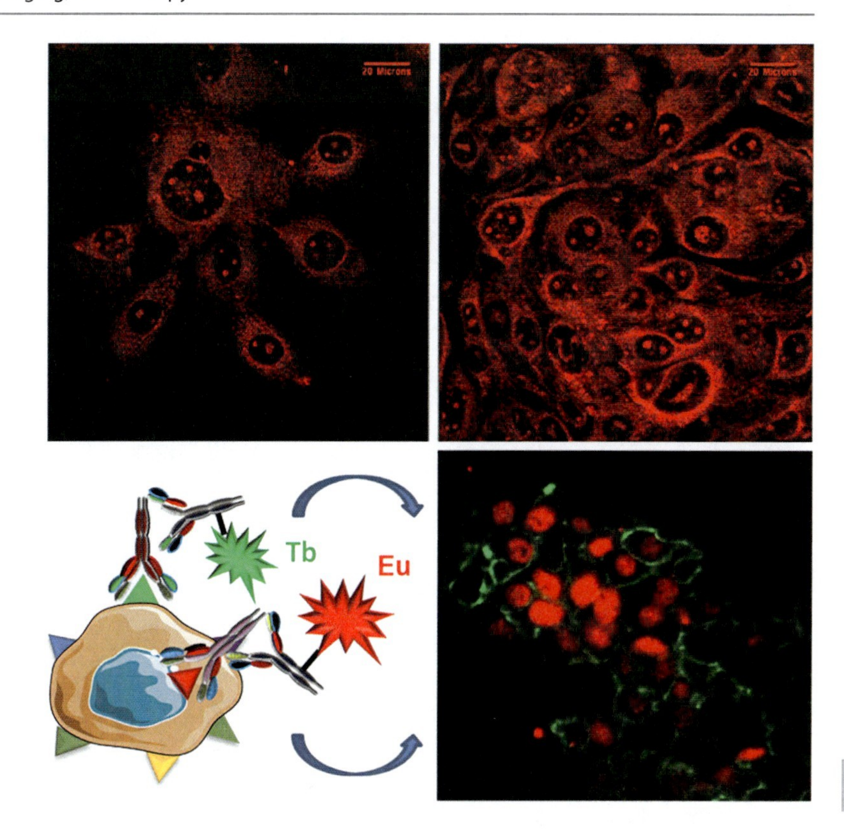

Lanthanides in Biological Labeling, Imaging, and Therapy, Fig. 5 Confocal microscopy images showing the intracellular localization profile of a Eu^{III} complex with a cyclen derivative bearing an azathiaxanthone chromophore (*top panel*) (Reproduced with permission from ref. (Montgomery et al. 2009) © American Chemical Society 2009). Principle of simultaneous analysis of two receptors expressed by cancerous cells in human breast tissue (*left*) and time-resolved luminescence microscopy image of these receptors (*right*) (*bottom panel*) (Reproduced with permission from ref. (Bünzli 2010). © American Chemical Society 2010)

($Q_{Ln}{}^{L} = 21\%$) and Tb^{III} (11%) but also Sm^{III} and Yb^{III}, thus allowing multiplex analyses. All of the chelates are noncytotoxic for several cancerous and nonmalignant cell lines, with $IC_{50} > 500$ μM. The cell uptake is slow and proceeds by endocytosis, with a localization in the endoplasmatic reticulum. Helicates are also easily amenable to bioconjugation, which eventually led to their application in the detection of specific markers expressed by cancerous cells. Using lab-on-a-chip technology, simultaneous analysis of two biomarkers expressed by human breast cancer cells, estrogen receptors (ER) and human epidermal growth factor receptors (Her2/*neu*), could be performed on tissue sections. The Ln^{III} helicates are coupled to two secondary antibodies: ER receptors are visualized by red-emitting Eu^{III} using goat anti-mouse IgG and Her2/*neu* receptors by green-emitting Tb^{III} using goat anti-rabbit IgG. The assay is much faster and requires far less reactants than conventional immunohistochemical assays. It is illustrated in Fig. 5 (bottom panel) (Eliseeva and Bünzli 2010; Bünzli 2010).

Recent developments of these imaging techniques have focused on multiphoton excitation in order to avoid phototoxicity and also to increase the examination depth as well as on improving the sensitivity by the use of nanoparticles and/or UCNPs. NIR-NIR imaging setups are now at hand, which combine both approaches (Bünzli 2010).

Magnetic Resonance Imaging (MRI) and Sensing

Magnetic resonance imaging is based on the fact that both spin–lattice (T_1) and spin-spin (T_2) proton relaxation times take different values depending on the tissue (Table 1). Contrast agents increase the proton relaxation rates of water in tissues in which they are distributed and therefore help making a better differentiation, enhancing the contrast of the MR image. In particular, Gd^{III} has a profound influence on the spin–lattice relaxation time, T_1 (up to 10^6-fold). When injected to the patient, gadolinium chelates distribute throughout the extracellular space and do not cross the blood–brain barrier; they are thus efficient for detecting pathologic abnormalities.

Lanthanides in Biological Labeling, Imaging, and Therapy, Table 1 Magnetic resonance relaxation times of water in selected human tissues

Tissue	T_1/ms	T_2/ms
Fat	150	150
Liver	250	44
Muscle	450	64
White matter	300	133
Gray matter	475	118
Spleen	400	107
Pancreas	275	43

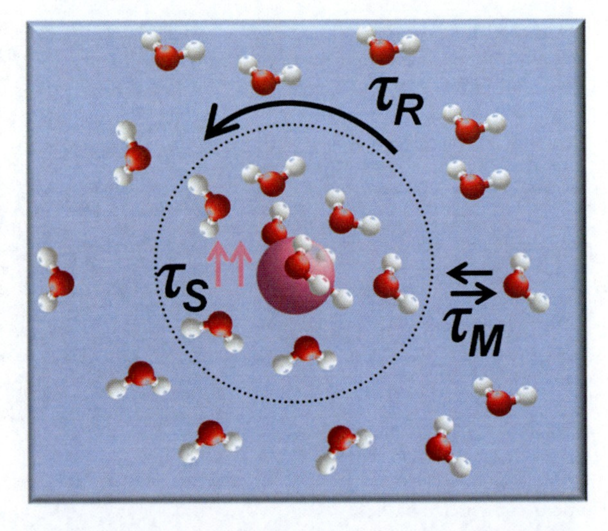

Lanthanides in Biological Labeling, Imaging, and Therapy, Fig. 6 Three factors influencing the relaxivity of a contrast agent

The effect of a CA is characterized by its relaxivity:

$$r_i = \frac{1}{\Delta T_i[\mathrm{CA}]} \quad (i = 1,\ 2)[\mathrm{mM}^{-1}\mathrm{s}^{-1}] \qquad (1)$$

The design of a CA must take into account several aspects. Firstly, free Gd^{III} is toxic, blocking calcium channels; it must therefore be bound to a strong chelating agent, and the resulting complex must not only have large thermodynamic stability (typically $\log K_{compl} > 22$) but also present kinetic inertness. In addition, it must be excreted from the body within a reasonable span of time (24–48 h). Relaxivity depends on three main factors (Fig. 6): (1) the exchange rate $1/\tau_M$ of the bound water molecules and of the water molecules hydrogen-bonded to those in the inner coordination sphere; diffusing molecules in the outer sphere also contribute to this parameter; (2) the rotational correlation rate of the molecule, $1/\tau_R$; and (3) the longitudinal electron spin relaxation time τ_S. For $[Gd(dota)]^-$, these values are equal to $4.1 \times 10^7\ s^{-1}$, $1.25 \times 10^{10}\ s^{-1}$, and 1 ns, respectively. The first-generation CA agents are sketched in Scheme 1. Their relaxivity is relatively small (3.4–4.6 $mM^{-1}s^{-1}$) so that sizeable concentrations have to be used (0.01–0.1 mM). This is a drawback for this imaging technique which otherwise provides a very good anatomical (three-dimensional) resolution in the submillimeter scale. But when it comes to molecular imaging, higher relaxivity is needed. One strategy to increase relaxivity is to couple the chelate to a large biocompatible macromolecule such as human (or bovine) serum albumin or a peptide. As a result, the CA conjugate rotates slowly and the correlation time goes up. However, the coupling often results in a decrease in the water exchange rate so that increases in relaxivity are less than expected. For a slowly rotating molecule ($\tau_R > 30$ ns), the theoretical limit for relaxivity is around 140 $mM^{-1}s^{-1}$. In recent time, BSA conjugates have been reported with relaxivities up to 90 $mM^{-1}s^{-1}$, while a commercially available (and approved) CA conjugate has $r_1 = 50\ mM^{-1}s^{-1}$ and works as a protein-targeted CA (Caravan 2009). Gd^{III} chelates can also be inserted into biological structures such as apoferritin, lipoproteins, micelles, liposomes, or microemulsions to enhance sensitivity.

The tendency now is to design responsive CAs ("smart" CAs), the relaxivity of which is modulated by interaction with a specific molecule or ion. Sensors for Ca^{II}, Zn^{II}, enzymes, or proteins have been proposed. Other types of lanthanide-containing CAs have emerged at the turn of the century: CEST agents, which are molecules containing labile protons. CEST CAs have the peculiarity of generating contrast only if the irradiation radiofrequency matches the absorption frequency of the labile protons. The recording of a preadministration image is therefore no more necessary. The so-called saturation transfer is not dependent on the concentration of the contrast agent, a feature which can be exploited for tailoring MRI-CEST agents responsive to various parameters such as pH, temperature, or metabolite concentrations. Despite moderate sensitivity, the potential of these new agents is high,

Lanthanides in Biological Labeling, Imaging, and Therapy, Scheme 1 First-generation contrast agents (trade names may change depending on the country)

$[Gd(dtpa)(H_2O)]^{2+}$ Magnevist® Log$K$ = 22.1 $r_1 = 3.7\ mM^{-1}s^{-1}$

$[Gd(dtpa\text{-}bma)(H_2O)]$ Omniscan® Log$K$ = 16.9 $r_1 = 4.6\ mM^{-1}s^{-1}$

$[Gd(dota)(H_2O)]^-$ Dotarem® Log$K$ = 25.8 $r_1 = 3.4\ mM^{-1}s^{-1}$

$[Gd(dota)(H_2O)]^-$ ProHance® Log$K$ = 23.8 $r_1 = 3.7\ mM^{-1}s^{-1}$

particularly for multiplex analyses, since each CA can be activated at will by judicious selection of the irradiation frequency (Aime et al. 2009).

Lanthanide in Therapeutical Treatment

Medicinal chemistry of rare earths has roots in the nineteenth century when cerium oxalate was widely prescribed as an antiemetic drug to cure sickness due to pregnancy. Moreover, the in vitro antimicrobial properties of several lanthanide complexes stimulated clinical trials in the treatment of tuberculosis and leprosy although their impact was minimal. Lanthanide ions have also anticoagulant properties, but their main therapeutic applications lie presently in the radioactive treatment of cancers.

Several Ln^{III} ions are β-emitters with a relatively long penetration depth between 2 and 12 mm, which makes them adequate for treating solid tumors with high heterogeneity. All of them but ^{90}Y have also a small γ emission which is useful for measuring the biodistribution of the radiopharmaceutical. ^{90}Y, which is produced from the decay of ^{90}Sr, has high-energy pure β-emission and is therefore well suited for systemic cancer therapy. ^{153}Sm is used to relieve pain when cancer has spread to bones; the most common cancers treated are lung, prostate, and breast cancers. ^{166}Ho emits β-rays having a penetration range of 9 mm and which are used for skin and hepatic cancer treatment. ^{177}Lu can be used similarly to ^{90}Y and since the β-ray penetration is smaller, it is provided for treating smaller tumors; it is presently being tested for the treatment of prostate cancer. Similarly to ^{166}Ho and ^{177}Lu, ^{149}Pm can be prepared with no added carrier and has similar biodistributions and properties.

Another emerging application of lanthanide ions in therapy is their implication as light transducer for the photodynamic treatment of cancer (PDT). The first step in this therapy is the selective uptake of a photosensitizer on the cancerous cells/tissue; subsequently, irradiation with predetermined doses of light activates the photosensitizer which generates reactive oxygen species themselves killing the tumor cells. Lutetium texaphyrins (a texaphyrin is an enlarged porphyrin containing five pyrrole units), for instance, display large efficiency for the production of singlet oxygen and have been tested for PDT of skin cancer and photoangioplasty. One problem in this therapy is the penetration depth of the excitation light. Therefore, UCNPs have been modified with a shell impregnated with the photosensitizer. In this way, deep-penetrating infrared light can be used to initiate upconversion, and the visible light emitted in situ by the UCNP then activates the photosensitizers. There are presently only a limited number of studies taking advantage of this technology, but it has a good prospect for the future.

Cross-References

▶ Lanthanides in Nucleic Acid Analysis
▶ Lanthanum, Physical and Chemical Properties

References

Aime S, Castelli DD, Crich SG, Gianolio E, Terreno E (2009) Pushing the sensitivity envelope of lanthanide-based magnetic resonance imaging (MRI) contrast agents for molecular imaging applications. Acc Chem Res 42:822

Bünzli J-CG (2010) Lanthanide luminescence for biomedical analyses and imaging. Chem Rev 110:2729

Caravan P (2009) Protein-targeted gadolinium-based magnetic resonance imaging (MRI) contrast agents: design and mechanism of action. Acc Chem Res 42:851

Cooper DE, D'Andrea A, Faris GW, MacQueen B, Wright WH (2007) Up-converting phosphors for detection and identification using antibodies, Ch. 9. In: Van Emon JM (ed) Immunoassay and other bioanalytical techniques. CRC Press/Taylor & Francis, Boca Raton, pp 217–247

Eliseeva SV, Bünzli J-CG (2010) Lanthanide luminescence for functional materials and bio-sciences. Chem Soc Rev 39:189

Frullano L, Meade TJ (2007) Multimodal MRI contrast agents. J Biol Inorg Chem 12:939

Ghose S, Trinquet E, Laget M, Bazin H, Mathis G (2008) Rare earth cryptates for the investigation of molecular interactions in vitro and in living cells. J Alloys Compd 451:35

Hemmilä IA (2008) Time-resolved fluorimetric immunoassays; instrumentation, application, unresolved issues, future trends. In: Resch-Genger U (ed) Standardization and quality assurance in fluorescence measurements II. Springer series on fluorescence, Wolfbeis OS, Hof M, Springer, Berlin, vol 3, pp 429–447

Montgomery CP, Murray BS, New EJ, Pal R, Parker D (2009) Cell-Penetrating metal complex optical probes: targeted and responsive systems based on lanthanide luminescence. Acc Chem Res 42:925

Overoye-Chan K, Koerner S, Looby RJ, Kolodziej AF, Zech SG, Deng Q, Chasse JM, McMurry TJ, Caravan P (2008) EP-2104R: A fibrin-specific gadolinium-Based MRI contrast agent for detection of thrombus. J Am Chem Soc 130:6025

Lanthanides in Nucleic Acid Analysis

Jean-Claude G. Bünzli
Center for Next Generation Photovoltaic Systems, Korea University, Jochiwon–eup, Yeongi–gun, ChungNam–do, Republic of Korea
École Polytechnique Fédérale de Lausanne, Institute of Chemical Sciences and Engineering, Lausanne, Switzerland

Synonyms

Lanthanide analytical probes

Definition

A lanthanide analytical probe is a lanthanide chelate, possibly its bioconjugate, which exhibits characteristic magnetic or luminescent properties allowing its easy detection, henceforth helping quantify the analyte.

Introduction

Most molecular biological and diagnostic applications require accurate quantification of nucleic acids extracted from various sources (e.g., blood, cells, bones), in particular, of DNA. Deoxyribonucleic acid (DNA) is a nucleic acid which contains the genetic instructions necessary to the development of all living organisms, with the exception of some viruses. Genetic information is carried by DNA at specific sequences. DNA consists of two long polymers built from simple units called nucleotides. The two strands lie antiparallel to each other in direction forming the double helix evidenced by Watson and Cricks in 1953 and which relies on hydrogen bonding between nucleobases belonging to each strand. These nucleobases are adenine (A), cytosine (C), guanine (G), and thymine (T). The GC base pair features three hydrogen bonds and the AT pair two (Fig. 1).

The simplest analytical method for the determination of DNA is based on its ultraviolet absorption band at 260 nm, with an estimated molar absorption coefficient of 50 $\mu g^{-1} \cdot cm^{-1} \cdot mL$. This method, however, bears several flaws. Firstly, the molar absorption coefficient may slightly vary from one source of DNA to the other. Secondly, large sample volumes are needed. Thirdly, the dynamic range is narrow and the method is sensitive to pH and salt content. Finally, contributions to the 260 nm absorbance by contaminating agents such as proteins and free nucleotides often lead to an overestimation of the DNA concentration.

DNA is very weakly luminescent and emission occurs in the UV, preventing practical applications. Thus, fluorometric commercial DNA assays have been developed using intercalating fluorescent reagents which display enhanced fluorescence intensity when interacting spontaneously with DNA. Some examples are rhodamine 800, Nile blue, oxazine, BODIPY™, ethidium bromide, bis(benzimidazole) dye Hoechst 33258, cyanine dyes PicoGreen®, and SYBR Green I®. Many of these dyes though suffer from one or

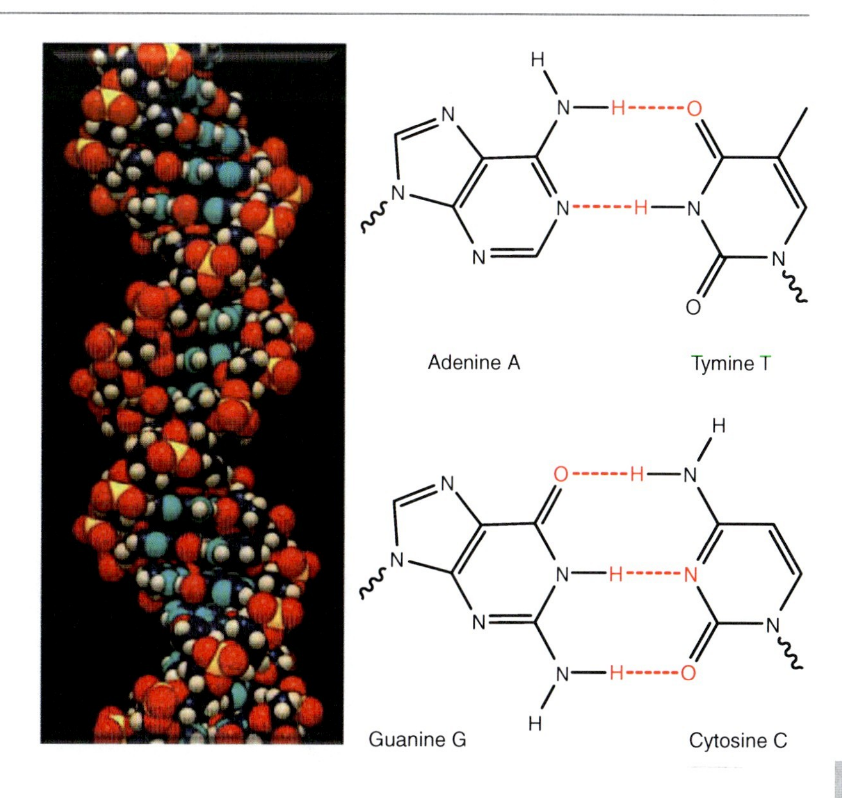

Lanthanides in Nucleic Acid Analysis, Fig. 1 DNA double-stranded structure (*left*). The two types of nucleobase pairs (*right*)

another drawback such as intrinsic luminescence limiting the sensitivity of the assays, selectivity depending on the salt concentration in the sample, association with a preferred DNA domain, reduced binding for small DNA fragments, and photobleaching. The specificity of DNA fluorescence analyses can be enhanced by bis (intercalating) dyes (popular ones are ethidium dimers or cyanine dye dimers such as TOTO®), which show linear responses in the 0.5–100 ng/mL^{-1} range (Lakowicz 2006). Methods using Förster resonance energy transfer (FRET) have also good sensitivity and are widely used in analysis relying on polymerase chain reactions (PCR) with technologies such as TaqMan®, Molecular Beacon®, or Scorpion®. Detection limit is always a crucial issue, especially in environmental samples. A natural limit of sensitivity is single molecule detection, which has been achieved for DNA by flow cytometry combined with correlated single photon counting. The price to pay for this sensitivity is a sophisticated instrumentation which is not easily handled in a routine way. Alternatively, one DNA molecule can be labeled with many chromophores; an efficient way of realizing such marking is to attach the chromophores on a biotinylated dendrimer.

In this context, it was interesting to take advantage of the ability of lanthanide luminescent probes (LLBs) to generate more sensitive and/or multianalyte detection analyses (Bünzli 2010; Nishioka et al. 2007). The principle of LLBs and associated time-resolved heterogeneous or homogeneous immunoassays (TR-FIA) is described in ► Lanthanides in Biological Labeling, Imaging, and Therapy. The most used ions are Eu^{III} and Tb^{III}, but multiplex analyses feature additionally Sm^{III} and Dy^{III}. For DNA (or RNA) assays, the lanthanide probe is often grafted onto a single-stranded DNA sequence so that the recognition step occurs via hybridization (Fig. 2). Once the lanthanide chelate is excited, its energy is transferred onto an organic chromophore grafted onto a complementary single-stranded DNA. This transfer can only occur if the two chromophores lie at a relatively close distance to each other, that is, if hybridization is completed. There is therefore no need to eliminate unreacted reagents.

Heterogeneous DNA Assays

Compared to homogeneous assays, heterogeneous analyses are time-consuming and labor-intensive since they necessitate thorough washing in order to separate the bound probes from the free probes, but their principle is simple, and so they are still performed

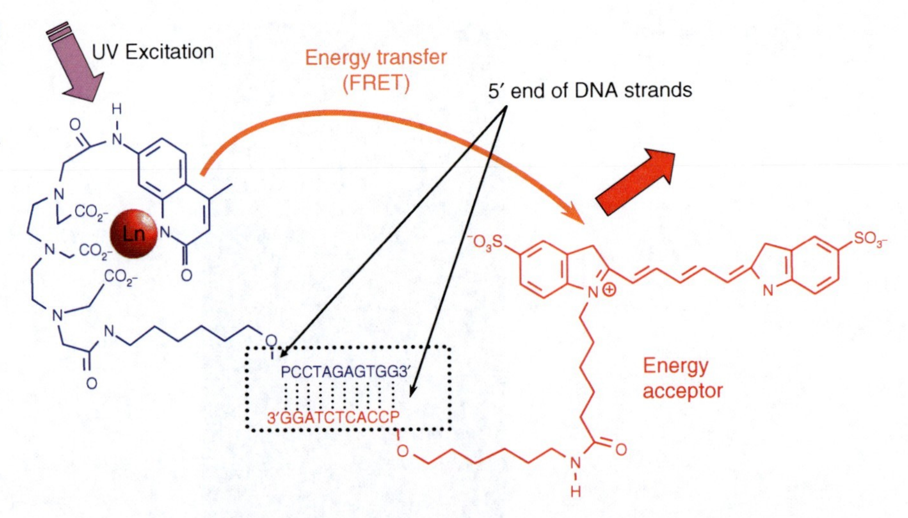

Lanthanides in Nucleic Acid Analysis, Fig. 2 Principle of hybridization recognition followed by FRET transfer. Used in homogeneous TR-FIA

even today since their sensitivity is large, thanks to time-resolved detection. For instance, a *Streptococcus pneumonia* DNA assay achieves sensitivity of 50 fg, corresponding to only 20 copies of the DNA. This sensitivity is obtained after PCR amplification with a primer labeled with biotin, followed by reaction with a Eu-labeled oligonucleotide complementary to the biotin-tagged DNA, and finally TR-FIA analysis.

Multicomponent analyses of PCR products also often rely on heterogeneous TR-FIA. An illustration is given on Fig. 3 for a seven-tag assay of the human papilloma virus associated with cervix cancer. The probes are seven oligonucleotides bearing different combinations of three lanthanide ions enabling therefore the simultaneous detection of up to seven target probes with a detection limit down to 10^{-16} mol (Samiotaki et al. 1997).

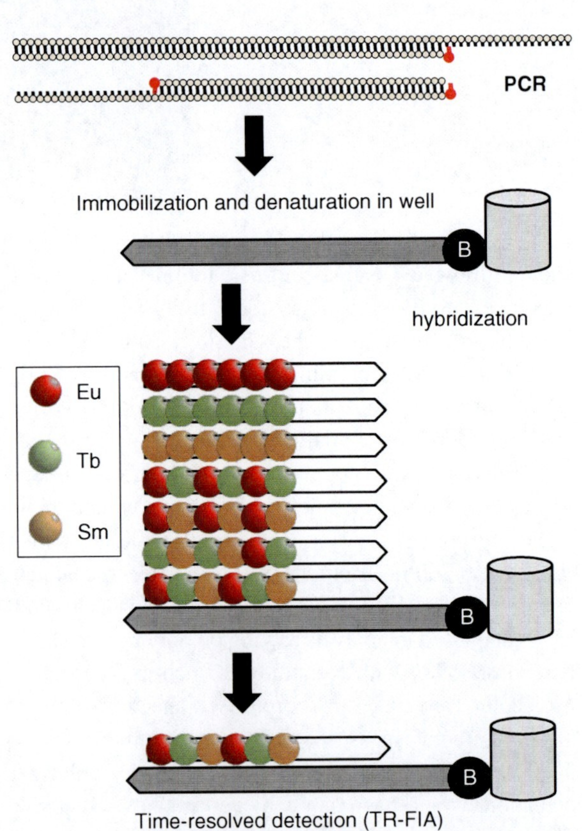

Lanthanides in Nucleic Acid Analysis, Fig. 3 Principle of multiplex analysis of PCR products using hybridization and TR-FIA. Redrawn from (Samiotaki et al. 1997)

Homogeneous Assays

These assays may be classified as follows (Nishioka et al. 2007).

Hybridization Assays

These analyses are designed for the detection of DNA (or RNA) sequences by hybridization as in the case of heterogeneous TR-FIA but without the time-consuming separation of the unreacted reagents. Oligonucleotide probes are an alternative too when allied to molecular beacons. The latter are hairpin-shaped nucleotides bearing a luminophore (L) and a quencher (Q) at their extremities and with a loop region complementary to the targeted nucleic acid. In absence of interaction, L and Q are in close proximity so that luminescence is switched off. When the loop of the beacon interacts with the targeted nucleic acid, emission is restored. Molecular beacons are efficient probes for PCR, genotyping, DNA binding, or cleavage assays of proteins and gene expressions. When assays are performed with live cells or tissue lysates, degradation may occur so that a more sophisticated technique is required with a pair of FRET probes each labeled with an acceptor or donor chromophore and which are complementary to neighboring regions of the analyte. When lanthanides are used as chromophores, improvement in sensitivity is regularly observed. In particular, DTPA-cs124 probes (Scheme 1) detected under time-resolved mode have a sensitivity of less than 1 pM (Krasnoperov et al. 2010).

An interesting variant of FRET analysis has been proposed by Laitala and Hemmilä based on nonoverlapping resonance energy transfer, nFRET. The lanthanide probe, e.g., a europium chelate such as the phosphoramidite depicted in Fig. 4 attached to the 3′ end of a 3′-TAC TTA TAT CTA TGT CTTC-5′ sequence, transfers energy onto Alexa Fluor® dyes conjugated to an amino-modified 3′-AAA TTA TAG TAA CCA CAAA-5′ sequence. Here, the underlined letters denotes bases which are noncomplementary to the target sequence and which prevent the probe of acting as primer during PCR. Changes in the lifetime of the acceptor upon hybridization with the target sequence make direct identification of the hybridized and nonhybridized probes feasible by time-resolved detection. The limit of detection for the DNA target is as low as 0.8 pM. The decay time of the organic fluorescent probe can be adjusted by changing the acceptor dye, which allows one-step multiplex

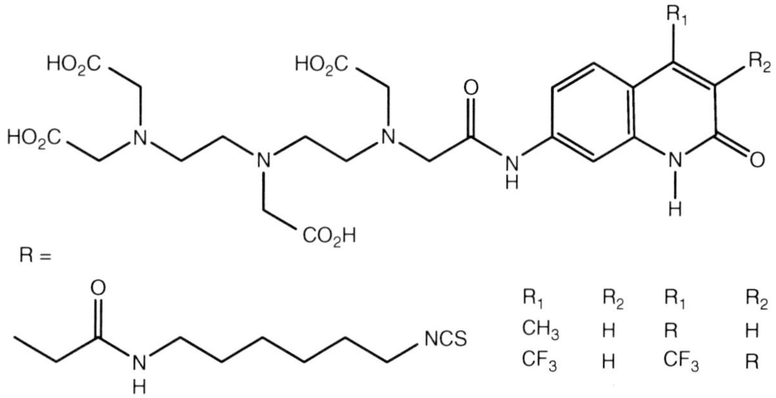

Lanthanides in Nucleic Acid Analysis, Scheme 1 Derivatized DTPA ligands for lanthanide DNA hybridization probes

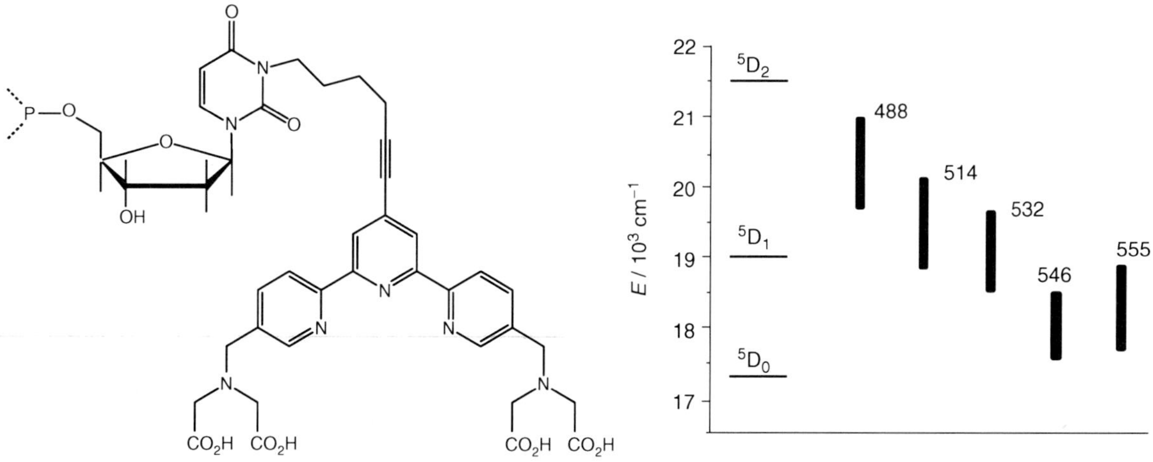

Lanthanides in Nucleic Acid Analysis, Fig. 4 Phosphoramidite ligand for nFret analyses (*left*). Comparison between the Eu^{III}-excited state energies and the absorption ranges of various Alexa dyes (*right*) (Redrawn from (Bünzli 2010))

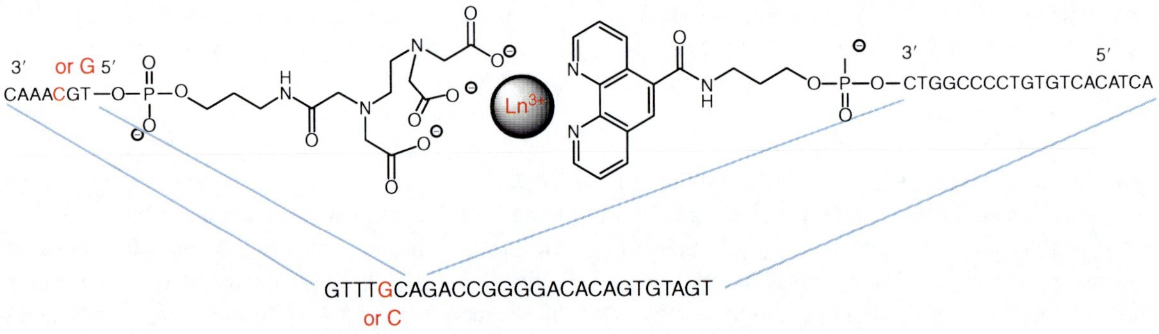

3′ or G 5′ O
CAAACGT—O—P—O ... N ... Ln3+ ... O—P—O—CTGGCCCCTGTGTCACATCA 3′ 5′

GTTTGCAGACCGGGGACACAGTGTAGT
or C

Lanthanides in Nucleic Acid Analysis, Fig. 5 DNA-templated lanthanide luminescent bioprobe and its target, a wild-type gene or mutant of the thiopurine-*S*-methyltransferase gene (Redrawn from (Kitamura et al. 2006))

homogeneous assays. It is noteworthy that the emission wavelength of the organic dyes is shorter than the Eu($^5D_0$) emission so that energy transfer occurs from upper excited levels of Eu^{III}. This is illustrated on the right part of Fig. 4 in which the energies of the Eu(^5D_J) levels are compared to the excited singlet state energies of a series of Alexa Fluor® dyes.

Intercalation Assays

These assays are designed to detect double-stranded DNAs; they have the disadvantage of nonspecificity but due to its simplicity, the method is rather widespread. Most of the intercalate probes are organic luminophores such as PicoGreen® (limit of detection: 0.25 ng/mL). In the past years, however, lanthanide complexes have been developed, e.g., based on quinoline derivatives or tetracyclines which give reasonable limits of detection although an order of magnitude larger than PicoGreen® (Nishioka et al. 2007). Alternatively, the disappearance of the quenching of a binuclear Eu^{III} helicate by acridine orange when it intercalates into DNA can be exploited to sense various types of DNAs and PCR products (Bünzli 2010).

Real-Time PCR Assays

PCR amplification is monitored at each cycle by either intercalations methods or by measuring the exonuclease activity of DN3A polymerases by the hybridization technique (TaqMan®). Some of these analyses have also been engineered with lanthanide luminescent chromophores.

Genotyping Assays

Cost-effective genotyping methods are presently in high demand in view of the needs for genome-based medicine ("personalized medicine"). Particular interest is focused on genotyping single nucleoside polymorphisms (SNPs) which are the most abundant for DNA sequence variation in the human genome and which are responsible for phenotypic diversity. Their large density, however, requires particularly powerful methods of analysis that are able to screen very large numbers of SNPs, and various approaches have been proposed based on hybridization of allele specific oligonucleotides and combining analytical methods such as aggregation of nanoparticles, electrochemical and microarray sensing, as well as FRET and enzymatic reactions. All these methods are multistep procedures, and some require sophisticated instrumentation or are limited by experimental factors such as temperature, pH, or ionic strength. Most of these drawbacks can be overcome with LLBs and highly sensitive "colorimetric" analyses similar to homogeneous immunoassays in their principle have been engineered. They involve DNA-templated cooperative complexation between a lanthanide luminescent ion and two oligodeoxyribonucleotide conjugates, one fitted with a polyaminocarboxylate (EDTA, DTPA) as chelating unit and the other decorated with an aromatic polyamine (1,10-phenanthroline, terpyridine, or dipyrido[3,2-a:2′,3′-c] phenazine) as sensitizing chromophore. The conjugates form a tandem duplex with the target in such a way that the auxiliary units face each other and build a coordinative environment for the Ln^{III} ions. Two situations are described in Fig. 5: the Ln^{III} conjugate with the C base recognizes the wild-type gene while the conjugate with the G base is specific for a mutant gene. It is noteworthy that only the combination of EDTA and 1,10-phenanthroline provides sufficient emission in the presence of the targets.

Outlook

In view of their specific photophysical properties and the ease with which lanthanide chelates can be linked to specific biomolecules, their use in DNA sensing and quantitation will undoubtedly increase in the future. Their introduction into nanoparticles with concomitant boost of the sensitivity has been demonstrated as well as their usefulness in combinatorial assays and in analyses based on microarrays.

Cross-References

- ▶ Lanthanides in Biological Labeling, Imaging, and Therapy
- ▶ Lanthanum, Physical and Chemical Properties

References

Bünzli J-CG (2010) Lanthanide luminescence for biomedical analyses and imaging. Chem Rev 110:2729–2755

Kitamura Y, Ihara T, Tsujimura Y, Osawa Y, Jyo A (2006) Colorimetric allele analysis based on the DNA-directed cooperative formation of luminous lanthanide complexes. Nucleic Acids Symp Ser (Oxf) 50:105–106

Krasnoperov LN, Marras SAE, Kozlov M, Wirpsza L, Mustaev A (2010) Luminescent probes for ultrasensitive detection of nucleic acids. Bioconjugate Chem 21:319–327

Lakowicz JR (2006) Principles of fluorescence spectroscopy. Springer Science, New York

Nishioka T, Fukui K, Matsumoto K (2007) Lanthanide chelates as luminescent labels in biomedical analyses. In: Gschneidner A Jr, Bünzli J-CG, Pecharsky VK (eds) Handbook on the physics and chemistry of rare earths, vol 37. Elsevier Science B.V, Amsterdam, pp 171–216

Samiotaki M, Kwiatkowski M, Ylitalo N, Landegren U (1997) Seven-color time-resolved fluorescence hybridization analysis of human papilloma virus types. Anal Biochem 253:156–161

Lanthanides, Basic Physico-chemical Properties

Fathi Habashi
Department of Mining, Metallurgical, and Materials Engineering, Laval University, Quebec City, Canada

Position in the Periodic Table

Lanthanides are a group of 13 metals that occur closely associated in nature with very similar chemical properties. To this group, an additional member, europium, does not occur in nature and the group adds to 14. Closely related to these 14 metals are scandium, yttrium, and lanthanum as shown in (Fig. 1). The lanthanides together with the other three members Sc, Y, and La are generally known as "rare earths". They are neither rare nor earths but this was a historical term that was introduced during the early years when they were discovered.

Physical Properties

See Table 1

Occurrence

While scandium and yttrium occur mainly in uranium ores the other members occur in monazite sand, a rare earth phosphate, $LnPO_4$, in bastnasite, a rare earth fluorocarbonate, $LnFCO_3$, and in small amounts in igneous phosphate rock. Promethium does not occur in nature – is separated from fission products produced in nuclear reactors. Ln is abbreviation for lanthanides.

Chemical Properties

Since the lanthanides have the same number of electrons in the two outermost shells their chemical properties will be nearly the same and it will be difficult to separate by chemical methods. That is why physical and physicochemical methods are used. Earlier methods were based on fractional crystallization. Recent methods are now based on ion exchange and solvent extraction. Since the outermost shell contains 2 electrons it would have been expected that the lanthanides have a valency of 2. This is, however, not the case – the valency is 3 and in the case of cerium it is 3 and 4.

Ce	Pr	Nd	Pm	Sm	Eu	Gd	Tb	Dy	Ho	Er	Tm	Yb	Lu
2	2	2	2	2	2	2	2	2	2	2	2	2	2
8	8	8	8	8	8	8	8	8	8	8	8	8	8
18	18	18	18	18	18	18	18	18	18	18	18	18	18
19	20	21	22	23	24	25	26	27	28	29	30	31	32
9	9	9	9	9	9	9	9	9	9	9	9	9	9
2	2	2	2	2	2	2	2	2	2	2	2	2	2

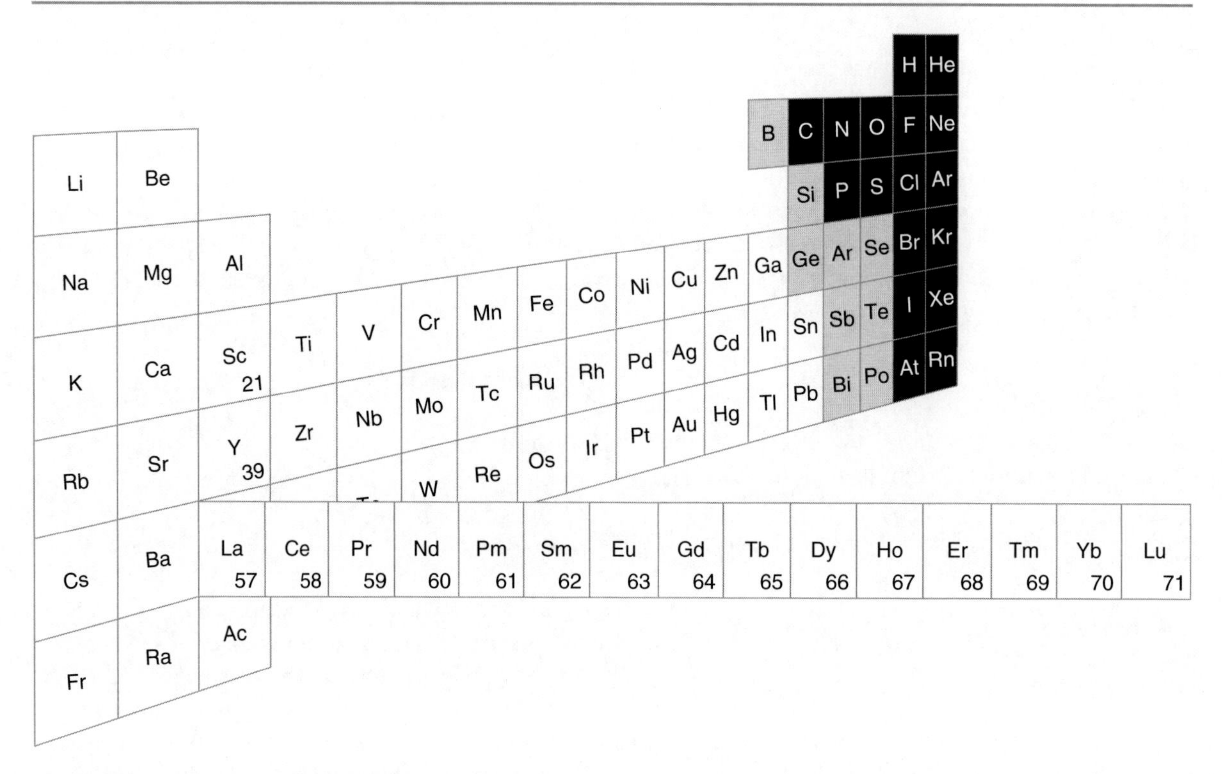

Lanthanides, Basic Physico-chemical Properties, Fig. 1 The periodic table of the elements showing the position of the lanthanides

Lanthanides, Basic Physico-chemical Properties, Table 1

Element	Symbol	Relative abundance, ppm	mp, °C	bp, °C	Crystal structure of stable phase at room temperature (high-temperature polymorphs)	Lattice constants, nm *a*	*c*	Density, g/cm³
Scandium	Sc	5–10	1,540	2,832	a (d)	0.33085	0.52683	2.989
Yttrium	Y	28–70	1,525	3,337	a (d)	0.36482	0.57318	4.469
Lanthanum	La	5–18	920	3,457	b (c, d)	0.3774	1.2159	6.145
Cerium	Ce	20–46	798	3,427	c (b, d)	0.51610		6.770
Praseodymium	Pr	3.5–5.5	931	3,512	b (d)	0.36721	1.18326	6.773
Neodymium	Nd	12–24	1,016	3,067	b (d)	0.36582	1.17926	7.007
Promethium	Pm	$<10^{-15}$	1,042	ca. 3,000	b (d)	0.365	1.165	7.26
Samarium	Sm	4.5–7	1,073	1,791	(a, d)	0.3621	2.625	7.536
Europium	Eu	0.14–1.1	822	1,597	d	0.45824		5.243
Gadolinium	Gd	4.5–6.4	1,312	3,267	a (d)	0.36360	0.57810	7.900
Terbium	Tb	0.7–1	1,357	3,222	a (d)	0.36055	0.56966	8.229
Dysprosium	Dy	4.5–7.5	1,409	2,562	a (d)	0.35915	0.56501	8.550
Holmium	Ho	0.7–1.2	1,470	2,695	a	0.35778	0.56178	8.795
Erbium	Er	2.5–6.5	1,522	2,862	a	0.35592	0.55850	9.066
Thulium	Tm	0.2–1	1,545	1,927	a	0.35375	0.55540	9.321
Ytterbium	Yb	2.7–8	816	1,194	c (d)	0.54862		6.965
Lutetium	Lu	0.8–1.7	1,663	3,397	a	0.35052	0.55494	9.840

a = hexagonal close packed (hcp); b = double hexagonal close packed (dhcp); c = face-centered cubic; d = body-centered cubic; r = rhombohedral.

References

Habashi F (2008) Researches on rare earths. History and technology. Métallurgie Extractive Québec, Québec, Distributed by Laval University Bookstore, www.zone.ul.ca
McGill I (1997) Rare earth metals General. In: Habashi F (ed) Handbook of extractive metallurgy. Wiley-VCH, Heidelberg/Germany, pp 1693–1741

Lanthanides, Luminescent Complexes as Labels

Loïc J. Charbonnière and Aline M. Nonat
Laboratoire d'Ingénierie Moléculaire Appliquée à l'Analyse, IPHC, UMR 7178 CNRS/UdS ECPM, Strasbourg, France

Synonyms

Biomarker; Lanthanide-bioconjugate; Luminescent tag

Lanthanide Label

"Lanthanide label" stands for lanthanide complexes which have been specifically designed to incorporate a labeling function for their conjugation to some biological material. The labeling function is a chemically reactive function, chosen to form a covalent link with residues contained into the biomaterial under mild conditions. In practice, two main families of residues will be targeted: amino acids (peptide, proteins, antibodies, etc.) and oligonucleotides (DNA, RNA, etc.).

The luminescence of trivalent lanthanide ions has found numerous applications in LASERs, optical telecommunications, medical diagnostics, and various other fields. The attention to luminescent lanthanide complexes for sensing and imaging started in the mid-1970s with the development of a family of Eu(III), Sm (III), Tb(III), and Dy(III) polyaminocarboxylate and β-diketonates complexes. Since then, luminescent lanthanide complexes have found applications in biology, biotechnology, and medicine with the development of lanthanide-based time-resolved fluoro-immunoassays, molecular probes for the quantification of pH, pO_2, and various simple anions, DNA targeting, and cell and tissue imaging (Eliseeva and Bünzli 2010).

All these bioanalysis require specific luminescent labels, i.e., luminescent lanthanide complexes which have been specifically modified to allow their covalent binding to antibodies, oligonucleotides, proteins, and other biomolecules of interest. This entry will first describe the prerequisites for these labels in terms of photo-physical and chemical properties, it will then introduce the chemistry of common bio-labeling reactions and we will finally present few applications of such labels for biomedical analysis.

General Properties of Luminescent Labels

In general, when exposed to a light, a sample can absorb photons at particular wavelengths corresponding to the differences between its ground state (S_0) and its various excited states (S_i, i = 1,2,. . ., Fig. 1). The representation of the intensity of light absorbed by the material over a range of incident wavelengths is called "absorption spectrum." The measurement of the absorbance A of a sample at a given wavelength λ (in nm), with an optical pathway l (in cm) and at a known concentration c gives access to the determination of the extinction coefficient $\varepsilon_{(\lambda)} = A_{(\lambda)}/lc$ (in M^{-1} cm^{-1}) according to the Beer-Lambert law.

Luminescence occurs when the deactivation of an excited state (S_i) is associated with the emission of a photon. Different types are distinguished depending on the source of excitation (chemoluminescence from a chemical reaction, bioluminescence by a living organism, thermoluminescence upon heating, etc.). In this entry, luminescence refers to photoluminescence, which occurs when the excited state is reached by absorption of a photon at a wavelength λ_{exc}, also called excitation wavelength. Upon relaxation to the ground state (S_0), the emission of photons at various wavelengths can be observed. Two mechanisms of luminescence can be distinguished and are usually represented by a Jablonski energy diagram (Fig. 1). Fluorescence happens when the emission of photons occurs directly from the electronic level which has been populated by the excitation. The average lifetime of a fluorescent excited state is between 10^{-12} and 10^{-6} s. Phosphorescence, however, implies a spin transition and emission arising from the transition between two states of

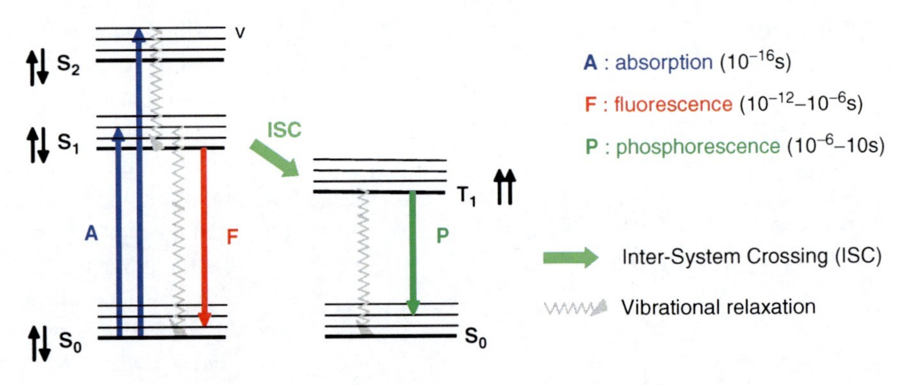

Lanthanides, Luminescent Complexes as Labels, Fig. 1 Jablonski energy diagram representing radiative and non-radiative processes between singlet state (S), triplet state (T), and their corresponding vibrational levels (v)

different multiplicities (e.g., from the triplet excited state to the fundamental singlet state). The lifetime of a phosphorescent excited state is significantly longer (from 10^{-6} to 10 s).

The plot of the emission intensity measured as a function of the emission wavelength is called "emission spectrum." The wavelength corresponding to the maximum of intensity in the emission spectrum is often referred to as the emission wavelength (λ_{em}). In addition to the excitation and the emission wavelengths, two other parameters are usually used to characterize luminescent compounds: The luminescence lifetime, τ, which is the characteristic time of the luminescent decay of an excited state and the luminescent quantum yield, which is a measure of the emission efficiency of the dye. The luminescent lifetime can be determined experimentally by fitting the time-decay of the emitted intensity according to the equation $I_{em}(t) = I_0 exp\left(\frac{-t}{\tau}\right)$ in which $I_{em}(t)$ is the emitted intensity at time t after a pulsed excitation and I_0 is the intensity at time 0. The luminescence quantum yield corresponds to the ratio of the number of emitted photons over the number of absorbed photons. Assuming that the samples are diluted (A < 0.05) and that the intensity of the lamp is corrected for its wavelength dependence, the quantum yield of a compound (x), ϕ_x, can be determined by comparison with a reference sample (ref) of known quantum yield, ϕ_{ref}, using the following equation: $\phi_x = \phi_{ref} \times \left(\frac{n_x}{n_{ref}}\right)^2 \times \left(\frac{\int I_x(\lambda)d\lambda}{\int I_{ref}(\lambda)d\lambda}\right) \times \frac{A_{ref}(\lambda_{exc.x})}{A_x(\lambda_{exc.ref})}$ in which n is the refractive index of the solvent used, $I(\lambda)$ is the luminescence spectra corrected for the wavelength dependence of the detector and $A(\lambda_{exc})$ corresponds to the absorbance at the excitation wavelength. Alternatively, direct methods exist to measure ϕ_x without the need of a reference sample but they require more sophisticated equipments such as an integrating sphere.

Photo-Physical and Chemical Properties of Lanthanide-Based Luminescent Labels

Luminescent lanthanide ions display narrow-line emission spectra that correspond to 4f-4f electronic transitions which are characteristic of each metal. Whereas all lanthanide(III) ions (except for La and Lu) are luminescent, visible emitters, such as Eu ($\lambda_{em} = 616$ nm) and Tb ($\lambda_{em} = 545$ nm) and to a lesser extent Sm and Dy, have received most attention for biomedical applications. Some other Ln(III), such as Nd and Yb, have relatively small gaps between the ground and the excited states, resulting in an emission signal in the near-infrared (NIR) region. Such NIR-emitting lanthanides are particularly attractive for probing biological interactions because biological tissues are transparent in this spectral range. Unfortunately, Nd and Yb complexes are only weakly emissive because radiative processes are in competition with non-radiative deactivation processes such as vibronic interactions with solvent molecules.

The low probability of the 4f-4f transitions results in long-lasting emission with luminescent lifetimes significantly longer than the organic dyes (in the order of milliseconds for europium and terbium, to be compared to nanoseconds for conventional fluorescent compounds). This allows the use of lanthanide-based labels for time-resolved detection. This technique consists in the implementation of a delay time between the pulsed photonic excitation of the sample and the integration of the luminescence signal by the detector.

During this short delay, all the short-lived luminescence arising from auto-fluorescence of a biological sample or from light scattering in the apparatus will vanish, so that the only remaining signal will originate from the long-lived luminophor and from the dark noise of the setup. By a careful choice of the delay, the integration time and the repetition rate of the pulsed excitation, time-resolved detection considerably enhances the signal-to-noise ratio with sensitivity gains of more than three orders of magnitude using long-lived luminescent lanthanide complexes.

Moreover, trivalent lanthanide ions display a high photostability, which is of great interest for their use in analysis and imaging. Unfortunately, the f-f transitions are characterized by very low extinction coefficients in the absorption spectra ($\varepsilon \leq 10\ cm^{-1}\ M^{-1}$). This drawback arises from the low probability of the f-f transitions which are generally forbidden by the selection rules. It can be overcome by using powerful excitation sources such as LASERs or by sensitizing the lanthanide ion via a chromophoric antenna. The indirect excitation by a bound chromophoric unit is referred to as "antenna effect." This sensitization mechanism involves four steps, which will be detailed later.

The other challenge of working with lanthanide ions for biomedical applications comes from their toxicity. For instance, lanthanide(III) has an ionic radius close to Ca(II) and can compete with endogenous calcium to produce a dysfunction of the calcium channels leading to neuromuscular disorders. Moreover, free Ln ions are also known to bind to several serum proteins and endogenous anions such as phosphates, carbonates, or hydroxides to form insoluble complexes which are then irreversibly fixed on the bones, liver, and bladder. However, when the Ln ions are complexed with appropriate ligands, they remain chelated in the body and are excreted intact. Such a strong chelation is achieved by combining a strong preorganization of the complexation cavity of the ligand and/or by the introduction of hard coordinating units such as carboxylate and phosphonate functions.

As mentioned earlier, the ligand should incorporate a chromophoric antenna, which will transfer its energy to the excited state of the lanthanide cation (Ln*, Fig. 2). In general, the population of the lanthanide excited state occurs via indirect energy transfer through the ligand centered triplet state ($^3\pi\pi^*$). Indeed, after the absorption of photons, the ligand reaches its singlet excited states which could then relax according to the three following mechanisms: emission of fluorescence F, non-radiative vibrational processes, or by energy transfer to the triplet state $^3\pi\pi^*$ through Inter-System Crossing (ISC). In the case of Ln complexes, ISC is favored by the presence of the heavy Ln(III) cation that favors vibronic coupling between the electronic states. The triplet state can then dissipate its energy by emission of phosphorescence P, by non-radiative pathways or by energy transfer through the lanthanide excited state Ln*. Finally, the lanthanide excited state returns to the ground state by emission of photons and/or by non-radiative processes.

From the above mechanism, it is foreseen that the antenna must be optimized in several ways in order to optimize the efficiency of the energy conversion processes and to sensitize efficiently the lanthanide excited state. First, the chromophoric unit should possess an extinction coefficient as high as possible and the excitation wavelength should be above ca. 300 nm in order to avoid the large absorption arising from the biological background below this limit and to minimize degradation of biological material. Importantly, the triplet state should match the excited state of the lanthanide in order to observe an efficient energy transfer. The design of well-suited antennae, which gives rise to stable complexes together with a quantitative sensitization of the Ln, is one of the key steps toward highly luminescent labels. In the case of Eu and Tb complexes, empirical rules have been established to predict the energy transfer efficiency from the triplet state of the antenna and therefore facilitate the ligand design (Latva et al. 1997). Actually, the energy gap between the ligand triplet state and the lanthanide excited state should be around 2,500–3,500 cm^{-1} in order to optimize the energy transfer. In fact, particularly in the case of Tb complexes, when the ligand-centered triplet state and the Tb($^5D_4$) emitting level at 20,600 cm^{-1} (ca. 485 nm) are too close in energy, back-energy transfer operates from Ln* to $^3\pi\pi^*$ with a concomitant loss of luminescence. This process is in general assisted by vibrations and is therefore strongly temperature dependent. For Eu, the energetic requirements of the chromophoric unit are less straightforward since three emitting levels ($^5D_2$, $^5D_1$ $^5D_0$) can be populated by energy transfer from the ligand triplet state. Nonetheless, an efficient sensitization is usually observed when the energy of the ligand triplet state is around 22,000 cm^{-1} (Latva et al. 1997).

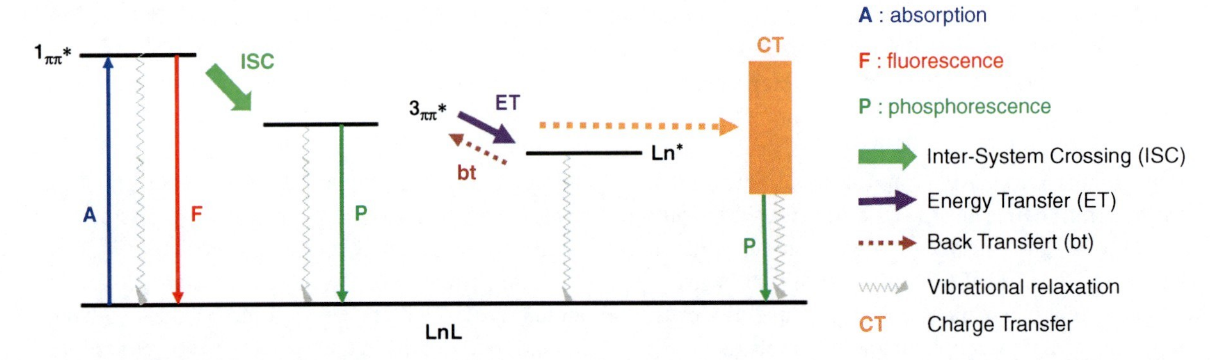

Lanthanides, Luminescent Complexes as Labels, Fig. 2 Principle of the antenna effect in a luminescent Ln(III) complex and competing energy pathways

Finally, the last requirement to achieve highly luminescent lanthanide complexes is to minimize the non-radiative processes. Deactivation through vibrations is especially effective and represents a major concern in the design of luminescent edifices. In aqueous solutions, interaction with water, both in the inner and outer coordination spheres of the Ln ion, leads to a significant quenching of the metal luminescence via O-H vibrations (Horrocks and Sudnick 1981). This non-radiative deactivation can be minimized by complexing the luminescent lanthanide ion with a rigid and saturated coordination sphere. Other deactivation processes occur through vibrations of bound ligands, particularly those with high-energy vibrations such as O-H, N-H, or C-H. This type of non-radiative processes are particularly harmful in the case of near-infrared emitting lanthanide(III). In a similar way, electronic deactivation processes have to be avoided by an adequate design of the ligand (Fig. 2). Another deactivation process is the photo-induced electron transfer (PET) leading to ligand-to-metal- or metal-to-ligand charge transfer states (LMCT or MLCT). PET processes are generally observed in the case of easily reducible Ln(III) such as Eu, Sm and Yb, and are due to the reduction of Ln(III) into Ln(II) upon photo-excitation and the concomitant energy transfer to low-lying ligand-to-metal charge transfer states.

Examples of Antennae

One of the simplest chromophoric units is the picolinate ligand (**1**) (Fig. 3), which has been extensively used for the sensitization of lanthanide emitters and particularly Eu and Tb. The bidentate ligand bipyridine (bipy = **2**) is also an efficient antenna for the sensitization of visible lanthanide luminescence. For instance, the tris-bipy cryptand molecules developed by Lehn and coworkers are routinely used for bioassays since more than 20 years. Regarding NIR emission, the incorporation of 8-hydroxyquinolate chromophore (**4**) often led to the formation of complexes with large stability in water and an efficient sensitization of Nd and Yb. Other well-studied chromophores are N-heterocyclic ligands including terpyridine (**5**), phenanthroline (**3**), azaxanthone (**7**), and hydroxyisophtalimide derivatives (**6**) (Fig. 3) (Bünzli 2010).

Two strategies are generally employed in order to afford highly stable complexes and to prevent non-radiative deactivation (Fig. 4). On the one hand, the chromophoric unit can be covalently linked to a multidentate ligand which bind the Ln(III) ion. Polyamine derivatives such as the linear diethylenetriaminepentaacetic acid (DTPA) and macrocyclic cyclen-based ligands (cyclen = 1,4,7,11-tetraazacyclododecane) are known to yield Ln complexes with high thermodynamic stability and kinetic inertness and they have been extensively used. To date, over 60 different cyclen-based Eu and Tb complexes have been developed with various antennae such as quinoline (**8**), phenanthroline, or azaxanthone derivatives (Parker 2000; Gunnlaugsson and Leonard 2005). With this strategy, it has to be noticed that, the chromophoric unit is not always coordinated to the Ln cation. Should that happen, ligand-to-metal energy transfer would be are less efficient. On the other hand, the chromophoric unit can be appended with strong coordinating units such as carboxylates, aminocarboxylates (**9**), or phosphonates. As such, the chromophoric unit is directly bound to the lanthanide cation, which is favorable to efficient energy transfer.

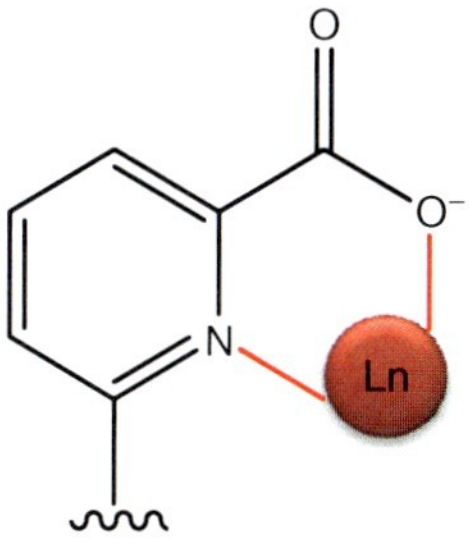

1 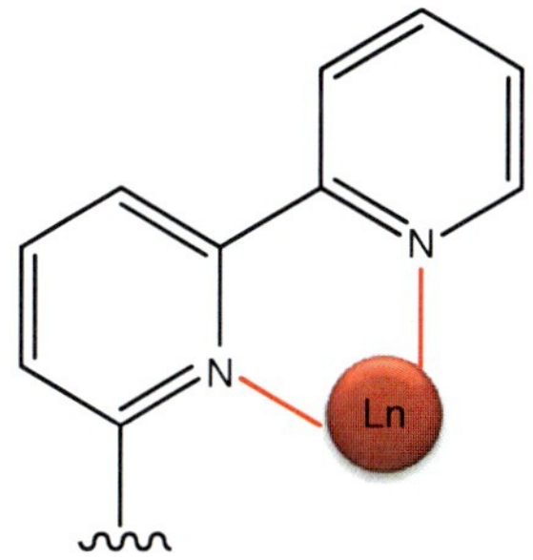 2 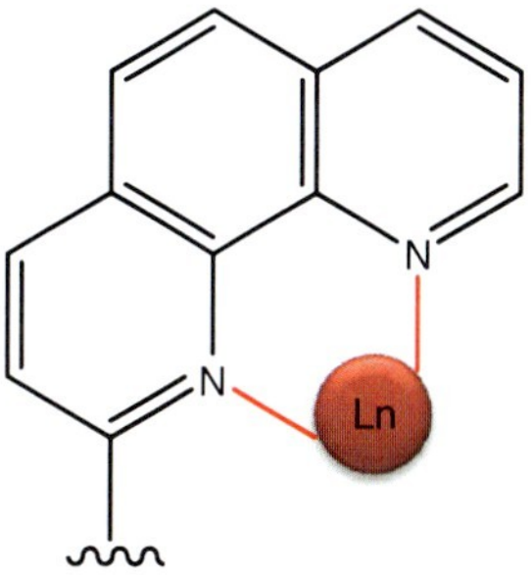 3 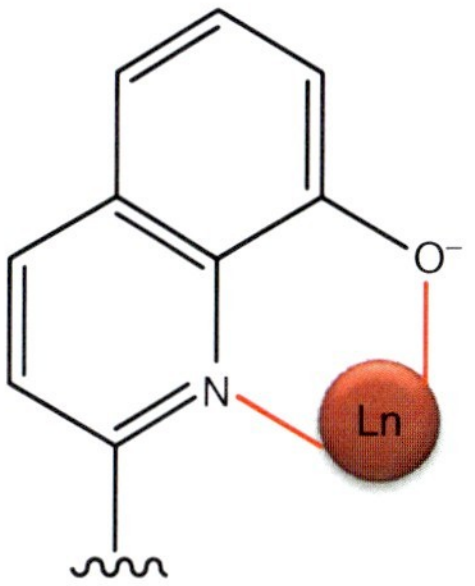 4

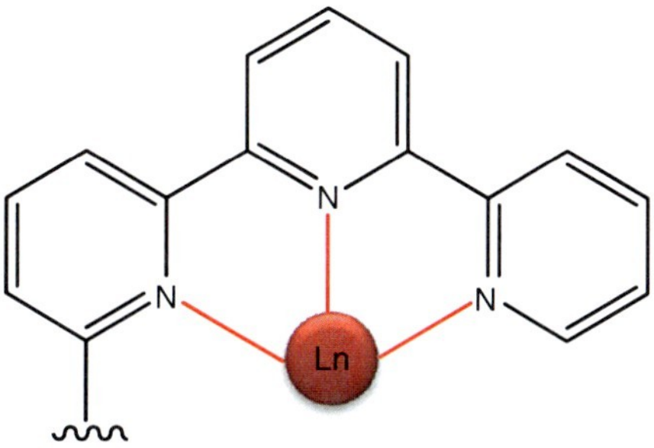

5 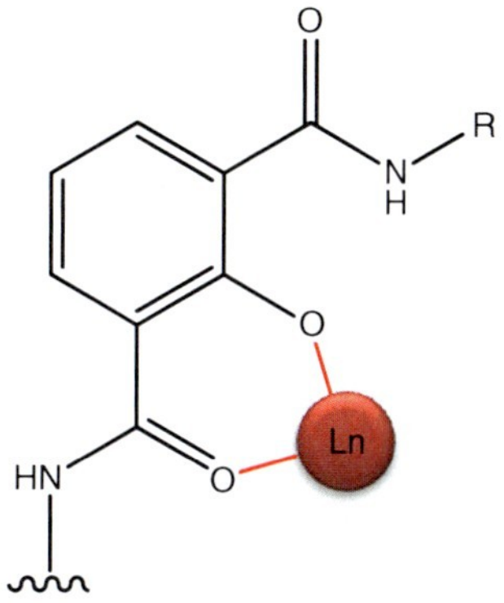 6 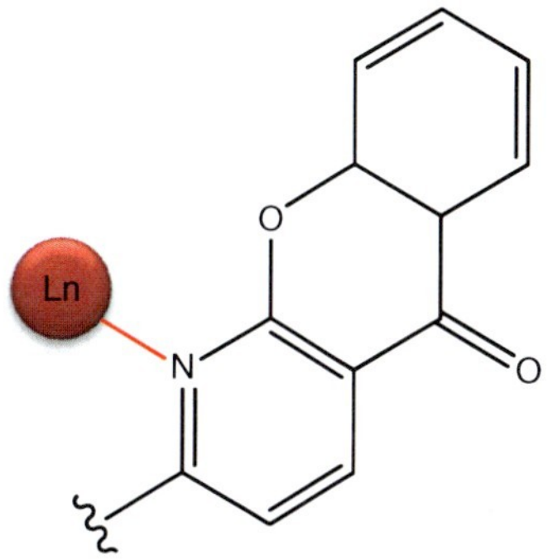 7

Lanthanides, Luminescent Complexes as Labels, Fig. 3 Examples of chromophoric units and their coordination to Ln(III) cations

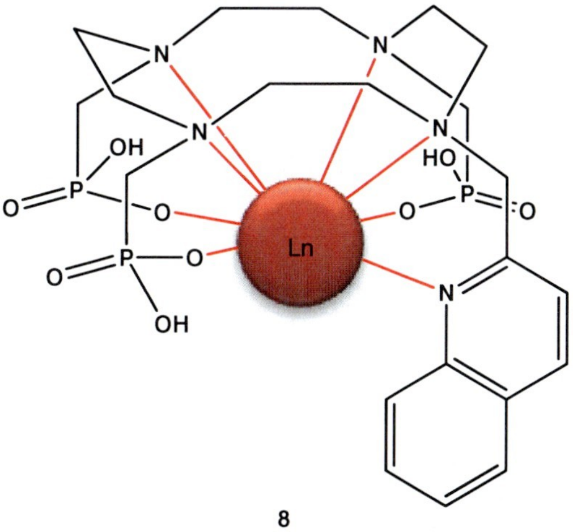

8 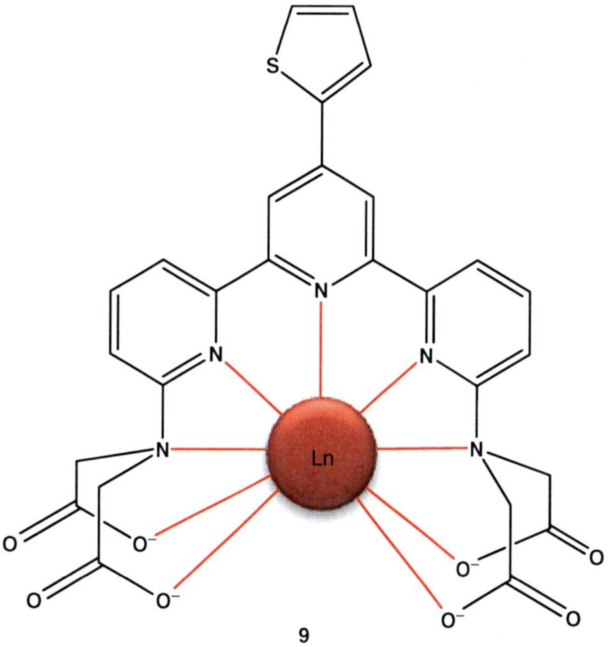 9

Lanthanides, Luminescent Complexes as Labels, Fig. 4 Example of cyclen derivative (**8**) and podant (**9**) for the complexation of Ln(III) ions (From Charbonnière (2011))

Lanthanides, Luminescent Complexes as Labels, Fig. 5 Schematic representation of a lanthanide label

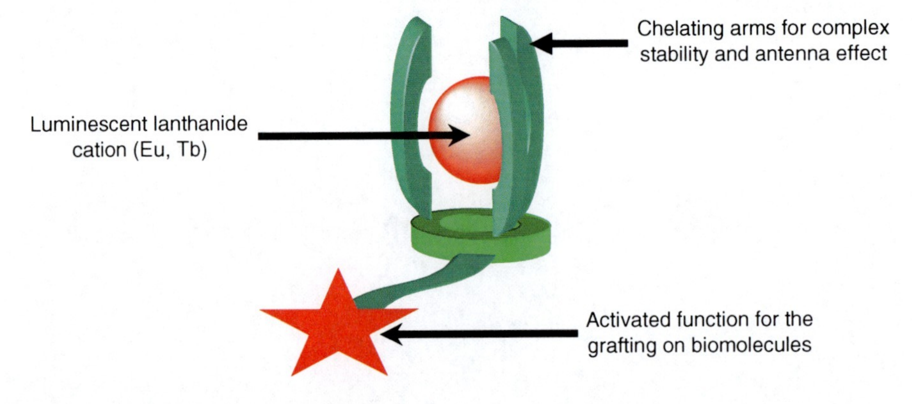

Other approaches have also been used for the synthesis of luminescent Ln complexes, as for example the formation of ternary complexes, in which the coordination sphere of the Ln cation is fulfilled by one or several β-diketonate antennae. However, also they have excellent photophysical properties; these complexes are generally less stable, particularly in aqueous solutions. Apart from the sensitization through organic chromophores, energy transfer from d-transition metal levels such as d-d transitions or MLCT states or from the excited state of another lanthanide has also been investigated. This strategy is particularly adapted for the sensitization of ions emitting in the near-infrared and will not be developed here (Eliseeva and Bünzli 2010).

The Labeling Function

Once the efficient design of the ligand has led to the formation of a lanthanide complex with interesting luminescent properties and suitable stability, its use as a labeling compound will require the introduction of a labeling function (Fig. 5). Indeed, this chemically reactive function will react with the material to be labeled by forming a strong covalent bond (Hovinen and Guy 2009; Charbonnière 2011). The labeling function being by essence a reactive entity, it is of prior importance to consider its introduction as late as possible in the retrosynthetic pathway leading to the final label in order to avoid its degradation before the final labeling.

Classical conditions for the bio-labeling reactions are aqueous buffered media at pH close to neutrality or slightly basic, and at temperatures ranging from 4°C to 40°C. Some organic solvents such as EtOH, DMSO, or DMF may also be used when necessary (often for solubility reasons), providing that they do not perturb the bioactivity of the labeled product. Considering that the material to be labeled is mainly concerned with peptides, proteins, or antibodies, the most popular activated functions are targeted toward accessible chemical functions on these residues: the amine function of lysine amino acids and the thiol function of cysteine residues. The following figure summarizes the main activated functions used for lanthanide labels targeted toward amine functions (Fig. 6).

The isothiocyanate function of **10** is a widely sprayed function obtained by reaction of thiophosgene on amino aromatic moieties. In the presence of amine, they react to form a thiourea bridge with the protein. When chlorocyanuric acid is reacted with an amine of the lanthanide complex, it readily forms the (4,6-dichloro-1,3,5-triazin-2-yl)amine **11**, which can be used to obtain a second substitution of a chloride atom to link the protein to the label. The chlorosulfonyl function of **12** was one of the first activated functions used to label lanthanide on biological compounds. It is obtained by the reaction of chlorosulfonic acid (HSO_3Cl) on the aromatic parts of the ligand. It is however to be noticed that in this case, the labeling is performed with the free ligand and that the lanthanide cation is added only after the labeling of the biomolecule. Finally, one of the most current activating functions currently used is the formation of activated esters, especially with *N*-HydroxySuccinimide (NHS), **13**. These activated esters are obtained from the reaction of a carboxylic acid contained on the complex, with NHS, in the presence of a coupling agent such as EthylDimethylaminopropylCarboDiimide (EDCI). In case of water solubilization troubles of the activated complex, the more hydrophilic Sulfo-NHS esters can be used, in which a sodium sulfonate function is introduced into the succinimidyl backbone. If one is

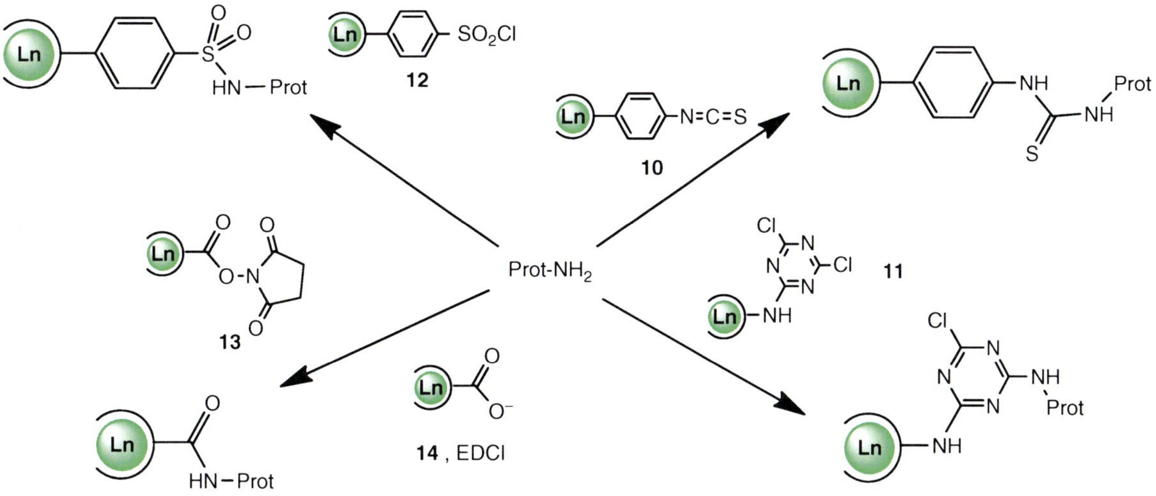

Lanthanides, Luminescent Complexes as Labels, Fig. 6 Activated functions for labeling lanthanide complexes on amine functions

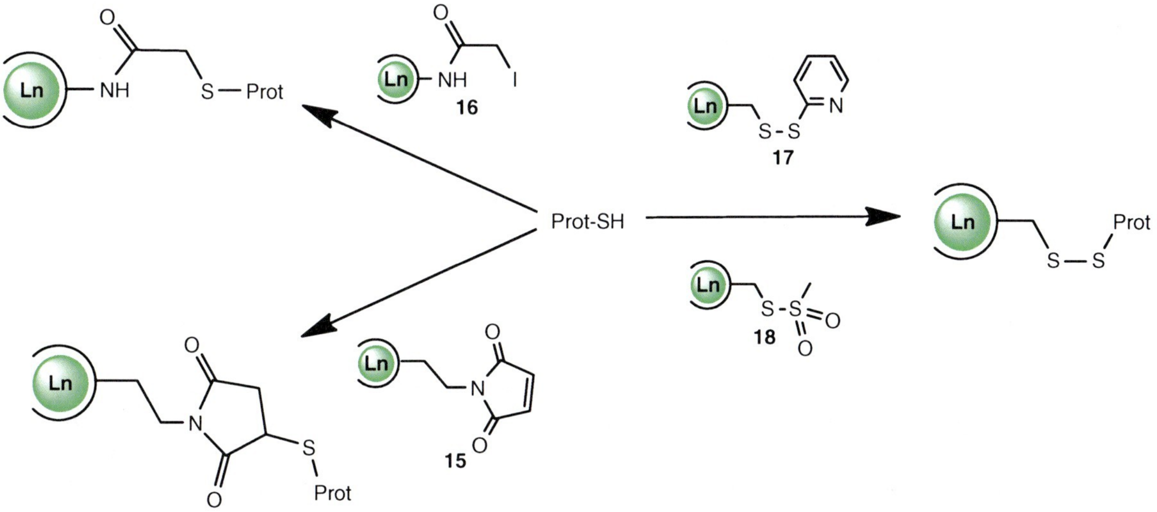

Lanthanides, Luminescent Complexes as Labels, Fig. 7 Activated functions for labeling lanthanide complexes on thiol functions

not interested in isolating the activated ester, it is also possible to couple directly the carboxylic acid function of the complex **14** on the biomolecule with EDCI. Nevertheless, most ligands used for a strong coordination of lanthanide cations are based on polyaminocarboxylic groups and the direct coupling may result in a difficult control of the reaction which can occur on more than one acid per ligand, leading to cross coupling reactions.

Thiol residues are another target of choice for the labeling of biomolecules (Fig. 7). Cysteine amino acids are far less present on proteins than lysines, resulting in a more selective coupling. Furthermore, disulfide bridges are present as cystine functions in the structure of whole antibodies and their reduction give access to thiols that can be labeled. Finally, the deprotonation of thiol functions to form RS^- nucleophiles happens at lower pH than that of ammonium

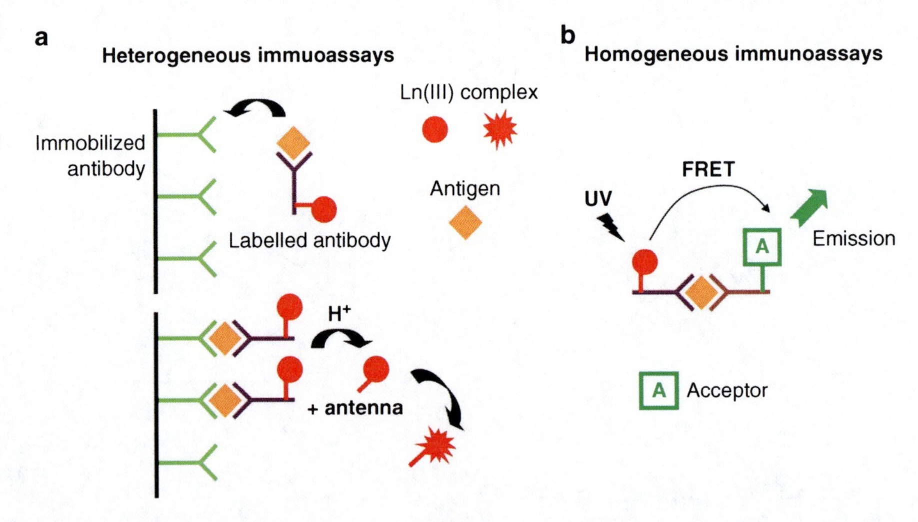

Lanthanides, Luminescent Complexes as Labels, Fig. 8 General principle of heterogeneous and homogeneous immunoassays

residues, affording a pH-dependent selectivity toward thiol residues. Coupling with thiol groups can be obtained with maleimide functions such as that of **15**, the addition of the thiol on the double bond resulting in a thioether link between the lanthanide tag and the biomolecule. Halogenoacetamides based on iodine (**16**) or bromide leaving groups are also good candidates, leading also to the formation of a thioether. For these two reactions based on nucleophilic reactions, caution must be taken to control the pH of the solution, as too basic conditions can lead to the concurrent reaction of amine functions of the protein. More selective, the reaction of thiols on disulfide-activated functions such as those of **17** or **18** allows for the selective bonding of thiol residues through the formation of disulfide bridges with the tag.

Applications : Fluoroimmunoassays, Microscopy, and Time-Resolved Luminescence

The earliest use of lanthanide labels for biomedical application arose from the need to develop high-sensitivity probes for use in bioassays. Nowadays, lanthanide-based assays are routinely used to detect analytes such as antigens, steroids, hormones, and nucleic acid, at nano- to picomolar concentrations (Hemmila and Laitala 2005). Using these techniques, the short-lived background fluorescence of the biological medium is allowed to fade of before measuring the long-lived lanthanide-centered luminescence, which greatly improves the signal-to-noise ratio. These assays are based on the use of antibodies labeled with lanthanide complexes. Heterogeneous luminescent immunoassays involve a two-step procedure in which the analyte is first coupled to an Ln-labeled antibody and immobilized on a solid surface (Fig. 8a). Then, the lanthanide complex is dissociated in acidic medium and the free lanthanide(III) ions are complexed by an antenna to give a highly luminescence complex, which emission signal is measured (Fig. 8a). Alternatively, homogeneous immunoassays involve two different antibodies: The first one is labeled with a lanthanide complex whereas the second one is conjugated to an organic fluorophore emitting at a wavelength distinct from the Ln emission. Luminescence of the organic dye is observed only when the analyte is coupled to the two antibodies simultaneously. Indeed, in that case, time-resolved energy transfer is observed from the lanthanide complex (which acts as an energy donor) to the organic dye (energy acceptor) (Fig. 8b). This non-radiative energy transfer process is called resonance energy transfer (RET) and occurs from a first luminescent compound in its excited state (i.e., the donor) to a second one (i.e., the acceptor), providing that (1) the two compounds are in a close spatial proximity and (2) the emission spectrum of the energy donor overlaps the absorption spectrum of the energy acceptor. RET can also be used to measure distances between the donor and the acceptor within the 1–10 nm range.

A similar technology has been developed for DNA targeting. A typical example is when a europium(III) complex is bound to the end of a single-stranded DNA or oligonucleotide, while an acceptor such as

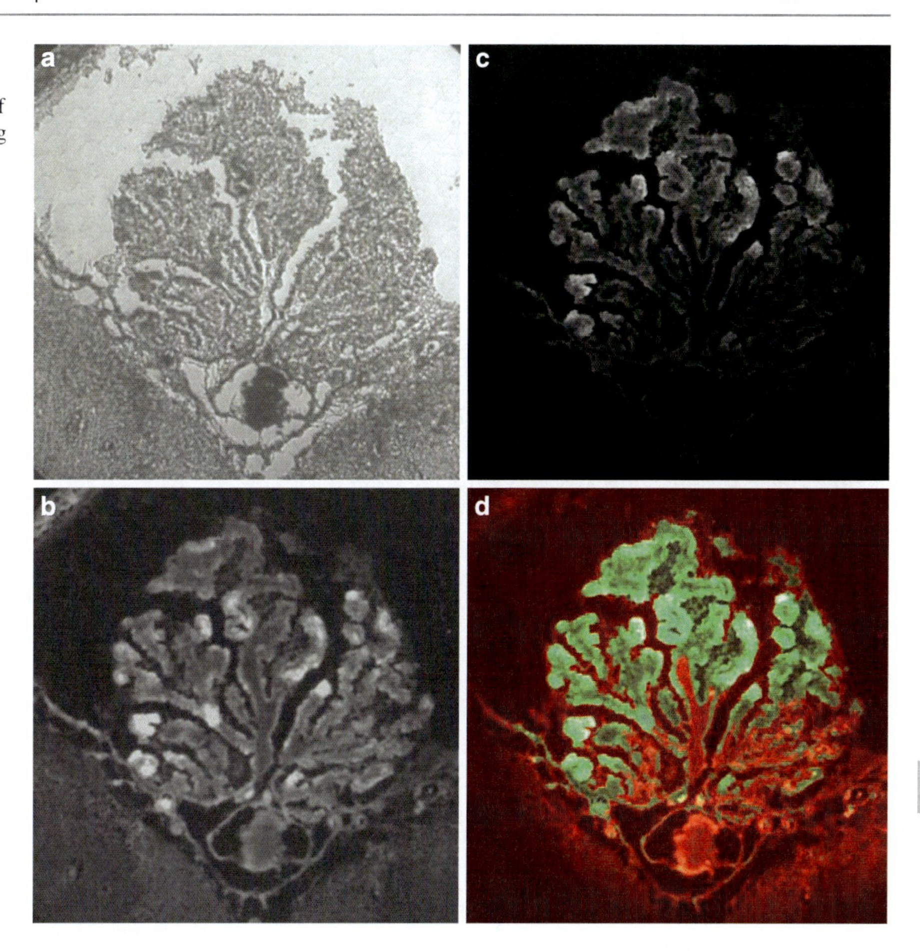

Lanthanides, Luminescent Complexes as Labels, Fig. 9 Microscope images of a rat brain slice corresponding to the region of the choroids plexus stained with a hydrophylic terbium complex. (**a**) Transmission. (**b**) Fluorescence. (**c**) Time-resolved emission ($\delta = 100$ μs). (**d**) Overlay of b and c (real size $= 1 \times 1$ mm^2)

a cyanine-5 dye is linked to an oligonucleotide with a complementary sequence. Upon hybridation, FRET occurs and emission of the acceptor is observed.

Among other interesting applications of luminescent lanthanide complexes, time-resolved microscopy is still poorly exploited, but should find a lot of applications. As for other time-resolved applications, the principle is simply to apply a delay time between the pulsed illumination of the sample and the detection of the fluorescence image. During this delay, all spurious signals originating from the fluorescence of the sample and light scattering in the apparatus will vanish, and the delayed acquisition will only reveal signals arising from the long-lasting luminescence of the lanthanide complex. As mentioned previously for time-resolved fluoroimmunoassays, this technique allows for a large improvement of the signal-to-noise ratio and of the limit of detection. Figure 9 displays microscopic images of a rat brain slice which has been stained by a hydrophilic terbium complex. As observed in the upper left part, the transmission image shows the region to consist of the choroids plexus, an important region of the blood-brain barrier at which exchanges occur between the blood and the brain. The central part of the images is constituted of hydrophilic epithelial cells, while the peripheral part shows the hydrophobic gray matter. In the fluorescence mode (Fig. 9b), the fluorescence signal can be observed in both regions, as a result of auto-fluorescence of the sample and luminescence from the Tb complex. Upon imposition of a delay (Fig. 9c), the short-lived signals disappeared and only the luminescence arising from Tb can be observed, principally in the hydrophilic regions as evidenced in Fig. 9d (Claudel-Gillet et al. 2008).

Although time-resolved luminescence microscopy clearly presents many advantages compared to conventional fluorescence microscopy, it requires the use of special optical components (excitation is mainly into the UV region with dedicated objectives), a pulsed light source, and a delayed acquisition setup, all of which contributing to largely increase the price of the apparatus. But with the development of improved lanthanide complexes excited at lower energy and of lower cost electronic devices (gated cameras), time-resolved microscopy may find a second wind.

Conclusion and Perspectives

With their sharp emission spectra, large effective Stokes shifts and long-lived luminescent lifetimes, lanthanide(III) labels are very powerful tools for biomedical diagnostics. In particular, they render possible both spectral and time-resolved discrimination of the luminescence signal from background autofluorescence and therefore have promoted the development of assays with improved detection sensitivities. However, efficient lanthanide labels must simultaneously meet very drastic photophysical, chemical, and biochemical criteria and remain a synthetic challenge. Over the past several years, considerable efforts have been made in understanding the relationship between the architecture of the ligand and the stability and the photophysical properties of the corresponding Ln complexes. Among the many compounds described in the literature, pyridine or bipyridine appended with anionic groups, such as carboxylates, aminocarboxylates, and phosphonates, have lead to highly luminescent complexes in the visible range with high stability. Chelates with hydroxyisophtalimide are also good candidates for the sensitization of visible emitters (Eu, Tb, and Sm), whereas podating ligands based on the hydroxyquinolate have been developed for the sensitization of NIR emitters (Nd, Yb, and Er).

Several bioconjugation reactions have now been developed in order to allow the covalent coupling of a lanthanide label to bioactive molecules such as proteins, nucleic acids, or peptide without altering their affinity for the target. A large variety of activated functions have been introduced on lanthanide complexes to allow the coupling reaction to the primary amines or thiol functions of peptidic residues in very mild conditions. These lanthanide bioconjugates are now used in a large number of bioanalysis for specific targeting such as time-resolved fluoroimmunoassays and time-resolved luminescence microscopy.

Recent works are focusing on improving immunoassays and imaging experiments sensitivity. For example, time-resolved luminescence resonance energy transfer assays enable to study molecular interactions with improved sensitivity and high throughput. Even better, single molecule detection would be possible in the near future with novel-imaging systems based on Ln-containing nanoparticles. Another focus is to shift the excitation wavelength in the visible range which would preserve the integrity of the biological material and would reduce the cost of excitation sources. For that purpose, ligands with extended conjugation as well as ligands for mutiphoton excitation are now being developed.

Cross-References

▶ Lanthanide Ions as Luminescent Probes
▶ Lanthanides in Biological Labeling, Imaging, and Therapy
▶ Lanthanides, Physical and Chemical Characteristics
▶ Lanthanum, Physical and Chemical Properties

References

Bünzli J-CG (2010) Lanthanide Luminescence for biomedical Analyses and Imaging. Chem Rev 110(5):2729–2755

Charbonnière LJ (2011) Luminescent lanthanide labels. Curr Inorg Chem 1:2–16

Claudel-Gillet S, Steibel J, Weibel N, Chauvin T, Port M, Raynal I, Ziessel R, Charbonnière LJ (2008) Lanthanide-based conjugates as polyvalent probes for biological labeling. Eur J Inorg Chem 18:2856–2862

Eliseeva SV, Bünzli J-CG (2010) Lanthanide luminescence for functional materials and bio-sciences. Chem Soc Rev 39:189–227

Gunnlaugsson T, Leonard JP (2005) Responsive lanthanide luminescent cyclen complexes: from switching/sensing to supramolecular architectures. Chem Commun 25: 3114–3131

Hemmila I, Laitala V (2005) Progress in lanthanides as luminescent probes. J Lumin 15(4):529–542

Horrocks WD Jr, Sudnick DR (1981) Lanthanide ion luminescence probes of the structure of biological macromolecules. Acc Chem Res 14:384–392

Hovinen J, Guy PM (2009) Bioconjugation with stable luminescent lanthanide(III) chelates comprising pyridine subunits. Bioconjugate Chem 20:404–421

Latva M, Takalo H, Mukkala V-M, Matachescu C, Rodríguez-Ubis JC, Kankare J (1997) Correlation between the lowest triplet state energy level of the ligand and lanthanide(III) luminescence quantum yield. J Lumin 75:149–169

Parker D (2000) Luminescent lanthanide sensors for pH, pO(2) and selected anions. Coord Chem Rev 205:109–130

Lanthanides, Physical and Chemical Characteristics

Jean-Claude G. Bünzli
Center for Next Generation Photovoltaic Systems, Korea University, Sejong Campus, Jochiwon–eup, Yeongi–gun, ChungNam–do, Republic of Korea
École Polytechnique Fédérale de Lausanne, Institute of Chemical Sciences and Engineering, Lausanne, Switzerland

Synonyms

4f Elements; Lanthanoids; Rare earths

Definition

According to IUPAC rules, elements 57–71 (Ln = La-Lu) should be called "lanthanoids," and when the series is completed by Sc (21) and Y (39), it takes the name of "rare earths" (Fig. 1).

The Elements and Their Uses

Despite the above definition, "lanthanides" is still the most used designation for the metallic elements 57–71 and their compounds (in >80% of the published papers). The first rare-earth element, yttrium, was isolated under the form of its oxide by the Finnish chemist Johan Gadolin in 1794 from a recently discovered mineral near Ytterby (Sweden). It took more than 100 years (1803–1907) to identify the remaining naturally occurring elements from minerals in which they appear as entangled mixtures, while radioactive Pm was synthesized in 1947. The mesmerizing story of the discovery of lanthanoids is full of incorrect claims and heated disputes among would-be discoverers whilst reflecting the developments in separation, analytical, and spectroscopic techniques which took place during this span of time.

The abundances of rare earths in the Earth's crust are relatively small, ranging between 0.5 (Tm, Lu) and 60 ppm (Y) and their concentration in minerals seldom exceeds 10–15%. Moreover, these minerals contain a range of rare earths and given very similar chemical properties, extraction and separation of these elements is difficult, time consuming, and polluting. Rare-earth resources are fairly well distributed among the continents, but their production is presently concentrated in China, the Chinese ores being easier to mine (Cotton 2006).

Nowadays, highly pure rare earths, obtained after lengthy separations, are at hand and are at the heart of high-technology applications. Rare earths and their compounds are vital to almost every aspects of our present world, being active components of catalysts, special alloys, magnets, rechargeable batteries, windmills, automotive vehicles, electronic devices, optical and telecommunication systems, lasers, economical lighting systems, magnetic resonance imaging, as well as security inks, to name a few. In addition, medical-oriented applications such as time- resolved luminescence immunoassays and bioanalytical sensing are now well established. Emerging photovoltaic technologies such as water splitting and solar-energy conversion are also turning to lanthanoids to improve their performances, while optical bioimaging taking advantage of time-resolved luminescence is gaining momentum. Less pure rare earth compounds or mixtures thereof are used as catalysts and components of many alloys and ceramics.

For commodity, the elements Ce-Lu are often listed separately in the periodic table, although it is usually assumed that they belong to the column Sc, Y, La, Ac. All rare-earth elements are highly electropositive (see Table 1) and their most stable oxidation state is +3, although some elements are stable under given conditions at +2 (Sm, Eu, Yb) or +4 (Ce, Pr, Tb) oxidation state. The field of low valence rare-earth compounds has well progressed during the past years and today there are six fully characterized Ln^{II} ions (Ln = Nd, Sm, Eu, Dy, Tm, Yb) and divalent reduction chemistry with organometallic compounds is known for almost all the others. Some mono-halides have also been characterized as well as Sc^{I} and a few Ln^{0} compounds (Nief 2010).

Photophysical Properties of Trivalent Ions

Trivalent lanthanoid ions have all a $[Xe]4f^n$ electronic configuration (n = 0–14); in this configuration, 4f electrons are shielded by the filled $5s^25p^6$ subshells, so that their interaction with orbitals of surrounding

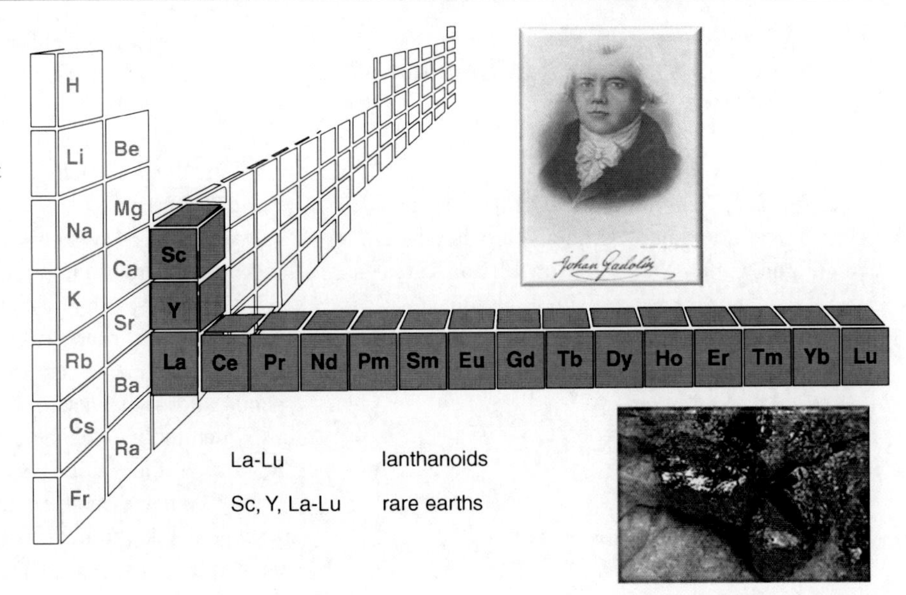

Lanthanides, Physical and Chemical Characteristics, Fig. 1 Position of the lanthanoids and rare earths in the periodic table, official IUPAC nomenclature, portrait of Johan Gadolin, and picture of the black stone of Ytterby from which yttrium was isolated

Lanthanides, Physical and Chemical Characteristics, Table 1 Lanthanoid elements and some of their properties

Element	Symbol	Z	Electronic configuration	Abund./ppm[a]	Principal isotopes, %	r_{at}/pm[b]	$\chi_{Pauling}$	I_{1-3}[c]
Lanthanum	La	57	$[Xe]4f^0 5d^1 6s^2$	32–39	^{139}La, 99.91	187.9	1.10	3,455
Cerium	Ce	58	$[Xe]4f^1 5d^1 6s^2$	60–68	^{140}Ce, 88.45	182.5	1.12	3,523
					^{142}Ce, 11.11			
Praseodymium	Pr	59	$[Xe]4f^3 6s^2$	8.7–9.5	^{141}Pr, 100	182.8	1.13	3,627
Neodymium	Nd	60	$[Xe]4f^4 6s^2$	33–41	^{142}Nd, 27.15	182.1	1.14	3,694
					^{144}Nd, 23.80			
Promethium	Pm	61	$[Xe]4f^5 6s^2$	0	–	181	1.13	3,738
Samarium	Sm	62	$[Xe]4f^6 6s^2$	6–7.9	^{152}Sm, 26.74	180.4	1.17	3,871
					^{154}Sm, 22.73			
Europium	Eu	63	$[Xe]4f^7 6s^2$	1.8–2.1	^{151}Eu, 47.81	204.2	1.2	4,035
					^{153}Eu, 52.19			
Gadolinium	Gd	64	$[Xe]4f^7 5d^1 6s^2$	5.2–7.7	^{158}Gd, 24.84	180.1	1.20	3,750
					^{160}Gd, 21.86			
Terbium	Tb	65	$[Xe]4f^9 6s^2$	0.9–1.2	^{159}Tb, 100	178.3	1.1	3,790
Dysprosium	Dy	66	$[Xe]4f^{10} 6s^2$	5.2–6.2	^{164}Dy, 28.26	177.4	1.22	3,898
					^{162}Dy, 25.48			
Holmium	Ho	67	$[Xe]4f^{11} 6s^2$	1.2–1.4	^{165}Ho, 100	176.6	1.23	3,924
Erbium	Er	68	$[Xe]4f^{12} 6s^2$	3.0–3.8	^{166}Er, 33.50	175.7	1.24	3,934
					^{168}Er, 26.98			
Thulium	Tm	69	$[Xe]4f^{13} 6s^2$	0.5	^{169}Tm, 100	174.6	1.25	4,045
Ytterbium	Yb	70	$[Xe]4f^{14} 6s^2$	2.8–3.3	^{174}Yb, 31.83	193.9	1.1	4,194
					^{172}Yb, 21.83			
Lutetium	Lu	71	$[Xe]4f^{14} 5d^1 6s^2$	0.35–0.5	^{175}Lu, 97.41	173.5	1.27	3,896

[a] Abundances in earth crust; ranges are indicated because values vary depending on sources
[b] For CN = 12 and the α-form at room temperature
[c] Sum of first three ionization potentials in $kJmol^{-1}$

Lanthanides, Physical and Chemical Characteristics, Table 2 Electronic properties of Ln^{III} ions (Bünzli and Eliseeva 2010)

f^n		Multiplicity	Number of terms	Number of levels	Ground level	
f^0	f^{14}	1	1	1	$^1S_0$	$^1S_0$
f^1	f^{13}	14	1	2	$^2F_{5/2}$	$^2F_{7/2}$
f^2	f^{12}	91	7	13	$^3H_4$	$^3H_6$
f^3	f^{11}	364	17	41	$^4I_{9/2}$	$^4I_{15/2}$
f^4	f^{10}	1,001	47	107	$^5I_4$	$^5I_8$
f^5	f^9	2,002	73	198	$^6H_{5/2}$	$^6H_{15/2}$
f^6	f^8	3,003	119	295	$^7F_0$	$^7F_6$
f^7		3,432	119	327	$^8S_{7/2}$	

ligands is minimum. Henceforth, ligand field effects are small (a few hundred of cm^{-1}) and electronic properties may therefore be adequately described within the frame of the ligand field theory; Russels-Saunders scheme for spin-orbit coupling is also usually assumed, so that the electronic configurations are described by the three quantum numbers S, L, and J. More detailed calculations though need to make use of the intermediate spin-orbit coupling scheme. Basic properties are summarized in Table 2 while a partial electronic level diagram is given in Fig. 2. It is noteworthy that presently this diagram has been experimentally extended up to about 70,000 cm^{-1} while calculations have been performed for levels up to 193,000 cm^{-1}. The splitting of the electronic levels under ligand field interaction is given in Table 3 and is useful for the determination of the site symmetry of the lanthanide ion either by absorption or by emission spectroscopy.

Absorption Spectra

Absorption of light is achieved through operators related to the nature of electromagnetic waves: the odd-parity electric dipole (ED), the even-parity magnetic dipole (MD), and the electric quadrupole (EQ) operators. Transitions between an initial state Ψ_i and a final state Ψ_f follow specific selection rules which depend on the nature of the operator and of the initial and final states, as well as on the symmetry of the chemical environment of the metal ion. Transitions involving lanthanoid ions are of three different kinds:

1. Sharp intraconfigurational 4f-4f transitions featuring rearrangement of the electrons within the $4f^n$ subshell. Electric dipole transitions are forbidden by both Laporte's and spin selection rules while magnetic dipole transitions are allowed, but are very weak. Henceforth, intensities of the 4f-4f transitions are rather weak, with molar absorption coefficients rarely larger than 10 and often smaller than 1 M^{-1} cm^{-1}. Some transitions are very sensitive to the metal-ion environment and are termed *hypersensitive* (or pseudo quadrupolar) transitions; intensity enhancements may reach 200-fold. A list of these transitions is presented in Table 4. They are very useful for analytical purposes.
2. Broader and more intense allowed 4f-5d transitions occurring when a 4f electron is moved into an empty 5d orbital. These transitions are energetic with $\lambda < 300$ nm, with the exception of Ce^{III} and sometimes Tb^{III}.
3. Broad allowed charge-transfer transitions during which one electron is transferred from the metal ion to the bonded ligands (MLCT) or vice versa (LMCT). MLCT transitions are very rarely identified in Ln^{III} spectra, except for Ce^{III} which can be readily oxidized into Ce^{IV}, the corresponding bands appearing up to 390 nm. There are also some reports of MLCT transitions for Tb^{III}, but at much higher energy and in the case of purely inorganic compounds. Ligand-to-metal charge transitions play an essential role in sensitizing the luminescence of phosphors used in modern lighting devices. As for 4f-5d transitions, LMCT transitions occur in the UV ($\lambda < 300$ nm), except for Eu^{III} and Yb^{III} for which they can extend up to 400 and 320 nm, respectively.

Emission Spectra

Barring La^{III} and Lu^{III}, all Ln^{III} ions are luminescent and their f-f emission lines cover the entire spectrum, from UV (Gd^{III}) to visible (e.g., Pr^{III}, Sm^{III}, Eu^{III}, Tb^{III}, Dy^{III}, Tm^{III}) and near-infrared (NIR, e.g., Pr^{III}, Nd^{III}, Ho^{III}, Er^{III}, Yb^{III}) spectral ranges. Several ions are simultaneously visible and NIR emitters (e.g., Sm^{III}, Eu^{III}, Er^{III}, Tm^{III}). Some ions are fluorescent ($\Delta S = 0$), others are phosphorescent ($\Delta S \neq 0$), and some are both. The 4f-4f emission lines are sharp because the electronic rearrangement consecutive to the promotion of an electron into a 4f orbital of higher energy does not perturb much the binding pattern in the molecules since 4f orbitals do not participate much in this binding (the covalency of Ln^{III}-ligand bonds is at most 5–7%).

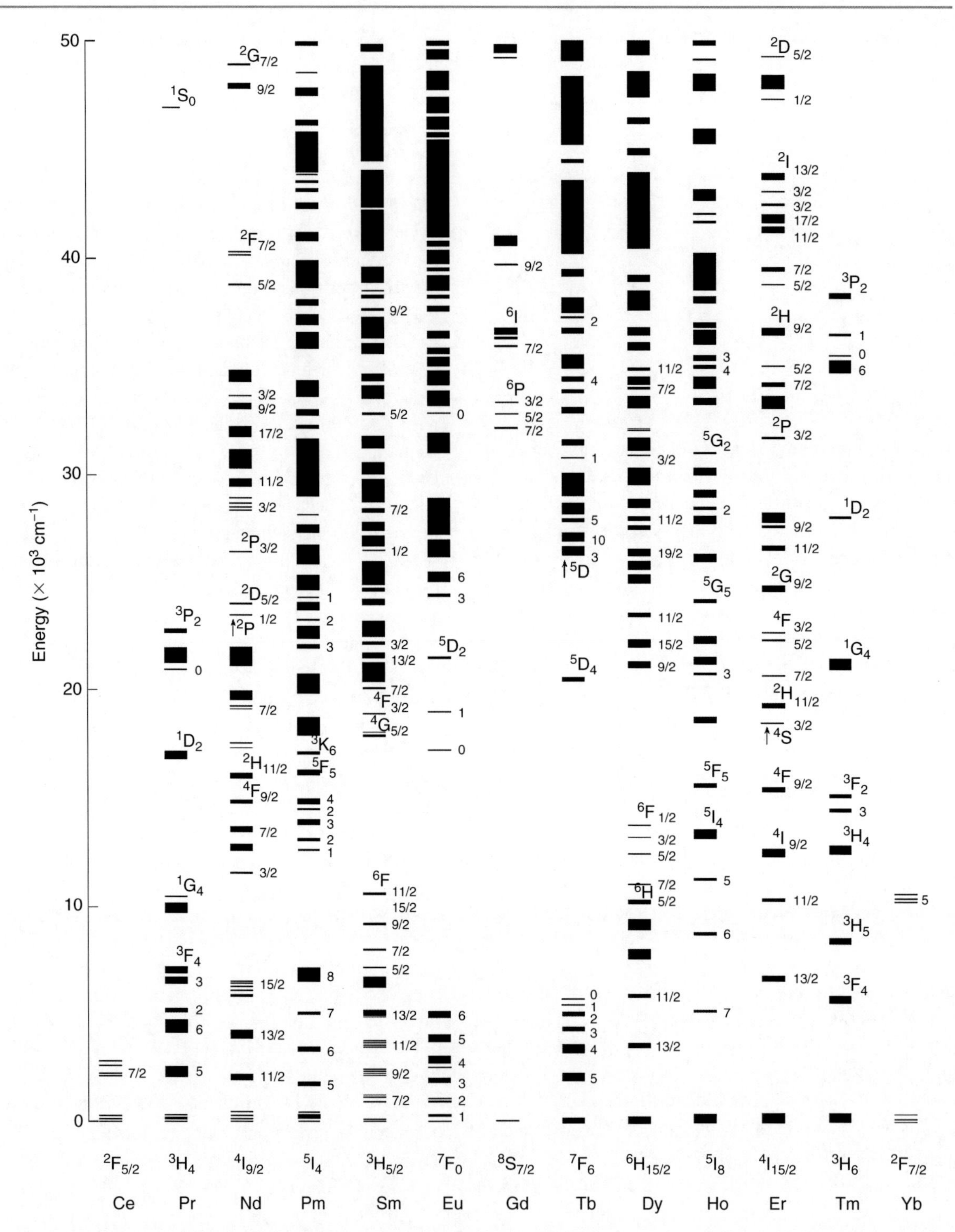

Lanthanides, Physical and Chemical Characteristics, Fig. 2 Partial energy diagrams for Ln^{III} ions doped into a low-symmetry crystal, LaF_3 (Reproduced with permission from reference Liu (2005) © 2005, Springer Verlag, Berlin)

Lanthanides, Physical and Chemical Characteristics, Table 3 Number of ligand-field sub-levels versus the value of the quantum number J

Symmetry	Site symmetry	Integer J								
Symmetry	Site symmetry	0	1	2	3	4	5	6	7	8
Cubic	T, T_d, T_h, O, O_h	1	1	2	3	4	4	6	6	7
Hexagonal	$C_{3h}, D_{3h}, C_6, C_{6h}, C_{6v}, D_6, D_{6h}$	1	2	3	5	6	7	9	10	11
Trigonal	$C_3, S_6, C_{3v}, D_3, D_{3d}$									
Tetragonal	$C_4, S_4, C_{4h}, C_{4v}, D_4, D_{2d}, D_{4h}$	1	2	4	5	7	8	10	11	13
Low	$C_1, C_S, C_2, C_{2h}, C_{2v}, D_2, D_{2h}$	1	3	5	7	9	11	13	15	17
Symmetry	Site symmetry	Half-integer J								
		1/2	3/2	5/2	7/2	9/2	11/2	13/2	15/2	17/2
Cubic	T, T_d, T_h, O, O_h	1	1	2	3	3	4	5	6	6
All others[a]	See above	1	2	3	4	5	6	7	8	9

[a]All sublevels are doubly degenerate (Kramer's doublets)

Lanthanides, Physical and Chemical Characteristics, Table 4 Identified hypersensitive transitions for Ln^{III} ions. Wavelengths are approximate (Bünzli and Eliseeva 2010)

Ln	Transition	λ/nm	Ln	Transition	λ/nm
Pr	$^3F_2 \leftarrow {}^3H_4$	1,920	Dy	$^6F_{11/2} \leftarrow {}^6H_{15/2}$	1,300
Nd	$^4G_{5/2}, {}^2G_{7/2} \leftarrow {}^4I_{9/2}$	578		$^4G_{11/2}, {}^4I_{15/2} \leftarrow {}^6H_{15/2}$	427
	$^2H_{9/2}, {}^4F_{5/2} \leftarrow {}^4I_{9/2}$	806	Ho	$^3H_6 \leftarrow {}^5I_8$	361
	$^4G_{7/2}, {}^2K_{13/2}, {}^2G_{9/2} \leftarrow {}^4I_{9/2}$	521		$^5G_6 \leftarrow {}^5I_8$	452
Sm	$^4F_{1/2}, {}^4{}_{3/2} \leftarrow {}^6H_{5/2}$	1,560	Er	$^4G_{11/2} \leftarrow {}^4I_{15/2}$	379
Eu	$^5D_2 \leftarrow {}^7F_0$	465		$^2H_{11/2} \leftarrow {}^4I_{15/2}$	521
	$^5D_1 \leftarrow {}^7F_1$	535	Tm	$^1G_4 \leftarrow {}^3H_6$	469
	$^5D_0 \rightarrow {}^7F_2$	613		$^3H_4 \leftarrow {}^3H_6$	787
Gd	$^6P_{5/2}, {}^6P_{7/2} \leftarrow {}^8S_{7/2}$	308		$^3F_4 \leftarrow {}^3H_6$	1,695
Tb	[a]				

[a]None identified positively, but the $^5D_4 \rightarrow {}^7F_5$ transition sometimes display ligand-induced pseudo-hypersensitivity

As a consequence, internuclear distances remain almost the same in the excited state, which generates narrow bands and very small Stokes' shifts when the ions are excited directly. A different situation prevails in organic molecules for which excitation leads frequently to a lengthening of the chemical bonds, resulting in large Stokes' shifts and since the coupling with vibrations is strong, in broad emission bands. Typical emission lines observed in Ln^{III} luminescence spectra are listed in Table 5, together with other key photophysical parameters.

The dipole strengths of 4f-4f transitions being very small, direct excitation into the Ln^{III} excited levels rarely yields strong luminescence, even if the intrinsic quantum yield is large, unless considerable excitation power is provided (by means of a laser, given the sharpness of the absorption bands). Therefore, an alternative solution is frequently used, which is called *luminescence sensitization* or *antenna effect*. The luminescent ion is imbedded into a matrix or an organic environment with good light-harvesting properties. Energy is then transferred from the excited surroundings onto the metal ion which eventually gives off its characteristic light. The following photophysical parameters are defined:

$$Q_{Ln}^{L} = \eta_{sens} \cdot Q_{Ln}^{Ln} = \eta_{sens} \cdot \frac{\tau_{obs}}{\tau_{rad}} \quad (1)$$

in which Q_{Ln}^{L} and Q_{Ln}^{Ln} are the overall (upon ligand excitation) and intrinsic (upon direct f-level excitation) quantum yields, η_{sens} the efficiency with which energy is transferred onto the metal ion, τ_{obs} and τ_{rad} the observed and radiative lifetimes. The intrinsic

Lanthanides, Physical and Chemical Characteristics, Table 5 Ground (G), main emissive (E), and final (F) states for typical 4f-4f emission bands for Ln^{III} ions with approximate corresponding wavelengths (λ), energy gap between the emissive state and the highest spin-orbit level of the receiving state, and radiative lifetimes. More NIR lines are listed in reference Comby and Bünzli (2007)

Ln	G	E	F	λ/μm or nm[a]	τ_{rad}/ms[a]
Ce	$^2F_{5/2}$	5d	$^2F_{5/2}$	Tunable, 300–450	–
Pr	$^3H_4$	$^1D_2$	$^3F_4$, $^1G_4$, $^3H_{4,5}$	1.0, 1.44, 600, 690	(0.05[b]–0.35)
		$^3P_0$	$^3H_{4–6}$	490, 545, 615	(0.003[b]–0.02)
		$^3P_0$	$^3F_{2–4}$	640,700, 725	
Nd	$^4I_{9/2}$	$^4F_{3/2}$	$^4I_{9/2–13/2}$	900, 1.06, 1.35	0.42 (0.2–0.5)
Sm	$^6H_{5/2}$	$^4G_{5/2}$	$^6H_{5/2–15/2}$	560, 595, 640, 700, 775, 910	6.26 (4.3–6.3)
		$^4G_{5/2}$	$^6F_{1/2–9/2}$	870, 887, 926, 1.01, 1.15	
		$^4G_{5/2}$	$^6H_{13/2}$	877	
Eu[c]	$^7F_0$	$^5D_0$	$^7F_{0–6}$	580, 590, 615, 650,720, 750, 820	9.7 (1–11)
Gd	$^8S_{7/2}$	$^6P_{7/2}$	$^8S_{7/2}$	315	10.9
Tb	$^7F_6$	$^5D_4$	$^7F_{6–0}$	490, 540, 580, 620, 650, 660, 675	9.0 (1–9)
Dy	$^6H_{15/2}$	$^4F_{9/2}$	$^6H_{15/2–9/2}$	475, 570, 660, 750	1.85 (0.15–1.9)
		$^4I_{15/2}$	$^6H_{15/2–9/2}$	455, 540, 615, 695	3.22[b]
Ho	$^5I_8$	$^5S_2$	$^5I_{8,7}$	545, 750	0.37 (0.51[b])
		$^5F_5$	$^5I_8$	650	0.8[b]
		$^5F_5$	$^5I_7$	965	
Er[d]	$^4I_{15/2}$	$^4S_{3/2}$	$^4I_{15/2,13/2}$	545, 850	0.7[b]
		$^4F_{9/2}$	$^4I_{15/2}$	660	0.6[b]
		$^4I_{9/2}$	$^4I_{15/2}$	810	4.5[b]
		$^4I_{13/2}$	$^4I_{15/2}$	1.54	0.66 (0.7–12)
Tm	$^3H_6$	$^1D_2$	$^3F_{4–2}$, $^3H_4$	450, 650, 740, 775	0.09
		$^1G_4$	$^3H_6$, $^3F_{4,5}$	470, 650, 770	1.29
		$^3H_4$	$^3H_6$	800	3.6[b]
Yb	$^2F_{7/2}$	$^2F_{5/2}$	$^2F_{7/2}$	980	1.3 or 2.0[e]

[a]Values for the aqua ions, otherwise stated, and ranges of observed lifetimes in all media, if available, between parentheses
[b]Doped in Y_2O_3 or in $YLiF_4$ (Ho), or in $YAl_3(BO_3)_4$ (Dy)
[c]Luminescence from $^5D_1$, $^5D_2$, and $^5D_3$ is sometimes observed as well
[d]Luminescence from four other states has also been observed: $^4D_{5/2}$, $^2P_{3/2}$, $^4G_{11/2}$, $^2H_{9/2}$
[e]Complexes in solution: 0.7–1.3 ms; solid-state inorganic compounds: ≈2 ms

quantum yield is difficult to measure in view of the weak intensity of the 4f-4f transitions, so that it is often calculated from the lifetimes. The radiative lifetime can be calculated from Einstein coefficients if Judd-Ofelt parameters are known (Bünzli and Eliseeva 2010) or from the absorption spectrum when the luminescence transitions terminate onto the ground level:

$$\frac{1}{\tau_{rad}} = 2303 \times \frac{8\pi c n^2 \tilde{\nu}^2 (2J+1)}{N_A(2J'+1)} \int \varepsilon(\tilde{\nu})\, d\tilde{\nu} \quad (2)$$

Here $\tilde{\nu}$ is the energy of the transition, n the refractive index, J and J' the quantum numbers of the initial and final state, c the velocity of light in vacuum and N_A Avogadro's number. In the particular case of Eu^{III}, τ_{rad} is accessible through the emission spectrum since the $^5D_0 \rightarrow {}^7F_1$ transition has a purely magnetic character:

$$\frac{1}{\tau_{rad}} = A_{MD,0} \cdot n^3 \left(\frac{I_{tot}}{I_{MD}}\right) \quad (3)$$

in which I_{tot} and I_{MD} are the total and $^5D_0 \rightarrow {}^7F_1$ emission area and $A_{MD,0} = 14.65\ s^{-1}$.

Lanthanide luminescence has found a wealth of applications in lighting (fluorescent lamps, light emitting diodes), electronic displays, fiber-optic telecommunications, sensitive analyses, security inks, various quality tags, and time-resolved bioanalysis and bioimaging (see "▶ Lanthanides in Nucleic Acid Analysis" and "▶ Lanthanides in Biological Labeling, Imaging, and Therapy").

Lanthanides, Physical and Chemical Characteristics, Table 6 Magnetic moments (B.M.) of trivalent 4f ions at 298 K

Ln	Conf.	n[a]	Ground state	1st exc. state	$\Delta E/cm^{-1}$	g_J	μ_{calc}[b]	μ_{calc}[c]	μ_{exp}[d]
Ce	f^1	1	$^2F_{5/2}$	$^2F_{7/2}$	2,200	0.86	2.54	2.56	2.5–2.8
Pr	f^2	2	$^3H_4$	$^3H_5$	2,100	0.80	3.58	3.62	3.2–3.6
Nd	f^3	3	$^4I_{9/2}$	$^4I_{11/2}$	1,900	0.73	3.62	3.68	3.2–3.6
Pm	f^4	4	$^5I_4$	$^5I_5$	1,600	0.60	2.68	2.83	n.a.
Sm	f^5	5	$^6H_{5/2}$	$^6H_{7/2}$	1,000	0.29	0.85	1.60	1.3–1.5
Eu	f^6	6	$^7F_0$	$^7F_1$	300	[e]	0	3.45	3.1–3.4
Gd	f^7	7	$^8S_{7/2}$	$^6P_{7/2}$	32,000	2.00	7.94	7.94	7.9–8.1
Tb	f^8	6	$^7F_6$	$^7F_5$	2,000	1.50	9.72	9.72	9.2–9.7
Dy	f^9	5	$^6H_{15/2}$	$^6H_{13/2}$	3,300	1.33	10.65	10.6	10.1–10.6
Ho	f^{10}	4	$^5I_8$	$^5I_7$	5,300	1.25	10.61	10.6	10.0–10.5
Er	f^{11}	3	$^4I_{15/2}$	$^4I_{13/2}$	6,500	1.20	9.58	9.6	9.2–9.6
Tm	f^{12}	2	$^3H_6$	$^3F_4$	5,800	1.17	7.56	7.6	7.0–7.3
Yb	f^{13}	1	$^2F_{7/2}$	$^2F_{5/2}$	10,000	1.14	4.54	4.5	4.3–4.6

[a]Number of unpaired electrons
[b]Calculated according to (4)
[c]Exact calculation with van Vleck formula, taking into account the population of excited states
[d]Ranges of experimentally observed values
[e]Formula (4) yields an undetermined value; g_J has been estimated to be <5

Magnetic Properties of Trivalent Ions

Effective magnetic moments for the free ions can be estimated within Russel-Saunders approximation with the following formula:

$$\mu_{eff} = g_J\sqrt{J(J+1)} \quad \text{with} \qquad g_J = \frac{J(J+1)+S(S+1)-L(L+1)}{2J(J+1)} \tag{4}$$

La^{III} ($4f^0$) and Lu^{III} ($4f^{14}$) ions have $^1S_0$ ground levels and are diamagnetic. For the other ions, (4) gives satisfying results except for Eu^{III} and, to a lesser extent, Sm^{III}. Indeed the first excited level of Eu^{III} is at low energy, $\approx$ 300 cm^{-1} leading to a $^7F_1$ population of about 5% at room temperature. For Sm^{III}, $^6H_{7/2}$ lies at and $\approx$1,000 cm^{-1} and its population is about 1%. The orbital contribution is often important. Indeed, the most magnetic ion is not the one with the largest number of unpaired electrons (Gd^{III}) (Table 6, Fig. 3).

The paramagnetism of Ln^{III} ions finds application in the design of high coercitivity magnets ($CoSm_5$ and $Fe_{14}Nd_2B$), magnetic resonance imaging (Gd complexes as contrast agents), shift reagents for interpreting nuclear magnetic resonance spectra, and magnetic refrigeration (Eliseeva and Bünzli 2011).

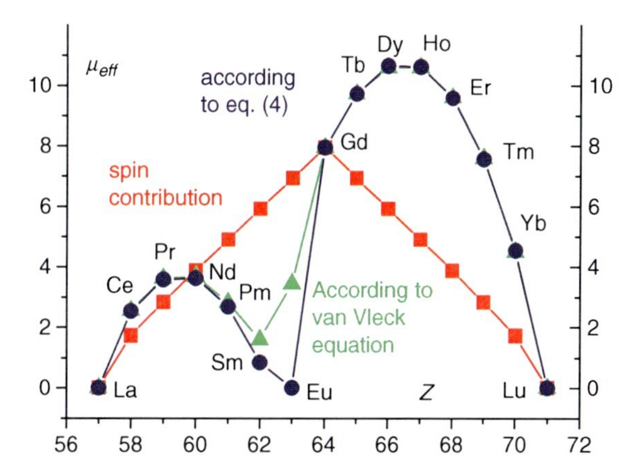

Lanthanides, Physical and Chemical Characteristics, Fig. 3 Paramagnetic effective moments of Ln^{III} ions (in Bohr magnetons)

Chemical Properties of Trivalent Ions

Coordination Chemistry

From the chemical point of view, Ln^{III} ions behave like hard Lewis acids and their bonding is essentially electrostatic, with small covalent contributions. Generally speaking, therefore, they prefer to form complexes with oxygen donors than with softer donors like nitrogen or sulfur. A recent survey of 1,391 crystal

Lanthanides, Physical and Chemical Characteristics, Table 7 Ionic radii (Å) for coordination numbers 6, 9, and 12, hydration enthalpy at 298 K ($kJmol^{-1}$), hydrolysis constants $-Log^{*}\beta_{11}$ for the formation of $[Ln(OH)]^{2+}$ from Ln_{aq}^{3+} at 298 K, $I = 0.3$ M ($NaClO_4$), and pH at which precipitation starts in $Ln(NO_3)_3$ solutions 0.1 M in water (Bünzli 1998)

Ln^{III}	Color	$r_i(6)$	$r_i(9)$	$r_i(12)$	$-\Delta H_h^0$	$-Log^{*}\beta_{11}$[a]	pH
La	Colorless	1.03	1.22	1.36	3,326	9.01	7.47
Ce	Colorless	1.01	1.20	1.34	3,380	10.6	7.10[b]
Pr	Green	0.99	1.18	1.32	3,421	8.55	6.96
Nd	Violet	0.98	1.16	1.30	3,454	8.43	6.78
Pm	Rose	0.97	1.14	1.28	3,482	n.a.	n.a.
Sm	Cream	0.96	1.13	1.27	3,512	8.34	6.65
Eu	Pale pink	0.95	1.12	1.25	3,538	8.31	6.61
Gd	Colorless	0.94	1.11	1.24	3,567	8.35	6.58
Tb	Colorless	0.92	1.10	1.23	3,600	8.16	6.47
Dy	Cream	0.91	1.08	1.22	3,634	8.10	6.24
Ho	Yellow	0.90	1.07	1.21	3,663	8.04	6.20
Er	Pink	0.89	1.06	1.19	3,692	7.99	6.14
Tm	Green	0.88	1.05	1.18	3,717	7.95	5.98
Yb	Colorless	0.87	1.04	1.17	3,740	7.92	5.87
Lu	Colorless	0.86	1.03	1.16	3,759	7.90	5.74

[a] $^{*}\beta_{11} = [LnOH^{2+}][H^{+}]/[Ln^{3+}]$
[b] $I = 0.005$ M

structures showed that 42% of the investigated complexes contain exclusively Ln-O bonds while 78% contain a least one Ln-O bond (Huang and Bian 2010). The ionic radius of Ln^{III} ions is large and depends on both the ion and the coordination number (CN). They do not vary much along the series, but the change is gradual and reflects the *lanthanide contraction*. Although coordination numbers between 3 and 12 are well documented, most complexes exhibit coordination numbers 8 (37%) or 9 (26%), with CN = 6 (9%), 7 (8%), and 10 (11%) being also well represented. When complexation occurs in water, complex formation has to overcome the large hydration energies of the Ln^{III} ions, so that chelating agents are usually preferred in order to get a large entropic stabilization. Hydrolysis may also be a problem when working at basic pH or around neutral pH. Key parameters are reported in Table 7.

Lanthanoid coordination compounds and associated coordination polymers are presently finding numerous applications as electroluminescent materials, magnetic and luminescent probes in biosciences (Eliseeva and Bünzli 2010), as well as in organic catalysis.

Organometallic Chemistry

The first organometallic compounds involving rare-earth ions, $Ln(Cp)_3$ (Ln = Sc, La, Pr, Nd, Sm, Gd), have been synthesized 2 years after the isolation of ferrocene and opened the way to a rich organolanthanoid chemistry. Presently, both alkyl, aryl and cyclopentadienyl compounds are known and this chemistry allowed the characterization of low-valence lanthanide compounds, as well as the activation of small molecules such as nitrogen and carbon monoxide (Nief 2010; Evans 2007).

Cross-References

▶ Lanthanides in Biological Labeling, Imaging, and Therapy
▶ Lanthanides in Nucleic Acid Analysis

References

Bünzli J-CG (1998) Coordination chemistry of the trivalent lanthanide ions: an introductory overview. In: Saez Puche R, Caro P (eds) Rare Earths. Editorial Complutense, Madrid, pp 223–259

Bünzli J-CG, Eliseeva SV (2010) Basics of lanthanide photophysics. In: Hänninen P, Härmä H (eds) Lanthanide luminescence: photophysical, analytical and biological aspects, vol 7, Springer series on fluorescence. Springer, Berlin, Ch. 1, pp 1–45

Comby S, Bünzli J-CG (2007) Lanthanide near-infrared luminescence in molecular probes and devices. In: Gschneidner, KA Jr, Bünzli J-CG, Pecharsky VK (eds) Handbook on the physics and chemistry of rare earths, vol 37. Elsevier Science BV, Amsterdam, Ch. 235, pp 217–470

Cotton S (2006) Lanthanide and actinide chemistry. Wiley, Chichester

Eliseeva SV, Bünzli J-CG (2010) Lanthanide luminescence for functional materials and biosciences. Chem Soc Rev 39:189

Eliseeva SV, Bünzli J-CG (2011) Rare earths: jewels for functional materials of the future. New J Chem 35:1165–1176

Evans WJ (2007) The importance of questioning scientific assumptions: some lessons from f-element chemistry. Inorg Chem 46:3435

Huang C, Bian Z (2010) Introduction, Ch. 1. In: Huang C (ed) Rare Earth coordination chemistry, fundamentals and applications. Wiley, Singapore, pp 1–40

Liu GK (2005) Electronic energy level structure, Ch 1. In: Liu GK, Jacquier B, Hull R, Osgood RMJ, Parisi J, Warlimont H (eds) Spectroscopic properties of rare earths in optical materials, vol 83, Springer series in materials science. Springer, Berlin, pp 1–94

Nief F (2010) Molecular chemistry of the rare-earth elements in uncommon low-valent States, Ch 246. In: Gschneidner KA Jr, Bünzli J-CG, Pecharsky VK (eds) Handbook on the physics and chemistry of rare earths, vol 40. Elsevier Science B.V, Amsterdam, pp 241–300

Lanthanides, Rare Earth Elements, and Protective Thiols

Petr Babula[1], Vojtech Adam[2,3] and Rene Kizek[2,3]
[1]Department of Natural Drugs, Faculty of Pharmacy, University of Veterinary and Pharmaceutical Sciences Brno, Brno, Czech Republic
[2]Department of Chemistry and Biochemistry, Faculty of Agronomy, Mendel University in Brno, Brno, Czech Republic
[3]Central European Institute of Technology, Brno University of Technology, Brno, Czech Republic

Synonyms

Thiols and lanthanides

Definition

The rare earth elements group comprises scandium and yttrium and the group of lanthanides, which includes elements with atomic numbers from 57 (lanthanum) to 71 (lutetium). They are widely used in agriculture, chemistry, and industry as well as medicine; so, they are able to enter the living environment and food chains. Ecological risk as well as effect of lanthanides on living organisms at different levels – molecular, cytological, histological – is still almost unknown. Lanthanides are supposed to affect different metabolic and biochemical processes. In animals, interference of lanthanum(III) and gadolinium(III) ions with calcium channels as well as inhibition of different enzymes was demonstrated. It was proved that lanthanide ions are able to regulate synaptic transmission with the effect to some receptors, especially their blockade (glutamate receptors). Connection of lanthanides with human health is still discussed. They are widely used in medicine, especially as contrast agents for magnetic resonance imaging, such as gadolinium compounds. Some newly synthesized complexes of lanthanum as well as other lanthanides demonstrated significant cytotoxic effect based on interactions with biomolecules. Negative effect of lanthanides on human health is still only little investigated. Oxidative damage of some tissues, including generation of reactive oxygen species with subsequent lipid peroxidation, has been evidenced after application of lanthanum(III) and cerium(III) ions to experimental animals. Relation between lanthanides and protective thiols including cysteine, reduced glutathione, phytochelatins, and metallothioneins is discussed.

Lanthanides, Rare Earth Elements, Their Classification and Chemical Properties with respect to Possible Biological Properties

The rare earth elements (REEs) consist of a chemically uniform group including scandium, yttrium, and the group of lanthanides, which includes elements with atomic numbers from 57 (lanthanum) to 71 (lutetium) – lanthanum (La) and 14 lanthanides – cerium (Ce), dysprosium (Dy), erbium (Er), europium (Eu), gadolinium (Gd), holmium (Ho), lutetium (Lu), neodymium (Nd), promethium (Pm), praseodymium (Pr), samarium (Sm), terbium (Tb), thulium (Tm), and ytterbium (Yb). REEs are usually divided into two groups – the light rare earth elements (LREEs), from lanthanum to europium, and heavy rare earth elements (HREEs), from gadolinium to lutetium. In nature, lanthanides are presented in different types of rocks, such as pegmatites and granites, especially in the form of phosphates, carbonates, fluorides, and silicates. Occurrence of lanthanides in phosphate rocks, sedimentary rock containing high levels of phosphate-bearing minerals has been reported too. Lanthanides are intensely used in agriculture, especially in Asian countries, as potential substitution for antibiotics medicine as imaging agents and probes and especially in industry.

Biologically Important Thiols

A thiol is an organo-sulfur compound that contains a carbon-bonded sulfhydryl (–C–SH). As the functional group of the amino acid cysteine (Fig. 1), the thiol group plays an important role in biology. When the thiol groups of two cysteine residues (as in monomers or constituent units) are brought near each other in the course of protein folding, an oxidation reaction can generate a cystine unit with a disulfide bond (–S–S–). Disulfide bonds can contribute to a protein's tertiary structure if the cysteines are part of the same peptide chain, or contribute to the quaternary structure of multi-unit proteins by forming fairly

Lanthanides, Rare Earth Elements, and Protective Thiols, Fig. 1 Model of cysteine molecule

strong covalent bonds between different peptide chains. In addition, the sulfhydryl group is highly reactive and is often found conjugated to other molecules, mainly to heavy metal ions. It is not surprising that biologically active molecules rich in –SH moiety are responsible for maintaining homeostasis of metal ions and/or their detoxifying (Fig. 2).

Reduced Glutathione

Reduced glutathione was discovered by M. J. de Rey Pailhade at the end of the nineteenth century as the substance "hydrogénant le soufre," which was renamed by F.G. Hopkins in 1921. Hopkins first characterized the compound as a dipeptide of glutamic acid and cysteine. Few years later, he suggested the correct structure to be a tripeptide which also contains glycine. Reduced glutathione (GSH, Fig. 3) as a ubiquitous tripeptide thiol is a vital intra- and extracellular protective antioxidant. It plays a number of key roles in the controlling of signaling processes, detoxifying of some xenobiotics and heavy metals.

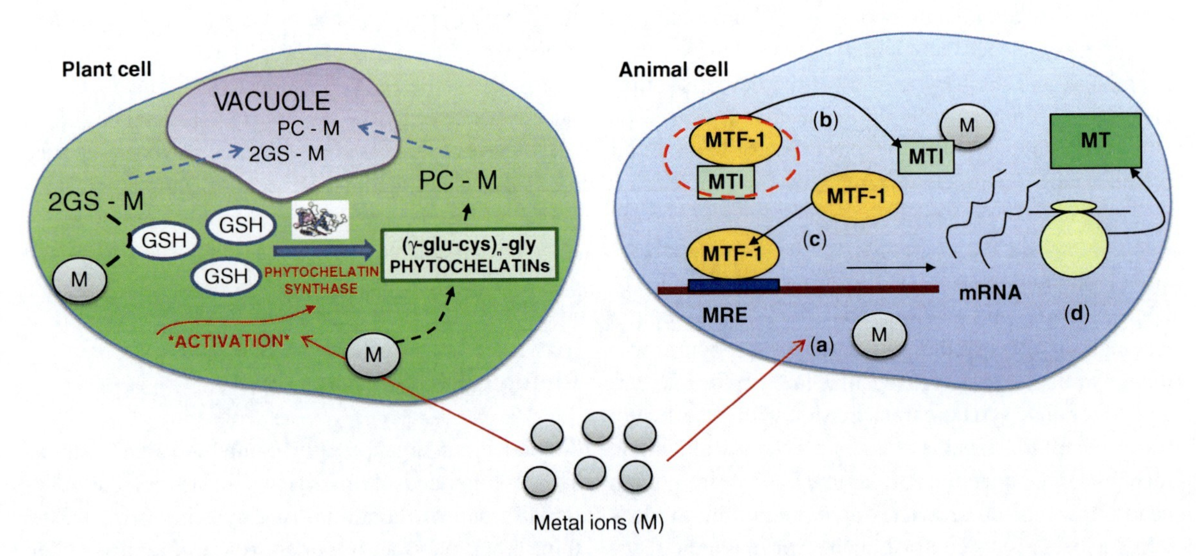

Lanthanides, Rare Earth Elements, and Protective Thiols, Fig. 2 *Plant cell*. Glutathione serves as a precursor of phytochelatins, which are composed of two or more repeating gamma-glutamylcysteine units with a terminal glycine residue; (gamma-glutamylcysteine)$_n$-gly, where $n = 2$–11. The enzyme responsible for the synthesis of these peptides is known as phytochelatin synthase (glutathione gamma-glutamylcysteinyltransferase or gamma-glutamylcysteine dipeptidyl transpeptidase), which is a constitutive enzyme that is activated by cadmium and other metal ions. *Animal cell*. A heavy metal ion enters through a cytoplasmic membrane of a cell via ionic channels or special transporters (**a**). After entering the cytoplasm, the ion interacts with a complex of metal-regulatory transcription factor-1 (MTF-1) and metal synthesis inhibitor (MTI) (**b**). The ion binds to MTI. Due to this, MTF-1 is released and can bind to a regulatory sequence of DNA called metal responsive element (MRE) (**c**). Then, the gene responsible for synthesis of metallothioneins is transcribed. The synthesized mRNA molecule is translated into MT (**d**). MT binds to the heavy metal ion

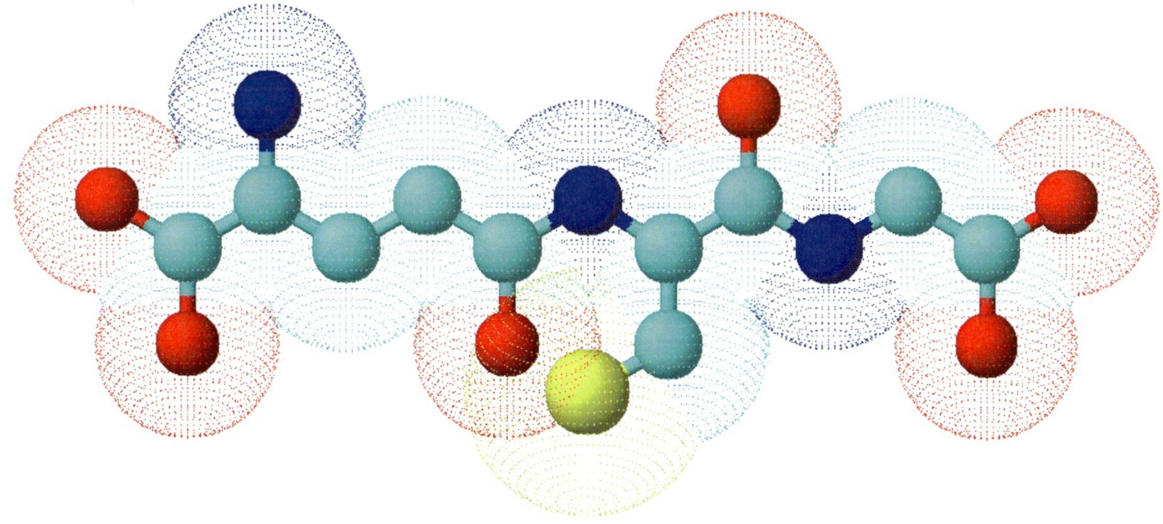

Lanthanides, Rare Earth Elements, and Protective Thiols, Fig. 3 Model of reduced glutathione molecule

Glutathione is found almost exclusively in its reduced form since the enzyme, which reverts it from its oxidized form (GSSG) called glutathione reductase, is constitutively active and inducible upon oxidative stress. The glutathione–ascorbate cycle is a metabolic pathway that detoxifies hydrogen peroxide (H_2O_2), which is a reactive oxygen species that is produced as a waste product in metabolism. The cycle involves the antioxidant metabolites – ascorbate, glutathione, and NADPH – and the enzymes linking these metabolites. In the first step of this pathway, H_2O_2 is reduced to water by ascorbate peroxidase using ascorbate as the electron donor. The oxidized ascorbate (monodehydroascorbate) is regenerated by monodehydroascorbate reductase. However, monodehydroascorbate is a radical and if not rapidly reduced it breaks down into ascorbate and dehydroascorbate. Dehydroascorbate is reduced to ascorbate by dehydroascorbate reductase at the expense of GSH, yielding oxidized glutathione (GSSG). Finally GSSG is reduced by glutathione reductase using NADPH as the electron donor. The reduction of dehydroascorbate may be nonenzymatic or catalyzed by proteins with dehydroascorbate reductase activity. In fact, the ratio of reduced to oxidized glutathione within cells is used as a marker of cytotoxicity. In connection with this, many fundamental events of cell regulation such as protein phosphorylation and binding of transcription factors to consensus sites on DNA are driven by physiological oxidant–antioxidant homeostasis, especially by thiol-disulfide balance. Therefore, endogenous glutathione and thioredoxin systems may be considered to be effective regulators of redox-sensitive gene expression (Meister and Anderson 1983).

Phytochelatins

Moreover, GSH can be used for synthesis of phytochelatins (a basic formula $(\gamma\text{-Glu-Cys})_n\text{-Gly}$ ($n = 2–11$), molecule of PC_2 is shown in Fig. 4) participating in the detoxification of heavy metals in plants, because they have the ability to bind heavy metal ions via SH groups of cysteine units and consequently transport them to vacuole, where there is no threat of immediate toxicity. Complex of PC and a metal ion is called a low-molecular-weight metal–phytochelatin complex (LMW M-PC). After transport of this complex through tonoplast to vacuole, low-molecular-weight complex is transformed to high-molecular-weight metal–phytochelatin complex (HMW M-PC) via –S–S– groups. The synthesis of PC itself involves the transpeptidation of the γ-Glu-Cys moiety of GSH into initially a second GSH molecule to form PC_2 or, in later stages of the incubation, into a PC molecule to produce an $n + 1$ oligomer. The reaction is catalyzed by γ-Glu-Cys dipeptidyl transpeptidase (EC 2.3.2.15), which has been called as phytochelatin synthase. In vitro the purified enzyme was active only in the presence of metal ions.

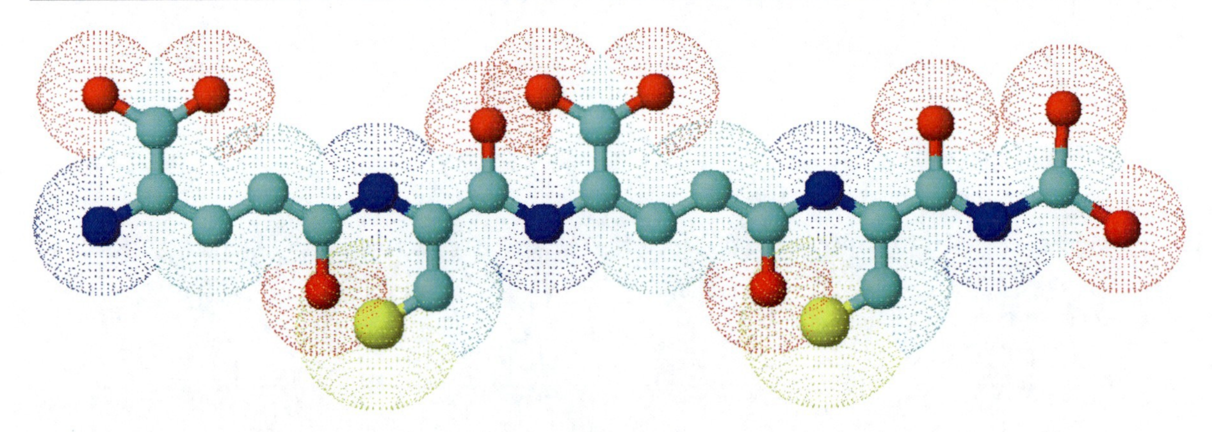

Lanthanides, Rare Earth Elements, and Protective Thiols, Fig. 4 Model of phytochelatin-2 molecule

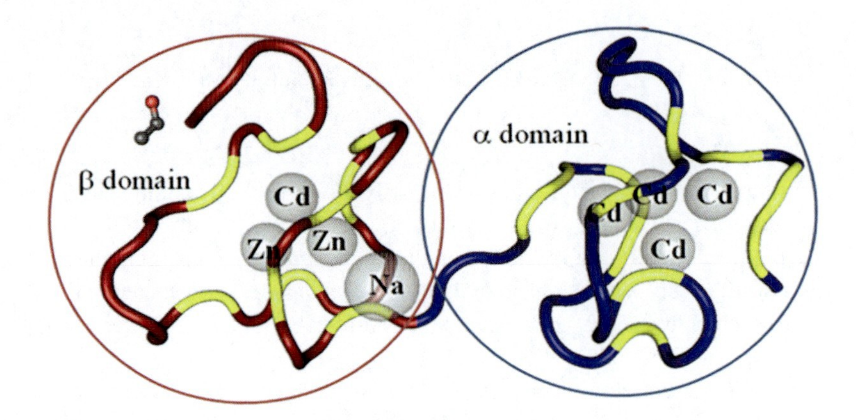

Lanthanides, Rare Earth Elements, and Protective Thiols, Fig. 5 Model of metallothionein molecule

Cadmium was the best activator of phytochelatin synthase followed by Ag, Bi, Pb, Zn, Cu, Hg, and Au cations (Cobbett and Goldsbrough 2002).

Metallothionein

Metallothioneins (MTs, Fig. 5) were discovered by Margoshes and Valee in 1957 as newly identified proteins isolated from a horse renal cortex tissue. These proteins occur across the animal kingdom with a high degree of homology. Similar proteins are expressed by bacteria, fungi, and even plants. MTs are low-molecular-mass (from 2 to 16 kDa) proteins with unique abundance of cysteine residues (more than 30% from all amino acids). Other interesting structural property is the lack of aromatic amino acids. However – as discovered recently – there is an exception: a group of certain yeast and bacterial species rarely containing histidine. MTs are single-chain proteins with amino acid numbers oscillating between approximately 20 and more than 100 residues according to organisms. Almost one third of this number is cysteine occurring in conserved sequences *cys*-**x**-*cys*, *cys*-**x**-**y**-*cys* a *cys-cys* where **x** and **y** represent other amino acids. Divalent metal ions bonded to sulfhydryl groups of cysteines are creating tetrahedric configuration of thiolate clusters. MT exhibits the highest affinity for Cu^{+} (stability constant 10^{17}–10^{19}), followed by Cd^{2+} (10^{15}–10^{17}) and Zn^{2+} (10^{11}–10^{14}); however, it is not capable of binding Cu^{2+}. Generally, 18 metal ions suitable to be bonded by MT are known but only Cu^{+}, Cd^{2+}, Pb^{2+}, Hg^{2+}, Ag^{+}, and Bi^{2+} can replace Zn^{2+} in MT structure. Binding capacity of MT is 7 and 12 atoms for divalent and monovalent ions, respectively. MT's tertiary structure consists of two domains: more stable α (C-terminal), containing four ion binding sites, and β (N-terminal) capable of incorporating three ions (Ryvolova et al. 2011).

Expression of MTs is started by binding of metal-regulatory transcription factor-1 (MTF-1) to the regulative region of MTs gene called metal responsive element (MRE). Transcription of MTs through the MRE may be initiated by several metal ions (Zn, Cd, Cu, Hg, Pb, Au, and Bi); however, only Zn can activate MTF-1. Moreover, MRE is capable of interacting with many proteins, which can regulate MT expression. Induction of MT expression by chemicals producing free radicals as well as various organic solvents has been shown. It has been found that the expression reaches the highest levels in the late G1 phase and during onset of the S phase. Nowadays, the attention is focused on MT's role in cancerogenesis and on its relation to cancer cell cycle (Eckschlager et al. 2009).

Due to the involvement in the wide range of cell processes and variety of vital functions, MT is considered an essential protein in living organisms. Also its potential use as an environmental pollution as well as cancer marker has been extensively studied. Moreover, the progress in the instrumental analytical techniques enables studying MT in detail and in the context of the whole organism. The interest of scientific society in this topic can be expressed in terms of the number of publications focused on this compound. It can be concluded that despite the abundance of applicable methods, immunoanalytical methods remain the key techniques employed in MT analysis. Also investigation of MT-mRNA is in the center of interest and rapid expansion of mass spectrometric detection is evident (Ryvolova et al. 2011).

Historically, MTs were classified into three groups. Class I comprised of all proteinaceous MT with locations of cysteine closely related to those in mammals. Some mollusks and crustacean MT belonged to this class, such as those characterized in mussels, oysters, crabs, and lobsters. Class II included proteinaceous MT that lacks this close similarity to mammalian MTs, while class III consisted of non-proteinaceous MTs, into which some authors included plant heavy-metal-binding peptides called phytochelatins. However, due to the enormous diversity of MT group, complex classification system has been introduced by Binz and Kägi. This system involves families, subfamilies, subgroups, and isoforms. Fifteen families include vertebrate, mollusk, crustacean, echinodermata, diptera, nematode, ciliate, fungi-I, fungi-II, fungi-III, fungi-IV, fungi-V, fungi-VI, prokaryota, and planta. The biggest family – vertebrate – is subdivided into 11 subfamilies including five mammalian, three avian, one each of batrachian, anura, and teleost subfamily. Probably the most widely studied group of MTs is the mammalian subfamily. Four mammalian MT isoforms (MT-1–MT-4) are known and 13 MT-like human proteins were identified. Eleven genes have been discovered for MT-1 isoform (MT-1A, B, E, F, G, H, I, J, K, L, and X) and one gene for each of other isoforms. Differences of constituent forms come mainly from posttranslational modifications, small changes in primary structure, type of incorporated metal ion, and speed of degradation. Despite the physical-chemical similarity of the forms, their roles and occurrence in tissues vary significantly. MT-1 and MT-2 are present almost in all types of soft tissues; MT-3 is expressed mostly in brain tissue but also in heart, kidneys, and reproductive organs; and MT-4 gene was detected in epithelial cells. MT without metal ion, apo-MT, is present in zinc-deficient cells. Recently, this form was discovered also in tumor cells (Ryvolova et al. 2011).

As mentioned previously, mussels belonged to the class I. However, in classification according to Binz and Kägi, there is a family devoted to mollusk, which is further subdivided into four subfamilies. Two of them are specifically dedicated to mussels. In contrast to the mammals, two groups of isoforms – MT-10 and MT-20 – were identified in this species. The most significant difference between MT-10 and MT-20 is the presence of two additional cysteine residues in the MT-20 isoforms. The exact role of each specific MT isoform or isoform group has not been clarified yet, but there are some indications that MT-20 might play a role in detoxification of cadmium in exposed mussels (Ryvolova et al. 2011).

Lanthanides and Thiols

Due to widespread application of lanthanides in agriculture, different branches of industry, and medicine, they represent important group of compounds causing acute as well as chronic effects on living organisms including human. Negative effects of REEs on human health have been reported. Nevertheless, uptake of lanthanides as well as mechanisms of toxic effect stays almost unknown. Differences in behavior of even rare elements chlorides (Y, Ce, Pr, Eu, Dy, Yb, and Lu) after intravenous application to rats were investigated. More than 78 % of applied lanthanides were administered into liver, bone, and spleen. In the case of Y, Eu, and Dy, application of high doses led to their accumulation in lungs and spleen and to their increased deposition of calcium in liver, spleen, and lungs. One day after application, REEs were not detectable in blood. The maximum percentages of REEs were found in liver from 8 hours to 48 hours. After that, there was a gradual decrease in their levels except Pr whose concentration remained still high. Significant hepatotoxicity (fatty liver, jaundice) was demonstrated for two REEs elements – Ce and Pr. Treatment of rats with subchronic dosage of terbium led to changes in the uptake of phosphorus and calcium. Complexes of lanthanides, which are intensely used in medicine as specific probes and stains, are after parenteral application taken up by macropinocytosis. However, toxic effects have not been recorded. Oxidative stress injury of spleen induced by lanthanides, particularly their chlorides – $LaCl_3$, $CeCl_3$, and $NdCl_3$ – led to induction of splenomegaly; oxidative stress was connected with increased lipid peroxidation and super oxide dismutase (SOD) activity. In addition, generation of nonenzymatic antioxidants (glutathione and ascorbic acid) was recorded. Compared to the above-mentioned three lanthanides, Ce^{3+} and Nd^{3+} exhibited higher oxidative stress than La^{3+} ions (Liu et al. 2010). Long-term application of lanthanides, particularly lanthanum chloride, to rats from day 0 to 6 months of age led to negative effects on central nervous system – neurotoxicity – changes in Ca^{2+}-ATPase activity, inhibition of activities of antioxidant enzymes, and subsequent cell damage. Antioxidant response of La^{3+}, Ce^{3+}, and Nd^{3+} in the form of chlorides was investigated on the lungs of mice. Authors detected increase of lipid peroxidation with subsequent reduction of antioxidative mechanism including superoxide dismutase, catalase, ascorbate peroxidase, and individual antioxidant molecules – glutathione and ascorbate – in the case of $LaCl_3$ treatment. Only minimal decrease of protective antioxidant mechanism has been detected for Ce^{3+}, medium for Nd^{3+}. Authors suggested a possible role of valence of individual lanthanides in Ln-induced toxicity (Li et al. 2010b).

As it is above concluded, lanthanides are able to induce reactive oxygen species, which are closely connected with mitochondrial dysfunction. Reactive oxygen species (ROS) are not necessarily connected with pathological processes, but also with important defensive processes including phagocytic cells that are responsible for killing of tumor cells and microorganisms. Due to ability of ROS to damage essential cellular structures, such as DNA and polyunsaturated lipid acids, there must be protective mechanisms to prevent the damage. These mechanisms include enzymes superoxide dismutase, catalase, glutathione peroxidase, and low-molecular-weight compounds, such as vitamin E and especially glutathione. Conjugation, nonenzymatic reactions of glutathione with some chemicals, has been described. Besides glutathione, phytochelatins and metallothioneins also play a role in detoxification of lanthanides. *Lactuca sativa* (*Asteraceae*, syn. *Compositae*) simultaneously treated by La^{3+} and Cd^{2+} ions demonstrates the ability of lanthanum ions to decrease Cd^{2+} ions and enhancement of phytochelatin synthase expression (He et al. 2005). Due to enhancement of phytochelatin synthase expression, La^{3+} ions may serve as modulators of heavy metals ions. In vitro as well as in vivo experiments demonstrated interactions between MTs and Eu^{3+} and La^{3+} ions (Huang et al. 1996). This effect is closely connected with accumulation of lanthanides in tissues of treated animals. Rats treated by lanthanum, cerium, neodymium, and praseodymium ions showed different abilities of accumulation in liver in the order La – Ce – Nd – Pr (Lu et al. 2002); this must be carefully considered in the light of the work of Huang et al. and the fact that the liver represents an important site of metallothionein metabolism and localization.

Connection between oxidative stress and inflammatory is well known and has been demonstrated. Ability of some lanthanides to induce oxidative stress may be connected with inflammatory response. One of the possible markers of inflammatory is enzyme alanine aminotransferase (ALT). Some lanthanides, especially Ce^{3+} as chloride, were found to provoke inflammatory responses in mice and rats. However, mechanisms of activity are still unknown. Possible role of cerium ions in alteration and remodeling of the structure of ALT in the active site was suggested (Li et al. 2010a). On the other hand, the effect of lanthanum chloride on lipopolysaccharide-challenged mice, particularly on proinflammatory factors, was demonstrated. They recorded inhibition of secretion of tumor necrosis factor-alpha (TNF-alpha) and interleukin-1-beta (IL-1-beta) as well as TNF-alpha mRNA expression due to inhibition of NF-kappa B activation.

Conclusion

The group of lanthanides comprises metals with atomic numbers from 57 (lanthanum) to 71 (lutetium). Due to their similarity with calcium ions, they are able to interfere with many Ca^{2+}-based cellular processes. One of the most important potency of lanthanides consists in ability to interfere with Ca^{2+} metabolism in mitochondria and endoplasmic reticulum, which may in certain concentrations result in the production of reactive oxygen species (ROS), which are closely connected with damage of crucial cellular structures and apoptosis or cell death, processes linked with elimination of damaged cells. ROS are responsible for induction of protective mechanisms to eliminate this damage. Generation of ROS under lanthanides treatment was demonstrated in some works; however, knowledge about close connection between lanthanides, ROS, and protective mechanisms – thiol compounds – is still missing and must be further investigated.

Acknowledgments Financial support from CEITEC CZ.1.05/1.1.00/02.0068 and NANIMEL GA CR 102/08/1546 is highly acknowledged.

Cross-References

- ▶ Lanthanides, Rare Earth Elements, and Protective Thiols
- ▶ Metallothioneins
- ▶ Thiols and Lanthanides

References

Cobbett C, Goldsbrough P (2002) Phytochelatins and metallothioneins: roles in heavy metal detoxification and homeostasis. Annu Rev Plant Biol 53:159–182

Eckschlager T, Adam V, Hrabeta J, Figova K, Kizek R (2009) Metallothioneins and cancer. Curr Protein Pept Sci 10:360–375

He ZY, Li JC, Zhang HY, Ma M (2005) Different effects of calcium and lanthanum on the expression of phytochelatin synthase gene and cadmium absorption in *Lactuca sativa*. Plant Sci 168:309–318

Huang ZX, Zheng Q, Gao HY, Gu WQ, Liu F (1996) The reactions between lanthanons and metallothionein. Chem J Chin Univ-Chin 17:190–192

Li N, Duan YM, Liu C, Hong FS (2010a) The mechanism of $CeCl_3$ on the activiation of alanine aminotransferase from mice. Biol Trace Elem Res 136:187–196

Li N, Wang SS, Liu J, Ma LL, Duan YM, Hong FS (2010b) The oxidative damage in lung of mice caused by lanthanoide. Biol Trace Elem Res 134:68–78

Liu J, Li N, Ma LL, Duan YM, Wang J, Zhao XY, Wang SS, Wang H, Hong FS (2010) Oxidative injury in the mouse spleen caused by lanthanides. J Alloy Compd 489:708–713

Lu R, Zhu YM, Chen HT, Zhao DQ, Ni JZ, Chen D, Nie YX (2002) Species of rare earth in rat liver. Chem J Chin Univ-Chin 23:1–5

Meister A, Anderson ME (1983) Glutathione. Annu Rev Biochem 52:711–760

Ryvolova M, Krizkova S, Adam V, Beklova M, Trnkova L, Hubalek J, Kizek R (2011) Analytical methods for metallothionein detection. Curr Anal Chem 7:243–261

Lanthanides, Toxicity

Milon Tichy and Marian Rucki
Centre of Occupational Health, Laboratory of Predictive Toxicology, National Institute of Public Health, Praha 10, Czech Republic

Synonyms

Biological activity; Biological effect; Lanthanides; Rare earth elements

Definition

Fifteen chemical elements with atomic number from 57 to 71 (lanthanum La, cerium Ce, praseodymium, Pr, neodumium Nd, promethium Pm, samarium Sm, europium Eu, gadolinium Ga, terbium Tb, dysprosium Dy, holmium Ho, erbium Er, thulium Tm, ytterbium Y band, lutecium Lu), sometimes chemical elements scandium (atomic number 21) and yttrium (atomic number 39) belong to the rare earth elements (REEs). They form the third subgroup of the periodic system of the elements. REEs were characterized from several points of view: occurrence, physicochemical properties, reactivity, biochemistry, toxicology, cell biology, application, or business.

Introduction

There are 15 chemical elements belonging to rare earth elements (REEs) with atomic number from 57 to 71 as lanthanides (lanthanum La, cerium Ce, praseodymium, Pr, neodymium Nd, promethium Pm, samarium Sm, europium Eu, gadolinium Ga, terbium Tb, dysprosium Dy, holmium Ho, erbium Er, thulium Tm, ytterbium Y band lutecium Lu), sometimes chemical elements scandium (atomic number 21) and yttrium (atomic number 39) are involved in the rare earth elements. They form the third subgroup of the periodic system of the elements.

REEs were studied from all viewpoints possible: occurrence, physicochemical properties, reactivity biochemistry toxicology, cell biology, application, or business. These elements and their salts or compounds have a wide application in our life and, actually, without them a common daily life today cannot be possible. Toxicity is mentioned in vitro, rarely toxicity in vivo for synthetic compounds mostly in a form of complexes with organic ligands. Distribution of the elements in body organs was studied for most of the REEs.

Occurrence

History of the lanthanides has started with a discovery of a black mineral "ytterbite" (later called gadolinite) by Carl Axel Arrhenius in 1787 in Ytterby in Sweden (Haley 1965; Hedrick 1995). Emission of lanthanides to the environment increased as a result of the growing industrial application of these elements. However, data evaluating their environmental fate are scarce (Weltje et al. 2002) and on health of nature including human beings as well. Various complexes with organic ligands were synthesized and tested.

It is suitable to recall the occurrence of REEs, because they are neither rare nor uncommon. Ironically, the REEs are relatively abundant in the Earth's crust despite their name (Zhou and He 2008), being abundant like copper or zinc. Radioactive promethium is an exception and is really rare. The REEs are often found together and it is always difficult to isolate them due to small atomic size. Lanthanides occur in various minerals. Despite their high relative abundance, it is difficult to mine them making the REEs relatively expensive. They are in monazite phosphate minerals, bastmasite as fluorocarbonate minerals, radioactive samarskite, euxenite, fergusinite, yttrofluorite, thalenite, or yttrialite as another mineral containing REEs. Their form sometimes has large crystals used in jewelry.

Human and Environmental Exposure

Men involved in mining and jewelry production are subject to constant inhalation of REEs.

Another source of inhalation of REEs is electronic waste or other wastes having REEs components. They are used in many devices that people use daily, such as car catalytic converters, magnets, rechargeable batteries, and more. Extensive application of fertilizers causes distribution of REEs in aquatic and terrestrial ecosystems.

Markers of chronic toxicity measured in the population of South Jiangxi region, China, indicate an intake of REEs through food chain (Zhang et al. 2000). Workers employed in mining of monazite are at risk of exposure to dust with expected adverse health effects; this prediction is an extrapolation of studies with rats exposed to REEs. There are many papers on the harmful effects of REEs. Some of experimental results are given below.

Toxicity

The REEs and their salts are generally said to be nontoxic, their biological activity can be detected, however.

No toxicity was detected even with nanoscaffolds taken up by cells (Delgado et al. 2011). Nanorod-treated liver, kidneys, spleen, or lungs showed no or very mild histological changes on exposure to higher doses of lanthanide nanorods (Patra et al. 2009).

However, REES find therapeutic applications for burn wounds and hyperphosphatemia (Zhang et al. 2010). The complex aquatrichlordobis (1,10-phenanthroline)cerium(III) shows promising anticancer properties in vitro by inhibiting proliferation of human cancer cell lines (Biba et al. 2009). Complexes of La(III) and Dy(III) with deprotonated 4-hydroxy-3(1-(4-nitrophenyl)-3-oxobutyl)-2H-1-benzopyran-2-one exhibit inhibitory activity against melanoma B16 and fibrosarcoma L929 (Kostova and Stefanova 2010).

The lanthanide complexes REEs ions may regulate plant growth by influencing the distribution of enzymes (Ye et al. 2008). Accumulation of terbium (III) in soil and plant leaves led to various toxic effects on the plant leaves (Wang et al. 2010). Significant accumulation of lanthanoids in the liver causes liver histopathological changes and, thus, liver malfunctions. The real-time quantitative RT-PCR and ELISA analyses showed the alternation in the mRNA and protein expressions of some inflammatory cytokines, tumor necrosis factor, interleukins, or cross-reaction proteins (Fei et al. 2011a). The oxidative stress in the liver caused by lanthanoids decreases in the order: Ce(III), Nd(III), La(III) (Fei et al. 2011b). Results obtained with sea urchin embryos suggest adverse effects in REE-exposed biota (Oral et al. 2010). Gadolinium causes neuron cell apoptosis especially by inhibiting mitochondria functions and inducing oxidative stress (Feng et al. 2010). Neurotoxicity was evaluated in Wistar rats exposed to lanthanum chloride by oral administration (He et al. 2008). Chronic exposure to lanthanum could possibly impair the learning ability and possibly disturb the homeostasis of trace elements, enzymes, and neurotransmitter systems in brain. Application of lanthanide in pharmacology should be exercised with caution in clinics, because of their dangerous effects (Feng et al. 2006).

Colloidal conjugates with scandium deposit in glomeruli and may reduce glomerular filtration rate (Tanida et al. 2009). The concentrations of cerium, lanthanum, and neodym were found to be significantly higher in *Nautilus pompilius* than in *Nautilus macromphalus* in localized enrichment of Vanatuatu waters due to environmental processes such as vulcanism or upwelling (Pernice et al. 2009).

References

Biba F, Groessl M, Egger A, Jakupec MA, Keppler BK (2009) A novel cytotoxic complex: aquatrichlordobis (1,10-phenanthroline)cerium(III). Synthesis, characterization, behaviour in H_2O, binding towards biomolecules, and antiproliferative activity. Chem Biodivers 6:2153–2165

Delgado PE, Albelda MT, Frías JC, Barreiro O, Tejera E, Kubicek V, Jiménez BLJ, Sánchez MF, Tóth E, Alarcón J, García EE (2011) Lanthanide complexes as imaging agents anchored on nano-sized particles of boehmite. Dalton Trans 40:6451–6457

Fei M, Li N, Ze Y, Liu J, Wang S, Gong X, Duan Y, Zhao X, Wang H, Hong F (2011a) The mechanism of liver injury in mice caused by lanthanoids. Biol Trace Elem Res 140:317–329

Fei M, Li N, Ze Y, Liu J, Gong X, Duan Y, Zhao X, Wang H, Hong F (2011b) Oxidative stress in the liver of mice caused by intraperitoneal injection with lanthanoids. Biol Trace Elem Res 139:72–80

Feng L, Xiao H, He X, Li Z, Li F, Liu N, Zhao Y, Huang Y, Zhang Z, Chai Z (2006) Neurotoxicological consequence of long-term exposure to lanthanum. Toxicol Lett 165: 112–120

Feng X, Xia Q, Yuan L, Yang X, Wang K (2010) Impaired mitochondrial function and oxidative stress in rat cortical neurons: implications for gadolinium-induced neurotoxicity. Neurotoxicology 31:391–398

Haley TL (1965) Pharmacology and toxicology of the rare earth elements. J Pharm Sci 54:663–670

He X, Zhang Z, Zhang H, Zhao Y, Chai Z (2008) Neurotoxicological evaluation of long-term lanthanum chloride exposure in rats. Toxicol Sci 103:354–361

Hedrick JB (1995) The global rare-earth cycle. J Alloys Comp 225:609–618

Kostova I, Stefanova T (2010) Synthesis, characterization and cytotoxic/cytostatic activity of La(III) and Dy(III) complexes. J Trace Elem Med Biol 24:7–13

Oral R, Bustamante P, Warnau M, D'Ambra A, Guida M, Pagano G (2010) Cytogenetic and developmental toxicity of cerium and lanthanum to sea urchin embryos. Chemosphere 81:194–198

Patra CR, Abdel MSS, Wang E, Dutta S, Patra S, Eshed M, Mukherjee P, Gedanken A, Shah VH, Mukhopadhyay D (2009) In vivo toxicity studies of europium hydroxide nanorods in mice. Toxicol Appl Pharmacol 240:88–98

Pernice M, Boucher J, Boucher RR, Joannot P, Bustamante P (2009) Comparative bioaccumulation of trace elements between *Nautilus pompilius* and *Nautilus macromphalus* (Cephalopoda: Nautiloidea) from Vanuatu and New Caledonia. Ecotoxicol Environ Saf 72:365–371

Tanida E, Usuda K, Kono K, Kawano A, Tsuji H, Imanishi M, Suzuki S, Ohnishi K, Yamamoto K (2009) Urinary scandium as predictor of exposure: effects of scandium chloride hexahydrate on renal function in rats. Biol Trace Elem Res 130:273–282

Wang L, Zhou Q, Huang X (2010) Effects of heavy metal terbium on contents of cytosolic nutrient elements in horseradish cell. Ecotoxicol Environ Saf 73:1012–1017

Weltje L, Brouwer AH, Verburg TG, Wolterbeck HT, De Goeij JJM (2002) Accumulation and elimination of lanthanum by duckweed (Lemna minor L.) as influenced by organism growth and lanthanum sorption to glass. Environ Toxicol Chem 21:1483–1489

Ye Y, Wang L, Huang X, Lu T, Ding X, Zhou Q, Guo S (2008) Subcellular location of horseradich peroxidise in horseradish leaves treated with La(III), Ce(III and Tb(III). Ecotoxicol Environ Saf 71:677–684

Zhang H, Feng J, Zhu W, Liu C, Xu S, Shao P, Wu D, Yang W, Gu J (2000) Chronic toxicity of rare-earth elements on human beings: implication of blood biochemical indices in REE-high regions, South Jingxi. Biol Trace Elem Res 73:1–17

Zhang J, Li Y, Hao X, Zhang Q, Yang K, Li L, Ma L, Li X, Wang S (2010) Recent progress in therapeutic and diagnostic applications of lanthanides. Mini Rev Med Chem 11: 678–694

Zhou M, He Q (2008) Synthesis, characterization, and biological properties of nano-rare earth complexes with L-glutamic acid and imidazole. J Rare Earth 26:473–477

Lanthanoid

▶ Lanthanides and Cancer

Lanthanoids

▶ Lanthanides, Physical and Chemical Characteristics

Lanthanum

Takashiro Akitsu
Department of Chemistry, Tokyo University of Science, Shinjuku-ku, Tokyo, Japan

Definition

A lanthanoid element of the f-elements block, with the symbol La, atomic number 57, and atomic weight 138.9055. Electron configuration $[Xe]5d^1 6s^2$. Lanthanum is composed of stable (^{139}La, 99.91%) and two radioactive (^{137}La; ^{138}La, 0.09%) isotopes. Discovered by C. G. Mosander in 1839. Lanthanum exhibits oxidation states III (and II); atomic radii: 187 pm, covalent radii 208 pm; redox potential (acidic solution) La^{3+}/La −2.522 V; La^{3+}/La^{2+} −3.1 V; electronegativity (Pauling) 1.1. Most stable technogenic radionuclide ^{138}La (half-life 1.05×10^{11} years). The most common compounds: La_2O_3, LaF_3, and $La(OH)_3$ and $[La(OH_2)_9]^{3+}$ (in aqueous solution). Biologically, lanthanum is of low to moderate toxicity, can cause granulomatous lesions in the lung or pneumoconiosis (higher concentration of La was observed than in normal lung) (Atkins et al. 2006; Cotton et al. 1999; Huheey et al. 1997; Oki et al. 1998; Rayner-Canham and Overton 2006).

Cross-References

▶ Lanthanide Ions as Luminescent Probes
▶ Lanthanide Metalloproteins
▶ Lanthanides and Cancer
▶ Lanthanides in Biological Labeling, Imaging, and Therapy
▶ Lanthanides in Nucleic Acid Analysis
▶ Lanthanides, Physical and Chemical Characteristics

References

Atkins P, Overton T, Rourke J, Weller M, Armstrong F (2006) Shriver and Atkins inorganic chemistry, 4th edn. Oxford University Press, Oxford/New York
Cotton FA, Wilkinson G, Murillo CA, Bochmann M (1999) Advanced inorganic chemistry, 6th edn. Wiley-Interscience, New York
Huheey JE, Keiter EA, Keiter RL (1997) Inorganic chemistry: principles of structure and reactivity, 4th edn. Prentice Hall, New York
Oki M, Osawa T, Tanaka M, Chihara H (1998) Encyclopedic dictionary of chemistry. Tokyo Kagaku Dojin, Tokyo
Rayner-Canham G, Overton T (2006) Descriptive inorganic chemistry, 4th edn. W. H. Freeman, New York

Lanthanum, Physical and Chemical Properties

Fathi Habashi
Department of Mining, Metallurgical, and Materials Engineering, Laval University, Quebec City, Canada

Lanthanum is a malleable, ductile, and soft metal. It is obtained from monazite and bastnäsite and is mainly used as an alloying element and in lighter flints. It belongs to Group 3 of the Periodic Table:

Al
Sc
Y
Lanthanum
Ac

Physical Properties

Atomic number	57
Atomic weight	138.91
Density, g/cm^3	6.145

(*continued*)

Relative abundance, %	1.7×10^{-3}
Melting point, °C	920
Boiling point, °C	3,457
Crystal structure	Hexagonal ⇆
Undergoes phase transformation at 310 °C and at 865 °C	Face centered cubic ⇆ Body centered cubic
Lattice constants, nm	
a	0.3774
c	1.2159

Chemical Properties

Lanthanum oxidizes rapidly when exposed to air. It burns readily at 150 °C to form lanthanum oxide:

$$4La + 3O_2 \rightarrow 2La_2O_3$$

It reacts slowly with cold water and quickly with hot water to form lanthanum hydroxide:

$$2La + 6H_2O \rightarrow 2La(OH)_3 + 3H_2$$

Lanthanum metal reacts with all the halogens vigorously at 200 °C. It dissolves readily in dilute sulfuric acid liberating hydrogen:

$$2La + 6H^+ \rightarrow 2La^{3+} + 3H_2$$

Lanthanum is precipitated from solution as an oxalate. The oxide is obtained from the oxalate by thermal decomposition. The metal combines with nitrogen, carbon, sulfur, phosphorus, boron, selenium, silicon, and arsenic at elevated temperatures, forming binary compounds (Gupta and Krishnamurthy 2005; Habashi 2003; McGill 1997; Vickery 1953).

References

Gupta CK, Krishnamurthy N (2005) Extractive metallurgy of rare earths. CRC Press, Boca Raton

Habashi F (2003) Metals from ores. An introduction to extractive metallurgy. Métallurgie Extractive Québec, Quebec City. Distributed by Laval University Bookstore. www.zone.ul.ca

McGill I (1997) Rare earth metals. general. In: Habashi F (ed) Handbook of extractive metallurgy. Wiley, Heidelberg, pp 1695–1741

Vickery RC (1953) Chemistry of the lanthanons. Butterworths, London

Large Surface-Area-to-Volume

▶ Porous Silicon for Drug Delivery

L-Ascorbate

▶ Ascorbate Oxidase

Last Step of the Denitrification Pathway

▶ Nitrous Oxide Reductase

Lead

▶ Polonium and Cancer

Lead (Pb^{2+}) Permeases/Porters

▶ Lead Transporters

Lead and Alzheimer's Disease

Fernando Cardozo-Pelaez and Jennene A. Lyda
Department of Pharmaceutical Sciences, Center for Environmental Health Sciences, University of Montana, Missoula, MT, USA

Synonyms

Metal-induced oxidative DNA damage; Metal neurotoxicity

Definition

Lead and Alzheimer's disease refers to the association, supported by epidemiological and animal models, between the onset of Alzheimer's pathology and environmental or occupational exposure to the heavy metal lead.

Alzheimer's disease (AD) is defined as an irreversible neurodegenerative disease characterized by an age-dependent decline in learning, memory, and other cognitive functions. Although some cases may be linked to genetic mutations at various loci, most cases appear to be sporadic with no link to the genetic families. It has been argued that familial cases of AD (FAD) account to not more of 10% of the total AD cases and usually FAD has an early onset as compared to sporadic cases. Some of the genetic mutations linked to AD development occur in genes that code for proteins linked to the pathological mechanism(s) and to the histological markers of the disease (Hardy 2006). On the other hand, there are certain genetic alterations that are considered to be a risk factor for the development of the pathology, impacting genes that are not directly linked to the major histopathology of AD. Along with aging, known to be the major risk factor for the development of AD, these other genetic components suggests that there may be one or a combination of environmental/occupational factors that play a role in the origin of AD cognitive deficiencies beyond the degree that is normal during the aging process.

Lead (Pb) is a heavy metal ubiquitously distributed, and environmental and occupational exposures have been linked to neurotoxicity. Upon acute or chronic exposure, Pb may lead to cognitive and behavioral deficits in children and adults. Environmental exposure to Pb has been suggested as a causative factor of neurodegeneration in AD. Indeed, several epidemiological studies have shown a correlation between developmental Pb exposure as well as occupational Pb exposure and late age onset of AD (Stewart et al. 2002) and (Zawia et al. 2009). These studies serve to support the commonly described Barker hypothesis, which outlines the concept of connecting developmental chemical exposures with adult disease onset (Tamashiro and Moran 2010).

AD neuropathology and diagnosis is characterized by the presence of senile plaques and neurofibrillary tangles in the hippocampus and neocortex regions of the brain. These proteinaceous aggregations correlate with regionally specific neuronal dysfunction, dystrophy, and neuronal loss. Senile plaques are mostly composed of β-amyloid (Aβ), a fragment cleaved by the sequential activity of β- and γ-secretases from the larger protein, amyloid precursor protein (APP) (Wu et al. 2008b). The endogenous function of APP is still not known. Thus far, it has been established that APP may play a role in neuronal plasticity (Gralle and Ferreira 2007). This is supported by findings of APP's trophic role for stimulating neurite outgrowth, synaptogenesis, and cell adhesion (Thinakaran and Koo 2008). The other major histological changes linked to AD, neurofibrillary tangles, are composed of highly phosphorylated and aggregated protein clusters of tau protein (Götz et al. 2011). Normally, tau is responsible for the stabilization of microtubules in the axon and without the correct localization can cause axon collapsing and subsequent downstream effects as well as accumulation of hyper-phosphorylated tau in neurons. The diagnosis for patients with AD is still a difficult issue. With recent advancements in positron emission tomography (PET), tracking the development of Aβ plaque deposits with the decline in cognition has improved. However, the absolute diagnosis of AD is still determined at autopsy, by the histological evidence of Aβ plaques and tau-neurofibrillary tangles (Götz et al. 2011).

The exact cause for the abnormal formation and accumulation of Aβ peptides as well as their correlation to the onset of dementia are not fully understood, but there are many hypotheses of mechanisms linked to their initial formation and associated toxicity. Aβ formation and accumulation can be explained by some of the genetic deficiencies linked to AD. For instance, mutations in APP have been associated with altered protein processing that leads to the elevated release of the Aβ fragment from APP (Thinakaran and Koo 2008). Also, mutations in presenilins protein, which form part of the γ-secretase complex, lead to dysregulated levels of Aβ (Edbauer et al. 2003). However, how Aβ plaque formation occurs in sporadic cases of AD is unresolved.

Recent studies linking developmental exposure to Pb have shed light into Pb-associated cognitive deficiencies and their possible role on AD onset. These studies have demonstrated that there is increased Aβ formation and senile plaque formation (similar to AD pathology in humans) in aged rodents and nonhuman primates after developmental exposure to Pb (Basha et al. 2005; Wu et al. 2008a). Low levels of inorganic Pb were used to treat the subjects in these studies yielding blood levels slightly higher than the 10 μg/dL levels considered safe for humans. Exposure scenarios in aged animals did not lead to similar AD-like pathology compared to exposures during stages of development only. Additionally, mRNA levels of APP and the immunohistological evidence of Aβ accumulation were increased in aged animals exposed to Pb early in life. Based on epidemiological and animal evidence of Pb's role in AD, several models have been proposed for the mechanism that may link the increase of APP expression and Aβ formation and accumulation to post developmental exposure from Pb. These theories include oxidative damage to bases in DNA (Amadoruge and Barnham 2011), abnormal nucleotide methylation (Wu et al. 2008b), and the possible interactions with genetic predisposition factors (i.e., ApoE) (Stewart et al. 2002) which could make certain populations more susceptible to developing early-onset AD following Pb exposure.

Oxidative Damage as a Result of Pb Exposure

In addition to the established neuropathology of AD, oxidative damage to cellular components has also been a constant finding in affected brain regions of AD sufferers. It has been proposed that some of the toxic effects of Aβ are mediated through generation of free radicals (Bush 2003). The proposed mechanism responsible for producing free radicals in AD involves Aβ catalyzing the reduction of metals such as Cu^{2+} and Fe^{3+} using O_2, by the Fenton reaction, resulting in the production of H_2O_2 (Bush 2003). Neurons harbor a diverse number of cellular defenses that evolved to reduce and prevent the potential damaging impact of reactive oxygen species (ROS). These defenses include small reactive molecules such as glutathione, and ubiquitously expressed detoxifying enzymes such as superoxide dismutase, glutathione peroxidase, and catalase. If these systems of detoxification are impaired or reduced, there is an imbalance between the formation of oxidative radicals and the cellular capacity to contain them, thus overwhelming the cell and leading to a variety of damaged targets. Despite Pb not being considered highly redox active in the cellular milieu, exposure to Pb during development has been linked to increased oxidative damage to DNA later in life. The specific type of DNA damage found after Pb exposure resembles damage seen in AD, where there are higher levels of 8-oxo-2′-deoxyguanosine (8-oxo-dG), the most common oxidative product in DNA. These increases in 8-oxo-dG are specifically increased in AD impacted areas of the brain such as the neocortex and hippocampus, both relevant to processes of learning and memory. In the case of developmental Pb exposure, 8-oxo-dG levels have been found to be higher in cerebral cortical tissue of aged-rats, when compared to unexposed animals (Bolin et al. 2006); it is worth noting that such increase was not evident after exposure to Pb later in life. Interestingly, the increased levels in 8-oxo-dG were paralleled by increases in APP expression and increased levels of Aβ, and in sync with DNA damage these two markers were not elevated with exposure to Pb later in life. These results suggest that there is a window of susceptibility during the development of neurons that may link developmental exposure and the latency of APP and Aβ accumulation in relationship to oxidative DNA damage as measured by 8-oxo-dG (Basha et al. 2005). Both oxidative damage to DNA and increased Aβ accumulation linked to developmental Pb exposure has also been documented in primates (Zawia et al. 2009). Since increased 8-oxo-dG levels were only found months after Pb exposure, this indicates that Pb is altering the cell's response at the developmental stages leading to oxidative stress conditions where 8-oxo-dG can be formed and accumulate. Although certain amounts of oxidative stress accumulation occurs as a part of the normal aging process, developmental exposure to Pb can facilitate the accumulation of more oxidative stress in the brain impacting selectively specific genes involved in learning, memory, and neuronal survival (Castellani et al. 2006). As stated above, Pb does not participate in the Fenton reaction, thus any evidence of free radical damage after exposure must be due to an indirect mechanism different from the effects of other heavy metals.

DNA Methylation as a Result of Pb

Other possible mechanism by which Pb can increase Aβ in the brain is through an increase in expression of the parent protein APP. Evidence suggests that Pb can interfere with gene expression via epigenetic modulation of the APP promoter. Thus, by altering DNA methylation patterns in the promoter region of the APP gene, especially during critical transcriptional event stages in development, Pb can alter the life-long expression of the protein. DNA methylation patterns on promoter regions are important for the regulation of gene expression during a cell's life span. DNA methylation typically takes place on the cytosines resulting in silencing (reduced expression) of the downstream genes. The enzyme DNA methyltransferase 1 (DNMT1) is selective for the methylation of cytosine in a CpG dinucleotide. After exposure to Pb, DNMT1 was found to be reduced suggesting Pb's role in reducing methylation patterns after early life exposure (Wu et al. 2008b), such reduction will lead to an upregulation of APP protein expression and consequently increased Aβ levels.

A novel idea linking epigenetic changes and oxidative stress from results obtained in rats and primates has gained additional support by studies using synthetic DNA oligonucleotides with CpG sites containing the oxidatively modified base 8-oxo-2dG. Presence of 8-oxo-dG at the CpG sites decreased cytosine methylation capacity in newly synthesized DNA (Turk et al. 1995). Thus, environmental exposure to inorganic lead may influence methylation patterns as a consequence of effects in the methylation process and/or increased oxidative damage to adjacent guanines in CpG regions, the net effect will be reduced methylation of the CpG regions in the promoter of the APP gene leading to upregulation of APP protein expression. Conversely, Pb exposure in adulthood had no effect on APP upregulation or alterations in the promoter region providing further evidence of the critical developmental exposure period and its latent effects for this heavy metal.

Genetics

Interaction between environmental exposure to Pb and genetic variants may play a role in increasing the risk factor for AD development. Carrying the allele ε4 of apolipoprotein E (ApoE) has been shown to be associated with an increased risk for the development of late-onset AD. The mechanism linking the presence of this allele and the formation of Aβ aggregates is unknown. An environment/gene interaction is supported by an epidemiological study that establishes a correlation between tibia bone Pb levels in organo-lead workers and presence of the ε4 allele with an increased risk to develop AD (Stewart et al. 2002). This study was based on a cohort of 544 individuals where their ApoE genotype was determined and bone tibia lead levels determined. Neurobehavioral analysis was then conducted for the direct correlation of tibia bone lead levels and a decrease in performance on behavioral exams. The results in this study provide strong evidence for the delayed effects of lead, given that the last time of exposure has occurred 16 years prior to the time of assessments. Animal and cell culture studies have been presented to uncover the possible mechanism involved in lead and organo-lead neurotoxicologic mechanisms. Some of the most common organo-lead compound (triethyl and trimetyl lead), used in manufacturing process, could lead to morphological changes of astrocytes in the hippocampus and neurofibrillary tangles in frontal cortex and cerebellum.

References

Amadoruge PC, Barnham KJ (2011) Alzheimer's disease and metals: a review of the involvement of cellular membrane receptors in metallosignalling. Int J Alzheimers Dis. doi:10.4061/2011/542043,542043

Basha MR, Wei W, Bakheet SA, Benitez N, Siddiqi HK, Ge Y-W, Lahiri DK, Zawia NH (2005) The fetal basis of amyloidogenesis: exposure to lead and latent overexpression of amyloid precursor protein and beta-amyloid in the aging brain. J Neurosci 25(4):823–829. doi:10.1523/JNEUROSCI.4335-04.2005

Bolin CM, Basha R, Cox D, Zawia NH, Maloney B, Lahiri DK, Cardozo-Pelaez F (2006) Exposure to lead and the developmental origin of oxidative DNA damage in the aging brain. FASEB J 20(6):788–790. doi:10.1096/fj.05-5091fje

Bush AI (2003) The metallobiology of Alzheimer's disease. Trends Neurosci 26(4):207–214

Castellani RJ, Lee H-G, Perry G, Smith MA (2006) Antioxidant protection and neurodegenerative disease: the role of amyloid-beta and tau. Am J Alzheimers Dis Other Demen 21(2):126–130

Edbauer D, Winkler E, Regula JT, Pesold B, Steiner H, Haass C (2003) Reconstitution of gamma-secretase activity. Nat Cell Biol 5(5):486–488. doi:10.1038/ncb960

Götz J, Eckert A, Matamales M, Ittner LM, Liu X (2011) Modes of Aβ toxicity in Alzheimer's disease. Cell Mol Life Sci 68(20):3359–3375. doi:10.1007/s00018-011-0750-2

Gralle M, Ferreira ST (2007) Structure and functions of the human amyloid precursor protein: the whole is more than the sum of its parts. Prog Neurobiol 82(1):11–32. doi:10.1016/j.pneurobio.2007.02.001

Hardy J (2006) A hundred years of Alzheimer's disease research. Neuron 52(1):3–13. doi:10.1016/j.neuron.2006.09.016

Stewart WF, Schwartz BS, Simon D, Kelsey K, Todd AC (2002) ApoE genotype, past adult lead exposure, and neurobehavioral function. Environ Health Perspect 110(5):501–505

Tamashiro KLK, Moran TH (2010) Perinatal environment and its influences on metabolic programming of offspring. Physiol Behav 100(5):560–566. doi:10.1016/j.physbeh.2010.04.008

Thinakaran G, Koo EH (2008) Amyloid precursor protein trafficking, processing, and function. J Biol Chem 283(44):29615–29619. doi:10.1074/jbc.R800019200

Turk PW, Laayoun A, Smith SS, Weitzman SA (1995) DNA adduct 8-hydroxyl-2′-deoxyguanosine (8-hydroxyguanine) affects function of human DNA methyltransferase. Carcinogenesis 16(5):1253–1255

Wu J, Riyaz Basha Md, Brock B, Cox DP, Cardozo-Pelaez F, McPherson CA, Harry J et al (2008a) Alzheimer's disease (AD)-like pathology in aged monkeys after infantile exposure to environmental metal lead (Pb): evidence for a developmental origin and environmental link for AD. J Neurosci 28(1):3–9. doi:10.1523/JNEUROSCI.4405-07.2008

Wu J, Riyaz Basha Md, Zawia NH (2008b) The environment, epigenetics and amyloidogenesis. J Mol Neurosci 34(1):1–7. doi:10.1007/s12031-007-0009-4

Zawia NH, Lahiri DK, Cardozo-Pelaez F (2009) Epigenetics, oxidative stress, and Alzheimer disease. Free Radic Biol Med 46(9):1241–1249. doi:10.1016/j.freeradbiomed.2009.02.006

Lead and Immune Function

Jane Kasten-Jolly[1] and David A. Lawrence[2,3]
[1]New York State Department of Health, Wadsworth Center, Albany, NY, USA
[2]Department of Biomedical Sciences, School of Public Health, State University of New York, Albany, NY, USA
[3]Laboratory of Clinical and Experimental Endocrinology and Immunology, Wadsworth Center, Albany, NY, USA

Synonyms

Caspase 1; CASP1, ICE (IL-1 beta-converting enzyme), IL-1 beta convertase
Interferon gamma; Type-2 interferon, antigen-induced interferon, immune interferon
Interleukin 1; BCDF/BDF (B-cell differentiation factor), PIF (proteolysis-inducing factor)
Interleukin 4; BCGF-1 (B-cell growth factor-1), IgE-EF (IgE-enhancing factor), IgG1-enhancing factor, MCGF-2 (mast cell growth factor-2), TCGF-2 (T-cell growth factor-2)
Interleukin 6; BCSF (B-cell stimulating factor), CDF (cytotoxic T-cell differentiation factor), DIF (differentiation-inducing factor), TAF (T-cell activating factor)
Interleukin 12; CLMF (cytotoxic lymphocyte maturation factor), TSF (T-cell stimulating factor)
Lead; Pb, a heavy metal toxicant
NOS2; nitric oxide synthase-2, iNOS

Definitions

Adaptive Immunity – The adaptive immune system provides the ability to recognize and remember specific antigens, e.g., different pathogens. This system prepares the organism for future challenges.

Antigen-Presenting Cell (APC) – These cells display peptides bound to major histocompatibility complex (MHC) class II molecules on their cell surface for recognition by CD4+ T-cells via their antigen-specific T-cell receptors. APC also express additional co-stimulatory molecules, including members of the B7 family (CD80 and CD86).

B Lymphocytes – Bone marrow-derived cells that produce antibodies.

Complement System – This is a biochemical cascade of factors synthesized mostly by the liver as part of the innate immune system that aids or "complements" the ability of antibodies to clear pathogens.

Cytokines – Growth or differentiation factors secreted by various cells that regulate activation of B-cells, T-cells, macrophages, and various other cells that have roles in immune response.

Cytotoxic T-cells – CD8+ T-cells that induce the death of cells infected with pathogens. These T-cells are activated by presentation of antigen peptide by MHC class I.

Innate Immunity – The innate immune system is comprised of cells and mechanisms that provide an immediate response to infection via nonspecific means.

Lymphocytes – A type of white blood cell in the vertebrate immune system. There are four types of lymphocytes: NK, NKT, B, and T-cells.

Macrophages – These are large cells that are part of the innate immune response because of their ability to engulf and internalize (phagocytosis) pathogens in order to destroy them.

NK Lymphocytes – Natural killer (NK) lymphocytes are part of the innate immune response. These cells recognize infected and tumor cells (cells lacking MHC class I expression), then kill them by releasing cytotoxic granules into the abnormal cell.

T-Helper cells – CD4+ T-cells that help other immune cells to do their tasks. The T-helper cells recognize peptide antigen presented on MHC class II. T-helper cell subsets are identified by the kind of cytokine that they secrete, for example, IFN-γ by Th1 and IL-4 by Th2, and IL-17 by Th17.

T Lymphocytes (thymus-derived cells) – These cells are responsible for T-cell-mediated immunity that is part of the adaptive immune response.

Introduction

The entry presented here is meant as an overview of knowledge gained over the last several decades regarding the impact of Pb exposure on the immune system. Lead's effect on immune system function has been extensively reviewed by Lawrence and McCabe (1995) and by Dietert and Piepenbrink (2006). Therefore, much of the information presented here will be referenced by these review articles, but this manuscript will contain some additional information and new findings not included in the more recent review by Dietert and Piepenbrink mentioned above.

Pb Lowers Host Resistance to Pathogens

Lead (Pb) lowers host resistance to infections by various pathogens, including bacteria and viruses. Evidence of this was first suggested with human studies, but the first controlled animal study of this was reported by Hemphill et al. (1971), who exposed mice to 100 μg or 250 μg of Pb nitrate per day for 30 days followed by infection with *Salmonella typhimurium*. Mice in the two Pb-exposed groups had a significantly higher mortality rate than those in the control group, and this mortality rate was Pb dose dependent. Another study performed with bacteria indicated that the mortality of mice infected with *Listeria monocytogenes* was increased after exposure to $\geq$0.4 mM Pb in the drinking water for 4 weeks prior to the infection. However, in the absence of infection, mice that were maintained in a pathogen-restricted facility and exposed to 10 mM Pb acetate in their drinking water for 1 year remained visibly unaffected by the high-dose exposure (Lawrence 1985). Other bacterial studies have included the following bacteria: *Escherichia coli, Staphylococcus epidermidis, Serratia marcescens, and Pasteurella multocida.* All of these studies reported that Pb lowered host resistance to infection by these agents. With respect to viral infections, mice exposed to Pb also had increased mortality (Gainer 1974). These experiments also gave the first indication that Pb has an effect on interferon activity. Mice exposed to 0.5 mM Pb acetate from post-natal-day (pnd) 0 to pnd 22 with a resulting blood Pb level of 17.4 ± 0.6 μg/dl, followed by infection with a sublethal titer of *Listeria monocytogenes* had decreased food and water intake and decreased weight gain compared with infected mice that did not receive Pb exposure. Pb-exposed mice also had higher serum levels of IL-1 on days 2–4 after *Listeria* infection than did mice receiving *Listeria* without Pb exposure (Dietert and Piepenbrink 2006). These reductions in resistance to infections occurred despite Pb's ability to induce lymphopoeisis or leukocytosis (Hogan and Adams 1979). Additionally, it was observed that Pb could increase in vitro cell proliferation of spleen cells in a dose response manner, and Pb could increase proliferation in leukocyte cultures stimulated with PHA or LPS (Lawrence 1981).

Pb Effect on B-cells and T-cell Help

One of the observations of the effects of Pb on the immune system was that Pb could skew T-cell help toward Th2-type responses. Pb could increase B-cell proliferation and differentiation, and Pb enhances B-cell expression of MHC class II (Dietert and Piepenbrink 2006). Since it was known that B-cells need T-cell help to differentiate into antibody-producing cells, experiments were performed using membrane-segregated cultures to determine if Pb could increase B-cell proliferation without direct contact between the B- and T-cells. The study found that Pb enhanced B-cell differentiation by T-B-cell interactions via the enhanced MHC class II expression on the B-cell surface. In addition, it was observed that Pb influenced the production of T-cell factors needed for B-cell differentiation. These factors were later identified as interleukins produced by T-helper cells of the Th2 subtype. Experiments performed in vitro with T-cell clones of the Th1 or Th2 subtype demonstrated that Pb increased production of the Th2 cytokine, IL-4, while decreasing production of the Th1 cytokine, IFN-γ. In vivo experiments showed that Pb exposure of BALB/cByJ mice increased serum IL-4 and IgE, suggesting that Pb was skewing the immune response toward the Th2 subtype (Dietert and Piepenbrink 2006). Although the mechanism by which Pb accomplishes this skewing is still uncertain, it has been shown that it may be associated with an increase in cAMP levels within the cell after Pb exposure (Dietert and Piepenbrink 2006). Skewing toward a Th2 subtype may be mediated through the antigen-presenting cells. It was found that Pb-exposed bone-marrow-derived dendritic cells had increased expression of MHC class II and produced IL-10. When these dendritic cells were placed in culture with antigen-specific T-cells, following antigen stimulation the cytokine profile generated was that of the Th2 subtype (Gao et al. 2007). The skewing effect on cell-mediated immunity has been postulated to be part of the reason for the lowered host resistance by Pb to sublethal doses of pathogens. In support of this hypothesis, interleukin-12, which promotes Th1 helper cell development, has the ability to reverse the inhibition of Pb on host defenses against *Listeria monocytogenes* (Dietert and Piepenbrink 2006). Interestingly, although Pb interferes with production of IFN-γ and nitric oxide (NO), which are two factors needed for host defenses against *Listeria*, Pb does not inhibit mRNA expression of IFN-γ or inducible NO synthase (iNOS) (Heo et al. 2007). In addition to Pb's ability to lower resistance to numerous pathogens, the Th2 skewing by Pb has been implicated in increased development of allergic reactions and exacerbation of the autoimmune disease systemic lupus erythematosis (SLE). Elevated serum IgE levels have been reported in workers at a plant manufacturing batteries. In regard to autoimmunity, occupational Pb exposure led to autoantibody deposition in the kidneys of the workers (Wedeen et al. 1978, 1979). More recent studies using a mouse model for SLE showed that Pb exacerbated symptoms in lupus-prone NZM mice (Dietert and Piepenbrink 2006).

Pb Effects on Delayed-Type Hypersensitivity (DTH)

Pb exposure has been reported to suppress DTH responses. Briefly, in DTH T-cells activated by antigenically modified skin constituents will promote an inflammatory reaction in the skin after reexposure to the antigen, such as occurs upon exposure to poison ivy. Studies performed on mice have indicated that Pb suppresses DTH responses with respect to administration of sheep erythrocytes or *Mycobacteria* (Lawrence 1985). The DTH response is highly dependent on Th1-type T-cells and various chemokines. Pb also depressed DTH responses to the purified protein derivative of tuberculin after developmental exposure during gestation, weaning, and direct exposure of rats to 6 weeks of age. DTH activity toward a relatively small molecule, picryl chloride, was found to be depressed in mice given 0.5 mg/Kg Pb by sc administration. In this study, it was determined that Pb depressed DTH responses regardless of the time of sensitization or Pb administration. The DTH response has been employed to measure the effects of Pb on the immune system during various stages of development (Dietert and Piepenbrink 2006). Studies performed during gestation in rats and chickens were in agreement that the Pb effect on DTH occurred between days 9 and 15 of gestation. It was suggested that thymic development and expression of the Th1 cytokine, IFN-γ, both had a role in Pb suppression of the DTH response.

Pb Effects on the Innate Immune Response

Pb can lower host resistance by having an impact on phagocyte activation and function. Production of NO by macrophages has an important role in the destruction and killing of intracellular pathogens. For this increased NO synthesis it is necessary to have activated T-cells present to synergistically help expression of iNOS present in the macrophages (Dietert and Piepenbrink 2006). Activated T-cells provide the cytokines and surface molecules (e.g., LFA-1 and CD40L) necessary to upregulate the iNOS activity (Tian et al. 1995). In cultures of splenocytes which contain macrophages, lymphocytes, and other cell types, it was observed that Pb decreased NO production. Decreased ability of splenic macrophages to destroy intracellular pathogens after Pb exposure was observed by Queiroz

L

et al. 1994. Here, it was thought that the impaired ability of macrophages and neutrophils to kill *Candida albicans* was due to Pb exposure leading to a deficiency in myeloperoxidase. Disabled macrophage function also was observed by Lee and Battles 1994; here Pb exposure of macrophages in culture displayed increased prostaglandin E2 (PGE2) in a dose-dependent manner. This suggested that the macrophage immune function was compromised, since functional macrophages have lowered PGE2. That macrophage function was indeed compromised by Pb in these experiments was demonstrated by stimulation of the macrophages with LPS followed by measurement of zymosan particle ingestion. Although the chemotactic activity of macrophages is an important aspect of their function, the Pb effect on this activity has not been well studied, but preliminary in vitro studies suggest that Pb inhibits macrophage chemotaxis.

Initial reports on Pb's impact on the immune system suggested that Pb could be interfering with the normal functioning of the complement system by binding directly to antibody, thereby decreasing antibody/complement interaction (Hemphill et al. 1971). Recent experimental evidence indicates that Pb does interfere with the function of the complement system, but not through binding to antibodies. Gene expression analysis performed on spleens of mice exposed to 0.1 mM Pb from gd8 to pnd21 indicated that Pb increased expression of C4bp by fivefold (Kasten-Jolly et al. 2010). The function of C4bp is to decrease the activity of the complement pathway in two ways: by blocking formation of C3/C5 convertase and by blocking lysis by complement of cells undergoing apoptosis. Therefore, Pb dampens complement activity during the innate immune response and in doing so may allow pathogens to escape destruction.

Gene expression studies on the spleens of Pb-exposed mice also revealed that Pb influenced the innate immune response by increasing expression of caspase-12 and indoleamine-pyrrole 2,3 dioxygenase (IDO) (Kasten-Jolly et al. 2010). Studies performed in mice have shown that increased presence of caspase-12 diminishes clearance of systemically and abdominally located bacterial pathogens. Further, caspase-12 blocks maturation of IL-1β and IL-18 to their active forms by directly inhibiting the activity of caspase-1. Conversely, Pb assisted in the innate immune response by increasing expression of IDO by threefold. The function of IDO is to deplete L-Trp from local cellular microenvironments and to promote the formation of kynurenine pathway metabolites. Interestingly, in accordance with the effect of Pb on T-helper cells, IDO inhibits Th1 activity and enhances Th2 activity (Kasten-Jolly et al. 2010).

Pb Effects on the Inflammatory Response

Recent studies performed in mice and rats employing developmental exposure by Pb have observed that Pb exposure generates responses similar to those observed during inflammation (Struzynska et al. 2007; Kasten-Jolly et al. 2010, 2011). In the spleen, Pb increased the expression of many catabolic enzymes, increased apoptosis, and increased expression of chemokines and chemokine receptors. Gene expression was upregulated by Pb for chymotrypsin, carboxypeptidases, trypsin, and elastase. This suggested that Pb was causing injury to the spleen through inflammation, since proteases are released from lysozymes during an inflammatory response. In the brain, upregulation of the inflammatory cytokine IL-6 was observed by Struzynska et al. 2007 and Kasten-Jolly et al. 2011. This along with increased expression of the glial cell marker, GFAP, suggested that Pb was causing inflammation in the brain, as well. Increased tissue damage through inflammation causes increased oxidative stress, which decreases the immune system's ability to fight off pathogens and increases the likelihood of developing an autoimmune illness.

Summary

Animal studies of bacterial and viral infections indicated that Pb lowered host resistance to the pathogen. Paradoxically, Pb promotes B-cell proliferation and differentiation and enhances T-cell proliferation induced by phytohemagglutinin (PHA). However, analysis of the antibody production profile and cytokine production profile in the presence of Pb indicated that Pb was skewing the T-cell helper cells toward the Th2 subtype. Therefore, expression of IgE and IL-4 were increased by Pb exposure. With respect to the innate immune response, macrophage cell function was compromised, complement function was diminished, and factors, such as caspase-12, that block bacterial clearance were increased. Various studies performed up to this time have not observed any Pb effect on natural killer cell (NK) or CD8 T-cell function (Mishra 2009). However, the findings on the effect of Pb on the humoral and innate immune responses provide some explanation with regard to the mechanism by which Pb lowers host resistance to pathogens, increases allergic reactions, and promotes autoimmune disease.

Cross-References

▶ Mercury and Immune Function

References

Dietert RR, Piepenbrink MS (2006) Lead and immune function. Critical Rev Toxicol 36:359–385

Gainer JH (1974) Lead aggravates viral disease and represses the antiviral activity of interferon inducers. Environ Health Perspect 7:113–119

Gao D, Mondal TK, Lawrence DA (2007) Lead effects on development and function of bone marrow-derived dendritic cells promote Th2 immune responses. Toxicol Appl Pharmacol 222:69–79

Hemphill FE, Kaeberle ML, Buck WB (1971) Lead suppression of mouse resistance to *Salmonella typhimurium*. Science 172:1031–1032

Heo Y, Mondal TK, Gao D, Kasten-Jolly J, Kishikawa H, Lawrence DA (2007) Posttranscriptional inhibition of interferon-gamma by lead. Toxicol Sci 96:92–100

Hogan GR, Adams DP (1979) Lead-induced leukocytosis in female mice. Arch Toxicol 41:295–300

Kasten-Jolly J, Heo Y, Lawrence DA (2010) Impact of developmental lead exposure on splenic factors. Toxicol Appl Pharmacol 247:105–115

Kasten-Jolly J, Heo Y, Lawrence DA (2011) Central nervous system cytokine gene expression: modulation by lead. J Biochem Mol Toxicol 25:41–54

Lawrence DA (1981) Heavy metal modulation of lymphocyte activities. Toxicol Appl Pharmacol 57:439–451

Lawrence DA (1985) Immunotoxicity of heavy metals. In: Dean J (ed) Immunotoxicology and pharmacology. Raven, New York

Lawrence DA, McCabe MJ Jr (1995) Immune modulation by toxic metals. In: Goyer RA, Klaassen CD, Waalkes MP (eds) Metal toxicology. Academic, San Diego

Lee JJ, Battles AH (1994) Lead toxicity via arachidonate signal transduction to growth responses in the splenic macrophage. Environ Res 67:209–219

Mishra KP (2009) Lead exposure and its impact on immune system: a review. Toxicol In Vitro 23:969–972

Queiroz ML, Costa FF, Bincoletto C et al (1994) Engulfment and killing capabilities of neutrophils and phagocytic splenic function in persons occupationally exposed to lead. Int J Immunopharmacol 16:239–244

Struzynska L, Dabrowska-Bouta B, Koza K, Sulkowski G (2007) Inflammation-like glial response in lead-exposed immature rat brain. Toxicol Sci 95:156–162

Tian L, Noelle RJ, Lawrence DA (1995) Activated T cells enhance nitric oxide production by murine splenic macrophages through gp39 and LFA-1. Eur J Immunol 25:306–309

Wedeen RP, Malik DK, Batuman V, Bogden JD (1978) Geographic lead nephropathy: case report. Environ Res 17:409–415

Wedeen RP, Malik DK, Batuman V (1979) Detection and treatment of occupational lead nephropathy. Arch Intern Med 139:53–57

Lead and Phytoremediation

Rupali Datta[1], Emily Geiger[1] and Dibyendu Sarkar[2]
[1]Department of Biological Sciences, Michigan Technological University, Houghton, MI, USA
[2]Earth and Environmental Studies Department, Montclair State University, Montclair, NJ, USA

Synonyms

Botano-remediation; Green remediation; Plant-mediated bioremediation

Definitions

Phytoremediation is a process that utilizes plants, algae, and fungi for the removal, stabilization, transfer, and/or destruction of toxic contaminants in the environment (McCutcheon and Schnoor 2003). Based on the plant physiological processes involved, phytoremediation is classified into (1) phytoextraction – the relocalization of contaminants from the environment into aboveground plant structures, (2) phytostabilization – the immobilization of contaminants within the environment, (3) phytodegradation – the breakdown of contaminants, (4) phytovolatilization – the release of contaminants taken up by plants into the atmosphere, (5) rhizodegradation – the breakdown of contaminants by the local microbial community, and (6) rhizofiltration – the uptake of contaminants from an aquatic environment (McCutcheon and Schnoor 2003).

Principles of Phytoremediation

Phytoremediation has been utilized for treatment of wastewaters for more than 300 years (McCutcheon and Schnoor 2003). Application of phytoremediation techniques in superfund sites started in the late 1990s, for the cleanup of metals and organics. Since then, various new applications for this technique have been developed. Further advances in the field of phytoremediation are being made as researchers study various aspects of the plant-contaminant relationship and the plant mechanisms responsible for the remediation activity. Understanding the physiological and molecular mechanisms involved in the phytoremediation processes will enhance the applicability of this technique to a wider range of contaminated systems. Phytoremediation is highly appealing when compared to other remediation techniques because it is environment-friendly, inexpensive, and applicable to a diverse range of contaminants.

Conceptualizing Phytoremediation of Nonphysiological Metals

The accumulation of nonphysiological metals such as lead (Pb) in the environment is a major public health concern. In human-impacted systems, the primary sources of Pb include industrial activities such as mining operations, vehicle emissions, coal burning, refuse incineration, pesticide applications, and Pb-based paints applied to structural surfaces. In the USA, in order to reduce human exposure to Pb, federal environmental standards for the maximum allowable level of Pb in gasoline were lowered from 0.78 g/L to 0.026 g/L in 1977. In 1976, the allowable level of Pb in residential paint was lowered to 0.06 %. However, a significant number of houses in every city were built prior to implementation of these policies, and Pb-based paint continues to be the principal cause of most of the Pb-poisoning cases in children reported in the USA (Sarkar et al. 2008).

Although many body processes can be severely affected by acute Pb exposure, Pb poisoning is dangerous mainly because chronic exposure to moderate or low levels of Pb affects the developing nervous system of young children. Hence, the USEPA has established a soil Pb hazard cut-off value of 400 mg/kg for soil in play areas of children and a house dust Pb hazard standard of 40 μg/ft^2 for floors (USEPA 2001). Since Pb is highly immobile in soils, concerns about human exposure persist despite the ban on Pb in residential paints approximately 25 years ago. Lead-based paint continues to be the root cause of the majority of the severe Pb-poisoning cases in children reported in the USA (Sarkar et al. 2008), since these paints have very high concentrations of Pb and is found in approximately 38 million pre-1978 homes (Sarkar et al. 2008). One leading option to prevent the risk of Pb toxicity is to reduce the amount of bioavailable Pb in the environment. Phytoremediation is a suitable option for remediation of sites contaminated with Pb, based on the ease, minimal environmental impact, cost, and effectiveness of the process. Plants that are considered suitable for use in phytoremediation have to be fast-growing, have high biomass, have extensive root systems, tolerate a variety of climate and soil conditions, be easy to harvest, and be able to tolerate and accumulate (via detoxification mechanisms) high levels of Pb. Certain plants that are capable of accumulating unusually high levels of heavy metals in their aboveground parts are called hyperaccumulators. In order to be classified as a hyperaccumulator, a plant should be able to accumulate a minimum shoot concentration of 1,000 mg kg^{-1} dry weight of Pb (Sarkar et al. 2008). Plants that have been successfully used for Pb phytoremediation include vetiver (*Chrysopogon zizanioides*), rattlebush (*Sesbania drummondii*), sunflower (*Helianthus annuus*), bent grass (*Agrostis castellana*), and Indian mustard (*Brassica juncea*).

Challenges of Lead Phytoremediation

The efficacy of phytoremediation is primarily dependent upon the properties of the contaminant. Uptake of metals in plants is selective; plants take up some metal ions preferentially over others, as these may serve as micronutrients. The lack of a biological role poses a challenge for the phytoremediation of Pb. The tendency to take up Pb in plant cells is limited. Moreover, Pb is also toxic to plant cells. Another challenge posed by Pb is its lack of soil mobility, due to precipitation of insoluble phosphates, carbonates, and (hydr)oxides, which limits plant availability.

Understanding the Dynamics of the Lead-Plant Relationship at the Molecular Level

The natural mechanisms underlying the ability of certain plants to take up and tolerate nonphysiological metal contaminants from the environment must first be understood before the potential efficacy of the phytoremediation process can be maximized. Research over the past two decades has shown that accumulator plant species possess additional detoxification mechanisms, which enable them to tolerate and accumulate large quantities of nonessential metals.

Phytochelatins

Several plant species possess varying degrees of innate metal tolerance. Studies show that phytochelatins, oligomers of glutathione (GSH), contribute to the ability of several plants to grow in the presence of nonphysiological metals, such as Pb. Phytochelatins (PCs) are made up of three amino acids: Glutamine (Glu), Cysteine (Cys), and Glycine (Gly) with the Glu and Cys residues linked through a γ-carboxylamide bond. A whole range of PCs of varying lengths are found in plants with increasing repetitions of the γ-Glu-Cys dipeptide followed by a terminal Gly. The structure of PCs can be represented as (γ-Glu-Cys)n-Gly, where n is usually in the range of 2–5, but PCs where n is as high as 11 have been reported (Cobbet and Goldsborough 2002). In some plants, the C-terminal glycine can be replaced by serine, glutamine, glutamate, or alanine, forming iso-phytochelatins (iso-PCs). Phytochelatins are produced by a biosynthetic pathway, with glutathione as the substrate. A γ-Glu-Cys moiety of GSH combines with a second GSH molecule to form PC2 by a transpeptidation reaction catalyzed by the enzyme phytochelatin synthase (PCS), which is activated by heavy metal ions (Cobbet and Goldsborough 2002).

Phytochelatins are rapidly synthesized in response to various metals, such as Cd, Zn, Ag, Pb, Cu, Sn, and Au, which are subsequently stored in vacuoles (Cobbet and Goldsborough 2002), enabling plants to tolerate these metals. Phytochelatins have been reported to transport metals from root to shoot, as well as cytosolic PC-metal complexes are transported into vacuoles, where the metal is sequestered away from crucial plant metabolic functions. Most reports that describe the induction of PCs in plants have focused on Cd exposure; only a handful of studies have described the induction of PCs in plants exposed to Pb. Andra et al. (2009) explored the role of PCs in plant Pb tolerance using vetiver grass. Vetiver plants grown hydroponically were exposed to varying concentrations of Pb, and root and shoot tissues were analyzed for PCs using high-performance liquid chromatography electrospray mass spectrometry (HPLC-ES-MS). While no PCs were detected in the absence of added Pb, four types of PCs ($n = 1$–4) were detected in vetiver roots and three types ($n = 1$–3) were detected in vetiver shoot tissues exposed to 1,200 mg/L Pb. Moreover, PC concentrations in the samples were related to the amount of Pb uptake by vetiver plants. The authors also reported the presence of higher-order PCs ($n = 5, 6$) in root but not in shoot tissues, and also higher concentration of Pb in the root tissues.

Previous to the above report, the presence of PC2 and PC3 was documented in aquatic plants exposed to Pb. Mishra et al. (2006) reported the presence of PC2 and PC3 in coontail (*Ceratophyllum demersum* L.) when exposed to Pb. Zhang et al. (2008) reported the presence of PC2, PC3, and PC4 in *Sedum alfredii* plants exposed to Pb, where PCs could be detected in root and stem tissues, but not in leaves, and the PC content was consistent with Pb content of the tissues.

Lead-Induced Protein Expression

Recent proteomic studies have revealed that expression of several proteins in plants is influenced by exposure to metal stress (Walliwalagedera et al. 2010). Changes in the expression of proteins in dwarf sunflowers (*Helianthus annuus*) were investigated after the plants were exposed to several metals including Pb to gain insight into the molecular effects of Pb toxicity in plants. Dwarf sunflowers grown hydroponically were subjected to the "typical" soil environment of industry-polluted areas in Northeast Ohio by exposing them to Cd, Ni and Cr (III) (40 μM of each metal). A set of plants were subjected to Pb (40 μM) in addition to the three above metals. Proteomic analysis of the leaf tissue showed that while several proteins were differentially expressed in the leaves, chitinase, chloroplastic drought-induced stress protein CDSP-34, a thaumatin-like protein, heat-shock cognate 70-1, and the large subunit of ribulose-1,5-bisphosphate carboxylase/oxygenase were substantially up-regulated in the dwarf sunflowers exposed to the three metals and Pb (Walliwalagedera et al. 2010). However, these proteins were not up-regulated in control plants not exposed to metals, or exposed to the three "typical" metals but not Pb, which provides evidence that these proteins have a role in plant defense mechanism under Pb stress.

Antioxidant System Induction

The presence of heavy metals in the soil environment of a living plant causes the production of reactive oxygen species (ROS) within the plants, a phenomenon known as oxidative stress. Plants have developed an antioxidant mechanism, induced in response to oxidative stress, which limits the damaging effects of ROS within the plant. In 2006, a study that directly focused on the induction of the antioxidant system in response to Pb exposure in coontails (*Ceratophyllum demersum*) was conducted (Mishra et al. 2006). The antioxidant enzymes analyzed included catalase, glutathione reductase, and superoxide dismutase (Mishra et al. 2006). Catalase enzymes breakdown hydrogen peroxide, which is produced by mitochondrial peroxisomes in response to a metal stressor. Lead exposure was found to induce catalase activity in coontails (Mishra et al. 2006). The maximum activity of catalase was observed 1 day after exposure to 50 μM Pb. Glutathione reductase enzyme functions to reduce the disulfide bonds in glutathione disulfide, to produce the sulfhydryl form of glutathione, which is a powerful antioxidant. Glutathione reductase activity was the highest when coontail plants were exposed to 25 μM Pb, after 1 day of exposure. In addition, the activity of superoxide dismutase enzyme, which converts superoxide radicals into oxygen and hydrogen peroxide also increased to a maximum after 1 day of exposure of coontail plants to 10 μM Pb (Mishra et al. 2006).

Metallothioneins

In addition to enzymatically produced PCs, some plant species produce metallothioneins (MTs), which are cysteine-rich, metal-binding polypeptides with two domains, that give them the shape of dumbbells (Cobbet and Goldsborough 2002). Metallothioneins are gene-encoded, and are hypothesized to help plants in metal tolerance and homeostasis (Cobbet and Goldsborough 2002). Metallothioneins are classified based on the arrangement of cysteine residues. While Class I MTs contain 20 highly conserved Cys residues and are widespread in vertebrates, MTs without this arrangement of cysteines are referred to as Class II MTs. Plants, fungi, and non-vertebrate animals have Class II MTs (Cobbet and Goldsborough 2002). A large number of MT genes have been identified in plants and have been classified into four categories (Cobbet and Goldsborough 2002). Type 1 MTs contain six motifs of Cys-Xaa-Cys (Xaa being another amino acid) distributed equally among the two domains, which are separated by approximately 40 amino acids, including aromatic amino acids. The presence of the long spacer is a typical characteristic of plant MTs. Most other organisms have spacers that are less than 10 amino acids long, and contain no aromatic amino acids.

Type 2 MTs are very similar to type 1 MTs in that they also contain two cysteine-rich domains separated by a 40 amino acid long spacer, but the spacer has variations between species. The difference lies in the fact that the first pair of cysteines is present as a Cys-Cys motif in positions 3 and 4, and a Cys-Gly-Gly-Cys motif is present at the end of the N-terminal cysteine-rich domain. The sequences of the N-terminal domain of Type 2 MTs are highly conserved, with a sequence of Met-Ser-Cys-Cys-Gly-Gly-Asn-Cys-Gly-Cys-Ser, and the C-terminal domain contains three Cys-Xaa-Cys motifs (Cobbet and Goldsborough 2002).

Type 3 MTs contain a highly conserved motif Gln-Cys-Xaa-Lys-Lys-Gly at the N-terminal domain. The N-terminal domain also consists of four Cys residues, with a consensus sequence of Cys-Gly-Asn-Cys-Asp-Cys. The fourth cysteine is a part of the conserved motif mentioned earlier. At the C-terminal end, six Cys residues are arranged in Cys-Xaa-Cys motifs. Similar to Type 1 and Type 2 plant MTs, the two domains are separated from each other by approximately 40 amino acid residues (Cobbet and Goldsborough 2002).

Type 4 MTs have three cysteine-rich domains, each with five or six conserved cysteine residues, separated by 10–15 residues. Most of the cysteines are present as Cys-Xaa-Cys motifs. Not many Type 4 MTs have been reported, but among the known Type 4 MTs, Type 4 MTs from dicots contain additional 8–10 amino acids in the N-terminal domain before the first cysteine residue, in contrast to those from monocots (Cobbet and Goldsborough 2002).

The differential expression of MTs in the presence or absence of metals hints at their metal-managing role. However, the specific roles of each of the MTs identified in plants are still unclear. Studies have been successful in identifying MTs involved in exposure to excess copper (Guo et al. 2008). Metallothionein gene expression has been reported to be strongly induced mainly by Cu, and to a lesser extent by Cd and Zn in rice, *Arabidopsis* and the hyperaccumulators, *Thlaspi caerulescens*, *Silene vulgaris*, and *Silene paradoxa* (Cobbet and Goldsborough 2002). Guo et al. (2008) studied the heterologous expression of *Arabidopsis* MT genes in Cu- and Zn-sensitive yeast mutants *Dcup1* and *Dzrc1 Dcot1* to determine their role as metal chelators. Six *Arabidopsis* MT genes (*MT1a, MT2a, MT2b, MT3, MT4a*, and *MT4b*) were expressed in the yeast mutants. Guo et al. (2008) reported that all four types of *Arabidopsis* MTs imparted similar levels of copper tolerance and accumulation in the *Dcup1* yeast mutant. Moreover, type-4 MTs (*MT4a* and *MT4b*) conferred greater Zn tolerance and higher accumulation of Zn than other MTs to the *Dzrc1 Dcot1* mutant.

Moreover, *Arabidopsis* plants lacking *MT1a* showed a 30% decrease in copper accumulation in roots of plants exposed to 30 mM $CuSO_4$. Ectopic expression of *MT1a* RNA in the mutant restored copper accumulation in roots. Interestingly, when MT deficiency was combined with PC deficiency, growth of the *mt1a-2 mt2b-1 cad1-3* triple mutant was more sensitive to Cu and Cd compared to the *cad1-3* mutant (Guo et al. 2008). Together these results show that MTs are involved in metal homeostasis in plants, and that MTs and PCs function cooperatively to protect plants from Cu and Cd toxicity. However, no such study has been reported so far in the case of Pb.

Localization of Lead in Plants

Knowledge of distribution of Pb in accumulating plant structures is useful in understanding the biochemical mechanisms, which enable it to accumulate Pb. Sahi et al. (2002) studied Pb accumulation and localization in *Sesbania drummondii* treated with 1,000 mg/L Pb $(NO_3)_2$. The Scanning Electron Microscopy (SEM) of root tissue showed concentric circles of Pb spots in the central region of vascular bundles, on the surface of root epidermis, in the parenchyma cells of the cortical region and outside the endodermis. Lead deposits in the form of clumps or sheets were observed in the basal part of the stem. Lower amounts of deposits were also observed in upper part of the stem and leaf. Sahi et al. (2002) also reported that microanalysis spectra through root section show a decreasing gradient of Pb from epidermis to the central axis, which indicates both apoplastic and symplastic modes of transport. Transmission electron microscopy and X-ray microanalysis of root sections demonstrated the localization of Pb granules in the plasma membrane, cell wall, and vacuoles (Sahi et al. 2002).

Analyzing cross-sections of the root, stem, and leaves of vetiver grass using SEM revealed pattered deposits of Pb in plants exposed to 1,200 mg L^{-1} of Pb (Andra et al. 2009). The roots had Pb deposits in the endodermis and cortex. The vascular bundles of the vetiver stems also contained Pb deposits. The leaves showed Pb deposits in its mesophyll cells. The specific locations of deposited Pb in the plant indicated the transport mechanism of Pb to be apoplastic as well as symplastic (Andra et al. 2009).

Transgenic Plants

The process of successful phytoremediation involves numerous genes for the tolerance and accumulation of heavy metals, including PC, MT, transport proteins, the antioxidant system, etc. Knowing the genes that are directly responsible for enhancing the uptake, translocation and accumulation potential of a plant species opens doors for the use of genetically engineered plants in phytoremediation. Once the genes involved in metal hyperaccumulation are identified, transferring these genes into other plants may be the way to create the ideal plant for phytoremediation (Meharg 2005).

Lead hyperaccumulator, *Brassica juncea*, has been genetically engineered to produce higher amounts of PCs by overexpressing genes for the PCS enzyme (Meharg 2005). The genetically engineered *B. juncea* takes up more Pb than its natural counterpart. Another possibility of genetically manipulating a plant species for hyperaccumulation of Pb is to overexpress the enzymes of the antioxidant system to increase the threshold for adverse effects due to oxidative stress. Genes that limit biomass have also been genetically engineered in hyperaccumulators to allow for more metal uptake (Meharg 2005).

Future Research

Although several Pb hyperaccumulators have been identified, the mechanism of Pb tolerance and accumulation in plants is not yet fully understood. Several recent studies have elucidated the roles of metal transporters, PC, and MT genes in Pb hyperaccumulation. The developing "omics" technology is expected to give us unprecedented insight into the genetic and biochemical basis of Pb hyperaccumulation in plants. The construction of transgenic mutants to overexpress some of these proteins may result in developing plants with improved phytoremediation capability for the cleanup of Pb from the environment.

Cross-References

▶ Lead Detoxification Systems in Plants
▶ Lead in Plants
▶ Metallothioneins and Lead

References

Andra S, Datta R, Sarkar D, Makris K et al (2009) Induction of lead binding phytochelatins in vetiver grass [*Vetiveria zizanioides (L.)*]. J Environ Qual 38:868–877

Cobbet C, Goldsborough P (2002) Phytochelatin and metallothioneins: roles in heavy metal detoxification and homeostasis. Annu Rev Plant Biol 53:159–182

Guo W-J, Meetam M, Goldsbrough PB (2008) Examining the specific contributions of individual *Arabidopsis* metallothioneins to copper distribution and metal tolerance. Plant Physiol 146:1697–1706

McCutcheon SC, Schnoor JL (2003) Overview of phytotransformation and control of wastes. In: McCutcheon SC, Schnoor JL (eds) Phytoremediation: transformation and control of contaminants. Wiley, New York, pp 53–58

Meharg AA (2005) Mechanisms of plant resistance to metal and metalloid ions and potential biotechnological applications. Plant Soil 274:163–174

Mishra S, Srivastava S, Tripathi RD et al (2006) Lead detoxification by coontail (*Ceratophyllum demersum L.*) involves induction of phytochelatins and antioxidant system in response to its accumulation. Chemosphere 65:1027–1039

Sahi SV, Bryant NL, Sharma NC et al (2002) Characterization of a lead hyperaccumulator shrub, *Sesbania drummondii*. Environ Sci Technol 36:4676–4680

Sarkar D, Andra S, Saminathan S et al (2008) Chelant-aided enhancement of lead mobilization in residential soils. Environ Pollut 156:1139–1148

U.S. EPA (2001) Lead: identification of dangerous levels of lead; Final Rule. 40CFR745, Fed Regist, 66, pp 6763–6765

Walliwalagedera C, Atkinson I, van Keulen H et al (2010) Differential expression of proteins induced by lead in dwarf sunflower *Helianthus annuus*. Phytochem 71:1460–1465

Zhang Z, Gao X, Qiu B (2008) Detection of phytochelatins in the hyperaccumulator *Sedum alfredii* exposed to cadmium and lead. Phytochem 69:911–918

Lead and RNA

Charles G. Hoogstraten and Minako Sumita
Department of Biochemistry and Molecular Biology, Michigan State University, East Lansing, MI, USA

Synonyms

Lead binding to RNA; Lead-RNA interactions; Leadzyme; Pb(II)-RNA coordination; RNA structure probing; Transfer RNA; tRNA

Definitions

Divalent lead ions, Pb(II), have no known role in the normal functioning of any organism. Nonetheless, they have played a disproportionately prominent role in laboratory studies of RNA since detailed studies of the role of the latter molecule in molecular biology began. This role arises from the unusually low pK_a value of waters of hydration bound to Pb(II), which makes them highly chemically active in the nonspecific degradation of RNA. Lead(II) hydroxide acts via deprotonation or activation of the 2′-hydroxyl group followed by nucleophilic attack of the activated oxygen on the adjacent phosphorus atom to yield a 2′,3′-cyclic phosphate product with concomitant strand cleavage. This reaction has led to the use of Pb(II) as a biochemical probe for the presence of secondary structure in RNA strands, as well as for electrostatic pockets with specific affinity for multivalent cations. A chemically and structurally specific Pb(II) cleavage site in the well-studied phenylalanine tRNA from yeast led to the derivation of the small lead-activated ribozyme, or leadzyme, motif, which has become an important model system for mechanistic studies of the small nucleolytic class of catalytic RNA molecules.

Degradation of Single-Stranded RNA by Pb(II)

Early studies of the degradation of RNA by various metal species identified a strong correlation between the pK_a of water molecules in the first hydration sphere of the metal ion being used and the activity of the ion in RNA degradation. In addition, such degradation was observed to be strongly pH dependent (Farkas 1968; Pan et al. 1993). These observations implicated a mechanism in which the metal hydroxide either acts as a general base by directly deprotonating a group on the RNA, presumably the ribose 2′-hydroxyl that distinguishes RNA from the more chemically stable DNA, or coordinates to and activates that group as a nucleophile. Hydrated Pb(II), with its unusually low pK_a of 7.7, has a high fraction of hydroxide ions present in its first coordination sphere and is particularly active in nonspecific depolymerization of RNA. In addition, the rate of RNA depolymerization by Pb(II) depends on the secondary structure, with double-stranded hybrids (e.g., poly-A hybridized with poly-U) substantially protected from degradation compared to sequences in which double-stranded structures cannot form (Farkas 1968).

In combination with methods for the sequence-specific identification of cleavage sites in RNA molecules of defined sequence, the latter observation opened the possibility for using Pb(II) ions as a chemical probe for the structure and dynamics of complex RNA sequences. Although the complicated aqueous chemistry of lead poses some experimental pitfalls for the unwary, the experiments are in concept simple: The RNA of interest is end-labeled and incubated with appropriate concentrations of Pb(II) under folding conditions for a time consistent with an average of one or fewer cleavages per molecule, and cleavage sites are visualized and quantitated using gel electrophoresis (Pan 2000). In single-stranded regions, a well-established background rate of Pb(II)-catalyzed hydrolysis will be observed within a very reasonable experimental time. By contrast, in RNA regions that have been protected by the formation of secondary or tertiary structure (or, in binding experiments, by the presence of proteins or other ligands), substantial decreases in the rate of hydrolysis can be observed. An efficient mapping of the presence or absence of RNA structure, or a footprint of a binding partner, can be obtained. In some experiments, the question is instead the presence and location of specific metal-ion binding sites formed by electrostatic pockets arising from multiple backbone phosphate groups. These pockets can be indicated by the enhancement of Pb(II)-catalyzed cleavage above the background rate at specific phosphodiester bonds (Pan 2000). As discussed below in the case of tRNA, however, not all metal-binding pockets will give rise to such enhancements, since structured regions may constrain the backbone away from the geometry necessary for cleavage.

A wide variety of chemical and enzymatic reagents and protocols are now available for more targeted probing of features of RNA secondary and tertiary structure. Pb(II) probing still finds use for the purpose, however, because it allows a rapid global assessment of the structural features present with readily available reagents and straightforward protocols, and because of the possibility of simultaneously probing for structural order and for individual metal-binding pockets.

Pb(II) Binding and Cleavage of Phenylalanine tRNA: Structural Analysis of Metal Sites

Various species of transfer RNA (tRNA) were early paradigms for analysis of the three-dimensional structure of RNA, and tRNA continues to be a rich lode of information on the stereochemistry, energetics, interactions, and chemical modifications of RNA in general. A variety of heavy-metal derivatives including Pb (II) were used to solve the phase problem of crystallographic analysis in these systems. In a comprehensive analysis of yeast phenylanine tRNA ($tRNA^{Phe}$) by Klug and coworkers (Brown et al. 1985, and references therein), three well-structured Pb(II) sites were observed. The lead ions showed close contacts consistent with direct coordination to RNA groups including nitrogen lone pairs and carbonyl oxygen atoms on pyrimidine and purine bases. Interestingly, all contacts with negatively charged backbone phosphate groups were at longer range and presumably mediated by hydration waters, although the resolution of this crystal structure was insufficient to directly reveal such water-mediated interactions. Figure 1 shows one lead ion and its surrounding groups from the pH 5.0 (chemically intact) structure, including apparent direct contacts to groups on uridine 59 and cytosine 60.

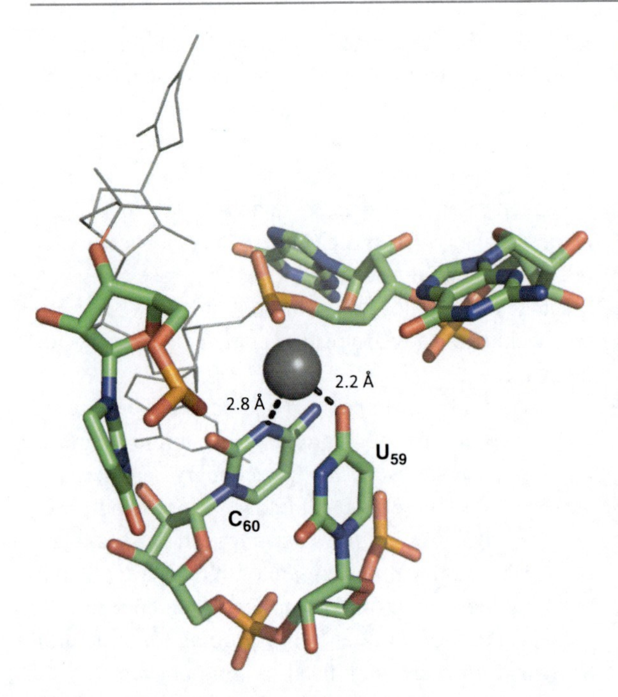

Lead and RNA, Fig. 1 The TΨC-loop binding site adjacent to the D17/G18 backbone cleavage site in yeast tRNA^{Phe} observed by Brown et al. (1985) at pH 5.0, PDB code 1TN2. The two atoms sufficiently close to the ion to represent inner-sphere ligands, N3 of Cyt-60 and O4 of Uri-59, are indicated (Figure prepared with PyMol)

The Pb(II) site shown in Fig. 1 is adjacent to a previously known site of enhanced Pb(II)-catalyzed hydrolysis between nucleotides 17 and 18, with the ion positioned approximately 6.0 Å from the nucleophilic 2′-hydroxyl on dihydrouridine-17. Further, in crystallographic experiments at pH 7.4, a loss of electron density corresponding to strand cleavage at that site was observed, and biochemical analysis of the resulting products revealed a 2′,3′-cyclic phosphate structure consistent with Pb(II)-activated cleavage (Brown et al. 1985). Thus, this ion in fact represents a binding geometry that leads to efficient cleavage at a nearby phosphodiester bond in solution. The other two Pb(II) ions in the tRNA^{Phe} structure show no such association with sites of cleavage. The cleavage-activating ion is located in the TΨC loop near the hinge region of the L-shaped tRNA molecule in a pocket of negative electrostatic potential created by numerous nearby phosphate groups. The combination of this potential pocket with the available Lewis-base coordination groups on Uri-59 and Cyt-60 serves to position Pb(II) in an appropriate location to activate the nucleophile for the geometrically favorable backbone cleavage at the 17–18 phosphodiester linkage.

In Vitro Selection Analysis of the Determinants of Specific Pb(II) Cleavage: The Lead-Dependent Ribozyme

Precise structural and biochemical analysis of the specific Pb-cleavage site in tRNA^{Phe} (see above) provoked the question of the sequence and structural determinants of the reaction, i.e., which of the surrounding nucleotides were obligatory for formation of the specific and reactive Pb(II) coordination site. In an attempt at an unbiased exploration of this question, the Uhlenbeck group applied the technology of RNA in vitro selection. In this technique, a fully or completely randomized library of up to 10^{14} DNA sequences is constructed and exposed to repeated cycles of transcription to RNA, selection for function (either binding to a particular target or catalysis of a given self-modifying reaction), reverse transcription to DNA, and amplification using PCR. In vitro selection can thus be considered either a technique of combinatorial chemistry applied to polynucleotides or a method of laboratory evolution with selection for a particular function, depending on the philosophy of the experimenter. In this case, DNA coding for the yeast tRNA^{Phe} sequence was randomized at either nine or ten nucleotides adjoining the cleavage-site lead ion binding pocket and exposed to selection for self-cleavage activity in the presence of divalent lead (Pan and Uhlenbeck 1992). The goal was to identify mutant tRNA sequences that maintained the Pb(II)-activated cleavage site. This experiment was among the very first applications of in vitro selection to the derivation of RNA catalytic activities, which has since become a major technique for exploring possible features of the putative "RNA World."

Unexpectedly, the selection resulted in the derivation of a small RNA motif that underwent highly site-specific lead-activated self-cleavage at a rate substantially higher than the parent tRNA sequence but showed no relationship to the tRNA secondary or tertiary structure (Pan and Uhlenbeck 1992). This motif, dubbed the lead-dependent ribozyme or leadzyme (Fig. 2), results in an intermediate 2′,3′-cyclic phosphate product followed by further

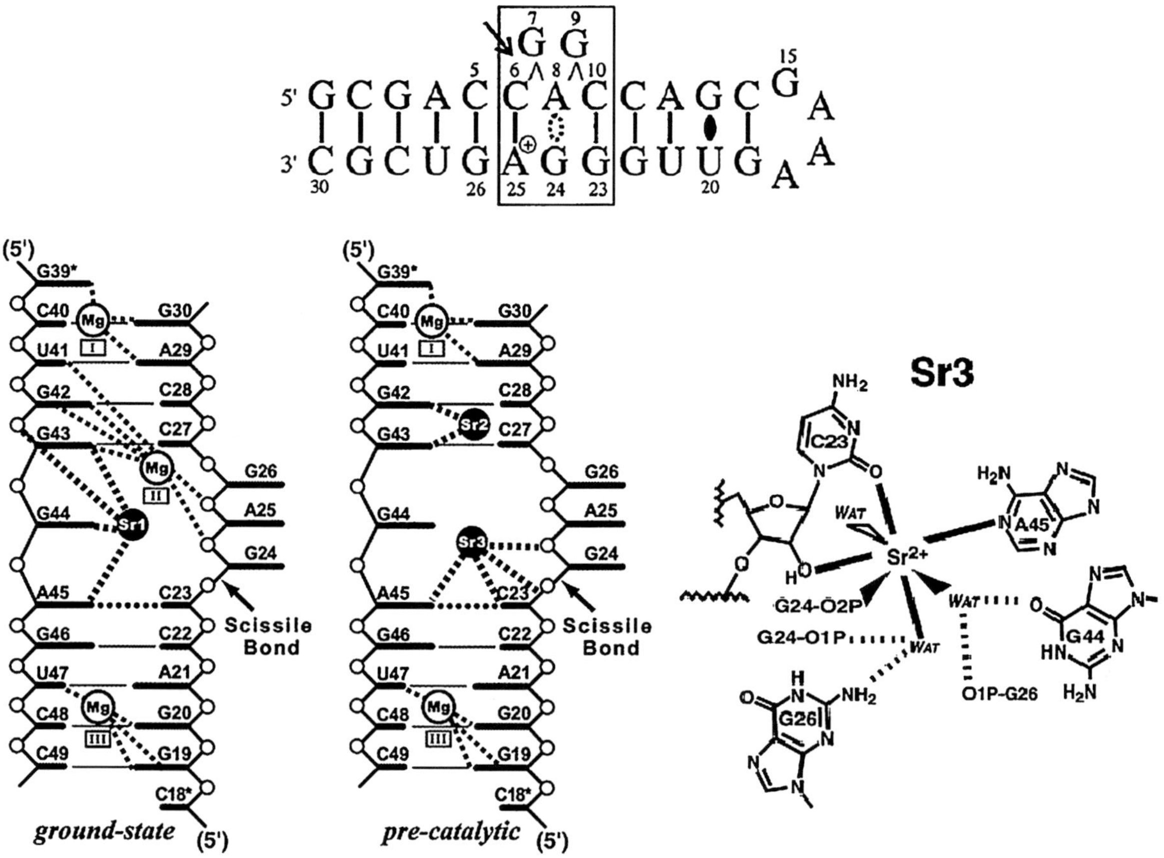

Lead and RNA, Fig. 2 The lead-dependent ribozyme. (*Top*) The leadzyme motif drawn as construct LZ2 from Pan and Uhlenbeck (1992) with structural features observed using solution NMR by Hoogstraten et al. (1998). The site of lead-activated cleavage is indicated with an *arrow*. In known active sequences, position 6 is always C and position 9 is always G, whereas at least some variation is tolerated at the other four sites within the internal loop. (*Bottom left*) Schematic structures of two crystallographically observed leadzyme molecules showing the binding of Sr^{2+} (a model ion for divalent lead) near the cleavage site. The "pre-catalytic" state more closely resembles the presumed catalytically active conformation than does the "ground-state." In comparison with the secondary structure shown at top, residues Cyt-23, Gua-24, Ade-25, Gua-26, Gua-44, and Ade-45 correspond to Cyt-6, Gua-7, Ade-8, Gua-9, Gua-24, and Ade-25, respectively. The Sr1 and Sr2 ions are fully hydrated in the crystal and take no inner-sphere ligands from RNA groups. (*Bottom right*) Detailed view of the coordination structure of Sr3 in the pre-catalytic state. *Bottom left* and *bottom right* figures (Reprinted with permission from Wedekind and McKay (2003), copyright 2003 American Chemical Society)

hydrolysis to a 3′-phosphate species and displays high specificity for Pb(II) as a cofactor. Divalent magnesium, the biological cofactor for naturally occurring self-cleaving ribozymes such as the hairpin and hammerhead motifs and the ribozyme from the human hepatitis delta virus, is actually a competitive inhibitor of the leadzyme. In short, the minimal sequence and structural requirements for site-specific lead-activated self-cleavage in RNA are much simpler than implied by the $tRNA^{Phe}$ structure, consisting only of a partially degenerate six-nucleotide asymmetric internal loop flanked by Watson-Crick double helices whose exact sequence is largely immaterial to catalytic function (Pan and Uhlenbeck 1992).

The prospect of a small motif that recapitulated many properties of both Pb(II)-activated RNA cleavage and naturally occurring self-cleaving ribozymes has attracted much attention to the leadzyme from chemists, biochemists, structural biologists, and calculational modelers (reviewed in Wedekind 2011). Structures of the leadzyme derived using solution NMR (Hoogstraten et al. 1998) and x-ray crystallography (Wedekind and McKay 2003 and references therein) show similar conformations in which the shorter strand of the internal loop closely resembles one strand of an A-form double helix, stabilized by a noncanonical base pair between Cyt-6 and Ade-25. NMR data provided strong support for a protonated adenosine at position 25, with the elevated pK_a of 6.5 of the imino nitrogen of the adenosine ring attributable to stabilization of the charged state via formation of a wobble base pair with Cyt-6. The remaining residues on the long strand of the internal loop display substantial conformational dynamics and, in the x-ray structure, engage in intermolecular base pairing with another leadzyme molecule within the crystallographic unit cell.

A refined (1.8 Å resolution) crystal structure determination of the leadzyme (Wedekind and McKay 2003) used Sr^{2+} ions, with an ionic radius nearly identical to that of Pb(II), as a structural mimic for lead cofactor binding. Interestingly, of the two different conformations of the leadzyme present in the crystal, it was the structure that most closely resembled the presumed catalytically active state (see below) that displayed a Sr^{2+} ion, denoted Sr3, bound directly at the self-cleavage site (Fig. 2, bottom). This ion showed direct coordination to multiple groups on the RNA, including carbonyl oxygen and ring nitrogen groups of the type observed in $tRNA^{Phe}$ as well as a single nonbridging phosphate oxygen at the cleavage site. The Sr^{2+} is also 3.8 Å from the nucleophilic 2′-hydroxyl of Cyt-6 (23 in the crystallographic numbering), consistent with the leadzyme in this conformation adopting a preformed metal-binding pocket structure capable of binding a Pb(II) ion in good position to catalyze backbone cleavage at the observed specific site. In support of the presence of a preformed metal-binding pocket, NMR titrations of various metal ions showed only minor perturbations of spectra obtained in the absence of divalent cations. The NMR solution structure itself, which was solved in the absence of divalent cations, showed general (although not perfect) agreement with that obtained in the crystal form.

One question left unresolved in this model is the role of the imino nitrogen on Ade-25 (45 in the crystallographic numbering). This nitrogen is one of the inner-sphere ligands that act to position Sr3 at the cleavage site in the "pre-catalytic" crystallographic structure (Fig. 2). Nevertheless, NMR data strongly indicate that this particular nitrogen is protonated and hydrogen bonded to the base of Cyt-6 at pH values below an observed pK_a of 6.5 in solution, apparently disrupting the Sr3 site, and pH-dependent kinetic studies of this system have displayed no evidence for a competition between protonation at this site and binding of the Pb(II) cofactor. In addition, an adenosine is not required at this site for function, and in fact an abasic substitution (a bare ribose with no nucleotide base present at all) at position 25 supports full leadzyme activity (Wedekind 2011). Some differences between the alkaline-earth ion Sr^{2+} and the relatively "soft" transition metal ion Pb(II) would not be unexpected; indeed, in comparative studies in tRNA, Pb(II) has shown a lesser tendency to coordinate to phosphate groups than does the alkaline-earth Mg^{2+}, even when bound at the same position (Feig and Uhlenbeck 1999). The precise structure of the leadzyme metal-binding pocket upon nucleophilic activation and transition-state approach, therefore, is unclear.

Conformational Dynamics and Metalloribozyme Catalysis

One overall goal of the detailed structural studies of the leadzyme was the mapping of the structural features that produced a lead-binding site and consequent backbone hydrolysis in comparison with the Mg^{2+}-activated sites in various naturally occurring ribozymes. The structures obtained, however, immediately posed a difficulty from this perspective in that a serious mismatch was observed between the NMR and crystallographic results and the extensive functional data in the system. Many apparently structurally critical nucleotide functional groups were not conserved in functional variants, whereas some absolutely conserved groups (notably Gua-9) appeared to be entirely irrelevant structurally. Further, the reactive groups at the scissile phosphodiester bond were in a conformation greatly differing from the "in-line attack" configuration consistent with reaction products, implying a substantial conformational rearrangement before a direct approach to the chemical transition state would be possible (Fig. 3a). Such inconsistencies may be best explained by an equilibrium between at least two conformations or ensembles of related conformations in the leadzyme, a "ground-state" (ES) form observed by structural studies but catalytically inactive, and a low-occupancy minor conformer ES* from which products may be formed (Fig. 3b).

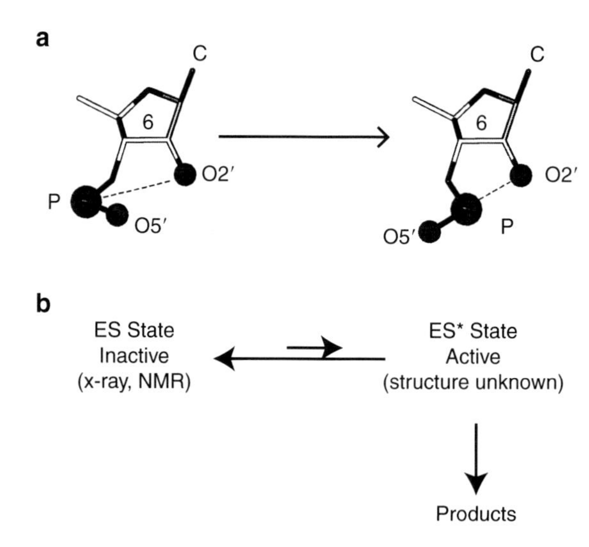

Lead and RNA, Fig. 3 (**a**) Comparison of the atomic configuration of the active-site nucleophile (O2′), reaction center (P), and leaving group (O5′) in the leadzyme in the NMR-derived solution structure (*left*) with that necessary for the in-line attack reaction that is necessary to yield the observed products (*right*) (Reprinted from Hoogstraten et al. (1998) with permission from Elsevier). (**b**) Kinetic scheme for the role of conformational dynamics in leadzyme catalysis, with only the minor conformer ES* shown as capable of reaching the chemical transition state and proceeding to products. The in-line attack configuration at *top right* is presumed to be a feature of ES*

Extensive calculational modeling studies in the leadzyme, reviewed by Wedekind (2011), resulted in models greatly diverging from experimental structures precisely because they took as input functional data that probed the ES* state rather than ES. This impression was strengthened by the observation that several active-site loop substitutions of chemically modified, conformationally restricted nucleotides designed to perturb the ground-state conformation in fact led to accelerated lead-activated hydrolysis, apparently by displacing the ES-ES* equilibrium of Fig. 3b to the right (Yajima et al. 2007; Julien et al. 2008). An implication of this model is that conversion to ES* could be thermodynamically linked to Pb(II) binding. In other words, detailed analysis of ES* using techniques such as functional probing, spectroscopic "invisible-state" analysis, and calculational modeling would be necessary to completely dissect the mode of catalytically active metal binding in the leadzyme. Efforts along these lines continue in several laboratories (Wedekind 2011).

The coupling between conformational dynamics and enzymatic function, as illustrated by the conformational-restriction studies mentioned above, is itself of no little interest. In both the hammerhead and hairpin ribozyme motifs, similar mismatches between structural and functional data were originally observed when structures were solved of the minimal catalytic core (in the hammerhead) or of one of two required internal loops (hairpin), but much better agreement was obtained when larger structures were solved that included additional motifs required for full tertiary structure formation. In other words, fluctuations to the active conformation similar to those schematized for the leadzyme in Fig. 3 are "captured" by the formation of additional structural interactions in these systems, allowing catalysis to occur with improved efficiency. In the leadzyme, no such larger structural context is present, and the activated conformation is only captured by autocatalytic cleavage itself.

A Proposed Role for RNA Degradation in Lead Toxicity

Since the unusually high activity of divalent lead in the depolymerization of RNA was first observed, there has been speculation about the potential role of this reaction in lead toxicity (Farkas 1968). More recently, the availability of large amounts of sequence data has enabled bioinformatic searches of human and other genomic sequences for the relatively well-characterized specific leadzyme motif, with at least one claim made for the potential presence of numerous active leadzyme sequences in human mRNAs (Barciszewska et al. 2005). Numerous difficulties plague such searches, including the difficulty of assessing whether a given RNA sequence is sequestered within the substantial amounts of secondary structure in messages rather than being exposed to solution reagents, as well as whether the discontinuous sequence requirements assemble properly in space. It should also be remembered that, even when engineered for full activity within in vitro constructs, the leadzyme motif results in Pb(II)-induced cleavage rates only approximately 1,000-fold higher than background (nonspecific) cleavage of single-stranded RNA; partially active leadzyme sequences would have to be common indeed to contribute substantially to any overall rate of cellular Pb(II)-induced RNA degradation. Perhaps more fundamentally, there is limited information available about the chemical state of Pb(II) within cells. The high levels of zinc- and calcium-binding proteins in most cells, including some with sulfur-rich metal sites that bind lead quite tightly, make it likely that much of the Pb(II) in a typical cell is bound up by proteins and/or small metabolites and unavailable for the relatively low-affinity sites on nucleic acids (Godwin 2001). As is described elsewhere in this volume, the free concentrations of many different metal ions in cells are much lower than their total concentrations as assessed by techniques such as atomic absorption would indicate. In short, although the potency of Pb(II) in both the nonspecific and site-specific degradation of RNA in vitro renders such degradation a chemically plausible mechanism for the biological effects of the ion, no convincing evidence that this effect is a significant contributor to physiologically relevant toxicity has appeared.

Conclusion

The unique physical and chemical properties of Pb(II) lead it to play a unique role in studies of RNA structure and function. Notably, the unusually low pK_a of Pb(II) waters of hydration tends to concentrate hydroxide ions near polyanionic RNA and to greatly accelerate both nonspecific and site-specific RNA hydrolysis reactions. The highly "soft" nature and flexible coordination sphere of lead result in a consistent pattern, within the limited high-resolution structural information available, of ionic binding within negative electrostatic pockets in folded RNA structures with direct coordination to nucleotide base groups rather than to negatively charged phosphate groups. Study of lead-activated RNA self-cleavage in the leadzyme has been complicated by the role of conformational dynamics, although interesting comparisons to the mechanisms of Mg^{2+}-activated ribozymes such as the hammerhead and hepatitis delta motifs can be made. Nevertheless, the interconnections among conformational dynamics, metal ion binding and catalytic function in this small and tractable system make the leadzyme a valuable model for RNA biophysical studies. The possible relevance of leadzyme-like mRNA motifs and/or nonspecific RNA cleavage by lead(II) hydroxide to lead toxicity in humans is also of continuing interest to researchers, although much remains to be done to establish RNA as an important target for the well-known deleterious effects of lead in vivo.

Cross-References

► Barium and Protein–RNA Interactions
► Chromium Binding to DNA
► DNA-Platinum Complexes, Novel Enzymatic Properties
► Lanthanides in Nucleic Acid Analysis
► Lead and Immune Function
► Lead, Physical and Chemical Properties
► Magnesium
► Mercury and DNA
► Metallothioneins and Lead
► Scandium, Interactions with Nucleotide and Nucleic Acids
► Strontium and DNA Aptamer Folding
► Zinc, Metallated DNA-Protein Crosslinks as Finger Conformation and Reactivity Probes

References

Barciszewska MZ, Szymanski M, Wyszko E, Pas J, Rychlewski L, Barciszewski J (2005) Lead toxicity through the leadzyme. Mutat Res 589:103–110

Brown RS, Dewan JC, Klug A (1985) Crystallographic and biochemical investigation of the lead(II)-catalyzed hydrolysis of yeast phenylalanine tRNA. Biochemistry 24:4785–4801

Farkas WR (1968) Depolymerization of ribonucleic acid by plumbous ion. Biochim Biophys Acta 155:401–409

Feig AL, Uhlenbeck OC (1999) The role of metal ions in RNA biochemistry. In: Gesteland RF, Cech TR, Atkins JF (eds) The RNA world, 2nd edn. Cold Spring Harbor Laboratory Press, Cold Spring Harbor

Godwin HA (2001) The biological chemistry of lead. Curr Opin Chem Biol 5:223–227

Hoogstraten CG, Legault P, Pardi A (1998) NMR solution structure of the lead-dependent ribozyme: evidence for dynamics in RNA catalysis. J Mol Biol 284:337–350

Julien KR, Sumita M, Chen PH, Laird-Offringa IA, Hoogstraten CG (2008) Conformationally restricted nucleotides as a probe of structure-function relationships in RNA. RNA 14:1632–1643

Pan T (2000) Probing RNA structure by lead cleavage. Curr Protoc Nucleic Acid Chem. Chapter 6:6.3.1–6.3.9

Pan T, Uhlenbeck OC (1992) A small metalloribozyme with a two-step mechanism. Nature 358:560–563

Pan T, Long DM, Uhlenbeck OC (1993) Divalent metal ions in RNA folding and catalysis. In: Gesteland RF, Atkins JF (eds)

The RNA world, 1st edn. Cold Spring Harbor Laboratory Press, Cold Spring Harbor
Wedekind JE (2011) Metal ion binding and function in natural and artificial small RNA enzymes from a structural perspective. Metal Ions Life Sci 9:299–345
Wedekind JE, McKay DB (2003) Crystal structure of the leadzyme at 1.8 Å resolution: metal ion binding and the implications for catalytic mechanism and allo site ion regulation. Biochemistry 42:9554–9563
Yajima R, Proctor DJ, Kierzek R, Kierzek E, Bevilacqua PC (2007) A conformationally restricted guanosine analog reveals the catalytic relevance of three structures of an RNA enzyme. Chem Biol 14:23–30

Lead Binding to RNA

▶ Lead and RNA

Lead Detoxification Mechanism

▶ Lead Detoxification Systems in Plants

Lead Detoxification Systems in Plants

Dharmendra K. Gupta[1] and Lingli Lu[2]
[1]Departamento de Bioquímica, Biología Celular y Molecular de Plantas, Estación Experimental del Zaidin, CSIC, Granada, Spain
[2]MOE Key Laboratory of Environment Remediation and Ecological Health, College of Environmental & Resource Science, Zhejiang University, Hangzhou, China

Synonyms

Lead detoxification mechanism; Lead remediation form environment; Lead toxicity

Definition

Lead (Pb) is a soft, blue-gray metal that is mined from the earth's crust and used for many industrial purposes for centuries and that was widely used in paint and gasoline. Lead is present in all parts of the environment, including inside homes. Most people, especially children, who suffer from lead poisoning, are exposed through lead-contaminated household dust or soil that gets into their mouths. Some people may be exposed to lead through working with or near lead industries. Lead poisoning is well known as plumbism.

Lead Contamination in the Environment

Heavy metal pollution is a widespread and important environmental concern. Lead is one of the most serious pollutants that affect plant productivity and causes health hazards to animals and in humans (Saifullah et al. 2009). Metals are introduced in the environment through natural means and anthropogenically. Industry and agriculture contributed considerably to the enhanced concentrations of heavy metals via waste disposal, smelter stacks, atmospheric deposition, fertilizer and pesticide use, and the application of sewage sludge in arable land (Saifullah et al. 2009); see Fig. 1. Many studies described that the food chain is the major pathway of Pb transfer from the environment to humans. It is well documented that many factors will affect their transfer through food chains (Saifullah et al. 2009). The transfer of metals from soils to plants depends on three factors: the total concentration (amount factor), the bioavailability of elements in the soil (intensity factor), and the rate of element transfer from solid to liquid phases and to plant roots. Excess metals in the contaminated soil were transferred to grasses and plants and, thus, to the food chain for cattle reared in the polluted areas. Once Pb contaminates the soil, it is available in the soil for a long time, and such soils, if ingested with food crops, may be a significant source of Pb toxicity to both humans and grazing animals (Saifullah et al. 2009).

Lead Toxicity in Plant and Humans

Excess Pb causes a number of toxicity symptoms in plants, e.g., stunted growth, chlorosis, and blackening of root system. Pb inhibits photosynthesis, upsets mineral nutrition and water balance, changes hormonal status, and affects membrane structure and permeability (Gupta et al. 2010). Pb caused changes in chloroplast ultrastructure inhibition of leaf/root growth and

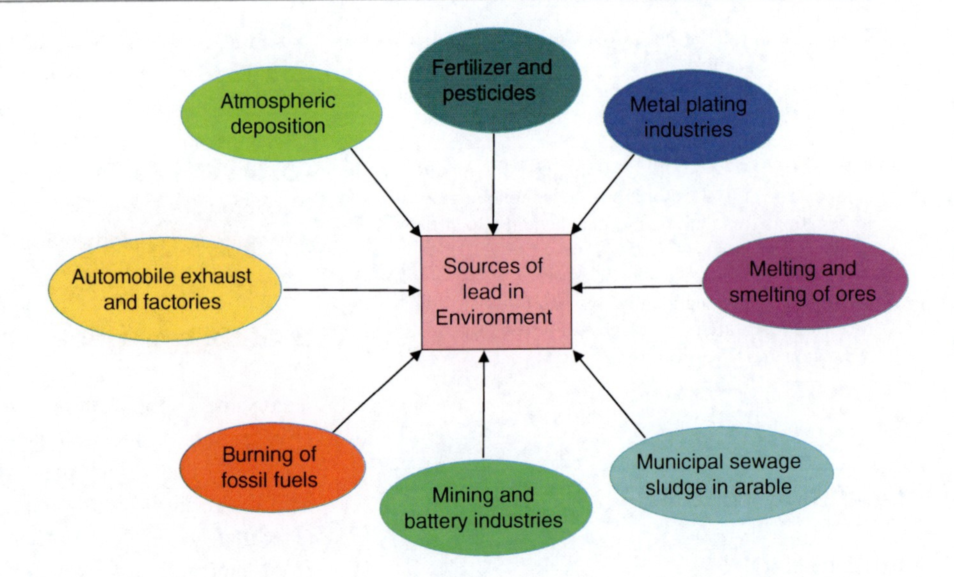

Lead Detoxification Systems in Plants, Fig. 1 Sources of lead pollution in environment

photosynthesis and induced the membrane damage of plants (Gupta et al. 2010). Pb exposure induced generation of reactive oxygen species (ROS) and increased the level of lipid peroxidation, accompanied by upregulation of activities of antioxidative enzymes (Gupta et al. 2009). Recent studies showed that primary site of Pb toxicity was localized in the cytoplasm (and hence transport across the plasma membrane (Kopittke et al. 2011)).

Common sources of lead exposure in children and adults include industrial and mining activities, paint, dust, soil, water, air, workplace, food, trinkets, ethnic folk remedies, and cosmetics. Symptoms and time to onset of symptoms postexposure may vary, and it can be difficult to identify the early, subtle neurologic effects of lead toxicity. The classic symptoms of lead toxicity generally correlate with blood lead concentrations of 25–50 μg/dL in children and 40–60 μg/dL in adults (Gracia and Snodgrass 2007). Symptoms of hemopoietic system involvement include microcytic, hypochromic anemia with basophilic stippling of the erythrocytes. Pb accumulates in the body organs (i.e., brain), which may lead to poisoning (plumbism) or even death. The gastrointestinal tract, kidneys, and central nervous system are also affected. Children suffer impaired development, lower IQ, hyperactivity, and mental deterioration. Adults experience loss of memory, nausea, insomnia, anorexia, and weakness of joints. Lead crosses the placenta during pregnancy and has been associated with intrauterine death, prematurity, and low birth weight (Papanikolaou et al. 2005). Currently approved clinical intervention methods are to give chelating agents, which bind and removed lead from lead burdened tissues. Studies indicate that there was a lack of safety and efficacy when conventional chelating agents are used (Ahamed et al. 2007).

Remediation of Pb from Environment by Plants

Heavy metals are kinds of toxicants that cannot be broken down to nontoxic forms. Once the heavy metals contaminate the ecosystem, they remain a potential threat for many years. Although traditional technologies for cleaning contaminated soils and waters have proven to be efficient, they are usually expensive, labor intensive, and in the case of soil, they produce severe disturbance (Huang et al. 2008). Phytoremediation has been suggested as an inexpensive, sustainable, in situ biotechnology to help rehabilitate soils contaminated with heavy metals without destructive effects on soil properties (Pilon-Smits 2005). The success of phytoremediation is largely determined by the amount of biomass and concentration of heavy metals in the aerial parts of plants. It is particularly difficult to remediate Pb-polluted soil using plants alone, as the metal is mostly concentrated in surface soil layers with only a small portion present in soil solution (Saifullah et al. 2009). Very rare plant species have been considered as potential hyperaccumulators and accumulators for Pb, which may benefit the extraction of Pb from contaminated soils or waters. *Sesbania drummondii*, a leguminous shrub occurring in the wild, was capable of accumulating greater than 1% Pb in shoots when grown in a Pb-contaminated nutrient solution (Sahi et al. 2002). Field surveys showed that *Buddleja asiatica* had an extraordinary accumulation capacity for Pb tolerance, it accumulates up to 4,335 mg kg^{-1} DW Pb in their shoots when grown in contaminated soils (Waranusantigul et al. 2008). Common buckwheat grown in Pb-contaminated soil was found to accumulate a large amount of Pb in leaves (8,000 mg kg^{-1} DW), stem (2,000 mg kg^{-1} DW), and in roots (3,300 mg kg^{-1} DW), without significant damage (Tamura et al. 2005). However, these metal hyperaccumulators/accumulators are of low biomass and growing slowly, while the high-biomass-yielding nonhyperaccumulator plants generally lack an inherent ability to accumulate unusual concentrations of Pb. Methods to facilitate the metal accumulation in the plants are therefore required to the success of phytoremediation.

Firstly, synthetic chelators have been widely applied to facilitate the phytoextraction of Pb from the contaminated sites (Saifullah et al. 2009). Among the chelates, EDTA (ethylenediaminetetraacetic acid) was the most widely used because of its high efficiency in extracting the metals (Komarek et al. 2007). Proper management of EDTA concentration can reduce Pb phytotoxicity and increase accumulation of low phytoavailable metal by plants (Komarek et al. 2007). In bean plants treated only with Pb, less than 6% of total lead accumulated was transported to the aboveground parts, while in the case of plants grown with Pb + EDTA, around 50% of total Pb was transported to shoots (Piechalak et al. 2008). However, controversial results were reported regarding the impact of EDTA on Pb accumulation in some plants. For instance, EDTA increased the Pb uptake but declined the biomass; subsequently, the total Pb accumulation was decreased in plant (Huang et al. 2008). Chin et al. (2009) reported that EDTA reduces lead accumulation in *Symphytum officinale* L. (comfrey) roots. In general, EDTA is usually suggested to increase the desorption of metals from the soil matrix to soil solution, thus enhancing their uptake by the plants. When plants were grown in soils, addition of EDTA may help to remobilize the fixed Pb in soils and, thus, enhances its uptake by plants (Huang et al. 2008) whereas application of EDTA to agar or liquid media was found to decrease Pb accumulation in some plants (Huang et al. 2008). It is suggested that metal-EDTA complex might enter the xylem vessels for translocation to aboveground harvestable plant parts (Saifullah et al. 2009) and, thus, increase the mobility of Pb within the plants. However, controversial results were reported whether the metal-EDTA complexes are directly absorbed by plant roots, or the dissociation of the complexes occurs prior to its uptake (Saifullah et al. 2009), and the stable Pb-EDTA complex in the soils were not in a plant-available form, and plant roots are inefficient in uptake of Pb-EDTA complex.

Besides, leaching of metals due to enhanced mobility during EDTA-assisted phytoextraction has been demonstrated as one of the potential hazards associated with this technology (Saifullah et al. 2009). Due to environmental persistence of EDTA in combination with its strong chelating abilities, the scientific communities turned to less aggressive alternative strategies, such as development of five types of controlled-release EDTA (polymer-coated EDTA) by coating the EDTA with a polyolefin polymer, polymer-coated EDTA has a potential for phytoextraction of Pb with a reduced environmental risk (Shibata et al. 2007). Addition of an electric field around the plants is an alternative approach to increase the uptake of lead, The accumulation of lead in the shoots using 0.5 mM kg^{-1} EDTA with electric potential increased by twofold to fourfold compared to EDTA only (Lim et al. 2004). The use of organic acids, more degradable chelates such as EDDS or APCAs (aminopolycarboxylic acids), and transgenic plants has considered to facilitate the efficiency of phytoextraction of Pb from contaminated soils. The soil amendment with biodegradable EDDS and organic acids may provide a good alternative to chelate-enhanced phytoextraction in enhancing metal uptake by plants and limiting metals from leaching out of soil (Luo et al. 2007). Promising role of plant hormones in translocation of lead was indicated in *Sesbania drummondii* plants, Pb accumulation in *Sesbania drummondii* shoots was enhanced by 654% and 415% in presence of 100 μM IAA(indole-3-acetic acid) and 100 μM NAA (naphthaleneacetic acid), respectively, compared to control plants (Pb alone). However, when IAA or NAA was added along with EDTA, Pb accumulation further increased in shoots by 1,349% and 1,252%, respectively (Israr and Sahi 2008). Coinoculation maize (*Zea mays* L.) with diozotrophs, which fix atmospheric nitrogen, was much better than single culture inoculations in Pb phytoextraction along with increase in plant growth and biomass. *Arbuscular mycorrhiza* (AM) fungi have been reported to improve Pb tolerance/accumulation in mesquite plants (Arias et al. 2010). Zaier et al. (2010) showed that Pb phytoextraction potential of halophyte *Sesuvium portulacastrum* is more efficient to extract Pb^{2+} in comparison with *Brassica juncea* commonly used in Pb phytoextraction, while seedlings of transformed plants grown in mining soils accumulated double concentration of heavy metal than wild type.

What Are the Mechanisms Involved in Detoxification of Lead?

Uptake and Distribution of Pb in Plants

Little is known regarding Pb uptake, localization, or the chemical forms in which Pb is found within plants, or indeed how plants tolerate elevated Pb in the environment.

One of the strategies to avoid Pb toxicity by plants is to restrict the metal in roots. The mobility of Pb in plants has been shown to be relatively low: most of the Pb taken up by plants is restricted to roots and only a small amount is transported to the shoots (Gupta et al. 2011).

Most of Pb absorbed by roots of plants was retained at root surface to prevent transport to vascular tissue for the xylem loading and subsequent translocation to shoots. Substantial and predominantly intracellular uptake of Pb was observed at the root tip of *Brassica juncea*, while sparse and predominantly extracellular uptake occurred at some distance from root tip (Meyers et al. 2008). X-ray microanalysis of *S. drummondii* root sections revealed a decreasing Pb gradient extending from the root epidermis to the axis (Sahi et al. 2002). Pb content in different part of 21 genotypes of rice grains changed as inner surface of rice glume>aleuronic layer>near aleuronic layer>surface of caryopsis>outer surface of rice glume>center of caryopsis (Chen et al. 2008). Histochemical methods (rhodizonate and dithizonate) of lead detection revealed significant accumulation of Pb on the root surface of *Dianthus carthusianorum*, followed by cell walls of pericycle (Baranowska-Morek and Wierzbicka 2004).

Compartmentation of Pb in Cell Walls and Vacuole

Compartmentation of heavy metals in cell wall and vacuole has been suggested to be an important detoxification mechanism in plants. At the ultrastructural level, Pb was localized mainly in the cell walls of plants, intercellular spaces and vacuoles, which may serve to minimize Pb toxicity to vital functions of cytosol (Phang et al. 2010). Transmission electron microscopy detected Pb(II) deposition in cell wall, plasma membrane, and vacuoles of root cells in *S. drummondii* (Sahi et al. 2002). Kopittke et al. (2008) suggested that the ability of signal grass *Brachiaria decumbens* Stapf and sequester insoluble Pb in cell wall represents an important mechanism of Pb tolerance. During transportation, Pb is usually retained in tracheid wall of vascular bundles. Recent investigation on Pb accumulator *S. alfredii* indicates that very low mobility of Pb out of vascular bundles, and the metal is largely retained in the cell walls (Tian et al. 2010).

Most of the Pb retained in the cell walls of the plants was either suggested to be in insoluble chemical forms or bound to pectin-like compounds. Kopittke et al. (2011) showed that a considerable quantity of Pb accumulated in the cell walls of signal grass roots as acicular deposits identified as chloropyromorphite ($Pb_5(PO_4)_3Cl$), as the same chemical form was reported in the roots of soil-grown *Agrostis capillaris* L. (Cotter-Howells et al. 1999). Lead in roots of *Arabidopsis thaliana* (L.) Heynh. showed a good affinity to galacturonic acid, the main component of two pectin domains homogalacturonan and rhanogalacturoman, suggestive of involvement of cell wall components in lead deactivation (Polec-Pawlak et al. 2007). Interestingly, some studies showed that Pb deposited in cell wall of plants (e.g., *Funaria hygrometrica* protonemata) is not stable, it can uptake or remobilized from the cell wall by internalization with low-esterified pectins (Krzeslowska et al. 2010).

Complexation of Pb in Plants

Although some reports indicated that the accumulation of Pb in plants was in the same oxidation state as supplied Pb(II), or as cerussite $PbCO_3$ (Sarret et al. 2001), chelating-bound Pb is considered as less toxic compared to free Pb ions and hence might induce less stress on plants. Phytochelatins (PC) are known as the most common peptides responsible for binding toxic metals in plants. Chelated Pb in conjunction with PCs synthesis and complexation reduces stress in lead-tolerant plants. The presence of Pb complexed by thiol groups, probably phytochelatins, has been reported in other plants (Gupta et al. 2010). The accumulation of PCs in Pb hyperaccumulator aquatic fern *Salvinia minima* has a direct response to Pb^{2+} accumulation, and phytochelatins do participate as one of the mechanism to cope with Pb^{2+} (Estrella-Gomez et al. 2009). Most probable mechanism for Pb detoxification in vetiver grass (*Vetiveria zizanioides* L.), can accumulate up to 19,800 and 3,350 mg kg^{-1} DW Pb in root and shoot tissues, respectively, is by synthesizing PCs and forming Pb-PCs complexes (Andra et al. 2010). Lead detoxification by coontail (*Ceratophyllum demersum* L.) involves induction of PCs and antioxidant system in response to its accumulation, PCs were synthesized to significant levels at 10 and 50 μM Pb with concomitant decrease in GSH levels (Mishra et al. 2006). Recent research also indicated that the detoxification of Pb in *Sedum alfredii* is not related to PCs but the GSH (Gupta et al. 2010), SH compounds may be partially involved in the Pb accumulation in *S. alfredii* (Tian et al. 2010).

The role of other bioligands, such as proteins, organic acids, flavonoids, and oligosaccharides, has received renewed attention. For instance, overexpression of membrane-associated acyl-CoA-binding protein ACBP1 enhances lead tolerance in *Arabidopsis* (Xiao et al. 2008). *Arabidopsis thaliana* ACBP1 binds lead [Pb(II)], its mRNA is induced by Pb (II)-treatment, and transgenic *Arabidopsis* overexpressing ACBP1 confer Pb(II) tolerance and accumulate Pb(II). *Sesbania drummondii* (rattlebush) has been reported to biotransform lead nitrate in nutrient solution to lead acetate and sulfate in its tissues, complexation with acetate and sulfate may be a lead detoxification strategy in this plant (Sahi et al. 2002).

Protective Effect of Cations

Cations, such as Ca and Mg, are generally thought to alleviate toxicities of heavy metals through site-specific competition. Calcium was found to alleviate Pb toxicity through a specific effect (Kopittke et al. 2011). The protective effect of Ca^{2+} on Pb^{2+} toxicity may involve multiple mechanisms including competition at entry level and that Pb^{2+} and Cd^{2+} may compete with divalent cations for transport into roots of rice plants (Kim et al. 2002). The involvement of TaLCT1 has been suggested in regulation of Ca-dependent Pb detoxification, and under conditions of low Ca in Pb uptake and distribution (Wojas et al. 2007).

Genes Involved in Tolerance of Pb

The response of a metal-tolerant plant to heavy metal stress involves a number of biochemical and physiological pathways. Molecular mechanisms of Pb detoxification in plants, however, were much less studied as compared to other metals. Up to date, several genes have been reported involved in tolerance of Pb in plants, mostly found in model plant *Arabidopsis*. The *Arabidopsis* ethylene-insensitive 2 (EIN2) gene mediates Pb(II) resistance, at least in part, through two distinct mechanisms, a GSH-dependent mechanism and a GSH-independent AtPDR12-mediated mechanism (Cao et al. 2010). AtPDR12, ATP-binding cassette (ABC) protein functions as a pump to exclude Pb (II) and/or Pb(II)-containing toxic compounds from the cytoplasm and contributes to Pb(II) resistance in *Arabidopsis*. AtHMA3 likely plays a role in the detoxification of biological (Zn) and nonbiological (Cd, Co, and Pb) heavy metals by participating in their vacuolar sequestration, an original function for a P1B-2 ATPase in a multicellular eukaryote (Morel et al. 2009). Analysis of transgenic *A. thaliana* plants overexpressing YCF1 showed that YCF1 is functionally active, and the plants have enhanced tolerance to Pb(II) and Cd(II) and accumulated higher amounts of these metals. These results suggest that transgenic plants expressing YCF1 may be useful for phytoremediation of lead and cadmium (Song et al. 2003). Cold treatment enhanced Pb(II) resistance in *Arabidopsis*, at least in part, by activating the expression of *AtPDR12* gene (Cao et al. 2010). In Pb hyperaccumulator *Sesbania drummondii*, which exhibits a significant level of tolerance to lead, several differentially expressed cDNA clones, including a type 2 metallothionein (MT) gene which is involved in detoxification and homeostasis shown to be differentially regulated in lead treated plants (Srivastava et al. 2007).

Conclusion

Since there are no known high-biomass plants, which could hyperaccumulate lead, researches are going on for genetically engineered plants having more biomass production in short time by incorporating genes from hyperaccumulators to achieve the goal. However, this is not so simple, and complete understanding of the mechanism is required for tolerance and detoxification. Tolerance requires lesser vulnerability to toxicity and, thus, quick defense system, more uptake then efficient translocation, binding with PC, and final sequestration to vacuole.

Cross-References

▶ Lead and Phytoremediation
▶ Lead in Plants

References

Ahamed M, Kaleem M, Siddiqui J (2007) Environmental lead toxicity and nutritional factors. Clinical Nut 26: 400–408

Andra SS, Datta R, Sarkar D, Makris KC, Mullens CP, Sahi SV, Bach SBH (2010) Synthesis of phytochelatins in vetiver grass upon lead exposure in the presence of phosphorus. Plant Soil 326:171–185

Arias JA, Peralta-Videa JR, Ellzey JT, Ren MH, Viveros MN, Gardea-Torresdey JL (2010) Effects of *Glomus deserticola* inoculation on prosopis: enhancing chromium and lead uptake and translocation as confirmed by X-ray mapping, ICP-OES and TEM techniques. Environ Exp Bot 68:139–148

Baranowska-Morek A, Wierzbicka M (2004) Localization of lead in root tip of *Dianthus carthusianorum*. Acta Bio Cracov Ser Bot 46:45–56

Cao SQ, Bian XH, Jiang ST, Chen ZY, Jian HY, Sun ZH (2010) Cold treatment enhances lead resistance in *Arabidopsis*. Acta Physiol Planta 32:19–25

Chen G, Sun GR, Liub AP, Zhou WD (2008) Lead enrichment in different genotypes of rice grains. Food Chem Toxicol 46:1152–1156

Chin L, Leung DWM, Taylor HH (2009) EDTA reduces lead accumulation in *Symphytum officinale* L. (comfrey) roots. Chem Ecol 25:397–403

Cotter-Howells JD, Champness PE, Charnock JM (1999) Mineralogy of Pb-P grains in the roots of *Agrostis capillaris* L-by ATEM and EXAFS. Mineral Mag 63:777–789

Estrella-Gomez N, Mendoza-Cozatl D, Moreno-Sanchez R, Gonzalez-Mendoza D, Zapata-Perez O, Martinez-Hernandez A, Santamaria JM (2009) The Pb-hyperaccumulator aquatic fern *Salvinia minima* Baker, responds to Pb^{2+} by increasing phytochelatins via changes in SmPCS expression and in phytochelatin synthase activity. Aquat Toxicol 91:320–328

Gracia RC, Snodgrass WR (2007) Lead toxicity and chelation therapy. Am J Health Syst Pharm 64:45–53

Gupta DK, Nicoloso FT, Schetinger MRC, Rossato LV, Pereira LB, Castro GY, Srivastava S, Tripathi RD (2009) Antioxidant defence mechanism in hydroponically grown *Zea mays* seedlings under moderate lead stress. J Hazard Mater 172:479–484

Gupta DK, Huang HG, Yang XE, Razafindrabe BHN, Inouhe M (2010) The detoxification of lead in *Sedum alfredii* H. is not related with phytochelatins but the glutathione. J Hazard Mater 177:437–444

Gupta DK, Nicoloso FT, Schetinger MRC, Rossato LV, Huang HG, Srivastava S, Yang XE (2011) Lead induced responses of *Pfaffia glomerata*, an economically important Brazilian medicinal plant, under *in vitro* culture conditions. Bull Environ Contam Toxicol 86:272–277

Huang HG, Li TX, Tian SK, Gupta DK, Zhang XZ, Yang XE (2008) Role of EDTA in alleviating lead toxicity in accumulator species of *Sedum alfredii* H. Bioresour Technol 99: 6088–6096

Israr M, Sahi SV (2008) Promising role of plant hormones in translocation of lead in *Sesbania drummondii* shoots. Environ Pollut 153:29–36

Kim YY, Yang YY, Lee Y (2002) Pb and Cd uptake in rice roots. Physiol Planta 116:368–372

Komarek M, Tlustos P, Szakova J, Chrastny V, Ettler V (2007) The use of maize and poplar in chelant-enhanced phytoextraction of lead from contaminated agricultural soils. Chemosphere 67:640–651

Kopittke PM, Asher CJ, Blamey FP, Auchterlonie GJ, Guo YN, Menzies NW (2008) Localization and chemical speciation of Pb in roots of signal grass (*Brachiaria decumbens*) and Rhodes grass (*Chloris gayana*). Environ Sci Tech 42: 4595–4599

Kopittke PM, Kinraide TB, Wang P, Blamey FPC, Reichman SM, Menzies NW (2011) Alleviation of Cu and Pb rhizotoxicities in cowpea (*Vigna unguiculata*) as related to ion activities at root-cell plasma membrane surface. Environ Sci Technol 45:4966–4973

Krzeslowska M, Lenartowska M, Samardakiewicz S, Bilski H, Wozny A (2010) Lead deposited in the cell wall of *Funaria hygrometrica* protonemata is not stable – a remobilization can occur. Environ Pollut 158:325–338

Lim JM, Salido AL, Butcher DJ (2004) Phytoremediation of lead using Indian mustard (*Brassica juncea*) with EDTA and electrodics. Microchem J 76:3–9

Luo CL, Shen ZG, Li XD (2007) Plant uptake and the leaching of metals during the hot EDDS-enhanced phytoextraction process. Int J Phytoremediation 9:181–196

Mishra S, Srivastava S, Tripathi RD, Kumar R, Seth CS, Gupta DK (2006) Lead detoxification by coontail (*Ceratophyllum demersum* L.) involves induction of phytochelatins and antioxidant system in response to its accumulation. Chemosphere 65:1027–1039

Meyers DER, Auchterlonie GJ, Webb RI, Wood B (2008) Uptake and localisation of lead in the root system of *Brassica juncea*. Environ Pollut 153:323–332

Morel M, Crouzet J, Gravot A, Auroy P, Leonhardt N, Vavasseur A, Richaud P (2009) AtHMA3, a P-1B-ATPase allowing Cd/Zn/Co/Pb vacuolar storage in *Arabidopsis*. Plant Physiol 149:894–904

Papanikolaou NC, Hatzidaki EG, Belivanis S, Tzanakakis GN, Tsatsakis AM (2005) Lead toxicity update. A brief review. Med Sci Monit 11:RA329–RA336
Phang IC, Leung DWM, Taylor HH, Burritt DJ (2010) Correlation of growth inhibition with accumulation of Pb in cell wall and changes in response to oxidative stress in *Arabidopsis thaliana* seedlings. Plant Growth Regul 64:17–25
Piechalak A, Malecka A, Baralkiewicz D, Tomaszewska B (2008) Lead uptake, toxicity and accumulation in *Phaseolus vulgaris* plants. Biol Planta 52:565–568
Pilon-Smits E (2005) Phytoremediation. Ann Rev Plant Biol 56:15–39
Polec-Pawlak K, Ruzik R, Lipiec E, Ciurzynska M, Gawronska H (2007) Investigation of Pb(II) binding to pectin in *Arabidopsis thaliana*. J Anal At Spectrom 22:968–972
Sahi SV, Bryant NL, Sharma NC, Singh SR (2002) Characterization of a lead hyperaccumulator shrub, *Sesbania drummondii*. Environ Sci Technol 36:4676–4680
Saifullah Meers E, de Qadir M, Caritat P, Du Tack FMG, Laing G, Zia MH (2009) EDTA-assisted Pb phytoextraction. Chemosphere 74:1279–1291
Sarret G, Vangronsveld J, Manceau A, Musso M, D'Haen J, Menthonnex JJ, Hazemann JL (2001) Accumulation forms of Zn and Pb in *Phaseolus vulgaris* in the presence and absence of EDTA. Environ Sci Technol 35:2854–2859
Shibata M, Konno T, Akaike R, Xu Y, Shen RF, Ma JF (2007) Phytoremediation of Pb contaminated soil with polymer-coated EDTA. Plant Soil 290:201–208
Song WY, Sohn EJ, Martinoia E, Lee YJ, Yang YY, Jasinski M, Forestier C, Hwang I, Lee Y (2003) Engineering tolerance and accumulation of lead and cadmium in transgenic plants. Nat Biotechnol 21:914–919
Srivastava AK, Venkatachalam P, Raghothama KG, Sahi SV (2007) Identification of lead-regulated genes by suppression subtractive hybridization in the heavy metal accumulator *Sesbania drummondii*. Planta 225:1353–1365
Tamura H, Honda M, Sato T, Kamachi H (2005) Pb hyperaccumulation and tolerance in common buckwheat (*Fagopyrum esculentum* Moench). J Plant Res 118:355–359
Tian SK, Lu LL, Yang XE, Webb SM, Du YH, Brown PH (2010) Spatial imaging and speciation of Lead in the accumulator plant *Sedum affredii* by microscopically focused synchrotron X-ray investigation. Environ Sci Technol 44:5920–5926
Waranusantigul P, Kruatrachue M, Pokethitiyook P, Auesukaree C (2008) Evaluation of Pb phytoremediation potential in *Buddleja asiatica* and *B. paniculata*. Water Air Soil Pollut 193:79–90
Wojas S, Ruszczynska A, Bulska E, Wojciechowski M, Antosiewicz DM (2007) Ca^{2+}-dependent plant response to Pb^{2+} is regulated by LCT1. Environ Pollut 147:584–592
Xiao S, Gao W, Chen QF, Ramalingam S, Chye ML (2008) Overexpression of membrane-associated acyl-CoA-binding protein ACBP1 enhances lead tolerance in *Arabidopsis*. Plant J 54:141–151
Zaier H, Ghnaya T, Lakhdar A, Baioui R, Ghabriche R, Mnasri M, Sghair S, Lutts S, Abdelly C (2010) Comparative study of Pb-phytoextraction potential in *Sesuvium portulacastrum* and *Brassica juncea*: tolerance and accumulation. J Hazard Mater 183:609–615

Lead in Plants

Stephan Clemens
Department of Plant Physiology, University of Bayreuth, Bayreuth, Germany

Synonyms

Pb in plants

Definition

The entry addresses plant exposure to lead, uptake and accumulation of lead in plants, lead toxicity and detoxification, as well as environmental and human health consequences of plant lead accumulation.

Introduction

Lead (Pb) has no biological functions in organisms including plants. Ever since the beginning of human civilization, however, plants have been exposed to a nearly exponentially increasing Pb emission closely linked with technological development. Pb ranks first among heavy metals with respect to tonnage produced and release into the environment. Because Pb is highly malleable and easy to extract and to smelt, its use has been widespread for thousands of years, dating back to the origins of metal technology (Nriagu 1998). Pb is highly toxic and therefore an important pollutant of worldwide concern. It is placed number two on the CERCLA priority list of hazardous substances, assembled by the US Environmental Protection Agency and the Agency for Toxic Substances and Disease Registry (www.atsdr.cdc.gov/cercla/07list.html). In spite of its relevance for human and environmental health, the knowledge about Pb uptake, accumulation, and detoxification by plants is very limited. This is largely due to serious experimental challenges.

Exposure of Plants to Pb

The average concentration of Pb in the Earth's crust is around 13 mg kg^{-1}. Pb does not occur as metallic lead but in a range of minerals such as PbS and $PbCO_3$. Most of the exposure of plants to Pb, however, is attributable to atmospheric deposition caused by human activities. The earliest releases into the environment were mainly due to ancient silver mining. Since then, the corrosion-resistant metal Pb has increasingly been used in construction (e.g., water pipes) and manufacturing. Nonferrous metal smelting is therefore an important Pb source. Addition of tetraethyl lead to gasoline in the 1950s caused further increases in the emission of lead to the atmosphere until the 1970s. In agriculture, the use of Pb in pesticides before the 1950s accounted for input mainly in fruit orchards. Sewage sludge can be another source, while, in contrast to Cd, little Pb is added through fertilizer (McLaughlin et al. 1999).

Most of the Pb present in the environment is in the inorganic form. Plants are exposed to Pb in the soil and to Pb particles and aerosols in the air. Low mobility of Pb in soil (see section on "Plant Uptake and Accumulation of Pb") results in accumulation predominantly in the top soil. The Pb content of noncontaminated soil usually varies between <10 and 70 mg kg^{-1} with an average at around 20 mg kg^{-1} (Kabata-Pendias and Pendias 2001). Close to sources of Pb emission values in excess of 1,000 mg kg^{-1} or sometimes even 10,000 mg kg^{-1} can be found. Elevated Pb levels are also consistently found along motorways.

Pb shows very low solubility at $pH > 5$. Specific adsorption to soil particles, for instance, to Fe, Al, and Mn oxides, is the strongest of all heavy metals. Organic matter reduces availability even further through the formation of highly stable metallo-organic complexes. In the presence of high phosphate levels, lead phosphates precipitate. Consequently, bioavailability of Pb is extremely low in most soils. Concentrations in the soil solution are usually in the subnanomolar to nanomolar range. For contaminated soil in Europe, concentrations of up to 1 μM Pb have been reported, for US soils up to 2 μM (McLaughlin et al. 1999).

Plant Uptake and Accumulation of Pb

The mechanistic understanding of Pb acquisition by plants is poor. Achieving realistic experimental conditions accounting for the very low solubility of Pb in soil has been difficult. Many studies employed cultivation of plants in common solution culture with a pH between 5.5 and 6.0 plus phosphate concentrations in the millimolar range. Under these conditions, Pb precipitates and only a minute, and in most cases not quantified, fraction is available for uptake. Accordingly, concentrations up to several millimolars were applied, i.e., at least three orders of magnitude higher than what can be found in soil solution (Kopittke et al. 2008). On the other hand, field studies mostly did not differentiate uptake of Pb into leaves via the roots from aerial deposition of Pb particles on the leaf surface. This can result in an overestimation of Pb long-distance transport within the plant.

Uptake of ionic Pb into plant roots is passive, i.e., driven by the negative membrane potential of root cells. Pathways for the transport across plasma membranes have so far not been identified molecularly. It is assumed that similar to other toxic nonessential ions such as arsenate or selenate, transporters mediating the uptake of essential ions do not perfectly discriminate between chemically similar ions and allow entry also of the toxic ions. In case of Pb^{2+}, transporters and channels mediating uptake of Ca^{2+} have been implicated. This assumption is largely based on the observed amelioration of Pb^{2+} toxicity by Ca^{2+} addition, which may suggest competition for uptake pathways. Alternatively, in analogy to presumed entry pathways in the human gut, Pb^{2+} may be taken up by transporters of other metal cations such as Fe^{2+} (Clemens 2006).

Available evidence shows that most of the root-associated Pb is in fact not transported into root cells but instead precipitates in the apoplast, i.e., the cell wall, of the outer cell layers. In addition, ionic Pb can strongly interact with the negative charges of cell wall components such as pectins (polygalacturonic acids). The minor intracellular Pb fraction could be sequestrated in vacuoles of root cells. As a consequence of this strong root retention, the rate of Pb translocation from roots to shoots is extremely small in most plants (<5%) and much lower when compared to metals like Cd (Pourrut et al. 2011). Most of the Pb detected in aboveground plant tissues of field-grown plants is therefore commonly attributed to deposition of lead particles or aerosols from the atmosphere onto the plant's surface. The speciation of translocated or otherwise cellular Pb inside plants is virtually unknown. Respective analyses such as X-ray fluorescence are challenging because the Pb signal is practically always dominated by Pb precipitates. Candidate chelators are thiols such as glutathione and phytochelatins (see section on "Pb Toxicity and Detoxification by Plants").

An extreme exception from the general trend of very low Pb translocation factors are plants that hyperaccumulate Pb. About 0.2% of all angiosperm species are known to hyperaccumulate certain metals, with the vast majority hyperaccumulating Ni. Zn hyperaccumulation has to date been found in 15 taxa. Several Zn hyperaccumulating species also show Pb hyperaccumulation, defined as a shoot Pb level in field samples of > 1,000 μg g^{-1} dry biomass (Krämer 2010). While field data have to be treated with caution because of possible aerial contamination, there is solid evidence for the existence of a few Pb hyperaccumulating plant species including *Noccaea praecox*, *Arabis paniculata*, *Sedum alfredii*, and *Viola baoshanensis*. Pb hyperaccumulation has been reported for plants growing in Pb/Cd/Zn-rich soil near former metal smelters or mining sites. *Noccaea praecox*, for example, was found to accumulate Pb to levels up to 0.4% of the leaf dry weight. As an integral component of Pb hyperaccumulation, efficient translocation of Pb via the xylem as part of the vascular system has to be postulated. This requires suppression of Pb precipitation (both extracellularly and intracellularly) and sequestration in root cells as well as radial transport across the root into xylem vessels. Unlike for Zn and Cd, which are normally hyperaccumulated by the same hyperaccumulating plant individuals, no mechanistic insight into these processes is available. There is, for instance, no evidence for an involvement of $P1_B$-type ATPases such as HMA4 in effluxing Pb (Krämer 2010). Also, it is unknown which low molecular weight chelators mobilize Pb into the vascular tissue. In leaves, Pb accumulates both inside cells and in the apoplast, indicating again strong interaction of Pb with cell wall polymers. Binding partners for efficient storage in leaf cells remain to be identified.

Several studies report enormous accumulation also in shoots when plants such as certain *Brassica* species are grown in solution culture or sometimes even in soil. However, most of the respective experiments employed treatment with ethylenediaminetetraacetic acid (EDTA) or other chelators to strongly enhance mobilization of Pb. This approach was widely propagated in the early days of phytoremediation (see section on "Plant Pb and Environmental and Human Health") to overcome the very limited bioavailability of Pb but is no longer encouraged as EDTA is toxic to plants and can cause leaching of the otherwise extremely immobile Pb into groundwater.

Pb Toxicity and Detoxification by Plants

Progress in understanding Pb toxicity in plants and mechanisms of tolerance has been severely hampered by the same experimental problems referred to above. Many studies are flawed because of cultivation conditions that result in Pb precipitation and thus unknown exposure levels. This is the reason why growth-inhibiting effects have been studied in solution culture at concentrations ranging from 0.1 μM to several millimolar. In carefully designed experiments employing low pH/low phosphate medium to ensure Pb solubility, symptoms of Pb toxicity are observed at concentrations below 1 μM, and together with Hg^{2+}, Pb^{2+} has the most pronounced growth-inhibiting effect among metal cations on plants in solution culture (Kopittke et al. 2010). Main site of toxicity is the root. Growth inhibition results from effects on both cell division and cell elongation. Primary targets of Pb have to date not unequivocally been identified. A possible exception may be acyl-CoA-binding proteins which appear to bind Pb with high affinity both in animals and plants (Xiao et al. 2008). Interference with Ca homeostasis or displacement of Ca from cell wall polymers has been proposed. Pb ions also strongly interact with thiols and may indirectly cause oxidative stress.

Like other groups of organisms, plants exhibit basal tolerance of Pb, meaning that a variety of processes and characteristics ensures survival up to a certain level of exposure (Pourrut et al. 2011). Furthermore, there is clear evidence for natural variation in Pb tolerance between species and possibly even within species. For instance, different grass species can vary in their growth response to Pb exposure by factor 10. The most extreme example is provided by Pb hyperaccumulators. The critical leaf tissue levels for Pb of around 20 μg g^{-1} dry biomass can be exceeded >100-fold.

Potential mechanisms underlying basal Pb tolerance and its variation include effective immobilization of Pb in the extracellular space, effluxing of Pb and intracellular sequestration (Pourrut et al. 2011). Candidate binding partners for sequestration in plant cells are glutathione and glutathione-derived peptides called phytochelatins. Phytochelatins are synthesized in response to exposure to various physiological and nonphysiological metal ions. They are peptides of the general structure (γ-Glu-Cys)$_n$-Gly ($n = 2$–11). They are nonribosomally synthesized from glutathione in a transpeptidase reaction, catalyzed by phytochelatin synthases. Phytochelatin synthases are activated by Cd^{2+}, Zn^{2+}, Cu^{2+}, Sb^{3+}, Hg^{2+}, Ag^{+}, $AsO_4{}^{3-}$, and also Pb^{2+}. Thus, in the presence of excess metals, phytochelatins are formed, and they effectively capture metals. Pb-activated PC accumulation has been observed in plants, but in contrast to Cd, Zn, As species, and Hg, proof for their role in detoxification is lacking (Clemens 2006).

The actual contribution of extracellular binding and precipitation of Pb has not been quantified either. However, virtually all studies addressing the question of Pb localization find a large fraction extracellularly. This emphasizes the role of the cell wall in Pb tolerance. Alternatively, secretion of ions and low molecular weight chelators could influence Pb adsorption to root tissue. Potential differences between plants in the ability to immobilize Pb have to date not been explained molecularly.

Evidence for active exclusion as a component of basal Pb tolerance has largely come from work on the model plant *Arabidopsis thaliana*. With the caveat that extreme Pb concentrations were applied in the assays with resulting uncertainty about bioavailable Pb, ATP-binding cassette (ABC) transporters appear to mediate the efflux of Pb resulting in detoxification. Respective mutants show stronger growth inhibition. Conversely, overexpression of these proteins enhances Pb tolerance relative to wild-type plants. The second transporter family implicated in Pb efflux is $P1_B$-type ATPases. Some of them confer Pb tolerance to bacteria. Overexpression of bacterial and plant $P1_B$-type ATPases yielded again transgenic plants with improved Pb tolerance.

Plant Pb and Environmental and Human Health

Pb is a potent neurotoxin for humans and animals and is classified as probably carcinogenic to humans. It accumulates both in soft tissues and the bones. Human exposure can occur via food, water, air, soil, and dust, with food as the major source of exposure. Thus, Pb accumulation in crop plants is a relevant health concern. Legislation has set maximum levels for plant-derived food in many countries. European Union examples are 0.2 mg kg^{-1} wet weight for cereals, legumes, and pulses and 0.1 mg kg^{-1} wet weight for vegetables and fresh herbs (EFSA Panel on Contaminants in the Food Chain 2010). International surveys showed that owing to the extremely low mobility of Pb within plant tissues, Pb concentrations of edible crop plant tissues are generally well below these thresholds. However, there is considerable variation in Pb translocation factors that is mechanistically not understood. In view of the fact that it is not possible to define a threshold level for which critical Pb effects on human health can be excluded, potential risks of Pb accumulation in plant-derived food could be reduced through better knowledge about environmental and genotypic factors influencing Pb accumulation and mobility. Also, Pb accumulation in aboveground tissues is likely to be different in natural habitats, which could affect wildlife. In contrast to an agricultural soil where a pH above 5 and phosphate fertilization strongly limit Pb bioavailability, an acidic forest soil can show much higher Pb concentration in the soil solution, leading potentially to much higher uptake. This context is relevant for the question of phytostabilization too. Principally desired revegetation of Pb-contaminated sites by grasses and other plants should aim to minimize transfer of Pb into the food chain.

Finally, the use of plants to remove metal pollutants from the environment in a cost-effective and sustainable manner (phytoremediation) depends on the availability of plants that can efficiently extract, translocate, and accumulate Pb in leaf tissue. The few demonstrated Pb hyperaccumulators show that this is possible. However, mechanistic insight is again extremely limited (Krämer 2010).

Cross-References

- ▶ Lead and Phytoremediation
- ▶ Lead Detoxification Systems in Plants
- ▶ Lead Transporters
- ▶ Lead, Physical and Chemical Properties
- ▶ Metallothioneins and Lead

References

Clemens S (2006) Toxic metal accumulation, responses to exposure and mechanisms of tolerance in plants. Biochimie 88:1707–1719

EFSA Panel on Contaminants in the Food Chain (2010) Scientific opinion on lead in food. EFSA J 8:1570

Kabata-Pendias A, Pendias H (2001) Trace elements in soils and plants. CRC Press, Boca Raton

Kopittke PM, Asher CJ, Menzies NW (2008) Prediction of Pb speciation in concentrated and dilute nutrient solutions. Environ Pollut 153:548–554

Kopittke PM, Pax F, Blamey C, Asher CJ, Menzies NW (2010) Trace metal phytotoxicity in solution culture: a review. J Exp Bot 61:945–954

Krämer U (2010) Metal hyperaccumulation in plants. Annu Rev Plant Biol 61:517–534

McLaughlin MJ, Parker DR, Clarke JM (1999) Metals and micronutrients – food safety issues. Field Crops Res 60:143–163

Nriagu JO (1998) Tales told in lead. Science 281:1622–1623

Pourrut B, Shahid M, Dumat C, Winterton P, Pinelli E (2011) Lead uptake, toxicity, and detoxification in plants. Rev Environ Contam Toxicol 213:113–136

Xiao S, Gao W, Chen Q-F, Ramalingam S, Chye M-L (2008) Overexpression of membrane-associated acyl-CoA-binding protein ACBP1 enhances lead tolerance in *Arabidopsis*. Plant J 54:141–151

Lead Nephrotoxicity

Alessandra Stacchiotti, Giovanni Corsetti and Rita Rezzani
Division of Human Anatomy, Department of Biomedical Sciences and Biotechnologies, Brescia University, Brescia, Italy

Synonyms

Lead-associated aggresome diseases; Lead-induced chronic renal failure

Definition

Lead (Pb) nephrotoxicity relates to experimental, epidemiologic, and clinical evidence of renal damage and hypertension associated to environmental or occupational Pb exposure. The biology of Pb is very complex because even if at low or occult and often underestimated doses, this poisonous heavy metal may exert chronic nephropathy that becomes evident many decades later from the first exposure period, occurred even during the pregnancy status or in fetal life. Recent proteomic studies and the use of transgenic rodent models provide new clues to address the kidney not simply as the main Pb excretion route but also as an important target for Pb-induced protein aggregates.

What Is Lead? Brief History

Pb is a ubiquitous heavy metal that man encountered thousands of years ago, called *plumbum* by Romans, from which was derived the symbol *Pb*; it was known also in ancient Egypt and cited in the Bible. It was ranked second among 275 toxic compounds on the Agency for Toxic Substances and Disease (ATSDR/EPA) Registry Priority List of Hazardous Substances since 2003 and still induces one of the most dangerous and persistent poisoning.

As a chemical element, Pb belongs to carbon family that includes five elements at the Group 14 (IVA) of the *periodic table*. This term indicates a chart that shows how chemical elements are related to each other. Although a member of the carbon family, Pb looks and behaves very differently from carbon. Pb is an easily worked metal, which means bending, cutting, shaping, pulling, and otherwise changing the shape. Its melting point is 327.4°C (621.3°F), and its boiling point is 1,750–1,755°C (3,180–3,190°F). Its density is 11.34 grams per cubic centimeter. Pb does not conduct an electric current, sound, or vibrations very well. It slowly dissolves in water but better in hot acids. It does not react with air oxygen, so does not burn. Pb is a metal without redox-reactive potential, and its induction of oxidative damage is probably mediated by other factors. Its peculiar chemical properties, such as corrosion resistance, density, and low melting point, make it a familiar metal in pipes, solder, weights, and storage batteries. Read more in http://www.chemistryexplained.com/elements/L-P/Lead and in http://www.osha.gov-Toxic Metals: Lead.

Epidemiologic Impact of Pb Nephrotoxicity: An Update Point

Even if the association between Pb poisoning and renal diseases in humans has been indicated since about a century ago when exposure rate was higher, actually, the lower exposure and its reduction due to public health measures have produced contrasting findings. However, it is important to mind that although Pb exposure decreased in population, the body burden of Pb still represents a dangerous endogenous source of exposure. In the human body, this pollutant metal accumulates mainly in blood and bone, and its biological half-life is in the order of decades (30 years). So, subjects exposed in the past still might develop diseases due to Pb deposition in their body (*Agency for Toxic Substances and Disease Registry*, ATSDR, 2007 in http://www.atsdr.cdc.gov-). This problem is particularly addressed in the renal pathology that represents a growing and important sanitary issue, often associated to aging and other chronic metabolic disorders such as hypertension, diabetes, and atherosclerosis.

Actually, an urgent problem that needs to be solved as soon as possible is the link between renal dysfunctions and Pb exposure in children or adolescents that often do not show any renal failure respect to adult or senescent population but might develop it in the future. In undeveloped countries and oriental area, comprising India, Taiwan, and China, there is an increasing alert for Pb intake and contamination in children. The main cause, in particular in China, was due to increasing industrialization and major pollution in rural areas associated to new factories, that escape surveillance from authorities, and Pb-contained paints in houses and toys.

Unfortunately, the risk assessment of Pb nephrotoxicity has been often underestimated probably due to the lack of suitable and sensible analytical methods and limited epidemiological studies on children and younger people, but the prevalence of occupational studies oriented on adult workers. In particular, in adults, the adverse association between Pb in the blood and kidney physiology has been documented even at low Pb blood levels (under 10 μg/dL), but this limit must be reduced in children and adolescents. Unfortunately, the index that mainly is associated to renal function, called *glomerular filtration rate* (GFR), measuring *serum creatinine* levels is unable to detect Pb below 10 μg/dL. So, despite the drastic reduction of Pb contamination, obtained in the USA by Federal regulations that required phaseout of painting for residential homes since 1978 and Pb prohibition in gasoline since 1996, general population still has detectable blood levels. Indeed, even Pb blood levels below 5 μg/dL were associated to increased chronic renal diseases in adults, and an update evidence suggests that no threshold for Pb toxicity exists. In particular, there is no remedy to remove Pb from the body at levels under 30 μg/dL, rendering treatment of blood Pb levels ineffective at this value. Moreover, blood Pb levels with a half-life of approximately a month reflect a recent exposure, whereas bone Pb levels with a half-life of 25–30 years represent accumulated and persistent exposure in population. So, the real and effective reaction against Pb contamination is the primary prevention and remotion from the environment before it acts on children. Recently, in adolescents from 12 to 20 years old, a link between Pb and development of chronic kidney diseases in the USA has been hypothesized (Fadrowski et al. 2010). Actually, newer and more accurate estimates of GFR have been introduced, using serum cystatin C instead of creatinine and specific equations to reveal the association between Pb in the blood (valuable even when Pb level was 3 μg/dL) and risk of chronic kidney diseases in clinical and epidemiologic studies. In the USA, the prevalence of chronic kidney disease affects about 26 million people, and their exposure to Pb must be underestimated. So, low-level environmental Pb contamination, mainly in selected social groups associated to lower socioeconomic status, might be strongly reconsidered. A recent reexamination of epidemiologic data obtained in adult people from 1999 to 2002 available by the US National Center for Health and Statistics (NCHS) suggested that there was a negative correlation between detectable Pb levels and obesity. However, Pb levels in blood were strictly related to a constant exposure and that their reduction, due to modern legal restrictions, did not necessarily mean that the body was devoid of this persistent toxic metal. Pb involvement in the onset of metabolic disorders such as diabetes, atherosclerosis, and chronic kidney diseases has been certainly underestimated.

Lead: Mechanism of Action and Its Main Binding Proteins

Environmental and occupational exposure to Pb contributes to several disorders in the hematopoietic, reproductive, gastroenteric, and nervous systems. Interaction of Pb cations (Pb^{2+}) with cellular fundamental proteins such as enzymes and high-affinity metal-binding proteins represents an important mechanism thought to be responsible for its toxicity.

Remarkably, the main danger of low doses of Pb^{2+} in the body is linked to its intracellular partitioning in target organs and its binding to soluble low molecular size proteins. Besides specific competition of Pb^{2+} with native metal ions in proteins/enzymes, this dangerous cation strongly bound in solution to amino acids in proteins that worked as carriers.

Proteins that formed complex with Pb^{2+}, called Pb-binding proteins, may fundamentally fall into two categories: proteins that naturally bind calcium (Ca) and others that bind zinc (Zn). If the Pb-protein complex is physiologically relevant, this may progressively induce a disease even after a long term.

The knowledge of soluble protein carriers may strongly influence the intracellular Pb bioavailability mainly when the metal is present at very low dosage in target organs. However, it is necessary that Pb binding is tight enough to proteins, and this is related to the *dissociation constant* for the Pb-protein complex (K_d^{Pb}). The dissociation constant is a specific type of *equilibrium constant* that measures the propensity of a larger object to separate (dissociate) reversibly into smaller components, as when a complex falls apart into its component molecules. This value indicates the stability or strength of a given multiple chemical complex, and more a binding grade is stable, less the dissociation constant will be. So, considering that the concentrations of bioavailable Zn or Ca cations (respectively, Zn^{2+} and Ca^{2+}) have been measured within 10^{-12} up to 10^{-6} M and 10^{-8} up to 10^{-6} M, respectively, it can be assumed that Pb divalent cations (Pb^{2+}) bind to proteins containing the above metals with a $Kd \leq 10^{-12}$ M. This value means that Pb^{2+} links cellular proteins at least one or three orders of magnitude more than other metal cations and may displace them from their natural sites. Obviously, these events deeply change the function of the proteins/enzymes involved. Both Pb^{2+} and Zn^{2+} ions are classified as borderline acids, showing properties of both hard and soft acids, and the binding coordination of Pb^{2+} based on number of ligands is more strictly related to that of Zn^{2+} than other cations.

One of the most recognized Pb targets is a Zn-containing enzyme called aminolevulinic acid dehydratase (ALAD). ALAD catalyzes the second step of heme synthesis, and its deficiency could determine a cutaneous disease called plumboporphyria with acute neurologic symptoms. The reduced activity of ALAD is strictly associated to Pb poisoning, even if inhibition of ALAD activity does not account for cognitive deficits in children with low blood Pb levels. As a consequence of ALAD activity reduction, Pb stimulated an auto-oxidation process that leaded to the production of reactive oxygen species (ROS).

Remarkably, the inhibition of ALAD is reversed in the presence of renal Pb-binding metallothionein, a kind of Zn-containing Pb-binding carrier, fundamental to regulate its excretion and detoxification.

However, even if Pb–Zn duel is more relevant in the biology of multiple metalloprotein clusters, Pb^{2+} may interfere also with Ca^{2+} cations in their flux and transport through membranes, accounting for its toxicity. A list of Pb^{2+}-protein complexes can be found in the Protein Data Bank (PDB) (in http://www.rcsb.org/pdb/) together with their models and crystallographic information.

In particular, calmodulin (CaM), an essential protein involved in Ca signaling but with high affinity to Pb^{2+}, has been extensively studied using different spectroscopic analytical methods. CaM contains Ca^{2+} in a helix-loop-helix structure with a pentagonal-bipyramid geometry and Pb^{2+} substituted Ca^{2+} from these loops. Pb^{2+}-CaM binding deactivated and modified the global conformation of the protein, independently of ion displacement but related to much stronger affinity for Pb^{2+} in regions of high surface negative charge potential (http://digitalarchive.gsu.edu/chemistry_diss/53). The implications of this mechanism on Pb^{2+} toxicity are several. First, Pb-binding proteins may result activated or inhibited according to metal concentration; second, nonspecific bindings may increase solubility and diffusion of proteins linked to toxic metals instead of native nontoxic ions; third and more dangerous, also many other nonmetalloproteins may opportunistically link Pb^{2+} and extend toxicity.

Pb^{2+} have high affinity for thiol groups (–SH) on proteins or enzymes, that blocks the activities in many enzymes where Pb^{2+} competes with other metal cations such as Zn^{2+} and iron (Fe^{2+}). This is very important when Pb^{2+} disrupts Zn-binding domain in DNA-regulatory proteins. Indeed, also in the heme synthetic pathway, a large number of sites for Pb activity are found. Moreover, even if the extent to which Pb^{2+} may directly substitute Fe^{2+} in biological effects is not easily valuable, it has been demonstrated to occupy the Fe site inside ribonucleotide reductase, an enzyme that catalyzes the formation of fundamental deoxyribonucleotides in cellular DNA.

Similar action is evident when Pb^{2+} substitutes Fe^{2+} in the structure of the divalent cation transporter-1, a specific transporter of this metal involved in the uptake. Fowler laboratory at the University of Maryland in the USA, well characterized many Pb-carrier proteins in the 1990s (Fowler. 1998). Remarkably, the main results obtained by Fowler's seminal studies were that individuals and different species significantly showed different reactions to Pb exposure and also different reactions in specific organs in the same subject. In particular, the equal Pb threshold detected in the blood did not prove the effective manifestation of human disease. When Pb^{2+} cations bound to cysteine residues in proteins, they induced a specific band in the ultraviolet region that might be easily registered.

A novel sensitive chemical analysis of albumin and Pb interactions using fluorescence and *Raman spectroscopy*, confirmed by circular dichroism absorption, indicated that Pb^{2+} cations linked to amino acids like tyrosine and phenylalanine in the albumin molecule and promoted the onset of multiple albumin clusters, due to specific dipole interaction.

Pb-Induced Renal Damage: Experimental Models and Clinical Studies

The kidney represents the main excretory route for absorbed Pb in human body, and so, this metal has been considered to be a risk factor for severe disorders, such as saturnine gout, hypertension cardiovascular diseases, and atherosclerosis. Acute Pb-induced nephropathy is associated to proximal tubules alterations and the development of a Fanconi-type disease, but chronic Pb damage results in irreversible disruption of glomerular activity and accelerated risk of renal cancer. Even if scientists are still debating on the real involvement or contributory role of Pb in the onset and progression of chronic renal failure in humans (Evans and Elinder 2011), chronic Pb nephropathy, associated to gout and hypertension, has been reported for moonshine consumption as end-stage renal disease when the common Pb-associated symptoms were absent.

Multiple effects induced by Pb poisoning, such as disruption of energy production and oxidative damage, altered calcium metabolism and glucose homeostasis, act on many different cellular targets, such as cytoplasmic enzymes, organelles like mitochondria, endoplasmic reticulum (ER), the nucleus, and plasma membrane.

In vitro, in a renal tubular cell line (Madin-Darby canine kidney-MDCK), Pb inhibited the mobilization of Ca ions from nonmitochondrial site such as ER. This fundamental organelle is crucial for cell survival and function like Ca ion homeostasis, protein production, and lipid-membrane biosynthesis. Pb exposure in renal cells has been demonstrated to stimulate the main Ca-receptor site in the ER, called phosphatidylinositol receptor (IP3R), responsible for the transport of the ion toward the cytoplasm. This event destabilizes Ca-dependent resident *molecular chaperones*, such as glucose-regulated protein, 78 kDa (Grp78 also known as BiP or HSPA5) and 94 kDa (Grp94 also known as HSP90B1). A molecular chaperone is called a protein that helps the correct folding of other client cellular proteins or enzymes so avoiding a nonfunctional status.

A fundamental aspect of Pb toxicity, firstly discovered in astrocyte-like C6 glioma cells by Tiffany-Castiglioni laboratory of Texas University (Qian et al. 2000), is that Pb targets the protein Grp78, the main resident ER chaperone. This strict interaction compromised protein secretion, exacerbated protein aggregation, and increased cellular sensitivity to oxidative damage. Using two-dimensional gel electrophoresis, the presence of a strong complex between Pb (used as acetate salt) and Grp78 was detected in nervous cells: this aggregation may represent a tolerance mechanism, because depletion of Grp78 protein sensitizes cells to Pb. However, even if different cell types may result in different Pb susceptibility and Grp78 expression, a strong Grp78 expression was detected also in a rat renal proximal tubular cell line (NRK-52E) exposed to Pb (used as chloride salt) and Grp78 immune-positive aggregates were observed within the nucleus after 24–48 h (Fig. 1) (Stacchiotti et al. 2009).

To address the mechanism implicated in Pb vasculopathy, a recent study suggests that Pb stimulates intense expression of resident ER chaperones in vascular endothelial cells, in a dose- and time-dependent manner. This condition is called ER stress that means that ER chaperones (Grp78 and Grp94 proteins) lose their ability to manage their client proteins and regulate Ca^{2+} homeostasis. These events

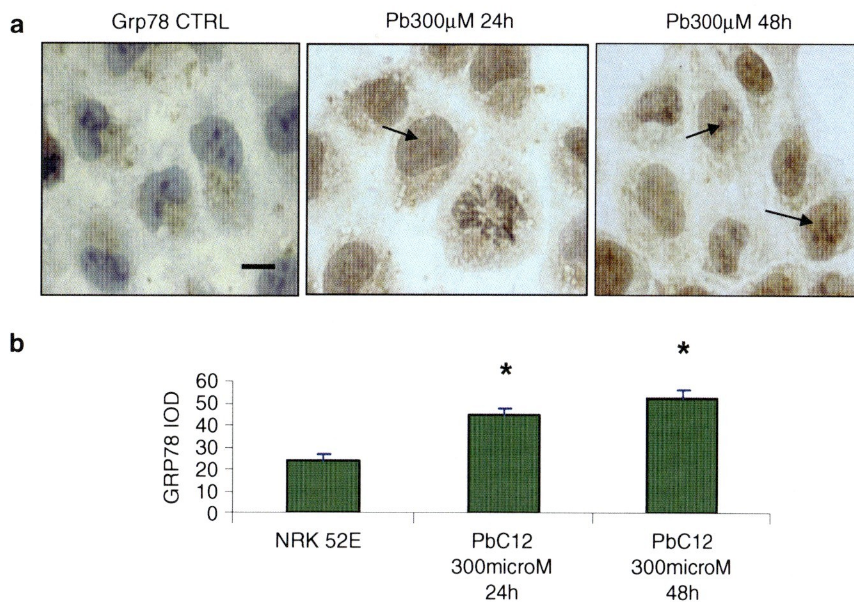

Lead Nephrotoxicity, Fig. 1 Immunohistochemical analysis of GRP78 in NRK52E proximal tubular renal cells. (**a**) Representative pictures from controls (ctrl) and cells treated with Pb chloride (PbCl2) at different times. Note brown aggregates in the cytoplasm that translocated in the nuclei after Pb intake. Bar = 50 μm. (**b**) Quantization of GRP78 immunostaining as integrated optical density (IOD) evidenced increased expression in a time-dependent manner (*significant at $p < 0.05$)

represent a specific and unique defense response of endothelial cells to Pb^{2+} and not other metal ions.

A remarkable characteristic of Pb intoxication is the presence in the cytoplasm and often the nucleus of poisoned cells in humans and animals of specific Pb-proteins complexes visualized by microscopy and called inclusion bodies (IBs). Since the early 1960s, many morphologic and ultrastructural studies characterized nuclear IBs in the kidney, mainly in proximal convoluted tubules and in other organs such as liver. They have been described as distinct from nucleolar components and sometimes able to shift from nucleus and cytoplasm and vice versa but always associated to specific proteins. However, if in renal cells exposed to Pb different inhibitors of protein synthesis were added, IBs disappeared or became very few respect to the same cultures exposed only to Pb. So, IBs require an active protein synthesis for their formation and persistence in vitro. Moreover, IBs, as a pathognomonic sign of Pb nephropathy, have been considered as a tool to render the metal toxicologically inert and to hamper its delivery to other intracellular targets. To address the specific protein complexed to form Pb-containing IBs in cells, different studies identified metallothioneins (MTs).

MTs belong to a class of low molecular cysteine-rich proteins found in most mammalian tissues that detoxificate several metal ions. Mammalian MTs contain 61–62 amino acids of which 20 are cysteine residues and therefore have high ability to bind different metal ions, such as Zn, cadmium (Cd), or Pb. The binding number of Pb atoms in MTs is 7, and two different species of Pb_7-MTs are formed at alkali/neutral or weak acid conditions. All Pb-MTs complexes have been demonstrated as spontaneous, exothermic, and entropy-increasing reactions, using microcalorimetry technique and a thermodynamic approach. Remember that *entropy*, in thermodynamic applied to chemistry, is defined as the tendency of a chemical reaction to be favored and determines the state of disorder and randomness obtained in the form of heat. Indeed, according to the second law of thermodynamics, entropy in an isolated system increases or remains constant.

Many MT isoforms have been discovered even if MT-I and MT-II isoforms are the most studied and ubiquitous. So MT-I and MT-II double knockout mice, called MT-null, is an animal model particularly useful to study Pb toxicity in vivo. This genetically modified model not present in nature does not express MT-I nor MT-II genes and, when exposed to Pb, showed evident carcinogenesis in the kidney with respect to the wild-type mice. Remarkably, the MT-null mice were unable to form Pb-induced IBs (Qu et al. 2002).

Besides MTs, alpha-synuclein (Scna) is another protein component of IB, that can be considered analogous to aggresomes, detected in vitro in renal

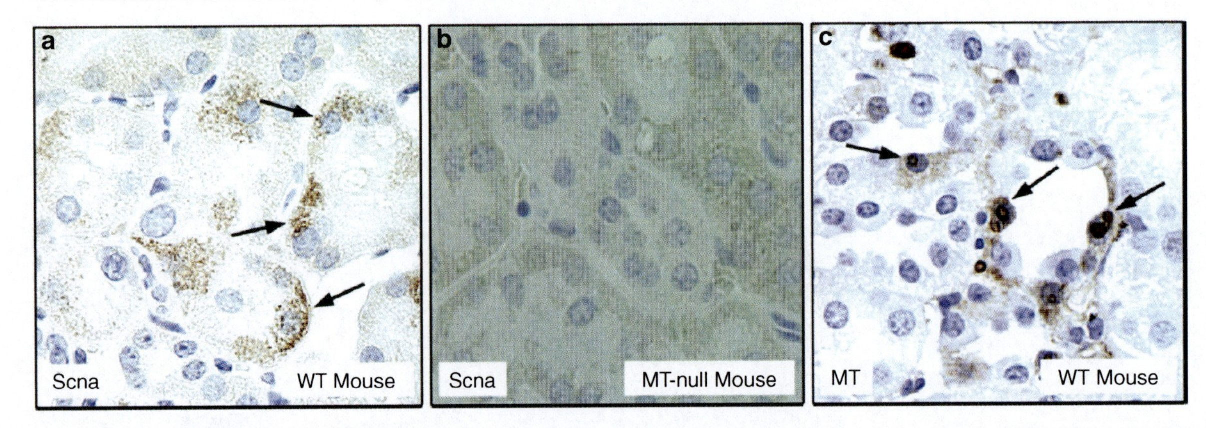

Lead Nephrotoxicity, Fig. 2 Immunohistochemical analysis of alpha-synuclein (Scna) and metallothioneins (MT) staining intensity and localization in Pb-treated wild-type (WT) mouse kidney. (**a**) Representative section from the kidney of a WT mouse chronically exposed to Pb in the drinking water showing intense and clustered staining (*brown*) for Scna that appeared to occur in cells with early, forming Pb-induced inclusion bodies (IBs) (*arrows*). (**b**) MT-null mice do not show staining for Scna. (**c**) A representative section of WT mouse kidney chronically exposed to Pb in the drinking water showing MT on the surface of several IBs (*arrows*) (From Zuo et al. license n° 2711850205991, July 18, 2011)

MT-null cells transfected with MTs exposed to Pb for 48 h. Moreover, this protein was strictly associated with MTs also ex vivo using a coprecipitation assay with specific MT antibody (Figs. 2 and 3) (Zuo et al. 2009).

Unfortunately, actually there is nothing comparable to MT-null animal model in human population, where a strict association between proteins-IBs and Pb is not so evident. However, it has been found a natural variability in MT expression in the relatively homogeneous ethnic groups, and subjects with low MT expression might be more susceptible to Pb poisoning and cancer.

It must be outlined that the characterization of specific proteins in aggresomes is a very hard work, because these proteins disappeared as IBs are formed and many data actually obtained are only inferred. Besides these considerations, it must be considered that intranuclear eosinophilic IBs are evident only during early years of exposure, so the clinical diagnosis of Pb nephropathy is usually obtained on a personal history of Pb environmental or occupational exposure associated with altered renal function test and disclosure of a nonimmunologically mediated interstitial nephritis. Usually in renal biopsies of chronic Pb nephropathy, nonspecific tubular atrophy and interstitial fibrosis with low inflammation were observed (Wedeen 2008).

Even if Pb is an unequivocal nephrocarcinogen in rodents even if associated with early life exposure (Tokar et al. 2010), also in humans, there is evidence that links cancer and Pb exposure.

International Agency for Research on Cancer (IARC) considered Pb as a possible human carcinogen (inserted in the group 2B), and its inorganic compounds have been classified as probable human carcinogens (inserted in the group 2A). However, if investigated in different biological systems, the genotoxic effects of Pb are still contradictory (recently reviewed by Garcia-Leston et al. 2010). However, for many years, a direct interaction of transcriptional regulator factors and Pb (where it substitutes Zn) has not been directly probed due to insufficient techniques and instrumentations that assumed Pb to be a spectroscopically silent metal. Actually, the genotoxic damage associated to Pb-proteins accumulation may be due to direct effect, that is, DNA structure damage, but also to indirect mechanisms. In this last case, Pb can substitute other metal ions in many DNA processing or repairing enzymes, so limiting or modifying DNA function. Even if in animal models Pb is a reproducible carcinogen, in humans, Pb was considered as a direct renal carcinogen only at high concentrations when it may interact with other genotoxic agents and so facilitate the onset of cancer. To support the "*facilitative role*" of Pb in carcinogenesis, that is, this metal by itself may not be both necessary and sufficient for the induction of cancer but may permit carcinogenic events. In particular, its direct interaction with DNA-surveillance proteins such as

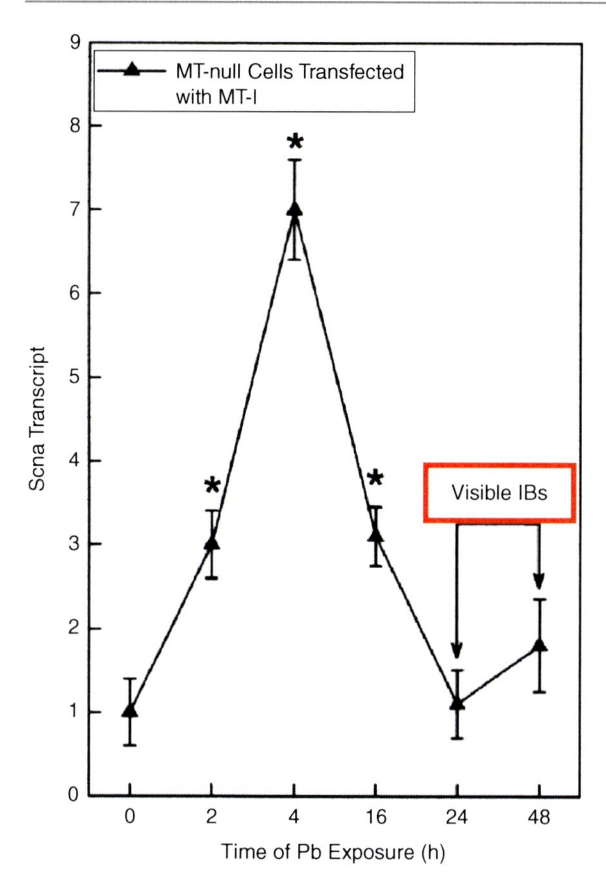

Lead Nephrotoxicity, Fig. 3 Alpha-synuclein (Scna) transcript in MT-null cells transfected with MT-I after Pb treatment. MT-null cells transfected with MT-I were exposed to 200 μM Pb for 0–48 h. Data are presented as the mean ± SEM. An *asterisk* (*) indicates a significant difference from untreated cells. The *arrows* indicate the approximate time Pb-induced inclusion bodies (IBs) become visible by light microscope. This suggest that both Scna and MT-I are strictly related in IBs formation (From Zuo et al. license n° 2711850205991)

protamines or histones reduced DNA protection. Moreover, Pb inhibits in vitro the repairing activity of endonuclease Apoe-1, a crucial enzyme that repairs mutagenic and cytotoxic basic sites in DNA, so acting as co-carcinogen. Indeed, Pb-protein interactions might affect posttranslational structure of specific proteins like the tumor suppressor protein p53, a Zn-binding protein, even if this mechanism has been clearly demonstrated for Cd but only hypothesized for Pb.

Another important indirect mechanism of Pb genotoxicity is related to MTs deficiency as evidenced in MT-null model as early preneoplastic renal lesions.

Several studies in human population have reported that delta-aminolevulinic acid dehydratase (ALAD), a gene located on chromosome 9q34, showed polymorphism related to Pb toxicity. Three different phenotypes have been characterized: ALAD1-1, ALAD1-2, and ALAD2-2. The ALAD2 mutations, the protective allele, in Asian and Caucasian populations are similar but absent in Africans. So, the presence of critical genes target in population and their polymorphism might potentially identify individuals at increased risk of Pb poisoning.

Pb is shown to alter antioxidant activities by inhibiting functional sulfhydryl (–SH) groups in several antioxidant enzymes such as superoxide dismutase (SOD), catalase, glutathione peroxidase, and glucose-6-phosphate dehydrogenase. This last enzyme contains many –SH group and supplies cells with most of the extramitochondrial NADPH through the oxidation of glucose-6-phosphate to 6-phosphogluconate and is inhibited by Pb. In a recent study performed on rat proximal tubular cells exposed to Pb acetate (0.25–1 μM), a dose-dependent alteration in mitochondrial membrane potential and depletion of intracellular *glutathione* were revealed, together with apoptotic changes as a consequence of oxidative damage.

Other enzymes linked to Pb-induced oxidative damage are delta amino levulinc acid synthetase (ALAS) and delta amino levulinic acid dehydratase (ALAD). In particular the last enzyme catalyzes the condensation of two molecules of alanine to form porphobilinogen in the heme biosynthetic pathway. When Pb is linked to ALAD, in the erythrocytes in the blood, it inhibits the enzyme activity and induces accumulation of its substrate delta amino levulinc acid (ALA). ALA can be easily transformed from a keto form into an enol form by auto-oxidation and produces superoxide ions. So, many free radicals or reactive oxygen species (ROS), like superoxide ions and hydrogen peroxide, have been originated in the blood and urine, with a crucial role in the progression of chronic renal failure.

In biological chemistry, ROS indicate different molecules with unpaired electrons on an open shell configuration that may have positive, negative, or zero charge. With some exceptions, the unpaired electrons cause radicals to be highly chemically reactive and, if allowed to run free in the body, are believed to be involved in overt oxidative damage. It has been experimentally demonstrated not only in vitro but

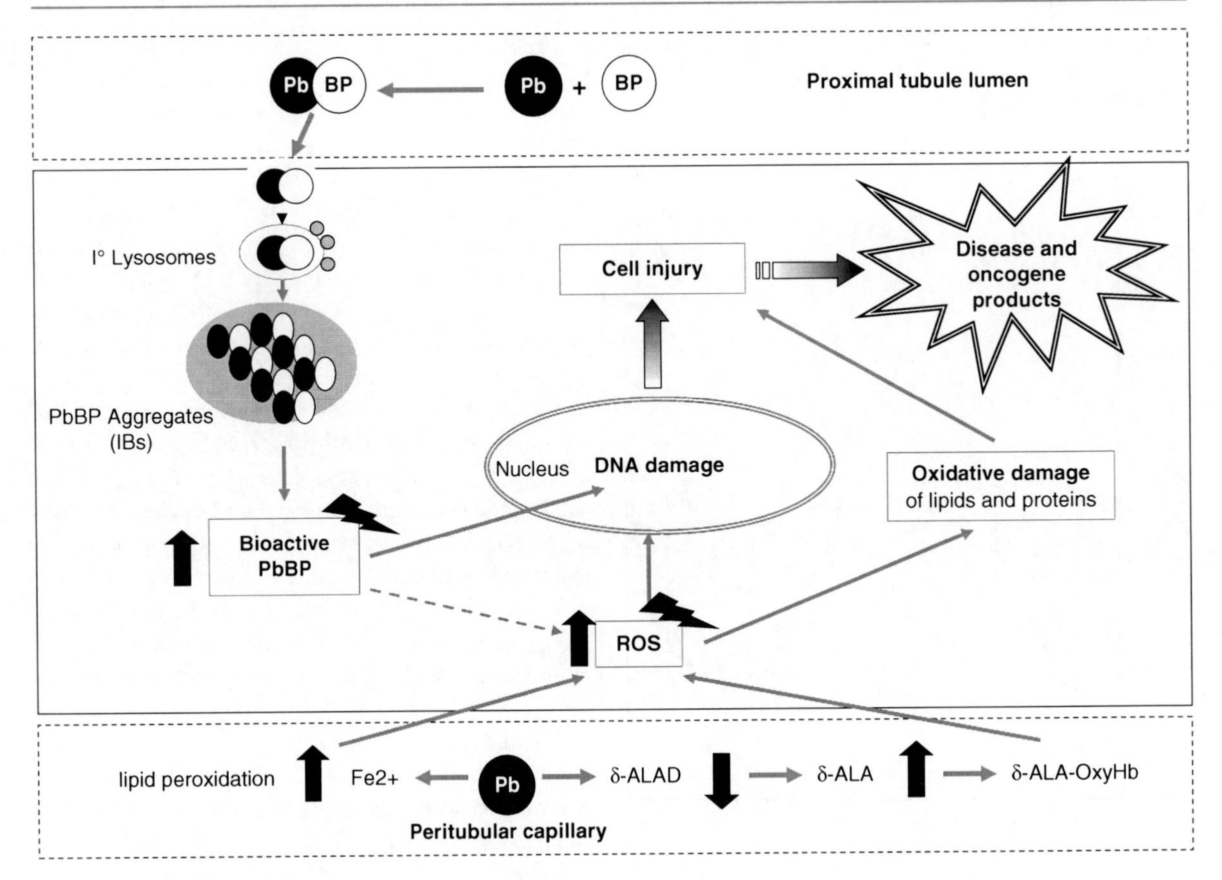

Lead Nephrotoxicity, Fig. 4 Multiple mechanisms for lead-induced oxidative damage to renal cells. Diagram of renal proximal tubule cell showing reabsorption of lead-proteins aggregates inside the tubule lumen and lead atoms inside blood. This event could alter DNA transcription, leading to oncogene expression and/or to cytoplasmatic oxidative damage by production of reactive oxygen species that induced cell injury. *Pb* lead, *BP* binding protein, *ROS* reactive oxygen species, *ALAD* aminolevulinic acid dehydratase, *ALA* aminolevulinic acid, *Oxy-Hb* oxidate hemoglobin, *IBs* inclusion bodies; arrows up, increase; arrows down, decrease

also in rats exposed to Pb (as Pb-acetate 100 ppm) in drinking water for 12 weeks, overt hypertension, and deregulated enzymatic activities of antioxidant markers, such as catalase, superoxide dismutase, glutathione peroxidase in the kidney, and thoracic aorta. Multiple hypotheses on Pb-induced renal toxicity are resumed in Fig. 4.

Novel Perspectives to Study Pb Nephrotoxicity

Novel perspectives in the actual research on Pb poisoning are probably opened by the application of proteomic techniques in the kidney exposed to subchronic and often asymptomatic doses (Chen et al. 2011), the study of genetic polymorphisms in occupationally exposed subjects, and the use of alternative in vivo models like *Drosophila* or Zebra fish (*Danio rerio*).

Toxic proteomics utilizes global protein measurement techniques to discover mechanisms associated to metal toxicity and eventually predict the onset of organ-specific damage. This new area of research even still at an early phase must be integrated with more classic and validated diagnostic tools, like chelation with ethylenediaminetetraaceticacid (EDTA) and dimercaptosuccinic acid (DMSA), but represents a challenge for the future.

Cross-References

▶ Calcium-Binding Protein Site Types
▶ Calcium-Binding Proteins, Overview

▶ Iron Homeostasis in Health and Disease
▶ Lead and RNA
▶ Lead, Physical and Chemical Properties
▶ Metallothioneins and Lead
▶ Zinc-Binding Proteins, Abundance
▶ Zinc-Binding Sites in Proteins

References

Chen J, Mercer G, Roth S, Abraham L, Lutz P, Ercal N, Neal RE (2011) Sub-chronic lead exposure alters kidney proteome profiles. Hum Exp Toxicol 30(10):1616–1625. doi:10.1177/0960327110396521

Evans M, Elinder C (2011) Chronic renal failure from lead: myth or evidence-based fact. Kidney Int 79:272–279

Fadrowski J, Navas-Acien A, Tellez-Plaza M, Guallar E, Weaver V, Furth S (2010) Blood lead level and kidney function in US adolescent. The third national health and nutrition examination survey. Arch Intern Med 170:75–82

Fowler B (1998) Roles of lead-binding proteins in mediating lead bioavailability. Environ Health Perspect 106: 1585–1587

Garcia-Leston J, Mendez J, Pasaro E, Laffon B (2010) Genotoxic effects of lead: an updated review. Environ Int 36:623–636

Qian Y, Harris E, Zheng Y, Tiffany-Castiglioni E (2000) Lead targets GRP78, a molecular chaperone, in C6 rat glioma cells. Toxicol Appl Pharmacol 163:260–266

Qu W, Diwan B, Liu J, Goyer R, Dawson T, Horton J, Cherian MG, Waalkes MP (2002) The metallothionein-null phenotype is associated with heightened sensitivity to lead toxicity and an inability to form inclusion bodies. Am J Pathol 160:1047–1056

Stacchiotti A, Morandini F, Bettoni F, Schena I, Lavazza A, Grigolato P, Apostoli P, Rezzani R, Aleo MF (2009) Stress proteins and oxidative damage in a renal derived cell line exposed to inorganic mercury and lead. Toxicology 264:215–224

Tokar E, Diwan B, Waalkes M (2010) Early life inorganic lead exposure induces testicular teratoma and renal and urinary bladder preneoplasia in adult metallothionein-knockout mice but not in wild type mice. Toxicology 276:5–10

Wedeen R (2008) Lead nephropathy. In: de Broe M, Porter G (eds) Clinical nephrotoxins – renal injury from drugs and chemicals, 3rd edn. Springer, Heidelberg, pp. 777–784

Zuo P, Qu W, Cooper R, Goyer RA, Diwan BA, Waalkes MP (2009) Potential role of alpha-synuclein and metallothionein in lead-induced inclusion body formation. Toxicol Sci 111:100–108

Lead Remediation Form Environment

▶ Lead Detoxification Systems in Plants

Lead Toxicity

▶ Lead Detoxification Systems in Plants

Lead Transporters

Maksim A. Shlykov, Henry Chan and Milton H. Saier Jr.
Department of Molecular Biology, Division of Biological Sciences, University of California at San Diego, La Jolla, CA, USA

Synonyms

Lead (Pb^{2+}) permeases/porters; Pb^{2+} excretion/efflux; Pb^{2+} resistance; Pb^{2+} uptake; P-type ATPase/ion pump

Definition

Transmembrane proteins that mediate the transport of Pb^{2+} and its derivatives into and out of the cell.

Introduction

Many transmembrane export systems are capable of acting on the lead ion, Pb^{2+}, but few of these appear to exhibit strict metal ion specificity. They provide protection against the toxic effects of high concentrations of Pb^{2+} and a variety of other metals. However, metal uptake porters, mostly secondary carriers, are also known. In this brief synopsis of lead transporters, a summary of what is known about these systems is presented.

Heavy-Metal-Ion-Transporting ATPases

P-type ATPases utilize the energy liberated in the exergonic ATP hydrolysis reaction to translocate positively charged substrates across membranes against their concentration gradients. For example, the animal Na^+, K^+-ATPases pump both Na^+ (out) and K^+ (in) against their concentration gradients.

These ATP-driven pumps form acylphosphate intermediates in a reaction cycle in which the γ-phosphate of ATP is transferred to an aspartyl residue within a well-conserved motif in the ATPase domains of the enzymes. This phosphorylation site can be found in all functional P-type ATPases (Chan et al. 2010). Most such enzymes are pumps that help to maintain the homeostasis of essential ions. They also mediate resistance to toxic concentrations of Pb^{2+}, Zn^{2+}, Cd^{2+}, Cu^{2+}, Cu^{+}, Ag^{+}, and other metal ions. Probably none of these enzymes/transporters are specific for lead. They catalyze Pb^{2+} expulsion from cells derived from tremendously varied organismal types.

Heavy-metal-ion ATPases are found ubiquitously in bacteria, archaea, and eukaryotes with a predominance in prokaryotes (Chan et al. 2010). The best characterized of these pumps are found in bacteria, but their functions and mechanisms of action are probably similar or the same in other types of organisms (Thever and Saier 2009; Chan et al. 2010). While the copper ATPases (TC subfamily 5; see the Transporter Classification Database; TCDB; www.tcdb.org) can function with either inwardly (uptake) or outwardly (efflux) directed polarity, almost all heavy metal ATPases (TC subfamily 6) function with outwardly directed polarity. Their primary function appears to be protection from toxic levels of these ions.

Rensing et al. (1998) reported that CadC of *Staphylococcus aureus* is a metal-responsive repressor that responds to metal ions in the order $Pb^{2+} > Cd^{2+} > Zn^{2+}$ and controls expression of the *cadA* gene which encodes a lead-, cadmium-, and zinc-expelling P-type ATPase. They showed that Zn^{2+} transport by both CadA (TC# 3.A.3.6.1) of *S. aureus* and ZntA (TC# 3.A.3.6.2) of *Escherichia coli* is inhibited by Pb^{2+} and that these pumps confer resistance to Pb^{2+}. Various metal-transporting P-type ATPases of TC families 5 (copper ions) and 6 (heavy-metal ions) exhibiting different characteristics have been described as reported in TCDB. These ATPases have metal binding domains in both their cytosolic N-terminal regions that often exhibit a GXXCXXC motif [G = glycine, C = cysteine, X = any amino acid] and in their sixth transmembrane domains that exhibit a conserved CPC motif [P = proline]. Each of these enzymes also displays a translocation domain and a highly conserved histidine-proline sequence 34 to 43 residues C-terminal to the CPC motif. These conserved residues are essential for function (Chan et al. 2010; Thever and Saier 2009).

ZntA of *E. coli* is a P-type ATPase that, similar to the *S. aureus* CadA, transports Zn^{2+}, Cd^{2+}, and Pb^{2+}. Its synthesis at the transcriptional level is regulated by ZntR, a MerR transcription factor homologue. Disruption of ZntA results in sensitivity to Zn^{2+}, Cd^{2+}, and Pb^{2+}. The unavailability of a Pb^{2+} radioisotope prevented a direct demonstration that ZntA transports Pb^{2+}, but ATP-dependent Zn^{2+} uptake was inhibited to the same degree by Pb^{2+} and Cd^{2+}, and the latter is a known substrate (Rensing et al. 1998).

The distinctive, highly polar, metal binding amino-terminal domains in both the Cu^{+}/Ag^{+} and the $Pb^{2+}/Zn^{2+}/Cd^{2+}$ ATPases contain 1–6 repeats of the conserved heavy-metal-binding domain (HMBD), each of about 70 amino acids (aas) in length, each including the GXXCXXC motif. The conserved CPC motif in the sixth transmembrane helix is believed to be part of the translocation pathway. The N-terminal domains are not absolutely required for transport, but they facilitate the process either by feeding the metal ion into the channel or by providing an activation function.

The substrate specificity of the *E. coli* ZntA has been examined using an assay which involves measurement of the formation of the aspartyl phosphate bond in the enzyme. ZntA formed the acylphosphate intermediate, not only with the known physiological substrates, Pb^{2+}, Zn^{2+}, and Cd^{2+}, but also with other divalent metal ions, including Co^{2+}, Cu^{2+}, and Ni^{2+}. No activity was observed with monovalent metal ions, Ag^{+} and Cu^{+}. ZntA is selective toward divalent metal ions that prefer sulfur as ligands. ZntA, and possibly other heavy-metal P-type ATPases, appears to recognize their substrates using charge and ligand preferences instead of ionic radii. Metal ion size cannot be significant in determining metal ion specificity or selectivity because Pb^{2+} and Cd^{2+} are much larger than Zn^{2+}, Ni^{2+}, Co^{2+}, and Cu^{2+}. Asp436 is the site of phosphorylation, and no intermediate could be detected in the presence of any of the metal ions for the D436N mutant in which the nonphosphorylatable asparagine (N) replaces the phosphorylatable aspartate (D) in the enzyme, confirming this prediction.

As noted above, the *cad* operon of *S. aureus* has two genes encoding the CadC transcriptional repressor and the CadA P-type ATPase (TC#3.A.3.6.1). The *cadA* gene complemented the $Zn^{2+}/Cd^{2+}/Pb^{2+}$-sensitive

Lead Transporters, Fig. 1 Six topological types of P-type ATPases. Heavy-metal ATPases comprising families 5 and 6 are unique in having the same 8 TMS topology which differs from all other types

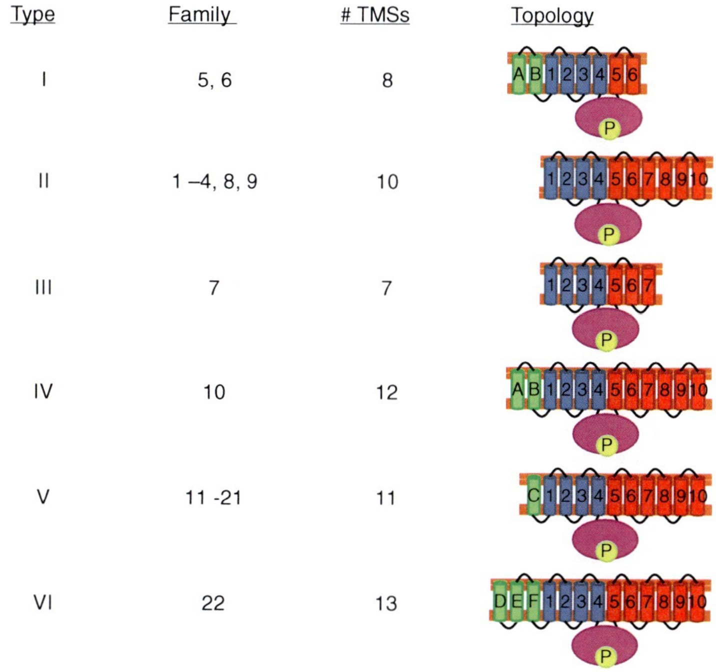

Type	Family	# TMSs	Topology
I	5, 6	8	AB123456 P
II	1 –4, 8, 9	10	12345678910 P
III	7	7	1234567 P
IV	10	12	AB12345678910 P
V	11 -21	11	C12345678910 P
VI	22	13	DEF12345678910 P

phenotypes of an *E. coli zntA*-disrupted strain. Thus, CadA, like ZntA, exports Zn^{2+}, Cd^{2+}, and Pb^{2+}. The topology of CadA has been experimentally determined using *phoA* and *lacZ* fusions. In contrast to ATPases outside of subfamilies 5 and 6, an 8 TMS topology was demonstrated as expected, based on hydropathy analyses (see Fig. 1; Chan et al. 2010).The recently elucidated X-ray structure of a copper-transporting P-type ATPase of the same topology confirmed this conclusion (Gourdon et al. 2011; see below).

CadA of *Listeria monocytogenes* (TC#3.A.3.6.8) also protects against heavy metal $Zn^{2+}/Cd^{2+}/Pb^{2+}$ toxicity. In some cases, the typical CXXC-motif-possessing N-terminal heavy-metal-binding domains include histidine-rich sequences, sometimes instead of the usual cysteine motif, and this may serve a comparable metal-binding function.

Over 30 mutations have been introduced into the *L. monocytogenes* heavy-metal ATPase *cadA* gene. Functional studies of the mutants showed that Cys354 and Cys356 in TMS 6 as well as Asp692 in TMS 8 and Met149 in TMS 3 participate in Cd^{2+} binding. In the canonical CPC motif, each of the two cysteines acts at distinct steps in the transport mechanism. Cys354 is directly involved in Cd^{2+} binding, while Cys356 seems to be required for Cd^{2+} occlusion within the enzymes. In TMS 4, Glu164, which is conserved among heavy-metal P-ATPases, may be required for Cd^{2+} release. Finally, analysis of the role of Cd^{2+} in the phosphorylation by ATP or inorganic phosphate suggested that two Cd^{2+} ions are involved in the reaction cycle of CadA. These observations are likely to be applicable to lead transport catalyzed by CadA as well as other heavy-metal-transporting ATPases.

It has been shown that the two metal-binding domains of the *Anabaena* $Pb^{2+}/Zn^{2+}/Cd^{2+}$ efflux pump, AztA (TC#3.A.3.6.13), are functionally nonequivalent with respect to zinc resistance. Thus, the first metal-binding domain is more critical for Cd^{2+}, Zn^{2+}, and Pb^{2+} resistance than the second. The two domains apparently function independently with Zn^{2+} but cooperatively with Cd^{2+} or Pb^{2+}. Each of these two metal ions is capable of bridging the two metal-binding domains, possibly accounting for the observed cooperativity.

Detailed high-resolution X-ray structures of heavy-metal P-type ATPases were not available prior to 2011 when Gourdon et al. reported the structure of CopA (TC# 3.A.3.5.30), a Cu^{+}-ATPase from *Legionella pneumophila* at 3.2 Å resolution (PDB code 3RFU; Fig. 2). A proposed three-stage copper transport pathway involves several well-conserved residues as noted above for the *Listeria* homologue.

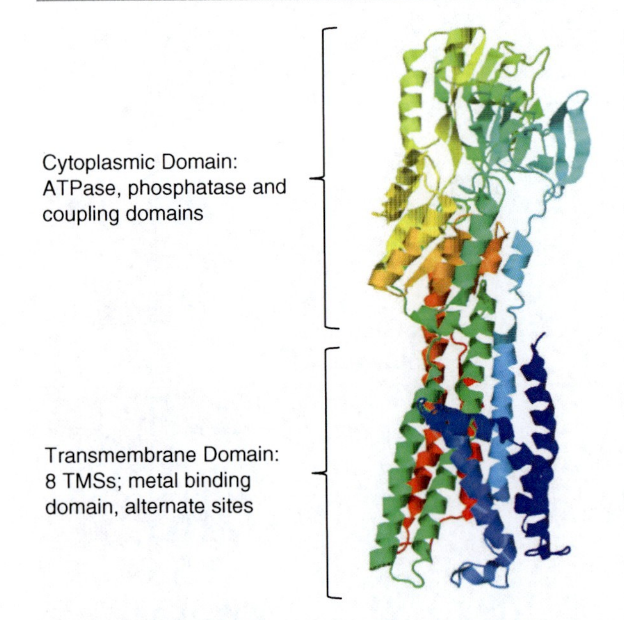

Lead Transporters, Fig. 2 3.2 Å X-ray structure of the CopA copper-transporting P-ATPase of *Legionella pneumophila*. An 8 TMS transmembrane topology is observed as shown also in Fig. 1. The cytoplasmic domains include the ATPase, phosphatase, and coupling domains. The structure was uploaded to the Protein Data Bank (PDB) by Gourdon et al. (2011)

An N-terminal transmembrane helix kinks at a double-glycine motif displaying an amphipathic helix that lines a putative copper entry point at the intracellular interface. Further, an ATPase-coupled copper release mechanism from the binding sites in the membrane via an extracellular exit site was apparent (Gourdon et al. 2011). The results provide the first in-depth description of a heavy-metal-translocating P-type ATPase.

Secondary Active Lead Transporters

P-type ATPases represent the class of lead and other metal ion transporters that function via a primary active transport mechanism, but transporters functioning by other mechanisms have been discovered as well. The Pb^{2+}-resistance determinant, *pbr*, on the plasmid, pMOL30, in *Ralstonia* (*Cupriavidus*) *metallidurans* strain CH34 has been studied in detail. The inducible *pbr* regulon is responsible for the uptake, efflux, and accumulation of Pb^{2+} (Taghavi et al. 2009). There are actually three operons: (1) *pbrRT* which encodes the Pb^{2+}-dependent regulator, PbrR, of the *pbr* regulon and a Pb^{2+} uptake protein, PbrT (TC# 9.A.10.2.1); (2) *pbrU*, encoding a defective major facilitator superfamily (MFS; TC# 2.A.1) porter of unknown substrate specificity; and (3) *pbrABCD*, transcribed divergently from *pbrRTU* and encoding a P-type Pb^{2+} efflux ATPase, PbrA (TC# 3.A.3.6.4), a phosphatase, PbrB (Hynninen et al. 2009), a predicted prolipoprotein signal peptidase, PbrC, and a cytoplasmic Pb^{2+}-binding protein, PbrD, that is essential for lead sequestration. Although originally predicted to be two separate ORFs, the *prbBC* gene codes for the PbrB/PbrC fusion protein (Monchy et al. 2007; Mergeay et al. 2009).

The *pbrU* gene was missed initially because it had been knocked out by a transposon. The MFS porter gene, *pbrU*, is incomplete, probably encoding an inactive transporter. The lead resistance transcriptional regulator, PbrR, binds to two promoters, one in front of *pbrTR* and one in front of *pbrU*. Binding to the first promoter leads to expression of *pbrTR*, while binding to the second promoter leads to the expression of *pbrU*. *pbrABCD*, also controlled by PbrR, is divergently expressed from the complementary strand.

Expression of PbrT in the absence of PbrABCD resulted in Pb^{2+} hypersensitivity, due to increased Pb^{2+} uptake into the cytoplasm (Taghavi et al. 2009). Once Pb^{2+} has entered the cell, it can be exported by the PbrA Pb^{2+}-efflux ATPase but can also be bound by the PbrD protein. PbrD is not absolutely required for Pb^{2+} resistance, but cells lacking it showed decreased accumulation of Pb^{2+} compared to wild-type cells.

The PbrA Pb^{2+}-efflux ATPase has been shown to be functional and able to counteract the consequences of PbrT-driven Pb^{2+} uptake. However, for full Pb^{2+} resistance, PbrB and PbrC are required. It has been proposed that PbrB, a phosphatase, cooperates with PbrA by sequestering the Pb^{2+} exported by PbrA as a phosphate salt. ATPase and phosphatase cooperativity may provide a dual mechanism of toxic metal detoxification. The presence of *pbrC*, encoding a predicted prolipoprotein signal peptidase, was the first prolipoprotein signal peptidase gene to be found within a heavy-metal-resistance operon.

Proteome and transcriptome analyses, combined with mutagenesis, were used to better understand the response of *R. metallidurans* CH34 to Pb^{2+} (Taghavi et al. 2009). Structural Pb^{2+}-resistance genes of the pMOL30-bearing *pbrUTRABCD* regulon formed the major line of defense against Pb^{2+}. Expression of the *pbrR*(2)-*cadA*-*pbrC*(2) operon of the CMGI-1 region and the chromosomally encoded *zntA* gene were clearly induced in the presence of Pb^{2+}. After inactivation of the *pbrA*, *pbrB*, or *pbrD* gene, the expression of the *pbrR*(2)-*cadA*-*pbrC*(2) operon increased, suggesting synergistic interactions between *pbrUTRABCD* and *pbrR*(2)-*cadA*-*pbrC*(2)to maintain a low intracellular Pb^{2+} concentration. *pbrR*(2) *cadA* and *pbrC*(2) gene functions can complement and compensate for mutations in the *pbrA* and *pbrD* genes. This ability of *zntA* and *cadA* to complement the loss of *pbrA* was confirmed by mutational analyses. The *pbrB*::Tn(Km2) mutation resulted in the most significant decrease of Pb^{2+} resistance, indicating that Pb^{2+} sequestration, avoiding reentry of this toxic metal ion, forms a critical step in *pbr*-encoded Pb^{2+} resistance (Taghavi et al. 2009).

Lead resistance has been reported in both Gram-negative and Gram-positive bacteria. *Pseudomonas marginalis*, a Gram-negative bacterium, shows extracellular lead exclusion, and *Bacillus megaterium*, a Gram-positive bacterium, catalyzes intracellular cytoplasmic lead accumulation. Pb^{2+}-resistant strains of *Staphylococcus aureus* (Gram +) and *Citrobacter freundii* (Gram −) have also been isolated.

Analyses of the OFeT (iron; TC# 9.A.10.1) and PbrT (lead; TC# 9.A.10.2) transporter families led to the formation of the iron/lead transporter (ILT) superfamily (TC# 9.A.10; Debut et al. 2006). Both families share a 7 TMS topology, and TMSs 1–3 and 4–6 are homologous in each due to an ancient intragenic duplication event. Members of the PbrT family appear to be ubiquitous, with putative archaeal members being very distant homologues. PbrT is the only functionally characterized homologue known to take up Pb^{2+}. The remaining members were predicted to be Fe^{2+}/Fe^{3+} uptake porters. The lead porter, PbrT (TC# 9.A.10.2.1), along with the iron porter, Ftr1 (TC# 9.A.10.2.4), possess N-terminal extensions of about 400 aas. The N-terminal hydrophobic end of this extension in both proteins is believed to represent a TMS. A search using CDD revealed that PbrT possesses the cytochrome C domain. It would be of interest to investigate the role that a heme-containing domain plays in the function of PbrT.

It is not clear whether a carrier or a channel mechanism is used by members of the two families of the ILT superfamily to transport Fe^{3+} and/or Pb^{2+}. Whereas members of the OFeT family are believed to utilize their oxidase domains to oxidize Fe^{2+} prior to transport, PbrT family proteins lack such oxidase domains. It is possible, but not essential, that all of the ILT superfamily members function via a secondary active transport mechanism, but the cotransported substrate has not been identified. Because the membrane potential in virtually all cells is negative inside, a channel mechanism would allow accumulation against large concentration gradients. However, any active cation efflux mechanism would require antiport (countertransport) against a proton or sodium ion.

Several secondary active lead transporters have been identified. One example is TC# 2.A.5.4.8 which has been shown to be a $Cd^{2+}:HCO_3^-/Zn^{2+}:HCO_3^-$ symporter. Evidence indicates that this carrier additionally takes up Cu^{2+}, Pb^{2+}, and Hg^{2+}, based on competitive inhibition studies (Liu et al. 2008). The pmf-driven Nramp2 (TC# 2.A.55.2.1) protein, similarly exhibits broad specificity, transporting Fe^{2+}, Zn^{2+}, Mn^{2+}, Cu^{2+}, Co^{2+}, Ni^{2+}, and Pb^{2+}. Pb^{2+} transporting channels have yet to be identified and characterized.

Conclusions

It has been seen that several types of transporters, belonging to different protein families, have evolved independently to protect cells against Pb^{2+} toxicity, but Pb^{2+} uptake porters (e.g., PbrT) have also been identified. Generally, the efflux systems exhibit broad specificity, transporting many heavy metal ions. Since lead is a toxic ion with no known cellular benefits, it seems likely that PbrT and other lead importers will prove to also transport other metal ions, thus providing benefit to the organism. Further studies will be required to test this postulate.

Acknowledgments Work in the authors' laboratory was supported by NIH grant GM077402.

Cross-References

- ▶ Lead and Alzheimer's Disease
- ▶ Lead and Immune Function
- ▶ Lead and Phytoremediation
- ▶ Lead and RNA
- ▶ Lead Detoxification Systems in Plants
- ▶ Lead in Plants

▶ Lead Nephrotoxicity
▶ Lead, Physical and Chemical Properties
▶ Mercury and Lead, Effects on Voltage-Gated Calcium Channel Function
▶ Metallothioneins and Lead

References

Chan H, Babayan V, Blyumin E et al (2010) The P-type ATPase superfamily. J Mol Microbiol Biotechnol 19:5–104

Debut AJ, Dumay QC, Barabote RD et al (2006) The iron/lead transporter superfamily of Fe^{3+}/Pb^{2+} uptake systems. J Mol Microbiol Biotechnol 11:1–9

Gourdon P, Liu XY, Skjørringe T et al (2011) Crystal structure of a copper-transporting P1B-type ATPase. Nature 475:59–64

Hynninen A, Touzé T, Pitkänen L et al (2009) An efflux transporter PbrA and a phosphatase PbrB cooperate in a lead-resistance mechanism in bacteria. Mol Microbiol 74:384–394

Liu Z, Li H, Soleimani M et al (2008) Cd^{2+} versus Zn^{2+} uptake by the ZIP8 HCO_3–dependent symporter: kinetics, electrogenicity and trafficking. Biochem Biophys Res Commun 365:814–820

Mergeay M et al (2009) Megaplasmids in the *Cupriavidus* genus and metal resistance. In: Schwartz E (ed) Microbial megaplasmids, vol 11. Springer, Berlin

Monchy S, Benotmane MA, Janssen P et al (2007) Plasmids pMOL28 and pMOL30 of *Cupriavidus metallidurans* are specialized in the maximal viable response to heavy metals. J Bacteriol 189:7417–7425

Rensing C, Sun Y, Mitra B et al (1998) Pb (II)-translocating P-type ATPases. J Biol Chem 273:32614–32617

Taghavi S, Lesaulnier C, Monchy S et al (2009) Lead(II) resistance in *Cupriavidus metallidurans* CH34: interplay between plasmid and chromosomally-located functions. Antonie Van Leeuwenhoek 96:171–182

Thever MD, Saier MH Jr (2009) Bioinformatic characterization of P-type ATPases encoded within the fully sequenced genomes of 26 eukaryotes. J Membr Biol 229:115–130

Lead, Physical and Chemical Properties

Fathi Habashi
Department of Mining, Metallurgical, and Materials Engineering, Laval University, Quebec City, Canada

Physical Properties

Lead is an ancient metal and was used in Roman times to make pipes for water transportation. It is a lustrous metal. When freshly cut, surfaces are bluish white but oxidize readily to gray color. Its relative abundance in the Earth's crust is $1.6 \times 10^{-3}\%$ or 16 g/t.

Atomic number	82
Atomic weight	207.21
Melting point, °C	327.4
Latent heat of fusion, J/g	23.4
Boiling point, °C	1,741
Latent heat of vaporization, J/g	862
Vapor pressure, kPa	
980°C	0.133
1,160°C	1.33
1,420°C	13.33
1,500°C	26.7
1,600°C	53.3
Density, g/cm^3	
20°C	11.336
327.4°C (solid)	11.005
327.4°C (liquid)	10.686
400°C	10.597
500°C	10.447
Mean specific heat, J g^{-1} K^{-1}	
−200–0°C	0.1202
0–100°C	0.131
0–200°C	0.134
0–300°C	0.136
Specific electrical resistance, Ω/cm	
20°C	20.65
100°C	27.02
200°C	36.48
300°C	47.94
Thermal conductivity, W m^{-1} K^{-1}	
−100°C	0.371
0°C	0.355
100°C	0.342
200°C	0.329
300°C	0.316
327.4°C (solid)	0.313
327.4°C (liquid)	0.155
Surface tension of liquid lead, mN/m	
327.4°C	444
350°C	442
400°C	438
500°C	431
600°C	424
800°C	410
Dynamic viscosity of liquid lead, mPa·s	
327.4°C	2.75
350°C	2.60
400°C	2.34

(continued)

450°C	2.12
500°C	1.96
550°C	1.70
Coefficient of linear expansion, K^{-1}	
At 20°C	29.1×10^{-6}
Mean 20–300°C	31.3×10^{-6}

Three of the natural isotopes of lead are decay products of radioactive elements: ^{208}Pb is the end product of thorium decay, while ^{206}Pb and ^{207}Pb are formed from the uranium series via actinium and radium, respectively. Lead crystals are face-centered and cubic.

Chemical Properties

Fresh cut or cast lead surfaces undergo oxidation and tarnish rapidly to form an insoluble protective layer of basic lead carbonate. Lead might be expected to dissolve in acids with liberation of hydrogen, but this is prevented by the high hydrogen overvoltage, and attack occurs only when the combination of oxidizing conditions and soluble salt species exists. Thus, the metal resists hydrochloric acid corrosion to quite high concentrations, sulfuric acid to ca. 13.3 mol/L, and hydrofluoric acid totally; however, it dissolves readily in warm, dilute nitric acid to form soluble lead nitrate. Lead is also corroded by weak organic acids, such as acetic or tartaric acid, in the presence of oxygen. The protective coating formed by oxidation is removed by alkali with formation of soluble plumbites and plumbates.

Lead displays valences of +2 and +4. Lead (II) compounds are ionic while those of lead (IV) are essentially covalent. The metal is amphoteric and forms lead salts, plumbites, and plumbates. When melted in air, lead oxidizes readily to its monoxide PbO (also called litharge). Other lead oxides are the dioxide PbO_2 (actually sesquioxide Pb_2O_4) and red lead Pb_3O_4 (the orthoplumbate $Pb{\cdot}Pb_2O_4$). Lead halides are considerably more soluble in hot water than in cold; they also form double chlorides with alkali metals and a series of oxyhalides. Other common compounds of lead are the soluble nitrate and acetate; lead chromate (chrome yellow) used as a pigment along with red lead; lead borate used along with oxides in glassmaking; lead sulfide (galena), which is the principal naturally occurring form of the metal; and lead sulfate that is readily formed by sulfating a soluble lead salt and by natural oxidation of galena.

Other Data

Lead is a preferred shielding material against gamma rays and X-rays because of its high density and atomic number. Lead is a toxic metal. Air quality criteria are set in the range of 0.5–2.0 $\mu g/m^3$. Lead-bearing dust can be especially dangerous. Lead (mainly organic compounds) can enter the human body through the skin, by inhalation, or by ingestion. Ingestion in food is generally the major source of lead intake. Lead in the diet can come from numerous sources such as uptake of lead into plants, deposition onto vegetables from soil or water, and from lead solder in canned goods.

Workers employed in primary and secondary lead smelters, as well as in the production of lead metal and lead compounds, are exposed to lead absorption hazards. Because of the fine particle size of lead fume and its solubility, it can readily enter the respiratory system and be absorbed in the blood stream.

References

Sutherland CA et al (1997) Lead. In: Habashi F (ed) Handbook of extractive metallurgy. Wiley, Weinheim, pp 581–639

Lead-Associated Aggresome Diseases

▶ Lead Nephrotoxicity

Lead-Induced Chronic Renal Failure

▶ Lead Nephrotoxicity

Lead-RNA Interactions

▶ Lead and RNA

Lead-Thioneins

▶ Metallothioneins and Lead

Leadzyme

► Lead and RNA

Leather

► Chromium and Allergic Reponses

Lectins

► Calnexin and Calreticulin

Leukotriene B_4

► Zinc Leukotriene A_4 Hydrolase/Aminopeptidase Dual Activity

LeuT

► Sodium-Coupled Secondary Transporters, Structure and Function

$LeuT_{Aa}$

► Sodium-Coupled Secondary Transporters, Structure and Function

LH2 Domain

► C2 Domain Proteins

Li

► Lithium in Biosphere, Distribution

Ligand, Inhibitor

► NMR Structure Determination of Protein-Ligand Complexes using Lanthanides

Ligand-Cr-DNA Cross-Link

► Chromium Binding to DNA

Lipases

Anita Sahu and Ruth Birner-Gruenberger
Institute of Pathology and Center of Medical Research, Medical University of Graz, Graz, Austria

Synonyms

Lipolytic enzymes; Triacylglycerol hydrolases

Definition

Lipases are water-soluble enzymes that catalyze the hydrolysis of ester chemical bonds of water-insoluble lipid substrates at the lipid/water interface.

Introduction

Lipids are a broad group of naturally occurring hydrophobic molecules which include triacylglycerols, diacylglycerols, monoacylglycerols, phospholipids, and others. The main biological functions of lipids include energy storage; they act as structural components of cell membranes, and as important signaling molecules. Their digestion, transport, formation, intracellular storage, and mobilization are tightly controlled processes to ensure overall energy balance. Therefore, a disturbed lipid metabolism can be associated with many metabolic pathophysiological conditions. Lipases perform essential roles in the digestion, transport, and mobilization of lipids in most, if not all, living organisms. They are widely distributed in

Lipases, Scheme 1 *Reaction catalyzed by triacylglycerol lipases*: triacylglycerol + water => diacylglycerol + free fatty acid

animals, plants, and prokaryotes, even viruses. Cofactors are generally not required for lipase activity, but divalent cations like calcium often stimulate enzyme activity.

Reaction Mechanism and Structure

Lipases are members of the serine hydrolase family. The most conserved structural feature in this family is the active site containing a nucleophilic serine residue which is typically coordinated to a histidine and an aspartic acid residue in a charge-dependent manner in a so-called catalytic triad. Such a region is also present in serine proteases and in acyltransferases.

The reaction mechanism involves a nucleophilic attack of the active serine of the lipase on the carbon of the carboxylic ester, resulting in the formation of a tetrahedral transition state of the enzyme-substrate complex whose hydrolysis leads to the cleavage products, that is, the free (fatty) acid and the respective alcohol (glycerol), and the free lipase. The reaction catalyzed by triacylglycerol hydrolases is depicted below (Scheme 1).

The 3D-structures of crystallized lipases show a common α/β hydrolase fold as well as a so-called nucleophilic elbow to which the catalytic serine is bound. Typically, the nucleophilic serine is contained in a G-X-S-X-G motif. One specific feature of lipases is the shielding of their catalytic site by a surface loop (lid), controlling the access of the substrate to the hydrophobic active site. Movement of the lid is a requirement for the lipase to adopt an active conformation and allowing its adsorption at the lipid/water interface by a mechanism called interfacial activation (Reis et al. 2009). Unfortunately, the classical lid definition is not applicable to all lipases, since several lipases exhibit no or only a very rudimentary lid, while their preferred substrates are still medium- to long-chain triacylglycerols. One remaining possibility to clearly classify lipases and carboxyl esterases is the K_m value for their natural substrates, which is in the low millimolar range for carboxyl esterases and in the range of 10^2 mM for lipases. In contrast to carboxyl esterase substrates, the substrate concentration for half-maximal activity of lipases is well above solubility limit, also reflecting the activation of lipases at water/lipid interfaces. Calcium-binding sites have been found in many solved lipase structures including human pancreatic lipase and several microbial lipases. However, calcium-binding sites and calcium requirement for activity are not common features of all lipases but rather appear to be specific for certain enzymes. Thus, location and function of calcium binding is discussed below for the individual proteins.

Lipases show various specificities for fatty acyl chain lengths, positional and optical isomers. They may also possess activity toward short-chain carboxylic esters and, thus, accept a wide range of substrates. The hydrolytic reaction is reversible, and if the amount of water is limiting, for example, in the presence of organic solvents, the enzymes effectively catalyze inter- and trans-esterification reactions. Thus, their enantio-, chemo-, and stereoselectivity make them important tools in organic synthesis. Development of new or improved lipolytic enzymes is mandatory for making progress in biocatalysis. In addition, lipases are used in the dairy industry for the hydrolysis of milk fat and applied in detergent formulations to remove fat-containing stains. Applications in biocatalysis and industrial processes prefer cheap enzyme preparations, which are often contaminated with other than the desired lipolytic activities, leading to undesired side reactions and lower enantioselectivity.

Determination of enzymatic activities is required during isolation, purification, and characterization of lipases, as well as for characterization of natural lipase sources originating from animals, plants, and microorganisms, or screening of recombinant proteins. Since lipases play an important role in human (patho-) physiology (obesity and cardiovascular disease), determination of lipase activities is also important in medical diagnosis and research. Moreover, their potential as drug target for treatment of metabolic disease has been increasingly recognized.

Substrates and inhibitors are useful tools for fingerprinting of lipases and esterases. These systems enable us to measure enzyme activity (using substrates) and determine the quality of an enzyme preparation (using inhibitors). Moreover, enzymes can be classified on the basis of structurally different substrates and inhibitors. Important factors for substrate and inhibitor design are the detailed chemical structure, overall polarity, and stereochemistry of the probe. In addition, the "quality" of the hydrophobic/hydrophilic interface critically influences the apparent activities. Typical systems for substrate or inhibitor solubilization include synthetic detergents, phospholipids, and/or proteins. Tagged covalent inhibitors, such as alkylphosphonates, reacting with the active serine of lipases, can be designed in such a way that they act as activity-based probes, which are suitable tools for discovery, profiling, and localization of lipolytic activities in complex proteomes and even in living cells (Schittmayer and Birner-Gruenberger 2009).

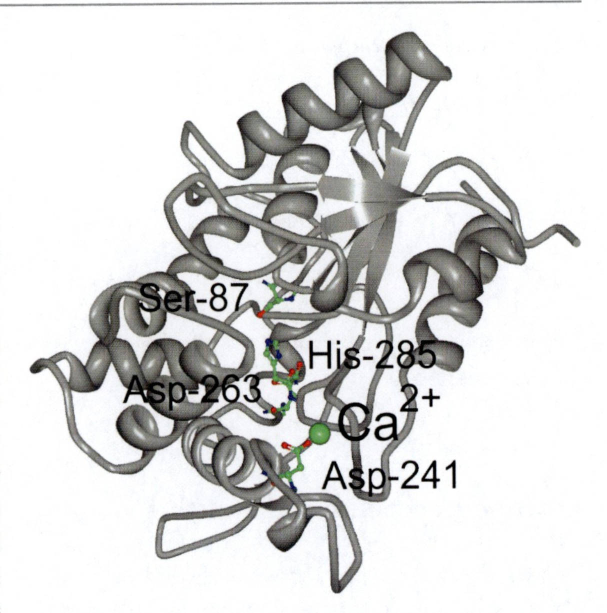

Lipases, Fig. 1 *Structure of Burkholderia glumae lipase depicting the active site and the calcium-binding site*: The *Burkholderia glumae* lipase entry 1TAH of Noble et al. (1993), was downloaded from the protein data bank (PDB) and modified using PDB protein workshop 3.9 to highlight catalytic triad residues Ser-87, His-285, and Asp-263, as well as the calcium-binding residue Asp-241

Microbial Lipases with Special Focus on Calcium-Binding Enzymes

Apart from their biological significance in lipid metabolism and in pathogenesis, such as skin colonization, microbial lipases are of major biotechnological interest. Virtually all of the technologically employed lipases are microbial due to their easy accessibility and efficient expression. For the same reasons, most of the structural information is available from lipases of microbial origin. In several cases, divalent calcium ions are required for enzymatic activity and/or protein stability.

Consequently, it is a lipase, namely the *Pseudomonas aeruginosa* lipase, which constitutes the best studied example of creating enantioselective enzymes by directed evolution. Interestingly, mutations near the surface remote from the active site but close to the calcium-binding site improve amidase activity despite that native lipases do not hydrolyze amides (Fujii et al. 2005). The structure shows a bound calcium ion coordinated with two aspartates, two carbonyl oxygens, and water molecules, reflecting a rather rigid conformation (Nardini et al. 2000). Calcium protects the protein from denaturation, suggesting a role in enzyme stabilization. The same was found for the homologous *Pseudomonas fluorescens* lipase, where the increased activity in the presence of calcium was correlated with secondary structural changes, namely, an increase in alpha-helix and beta-sheet content (Kim et al. 2004).

In *Burkholderia glumae* and *Pseudomonas cepacia* lipases, a single calcium site bridges the active site region containing the catalytic triad histidine to a second subdomain (Noble et al. 1993; Kim et al. 1997). Because of the distance of the calcium site to the active site, a direct involvement of calcium in catalysis is unlikely as can be seen in Fig. 1. Although in *Burkholderia glumae* lipase, one of the calcium ligands, an aspartate (Asp-241), is essential for protein activity since a site-specific mutation of the residue to alanine abolishes calcium binding as well as lipase activity. On the other hand, denatured protein can be refolded in vitro into an active enzyme in the absence of calcium, excluding an essential role for the ion in catalytic activity or protein folding (El Khattabi et al. 2003). However, removal of calcium by chelation induces protein unfolding, as well as reduced activity and increased protease sensitivity, but at elevated temperature only. Thus, the calcium site is required for protein stability and integrity of the local structure close to the active site under denaturing conditions.

Staphylococcal lipases also contain a single calcium-binding site required for activity, as determined for *Staphylococcus hyicus* lipase, which binds one calcium ion with a dissociation constant of 55 μM, for *Staphylococcus aureus* (dissociation constant of 250 μM), and for *Staphylococcus epidermis* lipases (dissociation constant of 300 μM) (Simons et al. 1999). Also for staphylococcal lipases, the calcium-binding site is important for structural stabilization and thermostability of protein activity. Interestingly, calcium-independent *Staphylococcus hyicus* lipase variants, which are nearly as active as the wild-type enzyme at room temperature but show reduced activity at elevated temperature, can be engineered by mutation of the calcium ligand aspartates 354 and 357. Similarly, thermostability of *Bacillus stearothermophilus* lipase is increased by calcium by about 8°: The enzyme starts to unfold at 66°C in the presence of calcium but at 58°C in the absence of calcium ions (Kim et al. 2000).

More than one calcium-binding sites are found in extracellular lipases of Gram-negative bacteria that are secreted via the type I secretion system. These proteins, in addition to the N-catalytic domain, have a C-terminal secretion signal and several copies of a GGxGxDxux (u: hydrophobic residue) calcium-binding tandem repeat motif. The structures of two homologous lipases containing these motifs were solved, namely LipA from *Serratia marcescens*, which has an important biotechnological application in the production of a chiral precursor for the coronary vasodilator and calcium channel blocker diltiazem, and *Pseudomonas* sp. MIS38 lipase. Interestingly, the structure of LipA from *Serratia marcescens* reveals a second helical hairpin additional to the putative lid that exposes a hydrophobic surface to the aqueous medium and thus might function as an additional lid. The C-terminal tandem repeats form two separated beta-roll domains that pack tightly against each other with asymmetric binding of six calcium ions only one side of the roll, which may have an intramolecular chaperone function for the N-catalytic domain, as reported for the *Pseudomonas* sp. MIS38 lipase, where knock-out of the tandem repeats leads to strongly reduced protein and secretion levels. LipA from *Serratia marcescens* contains in total eight calcium-binding sites (Meier et al. 2007). The N-terminal lipase domain shows a variation on the canonical alpha/beta-hydrolase fold in an open conformation, where the putative lid-helix is anchored by a calcium ion essential for activity. This essential calcium ion is coordinated by aspartic acids 153 and 157. Mutation of Asp-157 to alanine abolishes activity completely even in the presence of calcium. Similarly, a calcium-binding site in *Pseudomonas* sp. MIS38 lipase is formed only in an open conformation, in which the lid-helix is anchored by the calcium ion, and removal by site-specific mutation of Asp-157 exhibited no lipase but residual weak esterase activity (Kuwahara et al. 2008). This essential calcium site is buried deep inside the protein, suggesting that an open conformation is fairly unstable unless the calcium is bound to this site. Thus, this site is required to stabilize the open conformation and make the protein active. A second calcium ion is found in both enzymes in the same large loop as the active site residue aspartate and may thus play a stabilizing role for the conformation of the active site. In contrast to LipA from *Serratia marcescens*, the *Pseudomonas* sp. MIS38 lipase contains a third calcium site relatively close to the second site in the N-catalytic domain. Removal of this site by mutation of Asp-337 to alanine results in a less thermostable but equally active enzyme.

Mammalian Lipases

In mammals, lipases are found in blood, gastric juices, pancreatic secretion, and adipose and other tissues. Digestive lipases, that is, gastric/lingual (GL) (EC 3.1.1.3) and pancreatic lipase (PL) (EC 3.1.1.3), have the capacity to break down triacylglycerols (fat) into fatty acid and glycerol molecules, which are then absorbed by intestinal cells to fulfill the energy requirements of the body. Plasma lipases, such as lipoprotein lipase (LPL) (EC 3.1.1.34), hepatic lipase (HL) (EC 3.1.1.3), and endothelial lipase (EL) (EC 3.1.1.3), are involved in lipid transport via lipoproteins between tissues to ensure overall energy balance. Intracellular lipases [namely, adipose triglyceride lipase (ATGL) (EC 3.1.1.3), hormone-sensitive lipase (HSL) (EC 3.1.1.79), and monoglyceride lipase (MGL) (EC 3.1.1.23)] are responsible for mobilization of intracellular lipid stores (lipid droplets) in adipose and other tissues during times of starvation. In the following paragraphs, the mechanism, structure, function, and regulation of relevant individual lipases are discussed in more detail.

Digestive Lipases

Most lipids are taken up as triacylglycerol. For absorption across the intestinal epithelium, they are degraded into free fatty acids and monoacylglycerol by digestive lipases produced by the pancreas and emulsified by bile produced by the gall bladder. Drugs inhibiting digestive lipases, such as orlistat, are used to decrease fatty acid absorption, but are associated with severe gastrointestinal side effects.

Pancreatic Lipase: PL is a water-soluble enzyme secreted by the pancreas into the gut and has a central role in digestion of dietary fats. PL deficiency is associated with many pathophysiological conditions including reduced absorption of long-chain triacylglyceride fatty acids. Unlike other pancreatic digestive enzymes like trypsinogen and chymotrypsin, PL is secreted in pancreatic juice as an active enzyme and does not require active conversion to digest lipids.

Structural studies of porcine PL show that it possesses a two domain organization, the N-terminal domain bearing the catalytic triad (Ser-153, Asp-176, His-263) and the C-terminal domain devoted to colipase-binding site. Colipase is a protein coenzyme, which is required for optimal enzymatic activity of PL. Optimal activity also depends on the presence of calcium. The suggested mechanism is that divalent calcium ions are required for colipase binding (Hu et al. 2010). In the crystal structure, a calcium ion is located remotely from the active site, playing a purely structural role in stabilizing the conformation of a surface segment. Direct evidence for the function of the catalytic residues or of the lid and for the involvement of residues in colipase binding was provided by site-directed mutagenesis. In vitro, PL exhibits the maximum activity at pH 7.5–8.5, whereas, in vivo, bile salts shift the pH optimum to slightly acidic values (pH 6.5) that actually prevail in the upper intestine.

Similarly, the three-dimensional structure of human PL shows that the single polypeptide chain glycoprotein (449 amino acids) is folded into two domains: (1) a large N-terminal domain (residues 1–335), which is a typical α/β hydrolase fold containing the active site, and (2) a small C-terminal domain (residues 336–449), which resembles a double bound parallel β sheet sandwich type. A surface loop (from Cys-237 to Cys-261), the so-called lid or flap, surrounding the active site, prevents the access of substrate to the active site in the closed conformation. Human PL also requires colipase, a 12-kDa protein cofactor, to anchor the lipase to the surface of lipid micelles (triacylglycerol/water interface), counteracting the destabilizing influence of bile salts. Colipase binds to the C-terminal domain of PL and exposes the hydrophobic tips of its fingers at the opposite side of its lipase-binding site. The open lid and the extremities of the colipase fingers, as well as the β9 loop, form an impressive continuous hydrophobic plateau extending over more than 50 $A^{\circ 2}$, which may interact strongly with the lipid/water interface. These two different confirmations provide a convincing model to explain the preference of this enzyme for water-insoluble substrates over water-soluble substrates.

Some single nucleotide polymorphisms of the *PNLIP* gene, which encodes PL, have been associated with metabolic disorders. However, larger studies using more diverse study samples are required to confirm these results. Severe pancreatic exocrine insufficiency leading to malabsorption of nutrients is one of the most important late features of chronic pancreatitis; thus, through measurement of serum concentration of pancreatic lipase, acute pancreatitis can be diagnosed. Current standards, options, and future aspects of enzyme replacement therapy are still under discussion.

Gastric Lipase: Fat digestion in humans requires not only the classical pancreatic lipase but also GL, which is stable and active despite the highly acidic stomach environment (Roussel et al. 1999). Human GL is secreted by the chief cells in the fundic mucosa, which is a dome-like subdiaphragmatic portion of the stomach and is colocalized with pepsin, where it initiates the digestion of triacylglycerols. This lipase does not require any cofactor for optimal enzymatic activity. Presence of the signaling peptide gastrin in the plasma works as a trigger for secretion of this enzyme.

Human GL is a polypeptide of 379 residues in length. The structure of GL was determined using X-ray diffraction. The catalytic machinery consists of the catalytic triad (Ser-153, His-353, Asp-324) and the oxyanion hole, which stabilizes the oxyanion transition state via hydrogen bonds with two main-chain nitrogens (backbone NH groups of Gln-154 and Leu-67).

The nucleophilic serine belongs to the typical consensus sequence G-X1-S-X2-G (X1 and X2 being His, and Gln, respectively). The catalytic serine is deeply buried under a segment consisting of 30 residues, which can be defined as a lid and belonging to the cap domain (residues 184–308), an intricate mixture of eight helices turns, and random coils. The displacement of the lid is necessary for the substrates to have access to Ser-153.

GL plays a crucial role in neonates providing up to 50% of the total lipolytic activity because PL is not yet fully developed at this age. It successfully penetrates milk fat globules and facilitates digestion of milk fat. The products of gastric lipolysis were suggested to maintain the sterility of the gastrointestinal tract.

Various clinical studies have been conducted on both animals and humans to assess the efficacy of enzymatic replacement therapies using acid-resistant lipases to treat exocrine pancreatic insufficiency. A limitation of acidic lipases is that they remove only one fatty acid from triacylglycerol. The free fatty acid can readily cross the epithelial membrane lining the gastrointestinal tract, but the diacylglycerol cannot be transported across the membrane efficiently.

Plasma Lipases

After uptake of free fatty acids and monoacylglycerol by enterocytes located at the microvilli of the small intestine, triacylglycerols are reassembled and transported via the lymphatic system to the liver by chylomicrons. During the transport, chylomicrons encounter LPL, which hydrolyzes chylomicron triacylglycerol to free fatty acids and glycerol. The remnants of the chylomicrons are taken up by the liver, where triacylglycerols are stored in lipid droplets or secreted as very low density lipoprotein (VLDL). After loss of apolipoproteins and of a portion of triacylglycerol through the action of LPL, VLDL remnants are termed intermediate density lipoprotein (IDL). Approximately 50% of IDL is directly cleared by the liver; the rest is converted to low-density lipoprotein (LDL) by further reducing triacylglycerol content through LPL activity. The released free fatty acids are consumed by peripheral tissues to produce energy, as building blocks for phospholipids or are reassembled to triacylglycerol and stored in lipid droplets, for example, in adipocytes. The remaining LDL particles consist primarily of cholesteryl esters and are taken up by target cells via LDL receptors. Excess cholesterol is transported back to the liver by high-density lipoprotein (HDL), where it may be effluxed from the body by secretion via the bile.

Lipoprotein Lipase: LPL plays a major role in the metabolism and transport of lipids (Wang and Eckel 2009). It is the rate-limiting enzyme responsible for the hydrolysis of core triacylglycerol in chylomicrons and VLDL. LPL is a multifunctional enzyme produced by parenchymal cells in many tissues, including adipose tissue, cardiac and skeletal muscle, islets, and macrophages. After translocation into small capillaries of LPL secreting tissues, it docks onto heparan sulfate proteoglycans (HSPG) located in lipid rafts at the luminal endothelial cell surface. HSPG provide a platform for interaction with triacylglycerol-rich lipoproteins, that is, chylomicrons and VLDL, by facilitating binding through the presence of excess negative charge on the surface of the HSPG molecule. Besides its hydrolytic activity, LPL interacts with lipoproteins to anchor them to the vessel wall and facilitates lipoprotein particle uptake. Once triacylglycerol-rich lipoproteins are bound, LPL mediates triacylglycerol hydrolysis, causing the release of fatty acids that are subsequently taken up by fatty acid receptors such as CD-36, which are located on the plasma membrane of, for example, adipocytes and myocytes. Within these cells, the fatty acids are re-esterified and used for storage or provide energy through mitochondrial oxidation.

LPL is organized into two structurally distinct domains, an N-terminal domain and a smaller C-terminal domain with a flexible peptide connecting the two domains. The N-terminal domain contains the catalytic triad (Ser-132, Asp-156, His-241) responsible for lipolysis. The C-terminal domain contains the dominant heparin-binding domain and is thought to be important for binding lipoproteins. Native LPL monomers are arranged in a head-to-tail subunit orientation to form the noncovalent active homodimer.

LPL is regulated in response to energy requirements and hormonal changes at transcriptional, posttranscriptional, and posttranslational levels in a tissue-specific manner. It requires a specific cofactor, apolipoprotein C-II (apoC-II), to be fully active. The binding site for this physiological activator has been located to 11 amino acid residues in two different regions of the N-terminal domain of LPL, and these two regions appear to act cooperatively to enable the activation of LPL by apoC-II. Several other effectors of LPL that modulate the in vivo function of LPL have been identified. Lipase maturation factor 1 (LMF1) plays an essential role in the formation of catalytically active LPL from newly synthesized polypeptides in the endoplasmic reticulum (ER), a process called lipase maturation. Another factor, glycosylphosphatidylinositol-anchored high-density lipoprotein binding protein 1 (GPIHBP1), is critically involved in LPL-mediated triacylglycerol hydrolysis of triacylglycerol-rich lipoproteins.

About 20% of the patients with hypertriglyceridemia are carriers of common LPL gene mutations, namely, Asp-93Asn, Asn-2913Ser, Trp-863Arg, Gly-1883Glu, Pro-2073Leu, or Asp-2503Asn. The importance of the Asn-2913Ser gene variant to hypertriglyceridemia has been reviewed using meta-analysis. This variant also predisposes to more severe dyslipidemia with increasing age and weight gain. Two common LPL polymorphisms (HindIII and Ser447Ter) have also been shown to be associated with low HDL cholesterol levels and hypertriglyceridemia in Asian Indians. Variants in the promoter of the LPL gene have been associated with changes in lipid metabolism, leading to obesity and type-2 diabetes. The HindIII polymorphism is significantly associated with body mass index in obese people.

In recent years, LPL mass in preheparin serum (so-called preheparin LPL mass) has been suggested as a biomarker of the metabolic syndrome, a condition that is characterized by a combination of obesity, insulin resistance, and dyslipidemia.

Overall, LPL contributes in a pronounced way to normal lipoprotein metabolism, tissue-specific substrate delivery and utilization, and the many aspects of obesity and other metabolic disorders that relate to energy balance, insulin action, and body weight regulation. Elevated plasma triacylglycerol levels increase the risk for cardiovascular disease. A possibility to target LPL and its effector molecules for therapeutic use is under discussion and investigation.

Hepatic Lipase: HL is a lipolytic enzyme produced in hepatocytes (Perret et al. 2002). After secretion, the enzyme binds to cell surfaces in the liver sinusoidal capillaries in most animal species. Only small amounts of hepatic lipase can be detected in other tissues (e.g., macrophages, adrenal gland in rats). In humans, the enzyme is mostly bound onto HSPG at the surface of hepatocytes and also of sinusoidal endothelial cells. HL shares a number of functional domains with LPL and with other members of the lipase gene family. It is clear, however, that the two enzymes use different binding sites. It is secreted as a glycoprotein, and remodeling of the N-linked oligosaccharides appears to be crucial for the secretion process, rather than for catalytic activity. Systematic investigation of factors involved in maturation of HL in the ER underscores that folding is a major bottleneck for HL production. HL is also present in adrenals and ovaries, where it might promote delivery of lipoprotein cholesterol for steroidogenesis. However, evidence of a local synthesis is still controversial.

Hepatic lipase displays both triglyceride and phospholipase activities and participates in hepatic handling of chylomicron remnants, in conversion of IDL to LDL and in high-density lipoprotein (HDL) metabolism. High HL activity is often associated with low HDL, and this has been ascribed to HL action on phospholipids and triacylglycerol of HDL. The mechanism involves, most probably, that HDL binds the lipase, thereby extracting it from cell surfaces. In vivo, the extracted lipase would presumably transfer to binding sites on endothelial cells in the liver sinusoids and this could create a self-regulatory system: High HDL would stimulate HL production, which, in turn, would tend to lower HDL levels. HL activity is fairly regulated according to the cell cholesterol content and to hormonal status. Coordinate regulations have been reported for both HL and the scavenger-receptor B-I, suggesting complementary roles in cholesterol metabolism. Genetic variants of HL may contribute to development of a dyslipidemic phenotype in insulin-resistant subjects.

Endothelial Lipase: EL shares many properties with HL and LPL, but is mainly active on phospholipids of HDL (Huang et al. 2010). It shares considerable molecular homology with LPL (44%) and HL (41%). EL is synthesized by vascular endothelial cells and functions at the site where it is synthesized. The active form of EL was shown to be a head-to-tail-oriented homodimer, just as is the case for LPL. Studies in humans and mouse genetic models showed that high EL activity is associated with low HDL-cholesterol, and conversely low EL activity is associated with elevated HDL-cholesterol. Unlike LPL and HL, expression of EL is highly regulated by cytokines and physical forces. This, in turn, EL has the expected impact on risk for atherosclerosis, meaning that low EL activity is an atheroprotective factor.

Two interesting extensions of the relation of EL to clinical disease are (1) that in type 2 diabetic patients, EL activity was positively associated with the degree of inflammation (and insulin-lowered EL activity) and (2) that statins decrease the expression of EL, both in vitro and in vivo. It is apparent that specific inhibitors of EL would have a clinical potential as HDL-raising drugs. However, there is only a limited amount of information available about this enzyme.

Intracellular Lipases

Lipid droplets serve as intracellular storage organelles for energy substrates inside many different cell types and organisms. In mammals, the main site for intracellular storage is adipose tissue for long term and liver for short term storage. Lipid droplets consist of a neutral lipid core surrounded by a phospholipid monolayer which is coated with lipid droplet–associated proteins. Proteomics studies in adipocytes identified a surprising complexity of the lipid droplet coat proteome including proteins involved in signaling, cytoskeletal organization, intracellular trafficking, and lipid metabolism. Enzymatic degradation of lipid droplet-triacylglycerols involves the sequential formation of diacylglycerol, then monoacylglycerol and finally glycerol. Each step is accompanied by the release of one free fatty acid molecule. The ratio of these products depends on the substrate specificity of the responsible lipases, which have to localize to the lipid droplet to be able to access their substrates, as shown for ATGL in Fig. 2. Three enzymes appear to be required for complete hydrolysis of triacylglycerol of lipid droplets. First, ATGL acts on triacylglycerol to generate diacylglycerol. Diacylglycerol is hydrolyzed to monoacylglycerol by HSL. However, HSL possesses rather broad substrate specificity and can also use cholesteryl esters, triacylglycerol, and monoacylglycerol as substrates. Finally, MGL cleaves monoacylglycerol into glycerol and free fatty acid.

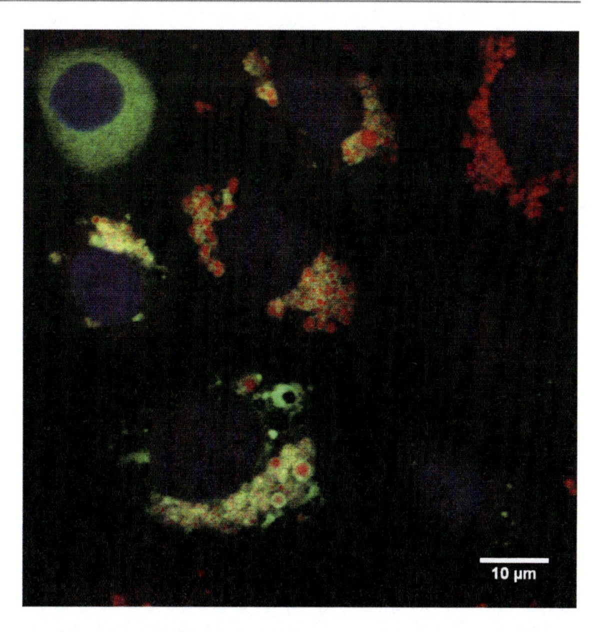

Lipases, Fig. 2 *Lipid droplet localization of the lipase ATGL*: The fluorescent lipase fusion protein YFP-ATGL (*green*) is localized preferentially on the surface of lipid droplets (*red*). Mammalian (COS7) cells transiently expressing murine YFP-ATGL were loaded overnight with 150 μM oleic acid. Nuclei (*blue*) and lipid droplets (*red*) were counterstained with Hoechst 33422 (0.1 μg/ml) and *nile red*, respectively. Images were acquired on a LSM510 meta confocal laser scanning microscope (Zeiss, Oberkochen, Germany) using a 63x oil immersion objective (NA 1.4)

Adipose Triglyceride Lipase: ATGL was discovered in 2004 by several groups as a lipid droplet–associated lipase (Lass et al. 2011). Studies in various cell culture and genetically modified animal models have shown that ATGL plays a rate-limiting role in both basal and stimulated lipolysis. This enzyme selectively hydrolyses the first ester bond from the triacylglycerol, leading to generation of diacylglycerol and free fatty acid. ATGL is highly expressed in adipose tissue, but has been also observed in other tissues, like cardiac muscle, skeletal muscle, testis, and liver. Its property to hydrolyze triacylglycerol is evolutionarily conserved between humans, mice, drosophila, *Saccharomyces cerevisiae*, and *Arabidopsis*. ATGL is encoded by *PNPLA2* in human and mouse. The proteins both have a molecular weight of approximately 56kDs and share 86% amino acid identity. The N-terminal half of ATGL comprises of the α/β hydrolase fold and an overlapping patatin-like domain. Although no 3D-structure of ATGL is available so far, multiple other analyses indicate that ATGL acts through a catalytic dyad consisting of Ser-47 present in a GXSXG motif and Asp-166 within a DXG motif. A hydrophobic C-terminal region of 45 amino acids may mediate ATGL localization to lipid droplets. Apart from it, like other extracellular lipases, ATGL activity is drastically affected by a co-activator protein, in this case CGI-58 (comparative gene identification-58, α/β hydrolase domain containing protein-5 or ABHD5). Even though much evidence supports the role of CGI-58 in ATGL activation, a defined mechanism for ATGL activation by CGI-58 remains to be clarified. In addition to CGI-58, recently, a protein called G0S2 has been brought into picture as a selective inhibitor of ATGL. However, detailed studies of CGI-58, S0G2, and ATGL will be required to elucidate their exact roles and their interplay in lipolysis. ATGL transcript levels in murine adipose tissue are altered with respect to fasting, feeding, and refeeding, indicating that regulation of ATGL is closely associated with the metabolic state of the organism. Moreover, ATGL deficiency in mouse is associated with reduced lipolysis causing disproportionate fat accumulation in many tissue types, leading to overall twofold deposition in body fat with enlarged adipose fat depots. On the other hand, mice overexpressing ATGL are leaner, triacylglycerol content is decreased specifically in adipose tissue, and they are resistant to diet-induced obesity with improved insulin sensitivity.

Hormone-Sensitive Lipase: Before discovery of ATGL, HSL was thought to be the rate-limiting enzyme in lipolysis. It was the first enzyme discovered to facilitate stimulated lipolysis, that is, hormone-induced catabolism of lipids. The gene for human HSL, *LIPE*, is located on chromosome 19q13.2. HSL is the major enzyme responsible for intracellular diacylglycerol and cholesteryl ester hydrolysis, but it also accepts triacylglycerol, monoacylglycerol, and retinyl esters as substrates. HSL is mainly expressed in adipose tissue but is also present at low level in many other tissues including heart, skeletal muscle, macrophages, and pancreatic β-cells. HSL knockout mice are not overweight or obese but accumulate diacylglycerol in several tissues, indicating that HSL is rate limiting for diacylglycerol hydrolysis.

No 3D structure is available yet, but based on the domain structure, the human enzyme can be subdivided into three to four functional regions: The N-terminal domain (residues 1–300) is thought to mediate enzyme dimerization, lipid binding, and interaction with fatty acid–binding protein (FABP4) known to enhance HSL enzymatic activity. The C-terminal domain (residues 301–768) contains the catalytic triad composed of Ser-424, Asp-693, and His-723, within a α/β hydrolase fold. The third domain corresponds to the regulatory module of enzyme: This loop region (residues 521–669) contains all the known phosphorylation sites of HSL. Finally, a putative lipid-binding domain is present at the C-terminal end of the protein. Human HSL is an 88 kD protein, which is the largest as compared to other species, for example, rat (84 kD) and mouse (82 kD), although the general structure and function are conserved. The amino acid sequences of human and rat HSL are 83% identical, and mouse and rat share 94% amino acid sequence similarity.

HSL activity is regulated by two major mechanisms: first phosphorylation by protein kinases, and secondly, by interacting with auxiliary proteins. The molecular mechanism to activate HSL by phosphorylation is well established. HSL-mediated lipolysis is strictly controlled by lipolytic (e.g., catecholamine, glucagon) and antilipolytic (e.g., insulin) hormones. The pathway of β-adrenergic receptor, a G-coupled protein receptor, activates adenylate cyclase, which increases cyclo-adenosine monophosphate (cAMP) levels activating protein kinase A (PKA). PKA phosphorylates HSL at Ser-563 and Ser-660, resulting in translocation of HSL from the cytosol onto lipid droplets. Ser-565 is considered to be the basal phosphorylation site while Ser-563 resembles the regulatory site. In addition to PKA, other kinases can phosphorylate HSL at Ser-565, such as glycogen synthase kinase (GSK), AMP-dependent protein kinase (AMPK), and Ca^{2+}/calmodulin-dependent protein kinase II (CaMK-II),which does not seem to activate the enzyme. Dysregulation of HSL leading to lipid accumulation and imbalance of fat mobilization may contribute to the development of metabolic disorders.

Monoglyceride Lipase: The last step of intracellular triacylglycerol hydrolysis is the breakdown of monoacylglycerol by MGL. In absence of MGL, monoacylglycerol accumulates. The MGL gene, *MGLL*, was mapped to human chromosome 3 and mouse chromosome 6. In humans, *MGLL* spans over 134kb and codes for a protein of 313 amino acid residues with a molecular weight of 34kD. Recently, the 3D structure of MGL was solved by X-ray crystallography. MGL is a member of the α/β hydrolase family and contains the classical GXSXG consensus sequence. The catalytic triad consists of Ser-122, Asp-239, and His-269. MGL mRNA was detected in many tissues including adipose tissue, kidney, and testis, suggesting a role for MGL in diverse tissues. The endogenous cannabinoid 2-arachidonoyl glycerol is an important substrate of MGL, and therefore, MGL plays a role in endocannabinoid signaling. Moreover, MGL activity is elevated in aggressive human cancer cells and primary tumors, where it regulates free fatty acid levels, which are educts of oncogenic signaling lipids, such as lysophospholipids, ether lipids, phosphatidic acid, and prostaglandin E2. MGL is therefore elucidated as potential therapeutic target for future cancer therapy.

Lysosomal Acid Lipase: Despite the close amino acid sequence similarities [59% of the amino acids are identical between GL and human lysosomal acid lipase (LAL, EC 3.1.1.3)], LAL hydrolyzes not only triacylglycerol delivered to the lysosomes by low-density lipoprotein receptor–mediated endocytosis but also cholesteryl esters (Roussel et al. 1999). The released cholesterol plays an important regulatory role in cellular sterol metabolism. Defective LAL activity is associated with two rare autosomal recessive diseases, Wolman disease and cholesteryl ester storage disease. Wolman disease is characterized by accumulation of cholesteryl esters and triacylglycerols in the lysosomes in most tissues, resulting in early hepatic and adrenal failure. In cholesteryl ester storage disease, which is less detrimental, residual LAL activity can be detected.

Acknowledgments Anita Sahu is supported by the FWF Austrian Science Funds project W1226: "DK-MCD: Doctoral school metabolic and cardiovascular disease." We are grateful to Martin Viertler for preparing Fig. 2 and to Rudolf Zechner, University of Graz, Austria, for providing the YFP-ATGL plasmid.

References

El Khattabi M, Van Gelder P, Bitter W et al (2003) Role of the calcium ion and disulfide bond in the *Burkholderia glumae* lipase. J Mol Catalysis B: Enzymatic 22:329–338

Fujii R, Nakagawa Y, Hiratake J et al (2005) Directed evolution of *Pseudomonas aeruginosa* lipase for improved amide-hydrolyzing activity. Prot Eng Des Sel 18:93–101

Hu M, Li Y, Decker EA et al (2010) Role of calcium and calcium-binding agents on the lipase digestibility of emulsified lipids using an *in vitro* digestion model. Food Hydrocolloids 24:719–725

Huang J, Qian HY, Li ZZ et al (2010) Role of endothelial lipase in atherosclerosis. Transl Res 156:1–6

Kim KK, Song HK, Shin DH et al (1997) The crystal structure of *Pseudomonas cepacia* reveals a highly open conformation in the absence of bound inhibitor. Structure 5:173–185

Kim M-H, Kim H-K, Lee J-K et al (2000) Thermostable lipase of *Bacillus stearothermophilus*: high-level production, purification, and calcium-dependent thermostability. Biosci Biotechnol Biochem 64:280–286

Kim KR, Kwon DY, Yoon SH et al (2004) Purification, refolding and characterization of recombinant Pseudomonas fluorescens lipase. Prot Expr Purif 39:124–129

Kuwahara K, Angkawidjaja C, Matsumara H, Koga Y, Takano K, Kanaya S (2008) Importance of the Ca^{2+}-binding sites in the N-catalytic domain of a family I.3 lipase for activity and stability. Prot Eng Des Sel 21:737–744

Lass A, Zimmermann R, Oberer M et al (2011) Lipolysis – a highly regulated multi-enzyme complex mediates the catabolism of cellular fat stores. Prog Lipid Res 50: 14–27

Meier R, Drepper T, Svensson V, Jaeger K-E, Baumann U (2007) A calcium-gated lid and a large beta-roll sandwich are revealed by the crystal structure of extracellular lipase from *Serratia marcescens*. J Biol Chem 282:31477–31483

Nardini M, Lang DA, Liebeton K et al (2000) Crystal structure of *Pseudomonas aeruginosa* lipase in the open conformation. J Biol Chem 275:31219–31225

Noble MEM, Cleasby A, Johnson LN, Egmond MR, Frenken LGJ (1993) The crystal structure of triacylglycerol lipase from *Pseudomonas glumae* reveals partially redundant catalytic aspartate. FEBS Lett 331:123–128

Perret B, Mabile L, Martinez L et al (2002) Hepatic lipase: structure/function relationship, synthesis, and regulation. J Lipid Res 43:1163–1169

Reis P, Holmberg K, Watzke H et al (2009) Lipases at interfaces: a review. Adv Colloid Interface Sci 147–148:237–250

Roussel A, Canaan S, Egloff MP et al (1999) Crystal structure of human gastric lipase and model of lysosomal acid lipase, two lipolytic enzymes of medical interest. J Biol Chem 274:16995–17002

Schittmayer M, Birner-Gruenberger R (2009) Functional proteomics in lipid research: lipases, lipid droplets and lipoproteins. J Proteomics 72:1006–1018

Simons J-WFA, van Kampen MD, Ubarretxena-Belandia I, Cox RC, Alves dos Santos CM, Egmond MR, Verheij HM (1999) Identification of a calcium site in *Staphylococcus hyicus* lipase: generation of calcium independent variants. Biochem 38:2–10

Wang H, Eckel RH (2009) Lipoprotein lipase: from gene to obesity. Am J Physiol Endocrinol Metab 297:E271–E288

Lipid Mediator

▶ Zinc Leukotriene A_4 Hydrolase/Aminopeptidase Dual Activity

Lipocortins

▶ Annexins

Lipolytic Enzymes

▶ Lipases

Lipoxygenase Homology Domain

▶ C2 Domain Proteins

Lithium as Mood Stabilizer

Janusz K. Rybakowski
Department of Adult Psychiatry, Poznan University of Medical Sciences, Poznan, Poland

Synonyms

Prevention of recurrences in bipolar disorders; Treatment of depression in bipolar disorders; Treatment of mania in bipolar disorders

Definition

The term "mood stabilizer" has been introduced in relation to the treatment of bipolar (manic-depressive) mood disorder. Mood-stabilizing property of a drug denotes its therapeutic and prophylactic action against both psychopathological poles of bipolar disorder (mania and depression). A prototype for such drugs is lithium ion where such properties have been demonstrated since the early 1960s. A proposed definition of mood stabilizer elaborated by the author of this entry would be: "A drug that if used as monotherapy: (1) acts therapeutically in mania or/and in depression; (2) acts prophylactically against manic or/and depressive episodes as demonstrated in a trial of at least one year's duration and (3) does not worsen any therapeutic or prophylactic aspect of the illness outlined above" (Rybakowski 2007a).

History of Mood Stabilizers

The introduction into psychiatric armamentarium of individual mood stabilizers fulfilling the criteria mentioned above started nearly a half century ago. The mood-stabilizing property of lithium was first suggested in the early 1960s (Hartigan 1963). This was followed by observations showing that some antiepileptic drugs may also exert a mood-stabilizing effect. The possibility of this for valproates was put forward at the turn of the 1960/1970s, when French psychiatrists coined a term "thymoregulatrice" (Lambert et al. 1971). Mood-stabilizing action of carbamazepine was proposed by Japanese psychiatrists in the early 1970s (Okuma et al. 1973). More than 20 years later, a suggestion that the atypical antipsychotic drug, clozapine, had a mood-stabilizing action was advanced (Zarate 1995). Following this, the evidence for such action of other atypical antipsychotic drugs such as olanzapine, quetiapine, aripiprazole, and risperidone has been accumulated in subsequent years. A proposal of mood stabilizer status for novel antiepileptic drug, lamotrigine, was made in the early 2000s. On the basis of such chronology, the author of this entry put in a claim to classify mood-stabilizing drugs into first- and second-generation ones. Therefore, a proposal has been made to name lithium, carbamazepine, and valproate first-generation mood stabilizers and atypical neuroleptics and lamotrigine second-generation mood stabilizers (Rybakowski 2007a).

History of Lithium as a Mood Stabilizer

Modern lithium pharmacotherapy begins in 1949, when the Australian psychiatrist John F. Cade (1912–1980) gave lithium carbonate to ten patients with acute and chronic manic states. The results were better than expected. Even though it was not possible to exclude spontaneous remission among some ill persons with acute symptoms of mania, the significant improvement observed in patients among whom manic symptoms had lasted for many months deserved attention. Cade's publication in the Medical Journal of Australia in 1949 titled "Lithium salts in the treatment of psychotic excitement," in which the psychiatrist described the results of his research, may be acknowledged as evidence of the introduction of lithium to modern psychiatric therapy (Cade 1949). It is also thought that this publication marked the beginning of modern clinical psychopharmacology because it preceded by 3 years the first French publication concerning the use of a neuroleptic (antipsychotic) drug, chlorpromazine (Rybakowski 2009).

The Danish psychiatrist Mogens Schou (1918–2005) in 1954 carried out research on lithium's effectiveness among patients in a manic state which, when it was performed over half a century ago, was unusual in two aspects. First, Schou used a neutral preparation (placebo) for comparative purposes to show the real effect of lithium, which was not well known at that time. The research included 38 patients in a manic state among whom 30 had "clear" affective symptoms. Among these patients, a spectacular improvement was noted in 12, improvement in 15, and a lack of effect in 3 of them. Second, within the therapy, measurements of concentration of the drug in blood serum were systematically made, and in six of them also in the cerebrospinal fluid. It was found that concentrations of lithium ion in the serum remained within 0.5–2 mmol/l, which was an important element for further findings of relations between concentration of lithium in serum and its clinical effectiveness and toxic symptoms (Rybakowski 2009).

A meta-analysis of lithium efficacy in mania was done by Poolsup et al. (2000) on the 50th anniversary of documenting antimanic activity of the drug. A total of 658 patients from 12 controlled trials were included. It was shown that bipolar manic patients were twice as likely to obtain improvement with lithium than with placebo. The antimanic efficacy of lithium as well as the side effects were similar to those of antipsychotics (haloperidol, risperidone) and other first-generation mood stabilizers (carbamazepine, valproates). Nowadays, after more than 60 years of Cade's observation, lithium is still considered as an antimanic drug of first choice and a monotherapy with lithium has been recommended, mainly for patients with moderate intensity of symptoms. Lithium is also a suitable drug for combination treatment of mania, especially with valproate and atypical antipsychotics.

From the beginning of the 1960s papers were published that changed the way of understanding the essence of the effect of lithium in mood disorders. It has appeared that the most important property of lithium salts may be not its antimanic effect but the possibility to prevent recurrences of mood disorders, both mania and depression, that is, the mood-stabilizing action. The first paper of this kind was published in the British Journal of Psychiatry in 1963. Its author was the British psychiatrist Geoffrey Hartigan, who presented his observation concerning chronic use of lithium for 3 years among 15 patients with affective disorder: 7 with bipolar affective disorder and 8 with recurrent depression. It resulted from these observations that among 6 persons from the first group and 6 from the other group recurrences of illness were not observed at that time. Hartigan (1963) suggested that long-term use of lithium has a "prophylactic" effect on recurrences of illness. A year later, similar observations were described by the Danish psychiatrist Paul Christian Baastrup. And in 1967, the first paper summing up the experiences of the use of lithium for a longer time (on average 6 years) among a large group of 88 patients with unipolar and bipolar mood disorder appeared. Its authors were the Danish psychiatrists Baastrup and Schou. All patients participating in the research started to use lithium during their hospitalization in the psychiatric hospital in Glostrup. The criterion for comparison was the total time of being in a state of disordered mood (mania or depression) within a year. Results presented in the paper showed that the average duration of mood disorders within a year before using lithium was 13 weeks, whereas using lithium it was 2 weeks. This demonstrated with great probability that lithium may have a beneficial prophylactic influence on the course of mood disorders (Rybakowski 2009).

In 1970–1973, the results of eight controlled studies including the use of placebo performed in Europe (in Denmark and the UK) and in the USA, researching prophylactic effectiveness of lithium, were published. In patients qualified for clinical observation in the period of the preceding 2 years at least two recurrences of illness occurred. Most of these studies employed a method comparing the course of illness in a group in which lithium was discontinued and replaced with a placebo in a group which continued to receive lithium (discontinuation design). Recurrence of illness was defined as a deterioration that would require psychiatric hospitalization or commencing regular antidepressive or antimanic treatment. Analysis of all research showed that the percentage of patients in whom recurrences of depression or mania occurred was significantly lower while receiving lithium (on average 30%) than while receiving placebo (on average 70%) (Rybakowski 2009).

A few years later, in Poland, a paper from a Poznań center concerning assessment of the prophylactic effect of lithium appeared in Psychiatria Polska (Polish

Psychiatry). A group of 61 patients was included in the analysis, among whom lithium was used for 5 years on average. Similar to the research described above, in all patients participating in the research carried out by the Poznań center, within 2 years before starting to give them lithium, at least two episodes of illness occurred. To assess the effect of lithium, a mirror image method was used, comparing the course of illness in the period of use of lithium with an analogical period before the beginning of giving lithium (i.e., if an ill person received lithium for 4 years, the course of illness was assessed in the period of 4 years before giving him/her lithium). Comparing the analysis of these periods, submitted to global inspection among all 61 patients, showed that in the period of using lithium the number of recurrences decreased by 71% and the number of hospitalizations decreased by 72%. Among 44% of patients, recurrences of illness while using lithium were not observed (Rybakowski 2009).

A meta-analysis of long-term lithium efficacy in bipolar disorder was performed by Geddes et al. (2004) on the 40th anniversary of documenting a prophylactic activity of the drug. Five randomized controlled trials with total of 770 participants were included. The analysis showed that lithium was significantly more effective then a placebo in preventing all relapses, with a relative risk of 0.65. Such relative risk for manic relapses amounted to 0.62 and was slightly less for depressive relapses (0.72). By this time, it also became evident that in one third of the patients with bipolar disorder, a continuous monotherapy with lithium can totally prevent the recurrence of affective episodes for 10 years and more (Rybakowski et al. 2001). Lithium has been also a most important component of polytherapy aimed to prevent manic or depressive recurrences of bipolar disorder being successfully combined with both first generation of mood-stabilizing drugs (carbamazepine, valproate) as well as second-generation ones (atypical antipsychotics, lamotrigine).

Starting from the 1970s, the papers showing the possibility of a therapeutic effect of lithium in a depressive episode in the course of mood disorder have appeared. Such effect was generally better in depression in the course of bipolar mood disorder than in periodic depression (unipolar mood disorder). In a number of current guidelines, lithium has still been recommended as an important treatment modality during depressive episode of bipolar mood disorder.

Lithium was also the first mood-stabilizing drug to provide evidence of being efficacious in the augmentation of antidepressant drugs in treatment-resistant depression. The initial paper on this topic was published in the early 1980s by Canadian researchers, and our own study, published a decade later, discussed 51 patients with treatment-resistant depression in whom the lithium augmentation effect was superior in bipolar depression (79%) but was also considerable in unipolar depressed patients (46%). The most recent review on this issue performed by a German group headed by Michael Bauer, current president of the International Group for the Study of Lithium-treated patients (IGSLI), found that lithium is an effective remedy reinforcing the effect of antidepressant drugs in treatment-resistant depression, and a successful effect may be expected in at least 50% of patients. In their analysis, more than 30 open-label studies and 10 placebo-controlled double-blind trials were included demonstrating the substantial efficacy of lithium augmentation in the acute treatment of depressive episodes. A meta-analysis of placebo-controlled trials revealed a mean response rate of 41.2% in the lithium group and 14.4% in the placebo group. According to the authors of the review, the augmentation of antidepressants with lithium is currently the best-evidenced augmentation therapy in the treatment of depressed patients who do not respond to antidepressants (Crossley and Bauer 2007). Indeed, in most current therapeutic standards, the strategy of augmenting antidepressant drugs by lithium is recommended in the first place.

The Status of Lithium Among Other Mood Stabilizers

Besides lithium, valproates and carbamazepine belong to the first generation of mood-stabilizing drugs. Initially, valproates were used as valproic acid amide (valpromide) and more recently as valproic acid and sodium valproate combination (divalproex). In the 1980s, Polish investigators observed the prophylactic effect of valpromide in two third of bipolar patients previously poorly responding to lithium. In the 1990s, an open 18-month comparative study of valpromide versus lithium performed by French researchers demonstrated the comparable effect of both drugs. In 2000, the first double-blind, randomized, placebo-controlled maintenance study was conducted in bipolar I disorder that compared divalproex and lithium over a 1-year period, showing similar results with both these drugs. However, 10 years later, the results of the BALANCE study demonstrated a distinct advantage of lithium over divalproex for prevention of the recurrence of affective episodes in bipolar mood disorder. In this study, 330 patients aged 16 years or older with bipolar I disorder were randomly allocated to open label lithium monotherapy (n = 110), valproate monotherapy (n = 110), or both agents in combinations (n = 110). The participants were followed up for up to 24 months. New intervention for an emergent mood episode was initiated in 54% of patients in the combination therapy group, in 59% of patients in the lithium group, and in 69% of patients in the valproate group. This study also showed that emerging depressive episodes were least frequent with lithium monotherapy (32%), compared with combination (35%) and valproate monotherapy (45%) (BALANCE Investigators and Collaborators et al. 2010).

In a meta-analysis performed in 1999, ten double-blind studies comparing carbamazepine to lithium in maintenance treatment for 1–3 years were included, suggesting that the overall prophylactic efficacy of both drugs may be similar. Interesting results were obtained in German MAP study (Multicentre study of long-term treatment of Affective and schizoaffective Psychoses) performed in late 1990s in which the prophylactic efficacy of lithium and carbamazepine was compared over the period of 2.5 years. Both drugs were similarly efficacious in bipolar II illness and slightly better effect of lithium was observed in bipolar I patients, the superiority of carbamazepine over lithium was found in patients with atypical features (Greil et al. 1997).

Several studies compared lithium with atypical antipsychotic drugs with mood-stabilizing properties assessing an efficacy for prevention of recurrences in bipolar mood disorders. Two double-blind controlled trials lasting about a year, in which olanzapine was compared with lithium or with divalproex did not demonstrate significant differences in relapse to an affective episode. However, the relapse into a manic episode was significantly lower with olanzapine than with lithium. Also, an open 12-month study did not show differences between quetiapine and classical mood stabilizers such as lithium and valproate for prevention of affective recurrences (Rybakowski 2007b). Furthermore, the prophylactic efficacy of olanzapine, quetiapine, aripiprazole, and risperidone when added to a first-generation mood stabilizer such as lithium or valproate was found to be better than the use of mood stabilizer alone (Altamura et al. 2008; Yatham et al. 2009).

New anticonvulsant drug, lamotrigine, has already been given a special category among mood stabilizers as a "mood stabilizer from below," since it has greater antidepressant than antimanic properties. Two placebo-controlled studies comparing the prophylactic efficacy of lamotrigine versus lithium over an 18-month period demonstrated that both lamotrigine (200 and 400 mg/day) and lithium were significantly superior to a placebo in delaying the time to intervention for any mood episode. Lamotrigine showed greater efficacy for prolonging the time to a depressive episode and lithium was superior at prolonging the time to a manic, hypomanic, or mixed episode (Fountoulakis and Vieta 2008; Yatham et al. 2009). An attempt was also made to compare clinical features of good prophylactic responders to lamotrigine or lithium. Lamotrigine responders had more frequently a chronic or rapid cycling course of illness, comorbidity of anxiety disorders (panic attacks), and a tendency to psychoactive substance abuse. Schizoaffective disorder, recurrent depression, or anxiety disorders often existed in their families. Lithium responders tended to have a classical form of bipolar disorder, with a periodic course and full remission, and pure bipolar illness dominated their family history. The antidepressant efficacy of lamotrigine has been observed when the drug is used as monotherapy in the treatment of bipolar depression as well as in brief recurrent depression believed to belong to the bipolar spectrum. An effective augmentation by lamotrigine of antidepressant drugs in treatment-resistant depression, with an efficacy comparable to that of lithium has also been reported.

Being the first member of the family of mood-stabilizing drugs, lithium still remains a cornerstone for the prophylaxis of a bipolar mood disorder. Such a statement formed the title of a recent meta-analysis carried out by Spanish investigators (Nivoli et al. 2010). The results of randomized controlled trials with long-term treatment, with at least 6 months of follow-up and involving 1,561 bipolar I and II patients, of whom 534 were randomized to lithium, were analyzed. The results point to a significant antimanic prophylactic efficacy of lithium and also to some efficacy in the prevention of depression. The efficacy of lithium was slightly less effective than lamotrigine in preventing depression, and less effective than olanzapine in the prevention of manic or mixed episodes. Making an extensive MEDLINE search of the past four decades brings about a conclusion that the evidence for bipolar maintenance efficacy has been overwhelmingly for lithium, better than that for any other drug. Therefore, the gold standard for treating bipolar disorder in 1970 was lithium, and the gold standard 40 years later remains lithium.

Excellent Lithium Responders

Those patients with bipolar mood disorder who can benefit most from long-term lithium treatment are described as "excellent lithium responders." This term was introduced by Canadian psychiatrist of Czech origin, Paul Grof, to describe patients in whom monotherapy with lithium can totally prevent the further recurrence of episodes for 10 years or more. The clinical profile of such patients may include complete remissions and other characteristics of an episodic clinical course, bipolar family history, low psychiatric comorbidity, and a characteristic presenting psychopathology, approximating to the classical description of a manic-depressive patient which was done in Kraepelin' textbook of psychiatry in 1899. Excellent lithium responders make up about one third of lithium-treated patients. In a previous study, the percentage of excellent lithium responders, defined as those having an absence of recurrence for 10 years of lithium monotherapy, remained similar in patients entering lithium treatment in the 1970s and 1980s (34% vs. 32%, respectively) (Rybakowski et al. 2001). Grof (2010) also presents interesting findings emerging from his prospective observations of the next generation of lithium responders and from their counterparts, that is, children of parents who did not respond to lithium and from control children. These findings indicate that parents and offspring suffer from a comparable brain dysfunction that manifests clinically in distinct stages.

The concept of excellent lithium responders resulted in a number of molecular-genetic and pharmacogenetic studies, where the degree of lithium response was regarded as an endophenotype, and excellent lithium responders made a clinical subgroup of patients with bipolar illness. In one of the genome-wide association studies (GWAS) in bipolar illness, the strongest association was related to a genetic variation in the diacylglycerol kinase (DGKH) gene, which encodes a key protein in the lithium-sensitive phosphatidyl inositol pathway. Among the candidate genes associated with lithium prophylaxis, interesting findings were obtained with the brain-derived neurotrophic factor gene (BDNF) and serotonin transporter gene as well as with an interaction of both. In past 2 years, following an initiative by the International Group for the Study of Lithium-Treated Patients (IGSLI) and the Unit on the Genetic Basis of Mood and Anxiety Disorder at the National Institute of Mental Health, lithium researchers from around the world have formed the Consortium on Lithium Genetics. Its main aim is to establish the largest sample to date for performing the genome-wide association study (GWAS) of lithium response in bipolar disorder, using a stringent phenotype definition of response. The sample currently comprises more than 1,200 patients characterized by response to lithium treatment, and the results of GWAS will appear in 2012 (Schulze et al. 2010).

Cross-References

▶ Lithium, Neuroprotective Effect

References

Altamura AC, Mundo E, Dell'Osso B (2008) Quetiapine and classical mood stabilizers in the long-term treatment of bipolar disorder: a 4-year follow up naturalistic study. J Affect Disord 110:135–141

BALANCE Investigators and Collaborators, Geddes JR, Goodwin GM, Rendell J et al (2010) Lithium plus valproate combination therapy versus monotherapy for relapse prevention in bipolar I disorder (BALANCE): a randomized open-label trial. Lancet 375:385–395

Cade J (1949) Lithium salts in the treatment of psychotic excitement. Med J Aust 2:349–352

Crossley NA, Bauer M (2007) Acceleration and augmentation of antidepressants with lithium for depressive disorders: two meta-analyses of randomized controlled trials. J Clin Psychiatry 68:935–940

Fountoulakis KN, Vieta E (2008) Treatment of bipolar disorder: a systematic review of available data and clinical perspective. Int J Neuropsychopharmacol 11:999–1029

Geddes JR, Burgess S, Hawton K et al (2004) Long-term lithium therapy for bipolar disorder: syatmatic review and

meta-analysis of randomized controlled trials. Am J Psychiatry 161:217–222
Greil W, Ludwig-Mayerhofer W, Erazo N et al (1997) Lithium versus carbamazepine in the maintenance treatment of bipolar disorders – a randomized study. J Affect Disord 43:151–161
Grof P (2010) Sixty years of lithium responders. Neuropsychobiology 62:8–16
Hartigan G (1963) The use of lithium salts in affective disorders. Br J Psychiatry 109:810–814
Lambert PA, Borselli S, Marcou G, Bouchardy M, Cabrol G (1971) Action thymoregulatrice a long terme de Depamide dans la psychose maniaco-depressive. Ann Med Psychol 2:442–447
Nivoli AMA, Murru A, Vieta E (2010) Lithium: still a cornerstone in the long-term treatment in bipolar disorder. Neuropsychobiology 62:27–35
Okuma T, Kishimoto A, Inue K (1973) Anti-manic and prophylactic effect of carbamazepine (Tegretol) on manic depressive psychosis. Folia Psychiatry Neurol Jpn 27:283–297
Poolsup N, Li Wan Po A, de Oliveira IR (2000) Systematic overview of lithium treatment in acute mania. J Clin Pharm Ther 25:139–156
Rybakowski JK, Chlopocka-Wozniak M, Suwalska A (2001) The prophylactic effect of long-term lithium administration in bipolar patients entering treatment in the 1970s and 1980s. Bipolar Disord 3:63–67
Rybakowski JK (2007a) Two generations of mood stabilizers. Int J Neuro Psychopharmacol 10:709–711
Rybakowski J (2007b) Long-term pharmacological treatment of bipolar disorder. Neuro Endocrinol Lett 28(Suppl 1):71–93
Rybakowski J (2009) The faces of manic-depressive illness. Termedia Wydawnictwa Medyczne, Poznan
Schulze TG, Alda M, Adli M (2010) The international consortium on lithium genetics (ConLiGen) an initiative by the NIMH and IGSLI to study the genetic basis of response to lithium treatment. Neuropsyhobiology 62:72–78
Yatham LN, Kennedy SH, Schaffer A et al (2009) Canadian Network for Mood and Anxiety Treatments (CANMAT) and International Society for Bipolar Disorders (ISBD) collaborative update of CANMAT guidelines for the management of patients with bipolar disorder: update 2009. Bipolar Disord 11:225–255
Zarate CA (1995) Is clozapine a mood stabilizer? J Clin Psychiatry 56:108–112

Lithium in Biosphere, Distribution

Gerhard N. Schrauzer
Department of Chemistry and Biochemistry, University of California, San Diego, La Jolla, CA, USA

Synonyms

Allocation on earth's surface; Li

Introduction

Lithium (Li) ranks 35th in abundance in the Earth's crust and with its atomic weight of 6.941 is the lightest of the alkali metals. It is recognized as a nutritionally essential trace element and is widely known for its use, as a carbonate, in the treatment of bipolar depression. Its occurrence in the biosphere, its uptake by plants, and its distribution in organs and tissues are delineated first, followed by a discussion of its nutritional essentiality and its toxicity at high dosages.

Lithium in Soils and the Water Supplies

Lithium was named after the Greek *lithos* because of its presence, in trace amounts, in virtually all rocks. In soils, it is found primarily in the clay fraction, to a lesser extent in the organic soil fraction, in amounts ranging from 7 to 200 μg/g. It is leached out by surface waters and ultimately ends up in the oceans. In some municipalities, Li is present in the tap water. In some Texas counties, for example, Li levels may reach 170 μg/L, adding about 340 μg to the daily Li intake from foods. In these counties, urinary Li excretions of local residents vary inversely with rainfall, reflecting the dilution of drinking water supplies. Significantly higher amounts of Li, from 1,508 to 5,170 μg Li/L, are found in the water from certain rivers in northern Chile. The consumption of 2 L of this water per day by adult residents in these regions provides an extra 10 mg of Li per day, or about five times the typical daily dietary intake of Li, without apparently producing any adverse effects (Weiner 1991).

Uptake by Plants

All plants take up Li, although they appear not to require it for their growth and development. However, this question is not yet completely resolved, since, in the ppb range, stimulatory effects of Li on plant growth have been observed (Schweigart 1962). At high levels in the soil, Li is toxic to all plants, causing a chlorosis-like condition. Uptake and sensitivity to Li are species dependent. Some plants, notably *Cirsium arvense* and *Solanum dulcamara*, accumulate Li three- to sixfold over other plants. Halophilic plants such as *Carduus arvense* and *Holoschoenus vulgaris* may reach Li contents of 99.6–226.4 μg/g (Tölgiesi 1983). Lithium is relatively toxic to citrus plants; nightshade species are remarkably lithium tolerant and may reach Li contents of up to 1,000 μg/g. Yeast (*Saccharomyces cerevisiae*) takes up limited amounts of Li, high levels (115–400 ppm) in the medium cause growth inhibition. In general, plants take up more Li from acidic than alkaline soils. Since soil acidity also increases the solubility of the heavier metallic elements, plant Li levels are directly and significantly correlated with those of iron, nickel, cobalt, manganese, and copper, and to some extent also to those of aluminum, lead, and cadmium.

Lithium in Biosphere, Distribution, Table 1 Sources of dietary lithium (According to Weiner 1991)

Food group	Quantity ingested kg food/day	Li Level mg/kg food	Total μg/day
Grains and vegetables	0.85	0.5–3.4	430–2,900
Dairy products	0.44	0.50	222
Meat	0.21	0.012	2.5
Total			650–3,100

Dietary Sources of Lithium

The U.S. Environmental Protection Agency (EPA) (Saunders 1985) estimated the daily Li intake of a 70 kg American adult to range from 650 to 3,100 μg, see Table 1.

According to Table 1, grains and vegetables contribute from 66% to more than 90% of the total Li intake; the remainder is from animal-derived foods. In general, diets rich in grains and vegetables may be expected to provide more Li than diets rich in animal proteins. However, due to the uneven distribution of the element in the Earth's crust, a predominantly vegetarian diet is not necessarily Li-rich. Accordingly, the estimated dietary Li intakes in populations of different countries vary over a wide range and exhibit large standard deviations from the means. For example, the average dietary Li intakes of adults in Stockholm (Sweden) were estimated to be 1,090 ± 324 μg/day and of adults in Galveston (Texas) and San Diego (California) to be 821 ± 684 μg/day and 429 ± 116 μg/day, respectively (Schrauzer 2002).

Lithium in Organs and Tissues

Ingested in the form of its soluble salts, Li is absorbed to virtually 100% from the small intestine via the Na+-channels and is uniformly distributed in body water, with only a small difference between the extra- and intracellular levels. Autopsy studies of adults revealed that the cerebellum retains more Li than other organs, followed by the cerebrum and the kidneys (Baumann et al. 1983). Organ Li levels showed some unexplained gender differences, with women exhibiting 10–20% more Li than men in the cerebellum, cerebrum, kidneys, and the heart, and 13% less Li in the pancreas; the Li concentrations in the liver, lungs, ribs, and thyroid were about the same for both genders. Li appears to play an important role especially during the early fetal development, since fetal organ Li levels reach maximal values in the first trimester of gestation and subsequently decline.

At the end of the third trimester, the Li concentration of the fetus is one third of that in the first. The Li contents of the kidneys, the liver, and the ribs continue to decline during the first 5–10 years of life, that of the prostate continues to decline over the entire lifespan. The serum Li concentrations are approximately proportional to the Li intakes. Baseline serum Li levels in adults typically range from 7 to 28 μg/L, corresponding to adult Li intakes of 385–1,540 μg/day. Scalp hair Li levels reflect the average dietary intakes of Li over a period of several weeks to months and represent a noninvasive means of determining the dietary Li intakes. Hair Li levels of adults from the New York area ranged from 0.009 to 0.228 μg/g (N = 206); the values were slightly higher for females than for males. This is also true for a number of other elements and was attributed to the generally higher inorganic (ash) content for hair from females. From these data, the medium Li intake of these adults was calculated to be 650 μg/day, range 100–2,645 μg/day (Schrauzer 2002). Hair Li levels increase in proportion to dose in human subjects receiving 1,000 and 2,000 μg of extra-dietary Li in a supplement, reaching a steady state after 3 months of supplementation. However, the proportionality does not extend to pharmacological Li intakes; hair Li thus cannot be used to monitor the compliance of patients on lithium carbonate.

Lithium as an Essential Trace Element

Essentiality for the Rat. In order to demonstrate the nutritional essentiality of Li for the rat it was necessary first to formulate appropriate low-Li diets. Patt, Pickett, and O'Dell (1978) succeeded in the mid-1970s in preparing a corn-casein diet for the laboratory rat whose lithium content was 5–15 ng/g (=0.005–0.015 μg/g). Growth rate and behavior of rats were found to be the same as in the controls maintained on a commercial feed whose lithium content was 350 ng/g or in animals receiving the low-Li corn-casein diet supplemented with Li to 500 ng/g. The lithium deficient rats required a longer conception time in the first, but not in the second generation. Litter size was 20–30% smaller in all generations, and survival to 1 week was only 53–60% of that of controls. A key result of this study was that the mature, Li-deficient rats retained Li at control levels in the pituitary and adrenal glands, suggesting that these organs require Li for some functions. In other tissues, including blood, cerebrum, liver, kidney, spleen, heart, and bone, a 20–50% reduction in lithium contents compared to the controls occurred, and all soft tissues contained less Li than the endocrine tissues. In a second study, retention of Li was again seen in adrenal and pituitary, but this time also in hippocampus, mammary gland, ovary, and thyroid. Retention in the thymus and the pancreas was less marked; the Li contents of other tissues were lower than in the controls. This study also revealed that Li is retained and released by bone. Whereas the femurs of the Li supplemented rats contained 0.21 ± 0.01 μg/g in the first and second generation, Li concentrations were below detection limit in the Li deficient rats. In a third study, rats were fed diets based on purified casein and corn containing only 2 ng Li/g and on purified casein and rice containing 0.6 ng Li/g. The controls received the same diets whose Li contents were increased to 500 ng/g. The dams and their offspring were maintained on these diets for five successive litters and three successive generations, respectively. A significant negative effect of Li deficiency on litter size and litter weight at birth was observed. In addition, evidence for the interaction of lithium with sodium was obtained: Litter size and litter weight at birth were significantly lower among low-Li dams consuming diets with normal or high levels of sodium than those consuming low sodium diets.

Essentiality for the Goat. Experiments to prove the essentiality of Li in goats were initiated by Anke and his school in 1976 and continued until 1988 (Anke et al. 1983). The animals were maintained on a semisynthetic feed; the feed of the controls contained 12.7 μg Li/g, that of the experimental group <1 μg Li/g. The Li deficient goats required repeated inseminations for conception and exhibited reduced conception rates and barrenness. Gravid Li deficient goats experienced a higher incidence of spontaneous abortions and produced kids with the female to male ratio of 1:1.9, significantly different from the ratio of 1:0.9 observed in the controls. In lactating Li deficient goats, milk production was not reduced during the first 5 weeks of lactation, but dropped below that of the controls in the sixth week. Although Li deficiency did not significantly alter the fat content of the milk, the overall fat production during the first 56 days of lactation was significantly reduced. These effects of Li deficiency were accompanied by atrophy of the spleen, lowered immunological status, chronic inflammations, hemosiderosis, and calcification of the blood vessels. The newborn kids had lower birth weights and showed slower weight gains than control kids during the subsequent lactation period. Lithium deficiency did not affect the mortality of the kids during the 8th and 91st day of life, but resulted in a significantly higher mortality of the adult goats during their 1st year. In the Li-depleted mature goats, the Li contents of skeletal muscle, pancreas, and of cardiac muscle were the same as in the normally fed controls, while the serum Li levels dropped to 19% of the controls. The Li contents of hair, lungs, and milk were reduced to 30%, of spleen, carpal bone to 40%, of rib to 42%, of ovary to 45%, and liver to 48%. The Li contents of the kidneys, uterus, aorta, and cerebrum remained at 48–69% of those of the controls. Compared to the organ Li contents of the Li-deficient rats, those of the Li-deficient kids were somewhat higher, indicating that their degree of Li deficiency was not as extreme.

Essentiality for Humans. Lithium was detected in human organs and fetal tissues already in the late nineteenth century, leading to the suggestion that the element might be needed in small amounts by the organism for specific functions (Schulz 1903). Newer studies revealed that low Li intakes cause behavioral defects, as evidenced by statistically significant inverse associations of tap water lithium contents in Texas counties with the rates of mental hospital admissions (Dawson et al. 1970), suicides, homicides, the arrest rates for forcible rape, drug addictions, and other crimes (Dawson et al. 1972; Schrauzer et al. 1992). Recent Chinese ecological studies associated low dietary Li intakes with shortened human life spans. A provisional RDA for Li for a 70 kg adult of 1,000 μg/day has been proposed (Schrauzer 2002).

Mechanisms of Action

The biochemical mechanisms of action of Li are multifactorial and are linked with the functions of several enzymes, hormones, and vitamins as well as with growth and transforming factors. Lithium furthermore appears to play a role in the expansion of the pluripotential stem cell pool to more mature progenitor cells. The neuroprotective and neurotrophic effects of Li, and its use in the prevention and therapy of Alzheimer's disease are currently being explored, as recent studies demonstrated that Li upregulates the concentrations of the cytoprotective protein bcl-2 in areas of rodent brains and in human neuronal cells (Manji et al. 2000). Presently, Li is being tested as a possible means to achieve neuronal regeneration in patients with spinal cord injuries. Lithium may even promote the development of new brain cells.

In one much publicized study with hospitalized patients suffering from bipolar mood disorders, an increase of gray brain matter was observed in eight out of ten patients, following 4 weeks of treatment with pharmacologic doses of lithium carbonate (Moore et al. 2000).

Lithium Toxicity

As lithium carbonate is used therapeutically for extended periods and at high dosages, for example, 600–900 mg per day, serum Li levels must be carefully monitored to safeguard against the development of Li toxicity. Patients remained free of symptoms of Li toxicity as long as their serum Li did not exceed 1.5 milli-equivalents per liter (MeQ/L). At serum Li levels from 1.5 to 2.5 MeQ/l, signs of Li toxicity appeared, characterized by neuromuscular changes (tremor, muscle hyperirritability, and ataxia), central nervous system changes (blackout spells, epileptic seizures, slurred speech, coma, psychosomatic retardation, and increased thirst), cardiovascular changes (cardiac arrhythmia, hypertension, and circulatory collapse), gastrointestinal changes (anorexia, nausea, and vomiting), and renal damage (albuminuria and glycosuria), deaths occurred at serum Li levels above 3.5 MEQ/l.

Summary

Lithium (Li) is a nutritionally essential trace element for which a provisional RDA of 1,000 mcg/day has been proposed. However, because of the uneven distribution of lithium in the Earth's crust, the human dietary Li intakes are frequently very low. As insufficient Li intakes have been associated in several studies with higher incidences of suicide and violent crimes, lithium supplementation, as well as the addition of Li in small amounts to the municipal water supplies, have been recommended to prevent the development of lithium deficiency in affected populations.

Note

This chapter is an updated version of a review by the author published in the *Journal of the American College of Nutrition*, Vol. 21, No. 1, 14–21 (2002). Reprinted with permission from the publisher.

Cross-References

▶ Lithium as Mood Stabilizer
▶ Lithium, Physical and Chemical Properties

References

Anke M, Groppel B, Kronemann H, Grün M (1983) Evidence for the essentiality of lithium in goats. In: Anke M, Baumann W, Bräunlich H, Brückner C (eds) Proceedings 4. Spurenelement-symposium, 1983, Jena. VEB Kongressdruck, Jena, pp 58–65

Baumann W, Stadie G, Anke M (1983) Der Lithiumstatus des Menschen. In: Anke M, Baumann W, Bräunlich H, Brückner C (eds) Proceedings 4th spurenelement symposium, 1983. VEB Kongressdruck, Jena, pp 180–185

Dawson EP, Moore TD, McGanity WJ (1970) The mathematical relationship of drinking water lithium and rainfall on mental hospital admission. Dis Nerv Syst 31:1–10

Dawson EP, Moore TD, McGanity WJ (1972) Relationship of lithium metabolism to mental hospital admission and homicide. Dis Nerv Syst 33:546–556
Manji HK, Moore GJ, Chen G (2000) Lithium upregulates the cytoprotective protein Bcl-2 in the CNS in vivo: role for neurotrophic and neuroprotective effects in manic depressive illness. J Clin Psychiatry 61(Suppl 9):82–96
Moore GJ, Bebchuk JM, Wilds IB, Chen G, Manji HK (2000) Lithium-induced increase in human brain grey matter. Lancet 356(9237):1241–1242, Erratum: Lancet (2000); 356(9247):2104
Patt EL, Pickett EE, O'Dell BL (1978) Effect of dietary lithium levels on tissue lithium concentrations, growth rate, and reproduction in the rat. Bioinorg Chem 9:299–310
Saunders DS (1985) Letter: United States Environmental Protection Agency, Office of Pesticide Programs
Schrauzer GN (2002) Lithium occurrence, dietary intakes, nutritional essentiality. J Am Coll Nutr 21:14–21
Schrauzer GN, Shrestha KP, Flores-Arce AF (1992) Lithium in scalp hair of adults, students and violent criminals. Biol Trace El Res 34:161–176
Schulz H (1903) Vorlesungen über Wirkung und Anwendung swe unorganischen Arzneistoffe. K.F. Haug, Berlin, p 195
Schweigart A (1962) Vitalstoff-Lehre. H. Zauner, Munich, p 73
Tölgiesi G (1983) Die Verbreitung des Lithiums in ungarischen Böden und Pflanzen. In: Anke M, Baumann W, Bräunlich H, Brückner C (eds) Proceedings 4th spurenelement symposium, 1983. VEB Kongressdruck, Jena, pp 39–44
Weiner ML (1991) Overview of lithium toxicology. In: Schrauzer GN, Klippel KF (eds) Lithium in biology and medicine. VCH, Weinheim, pp 83–99

Lithium, Neuroprotective Effect

Janusz K. Rybakowski
Department of Adult Psychiatry, Poznan University of Medical Sciences, Poznan, Poland

Synonyms

Anti-apoptotic; Nervous cell's resistance; Neuroplasticity; Neurotrophic

Definition

The term "neuroprotection" denotes the totality of mechanisms and strategies used to increase the nervous cell's resistance to damaging agents, to promote neuroplasticity as well as neurotrophic and anti-apoptotic mechanisms.

Introduction

Evidence for the neurotrophic and neuroprotective effects of lithium has accumulated over the last two decades. The neurotrophic and neuroprotective effects of this ion are now regarded as important therapeutic mechanisms of lithium action in mood disorders. They may also be responsible for the favorable influence of lithium on cognitive functions and for an increase in cerebral gray matter volume in lithium-treated patients with bipolar mood disorder. The evidence for a neuroprotective effect of lithium makes also this ion a possible candidate for use as a therapeutic drug in neurology, especially in neurodegenerative disorders (Rybakowski 2011).

In this entry, neurochemical mechanisms of neuroprotective action of lithium will be characterized. The evidence for favorable effect of lithium on cognitive functions will be presented as well as for effect of lithium on brain structures reflected in neuroimaging studies. A possibility of using lithium in neurodegenerative disorders, especially in Alzheimer's disease, will be discussed.

Neurochemical Mechanism of Neuroprotective Effect of Lithium

Several biochemical targets have been involved in neurotrophic and neuroprotective effect of lithium. They include increased expression of neurotrophins (mainly brain-derived neurotrophic factor – BDNF), the inhibition of glycogen synthase kinase-3 (GSK-3), modulation of cyclic adenosine monophosphate (c-AMP) – mediated signal transduction with special role of c-AMP response element binding (CREB) and of the phosphatidylinositide (PI) cascade, protein kinase C (PKC) inhibition, and increased B-cell lymphoma 2 (bcl-2) expression. Furthermore, an inhibitory action of lithium on such enzymes as biphosphate nucleotidase (BPN-ase), fructose 1,6-biphosphatase (FBP-ase), and phosphoglucomutase (PGM) may also play some role. As a result of such actions, lithium increases cell survival by promoting neurogenesis in the adult brain and by inhibiting cell death (apoptosis) cascades. Studies of the biochemical targets involved in the neuroprotective action of lithium were also connected with elucidating the neurobiological pathology of bipolar mood disorder and with proposals for new treatments (Quiroz et al. 2004, 2010).

BDNF is a member of neurotrophin family, which includes also nerve growth factor (NGF) and neurotrophin-3 (NT-3), NT-4, NT-5, and NT-6. BDNF and other neurotrophins are necessary for the survival and function of neurons. BDNF modulates the activity of such neurotransmitters as glutamate, gamma-aminobutyric acid, dopamine, and serotonin. A transcription of *BDNF* gene is activated by CREB. In experimental studies, it was demonstrated that lithium activates CREB and increases BDNF expression. Other studies demonstrated that chronic lithium treatment may also result in the modification of other neurotrophic factors (Quiroz et al. 2010). In clinical studies, lithium treatment results in an increase of blood level of BDNF. Polymorphisms of the *BDNF* gene were shown to be associated with a predisposition to bipolar mood disorder. We found that the Val66Met BDNF gene polymorphism is associated with the degree of prophylactic response to lithium (Rybakowski 2008).

GSK-3 is a serine/threonine kinase that regulates diverse cellular processes and specifically cell apoptosis.Therefore, GSK-3 inhibition directly influences gene transcription leading to anti-apoptotic effects and improved cell structural stability. Since the first studies in 1996, the evidence has been accumulating using various experimental model showing that lithium inhibits GSK-3 activity and that this enzyme can be regarded as one of the therapeutic target of lithium. Clinical studies demonstrated that GSK-3 activity is altered in mania and regulated after treatment with mood stabilizers, including lithium (Quiroz et al. 2010).

The PI pathway plays a key role in signal transduction pathways which are connected with receptors of various neurotransmitters. The effect of lithium on this pathway has long been considered as one of the main therapeutic mechanism of this ion in mood disorders and lead to a formulation of the inositol-depletion hypothesis of lithium action. Lithium exerts effect on many steps of PI pathways, the most important being inhibition of inositol monophosphatase (IMPase). In one of the genome-wide association studies (GWAS) in bipolar illness, the strongest association was related to a genetic variation in the diacylglycerol kinase *(DGKH)* gene which encodes a key protein in the lithium-sensitive phosphatidyl inositol pathway (Baum et al. 2008). An inhibition of the collapse of sensory neuron growth and an increase in growth cone area, the processes dependent on PI pathway, have been proposed as a common mechanism of therapeutic action of three classical mood-stabilizing drugs, that is, lithium, carbamazepine, and valproates (Williams et al. 2002). Recently, it has been found that lithium inhibits sodium myo-inositol transporter 1 (SMIT-1) and similar action was also attributed to carbamazepine and valproates. Homozygote knockout mice for the *SMIT1* gene behave similarly to lithium-treated animals in depression models. Clinical studies demonstrated that the content of SMIT1 mRNA in neutrophils of patients with bipolar mood disorder is altered and regulated after treatment with mood stabilizers (Quiroz et al. 2010).

Protein kinase C (PKC) is an enzyme closely connected with PI pathway which regulates both pre- and postsynaptic aspects of neurotransmission and several cellular processes. Lithium inhibits the activity of PKC and decreases the levels of phosphorylation of myristoylated alanine-rich C-kinase substrate, a major PKC substrate that has been implicated in signaling and neuroplastic events associated with cytoskeletal architecture (Quiroz et al. 2010). A study of the role of PKC in manic-like behavior, which showed an inhibition of PKC by lithium and valproates, lead to a successful trial of the estrogen antagonist, and fairly selective PKC inhibitor, tamoxifen, in manic patients (Yildiz et al. 2008).

Bcl-2 is important protein for cellular resilience and plasticity, exerting mostly anti-apoptotic effects. Lithium treatment results in an increase of bcl-2 in brain of experimental animals. Lithium also increases the expression of bcl-2 associated athanogene (bag-1) which is known to attenuate glucocorticoid receptor nuclear translocation, thus potentiating the anti-apoptotic effect. In patients with bipolar disorder, polymorphism of *bcl-2* gene is associated with an abnormality of PI system (Quiroz et al. 2010; Machado-Vieira et al. 2011).

Cognitive Effects of Lithium

The results of recent animal research point to lithium having a favorable effect on cognitive function as demonstrated on various models. Lithium treatment protects irradiated hippocampal neurons from apoptosis and improves cognitive performance in irradiated mice. Such pro-cognitive effect of lithium was associated with the inhibition of GSK-3b and an increase in Bcl-2 protein expression. In other study, using three different positive reinforcement spatial cognitive tasks in rats it was demonstrated that lithium magnifies learning in all three tasks which may be associated with enhancing hippocampal synaptic plasticity (Rybakowski 2011).

The results of clinical studies performed in the last 3 years show that lithium treatment may not affect previously impaired cognitive functions in bipolar patients negatively but may preserve or even augment them. Spanish researchers (Lopez-Jaramillo et al. 2010) found a lower performance on episodic verbal and visual verbal memory tests of bipolar patients, compared to the healthy subjects. However, they did not find any differences between lithium-treated and bipolar patients with no medication intake and concluded that lithium therapy had no deleterious effect on cognition.

In our own studies, we have attempted to correlate cognitive functions in lithium-treated patients with a quality of lithium prophylactic effect. We found that nonresponders to lithium had significantly worse performance on many domains of the Wisconsin Card Sorting Test, compared to excellent and partial responders (Rybakowski et al. 2009). In the most recent study, using neuropsychological tests from a CANTAB battery which measured spatial working memory and sustained attention, we demonstrated that bipolar patients who are excellent lithium responders have cognitive functions comparable to those of matched control subjects, thereby probably constituting a specific subgroup of bipolar patients in which long-term lithium administration can produce complete normality in this respect. As a decreased BDNF serum level was proposed as a marker of bipolar mood disorder, it should be noted that the excellent lithium responders in our study, concomitantly with their normal cognitive functions also had normal serum BDNF levels (Rybakowski and Suwalska 2010).

Neuroimaging Evidence of Lithium Neuroprotection

Could the neuroprotective effect of lithium in bipolar patients be reflected at the clinical level, as evidenced by neuroimaging studies? Following a provocative research letter in 2000 to the Lancet submitted by Moore and his colleagues, postulating a lithium-induced increase in human brain gray matter, the results of several studies on this issue have been reported mainly over the past 3 years. The target brain structures which could presumably be influenced by either short-term, or long-term, lithium treatment were the hippocampus and prefrontal cortex (including the anterior cingulated region). Moore et al. (2009) extended their results published 9 years earlier as they found that an increase in total gray matter volume in the prefrontal cortex of bipolar depressed subjects after 4 weeks of lithium administration was significant only in lithium responders.

American researchers focusing on prefrontal gray matter showed greater cortical gray matter density in lithium-treated patients with bipolar disorder, especially in the right anterior cingulate region. In the second study, they gave lithium in therapeutic doses to healthy individuals for 4 weeks. By conducting vaxel-based morphometry analysis, they showed that lithium caused a significant increase in gray matter in the left and right dorsolateral prefrontal cortices and the left anterior cingulated region (Monkul et al. 2007).

Canadian investigators demonstrated a bilateral increase in hippocampal volume after both short-term (up to 8 weeks) as well as in long-term (2–4 years) lithium administration in patients with bipolar disorder (Yucel et al. 2008). A similar study was that of Bearden et al. (2008) who, using three-dimensional mapping of hippocampal anatomy, demonstrated that total hippocampal volume in lithium-treated bipolar patients was significantly larger, compared to that of both unmedicated bipolar patients and healthy control subjects. The localized difference applied to the right hippocampus in regions corresponding primarily to cornu Ammonis subfields.

In 2010, the reports of two neuroimaging studies have appeared which compare the effects of lithium with those of anticonvulsants and antipsychotics possessing mood-stabilizing properties. Lyoo et al. (2010) compared the course and magnitude of cerebral gray matter volume changes in 22 bipolar patients treated with either lithium or valproate. They found that lithium caused an increase in gray matter volume, peaking at week 10–12 and maintained through 16 weeks of treatment, and that this increase was associated with a positive clinical response.

By contrast, valproate-treated patients did not show gray matter volume changes over time. Germana et al. (2010) performed a cross-sectional structural brain magnetic resonance imaging study of 74 remitted bipolar patients receiving long-term prophylactic treatment with lithium, valproate, carbamazepine, or antipsychotics. They found that volume of gray matter in the subgenual anterior cingulate gyrus on the right and in the postcentral gyrus, the hippocampus/amygdala complex, and the insula on the left was greater in patients on lithium treatment compared to all other treatments.

In summary, there is considerable evidence for lithium causing an increase in cerebral gray matter volume in patients with bipolar mood disorder which may reflect a neuroprotective effect of lithium at a clinical level. Such an effect has not been demonstrated for any other mood-stabilizing drug.

Recently, a study has also been performed in order to find whether lithium may counteract the microstructural and metabolic brain changes which occur in individuals at ultra-high risk (UHR) for psychosis. Hippocampal T2 relaxometry was performed prior to initiation and following 3 months of treatment in 11 UHR patients receiving low-dose lithium and 10 UHR patients receiving treatment as usual. The results of this pilot study suggest that low-dose lithium may protect the microstructure of the hippocampus in UHR states as reflected by significantly reduced relaxation time (Berger et al., unpublished).

Lithium in Neurodegenerative Disorders

The evidence for a neuroprotective effect of lithium makes this ion a possible candidate for use as a therapeutic drug in neurology, especially in neurodegenerative disorders. An inhibition by lithium of GSK-3, which is a key enzyme in the metabolism of amyloid precursor protein and in the phosphorylation of the tau protein, the main pathogenetic players in Alzheimer's disease (AD) has been shown in animal models. Lithium treatment causes a reduction of GSK3 mRNA in cultures of rat cortical and hippocampal neurons and may arrest the development of neurofibrillary tangles in mutant tau transgenic mice with advanced neurofibrillary pathology. In the *Drosophila* fly adult-onset model of AD, lithium may ameliorate amyloid beta pathology through an inhibition of GSK-3. In this fly, experimental studies of presenilin, the gene of which may be one of the determining factors in the familial form of AD were also performed. Reducing presenilin activity in *Drosophila* may contribute to cognitive impairment during aging, and treatment with lithium during the aging process prevented the onset of these deficits, and treatment of aged flies reversed the age-dependent deficits (Rybakowski 2011).

The results of two human studies which investigated the association between dementia and lithium consumption appeared in 2005–2007. The first study, which included 87 patients, suggested an increasing risk of dementia with increasing numbers of lithium prescriptions. On the other hand, in the second study, it was found in the group of 114 bipolar patients that those receiving long-term lithium therapy had a decreased prevalence of AD compared with patients not receiving recent lithium therapy (Rybakowski 2011).

Two papers from the University of Copenhagen were recently published, in which the Danish nationwide register of lithium prescriptions was used. In the first paper, a total of 16,238 persons who had purchased lithium at least once, and 1,487,177 persons from the general population who had not purchased lithium were compared as regards the diagnosis of dementia or AD during inpatient or outpatient hospital care. It was found that persons who had purchased lithium at least once had in 1.5-fold increased rate of dementia, compared with persons not exposed to lithium. For persons who continued to take lithium, the rate of dementia decreased to the same level as the rate for the general population. Such a phenomenon was unique to lithium since, in persons treated with anticonvulsant drugs, the risk of dementia increased with the duration of treatment (Kessing et al. 2008). In the second study, a total of 4,856 patients with a diagnosis of a manic or mixed episode or bipolar disorder at their first psychiatric contact were investigated over the study period of 1995–2005. Among these patients, 50.4 % were exposed to lithium, 36.7 % to anticonvulsants, 88.1 % to antidepressants, and 80.3 % to antipsychotics. A total of 216 patients received a diagnosis of dementia during follow-up (103.6/10,000 person-years). Analysis revealed that continued treatment with lithium was associated with a reduced rate of dementia in patients with bipolar disorder, in contrast to continued treatment with anticonvulsants, antidepressants, or antipsychotics (Kessing et al. 2010).

There has been one reported randomized lithium trial in patients with mild AD where 71 patients were randomized to receive either lithium (n = 33) or a placebo (n = 38) for 10 weeks. In those receiving lithium its serum level was targeted at 0.5–0.8 mmol/L. Lithium treatment did not result in changes in global cognitive performance, as measured by the ADAS-Cog subscale, or in depressive symptoms. Also, the treatment did not exert any effect on either the plasma activity of GSK-3 or disease biomarker concentrations in the cerebrospinal fluid. More promising results were, however, obtained by Brazilian researchers (Forlenza et al. 2011) who employed lithium in placebo-controlled randomized trial of 45 patients with amnestic mild cognitive impairment for 12 months. They found that lithium treatment was associated with significantly better performance on the cognitive subscale of the Alzheimer's disease Assessment Scale and with significant decrease of P-tau protein in cerebrospinal fluid.

The attempts to use lithium in another neurodegenerative disorder, Huntington's disease (HD), can be traced back to the 1970s. The rationale for such a use at that time was that the application of lithium resulted in improvement in some movement disorders, such as, tardive dyskinesia. However, the results of three controlled trials of lithium in HD, involving very small numbers of patients were negative. In recent years, a rationale for using lithium in HD has been based on a neuroprotective action of this ion, evidenced in animal studies. In a rat excitotoxic model of HD disease, in which quinolinic acid was unilaterally infused into the striatum, lithium suppressed striatal lesions and promoted neuronal survival and proliferation in this brain structure. Since HD is caused by a polyglutamine expansion, mutation in the huntingtin protein, which can be cleared by macroautophagy, a potential therapeutic mechanism of lithium in HD, has also been attributed to its property of augmenting this process by inhibiting inositol monophosphatase and GSK-3b activity. However, no systematic clinical trials with lithium in HD have been performed in recent years (Rybakowski 2011).

Amyotrophic lateral sclerosis (ALC) is a devastating neurodegenerative disorder that usually leads to death within 3–5 years from diagnosis. There is no effective treatment of ALC and all attempts are at the experimental stage. Some ALC cases are due to mutations of the gene coding for the enzyme copper-zinc superoxide-dysmutase (SOD1), therefore a mouse model of ALC (transgenic mice over expressing the human mutant SOD1) has been produced. When administering lithium concurrently with an antioxidant, Neu2000, an improvement in motor neuron survival, motor function, and reduced mortality in this model was observed. Also, using this model, some researchers observed a marked neuroprotective action of lithium with a delay of disease onset and augmentation of the life span, with evidence of lithium ability to activate autophagy and to reduce mitochondrial changes. However, such a beneficial effect of lithium was not confirmed in other studies. At the clinical level, Italian investigators were the first to suggest that giving lithium, in doses leading to plasma levels of 0.4–0.8 mmol/L may delay the progression of ALC. Unfortunately, another Italian group was not able to confirm such an effect of lithium when used in either therapeutic or subtherapeutic doses and a North American group found no evidence that lithium, in combination with riluzole, the drug used in ALC, slows progression of the disease more than riluzole alone (Rybakowski 2011).

The efficacy of lithium was also tested in a knock-in mouse model of spinocerebellar ataxia (SCA), the cause of which is a polyglutamine expansion of ataxin protein. Dietary lithium carbonate supplementation resulted in improvement of motor coordination and cognitive functions, attenuation of hippocampal changes, as well as in a decrease of a marker of ataxin toxicity. The authors postulate that lithium is a good candidate drug for human SCA patients (Rybakowski 2011).

In summary, the evidence for the neuroprotective effects of lithium gives rise to a number of speculations regarding the wider use of this ion in neurodegenerative disorders. Unfortunately, despite the promising results of lithium treatment obtained on animal models in these disorders, the reflection of these results in human studies of AD, HD, and ALS has been fairly weak. There is, however, a possibility of reducing the risk of AD with long-term lithium administration as well as a beneficial effect of this ion in patients with mild cognitive impairment. Clearly, the application of lithium in neurological disorders should be the subject of additional, well-designed, clinical studies in the future.

Cross-References

▶ Lithium as Mood Stabilizer

References

Baum AE, Akula N, Habanero M, Cardona I, Corona W, Klemens B et al (2008) A genome-wide association study implicates diacylglycerol kinase eta (DGKH) and several other genes in the etiology of bipolar disorder. Mol Psychiatry 13:197–207

Bearden CE, Thompson PM, Dutton RA, Frey BN, Peluso MA, Nicoletti M et al (2008) Three-dimensional mapping of hippocampal anatomy in unmedicated and lithium-treated patients with bipolar disorder. Neuropsychopharmacology 33:1229–1238

Forlenza OV, Diniz BS, Radanovic M et al (2011) Disease-modifying properties of long-term lithium treatment for amnestic mild cognitive impairment: randomized controlled trials. Br J Psychiatry 198:351–356

Germana C, Kempton MJ, Sarnicola A, Christodoulou T, Haldane M, Hadjulis M et al (2010) The effects of lithium and anticonvulsants on brain structure in bipolar disorder. Acta Psychiatr Scand 122:481–487

Kessing LV, Forman JL, Andersen PK (2010) Does lithium protect against dementia ? Bipolar Disord 12:87–94

Kessing LV, Sondergard L, Forman JL, Andersen PK (2008) Lithium treatment and the risk of dementia. Arch Gen Psychiatry 65:1331–1335

Lopez-Jaramillo C, Lopera-Vasquez J, Ospina Duque J, Garcia J, Gallo A, Cortez V et al (2010) Lithium treatment effects on the neuropsychological functioning of patients with bipolar I disorder. J Clin Psychiatry 71:1055–1060

Lyoo K, Dager SR, Kim JE, Yoon SJ, Friedman SD, Dunner DL et al (2010) Lithium-induced grey matter volume increase as a neural correlate of treatment response in bipolar disorder: a longitudinal brain imaging study. Neuropsychopharmacology 35:1743–1750

Machado-Vieira R, Pivovarova NB, Stanika RI et al (2011) The bcl-2 gene polymorphism rs956572AA increases inositol 1,4,5-triphosphate receptor-mediated endoplasmic reticulum calcium release in subjects with bipolar disorder. Biol Psychiatry 69:344–352

Monkul ES, Matsuo K, Nicoletti MA, Dierschke N, Hatch JP, Dalwani M et al (2007) Prefrontal gray matter increases in healthy individuals after lithium treatment: a voxel-based morphometry study. Neurosci Lett 429:7–11

Moore GJ, Cortese BM, Glitz DA, Zajac-Benitez C, Quiroz JA, Uhde TW et al (2009) A longitudinal study of the effects of lithium treatment on prefrontal and subgenual prefrontal gray matter volume in treatment-responsive bipolar disorder patients. J Clin Psychiatry 70:699–705

Quiroz JA, Gould TD, Manji HK (2004) Molecular effect of lithium. Mol Interv 4:259–272

Quiroz JA, Machado-Vieira R, Zarate CA, Manji HK (2010) Novel insights into lithium's mechanism of action: neurotrophic and neuroprotective effects. Neuropsychobiology 62:50–60

Rybakowski JK (2008) BDNF gene: functional Val66Met polymorphism in mood disorders an schizophrenia. Pharmacogenomics 9:1589–1593

Rybakowski JK (2011) Lithium in neuropsychiatry. A 2010 update. World J Biol Psychiatry 12:340–348

Rybakowski JK, Suwalska A (2010) Excellent lithium responders have normal cognitive functions and plasma BDNF levels. Int J Neuropsychopharmacol 13:617–622

Rybakowski JK, Permoda-Osip A, Borkowska A (2009) Response to prophylactic lithium in bipolar disorder may be associated with a preservation of executive cognitive functions. Eur Neuropsychopharmacol 19:791–795

Williams RS, Cheng L, Mudge AW et al (2002) A common mechanism of action for three mood-stabilizing drugs. Nature 417:292–295

Yildiz A, Guleryuz S, Ankerst DP, Obgur D, Renshaw PF (2008) Protein kinase C inhibition in the treatment of mania: a double-blind, placebo-controlled trial of tamoxifen. Arch Gen Psychiatry 65:255–263

Yucel K, Taylor VH, McKinnon MC, Macdonald K, Alda M, Young LT (2008) Bilateral hippocampal volume increase in patients with bipolar disorder and short-term lithium treatment. Neuropsychopharmacology 13:361–367

Lithium, Physical and Chemical Properties

Fathi Habashi
Department of Mining, Metallurgical, and Materials Engineering, Laval University, Quebec City, Canada

Lithium is an alkali metal but has properties more similar to the alkaline earth magnesium than to its group member sodium. For example, lithium carbonate is insoluble in water like $MgCO_3$ unlike Na_2CO_3. This is the basis of separating lithium from natural brines containing alkali metals for its recovery. This property is known as diagonal similarity (Fig. 1). Lithium-containing brines have become the most important raw materials for the production of lithium chemicals. Their industrial exploitation began at Searle's Lake, California; in Clayton Valley, Nevada; at the Salar de Atacama in Chile; and at the Salar del Hombre Muerto in Argentina.

Lithium salts have been widely used for the clinical treatment of manic-depressive illnesses. Therapeutic doses of 170–280 mg Li per day, mainly in the form of lithium carbonate, are administered over long periods. The lithium ion is considerably more toxic than the

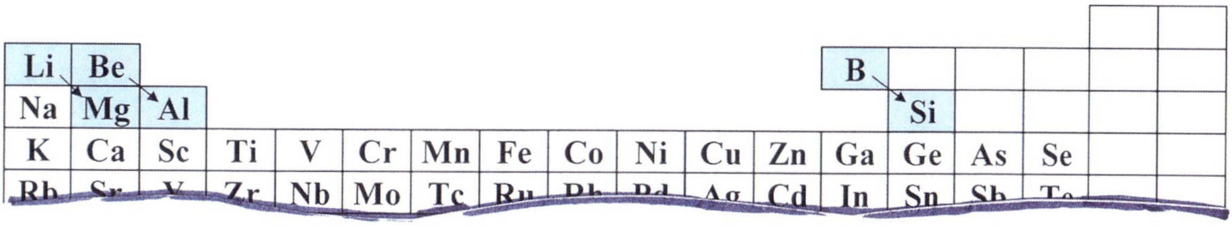

Li	Be											B					
Na	Mg	Al											Si				
K	Ca	Sc	Ti	V	Cr	Mn	Fe	Co	Ni	Cu	Zn	Ga	Ge	As	Se		
Rb	Sr	Y	Zr	Nb	Mo	Tc	Ru	Rh	Pd	Ag	Cd	In	Sn	Sb	Te		

Lithium, Physical and Chemical Properties, Fig. 1 Illustrating the diagonal similarity in the periodic table

sodium ion, for example, 5 g of lithium chloride can cause fatal poisoning. Since lithium is the least dense of all metals, then its alloys with aluminum are significantly less dense than aluminum. Commercial Al-Li alloys contain up to 2.5% lithium and are of interest to the aerospace industry. Lithium niobate is used extensively in telecommunication products. Lithium and its compounds are used in heat-resistant glass and ceramics, high strength-to-weight alloys used in aircraft, lithium batteries, and lithium-ion batteries. The transmutation of lithium atoms to tritium was the first man-made form of a nuclear fusion reaction, and lithium deuteride serves as a fusion fuel in staged thermonuclear weapons.

Physical Properties

Symbol	Li
Atomic number	3
Atomic weight	6.94
Density at room temp, g/cm^3	0.534
Melting point, °C	180.54
Heat of fusion, J/g	431.4
Boiling point, °C	1,342
Vapor pressure at the melting point, Pa	1.88×10^{-8}
Heat of vaporization, J/g	22,705
Dynamic viscosity at 180.5 °C, Pa·s	0.599×10^{-3}
Surface tension at the melting point, mN/m	398
Mean linear coefficient of thermal expansion between 273 and 368°K, K^{-1}	56×10^{-6}
Specific heat c_p between 20 and 180°C, J g^{-1} K^{-1}	3.3–4.2
Ionization energy, eV	5.37
Electrode potential of the half cell ($Li \rightarrow Li^{+} + e^{-}$), V	–3.024
Crystalline form	Body-centered cubic

Chemical Properties

Lithium is a reactive metal though considerably less so than other alkali metals. A freshly cut surface of lithium metal has a silvery luster. At room temperature in dry air with a relative humidity of less than 1%, the surface remains shiny for several days. Lithium metal can therefore be processed in dry air. However, in moist air, a gray coating, consisting mainly of lithium nitride, lithium oxide, and lithium hydroxide, forms within a few seconds. Even at room temperature, dry nitrogen reacts slowly with lithium metal.

Lithium burns with a luminous white flame, forming a dense white smoke consisting mainly of lithium oxide. Lithium reacts with water with formation of hydrogen, which ignites under normal conditions only if the metal is finely divided. Lithium reacts with hydrogen to form lithium hydride. This reaction is carried out on an industrial scale at 600–1,000°C. Lithium reacts with gaseous ammonia at elevated temperature to form lithium amide. The vigorous reaction with halogens produces incandescence. Lithium also reacts with boron, silicon, phosphorus, arsenic, antimony, and sulfur upon heating.

References

Habashi F (1999) A textbook of hydrometallurgy, 2nd edn. Métallurgie Extractive Québec, Québec City. Distributed by Laval University Bookstore. www.zone.ul.ca

Habashi F (2002) Textbook of pyrometallurgy. Métallurgie Extractive Québec, Québec City. Distributed by Laval University Bookstore. www.zone.ul.ca

Schreiter W (1961) Seltene metalle, vol 2. VEB Deutscher Velag für Grundstoffindustrie, Leipzig, pp 68–114

Wietelmann U, Bauer RJ (1997) Chapter 50. In: Habashi F (ed) Handbook of extractive metallurgy. Wiley-VCH, Heidelberg

Zelikman AN, Krein OE, Samsonov GV (1966) Metallurgy of rare metals. Israel Program for Scientific Translations,Jerusalem

L

Living System and Organisms

► Tungsten in Biological Systems

LLL, L-Leucyl-L-leucyl-L-leucine

► Zinc Aminopeptidases, Aminopeptidase from Vibrio Proteolyticus (Aeromonas proteolytica) as Prototypical Enzyme

Ln/An Adverse Effects

► Lanthanide/Actinide toxicity

Ln/An as Chemotherapeutics

► Lanthanide/Actinide in health and disease

Ln/An as Diagnostics

► Lanthanide/Actinide in health and disease

Loading and Release of Drugs/Proteins

► Porous Silicon for Drug Delivery

Lobaplatin

► Platinum Anticancer Drugs

Local Anesthetics

► Sodium Channel Blockers and Activators

Local Signaling

► Calcium, Local and Global Cell Messenger

Lou Gehrig's Disease

► Copper-Zinc Superoxide Dismutase and Lou Gehrig's Disease

Low Molecular Mass Aluminum Species in Serum

► Aluminum Speciation in Human Serum

Low Molecular Weight Protease

► Zinc-Astacins

Low-Grade Inflammation

► Magnesium and Inflammation

Low-Molecular-Weight Chromium-Binding Substance (LMWCr)

► Chromium(III) and Low Molecular Weight Peptides

LPA, L-Leucinephosphonic Acid

► Zinc Aminopeptidases, Aminopeptidase from Vibrio Proteolyticus (Aeromonas proteolytica) as Prototypical Enzyme

L-*p*NA, L-Leucine-p-nitroanilide

▶ Zinc Aminopeptidases, Aminopeptidase from Vibrio Proteolyticus (Aeromonas proteolytica) as Prototypical Enzyme

Luminescent Tag

▶ Lanthanides, Luminescent Complexes as Labels

Lung Cancer

▶ Polonium and Cancer

Lutetium

Takashiro Akitsu
Department of Chemistry, Tokyo University of Science, Shinjuku-ku, Tokyo, Japan

Definition

A lanthanoid element, the 14th element of the f-elements block (yttrium group), with the symbol Lu, atomic number 71, and atomic weight 174.967. Electron configuration [Xe] $6s^2 4f^{14} 5d^1$. Lutetium is composed of stable (^{175}Lu, 97.41%) and three radioactive (^{173}Lu; ^{174}Lu; ^{1766}Lu, 2.59%) isotopes. Discovered by G. Urbain in 1907. Lutetium exhibits oxidation state III; atomic radii 174 pm, covalent radii 178 pm; redox potential (acidic solution) Lu^{3+}/Lu −2.255 V; electronegativity (Pauling) 1.27. Ground electronic state of Lu^{3+} is $^1S_0$ with $S = 0$, $L = 0$, $J = 0$ with $\lambda = 0\ cm^{-1}$. Most stable technogenic radionuclide ^{176}Lu (half-life 3.78×10^{10} years). The most common compounds: Lu_2O_3, LuF_3, $LuCl_3$, $LuBr_3$, and LuI_3. Biologically, lutetium is of low to moderate toxicity, which indicates high affinity to cover Ca^{2+}-binding sites on surface of cells in smooth muscle (Atkins et al. 2006; Cotton et al. 1999; Huheey et al. 1997; Oki et al. 1998; Rayner-Canham and Overton 2006).

Cross-References

▶ Lanthanide Ions as Luminescent Probes
▶ Lanthanide Metalloproteins
▶ Lanthanides and Cancer
▶ Lanthanides in Biological Labeling, Imaging, and Therapy
▶ Lanthanides in Nucleic Acid Analysis
▶ Lanthanides, Physical and Chemical Characteristics

References

Atkins P, Overton T, Rourke J, Weller M, Armstrong F (2006) Shriver and Atkins inorganic chemistry, 4th edn. Oxford University Press, Oxford/New York
Cotton FA, Wilkinson G, Murillo CA, Bochmann M (1999) Advanced inorganic chemistry, 6th edn. Wiley-Interscience, New York
Huheey JE, Keiter EA, Keiter RL (1997) Inorganic chemistry: principles of structure and reactivity, 4th edn. Prentice Hall, New York
Oki M, Osawa T, Tanaka M, Chihara H (1998) Encyclopedic dictionary of chemistry. Tokyo Kagaku Dojin, Tokyo
Rayner-Canham G, Overton T (2006) Descriptive inorganic chemistry, 4th edn. W. H. Freeman, New York

LxVP

▶ Calcineurin

Printed by Books on Demand, Germany